AF411825

Nonlinear and Relativistic Effects in Plasmas

La Jolla International School of Physics
The Institute for Advanced Physics Studies

RESEARCH TRENDS IN PHYSICS

Nonlinear and Relativistic Effects in Plasmas

V. Stefan

Editor

American Institute of Physics   New York

Papers included in this volume are reproduced from camera-ready copy supplied by the contributors.

L.C. Catalog Card No. 91-55577
ISBN 0-88318-787-6
DOE CONF-900264

Printed in the United States of America.

Contents

Part I. INERTIAL CONFINEMENT FUSION

A. PARTICLE BEAM FUSION

Chairman: T. W. Mark (Lawrence Livermore National Laboratory,
Livermore, California)

B. LASER-DRIVEN FUSION

Chairman: K. A. Brueckner (University of California, San Diego,
La Jolla, California)

Part IV. COHERENT RADIATION GENERATION AND PARTICLE ACCELERATORS

Chairmen: S. L. Ossakow (Naval Research Laboratory, Washington, D.C.)
C. Pellegrini (University of California, Los Angeles, Los Angeles, California)

Part V. NONNEUTRAL PLASMAS

Chairman: T. M. O'Neil (University of California, San Diego, La Jolla, California)

PREFACE

The Topical Conference on Research Trends in Nonlinear and Relativistic Effects in Plasmas was held at the Catamaran Resort Hotel, San Diego February 5-8, 1990. The Conference was hosted by La Jolla Institute, La Jolla, California, in cooperation with General Atomics, San Diego; Jaycor, San Diego; Maxwell Laboratories, Inc., San Diego; the Institute for Advanced Physics Studies, La Jolla, California; and the Department of Physics of the University of California, San Diego. This meeting was an ambitious undertaking with the intent to bring together some of the most prominent plasma physicists in the country from various fields of applied plasma physics under the common denominator of applications based on nonlinear and/or relativistic effects in plasmas. Accordingly, this conference had the character of a cross-fertilization meeting. For example, the theory of Raman scattering, being relatively well developed in laser driven fusion research, can be applied to free electron laser and particle acceleration research and vice versa. In addition, the theory of Raman scattering is being used in tokamak research, and in space plasma research for interpretation of active experiments. A number of similar examples can be given for other plasma physics research areas.

The members of the Steering Committee were selected in a cumulative manner. The Committee was composed of the following members:

D. E. Baldwin
Institute for Fusion Studies
University of Texas at Austin
Austin, TX 78712

K. A. Brueckner
Department of Physics
University of California, San Diego
La Jolla, CA 92093

R. C. Davidson
Plasma Fusion Center
Massachusetts Institute of
Technology, Cambridge, MA 02139

J. M. Dawson
Department of Physics
University of California
Los Angeles, CA 90024

A. Hasegawa
AT&T Bell Laboratory
600 Mountain Avenue
Murray Hill, NJ 07974

A. M. Sessler
Lawrence Berkeley Laboratory
1 Cyclotron Road
Berkeley, CA 94720

N. A. Krall
Krall Associates
1070 America Way
Del Mar, CA 92014

W. L. Kruer
University of California
Lawrence Livermore National Laboratory
Livermore, CA 94550

K. D. Papadopoulos
Department of Physics and Astronomy
University of Maryland
College Park, MD 20742

M. N. Rosenbluth
Department of Physics
University of California, San Diego
La Jolla, CA 92093

N. Rostoker
Department of Physics
University of California, Irvine
Irvine, CA 92717

T. H. Stix
Plasma Physics Laboratory
Princeton University, P.O. Box 451
Princeton, NJ 08544

The members of the Steering Committee identified "hot" topics, suggested invited talks, and formed the final list of invited speakers. The Local Program Committee coordinated the decisions of the Steering Committee with the work of invited speakers. It consisted of:

V. Chan
General Atomics, Inc.
San Diego, CA

F. Felber
Jaycor, Inc.
San Diego, CA

P. Hammerling
Physical Dynamics, Inc.
La Jolla, CA 92037

J. Lovberg
Maxwell Laboratories, Inc.
San Diego, CA
Presently at
Directed Technologies, Inc.
San Diego, CA

V. Stefan, Conference Chairman
Institute for Advanced Physics Studies,
La Jolla, CA 92037 and Tesla
Laboratories, Inc., La Jolla, CA 92037

I am deeply indebted to all members of the Steering and Local Program Committees for their highly competent work and their efforts to make this meeting a success.

The major goal of the conference was to address the most important research problems and trends, as identified by the Steering Committee, in applied nonlinear and relativistic effects in plasmas. The emphasis was on unifying the principles and techniques relating to apparently disparate phenomena. The conference attracted a number of relatively young researchers and students enabling them to be in direct contact with mature researchers, learn insightful information about the addressed topics, and hopefully influence their future scientific activity. This meeting was organized through close and frequent contacts and exchanges of opinions between the members of the Steering Committee, Local Program Committee, and invited speakers and representatives of U. S. science supporting agencies.

The meeting dealt with the applications of nonlinear and relativistic effects in plasmas induced by external and/or self-consistent fields. Nonlinear wave-wave and wave-particle interaction within the plasma lead to a variety of effects such as plasma heating, current drive, stimulated radiation scattering, acceleration of particles, and stabilization of MHD instabilities in different plasma environments: magnetically confined plasmas, inertially confined plasmas, relativistic particle beams, nonneutral plasmas and upper atmospheric, space, and astrophysical plasmas.

The subject material was formatted under nine headings allocated into nine nonparallel sessions:

Particle Beam Fusion

Laser Driven Fusion

Upper Atmospheric, Space
and Astrophysical Plasmas

Magnetic Fusion

Coherent Radiation Generation
and Particle Accelerators, I and II

Relativistic Particle Beams

Nonneutral Plasmas

Nonlinear Electromagnetic Wave-
Plasma Interaction

This format allowed strong interaction among specialists in different areas of plasma research. The meeting purposely lacked the introductory and concluding talks. These are left to contributors and readers for evaluation. The conference was semiclosed. All participants were invited. The attendees of the conference were mainly invited speakers and/or their students or collaborators. The number of registered attendees was approximately 60.

On behalf of the Local Program Committee, I express my sincere thanks to those authors who submitted their well-prepared camera-ready manuscripts by the deadline. Sincere thanks are also due to those authors who cooperated with us in the course of time and managed to submit their manuscripts for publication, thus enriching the contents of the proceedings.

The Conference and the publication of the proceedings were supported by the following: La Jolla Institute, La Jolla, California; the Institute for Advanced Physics Studies, La Jolla, California; the National Science Foundation (Magnetospheric Physics, Division of Atmospheric Physics); and Naval Research Laboratory, Washington, D.C. I am especially grateful to Dr. V. Patel of the National Science Foundation (presently at NRL) and Dr. S. Ossakow of NRL for advice and encouragement. Special thanks are due to Dr. A. Hochstim, President of La Jolla Institute for his support. I am deeply indebted to Mrs. Rosalie Rocher and Dr. Charles Eminhizer for processing the proceedings manuscripts, and to Mrs. Grace Pitts for handling the meeting arrangements.

The "Research Trends in Physics" type of topical conferences in various areas of applied physics are now being regularly organized by the La Jolla International School of Physics.

Lastly, I would like again to express my gratitude to all those who contributed and/or attended the conference for their work and effort which made this meeting a success.

<table>
<tr><td>La Jolla, California
February 1991</td><td align="right">V. Stefan
Conference Chairman

La Jolla International
School of Physics
The Institute for Advanced
Physics Studies
La Jolla, California 92037 and
Tesla Laboratories, Inc.
La Jolla, California 92037</td></tr>
</table>

Yakov Alpert, Harvard-Smithsonian Center for Astrophysics, 60 Garden Street, Cambridge, Massachusetts 02138, (617) 495-7000

Ronald Blanken, U.S. Department of Energy, ER-542, Washington, D.C. 20545, (301) 353-3306

Alain Brizard, Lawrence Berkeley Laboratory, MS 4/230, 1 Cyclotron Road, Berkeley, California 94720, (415) 486-5928

K.A. Brueckner, Department of Physics, B-019, University of California, San Diego, La Jolla, California 92093, (619) 534-2892

Benjamin A. Carreras, Oak Ridge National Laboratory, P.O. Box 2009, Building 9201-2, Oak Ridge, Tennessee 37831-8070, (615) 574-1292

Anthony Chan, Plasma Physics Laboratory, P.O. Box 451, Princeton, New Jersey 08543, (609) 243-3653

Vincent S. Chan, General Atomics, P.O. Box 85608, Bldg. 13/311, San Diego, California 92138-5608, (619) 455-4162

Chiping Chen, Massachusetts Institute of Technology, Bldg. NW16-264, Cambridge, Massachusetts 02139, (617) 253-8506

Pisin Chen, Stanford Linear Accelerator Center, Stanford University, Stanford, California 94309, (415) 926-3384

Donald L. Cook, Sandia National Laboratories, P.O. Box 5800, Org. 1260, Albuquerque, New Mexico 87185, (505) 844-5429

Christopher Darrow, Lawrence Livermore National Laboratory, P.O. Box 808, L-473, Livermore, California 94550, (415) 423-7773

Ronald C. Davidson, Plasma Fusion Center, 167 Albany Street, NW 16-106, Cambridge, Massachusetts 02139, (617) 253-8102

Michael Desjarlais, Sandia National Laboratories, Division 1265, Albuquerque, New Mexico 87185-5800, (505) 846-6348

Patrick H. Diamond, Department of Physics, B-019, University of California, San Diego, La Jolla, California 92093, (619) 534-4025

Andris M. Dimits, LPR, University of Maryland, College Park, Maryland 20742, (301) 454-7104

Jack Dorning, Thornton Hall, University of Virginia, Charlottesville, Virginia 22903, (804) 924-6153

C.F. Driscoll, Department of Physics, B-019, University of California, San Diego, La Jolla, California 92093 (619) 534-2489

Daniel H.E. Dubin, Department of Physics, B-019 University of California, San Diego, La Jolla, CA 92093, (619) 534-4174

Don DuBois, Los Alamitos National Laboratory, MS B262, Los Alamos, New Mexico 87545, (505) 667-4135

Eduardo M. Epperlein, University of Rochester, Laboratory for Laser Energetics, 250 East River Road, Rochester, New York 14623-1299, (716) 275-5101

Eric Esarey, Naval Research Laboratory, Code 4791, Washington, D.C. 20375-5000, (202) 404-7720

Roger Falcone, Department of Physics, University of California, Berkeley, California 94720, (415) 642-8916

Franklin S. Felber, Jaycor, P.O. Box 85154, San Diego, California 92138, (619) 453-6580

Nat Fisch, Plasma Physics Laboratory, Princeton, New Jersey 08543, (609) 243-2643

Moshe Friedman, Naval Research Laboratory, Code 4754, 4555 Overlook Avenue SW, Washington, D.C. 20375-5000, (202) 767-3145

Akira Hasegawa, AT&T Bell Laboratory, 1E-351-600 Mountain Avenue, Murray Hill, New Jersey 07974, (201) 582-2886

Chan Joshi, Department of Electrical Engineering, University of California, Los Angeles, 7731 Boelter Hall, Los Angeles, CA 90024, (213) 825-7279

Rhon Keinigs, Los Alamos National Laboratory, Los Alamos, New Mexico 87545, (505) 667-9530

William L. Kruer, Lawrence Livermore National Laboratory, P.O. Box 5508, L-472, Livermore, California 94550, (415) 422-5437

Baruch Levush, Laboratory for Plasma Science, University of Maryland, College Park, Maryland 20742, (301) 454-7108

Chuan Sheng Liu, Department of Physics, University of Maryland, College Park, Maryland 20742, (301) 454-7483

John M.J. Madey, Department of Physics, Duke University, Durham, North Carolina 27706, (919) 684-8121

Robert L. McCrory, Laboratory for Laser Energetics, University of Rochester, 250 East River Road, Rochester, New York 14623-1299, (716) 275-4973

Dale M. Meade, Princeton University, Princeton Plasma Physics Laboratory, Forrestal Campus, P.O. Box 451, Princeton, New Jersey 08543, (609) 243-3301

Clifford W. Mendel, Sandia National Laboratories, Division 1263, Albuquerque, New Mexico 87185, (505) 844-6904

David Montgomery, Department of Physics and Astronomy, Dartmouth College, Hanover, New Hampshire 03755, (603) 646-3219

George Morales, Department of Physics, University of California, Los Angeles, 405 Hilgard Avenue, Los Angeles, California 90024, (213) 825-4318

Warren Mori, Department of Physics, University of California, Los Angeles, 405 Hilgard Avenue, Los Angeles, California 90024, (213) 825-7514

Sidney L. Ossakow, Naval Research Laboratory, Code 4700, 4555 Overlook Avenue SW, Washington, D.C. 20375, (202) 767-2723

Peter J. Palmadesso, Naval Research Laboratory, Code 4700.3, 4555 Overlook Avenue SW, Washington, D.C. 20375, (202) 767-6780

Dennis Papadopoulos, University of Maryland, Astronomy Program, College Park, Maryland 20742, (301) 454-6810

Vithal L. Patel, Naval Research Laboratory, Code 4701, 4555 Overlook Avenue SW, Washington, D.C. 20375-5000, (202) 767-2997

Claudio Pellegrini, Department of Physics, University of California, Los Angeles, 405 Hilgard Avenue, Los Angeles, California 90024-1547, (213) 825-4649

Donald Prosnitz, Lawrence Livermore National Laboratory, P.O. Box 808, L-626, Livermore, California 94550, (415) 422-7504

S. Rajagopalan, BIN 26, SLAC, PB 4349, Stanford University, Stanford, California 94309, (415) 926-2146

M.N. Rosenbluth, Department of Physics, B-019, University of California, San Diego, La Jolla, California 92093, (619) 534-3790

James Rosenzweig, Fermi National Accelerator Laboratory, Batavia, Illinois 60510, (208) 840-2981

Norman Rostoker, Department of Physics, University of California, Irvine, California 92717, (714) 856-6949

Andrew M. Sessler, Lawrence Berkeley Laboratory, University of California, Berkeley, California 94720, (415) 486-4992

Xiaowen Shan, Department of Physics and Astronomy, Dartmouth College, Hanover, New Hampshire 03755, (603) 646-2971

James P. Sheerin, KMS Fusion Inc., P.O. Box 1567, Ann Arbor, Michigan 48106-1567, (313) 769-8500 (ext.304)

Stephen A. Slutz, Sandia National Laboratories, P.O. Box 5800, Division 1265, Albuquerque, New Mexico 87185, (505) 844-1279

Phillip Sprangle, Naval Research Laboratory, 4555 Overlook Avenue SW, Code 4790, Washington, D.C. 20375-5000, (202) 767-3493

Leon A. Steinert, McDonnell Douglas Astronautics, P.O. Box 3973, Long Beach, California 90803, (714) 896-3874

J. Steward, Lawrence Livermore National Laboratory, P.O. Box 808, L626, Livermore, California 94550, (415) 422-0400

Erik Storm, Lawrence Livermore National Laboratory, P.O. Box 5508, L-481, Livermore, California 94550, (415) 422-0400

C.M. Surko, Department of Physics, B-019, University of California, San Diego, La Jolla, California 92093, (619) 534-6880

R.E. Waltz, General Atomics, P.O. Box 85608, Bldg. 13/303, San Diego, California 92138-5608, (619) 455-4584

David Whittum, Lawrence Berkeley Laboratory, MS 71-259, 1 Cyclotron Road, Berkeley, California 94720, (415) 486-4102

Scott C. Wilks, Lawrence Livermore National Laboratory, P.O. Box 5508, L-477, Livermore, California 94550, (415) 422-2974

Alfred Y. Wong, Department of Physics, University of California, Los Angeles, 405 Hilgard Avenue, Los Angeles, California 90024-1547, (213) 825-1642

Jonathan S. Wurtele, Massachusetts Institute of Technology, NW16-258, Cambridge, Massachusetts 02139, (617) 253-8445

Part I

INERTIAL CONFINEMENT FUSION

A. Particle Beam Fusion

Heavy Ion Two Stream

Keith A. Brueckner

Department of Physics, University of California, San Diego, La Jolla, CA 92092

A beam of heavy ions initially neutralized by co-moving electrons produces ionization in a low-density gas and also can have its charge state increased by ionizing collisions with the background gas. The ionization rate of the propagating beam is

$$\frac{dz_i}{dx} \cong 4\pi \frac{a_0 r_0}{\beta^2} z_g^2 \sum_B \frac{1}{\varepsilon_B (Ry)} n_g \tag{1}$$

$$a_0 = 0.521 \times 10^{-8} \, cm \, , \quad r_0 = e^2/mc^2$$

$\sum_B$ = sum over bound electrons in the beam

$\beta = v_{beam}/c$

ε_B = ionization energy of beam ions

z_i = beam ionization

z_g = atomic number of gas atoms

n_g = number density of gas

Eq. (1) omits the screening of the gas nuclei by the bound electrons, which is a reasonable approximation for lithium gas and heavy ions with charge state of three or more. The ionization of the gas by the beam ions is

$$\frac{dn_{e,g}}{dt} \cong 4\pi \frac{n_i z_i^2 a_0 r_0 c}{\beta} n_g \sum_g \frac{1}{\varepsilon_g (Ry)} \tag{2}$$

$\sum_g$ = sum over bound electrons in the gas

ε_g = ionization energyof gas atoms

n_i = beam number density

For $z_g = 3$, $n_g = 10^{14}/cm^3$, $\beta = 1/3$,

$$\frac{dz_i}{dx} \cong 1.5 \times 10^{-4} \sum_B \frac{1}{\varepsilon_B (Ry)} \, . \tag{3}$$

For lead with an initial ionization $z_i = 3$, the sum over the first 12 bound states gives the sum in Eq. (3) of the order of unity, and the ion charge increases only slightly in a propagation distance of 600 cm.

The gas ionization rate depends on the beam density, which is

$$n_i = \frac{P_B}{\pi r^2 \varepsilon_i v_B N_{beam}}$$

(4)

P_B = beam power

ε_i = ion kinetic energy

N_{beam} = number of separate ion beams carrying total beam power.

For $\varepsilon_i = 10\,GeV$, $v_B = 10^{10}\,cm/sec$, $P_B = 100\,TW$,

$$n_i = 1.99 \times 10^{12}/(N_{beam} R^2) .$$

(5)

The ionization rate for $n_g = 10^{14}/cm^3$ and $z_i = 3$ is

$$\frac{dn_{e,g}}{dt} = \frac{2.74 \times 10^9}{R^2 N_{beam}} \sum_g \frac{1}{\varepsilon_g (Ry)}/nsec .$$

(6)

The sum over bound states gives about 5, and

$$n_{e,g} \approx \frac{1.4 \times 10^{10}\, t\,(nsec)}{R^2 N_{beam}} .$$

The two-stream growth rate for $\omega_p < \omega_b$ is

$$\omega_{Im} = \left[\frac{\omega_p^2 \omega_b}{2} \right]^{1/3}$$

$$= \left[\frac{4\pi e^2 n_i}{m} \right]^{1/2} \left[\frac{2\pi a_0 r_0 c t n_g}{\beta} \right]^{1/3} z_i^{5/6} .$$

(7)

For a pulse length of 10 nsec, the growth is sufficiently rapid for the nonlinear saturation level to result. This is at

$$(k\,\Delta v)^2 \approx \omega_{Im}^2$$

(8)

or

$$(\Delta v)^2 \approx \omega_{Im}^2 R^2 .$$

(9)

The transverse motion of the electrons is contained by the electrostatic attraction of the ions, exerting a pressure of approximately

$$p_e \approx z_i n_i \frac{m (\Delta v)^2}{2} \ . \tag{10}$$

An approximate equation for the beam expansion is

$$\beta^2 n_i m_i c^2 \pi R^2 \frac{d^2 R}{dx^2} = 2\pi R p_e$$

or

$$\frac{d^2 R}{dx^2} = \frac{z_i}{\beta^2 m_i R c^2} m (\Delta v)^2$$

$$= \frac{z_i^{8/3}}{\beta^2 c^2} \frac{4}{R} \frac{e^2 N_i}{m_i} \left[\frac{2\pi a_0 r_0 c t n_g}{\beta} \right]^{2/3} . \tag{11}$$

Eq. (11) gives

$$\left[\frac{dR}{dx} \right]^2 = \left[\frac{dR}{dx} \right]_0^2 + \frac{8 z_i^{8/3} e^2 N_i}{\beta^2 c^2 m_i} \left[\frac{2\pi r_0 a_0 c t n_g}{\beta} \right]^{2/3} \ln \frac{R}{R_0} . \tag{12}$$

The minimum radius is at $dR/dx = 0$. For this radius to be less than a target radius R_t, the condition is

$$\frac{8 z_i^{8/3} e^2 N_i}{\beta^2 c^2 m_i} \left[\frac{2\pi r_0 a_0 c t n_g}{\beta} \right]^{2/3} \ln \frac{R_0}{R_t} < \left[\frac{dR}{dx} \right]_0^2 . \tag{13}$$

For $R_0/R_t = 100$, $n_g = 10^{14}/cm^3$, $t = 10 \ nsec$, $\beta = 1/3$, $m_i = 208 \ m_p$, $(dR/dx)_0 = 1/200$, and Eq. (5) for n_i, this condition is

$$N_B > 0.55 \ z_i^{8/3} \tag{14}$$

which is 10 for $z_i = 3$. If the ionization state is higher, however, due to gas stripping or photo ionization, the number of beams can become prohibitively large.

These results have also been studied with a simulation code in which the beam and plasma ions and electrons are represented by cylinders of charge interacting only transversely. The displacing beam ions and electrons are constrained to move along the axis at the constant beam velocity and the plasma electrons and ions move only transversely. For a typical case studied, the charge cylinders were 0.15 cm in length and the time step in the calculation chosen so the beam cylinders move 0.15 cm per time step. The number of propagation steps was 1000 and the number of beam pulses 100. The pulse duration was 0.5 nsec and the propagation distance of 150 cm. The number of components in the beam and plasma was 10 for both electrons and ions. The beam power was 100 TW carried by 30 beams of 10 GeV ions. The initial charge state for the beam ions was 3 and the atomic number of the plasma 3. For the propagation distance considered the initial beam radius was 0.5 cm. The plasma electrons were produced by the beam ionization computed classically. About 10-15% of the beam electrons are ejected by the transverse two-stream instability and the remaining trapped electrons quickly reach the saturation velocity which is about equal to $R\,\omega_{Im}$. The beam then expands in good agreement with the analytic result given above.

INERTIAL CONFINEMENT FUSION USING LIGHT ION BEAMS[*]

D. L. Cook, J. E. Bailey, K. W. Bieg, D. D. Bloomquist, R. S. Coats, G. C. Chandler,
M. E. Cuneo, M. S. Derzon, M. P. Desjarlais, P. L. Dreike, R. J. Dukart,
R. A. Gerber, D. J. Johnson, R. J. Leeper, T. R. Lockner, D. H. McDaniel,
J. E. Maenchen, M. K. Matzen, T. A. Mehlhorn, L. P. Mix, A. R. Moats,
W. E. Nelson, T. D. Pointon, A. L. Pregenzer, J. P. Quintenz, T. J. Renk,
S. E. Rosenthal, C. L. Ruiz, S. A. Slutz, R. W. Stinnett, W. A. Stygar, G. C. Tisone,
J. R. Woodworth, and J. P. VanDevender.
Sandia National Laboratories
Albuquerque, New Mexico 87185

ABSTRACT

Advances in ion beam theory, diagnostics, and experiments in the past two years
have enabled efficient generation of intense proton beams on PBFA II, and focusing
of the beam power to 5.4 TW/cm^2 on a 6-mm-diameter target. Target experiments
have been started with the intense proton beams, since the range of protons at
4-5 MeV is equivalent to that of lithium at 30 MeV. Three series of experiments
have been conducted using planar, conical, and cylindrical targets. These tests have
provided information on ion beam power density, uniformity, and energy deposition.
In order to increase the power density substantially for target implosion experiments,
we are now concentrating on development of high voltage lithium ion beams.

INTRODUCTION

Pulsed power technology offers an efficient, low-cost, and potentially repetitive
means for generating intense light ion beams for Inertial Confinement Fusion (ICF).
The technology produces a beam that couples well to matter and is scalable to very
high power levels. Considerable progress in developing this technology for ICF has
been made on PBFA II (the Particle Beam Fusion Accelerator II) within the last two
years. Advances in ion beam theory, diagnostics, and experiments have enabled
efficient generation of intense proton beams on PBFA II, and focusing of the beam
power to 5.4 TW/cm^2 on a 6-mm-diameter target.[1] At the present "3/4" energy
operating point of PBFA II (Marx generators charged to about 9 MJ, or 3/4 of the
nominal 12 MJ possible at $^+/_-$ 95 kV), this experimental achievement approaches the
theoretical limit with the measured 17 mrad divergence and a 50% source purity for
protons. Target experiments have been started with the intense proton beams, since
the range of protons at 4-5 MeV is equivalent to that of lithium at 30 MeV.

For PBFA II to drive implosion targets effectively, it must provide an ion beam
with adequate power, adequate power density, and proper range in the target.
Adequate power (50-100 TW) on PBFA II requires the plasma opening switch and full
energy operation. Adequate power density (50-100 TW/cm^2) requires an ion beam
with high magnetic stiffness. Proper range in the target (30-40 mg/cm^2) requires the
plasma opening switch (POS) and a magnetically stiff ion beam. In order to increase
the power density for implosion target experiments while maintaining an adequately
short ion range in the target, we are developing high voltage lithium ion beams. Our
goal on PBFA II, shown in cross-section in **Figure 1**, is to provide 1 MJ of energy in
a 30 MeV, 15 ns, 50-100 TW/cm^2 lithium ion beam to an ICF target for studying
implosion hydrodynamics and investigating ignition scaling.

---

*Major collaborators in this research include Cornell University and the Naval
Research Laboratory. Funding has been provided by the U. S. Department of Energy
under contract DE-AC04-76-DP00789.

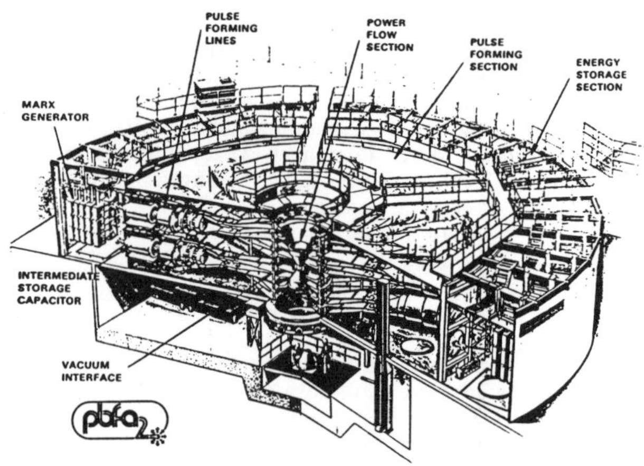

Figure 1. Cross-sectional drawing of PBFA II.

The key advantage in increasing the ion mass from protons to lithium is that the optimum range in the target is obtained at higher ion energy and, consequently, increased beam stiffness. The ion energy which provides the optimum range in the target is given approximately by

$$E_{opt} = 4\ Z(A)^{1/2}$$

where E_{opt} is in MV, and Z and A are the atomic number and atomic mass of the beam ions striking the target. High ion source species purity can also be obtained with lithium in principle, since singly ionized lithium has a helium-like closed electronic shell configuration. This results in a large difference between the first ionization potential (5.4 eV) and the second ionization potential (75.6 eV). In going from protons to lithium on PBFA II, at constant ion beam divergence, we should be able to increase the ion beam focal power density by a factor of about nine to approximately 50 TW/cm^2, by increasing the ion source purity (from 50% for protons from a hydrocarbon source to 90% for lithium), and by increasing the voltage using a plasma opening switch. We expect to be able to maintain a shot rate of one/day at the full energy level of PBFA II.

PHYSICS OF APPLIED-B DIODES

In order to meet the requirements for driving ICF targets, ion beam generation must be highly efficient, and the resulting ion beam must be highly focusable.[2] In Applied-B ion diodes, the efficiency of beam generation is sensitively dependent on the strength and uniformity of the magnetic field, on having an ion source capable of providing high current densities, and on the diode having a reasonably constant impedance. Factors which determine the power brightness of ion diodes include ion source spatial variations, the magnitude of the applied voltage, the uniformity and magnitude of magnetic insulation, the ion mass, and the charge-state purity of the ion source. Factors which determine the limits to ion beam focusing upon a target over the duration of the accelerator power pulse include the source power brightness, anode plasma motion, time-dependent magnetic bending of ions in the applied and self-magnetic fields due to the changing diode impedance, electromagnetic fluctuations in the acceleration gap, and beam emittance growth due to scattering and instabilities in the beam transport region. These elements of importance to diode behavior are

operation: (1) prior to arrival of the power pulse, (2) the early part of the power pulse, (3) near peak power, and (4) after peak power.

Prior to arrival of the power pulse, the magnitude and shape of the applied magnetic field are most important. The magnitude determines the critical insulation voltage, V_{crit}, the level of voltage beyond which most electrons are lost to the anode. The shape of the magnetic field determines both the origin of electron flow into the diode and the loss region for electron flow out from the diode. **Figure 2** depicts a typical shape of the insulating magnetic field used in the PBFA II diode. Since magnetic flux is provided by coils in both the anode and the cathode, there is a magnetic separatrix. The radial location of the separatrix can be chosen to be on the anode for optimal focusing of protons, or near the cathode tips for optimal focusing of lithium ions which are accelerated as Li^+, but are stripped to Li^{+3} at the gas cell boundary.[3]

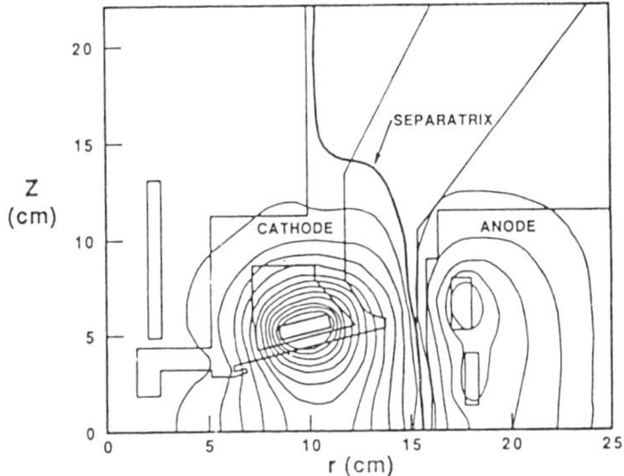

Figure 2. Configuration of the applied magnetic field in the PBFA II ion diode.

In the early part of the power pulse, electron flow into the diode is important in determining diode impedance. As the voltage pulse builds at the diode, electrons begin to move into the AK gap. These electrons can come either from the power feeds or from the cathode tips. With Gauss' law applied to the AK gap[4], the total charge in the gap $Q = Q_i + Q_e$ (ions and electrons) must remain zero for space-charge-limited emission surfaces. There is, therefore, a "bootstrap" behavior in the gap, with virtual cathode formation resulting in, and dependent upon, the emission of ions. An analytic model of diode behavior based upon this principle[5] predicts many of the features seen in experiments. Ion emission delay, i.e., the time from arrival of voltage at the diode to the time of significant ion emission, consists of two parts. The first is the time required to generate an ion source, and the second is the time required to "fill" the AK gap with electrons which create the virtual cathode. Flashover ion sources require deposition of low-energy electrons (200-400 J/gm) to create the anode plasma[6], before they become space-charge-limited ion emitters. Substantial research efforts are aimed at development of an active plasma ion source, which is fully-ionized before the voltage pulse arrives at the diode[7,8], or rapid formation of field-enhanced ion emission sites produced by an electrohydrodynamic (EHD) instability[9] in order to reduce this component of ion "turn-on" time. The second component of the ion emission delay is due to the formation time of the virtual cathode. Simulations using the MAGIC code have shown that this delay time depends on the shape of the applied magnetic field and can be reduced by using a low-density plasma fill in the diode.

Near peak power, the electron sheath moves closer to the anode. An analytic theory of ion diodes, based upon a self-consistent calculation of the diamagnetic effect on virtual cathode location, coupled with the assumption of uniform electron density in the sheath[10,11] has been successful in predicting the operating point for ion diodes; i.e., voltage and current, near peak power for many different experiments. The agreement between experiment and theory using this model is shown in **Figure 3** for diode experiments on the Proto I, Proto II, PBFA I, and PBFA II accelerators.

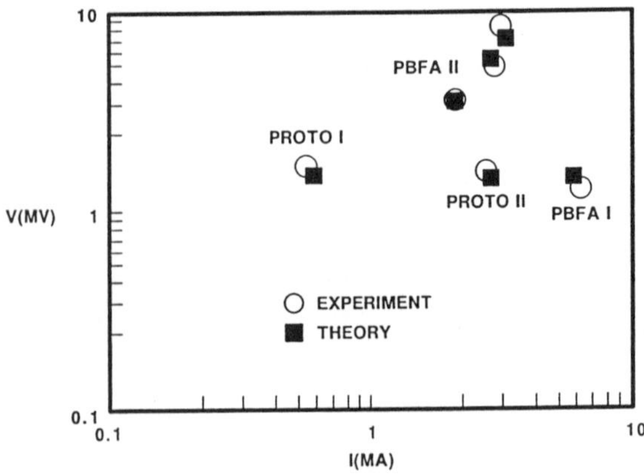

Figure 3. Agreement between the analytic diode theory and experiments on Proto I, Proto II, PBFA I, and PBFA II.

After peak power, penetration of magnetic field into the anode plasma appears to be a predominant cause of falling diode impedance. The large skin current driven in the anode plasma during compression of the magnetic field by the electron sheath can be the source of instabilities in the anode plasma. For the most likely instabilities, including the Buneman, ion acoustic, and lower hybrid drift, the effective anomalous collision frequency ranges from a few tenths to a few times the ion plasma frequency, allowing a more rapid penetration by the magnetic field than Spitzer resistivity would allow. The consequence of this penetration is to decrease the magnetic flux between the ion-emitting surface and the virtual cathode thereby decreasing the critical insulation voltage. Since the diode voltage is limited by the flux between the ion emission surface and the virtual cathode, this flux penetration will cause a corresponding decrease in the diode voltage.

RESULTING ION BEAM PROPERTIES

Measurements of ion divergence at the target location on axis in PBFA II have been made using an energy-resolved ion pinhole camera[12]. During testing at the one-quarter energy level of PBFA II, the best ion divergence at the target was measured to be 19 milliradians at 3.3 MV. In testing at the one-half energy level[13], ion divergence varied linearly as the size of the dynamic acceleration gap for voltages up to about 6 MV and for insulation magnetic field changes of less than 30% over the range of diode operating voltages from 4-6 MV. The dynamic acceleration gap is the distance between the virtual cathode and the ion-emission surface near the anode. For a given ion species, the dynamic gap, g, depends on V_{crit}, the mechanical gap d, the anode area, and the accelerator load time. It is given by $g = d(5.55 \, J_{CL}/J_i)^{1/2}$, where J_{CL} is the calculated Child-Langmuir current density based upon the geometric gap distance, d. The scaling of focal spot full-width at half-maximum intensity (FWHM) with the calculated dynamic gap is shown in **Figure 4**. The linear scaling of horizontal divergence with dynamic gap suggests the presence of instabilities in the electron sheath.

A number of investigators have assessed the physical models and the conditions for development of instabilities in the electron sheath, both without ions[14] and in the presence of ions[15]. The most serious of these appear to be the diochotron, the electron-ion two stream, and the ion transit-time instabilities. Experimental tests and 3-D PIC simulations are now in progress to search for and differentiate between these instabilities in intense ion diodes. Understanding this effect is therefore of fundamental importance to improving ion beam focusability.

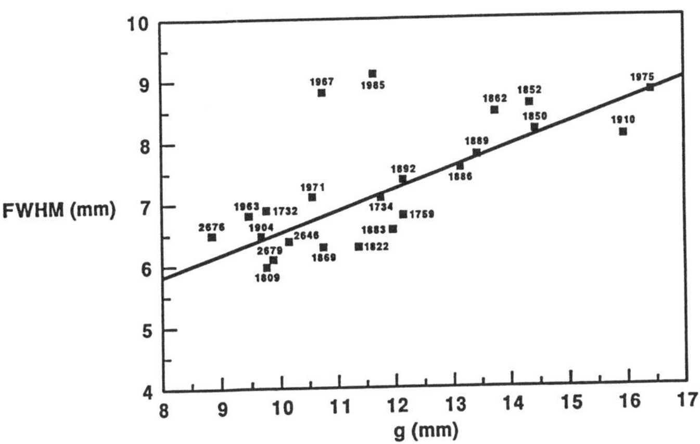

Figure 4. Scaling of focal spot size with dynamic gap.

In experiments at the one-half energy level of PBFA II using an integrated diagnostic package[16] for resolution of ion beam properties during transport and focusing, an energy spread of about 15% FWHM has been observed in the ion beam on axis[13]. A similar energy spread was seen earlier in 1984 in the Proto-I Applied-B experiment[17]. Whether this spread is due to charge-exchange or interaction of the ion beam with instabilities in the electron sheath can be resolved by accurate measurement of the center of the energy spread. If the energy spread lies entirely below the accelerating potential, instabilities are unlikely, but if the energy spread is centered on the accelerating potential, instabilities are the most probable explanation.

TARGET EXPERIMENTS

Three series of target experiments have been completed on PBFA II so far. The first series used large (15-20 mm midplane diameter) range-thick cones for development of target diagnostics. The second series used smaller cones (7.5-10 mm midplane diameter) for measurement of azimuthal beam uniformity and correlation of diagnostic data obtained at large radius with that obtained on the diode axis. The third series began the study of intense ion beam deposition in a low-density foam seeded with chlorine for x-ray imaging.

The configuration of the target experiments is shown in **Figure 5**. The ion beam, created at the 15 cm anode radius, is accelerated through the 1-2 cm anode-cathode gap, and passes through the 2 micron mylar gas cell window and into a few-torr Ar-filled propagation region. The beam is focused in the vertical direction from its initial 5-10 cm height to its 5-6 mm height at the target by bending in the self-magnetic field, bending in the applied magnetic field, and vertical shaping of the

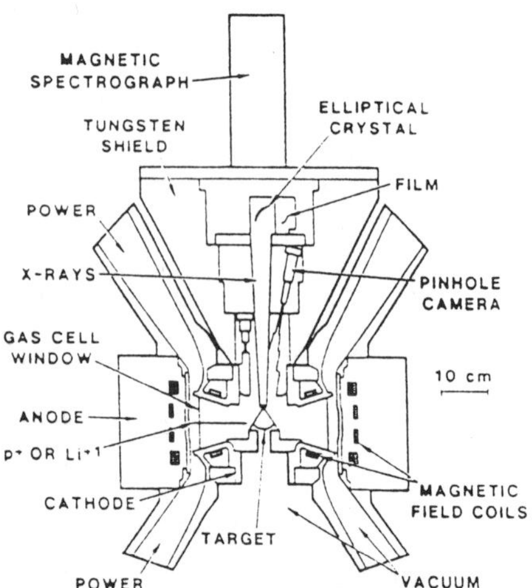

Figure 5. Configuration for target experiments on PBFA II showing upper diagnostics, upper shield, and diode.

anode profile. Power feeds to the diode are symmetric about the diode midplane. Tungsten shields on the top and bottom of the diode attenuate bremsstrahlung by a factor of about 2500 for diagnostics contained in within the shield cone. The upper diagnostics include multi-position x-ray pinhole cameras for observing inner shell excitation (e.g., K-, L-, or M-alpha) radiation from the target, an elliptic crystal spectrometer for obtaining x-ray spectra from the target, and a one-dimensional slit imaging magnetic spectrograph for obtaining Rutherford-scattered ion images, ion momenta, and ion power densities. A lower package includes both distant (3 m) and close-in diagnostics. The diagnostics mounted at about 3 meters from the target include on-axis x-ray pinhole cameras, a grazing incidence spectrometer, and a multichannel x-ray diode (XRD) detector array for obtaining energy cuts of the target x-ray spectrum. The close-in diagnostics include four more x-ray pinhole cameras, a streaked x-ray imaging camera, and a convex crystal spectrometer. The target configuration with a small titanium cone target mounted in the gas cell is shown in **Figure 6.**

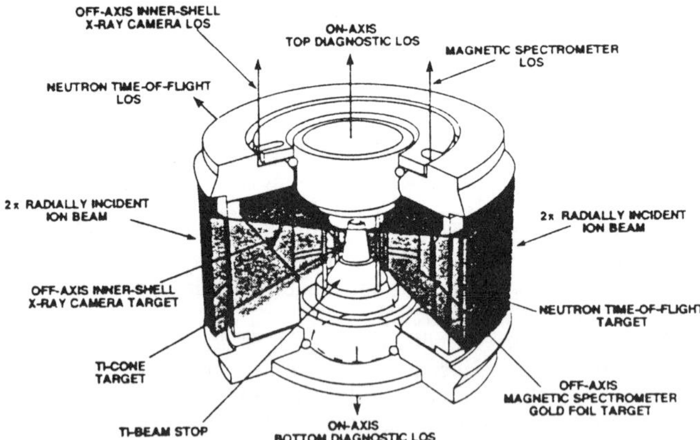

Figure 6. Configuration for target experiments showing small cone mounted in the gas cell.

In the first series of experiments, large conical targets (15-20 mm midplane diameter) were used to develop and characterize target diagnostics. The cones had a 30° half-angle, and were about 40-mm tall. Materials used for different shots included teflon (CF_2), aluminum, titanium, tantalum, and gold. The beam used for the experiments was a proton beam with a power intensity of about 1 TW/cm². An example of data obtained with a teflon target, shot in order to develop a grazing incidence spectrometer, is given in **Figure 7**. Most of the spectral lines have been identified, but some uncertainties remain. A collisional-radiative-equilibrium (CRE) code calculation[18] for the line ratios for carbon provided an estimate of electron temperature in the range of several tens of eV (50-75 eV) for the feature being observed by the spectrometer. From observing shots with titanium cones, we believe the spectrometer was viewing a heated stagnation mass on the axis of the cone.

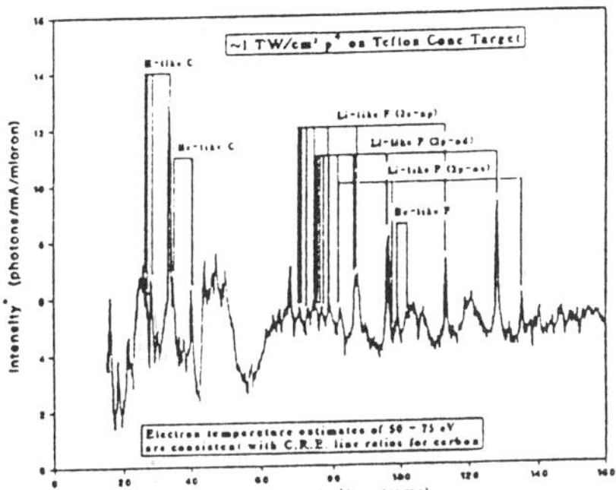

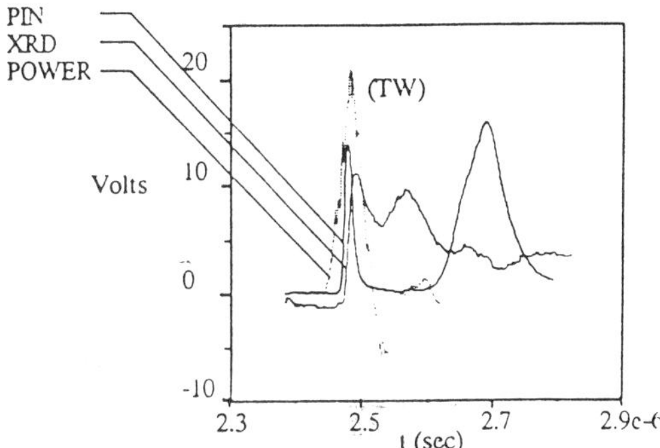

Figure 7. Grazing incidence spectrometer data obtained on a teflon cone target shot in the first target series.

Figure 8. Ion power, PIN diode, and XRD traces from a small titanium cone shot.

Clear x-ray pinhole camera images of the smaller titanium cones shot during the second series (with proton beam intensities of about 2-3 TW/cm²) were obtained. The hard (>600 eV) x-rays are produced by beam-induced inner shell (K-alpha) excitation of the target. The nonuniform brightness of the emission provides information on vertical and azimuthal variations in beam intensity. Images in the 100-300 eV band show stagnation of the cone material on the axis, expected to occur to some degree since the target thickness is about one-half of the ion range. The size of this feature may also be indicative of the beam azimuthal uniformity. The very soft x-rays (20-80 eV) show more symmetry than either of the harder x-ray images. Analysis of the PIN diode detector and XRD traces shown in **Figure 8** provides corroboration for the images. As the ion power pulse inferred from diode electrical monitors rises, the PIN trace shows the production of K-alpha radiation from the beam-target interaction. Direct heating of the target by the beam is seen in the XRD trace. A stagnation occurs roughly 100 ns after the power pulse, indicating inward movement of target material at a velocity of about 5 cm/microsecond. Late in time, ions which were Rutherford scattered by the cone arrive at the PIN detector. Spectroscopic data from one of the aluminum cone shots is shown in **Figure 9**. These data show the first observation of K-alpha satellites produced by an ion beam heated target[18]. As electrons are removed from the outer shells of the aluminum ions in the target, the energy levels of the K-alpha radiation, generated as electrons fill the innershell vacancy, is changed. The data indicate that boron-like aluminum (eight times ionized) was produced. CRE modeling indicates that this is consistent with electron temperatures from 20-60 eV.

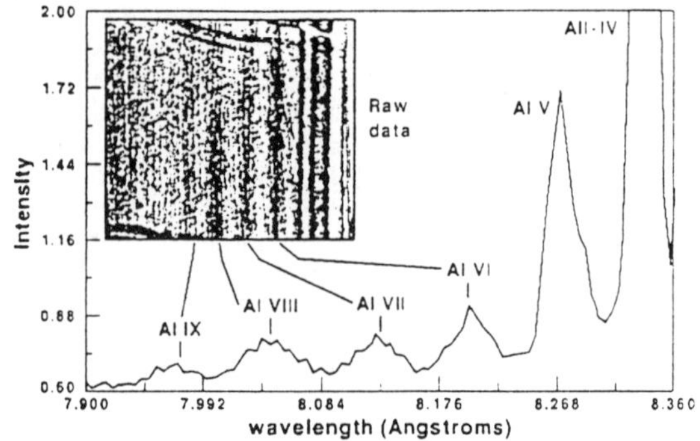

Figure 9. Spectroscopic data showing K-alpha satellites from a small aluminum cone target shot.

In the third series of experiments, more complex configurations were tested. The progression in this latest series is shown in **Figure 10**. In order to prepare for targets which preclude the use of a Rutherford scattering foil on axis, planar targets were used to obtain a correlation of on-axis data with off-axis data. The planar target provided the on-axis data via Rutherford scattering, while sectors of a conical target with 3.8 cm radius K-alpha target provided the off-axis data. A key element in obtaining the correlation was an ion propagation code, PICDIAG,[19] which provided information on ion orbits from the 3.8 cm radius position to the axis. Later, the "apron" targets were tested to provide ion pinhole and K-alpha images of the ion beam to get near-axis beam uniformity and intensity. The apron targets are thin-walled targets meant to give the "footprint" of the ion beam as it moves inward, with little effect on the beam itself. This ability to directly measure the beam footprint as it propagates to the central target is an important advantage of the "in-depth" nature of ion beam energy deposition in matter. Data from a recent apron target shot indicate proton beam azimuthal non-uniformity which is smaller than 20%. Three of the target experiments used a central cylinder of lithium in order to get total proton energy striking the lithium target from the p(Li, Be)n nuclear reaction. In this configuration, neutron activation of a praseodymium sample provides a measurement of total proton energy. On two out of the three tests, good beam quality was achieved, and total proton energy striking an 8-mm-diameter, 8-mm-tall lithium cylinder was measured to be 180-200 kJ. In the last configuration shown in **Figure 10**, a doped foam is used to measure the ion deposition profile. Image data from the only shot of this type, using a 15 mg/cm^3 chlorine dopant uniformly dispersed throughout a 40 mg/cm^3 carbon foam, show rather uniform deposition of the ion energy throughout the foam. Since the foam target was 10-mm in diameter, its diameter is equivalent to the ion range. Since this experiment provided good data, we have planned follow-on tests with a higher density foam, where the radius will be equal to the ion range, to resolve any nonuniformities in the deposition more clearly.

Future target experiments will concentrate upon diagnosis and improvement of ion beam deposition uniformity, measurement of the uniformity of the pressure pulse for driving implosion targets, and the hydrodynamics of target implosions. These target experiments will be more interesting if we are successful in developing intense lithium beams. Our strategy is to increase the energy we can deliver to a lithium beam by increasing the beam purity and increasing the diode voltage. Although an increase in the ion current density or a reduction in the ion divergence would each improve the ion power brightness at the anode, we have not assumed that the current density would exceed 5 kA/cm^2 or that the ion divergence would be smaller than 14 mrad. Since power brightness is given by[20]

$$P_B = (J)(V)(f_i)(f_\theta)/(\theta)^2,$$

where

J is the ion current density at the anode,
V is the diode voltage,
f_i is the fraction of ions in the focusable charge state, and
f_θ is the fraction of ions within the full width at half maximum
(= 0.5 for a gaussian beam),

we can increase the ion beam focal power intensity by about a factor of 9 from the 5 TW/cm^2 level we have achieved using protons if we can go from the present 50% proton purity with hydrocarbon sources to a 90% pure lithium beam, and increase the ion energy from 6 MV to 30 MV.

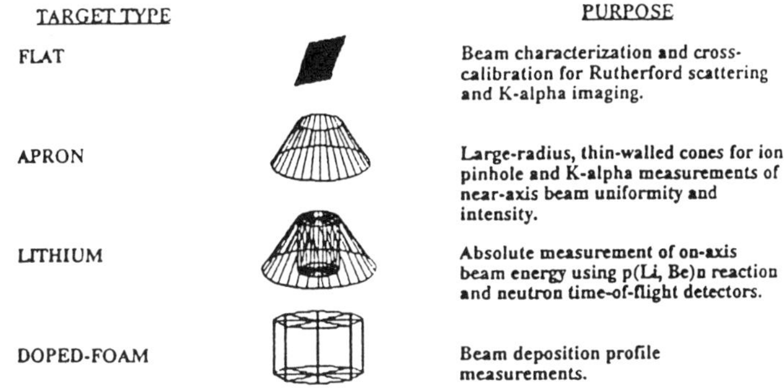

Figure 10. Progression of target configurations for the third series of target experiments on PBFA II.

LITHIUM SOURCE EFFORTS

Our present efforts to develop a lithium ion source include three approaches: (1) LEVIS,[7] a laser-heated, laser-ionized source; (2) EHD[9], an electrohydrodynamic liquid lithium ion source; and (3) BOLVAPS/LIBORS[8], an electrically heated, laser-ionized source. At present, we are concentrating most of our time on PBFA II on the LEVIS source, with the EHD source pursued as a backup. We believe we can increase the lithium fraction of the beam energy by either providing a clean liquid lithium surface, such as is done in the EHD fast-wicking liquid lithium source, or by removing hydrogen from the surface of a lithium-containing thin film. Optimization of the lithium source behavior for target experiments is the highest priority in our program.

SUMMARY

The progress in understanding the physics of intense light ion beam diodes has been impressive, but much remains unknown. The development and application of the analytic theory for Applied-B diodes has provided real insight into the behavior of these diodes. The shot rate now being obtained on PBFA II (1/day) and the application of the best available diagnostics to diode experiments is helping to accelerate the progress. The causes for the observed scaling of ion divergence with dynamic acceleration gap, and the energy spread of the ion beam have yet to be fully explained. The divergence increase with dynamic gap could be due to poorer uniformity of flashover ion sources with larger physical AK gaps. The development of uniform, large-area non-flashover ion sources is a high priority. In addition, if the hypothesis of magnetic field diffusion into a resistive anode plasma as the source of impedance collapse is correct, the development of liquid metal EHD ion sources with highly-conductive anode surfaces could reduce the impedance drop. There are several possible explanations for the divergence scaling with dynamic gap, but one of the most likely ones is an instability in the electron sheath coupling to the ion beam. Such an instability could grow to saturation very rapidly and could produce both the ion beam energy spread observed experimentally and a linear scaling of ion divergence with ion transit time through the sheath. Development of a satisfactory theory for ion divergence and its scaling with diode parameters is a high priority. Use of the recently completed QUICKSILVER[21], a 3D EM PIC code which can simulate the correct geometry, will be an important part of this effort. Finally, experimental tests which can differentiate between the various proposed sources of energy spread and ion divergence are needed. Further refinement of intense beam diagnostics to make more of them time resolved and useful at multiple radii within the beam transport region is needed. Broader application of spectroscopic diagnostics techniques[22] to exploration of the physics of AK gaps in intense ion-beam diodes is proving to be especially fruitful.

Advances in ion beam theory, diagnostics, and experiments in the past two years have produced a 5 TW/cm^2 proton beam. Target experiments using this beam have produced data on total proton energy, proton beam uniformity, and beam-target interaction physics. Experiments are being planned to measure ion beam deposition in heated material quantitatively. With success in lithium beam production, we will be able to begin implosion experiments driven by intense lithium beams, and demonstrate the utility of light ion beams for ICF.

REFERENCES

1. D. J. Johnson, T. R. Lockner, R. J. Leeper, J. E. Maenchen, C. W. Mendel, G. E. Rochau, W. A. Stygar, R. S. Coats, M. P. Desjarlais, R. P. Kensek, T. A. Mehlhorn, W. E. Nelson, S. E. Rosenthal, J. P. Quintenz, and R. W. Stinnett, Proc. 7th IEEE Pulsed Power Conference, Monterey, CA, June 11-14, 1989, IEEE Cat. No. 89CH2678-2.

2. J. P. VanDevender and D. L. Cook, Science, 232, 831 (1986).

3. P. L. Dreike and P. A. Miller, J. Appl. Phys. 57, 1589 (1985).

4. S. A. Goldstein and R. Lee, Phys. Rev. Lett. 35, 1079 (1975).

5. P. A. Miller and C. W. Mendel, Jr., J. Appl. Phys. 61, 529 (1987).

6. D. J. Johnson, J. P. Quintenz, and M. A. Sweeney, J. Appl. Phys. 57, 794 (1985).

7. G. C. Tisone, K. W. Bieg, and P. L. Dreike, Rev. Sci. Instrum. 61, 562 (1990).

8. P. L. Dreike, and G. C. Tisone, J. Appl. Phys. 59, 371 (1986).

9. A. L. Pregenzer, J. Appl. Phys. 58, 4509 (1985).

10. M. P. Desjarlais, Phys. Rev. Lett. 59, 2295 (1987).

11. M. P. Desjarlais, Phys. Fluids B 1, 1709 (1989).

12.   W. A. Stygar, R. J. Leeper, L. P. Mix, E. R. Brock, J. E. Bailey, D. E. Hebron, D. J. Johnson, T. R. Lockner, J. E. Maenchen, and P. Reyes, Rev. Sci. Instr., $\underline{59}$, 1703 (1988).

13.   T. R. Lockner, D. J. Johnson, R. J. Leeper, J. E. Maenchen, C. L. Ruiz, W. A. Stygar, T. A. Mehlhorn, J. E. Bailey, S. E. Rosenthal, R. S. Coats, J. P. Quintenz, M. P. Desjarlais, and R. P. Kensek, "Ion Focusing Experiments on PBFA II," Proc. 8th Int'l. Conf. on High Power Electron and Ion Beams, Karlsruhe, FRG, July 4-7, 1988.

14.   R. C. Davidson, K. T. Tsang, and J. A. Swegle, Phys. Fluids $\underline{27}$, 2332 (1984); H. S. UHM and R. C. Davidson, Phys. Rev. $\underline{A31}$, 2556 (1985).

15.   E. Ott, T. M. Antonsen, Jr., C. L. Chang, and A. T. Drobot, Phys. Fluids $\underline{28}$, 1948 (1985); R. C. Davidson, K. T. Tsang, and H. S. Uhm, Phys. Rev. $\underline{A32}$, 1044 (1985).

16.   R. J. Leeper, W. A. Stygar, J. E. Maenchen, C. L. Ruiz, R. P. Kensek, D. J. Johnson, T. R. Lockner, J. E. Bailey, G. W. Cooper, J. E. Lee, T. A. Mehlhorn, L. P. Mix, and R. W. Stinnett, Rev. Sci. Instr., $\underline{59}$, 1860 (1988).

17.   D. J. Johnson, R. J. Leeper, W. A. Stygar, R. S. Coats, T. A. Mehlhorn, J. P. Quintenz, S. A. Slutz, and M. A. Sweeney, J. Appl. Phys. $\underline{58}$, 12 (1985).

18.   J. E. Bailey, A. L. Carlson, G. Chandler, M. S. Derzon, R. J. Dukart, B. A. Hammel, D. J. Johnson, T. R. Lockner, J. E. Maenchen, E. J. McGuire, T. A. Mehlhorn, W. E. Nelson, L. E. Ruggles, W. A. Stygar, and D. F. Wenger, Proc. 5th Int'l. Workshop on Atomic Physics for Ion Driven Fusion, Schliersee, FRG, January 29 - February 2, 1990.

19.   T. A. Mehlhorn, W. E. Nelson, J. E. Maenchen, W. A. Stygar, C. L. Ruiz, T. R. Lockner, and D. J. Johnson, Rev. Sci. Instrum. $\underline{59}$, 1709 (1988).

20.   J. P. VanDevender, et al., Lasers and Particle Beams, $\underline{3}$, 93 (1985).

21.   D. B. Seidel, M. L. Kiefer, R. S. Coats, A. L. Siegel, and J. P. Quintenz, Proc. 12th Conf. on the Numerical Simulation of Plasmas, paper PT24, Sept. 20-23, 1987, San Francisco, CA.

22.   Y. Maron, M. D. Coleman, D. A. Hammer, and H. S. Peng, Phys. Rev. A $\underline{36}$, 2818 (1987).

Impedance Characteristics of Terawatt Ion Diodes

C. W. Mendel, Jr., M. P. Desjarlais, T. D. Pointon,
J. P. Quintenz, S. E. Rosenthal, D. B. Seidel, S. A. Slutz
Sandia National Laboratories,
Albuquerque, NM 87185

ABSTRACT

Light ion fusion research has developed ion diodes that have
unique properties when compared to other ion diodes. These diodes
involve relativistic electrons, ion beam stagnation pressures that
compress the magnetic field to the order of 10 Tesla, and large
space–charge and particle current effects throughout the
accelerating region. These diodes have required new theories and
models to account for effects that previously were unimportant.
One of the most important effects of the magnetic field compression
and large space–charge has been impedance collapse. The impedance
collapse can lead to poor energy transfer efficiency, beam
debunching, and rapid change of the beam focus. This paper
discusses our current understanding of these effects, some of the
methods we are using to ameliorate them, and the future directions
our theory and modeling will take.

INTRODUCTION

Ion diodes used at the terawatt level[1–14] are different from
other ion diodes for several reasons. Because all negative
electrodes are space–charge–limited electron emitters at the >5
MV/cm fields in the diodes, a strong magnetic field is imposed
transverse to the electric field to insulate the electrons from the
anode. The most significant differences between these diodes and
the more common low powered diodes are that collective effects are
more important, and that the ion beam reaction causes a large
compression of the diode–insulating magnetic field.[15] These
effects result in non–linear behavior that requires new and more
complete theories that include relativistic effects for electrons
and self–consistent treatment of electric and magnetic fields.[16–21]
The regions considered here are not quasi–neutral; moreover, finite
orbit effects require finite orbit theories where guiding center
models are of limited use because the electron orbits may be
comparable in size to the field gradient scale length.

Nevertheless, there are some simplifying assumptions that can
be justified. The lifetime of ions and the gyroperiods of
electrons in these systems are so short that pressure balance
across the flow can be assumed.[15,18–21] Moreover, the spatial
variations of fields in the direction parallel to the anode surface
are slow compared to gyrolengths, and temporal variations are slow
compared to gyroperiods. The dynamics of electrons, therefore can
be treated using a conserved action variable.

The subject treated in this paper is the collapse of the diode
impedance. This impedance collapse is complicated because it
involves several mechanisms causing two different effects. There

are effects that cause very non-linear impedance characteristics, and effects that cause an irreversible collapse of impedance. We will discuss our current understanding of the problem, the theories and models used to further that understanding, and some strategies we are pursuing experimentally.

Impedance collapse is very common in pulsed power devices; the only exceptions have been devices with an initial plasma fill. A plasma-filled diode[14,21,22] has many advantages, but the allowable density variation in the plasma fill is severally constrained. For this and other reasons they have not been used in light ion inertially confined fusion (LIICF) systems. Another plasma-filled device, the plasma opening switch[23-25] (POS), has been used extensively.

The falling impedance of the applied-B diode used in LIICF is problematic. The impedance can fall so rapidly with time that the pulser cannot deliver much of its electromagnetic energy to the diode. Additionally, when the voltage on the diode falls with time, the ion beam is debunched during its transit from the diode to the target. Perhaps most damaging of all, the beam focus changes rapidly through the power pulse. The latter effect causes the power pulse on target to be much shorter than the diode pulse because, much of the beam misses the target. Other more subtle, (but usually less damaging) effects can also be attributed to the falling impedance.

Early electron diodes showed falling impedance characteristics that were generally attributed to plasmas driven from the anode that reduce the anode-cathode gap with velocities of a few cm/μs. This was called "closure," and the interpretation is still believed to be correct. Holographic and Schlieren data confirmed the existence and expansion of these plasmas. Development of the Applied-B ion diode was begun in the mid 1970's,[3] and these diodes also were found to have falling impedance which could reasonably be attributed to closure. A series of Applied-B diodes was tested at Sandia National Laboratories, particularly by Johnson,[4-6,9] progressing from Proto-I[5] (1.8 MV, 0.6 MA), to Proto-II[6] (2 MV, 2 MA), to PBFA-I[7] (2 MV, 4 MA), and finally to PBFA-II[9] (6-9 MV, 3 MA). Miller,[26] who studied the set of all of these experiments together, observed that the same diode seemed to be able to work on all of these machines with only small changes in the diode setup parameters. The only significant change was to increase the insulating magnetic field (the applied B) in proportion to the operating voltage. The diode impedance always seemed to fall to a value determined by the power source. Closure was assumed at first, but it became clear that unrealistically high closure velocities of several tens of cm/μS were required. In addition, he noticed that there seemed to be a critical voltage where the diode impedance fell to zero,[26] i.e. the current diverged. Early theories and models[16-17] were usually related to the operating impedance, but did not explain falling impedance.

Miller proposed a theoretical model,[18] having to do with electron charge flowing into the diode, to explain the impedance collapse which allowed the diode to operate on pulsers with highly disparate currents. Goldstein and Lee[27] had pointed out that since the electric field normal to all boundaries of the diode was zero due to either space—charge—limited electron or ion emission, the net charge in the diode was zero, and the ratio of ion current to electron current was therefore equal to the ratio of electron to ion lifetimes in the diode. Miller reasoned that electrons emitted in the cathode region, and drifting into the diode due to E×B drift would not only raise the electron charge, but also the ion charge and therefore the ion current.[18] This model is still believed to partially explain the impedance collapse, but not the existence of a critical voltage. Desjarlais[19,20] was able to explain this in terms of the compression of the magnetic field near the anode by the momentum of the ion beam. Miller's model involved an irreversible collapse, whereas Desjarlais' model involves a very non—linear impedance character.

Theories of these devices including electrons with finite orbits show that electron pressure can be neglected even when orbit size cannot. In the model of Desjarlais, electron pressure was neglected and the electron distribution in the diode was assumed. Using that model to study data indicates that the distribution varies in time. The distribution also varies in space, but this is expected, since magnetic flux is compressed more and more near the anode with time. However, data also indicate that electrons move across magnetic flux surfaces with time. This has implications about momentum and energy transfer out of the cloud which are not yet understood. This effect is being studied with two— and three—dimensional, time—dependent, particle—in—cell simulations and analytic theory.

To understand the problem of coupling energy to an ion diode, a model of a magnetically insulated transmission line is needed. Because the cathode surfaces of these transmission lines are also space—charge—limited electron emitters, the characteristics of the source as seen by the load are not linear. It can be modeled with a linear source and a non—linear section with a flow impedance,[28] as shown in Fig. 1. The concept of flow impedance is valuable when electrons flow parallel to the electrodes, but also provides qualitatively correct results when electrons are being lost directly across the line. Flow impedance is defined as

$$Z_{f1} = \frac{V}{(I_a^2 - I_c^2)^{1/2}} = \frac{V}{(I_i^2 - I_o^2)^{1/2}}$$

where V is the line voltage, and I_a, I_c are the anode and cathode line currents. The difference between the I_a and I_c is the electron current flowing in the line gap. This can be a major fraction (>50%) of the multi—megampere currents involved when load impedance is comparable to transmission line vacuum impedance, or downstream of a transmission line obstruction. When this electron current reaches the end of the transmission line it cannot flow into an ion diode in most cases. Moreover, the electrons are energetically unable to reach the cathode, so they must be lost to the anode. For this reason the transmission line can be viewed as a lossy line element that has $I_i = I_a$ input and $I_o = I_c$ output currents. Ordinarily the flow impedance of a transmission line is a little below its vacuum impedance. It can never be as large as the vacuum impedance, but in conservatively designed systems it is very close to the vacuum impedance if there are no line obstructions. Line obstructions, e.g. the POS, reduce the flow impedance because they launch electron flows which cannot return to the cathode after they are past the obstructed region. A prescribed Z_{fl} that rises rapidly with time is a good model of a POS. Typically, the flow impedance reaches a little over half the line vacuum impedance. For PBFA-II at full power U = 15 MV at peak, Z_o = 3.2 Ω, and Z_{fl} = 11 Ω. These numbers are for two sides in parallel, there are actually two sides with Z = 5.0 (Ω·side) and Z_{fl} = 22 (Ω·side). When a POS is used, Z_{fl} = 4–8 Ω (two sides in parallel) are observed once the POS has opened.

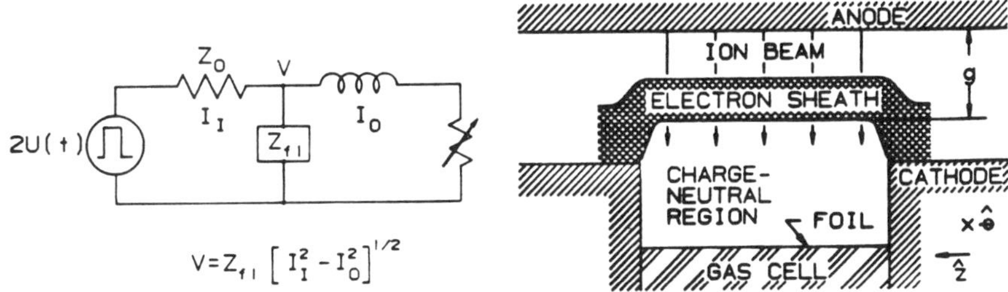

Fig. 1. Model of a pulsed-power source and magnetically insulated transmission feed. Z_o is the impedance of the source, U(t) the voltage of the forward wave.

Fig. 2. Schematic of the Applied-B diode as modeled. The diode is invariant to translations in the direction normal to the figure. Power is fed in symmetrically from both sides of the diode.

PRESENT STATUS OF APPLIED-B DIODE THEORY

Figure 2 shows a simplified schematic of an Applied-B ion diode. The applied-magnetic field (in the axial direction) in the PBFA-II diode is of order 2 T, which is comparable to the self-field (in the azimuthal direction) due to the electromagnetic power flowing into the diode. When the diode current rises, however, the axial component of the magnetic field is compressed to 5 to 15 T at the anode, whereas the azimuthal field remains about 2 T.

Over most of the anode the field gradients are in the ion flow direction, and are normal to the magnetic field. Moreover, the scale lengths of the gradients are small compared to the radius of the anode. For these reasons, a planar pressure balance

relationship accurately reflects the force balance in the diode. This can be written

$$\frac{B_z^2 + B_\theta^2}{2\mu_0} - \frac{\epsilon_0 E^2}{2} + J_i \left[\frac{2m_i(V-\phi)}{q_i}\right]^{1/2} = \text{constant}$$

(1)

$$= \frac{B_{cz}^2 + B_{c\theta}^2}{2\mu_0} + J_i\left[\frac{2m_i V}{q_i}\right]^{1/2} = \frac{B_c^2}{2\mu_0} + J_i\left[\frac{2m_i V}{q_i}\right]^{1/2}$$

where ϕ is the potential as a function of position across the diode. Using Eq. 1 at the anode (i.e. at $\phi=V$, $E=0$)

$$B_{az}^2 - B_{cz}^2 + B_{a\theta}^2 - B_{c\theta}^2 = 2\mu_0 J_i\left[\frac{2m_i V}{q_i}\right]^{1/2} \quad .$$

Electron pressure has been neglected since it is small when compared to field and ion pressure.

Since the ratio of cathode to anode azimuthal magnetic field is the same as the ratio of ion current to total diode current the efficiency of the diode[15] is equal to $B_{c\theta}/B_{a\theta}$. For diode voltages, currents, and dimensions of interest (10 MV, 15 cm radius, and 5 cm anode height) and for reasonable diode efficiency ($\epsilon\approx70$ to 80%), the pressure terms, written as magnetic fields squared, are

$$\underbrace{B_{az}^2 - B_{cz}^2 + B_{c\theta}^2\frac{1-\epsilon^2}{\epsilon^2}}_{\approx 2\ T^2} = \underbrace{2\mu_0 J_i\left[\frac{2m_i V}{q_i}\right]^{1/2}}_{\approx 100\ T^2} \quad .$$

It is clear that pressure balance primarily involves the z component of the magnetic field, and that the theta field contributes little to the pressure in the diode.

Poisson's equation for the diode is

$$\epsilon_0 \phi'' = n_e e - \frac{J_i}{[2q_i(V-\phi)/m_i]^{1/2}} \quad .$$

If the density is written as a function of potential[13] ($n_e = n_e(\phi)$) the above can be multiplied by ϕ' and all terms are integrable. This yields an expression with ϕ' on the left and a function of ϕ on the right. If both sides are divided by the right side the resultant expression is again integrable. The result is

$$\frac{\epsilon_0 E^2}{2} = J_i \left[\frac{2m_i V}{q_i}\right]^{1/2} [(1-\phi/V)^{1/2} - 1 + f(\phi/V)]$$

$$(2)$$

$$J_i = \left|\frac{\chi}{2}\right|^2 \left[\frac{2q_i \epsilon_0^2}{m_i}\right]^{1/2} \frac{V^{3/2}}{g^2}$$

where g is the distance between the virtual cathode and the anode, and χ is defined by

$$f(\eta) = \frac{1}{J_i (2m_i V/q_i)^{1/2}} \int_0^{V\eta} e\, n_e(\phi)d\phi \quad , \quad f(1)=1$$

$$\chi = \int_0^1 \frac{d\eta}{\{[1-\eta]^{1/2} - [1-f(\eta)]\}^{1/2}} \quad .$$

For a thin layer of electrons near the anode f=1 and χ=4/3, and for a uniform electron density f=η and χ=π.

Combining the above with Eqs. 1 and 2

$$E^2 = (\chi V/g)^2 [(1-\phi/V)^{1/2} - 1 + f(\phi/V)]$$

$$B^2 = B_c^2 + (\chi V/cg)^2 f(\phi/V) \qquad (3)$$

$$B_a^2 - B_c^2 = \left[\frac{\chi V}{cg}\right]^2 \quad .$$

Since the magnetic field in the gap is primarily in the z direction, and since the flux in the z direction is conserved because it is trapped between two closed conductors (the anode and cathode), we can approximate

$$B_a g = B_0 g_0$$

where g_0 is the gap between the real cathode and the anode, and B_0 is the initial magnetic field. Combining this with the pressure equation (third of Eqs. 3)

$$g^2 = \frac{(B_0 g_0)^2 - (\chi V/c)^2}{B_c^2} = \chi^2 \frac{V_*^2 - V^2}{B_c^2 c^2} \qquad (4)$$

where $V_* = cB_0 g_0/\chi$ is the limiting voltage at which the current diverges in equilibrium found by Desjarlais.[19,20]   Since the cathode field, B_c, can be no lower than the value of B_θ due to the ion beam current, the equilibrium gap has to go to zero at $V=V_*$.

Although the above expression for the gap involved an approximation for flux conservation, it is easily shown that there is a similar V_* when flux conservation is done in general,[20] and both the z and θ components of the field are retained. Using Eqs. 2 and 3, the conservation of flux can be expressed as

$$B_0 g_0 = \int_0^V d\phi \, \frac{B_z}{E} = \int_0^V d\phi \left[\frac{B_c^2 - B_\theta^2 + (\chi V/cg)^{1/2} f(\phi/V)}{(\chi V/g)^2 \, [(1-\phi/V)^{1/2} - 1 + f(\phi/V)]} \right]^{1/2}$$

$$= \frac{V}{c} \int_0^1 d\eta \left[\frac{f(\eta) + (cg/\chi V)^2 (B_c^2 - B_\theta^2)}{(1-\eta)^{1/2} - 1 + f(\eta)} \right]^{1/2} . \tag{5}$$

In this case the critical voltage is

$$V_* = cB_0 g_0 / \Lambda \tag{6a}$$

where

$$\Lambda = \int_0^1 \frac{f^{1/2}(\eta) \, d\eta}{[(1-\eta)^{1/2} - 1 + f(\eta)]^{1/2}} . \tag{6b}$$

The $B_c^2 - B_\theta^2$ term inside the radical and inside the integrand complicates the calculation since B_θ varies as ϕ varies in general.

If the diode were 100% efficient then $B_c^2 - B_\theta^2 = B_{cz}^2$ is constant. Otherwise the integral depends upon the precise way electrons flow through the diode before going to the anode. If the flux condition above is approximated

$$cB_0 g_0 \approx V \, [\Lambda^2 + (cg/V)^2 \, B_{cz}^2]^{1/2}$$

then

$$g^2 \approx \Lambda^2 \, \frac{V_*^2 - V^2}{B_c^2 \, c^2}$$

with V_* given by Eq. 6a. This expression for g has been satisfactory for interpreting PBFA-II data. For the distributions mentioned above, $\Lambda=4/3$ and 1.68 for $\chi=4/3$ and π, respectively.

Figure 3 shows diode V vs. I characteristic for $\chi= 4/3$ and π. The case of $\chi = 4/3$ is the minimum possible value, corresponding to

a zero thickness layer of electrons at the anode. The $\chi=\pi$ case
corresponds to uniform electron density across the gap. Higher
values of χ can occur. Early in time the diode is well described
by the $\chi=4/3$ characteristic. Also shown on Fig. 3 is the input
current to a section of transmission line of 10 Ω flow impedance
feeding a diode with a $\chi=4/3$ characteristic. This corresponds to
the transmission line impedance of PBFA-II. Notice that early in
the pulse (near the origin) there is a large difference between the
diode current and the input current. This difference is the
electron flow lost in the feed. This is wasted current, and it
holds the voltage down so that electrons are feeding into the diode
more slowly than desired. This problem is much worse with a POS
where the flow impedance might be 5 to 8 Ω.

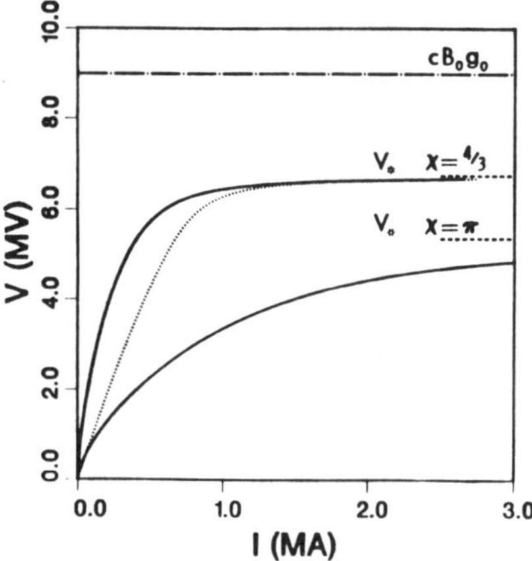

Fig. 3. I,V characteristics of an
ion diode for $\chi=4/3$ and $\chi=\pi$. The
dashed line is the characteristic
seen by the source for $\chi=4/3$ and
$Z_{fl}=10\Omega$. The horizontal
difference between the solid and
the dashed lines is the electron
current lost before the diode.

Fig. 4. Simulation data, for
four diode initial plasma fills,
showing V/I_{ion} vs. voltage. An
electron density of $4\cdot10^{12}$cm^{-3}
results in constant V/I without
reducing peak voltage.

STRATEGIES FOR IMPROVING IMPEDANCE BEHAVIOR

One strategy for improving the impedance character early in
the power pulse is now being implemented. By pre-filling the diode
with a tenuous plasma before power is applied to the diode, the
impedance can be made to start low and rise during the early part
of the power pulse. In addition, while the electric field will
remove the ions of the initial plasma fill in a short time (if the
density is low), the electrons will remain in the diode.
Therefore, by putting the plasma fill in carefully, the diode can
begin with χ near π instead of near 4/3. This will reduce the
initial high impedance, which reflects a great deal of the forward
power pulse, and can damage the transmission lines, particularly

when a POS is used. Figure 4 shows the effect of an initial plasma
fill in simulations done by Rosenthal. The figure shows diode
impedance plotted versus diode voltage. By using the optimum
initial fill density the impedance character can be greatly
improved, thereby reducing electron loss and damage in the
transmission lines, and improving initial diode performance as
discussed above.

Prefilling with electrons to a reasonable density will
eliminate the initial delay in the diode and ion current. Another
way to improve the diode impedance behavior would be to raise the
value of V_* so that the diode voltage would never approach it.

This might be done by raising the initial magnetic field. This can
be done to some degree, but the field coil and capacitor bank
designs have reached a level where further increases in field
energy are becoming difficult and very expensive.

COLLISIONAL EFFECTS IN THE ANODE PLASMA

Because the insulating magnetic field is very high at the
anode, a small layer of flux penetration into the anode is very
damaging[20] to diode impedance and ion beam focusing. The two main
problems are a reduced value of V_*, and a change in the canonical

angular momentum of the ion beam sufficient to cause much of the
beam to miss the target. In present flashover anodes this layer is
believed to be due to ionization of neutrals. The new preformed
anodes soon to be tested should overcome the ionization problem.
In the absence of ionization the remaining concern will be anode
plasma resistance due to turbulence. A small resistive skin depth
of a millimeter or less will allow enough flux to penetrate to

cause problems. Krall[29] has begun to study penetration of highly
ionized plasmas due to the lower hybrid instability. These
problems will not be discussed further here, except to say that
much more can be done.

FUTURE STUDIES:. ELECTRON MOTION ACROSS THE DIODE

It may be possible to increase V_* by changing the way

electrons are distributed in the diode. To accomplish this we need
to understand how electrons move across flux surfaces in these
diodes. At present, models which assume a family of electron
spatial distributions are being used to study the way electrons
flowing into the diode change the value of χ. This model connects
the theories of Miller and Desjarlais. It is desirable to know how
the electron spatial distribution develops.

A basic theory of relativistic electron flows in crossed
electric and magnetic fields has been developed.[30,31] In this theory
the electron distribution is expressed in terms of the electrons
canonical momentum, P, and total energy, W. These are given by
$P=\gamma mv/c - eA$, $W=(\gamma-1)mc^2 - e\phi$. We use y as the dimension in the
electric field (E) direction and z as the magnetic field (B)

direction. The vector potential, A, is therefore in the direction perpendicular to both. Using $v_y^2 = v^2 - v_x^2$ and $\gamma^2 = 1/(1-v^2/c^2)$,

$$\gamma m v_y / c = \pm Q^{1/2}(y,P,W) \quad ,$$

where $Q(y,P,W) = [mc^2 + e\phi(y) + W]^2/c^2 - (mc)^2 - [eA(y) + P]^2 \quad .$

Many features of the flow can be studied by considering the function Q. Any electron with P,W obeying $Q(y_c,P,W)=0$, where $y=y_c$ is the position of the cathode, will have a turning point at the cathode. Likewise those with $Q(y_a,P,W)=0$ will have a turning point at the anode. These equations form hyperbolae in P,W space which partially bound the region where electrons confined to the flow must reside. The final boundary is found by the parametric equations $Q(y,P,W)=0$ and $dQ(y,P,W)/dy=0$, or

$$\frac{P_L}{mc} = - \frac{eA_x(y)}{mc} + \frac{E(y)}{[c^2B^2(y) - E^2(y)]^{1/2}}$$

$$\frac{W_L}{mc^2} = -1 - \frac{e\phi(y)}{mc^2} + \frac{c\,B(y)}{[c^2B^2(y) - E^2(y)]^{1/2}} \quad .$$

The three boundaries in P,W space are shown in Fig. 5. The third boundary is called the laminar line because all orbits of electrons on this line are straight line orbits, and if all electrons in the flow lay on this line the flow is laminar, and vice versa. The laminar line actually consists of two parts: the stable and the unstable laminar lines. The unstable part is the small section that curves upward from the origin into the first quadrant. Orbits on this line are unstable. However, there are stable orbits of long gyro-length in the region near this line. Whether there is an unstable laminar line depends upon the local charge density.

An interesting, and very important action variable, H, can be defined everywhere in the allowed region of P,W space

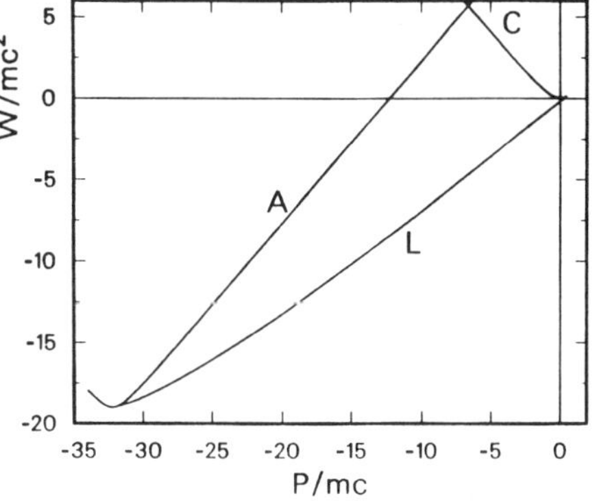

Fig. 5. Diagram of P,W space for the $\chi=\pi$ case. The anode voltage and magnetic flux in the diode determine the anode and cathode lines, A and C. The laminar line, L, is has zero slope at the origin and $(-B_o g_o, -V)$ for ion diodes.

$$H(P,W) = \int_{y_-}^{y_+} Q^{1/2}(y,P,W)\, dy \quad .$$

The limits of the integral, $y_{\pm}$, are two zeros of Q, i.e. $Q(y_{\pm},P,W)=0$, such that Q is positive between them. These are the turning points of electrons with P,W. The dynamics of an electron can be calculated using H because it is conserved if fields vary slowly over the time and length of an electron gyration. These are given by

$$\frac{dx}{ds} = -\frac{\partial H}{\partial P}\ , \qquad \frac{dt}{ds} = \frac{\partial H}{\partial W}\ , \qquad \frac{dP}{ds} = \frac{\partial H}{\partial x}\ , \qquad \frac{dW}{ds} = -\frac{\partial H}{\partial t}$$

where s is a 'phase' variable. Although not needed in the theory because of assumed random phase in the charge and current source integrals, the y position is given by $dy/ds=\pm(y_+ - y_-)$.

The assumption of slow variation over a gyration is a good approximation for many of the problems we are considering. Using this theory one can derive integral equations for the eigenfunctions of waves in these flows.[32] The equations are necessarily integral equations in general because of the finite orbits of the electrons which result in action-at-a-distance. It has been shown that waves in these flows are unstable if the density in P,W space is increasing in the direction down along the laminar line. The unstable waves have phase velocity near the drift velocity in the region of the reversed density gradient.

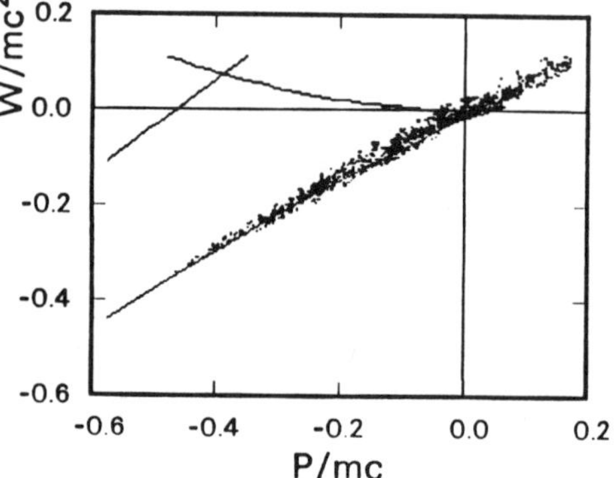

Fig. 6. P,W space for a sim—ulation of a magnetically insulated transmission line, showing the macro—particles. There are no ions in this simulation so the laminar line does not go to the $(-B_o g_o,-V)$, but meets A tangentially.

Constant H lines in P,W space are roughly parallel to the laminar line. Indeed, the laminar line is the H=0 line. Since electrons conserve H they tend to lie along the laminar line. This can be seen in Fig. 6 which is taken from a simulation of a magnetically insulated transmission line (electrons only) of Rosenthal.[33] The actual spread in simulations is higher than is seen experimentally,[34] but the tendency to lie along the laminar line is correct.

An interesting and important question arises. Fluctuations in these flows are believed to be slow. They change P and W, but in

such a way as to conserve H. Since all electric fields are normal
to electrode surfaces, all magnetic fields are parallel to
electrode surfaces in the symmetric direction, and all electrodes
are symmetric in that direction, there can be no torque on the
electron and ion cloud as a whole except by collecting electrons or
ions with non-zero angular momentum. Therefore, for the electron
cloud to have net angular momentum, it either must transfer angular
momentum to the ions, or to other electrons that subsequently leave
the system. In the simulation shown in Fig. 6, which had no ions,
the latter is what happened. In these flows electrons evaporate to
the cathode. Fluctuations transfer net energy and momentum to some
electrons at the expense of others. Those that gain energy and
momentum move up and to the right in P,W space, and hit the
cathode, thus heating and transferring forward momentum to it. The
electrons that lose energy and momentum move down to the left, and
are close to the anode, and cannot all move back to the cathode
without some external energy source.

Considering first the transfer of momentum to the ions, we can
put some limits on this transfer from the experiments. If
electrons travel down the laminar line to reach the anode, they
will have lost momentum $eB_o g_o$ by the time they reach the anode. If
all of this momentum went to the ions they would gain $eB_o g_o (1-\varepsilon)/\varepsilon$
transverse momentum on the average, since the electron loss rate is
$(1-\varepsilon)/\varepsilon$ times the ion loss rate. The maximum deflection angle is
thus

$$\left[\frac{m_e e}{m_i q_i}\right]^{1/2} cB_o g_o \left[\frac{e}{2Vm_e c^2}\right]^{1/2} \frac{(1-\varepsilon)}{\varepsilon}$$

which is of order $[m_e/m_i]^{1/2}$ or 40 mRad for protons. This is
larger than is seen, except for late in the pulse when the voltage
is low. If the particles move across the allowed region of P,W
space due to wave motion for waves moving more slowly than E/B they
might reach the cathode without losing as much momentum as $cB_o g_o$.

For the waves seen in 3-D, time-dependent simulations by Pointon
and Seidel[35] using the Quicksilver[36] code, which move at about c/3,
the ions would need to be deflected by 25 mR, still higher than is
observed. Waves high enough to get electrons to the anode by this
path would need to be very large if H is conserved, but large
amplitude waves are seen in Quicksilver simulations. Particle
motion in assumed wave fields[37] indicate that waves large enough to
give the experimentally observed beam divergence would be large
enough to carry the electrons to the anode through stochastic
effects that may not conserve H. Since the observed beam
deflection is smaller than would be expected from the electron
loss, transfer of momentum to the ions is not the entire source of
drift across the diode gap, but it might be a contributor.

A number of researchers, particularly Antonsen, Ott, Chang,
and Drobot,[38,39] have found instabilities and derived dispersion

relationships for these diodes. These waves involve the transit
time instability modified by the electron space charge. It has
been known for some time that a diode has negative resistance when
the particle transit time in the diode is about n+1/4 cycles of the
oscillation frequency. In two or more dimensions this leads to
instabilities connected with waves normal to the particle motion.
In addition to waves, the Quicksilver simulations show an electron
density distribution that is higher near the anode. It may be that
the wave motion due to the instability described above causes the
electron distribution to be unstable which causes more spreading of
the electron cloud due to electron instabilities. Since the
electron distribution makes a difference in the value of V_*, we

would like to understand how electrons reach a given distribution.
It might be possible to improve the impedance characteristics by
reducing the electron density near the anode. This could be done
either by keeping them from moving so far down the laminar line, or
by driving them into the anode once they are too close to the
anode.
    These effects can be studied using the general theory of
electron flow presented in Ref. 31. In this theory the source
terms for Maxwell's equations are

$$\rho = -\int dP \int dW \; Q^{1/2}(y,P,W) \; \frac{\partial J(P,W)}{\partial W}$$

$$j = \int dP \int dW \; Q^{1/2}(y,P,W) \; \frac{\partial J(P,W)}{\partial P}$$

where the integrals are carried out over the region of P,W space
where Q is positive. In the problems to be studied ρ, j, Q, and J
are also functions of x, and t, but the variation in those
directions are slow compared to gyrolengths and gyroperiods. To
make the solutions tractable the P,W integrals are transformed to
integrals of H and an orthogonal variable, y_0, related to the mean
position of the electron orbits in y. The Jacobian of the
transformation from (P,W) to (H,y_0) is approximated by the Jacobian
on the laminar line which is a function only of y_0. In addition,
by approximating Q by a parabola the theory is drastically
simplified. This theory is valid for electron distributions near
the laminar line. Because the orbit gyro-height $(y_+ - y_-)$, which is
zero on the laminar line, varies as the squareroot of the distance
from the laminar line, gyro-heights can be quite high. Using this
theory it has been found that electron orbits whose height is
comparable to the density scale length give electron pressures of
only one tenth the electro-magnetic pressure for fields comparable
to those in contemporary diodes.
    This theory can be used to get dispersion relationships with
finite orbits, so that the action-at-a-distance involved in such
orbits is retained. Following that, wave-particle and wave-wave
coupling should allow it to predict spreading of the distribution
in much the same way that Drummond and Pines[40] did for plasmas.

CONCLUSIONS

The pressure balance of ion diodes has proven a valuable tool for understanding their operation. It has explained the experimentally observed limiting voltage where the diode impedance goes to zero in equilibrium. Because of the high insulating magnetic fields used in these diodes, increasing the limiting voltage by increasing the field cannot be carried much further. For this reason it is desirable to improve our understanding of electron drift counter to the direction of the ion beam, i.e. toward the anode. The basic theory for doing this is available, and recent simplifications of the theory make the problems tractable by considering a reduced domain of the allowed phase space available to electrons.

REFERENCES

1. P. Dreike, C. Eichenberger, S. Humphries, and R. N. Sudan, J. Appl. Phys. **47**,85(1976).
2. S. Humphries, Jr., R. N. Sudan, and L. Wiley, J. Appl. Phys. **47**,2382(1976).
3. S. Humphries, Jr., Nucl. Fusion **20**,1549(1980).
4. D. J. Johnson, G. W. Kuswa, A. V. Farnsworth, J. P. Quintenz, R. J. Leeper, E. J. T. Burns, and S. Humphries, Jr. Phys. Rev. Lett. **42**,610(1979).
5. D. J. Johnson, E. J. T. Burns, A. V. Farnsworth, R. J. Leeper, J. P. Quintenz, K. W. Bieg, P. L. Dreike, D. L. Fehl, J. R. Freeman, and F. C. Perry, J. Appl. Phys. **53**,4579(1982).
6. D. J. Johnson, P. L. Dreike, S. A. Slutz, R. J. Leeper, E. J. T. Burns, J. R. Freeman, T. A. Mehlhorn, And J. P. Quintenz, J. Appl Phys. **54**,2230(1983).
7. P. L. Dreike, E. J. T. Burns, S. A. Slutz, J. T. Crow, D. J. Johnson, P. R. Johnson, R. J. Leeper, P. A. Miller, L. P. Mix, D. B. Seidel, and D. F. Wenger, J. Appl. Phys. **60**,878(1986).
8. J. N. Olsen, S. E. Rosenthal, L. P. Mix, D. B. Seidel, R. J. Anderson, P. L. Deike, and R. J. Leeper, J. Appl. Phys. **55**,1254(1984).
9. D. J. Johnson, T. R. Lockner, R. J. Leeper, j. E. Maenchen, C. W. Mendel, G. E. Rochau, W. A. Stygar, R. S. Coats, M. P. Desjarlais, R. P. Kenesk, T. A. Mehlhorn, W. E. Nelson, S. E. Rosenthal, J. P. Quintenz, and R. W. Stinnett., Proceedings of the 1989 IEEE Pulsed Power Conference (IEEE New York, 1989).
10. S. A. Goldstein, G. Cooperstein, R. Lee, D. Mosher, and S. Stephanakis, Phys. Rev. lett. **33**,1471(1974).
11. P. A. Miller, P. L. Dreike, J. P. Quintenz, R. J. Anderson, J. T. Crow, C. W. Mendel, Jr., G. S. Mills, L. P. Mix, S. E. Rosenthal, D. B. Seidel, and J. P. VanDevender, Laser and Particle Beams **2**,153(1984).
12. C. W. Mendel, Jr., J. P. Quintenz, D. M. Zagar, P. R. Johnson, R. J. Anderson, and M. M. Widner, J. Appl. Phys. **56**,637(1984).
13. C. W. Mendel, Jr., J. P. Quintenz, L. P. Mix, D. M. Zagar, R. L. Noack, T. Grasser, and J. A. Webb, J. Appl. Phys. **62**,3522(1987).

14. C. W. Mendel, Jr. and G. S. Mills, J. Appl. Phys. **53**,7265(1982).
15. C. W. Mendel, Jr. and J. P. Quintenz, Comments Plasma Physics Controlled Fusion **8**,43(1983).
16. T. M. Antonsen, Jr. and E. Ott, Phys. Fluids **19**,52(1976).
17. K. D. Bergeron, Phys. Fluids **20**,688(1977).
18. P. A. Miller and C. W. Mendel, Jr., J. Appl. Phys. **61**,529(1987).
19. M. P. Desjarlais, Phys. Rev. Lett. **59**,2295(1987).
20. M. P. Desjarlais, Phys. Fluids B **1**,1709(1989).
21. C. W. Mendel, Jr., Phys. Rev. A **27**,3528(1983).
22. J. Greenly, M. Ueda, G. D. Rondeau, and D. A. Hammer, J. Appl. Phys. **63**,1872(1988).
23. C. W. Mendel, Jr. and S. A. Goldstein, J. Appl. Phys. **48**,1006(1977).
24. R. A. Meger, R. J. Commisso, G. Cooperstein, and S. A. Goldstein, Appl. Phys. Lett. **42**,943(1983).
25. Special issue on fast opening switches, IEEE Trans. Plasma Sci., **PS-15**, (1987).
26. P. A. Miller, J. Appl. Phys. **57**,1473(1985).
27. S. A. Goldstein & R. Lee, Phys. Rev. Lett. **35**,1079(1975).
28. C. W. Mendel, Jr., J. P. Quintenz, S. E. Rosenthal, D. B. Seidel, M. P. Desjarlais, R. Coats, and M. E. Savage, Proceedings of the 7th Int'l. Conf. on High-Power Particle Beams, Kernforschungszentrum Karlsruhe Gmbh, Literaturabteilung, Postfach 3640, D-7500 Karlsruhe 1 (1988), W. Bauer and W. Schmidt Editors.
29. N. A. Krall, private communication.
30. C. W. Mendel, Jr., J. Appl. Phys. **50**,3830(1979).
31. C. W. Mendel, Jr., D. B. Seidel, and S. A. Slutz, Phys. Fluids **26**,3635(1983).
32. C. W. Mendel, Jr., J. A. Swegle, and D. B. Seidel, Phys. Rev. A. **32**,1091(1985).
33. S. E. Rosenthal, D. B. Seidel, and C. W. Mendel, Jr., Bul. Am. Phys. Soc. **31**,1398(1986).
34. C. W. Mendel, Jr., J. P. Quintenz, S. E. Rosenthal, D. B. Seidel, R. Coats, and M. E. Savage, IEEE Trans, on Plasma Sci., **17**,797(1989).
35. D. B. Seidel, T. D. Pointon, M. P. Desjarlais, R. S. Coats, M. L. Kiefer, J. P. Quintenz, and W. A. Johnson, Bull. Am. Phys. Soc. **34**,2064(1989).
36. D. B. Seidel, M. L. Kiefer, R. S. Coats, A. L. Siegel, and J. P. Quintenz, Proc. of the 12th Conf. on Numerical Simulation of Plasmas, Meeting of the Topical Group on Computational Physics, Am. Phys. Soc., 1987.
37. S. A. Slutz, Bull. Am. Phys. Soc. **34**,2063(1089).
38. T. M. Antonsen, Jr., W. H. Miner, E. Ott, and A. T. Drobot, Phys. Fluids **27**,1257(1984).
39. E. Ott, T. M. Antonsen, Jr., C. L. Chang, and A. T. Drobot, Phys. Fluids **28**,1948(1985).
40. W. E. Drummond and D. Pines, Nuclear Fusion, 1962 Supplement, Part 3, 1049.

B. Laser-Driven Fusion

Relativistic Plasma Physics with Ultra-High-Brightness Lasers

C.B. Darrow, E.M. Campbell, M.D. Perry
Lawrence Livermore National Laboratory
Livermore, CA 94500

and

C. Joshi, C. Clayton, T. Katsouleas, W.B. Mori, K. Marsh
Department of Electrical Engineering
University of California, Los Angeles

In this paper we consider several nonlinear processes which can be driven when a short-pulse, high-energy laser interacts with an appropriate plasma. Such a laser will be referred to hereafter as a "high-brightness laser" or "HBL". A more precise definition of an HBL will be given below but typically one can think in terms of laser pulses with (1) pulsewidths short compared to the growth times of the usual instabilities associated with laser-plasma interactions (e.g. SRS, SBS) and (2) sufficiently high intensities to drive relativistic electron quiver motions.

The notion of relativistic nonlinear laser-plasma interactions is by no means new and many interesting physical processes have been considered on a theoretical level[1-4]. The notion of relativistic nonlinear laser-plasma interaction experiments *is* relatively new and has been fostered by rather recent advancements in high-bandwidth, high-peak-power laser systems[5]. These advancements have been motivated by the quest for high-brightness lasers for other applications in physics such as recombination x-ray lasing schemes, multiphoton excitations of atomic systems, efficient high-order optical harmonic generation, and novel particle accelerator technologies[6].

"High-brightness" might be defined differently in any one of the laser-applications areas mentioned above. In most cases, however, the ultimate goal is neither the shortest possible pulse, nor the largest pulse energy. Rather, a laser system is designed, within the constraints of available technology and materials, to strike an optimized balance of pulse width and energy which pushes the physics for the particular application into new parameter regimes. In the case of plasma physics, two conditions may be thought of as providing a working definition of an HBL. First, the laser pulse width should be short compared to a plasma oscillation period,

$$(1) \qquad \tau_{laser} \ll \left(\frac{m}{4\pi n e^2} \right)^{\frac{1}{2}}$$

Second, motions of the plasma electrons quivering in the field of the incident laser should be relativistic,

$$(2) \qquad \frac{v_{osc}}{c} = 0.85 \sqrt{I_{18} \, \lambda_{laser}^2} \geq 1$$

These conditions, to some extent, represent conflicting goals since condition (1) is most easily satisfied for shorter laser wavelengths, λ, while condition (2) seeks to maximize λ. However, given that the focal-spot diameter scales as λ under Gaussian optics, condition (2) may be thought of as a power requirement since, with this scaling,

$$(3) \qquad \frac{v_{osc}}{c} \approx \sqrt{\frac{E_{laser}}{\tau_{laser}}}$$

Assuming the laser pulse width, τ_{laser}, has been made as short as necessary, condition (3) strives to maximize the energy extracted in the pulse. This is precisely the area in which efforts have been concentrated and progress has been made in recent years in high-brightness laser development, particularly in the case of glass lasers[7]   . We defer further discussion of laser physics to a later section on the LLNL high-brightness laser system. However, in answer to the question, "how bright are high brightness lasers?", one might characterize typical HBL's as having pulse widths from 0.5 - 1.0 ps, powers of order 10^{13} W or greater, and focal intensities of order 10^{19} W/cm^2 or greater. In terms of the the normalized oscillatory velocity, v_{osc}/c, which for such high intensities is nearly numerically equal to the relativistic factor, γ_{osc}, HBL's attain focal intensities in the range $v/c \approx 1 - 10$.

Non-linear Laser-Plasma Interactions

Not surprisingly, nearly every aspect of this broad topic is considered somewhere in the literature. What can be less certain from an experimental point of view is the connectivity between formalism and experiment. Here we present one possible simplified view of non-linear laser-plasma interactions with the intent of clarifying the niche of HBL's in laser-plasma interaction physics.

Beginning with Maxwell's equations and the equations of conservation of momentum and energy for an electron fluid, one can write a driven wave equation for the electromagnetic vector potential,

$$(4) \qquad \left(\partial_t^2 - c^2\,\nabla^2\right)\vec{A}_\perp = \omega_{po}^2\left(\frac{n}{n_o}\right)\frac{\vec{A}_\perp}{\gamma}$$

where γ is the relativistic factor,

$$(5) \qquad \gamma = \frac{1}{\sqrt{1 - \dfrac{v_{osc}^2}{c^2}}}$$

and ω_{po} is the plasma frequency of the unperturbed electrons,

$$(6) \qquad \omega_{po} = \sqrt{\frac{4\pi n_e e^2}{m_e}}$$

This differential equation is inherently nonlinear since the driving term on the RHS is the product of three components (n, A, and γ^{-1}) which are all at least first order the perturbation amplitude. For small perturbation amplitudes, a.k.a. the linear approximation, the departures of n/n_o and γ from unity are neglected and the usual dispersion relation for electromagnetic wave propagation in a plasma is obtained.

Because γ , given in eq .(5), is at least second order in the perturbation, the next highest order term in the driver neglects relativistic effects and results in a second-order non-linear driver,

$$(7) \qquad \left(\partial_t^2 - c^2\,\nabla^2\right)\vec{A}_\perp = \omega_{po}^2\,\vec{A}_\perp$$

This form of the wave equation governs processes in the plasma which might be characterized as ponderomotive effects, several examples of which Bill Kruer and others have discussed during this conference. The ponderomitive effects are the mainline of laser-plasma interactions in ICF plasmas and in microwave-plasma interactions in ionospheric heating and other smaller-scale laboratory experiments.

Third order contributions to the driver require that both the density-perturbation and the relativistic terms be kept:

$$(8) \qquad \left(\partial_t^2 - c^2 \nabla^2\right) \vec{A}_\perp = \omega_{po}^2 \left(1 + \frac{\delta n(A)}{n_0}\right)\left(1 + \frac{1}{2}\frac{v_{osc}^2}{c^2} + \ldots\right)\vec{A}_\perp$$

This form leads to mixed ponderomotive and relativistic effects such as relativistic self focussing[1,3,4,8,9], photon acceleration, and deceleration[12], and plasma wake generation[2-4]. Much will be said by other speakers concerning the details of these three effects in other talks during this conference. Here we discuss only briefly their significance in experimental relativistic laser-plasma interactions.

When a light pulse from an HBL passes through a plasma, its ponderomotive force establishes a stationary plasma electron density depression in the group-velocity frame of the pulse. In the lab frame, the passage of the light pulse and its associated density depression sets up a "wake" oscillation as ambient electrons respond to fill the depression trailing behind the optical pulse. The resulting electron plasma wave, or "laser wakefield", can have very large electrostatic fields,

$$E_{es}(V/cm) = \sqrt{n_e(cm^{-3})}$$

These large fields can then interact with electrons (laser wakefield accelerator[2,3,11,12]) or photons (photon acceleration or deceleration[10]). The latter is of interest for the development of tunable radiation sources, while the former holds promise for ultra-high gradient electron acceleration.

In either case, the "particles" must interact with a sufficiently strong electron plasma wave over a sufficiently long interaction length along the direction of propagation. While available lasers are certainly capable of driving strong electron plasma waves, diffraction allows this to happen only over a finite focal depth. It is possible that the interaction length can be extended through "optical guiding"[12,13] of the driver laser. Optical guiding occurs when the tendency of a laser beam to self focus (due to relativistic self-focusing) balances its tendency to defocus (due to diffraction). Under such conditions, the effective interaction length is dictated not by the focal depth, but rather by the pump depletion length (typically many times longer?). The topic of optical guiding in the laser wakefield accelerator is discussed in depth by P. Sprangle elsewhere in these proceedings.

HBL Development At LLNL

Our laser development work in the past has been based on both solid state and dye laser technology. Our current design for an HBL is based on the extrapolation of proven solid state laser technology to new parameter regimes. Several factors lead to the decision to proceed with a Nd:Glass based system. First, the candidate amplifier medium must have sufficient bandwidth to support the amplification of a short pulse. Second, the amplifier must have a high saturation fluence and a high energy extraction efficiency if the final design is to be practical and affordable. Finally, the beam intensity in the amplifier must be held sufficiently low so as to avoid the introduction of phase perturbations on the beam by intensity-dependent nonlinear optical effects (i.e. B-integral effects, for example self focussing of the beam in the amplifier glass). . Our expertise in Nd:Glass systems and its favorable properties as a large-bandwidth, high saturation fluence amplifier medium make it a natural choice for the needs of an HBL.

Despite the relative resilience of Nd:Glass to B-integral effects, the brighteness of an amplified short pulse for any reasonably apertured system would still exceed the nonlinear limits. This difficulty will be overcome through the use of the chirped pulse amplification (CPA) technique[14]. With CPA, the maximum light intensity at any time in an amplifier is held at a safe level by stretching the laser light pulse in time prior to its entering the amplifier chain as depicted in Fig. 1. A short oscillator pulse (several tens of picoseconds) is first frequency chirped by self phase modulation (SPM) in a long single-mode optical fiber. Because SPM is an intensity-sensitive process, the imposed bandwidth can be varied in a controlled manner by varying the oscillator pulse intensity. This chirped pulse is then passed through a grating pair which stretches the pulse in time. This stretching allows efficient extraction of energy from the amplifier while maintaining the peak intensity at any time during the pulse below that at which B-integral effects become significant. In the final stage the amplified, chirped pulse passes through a second grating pair arranged such that the pulse is compressed in time to the final desired transform-limited pulsewidth.

Laser and Experiments Status

Presently our 10 TW laser system is in its final stages of completion. This Nd:Glass based, CPA system has achieved 5 J in a 0.8 ps pulse. with a prepulse level below 10^{-3}. The experiments area includes two interaction vacuum chambers, one of which is dedicated primarily to HBL-plasma interactions. The first experiments will be a collaborative effort between LLNL and the UCLA laser accelerator group. These experiments will include investigations of laser wakefield excitation, photon acceleration and deceleration, and relativistic self-focusing. In brief, diagnostics will be configured to collect and diagnose the transmitted and scattered 1 μm light and energetic electrons. Where applicable, both optical and x-ray imaging diagnostics will be fielded. The results from these experiments will be reported as they are obtained

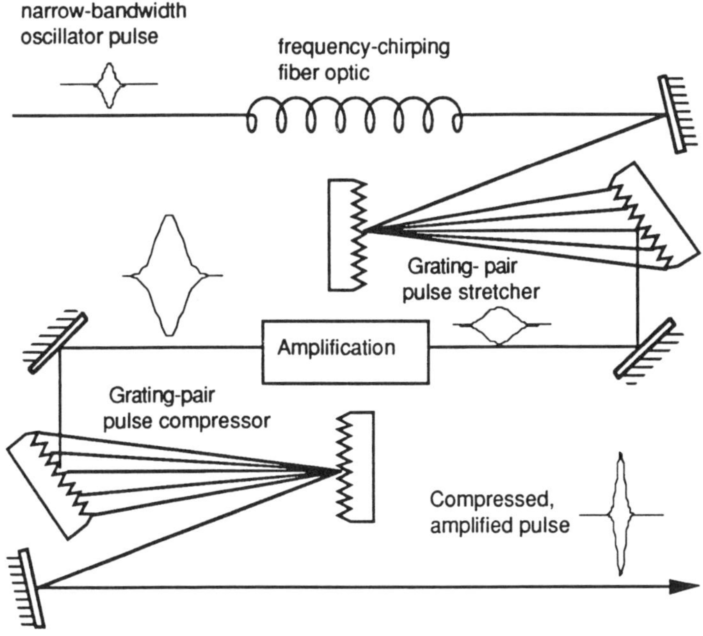

Fig. 1 Schematic diagram of the chirped-pulse amplification ("CPA") technique of producing short, high-peak-power optical pulses.

References

(1)   C. E. Max, Jonathan Arons, and A. Bruce Langdon, Phys. Rev. Lett. **33**, 209 (1974).

(2)   Tajima and J. M. Dawson, Phys. Rev. Lett **43** ,267 (1979)

(3)   W. B. Mori, UCLA Masters Thesis, Los Angeles, CA (1984).

(4)   G. Schmidt, and W. Horton, Comm. on Plas. Phys. and Contr. Fusion **9**, 85 (1985).

(5)   M. D. Perry and F. G. Patterson, Opt. Lett., **15**, 381 (1990).

(6)   *Advanced Accelerator Concepts*, American Institute of Physics Conference Proceedings Number 156, Madison,WI (1986)

(7)   M. D. Perry and E. M. Campbell, LLNL internal report.#

(8)　P. Sprangle, C. -M. Tang, and E. Esarey, IEEE Transactions on Plasma Science **2**, 145 (1987).

(9)　W. B. Mori, et al., Phys. Rev. Lett **60**, 1298 (1988).

(10)　S. Wilks, J. M. Dawson, and W. B. Mori, Phys. Rev. Lett **61**, 337 (1988).

(11)　D. J. Sullivan and D. B. Godfrey, *Laser Acceleration of Particles*, AIP Conference Proceedings #91, Los Alamos NM (1982), P. Channel ed.

(12)　P. Sprangle, E. Esarey, A. Ting, G. Joyce, Appl. Phys. Lett. **53**, 2146 (1988).

(13)　P. Sprangle, 1989 APS - DPP contributed paper, Anaheim, CA.

(14)　D. Strickland and G. Mourou, Opt. Comm., **56**, 219 (1985).

ELECTRON KINETICS IN LASER-DRIVEN INERTIAL CONFINEMENT FUSION

E. M. Epperlein

Laboratory for Laser Energetics, University of Rochester
250 East River Road, Rochester, NY 14623-1299

ABSTRACT

The validity of fluid models of electron transport has been investigated in the context of laser fusion. A brief derivation of the classical heat flow formula is presented. Comparisons with numerical solutions of the electron Fokker-Planck equation confirm the existence of flux inhibition for $(Z + 1)^{1/2}\lambda_t > 0.005l$, where λ_t is the scattering mean free path of a thermal electron and l is the appropriate temperature scale length. The applicability and breakdown of the "flux limiter" and delocalization formulae for classical heat conduction are discussed. Two-dimensional Fokker-Planck simulations of heat transport under nonuniform laser illumination show reduced smoothing in the corona and enhanced smoothing at the ablation surface, when compared with fluid predictions.

INTRODUCTION

One important requirement for the modeling of the dynamics of laser-driven inertial confinement fusion (ICF) is a proper understanding of electron transport.[1] It is electron transport from the critical surface, where most of the laser energy is absorbed, to the cold overdense plasma that gives rise to the ablatively driven implosion necessary for fusion. The presence of lateral transport is responsible for smoothing out short-scale thermal modulations that may arise as a result of nonuniform laser illumination, thereby reducing the "seed" for instabilities which are detrimental to efficient target compression.

All processes associated with thermal electron transport are normally studied by solving the fluid equations using classical heat conduction $\mathbf{q}_c = -\kappa_c \nabla T$, where κ_c is the thermal conductivity (also known as the Spitzer-Härm conductivity),[2] and T is the temperature in energy units. However, typical laser plasmas involve high temperatures and short scale lengths, where the mean free path of a heat carrying electron (with energies of about $7T$) may be comparable to $l = T/|\nabla T|$.[3] Under such conditions the electron transport becomes nonlocal, i.e., cannot be adequately described in terms of a local ∇T, and fluid theory breaks down. The need for a kinetic treatment of the electrons has thus resulted in many transport studies based on numerical solutions of the Fokker-Planck (FP) equation.[4-6] Figure 1 shows schematically the consequences of highly nonlocal transport in a laser-produced plasma.

A characteristic feature of nonlocal transport lies in the ability of coronal electrons to deposit their energy ahead of the main heat front, thereby undesirably preheating the fuel. Also, if electron thermalization is not sufficiently strong to maintain a Maxwellian distribution, the hot corona becomes partially depleted of heat-carrying electrons. This has the effect of reducing their phase-space density gradient, which in turn reduces their diffusive flow, thereby lowering the effective heat flow. This phenomenon, known as flux inhibition, is commonly identified with a flux-limiting parameter f, which when multiplied by the "free-streaming" heat flux $q_f = nmv_t^3$ (where $v_t = \sqrt{T/m}$) provides an upper bound to q_c.[7] The resultant heat front is correspondingly modified as shown qualitatively in Fig. 1. Another manifestation of the departure from classical heat flow arises in multidimensional transport where $\mathbf{q}$ is not necessarily parallel to $-\nabla T$.[5,6] In fact, it has been demonstrated that it is possible even in 1-D for $\mathbf{q}$ to be directed up a temperature gradient.[4]

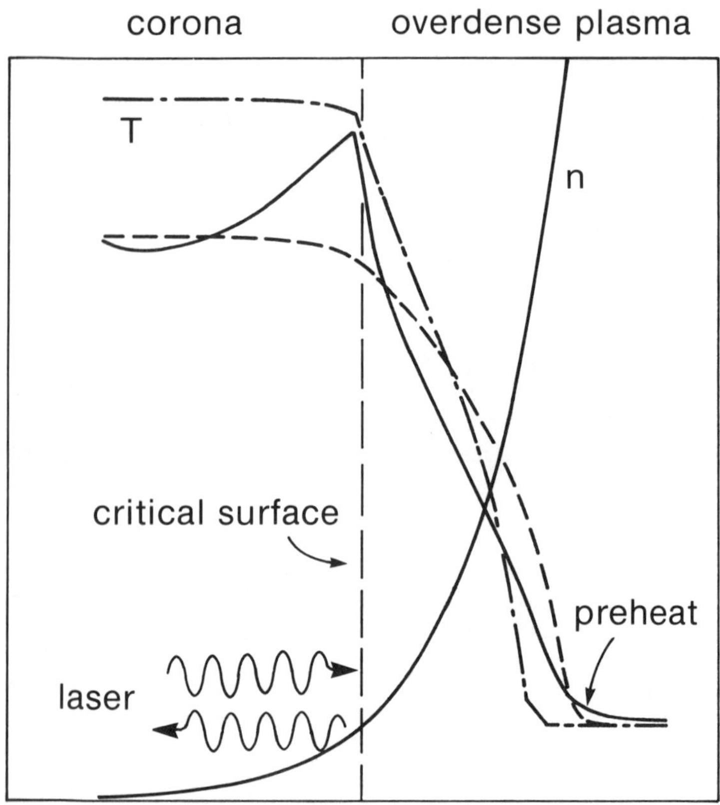

Fig. 1   Schematic of density (n) and temperature (T) profiles in a laser-produced plasma where kinetic effects are assumed dominant. Solid curves refer to FP heat flow, dashed curves to classical heat flow, and dash-dotted curves to flux-limited heat flow. For simplicity, n is assumed invariant.

The purpose of this article is to address these and other issues associated with nonlocal transport in ICF. Our main approach is based on comparisons between fluid and FP results. For a review of experimental investigations of heat transport we refer to an article by Delettrez (1986).[8] Other physical processes that may also affect the heat flow, such as magnetic field generation, ion-acoustic instability and strong inverse-bremsstrahlung heating, are reviewed by Kruer (1988).[9] In the following sections we start with a brief derivation of the classical heat flow formula, identify some of the reasons for its breakdown and discuss the physical motivation behind the flux limiter. We then provide a more detailed analysis of nonlocal transport by solving the FP equation. In particular we demonstrate a case where the simple flux limiter fails. This is followed by a section devoted to simulations more closely applicable to ICF conditions. We close by discussing the results and what further investigations are needed on the subject of nonlocal transport.

All of the transport simulations presented in this article are obtained using the 2-D FP code SPARK.[5,6] The code is Eulerian (in cartesian or cylindrical geometries), and is coupled to a fluid ion package. Magnetic field effects are not presently included.

CLASSICAL HEAT FLOW AND THE FLUX LIMITER

We start with the electron FP equation for a fully ionized plasma, where ions are assumed to be cold and at rest and magnetic field effects are neglected.

$$\frac{\partial}{\partial t}f + \mathbf{v} \cdot \nabla f - \frac{e}{m}\mathbf{E} \cdot \nabla_v f = C .\tag{1}$$

The electron distribution function at a spatial point $\mathbf{r}$, velocity $\mathbf{v}$ and time t is given by $f(\mathbf{r},\mathbf{v},t)$, e is the electron charge, m is the electron mass, $\mathbf{E}$ is the electric field, and C is the FP collision operator. The properties and form of the latter are described in detail elsewhere.[10]

The basis for calculating the heat flow from Eq. (1) involves assuming that the plasma is close to thermodynamic equilibrium, so that we may expand f as[10]

$$f = f_0 + \mathbf{v} \cdot \mathbf{f}_1/v,\tag{2}$$

where $f_0(\mathbf{r},\mathbf{v},t)$ is the isotropic part, taken to be a Maxwellian, and $\mathbf{f}_1$ is the anisotropic part describing the transport. Substituting (2) into (1) and dropping the time derivative we obtain,

$$\mathbf{f}_1 = -\lambda(v)\left[\nabla f_0 - \frac{e\mathbf{E}}{mv}\frac{\partial}{\partial v}f_0\right],\tag{3}$$

where $\lambda(v) = v^4/[n(Z+1)\,4\pi(e^2/m)^2 ln\Lambda]$ is the velocity-dependent scattering mean free path. In deriving (3) we have neglected electron-electron momentum exchange, an assumption that is strictly valid for high Z plasmas. However, for the purpose of comparing classical with nonlocal transport, this assumption is relatively unimportant.[4]

By invoking quasi-neutrality and hence zero current $\mathbf{j} = -(4\pi/3)\int dv v^3\mathbf{f}_1$ we obtain $e\mathbf{E}/m = -v_t^2(5\nabla T/2T + \nabla n/n)$ and

$$\mathbf{f}_1 = \lambda_t\left(\frac{v}{v_t}\right)^4\left[\frac{1}{2}\left(\frac{v}{v_t}\right)^2 - 4\right]\frac{\nabla T}{T}f_0 ,\tag{4}$$

where, $\lambda_t = \lambda(v_t)$. We now calculate the heat flow using the definition $\mathbf{q} = (2\pi/3)\int dv v^4\mathbf{f}_1$ and obtain $\mathbf{q}_c = -\kappa_c\nabla T$, where $\kappa_c = \gamma n v_t\lambda_t$, and $\gamma = 64(2/\pi)^{1/2}$. A more detailed calculation of transport coefficients for arbitrary Z and magnetic fields is given by Epperlein and Haines (1986).[11]

The first problem with the present calculation for $\mathbf{q}$ lies in the fact that the perturbation analysis breaks down, i.e., $|\mathbf{f}_1/f_0| \gg 1$, for large v thus giving rise to a negative distribution function. This unphysical behavior is most likely to occur for small values of l, as seen from Eq. (4). More precisely, since the main contribution to the heat flow integral comes from $v \sim v^* = 3.7\,v_t$ the validity criterion becomes $\lambda_t < 0.002l$, by requiring that $|\mathbf{f}_1/f_0| < 1$ at v^*.[3,4,12] However, as will be demonstrated in the next section, a more accurate criterion should also involve the electron-electron energy loss mean free path.

Another limitation of the classical heat flow formulation comes from the prediction of infinite energy flux as $l \rightarrow 0$. On physical grounds we might expect the energy flow to be limited to some fraction f of $q_f = nmv_t^3$, the "free-streaming limit". We may, therefore, arbitrarily define an effective heat flow as either $q^* = \min{(q_c, fq_f)}$, or $q^* = q_c/(1+q_c/fq_f)$,[7] based on the "sharp cut-off" or "harmonic mean" method, respectively. Both methods yield similar results, though the "harmonic mean", which is the one adopted in this article, gives a stronger reduction for a given value of f. Currently acceptable values for f range from about 0.03 to 0.2 (larger values essentially give $q^* = q_c$).[8] A priori, based on the simple arguments presented above there is no simple way to predict a unique value of f. Also, because of its local character, one cannot expect the flux limiter to be applicable in situations where nonlocal transport effects are dominant. In particular, it fails to account for the possibility of preheat.

Although the flux limiter is a very useful tool in many situations one must exercise caution when using it. As will be demonstrated in the following sections there are certain cases in which it is totally inadequate.

HEAT FLOW IN A HOMOGENEOUS PLASMA

In order to more accurately investigate nonlocal transport we solve the FP equation numerically. Our starting assumption is once again the expansion defined by Eq. (2), which is sufficient for demonstrating the major features of nonlocal transport and is probably adequate for most transport problems relevant to ICF.[4] The contributions from other terms in the expansion become necessary when one is interested in nearly collisionless phenomena and electromagnetic instabilities.[13]

Unlike in the previous section we do not assume a Maxwellian form for f_0, but rather solve for it using the equation

$$\frac{\partial f_0}{\partial t} + \frac{v}{3}\nabla\cdot\mathbf{f}_1 = \frac{1}{v^2}\frac{\partial}{\partial v}\left[\frac{v^2}{3}\mathbf{a}\cdot\mathbf{f}_1 + Y\left(C_0 f_0 + D_0\frac{\partial}{\partial v}f_0\right) + \frac{YnZ}{6v}v_0^2\frac{\partial}{\partial v}f_0\right], \qquad (5)$$

where $\mathbf{a} = |e|\mathbf{E}/m$, $Y = 4\pi(e^2/m)^2 \ln\Lambda$, and $C_0(f_0)$ and $D_0(f_0)$ are integral operators, as defined by Shkarofsky et al.[10] The last term in the equation is the inverse-bremsstrahlung operator, with v_0 being the electron quiver velocity in the laser field.[14] Equations (3) and (5), coupled with the quasi-neutrality condition and a few modifications to account for ion motion, are solved in the 2-D code SPARK, the details of which are described in Refs. 5 and 6.

As a first illustration of kinetic transport let us consider a homogeneous plasma of temperature T_0 where we apply a perturbation of the form $\delta T(t)e^{ikx}$, such that $\delta T \ll T_0$. Using the energy conservation equation $(3/2)n\partial_t T + \nabla\cdot\mathbf{q} = 0$, it is easily shown that $\delta T_c(t) = \delta T_c(t=0)e^{-t\alpha}$, where $\alpha = 2k^2\kappa_c/3n$ is the classical decay rate. By numerically calculating the temperature decay using SPARK, and defining $\alpha_{FP} = 2k^2\kappa_{FP}/3n$, it is then possible to obtain κ_{FP}/κ_c. This gives us a measure of departure from classical transport. By varying k we obtain different values of κ_{FP}/κ_c, which are then plotted in Fig. 2 as a function of $(k\lambda_d)^{-1}$, where $\lambda_d = (\lambda_e\lambda_t)^{1/2} = (Z+1)^{1/2}\lambda_t$ is an effective delocalization length (or stopping length), and $\lambda_e = T^2/4\pi ne^4 \ln\Lambda$ is the energy loss mean free path.[15] The origin λ_d may be traced back to Eqs. (3) and (5) by estimating $v^{-2}\partial(YC_0 f_0)/\partial v \sim (v_t/\lambda_e)f_0$ as the electron-electron energy loss term, and comparing it to the spatial diffusion term $v\nabla\cdot\mathbf{f}_1 \sim v_t\lambda_t\nabla^2 f_0 \sim v_t\lambda_t f_0/\lambda_d^2$, assuming $v = v_t$. In effect, λ_d provides a better measure of delocalization than λ_t alone (with some appropriate weighting to account for the higher velocity heat-carrying electrons), since it also depends on the electron-electron thermalization strength.

In Fig. 2 we observe a clear departure from classical heat flow, in the form of flux inhibition, for scale lengths $k^{-1} < 200\lambda_d$. Such an effect, which has been predicted by Bell in the context of ion waves,[16] clearly demonstrates a situation where nonlocal electron transport is present and yet $q_c \ll fq_f$, i.e., the heat flow is unsaturated. In fact, since the heat flow is arbitrarily small, there is no unique value of f that will give $q_{FP} = q_c$. Moreover, the breakdown due to negative distributions, mentioned in the previous section, occurs only for $v \gg v^*$. The explanation for the flux inhibition comes from the fact that, for $l \ll 200\lambda_d$, the heat carrying electrons are able to diffuse across many wavelengths thereby reducing their density gradient in phase-space. Since the modulation e^{ikx} refers to the thermal electrons (with energies $\sim T$) and the electron thermalization to higher energies ($E^* \sim 7T$) is not instantaneous, the effective heat flux is reduced. This is shown schematically in Fig. 3.

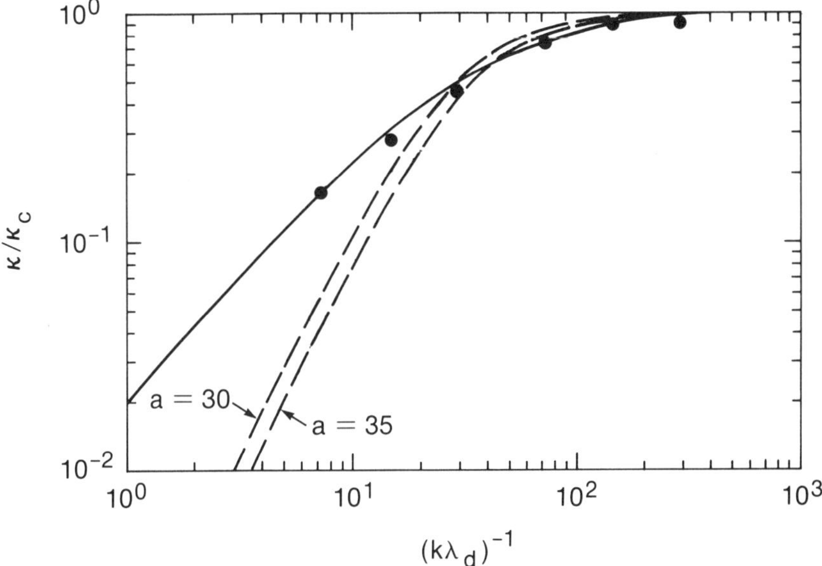

Fig. 2   Plot of κ_{FP}/κ_c (solid curve) and κ_d/κ_c (dashed curves) as functions of $(k\lambda_d)^{-1}$. Circle data points are obtained from a 2-D FP simulation.

It is interesting at this point to investigate the properties of so-called delocalization models of heat flux, which were designed to bridge the gap between fluid and kinetic theories. The simplest one, first proposed by Luciani et al.,[15] is given by:

$$q_d(x) = \int_{-\infty}^{\infty} dx' \frac{q_c(x')}{2\lambda_0(x')} \exp\left[-|x - x'|/\lambda_0(x')\right]$$

for a homogeneous plasma. Here $\lambda_0 = a\lambda_d$ and a is a free parameter. By Fourier analyzing the above equation it is straightforward to show that $q_d/q_c = \kappa_d/\kappa_c = (1+a^2k^2\lambda_d^2)^{-1}$ so that at first sight it appears to have the desired property that κ_d/κ_c decreases for large $k\lambda_d$. However, plotting this function in Fig. 2 using $a = 30$–35 (which covers the range of values quoted in the literature)[15] gives poor agreement with the FP data. The main reason is that for large $k\lambda_d$, $\kappa_{FP}/\kappa_c \sim (k\lambda_d)^{-1}$, whereas $\kappa_d/\kappa_c \sim (k\lambda_d)^{-2}$. Such a discrepancy may explain the somewhat limited success enjoyed by delocalization models in ICF-related transport simulations.[8,17]

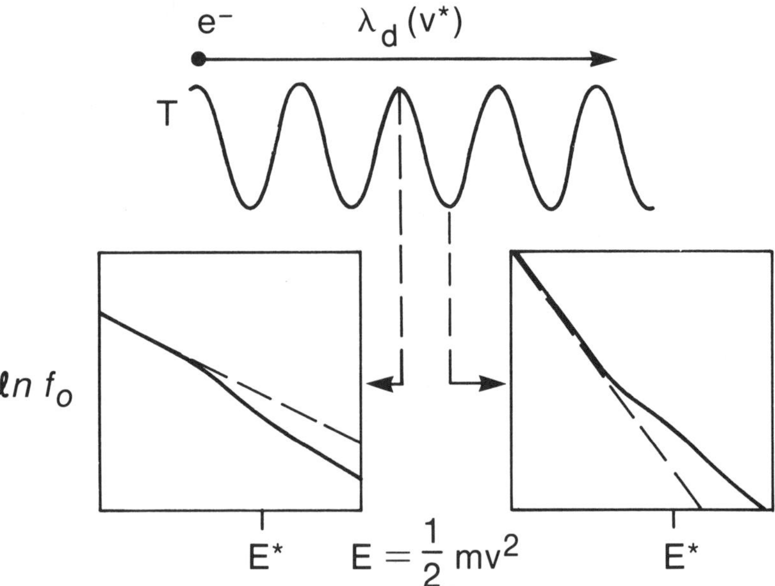

Fig. 3   Qualitative description of nonlocal transport. In the distribution plots, solid curves are based on a kinetic calculation, whereas dashed curves are obtained assuming a Maxwellian distribution at the local temperature.

HEAT FLOW IN A LASER-HEATED PLASMA

In this section we review the results of 2-D transport simulations that illustrate some of the subtleties associated with nonlocal transport. We investigate the thermal response of a CH target that has been subjected to nonuniform laser irradiation. It is particularly important to predict the level of thermal smoothing at the beginning of the laser pulse since the effectiveness of the implosion may depend on it.

As it is not yet feasible to run a full 2-D laser-ablative implosion using SPARK we restrict ourselves to the early time response by "freezing" the hydro. The background density profile, shown in Fig. 4, is obtained using the 1-D fluid code *LILAC*.[18] The conditions correspond to that of a 300-μm-diameter, 10-μm-thick CH target irradiated by a 0.35-μm laser of peak intensity 5 x 10^{14} W/cm^2 and a Gaussian temporal pulse of 600 ps FWHM. The plot is taken at 600 ps before the peak, such that the intensity is ~3 x 10^{13} W/cm^2. SPARK is initialized at this intensity (kept constant) and at a temperature of 50 eV. After a few tens of picoseconds the coronal temperature reaches approximately 300 eV and the thermal front evolves in a quasi-steady fashion. Figure 4 shows the FP and fluid results at 100 ps. The very slight flux inhibition (not noticeable in the figure) is well modeled by using $f = 0.1$ (harmonic). We note that the coronal temperature increases with flux limitation (as less heat is allowed into the overdense plasma) so that we also obtain better agreement for the laser absorption fraction φ, i.e., $\varphi_{FP} = 0.74$, $\varphi_C = 0.78$ and $\varphi^*(f = 0.1) = 0.73$. Such a result is typical for simulations with short wavelength lasers (<1 μm) at moderate intensities (<10^{15} W/cm^2).[4,15]

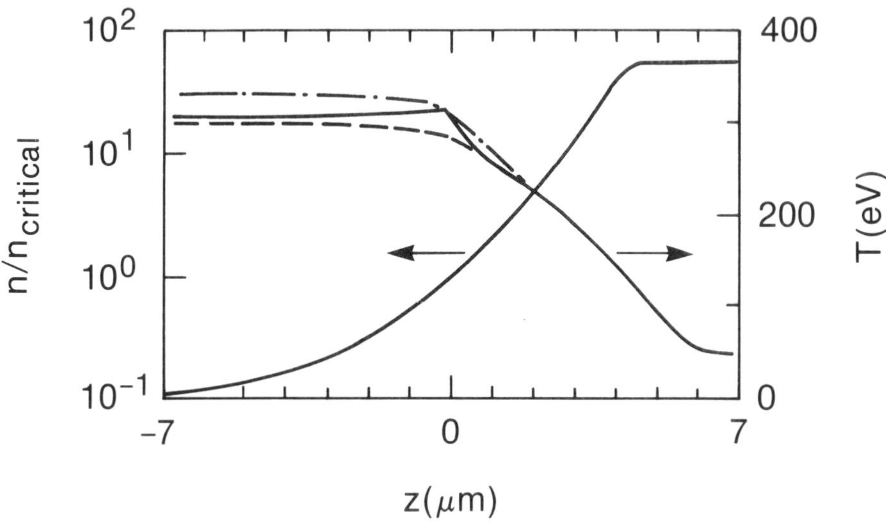

Fig. 4   Density n (normalized to the critical density value) and temperature T (in eV) profiles after a 100-ps laser pulse. As in Fig. 1.

The above simulation has been repeated by applying a small modulation in the x direction to the incident laser intensity of the form $\mathrm{Sin}(2\pi x/\lambda_\perp)$, where $\lambda_\perp = 10$ µm. Figure 5 shows the calculated root mean square (RMS) temperature deviation at 100 ps defined by

$$\sigma = \left\{ \left[\int dx \, (T - \langle T \rangle)^2 \right] \Big/ \int dx \right\}^{1/2} \langle T \rangle^{-1}$$

where $<T>$ is the average temperature. All plots have been normalized to the RMS laser intensity. Axial and transverse heat flow, normalized to q_f, are shown in Figs. 6(a) and 6(b), respectively. Here, q_c is calculated using the FP temperature profiles.

Despite the good agreement between fluid and kinetic results for the 1-D temperature profiles the same is not true for the transverse spatial modulation. The nonlocal transport gives rise to worse smoothing in the corona and an improvement at higher densities. By calculating σ_{FP}/σ_c at the critical surface and varying $\lambda_\perp$ we obtain the plot shown in Fig. 7. The increase of σ_{FP}/σ_c as $\lambda_\perp \to 0$ resembles the flux inhibition effect demonstrated in the previous section for the special case of a homogeneous plasma. To show that this is indeed the case we first make the assumption that $S \sim ik(q_{FP})x \sim ik(q_c)x$, in the corona, where S is the laser energy deposition rate. By defining $\kappa_{FP} = -(q_{FP})x/ik\delta T_{FP}$, we find that $\kappa_{FP}/\kappa_c \sim \delta T_c/\delta T_{FP} \sim (\sigma_{FP}/\sigma_c)^{-1}$, which then allows us to plot κ_{FP}/κ_c as a function of $\lambda_\perp/(2\pi\lambda_d)$ in Fig. 2, where $\lambda_d = 0.75$ µm at the critical surface. As observed, the agreement is excellent. The occurrence of reduced coronal thermal smoothing in longer scale length plasmas irradiated by higher intensities has been previously reported by Epperlein et al.[5]

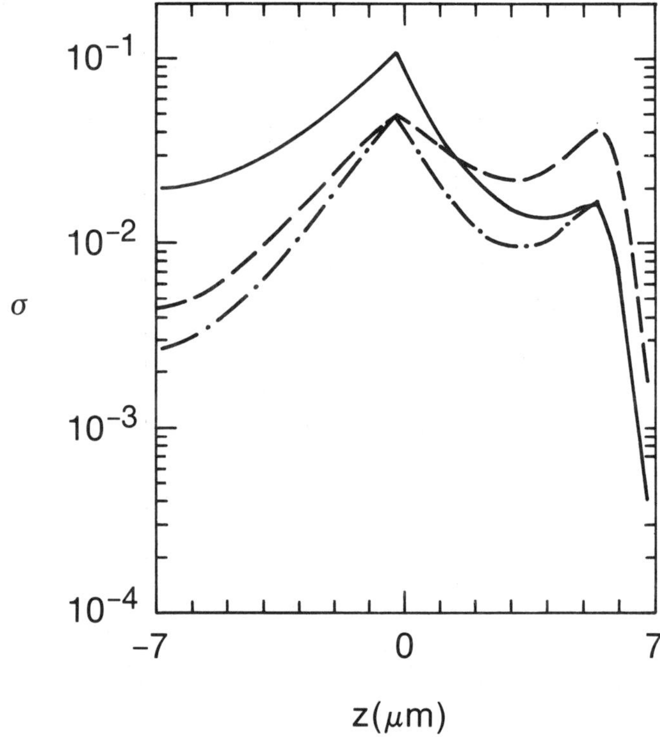

Fig. 5   Plot of the RMS temperature deviation (normalized to the RMS laser intensity deviation) as a function of z. As in Fig. 1.

The clue for the enhanced smoothing at high densities lies in the axial flux inhibition near critical [see Fig. 6(a)]. This acts to reduce the effective flow of thermal modulation away from the critical surface to the overdense plasma. By applying a flux limiter we can reproduce this effect to some extent as shown in Fig. 7. The enhancement of lateral thermal smoothing due to axial flux inhibition was first predicted by Skupsky,[19] using a crude steady-state fluid model. More recently Rickard et al.[20] used a FP code to predict a large increase in thermal smoothing in the long wavelength ($\lambda_\perp$) limit. They used, however, a much higher laser intensity (10^{15} W/cm^2) which gave rise to a much stronger axial flux inhibition and hence reduced thermal modulation at high densities. Here we calculate σ_{FP}/σ_c where pressure is a maximum p_{max} (~4.4 μm away from the critical surface) and obtain only a modest enhancement in smoothing at large wavelengths, as seen in Fig. 5. We do, however, predict that for a wide range of $\lambda_\perp$ (6–40 μm), $\sigma_{FP}/\sigma_c \sim 0.5$, which represents a factor of ~2 improvement in pressure uniformity at p_{max}. At smaller wavelengths the reduced lateral smoothing at critical starts to dominate and the value of σ_{FP}/σ_c increases again. (In practice this is relatively unimportant since, at p_{max}, we find that $\sigma < 10^{-2}$ for $\lambda_\perp < 6$ μm.) A failed attempt at accurately modeling this effect with flux limited classical transport is shown in Fig. 7 by plotting $\sigma^*(f = 0.1)/\sigma_c$.

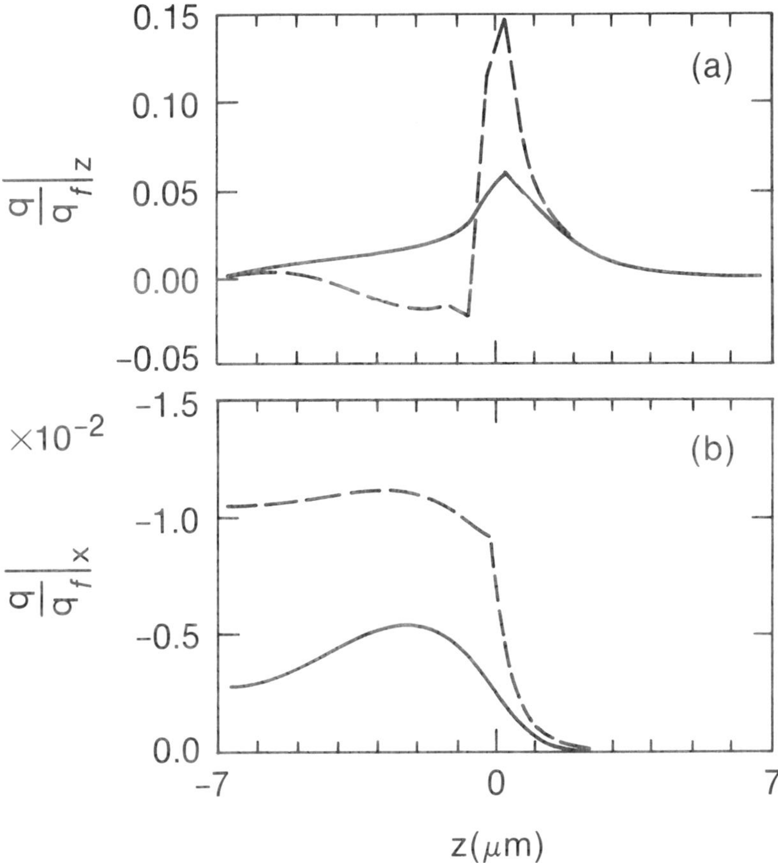

Fig. 6   Plot of heat flow (normalized to q_f) in the (a) z direction (axial) and in the (b) x direction (transverse) at $x = 0$, as functions of z. Here, solid curves refer to q_{FP}, whereas dashed curves refer to q_c using FP temperatures.

CONCLUSIONS

As shown in this article one has to be careful when interpreting transport data using flux-limited classical heat conduction. For short wavelength lasers (<1 μm) at moderate intensities (<10^{15} W/cm^2) a fluid model provides an adequate description of the transport under uniform illumination conditions. A modest amount of flux limitation ($f \sim 0.1$) is then sufficient to "fine tune" the results. One must be aware, however, that classical transport (flux limited or otherwise) cannot model preheat due to the long mean free path electrons. In the FP simulations presented in this article this issue was not addressed satisfactorily because of the relatively high initial plasma temperatures. Mima et al.[21] have recently shown using a 1-D FP code coupled to a Lagrangian fluid solver that preheat can significantly degrade the performance of thin low-Z shell targets irradiated by 0.53-μm laser light. The magnitude of this

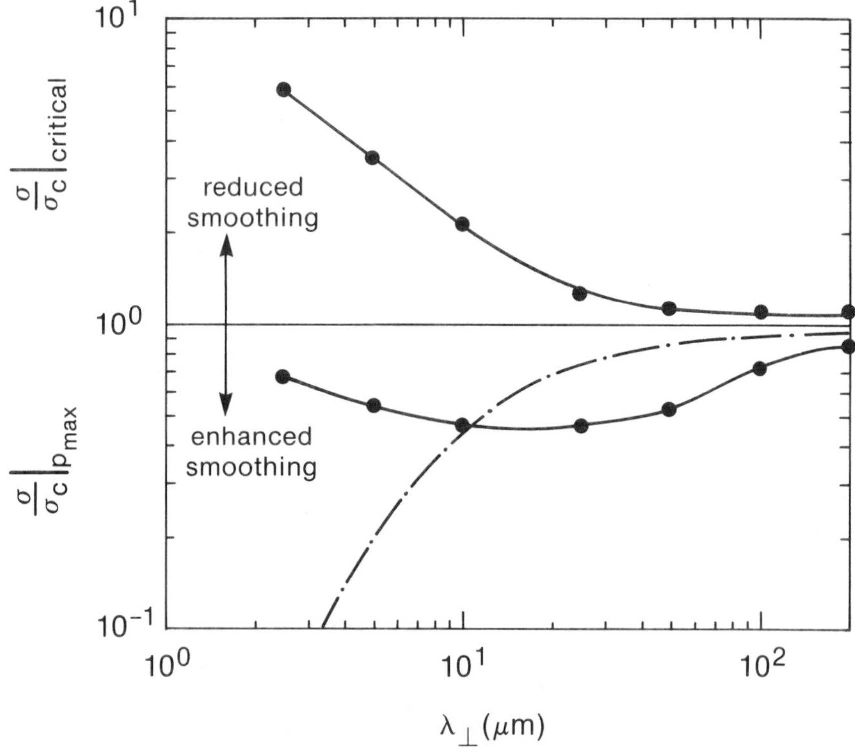

Fig. 7   Plot of σ_{FP}/σ_c (solid curves) and σ^* ($f = 0.1$)/σ_c (dash-dotted curve) as functions of z, calculated at $n_{critical}$ and at maximum pressure.

effect is expected to be less for shorter wavelength lasers, by virtue of the higher collisionality of the plasma, though the exact extent remains to be investigated.

When simulating 2-D transport, which may arise as a result of nonuniform laser irradiation, nonlocal effects can drastically alter the results, even when the average 1-D transport appears well modeled by fluid theory. The main consequences of nonlocal transport are given as follows:

1. A severe reduction in the coronal thermal smoothing for spatial modulations less than about $200\,(Z + 1)^{1/2}\lambda_t$.

2. An enhancement in the smoothing at high densities, especially near the pressure maximum, for all spatial modulations.

The first consequence is well explained in terms of flux inhibition of short scale modulations in a homogeneous plasma. Its main impact would be to reduce the threshold for thermal self-focusing instabilities[22] or any other instabilities that rely on coronal temperature modulations.

The second result arises from the axial flux inhibition from the critical density to higher densities, which acts to reduce the propagation of thermal modulations. The scaling of this phenomenon with plasma and laser conditions has not been fully investigated, nor has the issue of hydrodynamic feedback been addressed. However, the simulations presented here suggest the possibility of a two-fold increase in smoothing (when compared with classical heat flow modeling) over a wide range of modulation wavelengths. Such an outcome is potentially very beneficial to ICF since, for a given nonuniformity in the incident laser, it predicts a smoother ablation pressure and less chance for seeding hydrodynamic instabilities.

ACKNOWLEDGMENT

This work was supported by the U.S. Department of Energy Division of Inertial Fusion under agreement No. DE-FC03-85DP40200 and by the Laser Fusion Feasibility Project at the Laboratory for Laser Energetics which has the following sponsors: Empire State Electric Energy Research Corporation, New York State Energy Research and Development Authority, Ontario Hydro, and the University of Rochester.

REFERENCES

1.   J. H. Nuckolls, L. W. Wood, A. Thiesen, and G. Zimmerman, Nature **239**, 139 (1972).
2.   L. Spitzer and R. Harm, Phys. Rev. **89**, 977 (1953).
3.   D. R. Gray and J. D. Kilkenny, Plasma Phys. **22**, 81 (1980).
4.   A. R. Bell, R. G. Evans, and D. J. Nicholas, Phys. Fluids **46**, 243 (1981); J. P. Matte and J. Virmont, Phys. Rev. Lett. **49**, 1936 (1982); J. R. Albritton, Phys. Rev. Lett. **50**, 2078 (1983); T. H. Kho, D. J. Bond, and M. G. Haines, Phys. Rev. A **28**, 3156 (1983); J. P. Matte, T. W. Johnston, J. Delettrez, R. L McCrory, Phys. Rev. Lett. **53**, 1461 (1984); J. F. Luciani, P. Mora, and R. Pellat, Phys. Fluids **28**, 835 (1984); A. R. Bell, Phys. Fluids **28**, 2007 (1985); and S. Jorna and L. Wood, J. Plasma Phys. **38**, 317 (1987).
5.   E. M. Epperlein, G. J. Rickard, and A. R. Bell, Phys. Rev. Lett. **61**, 2453 (1988).
6.   E. M. Epperlein, G. J. Rickard, and A. R. Bell, Comput. Phys. Commun. **52**, 7 (1988); and G.J. Rickard, A. R. Bell, and E. M. Epperlein, Phys. Rev. Lett. **62**, 2687 (1989).
7.   R. C. Malone, R. L. McCrory, and R. L. Morse, Phys. Rev. Lett. **34**, 721 (1975).
8.   J. Delettrez, Can. J. Phys. **64**, 932 (1986).
9.   W. L. Kruer, The Physics of Laser Plasma Interactions (Addison-Wesley, 1988).
10.   I. P. Shkarofsky, T. W. Johnston, and M. A. Bachysnky, The Particle Kinetics of Plasmas (Addison-Wesley, London, 1966).
11.   E. M. Epperlein and M. G. Haines, Phys. Fluids **29**, 1029 (1986).
12.   D. Shvarts, J. Delettrez, R. L. McCrory, and C. P. Verdon, Phys. Rev. Lett. **27**, 247 (1981).
13.   E. M. Epperlein, Plasma Phys. & Controlled Fusion **27**, 1027 (1985).
14.   A. B. Langdon, Phys. Rev. Lett. **44**, 575 (1980).
15.   J. F. Luciani, P. Mora, and J. Virmont, Phys. Rev. Lett. **51**, 1664 (1983); J. R. Albritton, E. A. Williams, I. B. Bernstein, and K. P Swartz, Phys. Rev. Lett. **57**, 1887 (1986); and P. A. Holstein, J. Delettrez, S. Skupsky, and J. P. Matte, J. Appl. Phys. **60**, 2296 (1986).
16.   A. R. Bell, Phys. Fluids **26**, 279 (1983).
17.   M. K. Prasad and D. S. Kershaw, Phys. Fluids B **1**, 2430 (1989).
18.   J. Delettrez (private communication).
19.   S. Skupsky, University of Rochester, Report LLE 131 (1982).
20.   G. J. Rickard, I. R. G. Williams, A. R. Bell, contribution to the 19th Annual Anomalous Absorption Conference, Durango, CO, June 1989 (unpublished).
21.   K. Mima, H. Takabe, A. Nishiguchi, Y. Kihara, and S. Nakai, Laser and Part. Beams **7**, 487 (1989).
22.   E. M. Epperlein, contribution to the 19th Annual Anomalous Absorption Conference, Durango, CO, June 1989 (unpublished).

LASER-DRIVEN INSTABILITIES IN INERTIAL CONFINEMENT FUSION

W. L. Kruer

University of California
Lawrence Livermore National Laboratory
Livermore, CA 94550

ABSTRACT

Parametric instabilities excited by an intense electromagnetic wave in a plasma is a fundamental topic relevant to many applications. These applications include laser fusion, heating of magnetically-confined plasmas, ionospheric modification, and even particle acceleration for high energy physics. In laser fusion, these instabilities have proven to play an essential role in the choice of laser wavelength. Characterization and control of the instabilities is an ongoing priority in laser plasma experiments. Recent progress and some important trends will be discussed.

INTRODUCTION

The interaction of intense electromagnetic waves with plasmas is a basic problem with widespread applications. The use of high power lasers to heat plasmas is a critical issue for laser fusion.[1] Laser-heated plasmas play a key role in the ongoing development of laboratory x-ray lasers.[2] Laser beat work and laser wakefield accelerators[3] have potential applications for high energy physics. Short pulse, high brightness lasers promise access to novel plasma regimes.[4] In magnetic fusion, intense waves can be used for both plasma heating and current drive in Tokamaks.[5] Powerful radio-frequency waves continue to be used in studies of ionospheric modification.[6] Many of the same physical processes have been studied in these varied applications, with the wavelength of the light ranging from a fraction of a micron to meters.

A variety of instabilities can be excited by an intense electromagnetic wave in a plasma. Most of these instabilities represent the resonant coupling of the intense light wave into two other waves. There are many possibilities even in a plasma without large DC magnetic fields. Near the critical density n_{cr} (where the electron plasma frequency equals the light wave frequency), the wave can decay into electron plasma waves plus ion waves. Near $1/4\, n_{cr}$, the decay into two electron plasma waves can occur. For $n \lesssim 1/4\, n_{cr}$, the Raman instability can excite a scattered light wave plus an electron plasma wave. In the Brillouin instability, a scattered light wave plus an ion acoustic wave are generated. This instability can occur for $n \lesssim n_{cr}$.

When the electrostatic waves become heavily damped, these instabilities are transformed into much weaker nonresonant instabilities. For example, the

Raman instability goes over to stimulated Compton scattering on the electrons. In this process the light wave is scattered by a driven beat wave which is not a collective mode of the plasma. Another important example of a nonresonant instability is filamentation. In this process, local enhancements in the radial intensity profile create density depressions via either the ponderomotive force or enhanced local heating and subsequent plasma expansion. Refraction of light into these depressions enhance the intensity perturbations, leading to a potential break-up of the laser beam into intense filaments.

An illustration of the feedback leading to instability is shown in Figure 1. Let there be a small (thermal-level) density fluctuation with amplitude δn and an intense pump wave with electric field amplitude E_0 and frequency ω_0. By oscillating the electrons with velocity $v_{os} = eE_0/m\omega_0$, the pump wave generates a current $\delta J = -e\ \delta n\ v_{os}\ \sim\ \delta n\ E_0$. This current generates a wave (either electrostatic or transverse) with electric field amplitude δE. In turn, this field beats with the pump wave to generate a density fluctuation via the variation in the total electric field pressure; i.e., $\delta n\ \sim\ E_0\delta E$. If the frequencies and wavenumbers are properly matched, the density fluctuation δn and the field δE both exponentiate.

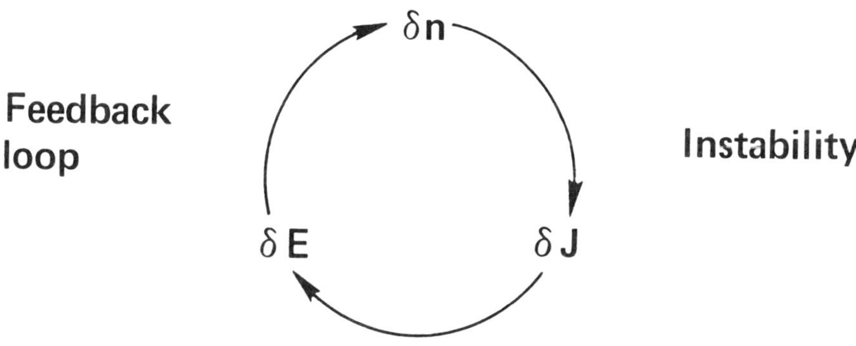

Figure 1.   A schematic illustrating the feedback loop for instability.

Laser-Driven Instabilities

A heuristic derivation of the dispersion relation describing the growth is instructive. In particular, let's consider stimulated scattering instabilities. Coupling of the pump field with amplitude $\mathbf{E_0}\cos(\mathbf{k_0}\ x - \omega_0 t)$ with the density fluctuation associated with either an electron plasma wave or an ion wave at frequency ω and wavenumber k drives electromagnetic waves at $k \pm k_0, \omega \pm \omega_0$:

$$D(k \pm k_0, \omega \pm \omega_0)\delta\mathbf{E}(k \pm k_0, \omega \pm \omega_0) \simeq \alpha\mathbf{E}_0\delta n(k,\omega). \tag{1}$$

Here $D(k,\omega)$ is the dispersion relation for light waves and α is a coefficient. In turn, beating of the pump field with $\delta\mathbf{E}$ drives the density fluctuation, giving

$$\epsilon(k,\omega)\delta n(k,\omega) \simeq \beta\mathbf{E}_0[\delta\mathbf{E}(k - k_0, \omega - \omega_0) + \delta\mathbf{E}(k + k_0, \omega + \omega_0)]. \tag{2}$$

Here $\epsilon(k,\omega)$ is the dispersion relation for the electrostatic waves and β is another coefficient. Substituting Eq. (1) into Eq. (2), we obtain the dispersion relation:

$$\epsilon(k,\omega) = \alpha\beta E_0^2 \left[\frac{1}{D(k - k_0, \omega - \omega_0)} + \frac{1}{D(k + k_0, \omega + \omega_0)}\right]. \tag{3}$$

Here $\alpha\beta$ is proportional to k^2 and the plasma density.

For the Raman and Brillouin instabilities, both electrostatic and electromagnetic waves are resonant. Then $\epsilon \simeq 0$ and $D \simeq 0$, and the growth rate varies as $\sqrt{\alpha\beta E_0^2}$. For stimulated Compton scattering, only the electromagnetic wave is resonant. The growth rate then varies as $\alpha\beta E_0^2$.

LASER PLASMA INSTABILITIES: STATUS AND UPDATE

Virtually all these instabilities have now been identified and studied in laser plasmas.[7] Stimulated Raman Scattering (SRS) has been the most studied instability in recent years. This process is easily identified by the frequency of the scattered light, which is down-shifted by an electron plasma frequency. Since the instability can occur for density $n \lesssim 1/4\ n_{cr}$,

$$\lambda_0 \lesssim \lambda_{sc} \lesssim 2\lambda_0, \tag{4}$$

where $\lambda_{sc}(\lambda_0)$ is the free-space wavelength of the scattered (incident) light. A typical scattered light spectrum in the backward direction is shown in Fig. 2. In these experiments,[8] CH targets were irradiated with 0.35 μm light. Most of the Raman scattering comes from $0.1 \lesssim n/n_{cr} \lesssim 0.2$. There is also a gap in the spectrum corresponding to scattering from densities $0.2 \lesssim n/n_{cr} \lesssim 0.25$. More will be said about this gap later.

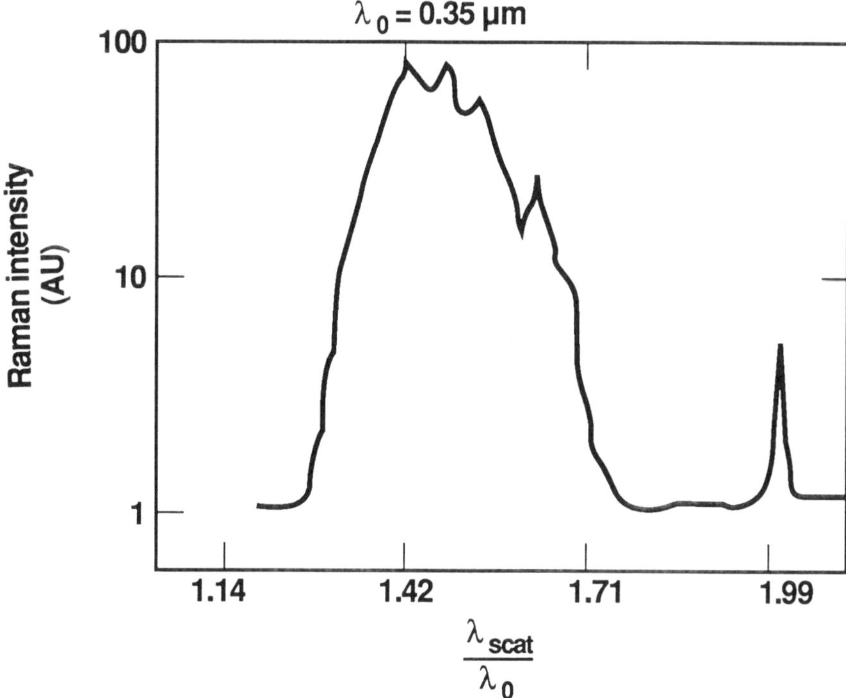

Figure 2.   An example of the wavelength spectrum from a CH target irradiated with 0.35 µm laser light (Seka, et al.,1984).

A key question is how efficient the SRS becomes in large, strongly driven plasmas. Fig. 3 shows the fraction of the incident light Raman scattered as a function of the estimated plasma scalelength L normalised to the free-space wavelength of the light.[9,10] In these experiments CH foils were irradiated with $I \gtrsim 10^{15}$ W/cm^2 using either 0.53 µm or 0.35 µm light as indicated. Plasma sizes $L/\lambda_0 \gtrsim 10^3$ were accessed using exploding foil targets. The results with $L/\lambda_0 \sim 5000$ were recently obtained[10] using the Nova laser. Note that SRS reflectivities $\gtrsim 10\%$ have been observed. The saturation of the reflectivity with L/λ_0 is probably only apparent. For the largest plasma, two-thirds or more of the reflected light is estimated to be collisionally absorbed before reaching the detector.

In a low density plasma, the electron plasma wave associated with Raman back scatter becomes heavily Landau damped. The resonant instability then goes over to a much weaker process called stimulated Compton scattering on the electrons (SCS-e). This heavy damping onsets when $k\lambda_{\mathrm{De}} \gtrsim 0.3$, where k is the wave number of the plasma wave and λ_{De} is the electron Debye length.

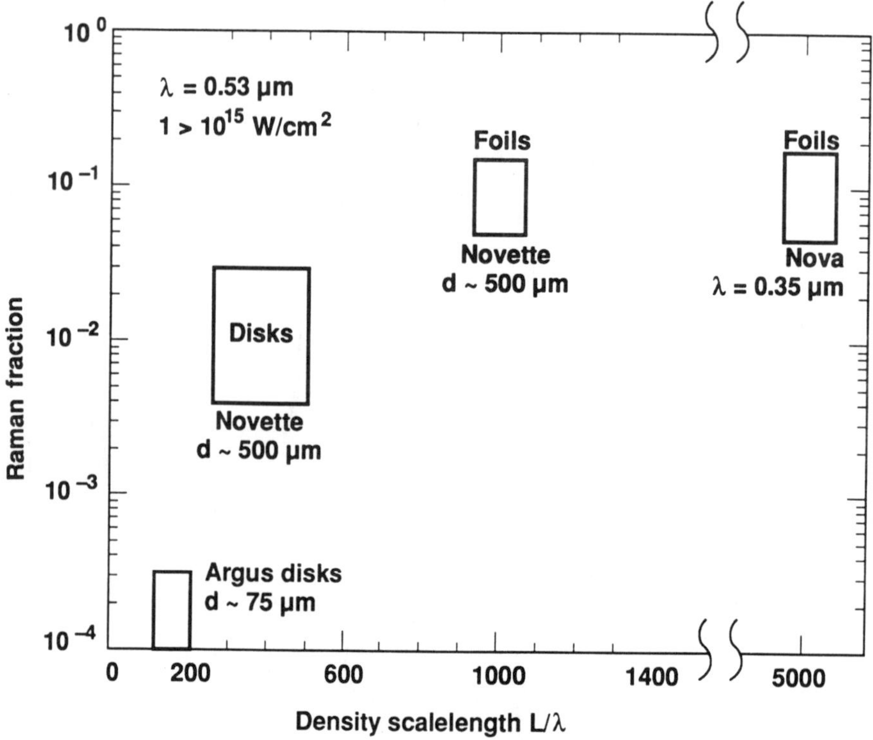

Figure 3.   The fraction of the laser light energy which was Raman-scattered versus the estimated plasma density scalelength (Drake, 1988 and Darrow et al.,1990).

If we estimate $k \sim 1.5\omega_0/c$ and $\theta_e \sim 2$ keV, this condition becomes $n/n_{cr} \lesssim 0.1$. SCS-e has been recently measured[11] in exploding foil plasmas irradiated with 0.35 μm light. Fig. 4 shows the spectral power near $\omega_{sc} = 0.82\omega_0$ as a function of the nominal laser intensity. The solid line is a theoretical estimate obtained by integrating the predicted gain over the calculated plasma density profile and includes hot spots.

Stimulated Raman forward scattering has also been observed. In very underdense plasmas, this process has a plasma wave with a small wavenumber ($k \sim \omega_{pe}/c$) and is much weaker than back or side scattering. However, the plasma wave is not suppressed by Landau damping and could produce very energetic electrons. Fig. 5 shows a measured forward scattered light spectrum in an experiment[12] in which a CH exploding foil was irradiated with 0.53 μm light. Both upshifted and downshifted components were observed, as expected for $k \ll \omega_0/c$. The level of the scattering was quite small ($r \lesssim 0.1\%$). Theoretical estimates compared reasonably well with the data. Further experiments to test the scaling are ongoing.

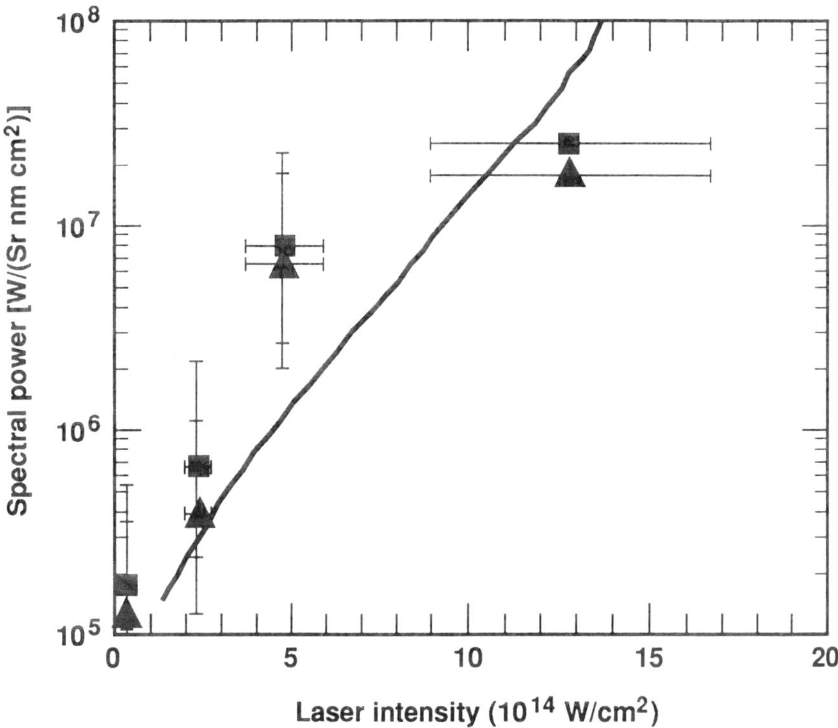

Figure 4.   The spectral power of scattered light near 418 nm as a function of the nominal laser intensity of the 0.35 μm incident laser light (Drake et al., 1990).

Filamentation is the least studied of the instabilities. The presence of filamentation is often difficult to assess, since current laser beams typically have large modulations in their radial intensity profile. Ponderomotive filamentation has been directly observed[13] by irradiating a preformed plasma with a laser beam with an imposed intensity modulation. Growth of the corresponding density perturbations was then directly observed by dark-field imaging. As shown in Fig. 6, significant growth of the density perturbations are observed when the intensity and modulation wave length are chosen properly. The threshold for growth is in reasonable agreement with a simple theory.

The consequences of filamentation remain poorly known. No direct correlation between filamentation and large levels of SRS and SBS has yet been established. A direct correlation between filaments and $2\omega_0$ emission in plasma with $n/n_{cr} \ll 1$ has been recently demonstrated in experiments[14] with 1.06 μm laser light. Figure 7 shows the spatially-resolved $2\omega_0$ emission as well as the size of the ponderomotively-driven density perturbation as inferred from interferom-

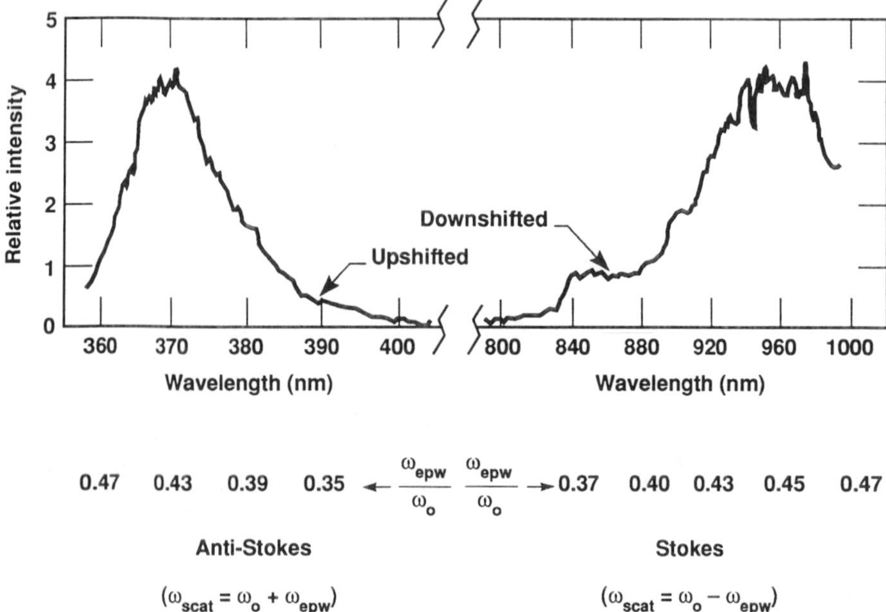

$$\leftarrow \frac{\omega_{epw}}{\omega_o} \quad \frac{\omega_{epw}}{\omega_o} \rightarrow$$

$$(\omega_{scat} = \omega_o + \omega_{epw}) \qquad (\omega_{scat} = \omega_o - \omega_{epw})$$

Figure 5. An example of the upshifted and downshifted spectra of Raman forward-scattered light. The intensity of the upshifted light is about ten times less than that of the downshifted light (Turner et al., 1986).

etry. The generation of $2\omega_0$ emission indicates that the filaments become quite intense and narrow. However, these filaments do not appear to persist to higher density plasma in these experiments.

Finally, there can be a rich nonlinear interplay among the various instabilities. Fig. 8 shows a time-dependent SRS spectrum on the left and the time history of the stimulated Brillouin back scatter (SBS) on the right. In this experiment,[15] a thin CH foil was irradiated by 0.35 μm light with an intensity $\sim 10^{15}\,\mathrm{W/cm^2}$. Note that during the period of significant SBS, there is very little SRS coming from the plasma with density $\gtrsim 0.2\ n_{cr}$. This suggests that ion fluctuations associated with SBS can efficiently suppress SRS at these densities. This is a possible explanation for the well-known gap in the Raman spectrum.

CONTROL OF LASER PLASMA INSTABILITIES

For laser fusion applications, the instabilities are generally deleterious. They can produce energetic electrons which preheat capsules or can scatter light which degrades efficiency and symmetry of irradiation. Hence, one wishes to control these instabilities.

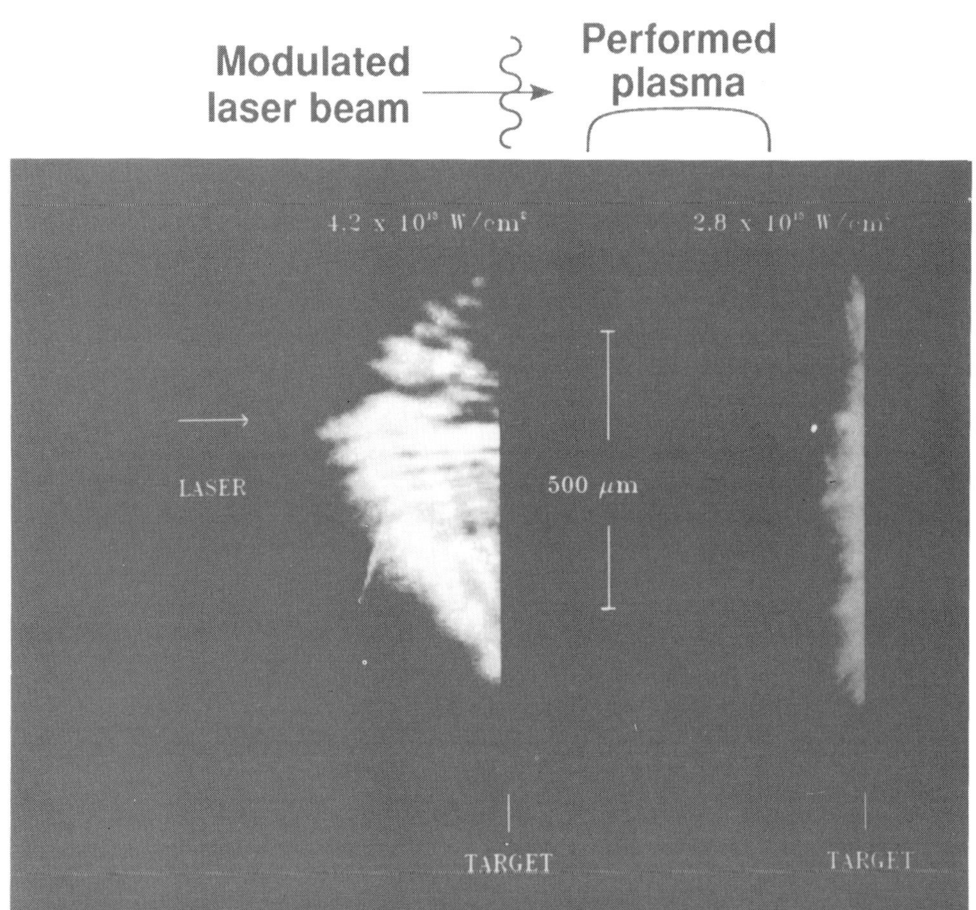

Figure 6.   A dark-field image of a plasma irradiated with a modulated laser beam (Young et al., 1988).

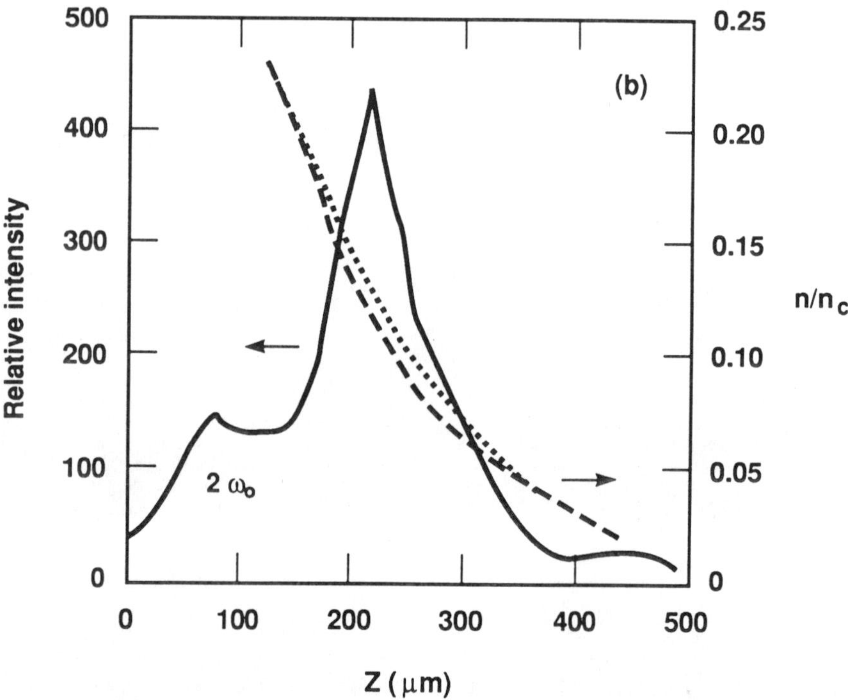

Figure 7.   A plot showing the correlation of the $2\omega_0$ emission with the density perturbation associated with ponderomotive filamentation. The broken lines are density profiles along the laser axis with and without the perturbation by the interaction beam (Young et al., 1989).

Various techniques for instability control are being exploited. The most important is the use of shorter wavelength laser light. As λ_0 decreases, the plasma accessed by the laser light is much more collisional and instabilities are more weakly driven. The more efficient collisional absorption and weaker instability generation was well documented in numerous experiments in the 1980's. Now implosions are driven not by 1.06 μm or 10.6 μm laser light but by 0.35 μm light. Current implosion experiments show very little instability generation.

Other techniques for instability control are also being explored. These include suppression by strong collisional damping - available with high Z plasmas in indirect drive applications - and the use of laser beams with temporal and spatial incoherence. Let us now briefly consider these techniques.

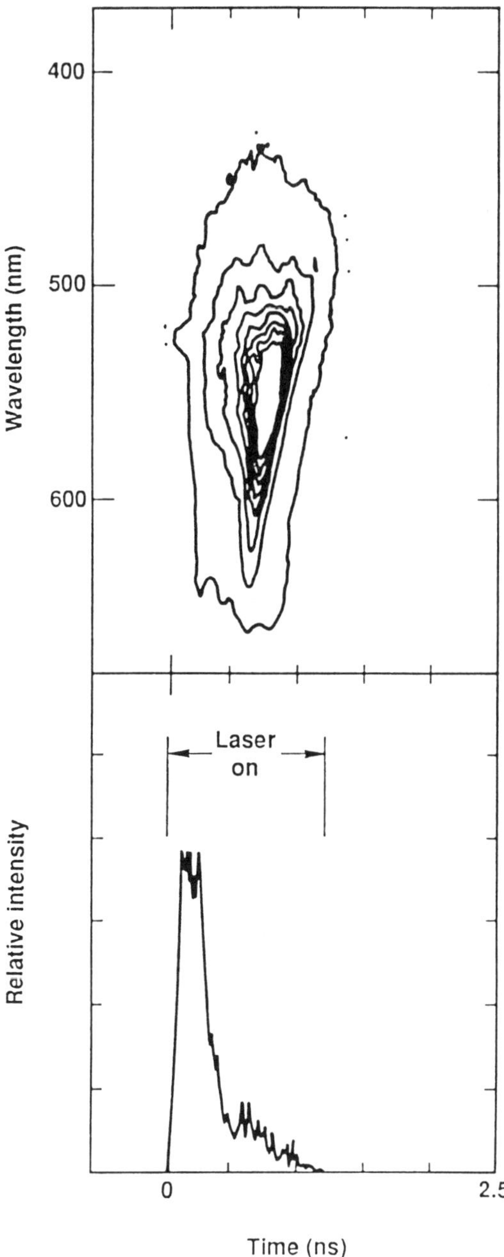

Figure 8.   A contour plot of the time and wavelength resolved Raman-scattered intensity and a plot of the time resolved Brillouin backscattered intensity (Baldis et al.,1989).

In high Z plasma irradiated with short wavelength light, collisional damping of both light waves and electron plasma waves can be very potent. Collisional stabilization of SRS requires $\gamma < (\nu_{ei}/2)\, \omega_{pe}/\omega_{sc}$, where γ is the homogeneous growth rate and ν_{ei} the electron ion collision frequency. In general, Landau damping of the electron plasma wave will also contribute to the stabilization. This stabilization is estimated to occur for $I \lesssim 5 \times 10^{14}$ W/cm^2 for Au targets irradiated with 0.35 μm light. Suppression of SRS by collisions has been demonstrated in experiments.[16,17] Figure 9 shows the measured fraction of the light which is Raman scattered versus the estimated ratio of the collision frequency to the growth rate. The boxes are data from gold targets irradiated with 0.35 μm light at various intensities. Note that the Raman fraction abruptly drops when $\nu_{ei}/\gamma \gtrsim 3-5$.

Finally, the introduction of temporal and spatial incoherence into the laser beam is a promising but still relatively untested technique for instability control. Laser beam incoherence has primarily been introduced[18−20] in order to address the stringent requirements for irradiation symmetry in direct drive. Temporal

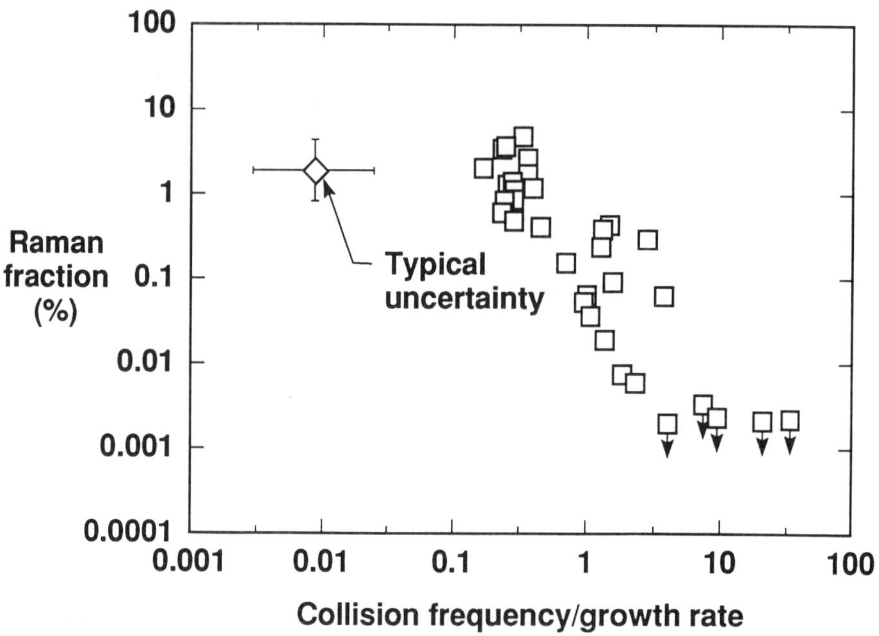

Figure 9.   The measured fraction of the laser light energy which is Raman-scattered versus the estimated ratio of the electron-ion collision frequency to the homogeneous growth rate (Drake,1989).

incoherence corresponds to a broadening of the frequency spectrum of the light, whereas spatial incoherence represents a broadening in angle of the wavenumber spectrum. One beam smoothing technique[18] is to use random phase plates to replace long scalelength hot spots in a beam with small scalelength ones which can be thermally smoothed. In the induced spatial incoherence (ISI) technique,[19] both echelons and bandwidth ($\Delta\omega$) are used to produce a very smooth beam on time scales $\gg 1/\Delta\omega$.

From the early 1970's, there have been calculatins[21] of how laser bandwidth reduces the growth of resonant instabilities. Spatial incoherence gives a similar effect[22] by disrupting the wavenumber matching. Bandwidth reduces the growth rate provided $\Delta\omega > \gamma$. To illustrate the numbers, consider 0.35 μm light with an intensity of 10^{15} W/cm^2 in a plasma with a temperature of 2.5 keV. For the Raman and $2\omega_{pe}$ instabilities near $n_{cr}/4$, $\gamma/\omega_0 \sim 2 \times 10^{-3}$. For SBS near 0.1 n_{cr}, $\gamma/\omega_0 \sim 5 \times 10^{-4}$. Hence, significant bandwidths would be needed to strongly affect the instabilities, unless they were already near threshold. It should be noted that bandwidth is less effective for controlling Raman forward scattering, since the incident and scattered waves then travel together. On the other hand, this weak instability is easily disrupted by gentle variations in the plasma density.

Nonresonant instabilities such as ponderomotive filamentation are not directly affected by bandwidth. The growth of this filamentation is reduced by spatial incoherence. Picture the beam as composed of overlapping beamlets. Beamlets whose wavevectors are at too large an angle not only do not contribute to but oppose ponderomotive filamentation of the beam. The angular spread needed for suppression has been calculated to be[23]

$$(\Delta\Theta)^2 = \left(\frac{\Delta k_0}{k_0}\right)^2 \gtrsim \frac{1}{2}\,\frac{n}{n_{cr}}\,\frac{I_{16}\lambda_{\mu m}^2}{\Theta_{keV}}, \tag{5}$$

where I_{16} is the intensity in units of 10^{16} W/cm^2, $\lambda\mu$m its wavelength in μm, and Θ_{keV} the electron temperature in keV. Physically, this equation corresponds to the condition that the spatial coherence length $(1/k_0\Delta\Theta)$ be less than the most unstable wavelength. Bandwidth helps indirectly by smoothing the interference pattern; i.e. by moving the hot spots around on a time scale $1/\Delta\omega < 1/\gamma$. This reduction of filamentation has been found in simulations[24] but has not yet been directly observed in experiments.

Experiments[25-27] have shown instability reduction by laser beam incoherence. The use of both spatial and temporal incoherence gives the most reductions. Figure 10 shows the measured Raman backscattered signal versus the average intensity in NRL experiments[25] using 1.05 μm light. Note the dramatic reduction by the use of broadband ISI. Either random phase plates or ISI have also been used in a number of other experiments.[26-27] As expected, the use of a modest bandwidth alone does not help much ($\Delta\omega/\omega_0 \lesssim 0.1\%$). Ob-

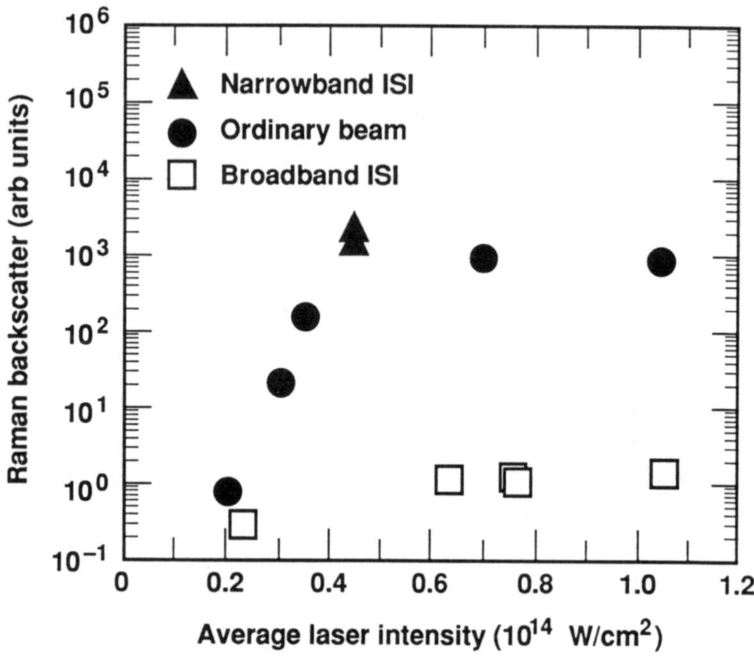

Figure 10.   The variation with intensity of the peak Raman backscatter in the wavelength band 1350-1750 nm for broadband ISI, narrowband ISI, and an ordinary laser beam (Obenschain et al., 1989).

served reductions using ISI (and sometimes random phase plates) are not yet well understood. Possibilities include the reduction of filamentation and/or the generation of smoother plasma density profiles. Defining the benefits of laser beam incoherence in long scale scalelength plasmas is a major thrust of current activity.

In summary, the interaction of intense electromagnetic waves with plasmas remains a very fertile area of research. Virtually all the instabilities have now been identified in laser plasma experiments, which are becoming increasingly well-diagnosed. These experiments are providing a crucial test bed for understanding and controlling any potential deleterious effects. Current research topics include understanding the consequences of filamentation, the nonlinear interplay between instabilities, and the extension to large plasmas. A major thrust is to define and understand the benefits of laser beam incoherence for laser plasma coupling. This is an important issue for the choice of a new laser driver. Finally, the advent of short pulse, high brightness lasers is allowing experiments on new topics, including acceleration by laser wakefields, optical guiding by plasmas, relativistic effects, and photon acceleration.

ACKNOWLEDGEMENTS

In preparing this mini-review and update, I am grateful to the many authors who have allowed me to quote their published data. Particular thanks goes to my colleagues in the Laser Experiments Program at the Lawrence Livermore National Laboratory, including E.M. Campbell, R.P. Drake, P.E. Young, C. Darrow, D.S. Montgomery, and R.E. Turner. There were also many valued discussions with H. Baldis, B. Langdon, E. Williams, R. Berger, K. Estabrook, and B. Lasinski.

Work performed under the auspices of the United States Department of Energy by the Lawrence Livermore National Laboratory under contract number W-7405-ENG-48.

REFERENCES

1. J. H. Nuckolls, L. Wood, A. Thiessen, and G. Zimmerman, Nature **239**, 139 (1972).

2. R. Elton, *X-Ray Lasers* (Academic Press, Mass., 1990).

3. See many articles in *Advanced Accelerator Concepts*, edited by C. Joshi (American Institute of Physics, NY, 1989).

4. M. M. Murname, H. C. Kapteyn, and R. W. Falcone, Phys. Rev. Lett. **62**, 155 (1989) and references therein.

5. V. Stefan and N. A. Krall, Phys. Fluids **28**, 2937 (1985).

6. A. Y. Wong, P. Y. Cheung, M. J. McCarrick, J. Stanley, R. F. Wuerker, R. Close, B. S. Bauer, E. Fremouw, W. Kruer, and B. Langdon, Phys. Rev. Lett. **63**, 271 (1989).

7. H. A. Baldis, E. M. Campbell, and W. L. Kruer, in Handbook of Plasma Physics, **Vol. 3** (to be published).

8. K. Tanaka, L. Goldman, W. Seka, M. Richardson, J. Soures, and E. Williams, Phys. Rev. Lett. **48**, 1179 (1982).

9. R. B. Drake, Laser and Particle Beams **6**, 235 (1988).

10. C.B. Darrow, R.P. Drake, D.S. Montgomery, P.E. Young, K. Estabrook, W.L. Kruer, and T.W. Johnston, Phys. Fluids (1990).

11. R. B. Drake, H. Baldis, R. Berger, W. Kruer, E. Williams, K. Estabrook, T. Johnston, and P. Young, Phys. Rev. Lett. **64**, 423 (1990).

12. R. E. Turner, K. Estabrook, R. B. Drake, E. A. Williams, H. N. Kornblum, W. L. Kruer, and E. M. Campbell, Phys. Rev. Lett. **57**, 1725 (1986).

13. P. E. Young, H. A. Baldis, R. P. Drake, E. M. Campbell, and K. Estabrook, Phys. Rev. Lett. **61**, 2336 (1988).

14. P. E. Young, H. A. Baldis, T. W. Johnson, W. L. Kruer, and K. G. Estabrook, Phys. Rev. Lett. **63**, 2812 (1989).

15. H. A. Baldis, P. E. Young, R. P. Drake, W. L. Kruer, Kent Estabrook, E. A. Williams, and T. W. Johnston, Phys. Rev. Lett. **62**, 2829 (1989).

16. R. E. Turner, K. Estabrook, R. L. Kauffman, D. Bach, R. P. Drake, D. Phillion, B. Lasinski, E. M. Campbell, W. L. Kruer, and E. A. Williams, Phys. Rev. Lett. **54**, 189 (1985).

17. R. P. Drake, Comm. Plasma Phys. Controlled Fusion XII, 189 (1989).

18. Y. Kato and K. Mima, Appl. Phys. B **29**, 186 (1982).

19. R. H. Lehmberg and S. P. Obenschain, Opt. Commun. **46**, 27 (1983).

20. S. Skupsky, R. W. Short, T. Kessler, R. S. Craxton, S. Letzring, and T. M. Soures, J. Appl. Phys. **66**, 3456 (1989).

21. J. J. Thomson, Nuclear Fusion **15**, 237 (1975).

22. E. A. Williams, R. Berger, and A. Bourdier, *Laser Program Annual Report 87*, Lawrence Livermore National Laboratory, Livermore, Calif., UCRL-50021-87 (1988), pp. 2-48 to 2-51.

23. A. B. Langdon, UCRL-89027, submitted to Phys. Fluids (1989).

24. A. J. Schmitt, Phys. Fluids **31**, 3079 (1988).

25. S. P. Obenschain, J. A. Stamper, C. J. Pawley, A. N. Mostovych, J. H. Gardner, A. J. Schmitt, and S. R. Bodner, Phys. Rev. Lett. **62**, 768 (1989).

26. Y. Kato, K. Mima, N. Miyanaga, S. Arinaga, Y. Kitagawo, M. Nakatsuka, and C. Yamanaka, Phys. Rev. Lett. **53**, 1057 (1984).

27. S. Coe, T. Afshar-rad, M. Desselberger, F.Khattak, O. Willi, A. Givlietti, Z. Q. Lin, W. Yu, and C. Danson, Europhys. Lett. **10**, 31 (1989).

Recent Trends in Inertial Confinement Fusion

Robert L. McCrory, Jr.

University of Rochester
Laboratory for Laser Energetics
250 East River Road
Rochester, NY 14623-1299

Inertial confinement fusion (ICF) requires high compression of fusion fuel, to densities approaching 1000 times liquid density of deuterium-tritium (DT), at central temperatures in excess of 5 keV. The direct-drive approach to ICF is one energy efficient than indirect drive if the stringent drive symmetry and hydrodynamic stability requirements can be met by a suitable laser irradiation and target design. Experiments using cryogenic fuel capsules in conjunction with distributed phase plates (DPPs) on the frequency-tripled OMEGA laser system have achieved compressed DT fuel densities in the 100-200 times liquid density regime, but the experiments exhibited deviations from one-dimensional performance. The deviations are believed to result from nonuniform implosion of fuel and shell material due to irradiation nonuniformities not removed by the DPPs. Improvements in irradiation uniformity through the use of a new technique, SSD, may lead to reduced hydrodynamic instability growth and early one-dimensional capsule performance. SSD allows high-frequency tripling in a solid-state laser system.

Elements of the LLE Inertial Fusion Research Program

- **Drive Uniformity Program**

- **Implosion Experiments on the OMEGA Facility**

- **Direct-Drive Target Design**

- **Plasma Physics**

SSD (Smoothing by Spectral Dirsion) Provides Uniform
Target Illumination for Frequency Converted Laser Systems

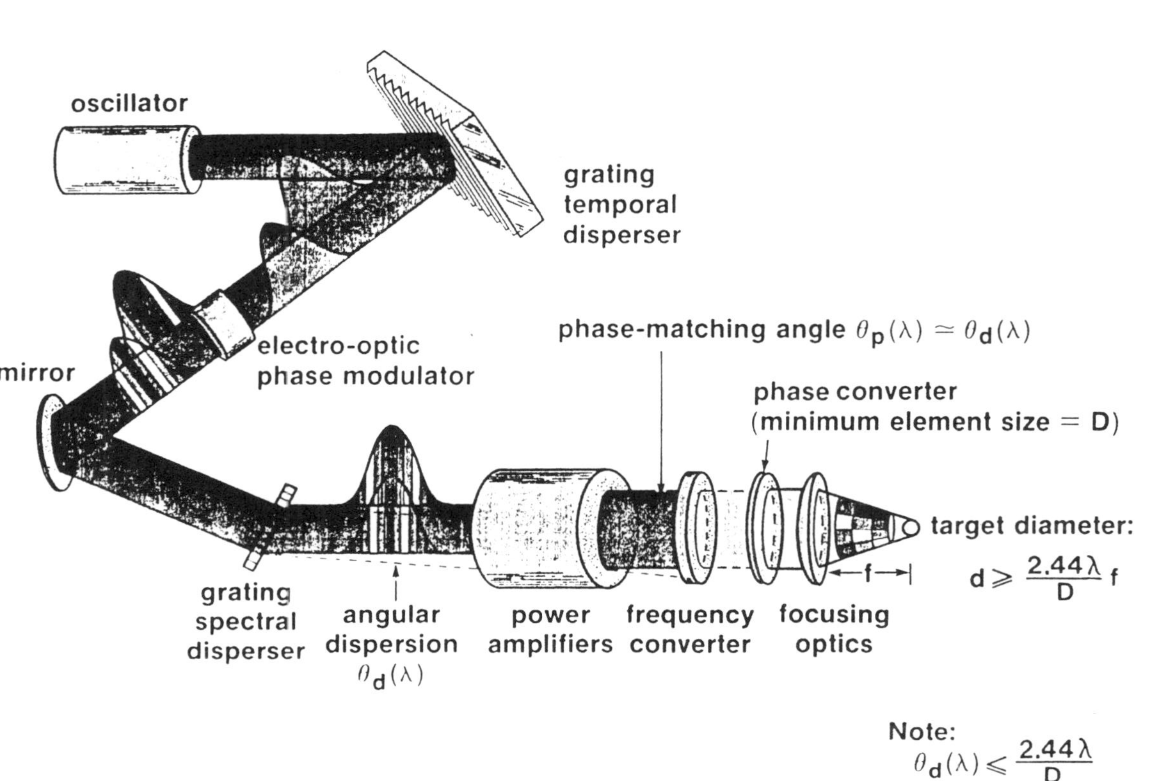

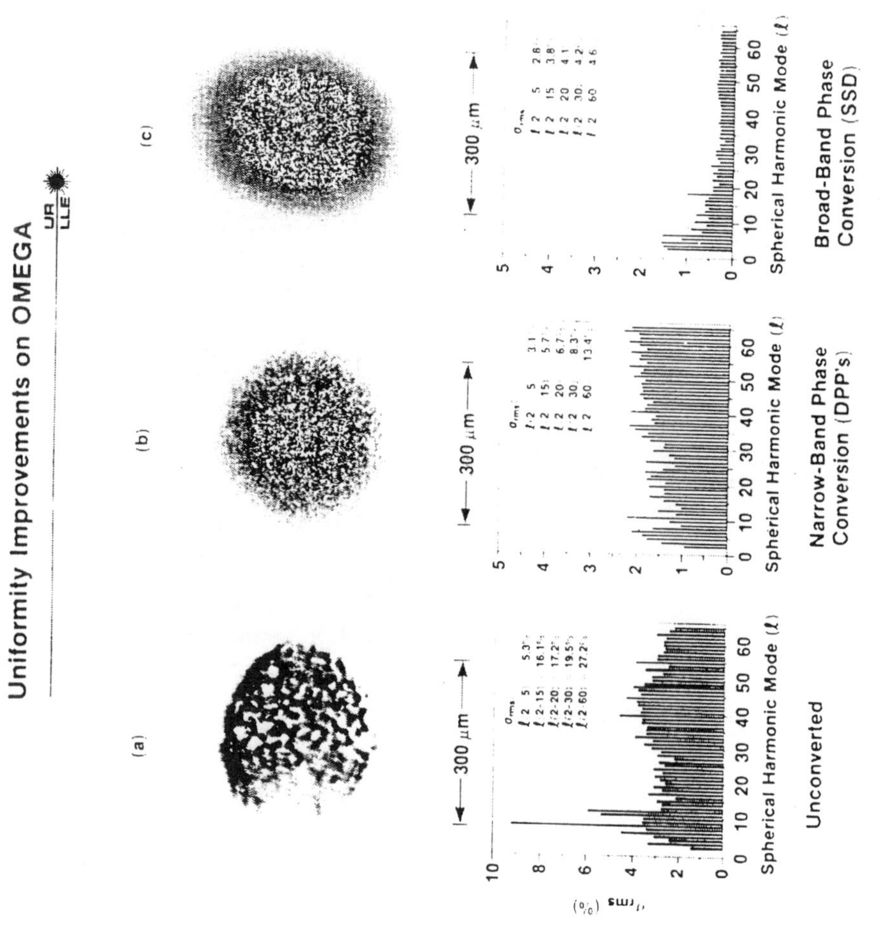

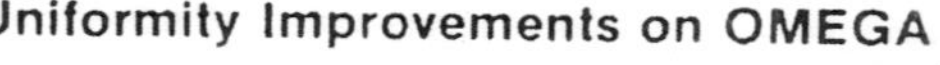

Uniformity Improvements on OMEGA
UR
LLE
(a)
(b)
(c)
300 μm
300 μm
300 μm
Spherical Harmonic Mode (ℓ)
Spherical Harmonic Mode (ℓ)
Spherical Harmonic Mode (ℓ)
Unconverted
Narrow-Band Phase Conversion (DPP's)
Broad-Band Phase Conversion (SSD)

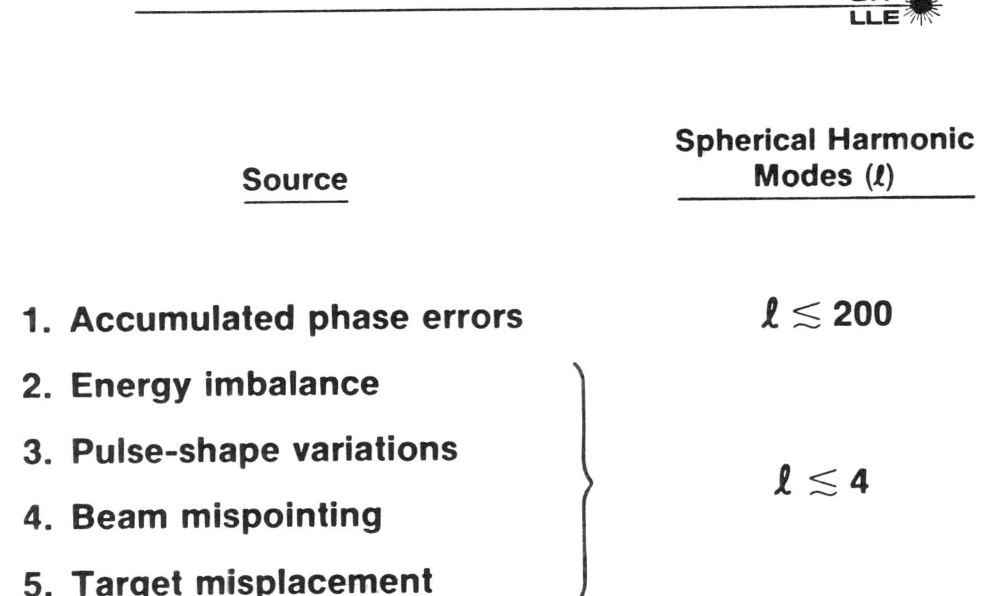

Sources of Irradiation Nonuniformity
UR
LLE
Source
Spherical Harmonic Modes (ℓ)
1. Accumulated phase errors
2. Energy imbalance
3. Pulse-shape variations
4. Beam mispointing
5. Target misplacement
$\ell \lesssim 200$
$\ell \lesssim 4$

Nonuniformities Due to Power Imbalance

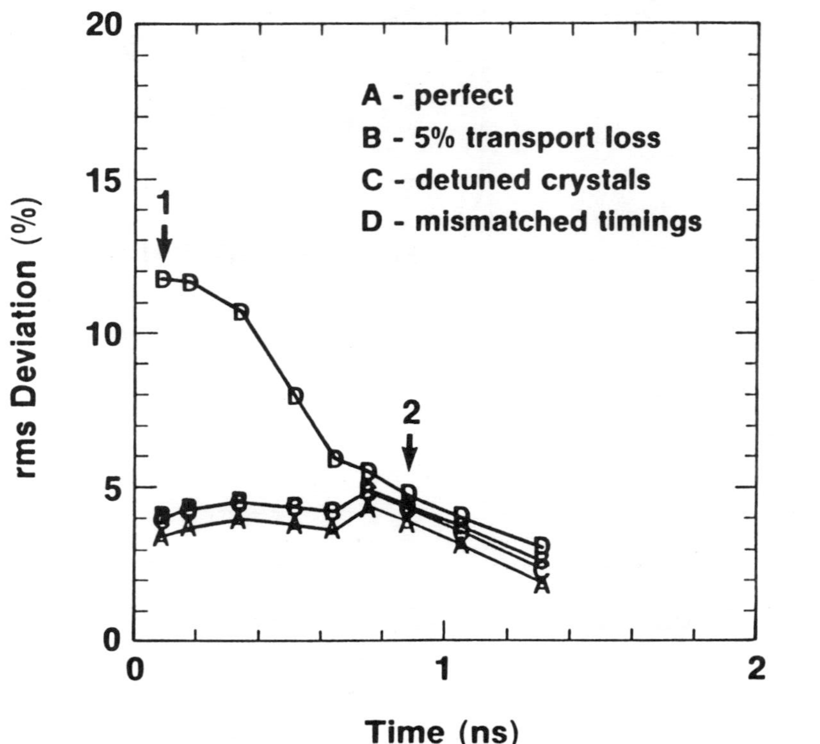

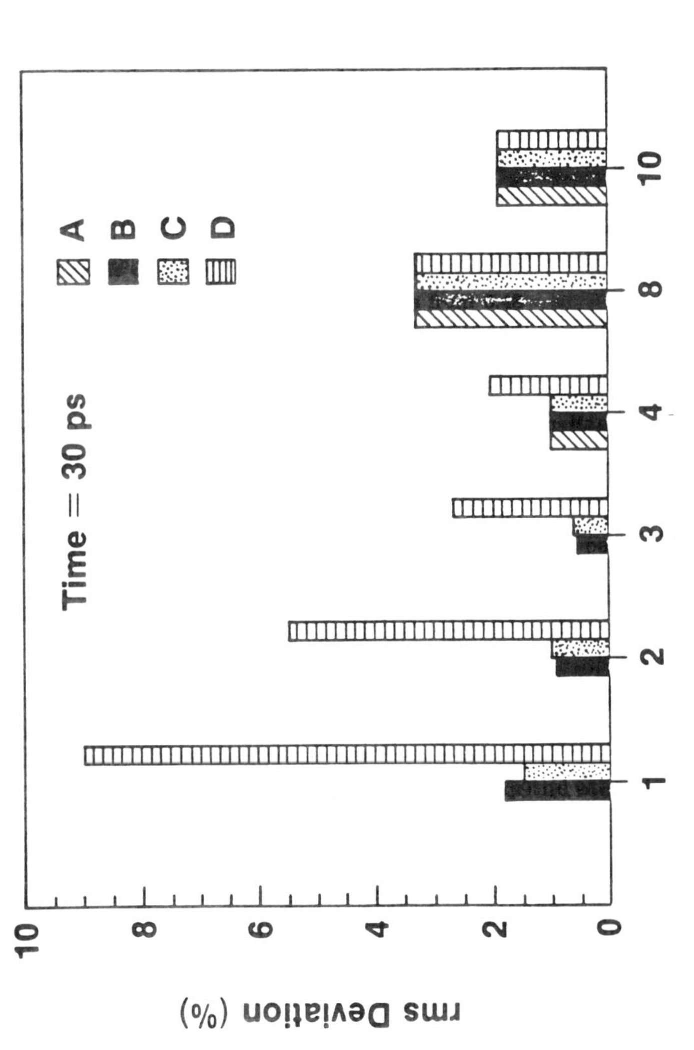

rms Deviation Due to Applied Levels of Power Imbalance at Times Early in the Implosion
UR
LLE
Time = 30 ps
A
B
C
D
rms Deviation (%)
ℓ-Mode Number
10
8
6
4
2
0
1
2
3
4
8
10

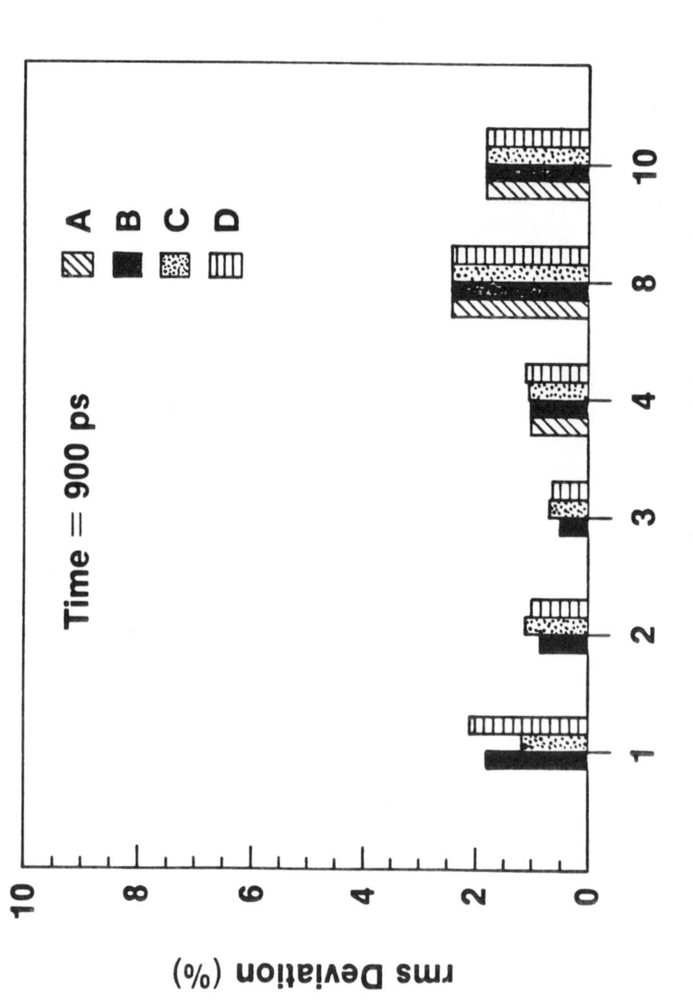
rms Deviation Due to Applied Levels of Power Imbalance
at Times Late in the Implosion
UR
LLE
Time = 900 ps
A
B
C
D
10
8
6
4
2
0
rms Deviation (%)
1
2
3
4
8
10
ℓ-Mode Number

Serpentine X-Ray-Framing Camera

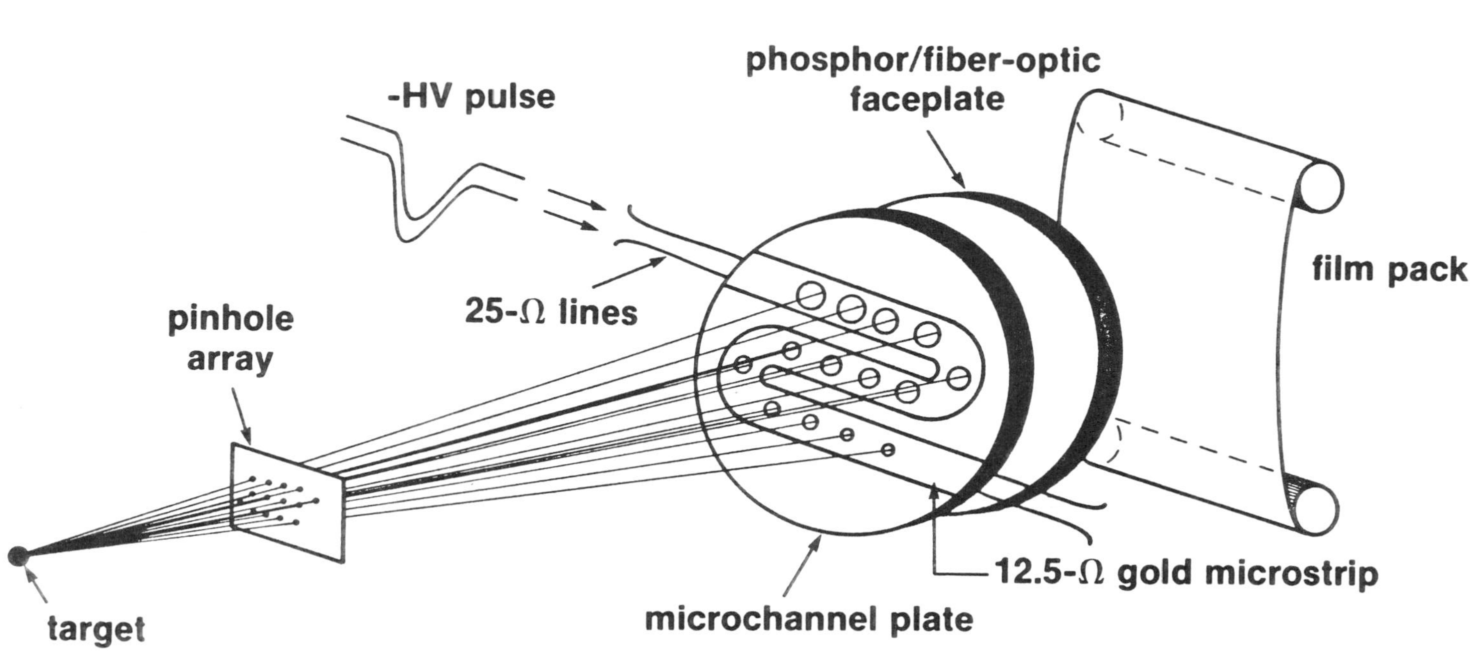

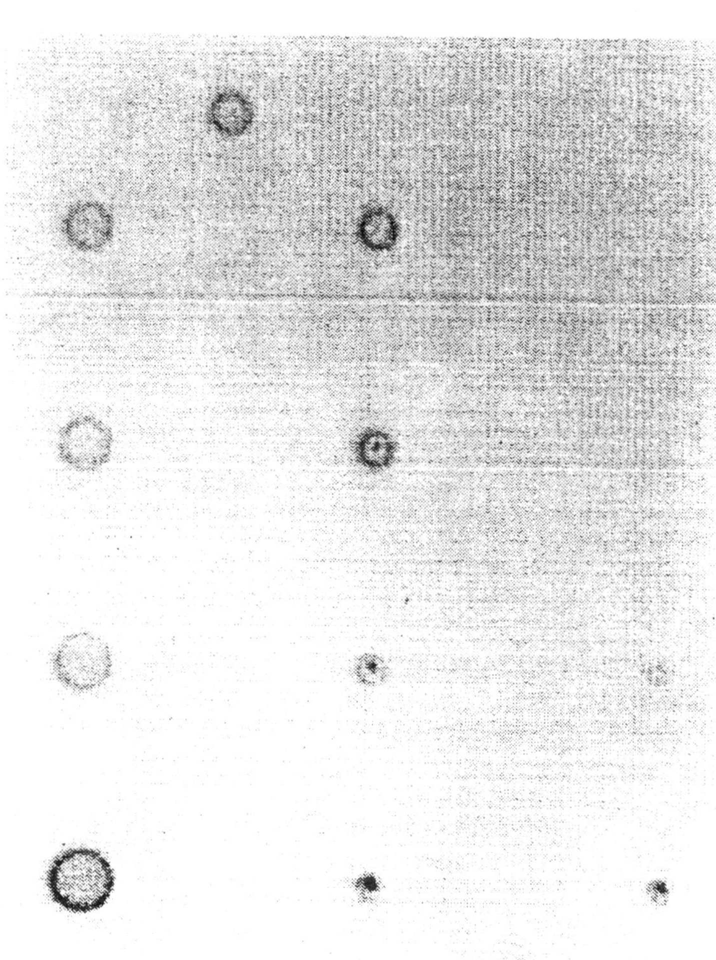

Multiframe, High-Speed X-Ray-Framing Camera Images of OMEGA Shot #18682

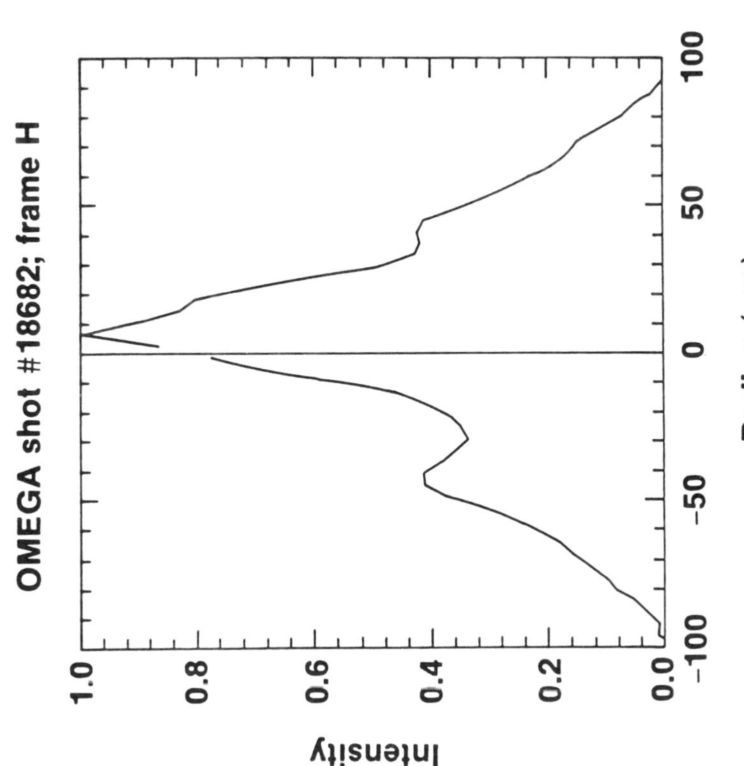
Azimuthally Averaged Intensity Profiles
UR
LLE
OMEGA shot #18682; frame H
Intensity
1.0
0.8
0.6
0.4
0.2
0.0
-100
-50
0
50
100
Radius (μm)

2-D *ORCHID* Simulation
of a Deliberate Power Imbalance Experiment

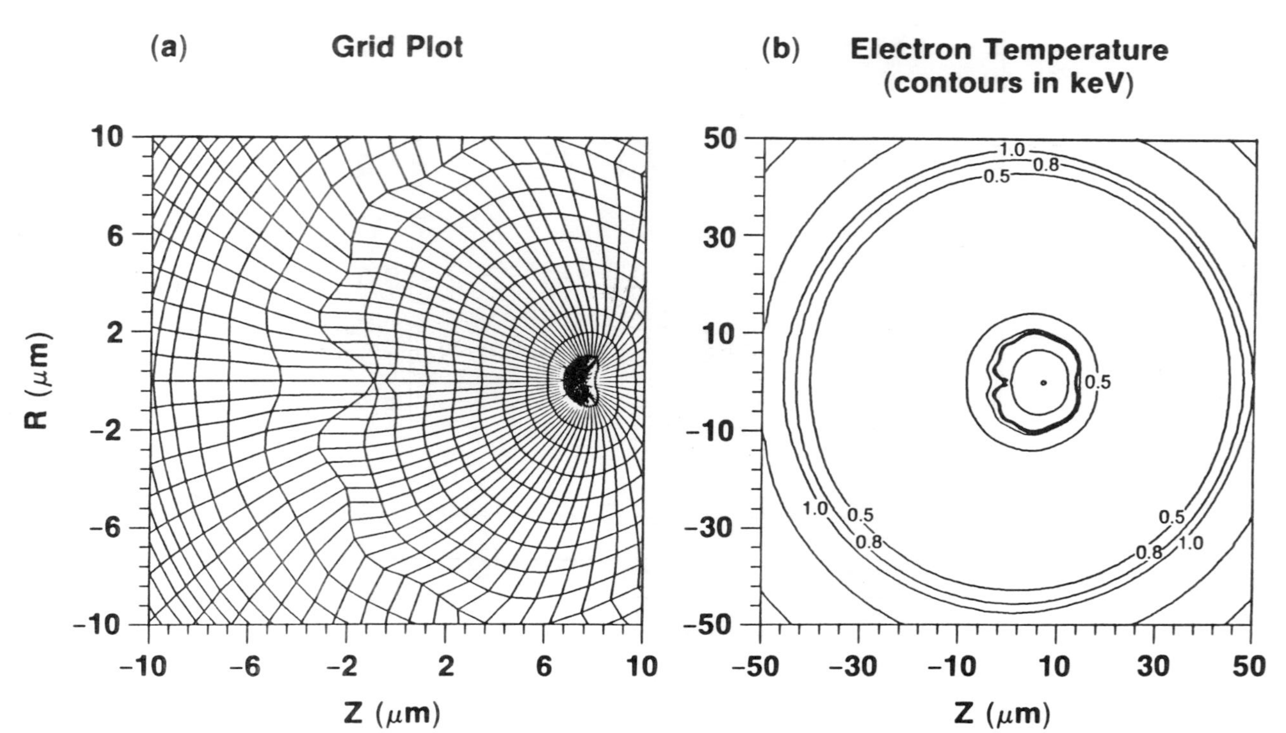

Summary
Initial Power Balance Experiments on OMEGA

- **Various sources of power imbalance on OMEGA have been characterized and modeled numerically.**

- **Frequency-conversion crystal detuning and UV transport losses make only a modest contribution to overall nonuniformity.**

- **Significant nonuniformity can exist on the target early in time due to pulse misshaping.**

- **Good correlation between theory and experiment is obtained for initial ($\ell = 1$) power imbalance experiments.**

Elements of the LLE Inertial Fusion Research Program

- **Drive Uniformity Program**

- **Implosion Experiments on the OMEGA Facility**

- **Direct-Drive Target Design**

- **Plasma Physics**

OMEGA Laser Parameters Are Optimized for Direct-Drive Target Experiments

- Beam-energy balance of 2%–3% rms is routinely obtained.

- Total on-target energy $\approx$ 1.5 kJ at 0.35 μm.

- Beam-pointing error $\lesssim$ 10 μrad.

- Phase conversion provides large depth of focus and reproducible beam profiles.

- Time-integrated irradiation uniformity of $<$ 5% rms is obtained with broad-bandwidth phase conversion (SSD).

- Synchronized short-pulse probe beam; synchronized high-energy beam (GDL).

OMEGA and GDL have provided more than 10,000 target shots from 1981–1989.

OMEGA Implosion Glass-Capsule Parameters

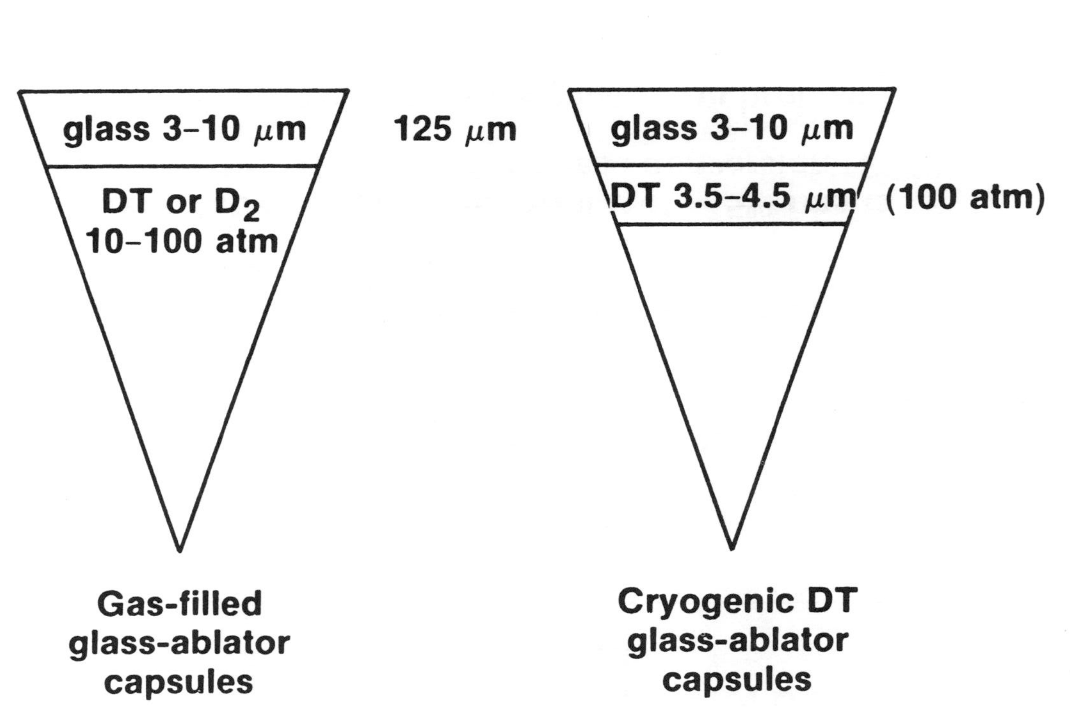

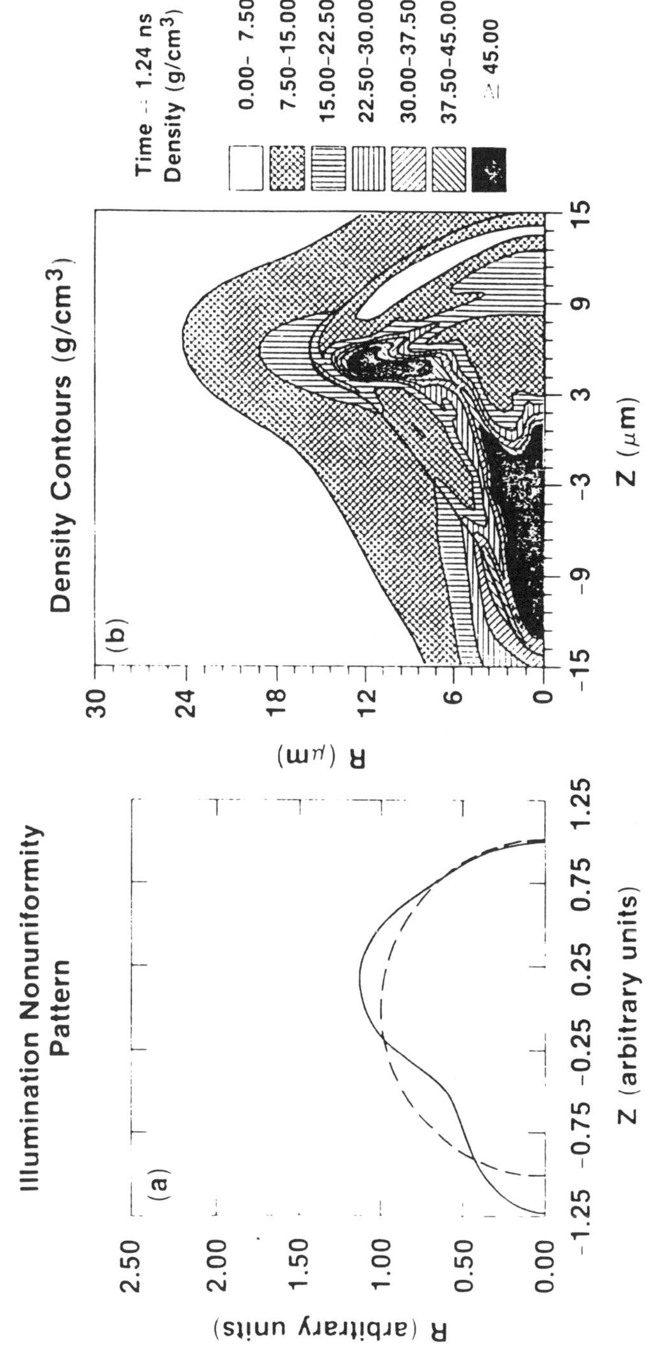

Effects of ~44% Peak-to-Valley Illumination
Nonuniformities (1 ≤ ℓ ≤ 4) on 5-μm Glass
Ablator (100 atm) Cryogenic Pellet Implosion
(time shown is for $Y_{exp} = Y_{simulation}$)
UR
LLE
Illumination Nonuniformity
Pattern
(a)
R (arbitrary units)
Z (arbitrary units)
-1.25 -0.75 -0.25 0.25 0.75 1.25
0.00 0.50 1.00 1.50 2.00 2.50
Density Contours (g/cm³)
(b)
Z (μm)
R (μm)
-15 -9 -3 3 9 15
0 6 12 18 24 30
Time = 1.24 ns
Density (g/cm³)
0.00- 7.50
7.50-15.00
15.00-22.50
22.50-30.00
30.00-37.50
37.50-45.00
≥ 45.00

Hydrodynamic Stability

UR
LLE

1. Two-dimensional *ORCHID* simulations are used to determine $\gamma(t)$ for a number of ℓ-modes (10–20 per capsule design).

2. *ORCHID* simulations show the Rayleigh-Taylor growth rate can be expressed in a form:

$$\gamma(t) = \sqrt{\frac{A(t)k\,g(t)}{1 + kL(t)}} - \beta k V_a(t)$$

where

$A(t)$ = Atwood number
$g(t)$ = acceleration
k = unstable wave number
$L(t)$ = ablation surface density scale length
$V_a(t)$ = ablation velocity
β = constant; depends on determination of $A(t)$, $L(t)$, and $V_a(t)$

3. Growth-rate spectrum is determined by interpolating between ℓ's found with ORCHID and/or fitting between ℓ's using expression above.

4. Growth evolution is determined using a model based on Haan's[1] model.

[1]S. W. Haan, Phys. Rev. A 39, 5812 (1989).

Hydrodynamic Stability
(continued)

1. Using "best" current theoretical understanding, hydrodynamic stability will place severe constraints on target fabrication and laser quality for CH base-line designs.

2. A number of different direct-drive designs are being examine to reduce the target fabrication/illumination uniformity constraints:

(a) High-Z-doped CH ablators

- change the isentrope of ablator by radiation preheat
  (increase density scale length and ablation velocity)
- some increase in radiation preheat of main fuel layer

(b) Different pulse shapes

- change the isentrope of imploding capsule by shock heating
  (increase density scale length and ablation velocity)
- results in increased preheat for main fuel layer

Measured Fuel Areal Densities and Inferred Fuel Densities
100-atm-DT, Cryogenic, Glass-Ablator Implosions

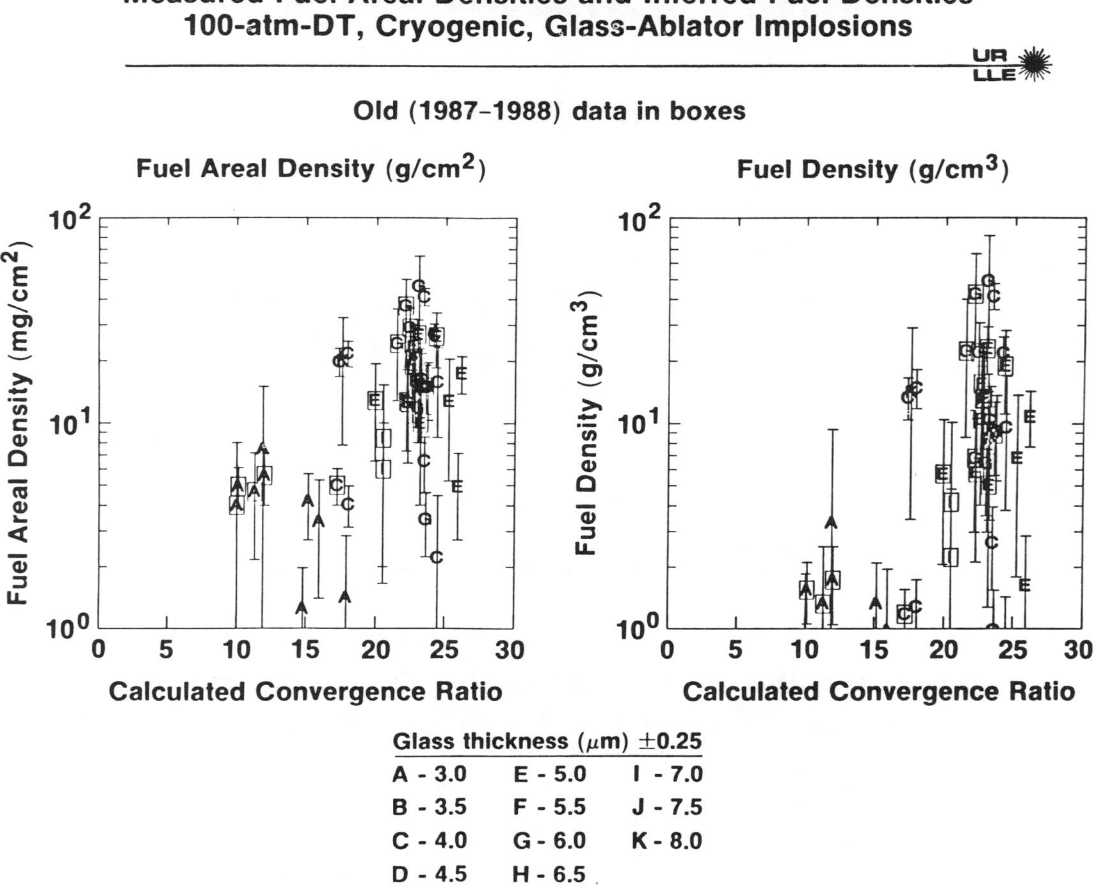

Pressure and Density Profiles at Various Times
100-atm-DT, 5-μm-Glass-Ablator Cryogenic Implosion

(1.25-kJ, 625-ps Gaussian; $R_{ID} = 125\ \mu$m)

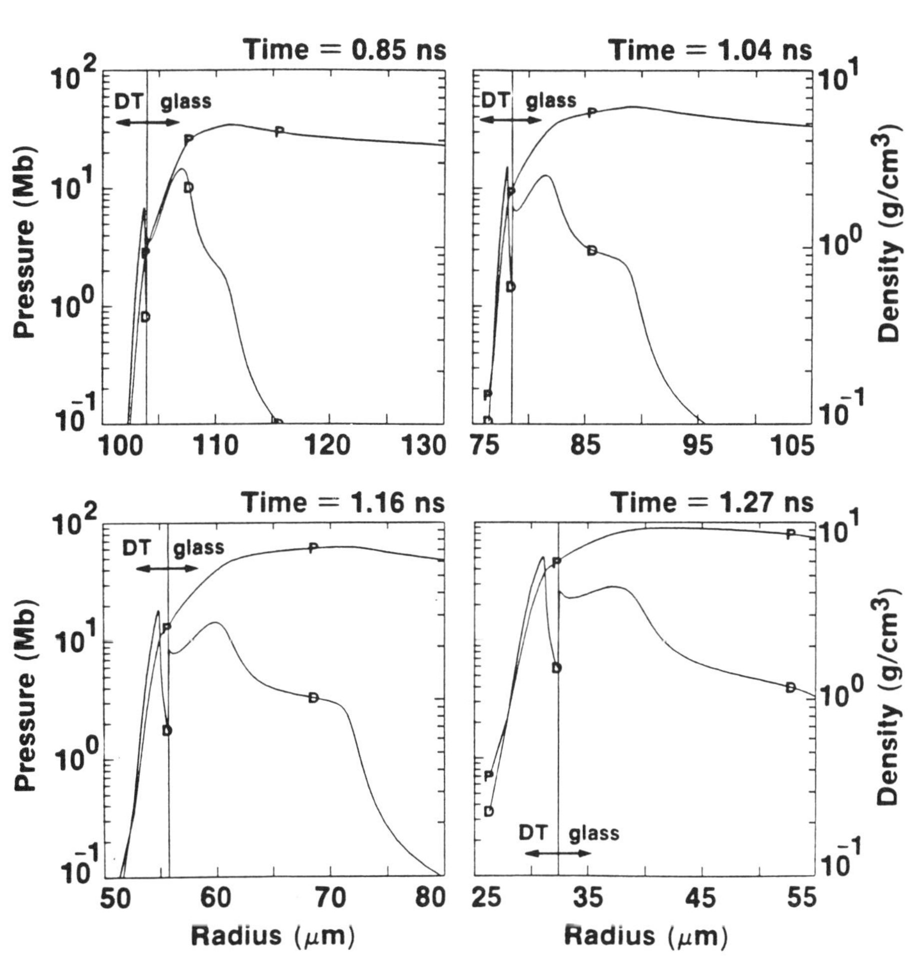

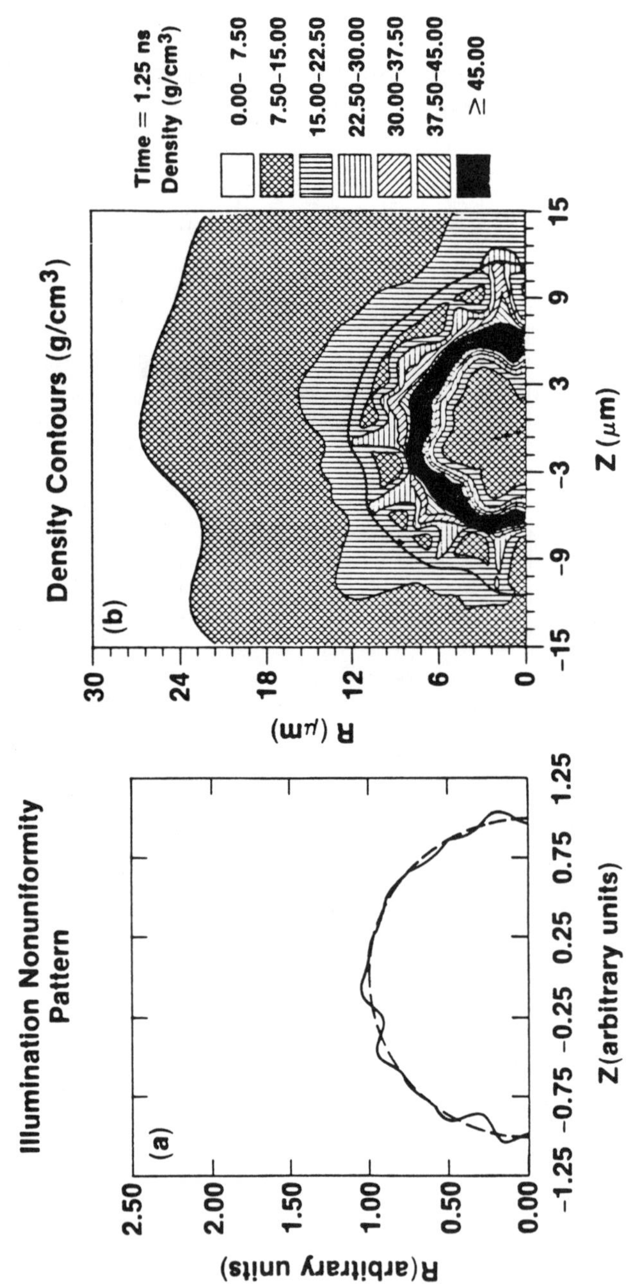
Effects of ~15% Peak-to-Valley Illumination Nonuniformities ($1 \leq \ell \leq 20$) on 5-μm Glass Ablator (100 atm) Cryogenic Pellet Implosion (time shown is for $Y_{exp} = Y_{simulation}$)
UR
LLE
Density Contours (g/cm^3)
Time = 1.25 ns
Density (g/cm^3)
0.00– 7.50
7.50–15.00
15.00–22.50
22.50–30.00
30.00–37.50
37.50–45.00
$\geq$ 45.00
(b)
R (μm)
30
24
18
12
6
0
Z (μm)
-15
-9
-3
3
9
15
Illumination Nonuniformity Pattern
(a)
R(arbitrary units)
2.50
2.00
1.50
1.00
0.50
0.00
Z(arbitrary units)
-1.25
-0.75
-0.25
0.25
0.75
1.25

One-Dimensional Simulations of Effects of Complete Mixing of Cryogenic Layer with Various Amounts of Glass
(100-atm-DT, glass-ablator cryogenic simulations)

C - clean; 1 - 14% by atom SiO_2; 2 - 25% by atom SiO_2

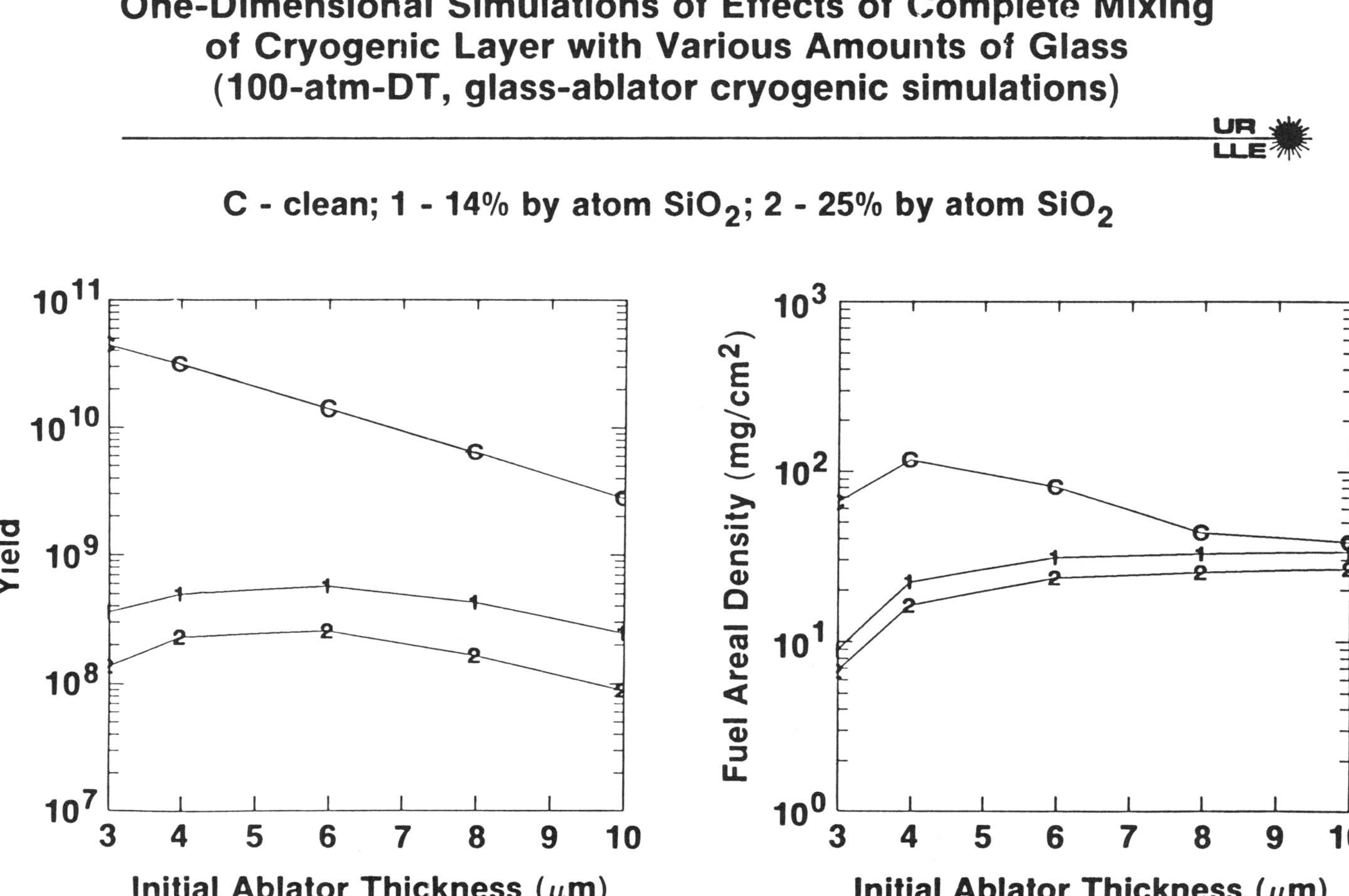

Summary
(Cryogenic-Fuel, Glass-Ablator Capsule Implosions)

- 100–200 XLD final-fuel densities have been obtained with cryogenic-DT, glass-ablator implosions.

- The effects of long-wavelength distortions and mixing of glass-ablator material into the fuel during the acceleration phase (due to unstable region in the DT near the DT glass interface) may be responsible for the observed departure from "1-D" performance.

High-Gain Pellet Physics Issues Could Be Addressed with a 30-kJ Direct-Drive Laser System

- **Energy-scaled MJ pellets could be used to examine**
    - illumination uniformity requirements
    - energy coupling and transport
    - hydrodynamic stability
    - high-density hot-spot physics

- **Long-scale-length plasma experiments could examine plasma physics issues using both single-beam and multiple-beam overlap configurations under conditions of optimal beam uniformity.**

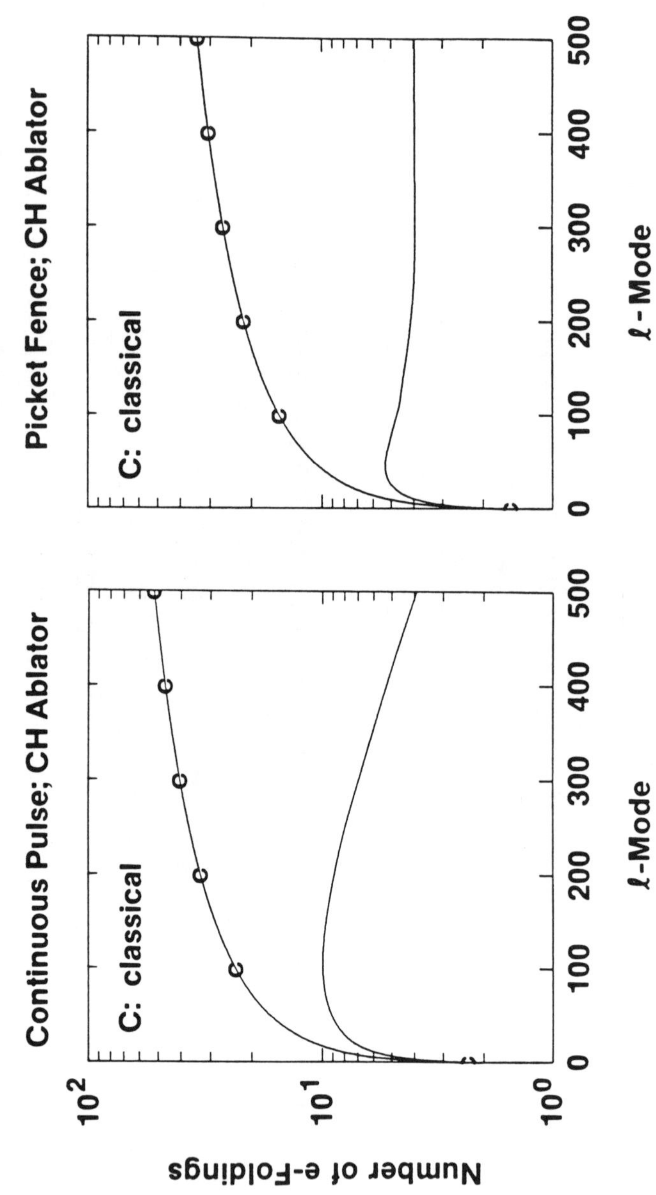

Number of e-Foldings per Mode
for 30-kJ OMEGA Upgrade Capsule Designs
(fuel isentrope changed by pulse shaping)
UR
LLE
Continuous Pulse; CH Ablator
C: classical
Picket Fence; CH Ablator
C: classical
Number of e-Foldings
ℓ-Mode
ℓ-Mode

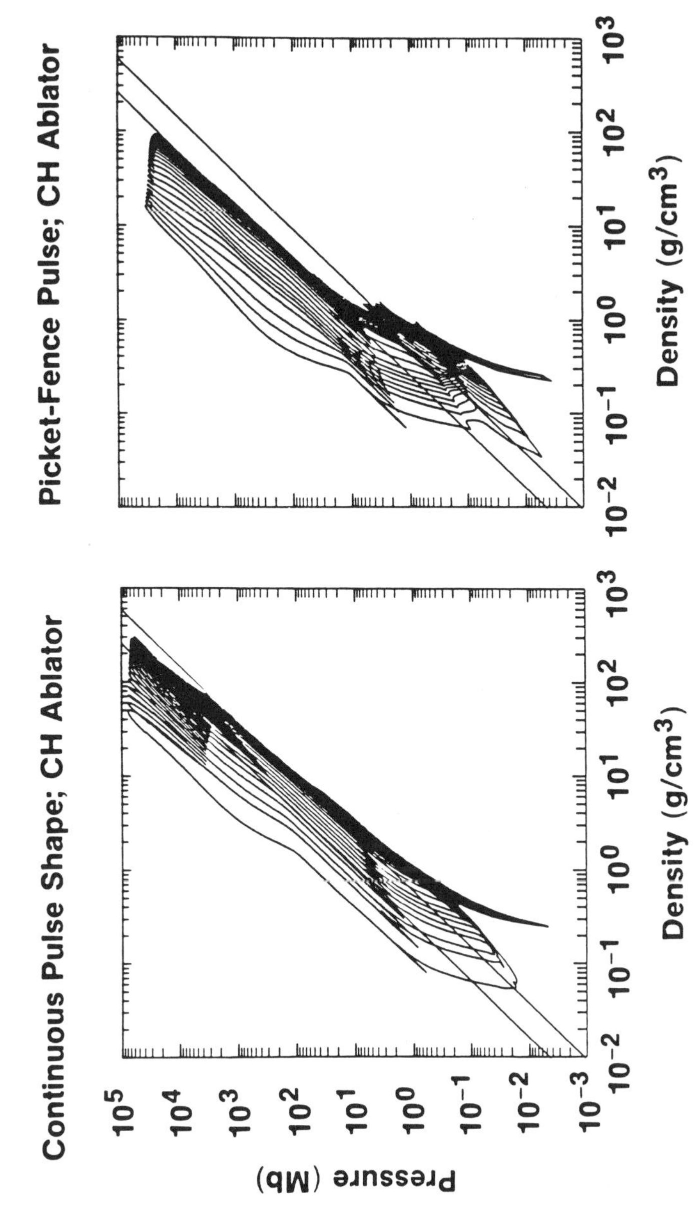

Performance of 30-kJ OMEGA Upgrade Capsule Designs
(fuel isentrope changed by pulse shaping)
(fuel pressure versus density)
UR
LLE
Continuous Pulse Shape; CH Ablator
Picket-Fence Pulse; CH Ablator
Density (g/cm³)
Density (g/cm³)
Pressure (Mb)

Performance of 30-kJ OMEGA Upgrade Capsule Designs
(fuel isentrope changed by pulse shaping)

Clean 1-D Results

	Continuous Pulse	Picket-Fence Pulse
Yield	4.4×10^{13}	4.6×10^{13}
Peak areal density (ρR)	0.52 g/cm^2	0.22 g/cm^2
Neutron-weighted ρR	0.43 g/cm^2	0.18 g/cm^2
Peak density (ρ)	310 g/cm^3	95 g/cm^3
Neutron-weighted Ti	2.80 keV	3.49 keV
Convergence ratio ($\rho R = 0.2$)	18	8.5

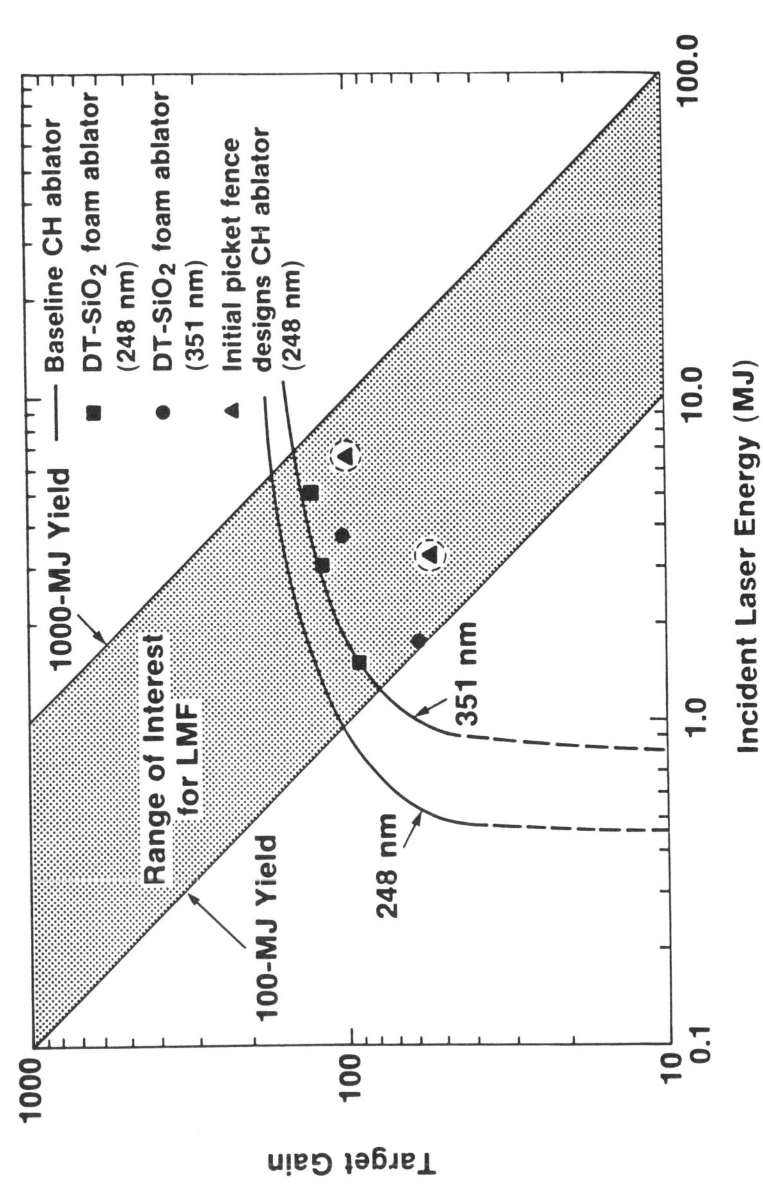
Target Gain versus Incident Laser Energy
UR
LLE
Target Gain
Incident Laser Energy (MJ)
1000
100
10
0.1
1.0
10.0
100.0
1000-MJ Yield
100-MJ Yield
Range of Interest
for LMF
351 nm
248 nm
Baseline CH ablator
DT-SiO2 foam ablator (248 nm)
DT-SiO2 foam ablator (351 nm)
Initial picket fence designs CH ablator (248 nm)

Summary

- **Further improvements to SSD and power balance are now underway on OMEGA.**

- **Glass-ablator targets will be used to study improvements.**

- **Polymer-ablator capsule implosions will be started in the near future on OMEGA.**

    - **Target fabrication and illumination uniformity will initially limit these experiments.**

- **30-kJ OMEGA Upgrade designs are being conducted to address concerns associated with hydrodynamic stability issues of high-gain direct-drive capsule implosions.**

THE LLNL ICF PROGRAM:
RECENT PROGRESS TOWARD INERTIAL CONFINEMENT FUSION

Erik Storm, Program Leader
ICF Program, LLNL

The Nova Program at the Lawrence Livermore National Laboratory has made substantial progress in inertial confinement fusion (ICF) target physics, target diagnostics, and laser science and technology. In each of these three areas, progress required the development of experimental techniques and computational modeling. Recent experiments and modeling have established much of the physics necessary to validate the basic concept of ignition and gain (fusion energy released/driver energy delivered) in the the laboratory.

Target Physics

The objectives of the target physics experiments on Nova are to address and understand critical physics issues that determine the conditions required to achieve ignition and gain of an ICF capsule. The LLNL experimental program primarily addresses indirect drive implosions in which the capsule is driven by x-rays that are produced by the interaction of the laser with a high Z plasma. Experiments address both the physics of generating the radiation environment in a laser driven hohlraum and the physics associated with impolding ICF capsules to ignition and high gain conditions.

Imploding ICF targets to high density on Nova requires shaped laser drive pulses which are temporally tailored to reduce shock heating in the target, instantaneous power balance among the ten beams at the target of about 5 percent, and placement of the beams on the target with high precision. We have made recent progress toward these goals by starting target irradiation experiments with shaped laser pulses, assessing percent energy balance, and beginning improvements in our diagnostics. Images taken of the x-ray emission from Argon doped fuel show that a shaped laser pulse drives similar targets to higher convergence. More recent experiments have produced fuel densities in excess of 40 g/cm^3 and have observed pusher areal density increases by greater than a factor of 3 when compared to non-pulse shaped implosions.

Target Diagnostics

Progress in target diagnostics has led to multiple frame (>10) x-ray and optical gated framing cameras with effective shutter times <100 ps. Prototype x-ray cameras have demonstrated 50 ps shutter times, and we expect to achieve ≤30 ps in the near future. LLNL is actively developing neutron spectroscopy, imaging, and time-resolving diagnostics that rival those based on x-ray detection. The diagnostics on Nova allow for an unprecedented study of target performance. Observables that are measured using Nova diagnostics include: a) hohlraum temperatures and temperature gradients, b) yield, c) fuel temperature, d) bang time, e) burn width, f) fuel ρr and pusher ρr , g) high energy x-ray image of core, h) neutron image of core and, i) implosion symmetry.

Laser Science and Technology

We have continued to increase Nova's productivity and flexibility as an experimental facility. The range of energies at 1.05 µm wavelength reliably provided by the laser now exceeds more than 120 kJ, a new performance record. Pulse durations varied from 10 picoseconds to over 5 nanoseconds. Temporal pulse shapes included square, Gaussian, and specially shaped pulses designed to improve target performance. We are now using the platinum-particle-free, high damage threshold laser glass in demonstrating record output energy and power at both 1.05 µm and its third harmonic wavelength at 0.35 µm. To reduce the risk of optical damage to components when the laser is operated at high output energy, we have improved the spatial uniformity of the beam. We have also begun a two-year program to increase control of the output power, pulse shape, and focusing of the ten beams.

We have performed a series of experiments to demonstrate the capabilities of the laser system at 1.05 µm at three pulse durations of interest — at 2.5 ns where damage to the output optics is of concern, at 1.0 ns which is the pulse of greatest interest for experiments, and at 0.1 ns where output power is limited by beam break-up from self-focusing. In a 2.5 ns square temporal pulse we obtained up to 125 kJ of output energy, in a 1.0 ns square pulse we obtained up to 82 kJ, and in a 0.1 ns Gaussian temporal pulse we obtained up to 125 TW of output power, all with 2% or less r.m.s. spread in energy among Nova's ten beams. The tests demonstrated that Nova can safely exceed its original 1.05 µm performance goals (80 to 120 kJ). A recent series of laser shots has demonstrated that Nova can deliver more than 40 kJ of 0.35 µm light into its target chamber.

Work performed under the auspices of the U.S. Department of Energy by the Lawrence Livermore National Laboratory under Contract Number W-7405-ENG-48.

Part II

MAGNETIC FUSION

A. Toroidal Magnetic Confinement

Radiation from Mildly Relativistic Electrons in Tokamaks

N. J. Fisch

Plasma Physics Laboratory, Princeton University, Princeton, NJ 08543

Abstract

As tokamak experiments progress towards the reactor regime, the hotter temperatures promote to greater prominence the role played by mildly relativistic electrons. These electrons dominate the plasma synchrotron radiation, and they can carry enormous electric current. We survey the means of describing mathematically these energetic electrons, the experimental evidence for this description, and several of the processes in which these electrons are expected to play a dominant role.

Introduction

Tokamak reactors will operate in a far different temperature regime than have nearly all of the experimental devices that are the basis of our current understanding of tokamak operation. The new temperature regime means that a large number of mildly relativistic electrons will contribute to power loss through intense synchrotron radiation. These power losses become more severe if the tokamak is subjected to wave heating of electrons for the purpose of ignition or current drive.

A number of issues, therefore, deserve close scrutiny: Can this intense radiation be reflected back into the tokamak, either to preserve heat or redirect current? Do radiative transfer processes dominate conventional particle and heat transport? Can the intense synchrotron radiation be employed to inform on tokamak parameters, particularly near the tokamak center, where other means of measurement are unavailable or difficult?

Central to our answers to these questions is the physics of these mildly relativistic, radiating electrons. By mildly relativistic we mean those electrons in the tail of the electron distribution function, possibly with velocities 4 to 10 times the thermal velocity and with energies of perhaps 50 keV to 1 MeV. These electrons often are found in an enhanced tail or plateau of the electron parallel velocity distribution function. By parallel we mean the direction of the largely toroidal dc magnetic field in the tokamak. The density of these plasmas range from 10^{12}cm^{-3} to 10^{14}cm^{-3}. The time it takes the energetic electrons to thermalize by collisions might be from 1 to 100 ms. Typically, the number of such electrons might be less than a percent of the total number of electrons, but when the heating of the tokamaks is accomplished by radio frequency (rf) waves, the superthermal component of the electron population is substantially enhanced.

Despite their small number, these electrons dominate several important physical processes, including the synchrotron radiation. For a reactor this can be a serious cooling problem, unless the radiation is reflected efficiently by the tokamak chamber walls. During tokamak start-up, mildly relativistic electrons might be produced through electron runaway, when the acceleration by the dc toroidal electric field in the tokamak overcomes collisional frictional forces. Eventually, the runaway electrons become relativistic. The current-drive effect, where rf waves produce an asymmetry in the superthermal electron parallel velocity distribution, might be the eventual way to achieve steady state tokamak operation. Indeed, the heating, in any event, of a tokamak to ignition, might be accomplished by the absorption of rf waves by the superthermal electrons; in the event that the absorption is asymmetric, the current-drive effect can occur. It should be noted, too, that the conductivity of the plasma itself can be significantly altered in the presence of a significant nonthermal population of electrons, such as is produced during rf heating.

It is remarkable that the description of these electrons is relatively manageable; the equations governing these electrons are often mathematically more tractable than those governing the bulk electrons and ions. The interaction with the rf waves is described well by quasilinear theory. The velocity space dynamics is also governed by acceleration possibly due to a dc electric field, and by classical collisions which for fast electrons take a particularly simple form. The fast electrons are affected only through the number of bulk electrons and ions, and are not affected by the bulk ion or electron temperature. Since the slowing down is over milliseconds, a process slower than even the low frequency turbulence that permeates the dynamics of the bulk electrons and ions, the fast electrons do not respond to even the turbulent fluctuations in the background plasma density. Since the collisional slowing down of the fast electrons is a process with so few parametric dependencies, since these electrons can be observed through theier copious synchrotron emission, it might, in fact, be feasible to deduce the plasma parameters on which the dynamics does depend by employing both knowledge of the electron dynamics and the emission data.

The increasing importance of the dynamics of these electrons naturally raises the question of how firmly based on experimental observation is the theory of these electrons. This paper will argue that our understanding of these electrons enjoys an unusually firm experimental basis, at least for the parallel dynamics. The perpendicular transport of these electrons is less certain, but certain conclusions can be drawn. The detailed evidence for our understanding comes primarily from experiments designed to elucidate the important physical processes in which the mildly relativistic electrons are involved. Here, we review both these processes and the experimental evidence for the transport equations that describe these processes. We indicate also how further measurements, particularly of the copious synchrotron radiation, can further elucidate the physics relevant to these electrons.

DESCRIPTION OF THE FAST ELECTRONS

To describe the velocity distribution function f of the fast electrons, separate $f = f_M(1 + \phi)$, where f_M is a Maxwellian distribution and ϕ describes the a time-dependent perturbation brought about by rf heating. In terms of contributing to the collision integral, ϕ may be treated as small, so that f obeys the linearized Fokker-Planck equation. The evolution of ϕ may then be written as

$$f_M \partial\phi/\partial t + q\mathbf{E} \cdot \nabla_\mathbf{p} f_M \phi - C(\phi) = -\nabla_p \cdot \mathbf{S} \tag{1}$$

where $\mathbf{p}$ is the electron momentum, C is a collision term and $\mathbf{S}$ is the rf induced flux. In the case of quasilinear theory, we can write

$$\mathbf{S} = -D_{QL}(\mathbf{p}) \cdot \nabla_p f_M \phi, \tag{2}$$

where $D_{QL}(\mathbf{p})$ is the rf diffusion tensor.

There are two important and excellent approximations that lead to considerable simplification of the collision integral. Most important is the the linearization of the collision integral, which is an excellent approximation. Physically, this corresponds to tail electrons colliding off of bulk electrons and ions, but not with each other. It can immediately be recognized to be a good approximation, that is at least to a few percent, because the tail electrons make up only a few percent at most of the total number of electrons. In fact, the approximation is even better than one might suppose from this argument alone, since collisions between tail electrons in any event conserve momentum and energy, hence these collisions in themselves cannot decrease the current carried by the energetic electrons, nor their average energy which governs the collision rate with the bulk. So ignoring these collisions is even better for problems of interest. The approximation not only leads to an accurate answer, but the mathematics is vastly simplified, since, once the collision operator is linearized, Eq.(1) becomes linear in ϕ, so a Green's function response can be constructed.

The Green's function response is even more powerful when we consider that, for many problems of interest, the rf induced flux can be considered to be at least partially known, and might be treated as an inhomogeneous term. This can be done because the direction of the flux is given by the nature of the rf waves; for example, for electrostatic waves the flux is entirely in the parallel direction. Moreover, the resonance condition very much localizes $\mathbf{S}$ in velocity space too. Therefore, the flux is often known up to a multiplicative constant. This multiplicative constant is the same, however, for all quantities of interest, so the *ratio* of quantities of interest, such as the ratio of current generated to power dissipated, is determined.

The second important, but less essential, approximation to the collision term is the so-called high-velocity limit of the linearized collision integral. Physically this also corresponds to the tail electrons colliding only off of bulk electrons and ions, but these bulk electrons and ions are now assumed to be at zero temperature. This approximation, which ignores the spread in the speeds of the scatterer particles is always good for ions and is good for electrons to the extent that the scattered electrons are highly superthermal. The errors incurred due to this approximation are formally of order v_T^2/v^2 — typically at three times thermal the errors are less than ten per cent. Actually, this approximation is often much better than that since the quantities that we are interested in such as current or radiation are themselves small at small velocities where the error is large. Thus, the error in the quantity of interest, which is a velocity-weighted average of an error of order v_T^2/v^2, can be far smaller than the average error.

The utility of the high-speed limit is that analytic solutions can be constructed at high accuracy. For example, it is far easier to construct the hot conductivity than it is to construct the Spitzer conductivity.

CURRENT DRIVE EFFICIENCY
AND OTHER NONTHERMAL TRANSPORT QUANTITIES

When the superthermal component of the electron distribution function is enhanced, as through rf heating, then the plasma is necessarily so far from thermal equilibrium that a hydrodynamic description of transport processes is insufficient. New, kinetic effects come into play, and to describe these effects new transport quantities must be introduced. Among the most important such quantities are the current drive efficiency, the enhanced conductivity, and the runaway probability. All of these quantities are defined here for an infinite homogeneous plasma, but the appropriate generalization to toroidal geometry can be made without conceptual difficulty. These quantities are all reviewed in detail elsewhere;[1] here we merely outline the major processes at play.

The kinetic processes which superthermal electrons dominate can be arrived at by considering the cumulative effect of individual electrons during the time in which they remain superthermal. To be specific, consider, for example, the current drive effect in the absence of a dc parallel electric field. An electron traveling in the parallel direction is subject to direction-randomizing and speed-decreasing collisions. Until its velocity is randomized or its speed is significantly decreased., however, the electron tends to carry current associated with its initial velocity. Eventually, the collisions dominate and the individual electron, on average, carries no directed current. Suppose that the initial electron velocity is $\mathbf{v}$ and the parallel velocity as a function of time and the initial velocity is $w_\parallel(\mathbf{v},t)$. One can define then the quantity

$$j_\parallel(\mathbf{v}) = \int_0^\infty w_\parallel(\mathbf{v},t)dt, \tag{3}$$

which is the total current that the electron carries integrated over all time. This quantity is a property of the initial velocity space coordinates only. The significance of pushing an electron in velocity space from one location to another can now be appreciated — the total current integrated over time may change even if the push itsef is not in the parallel direction.

One can calculate the current drive efficiency, the amount of power necessary to drive a given current. Consider the energy it takes to push an electron from one velocity space location (1) to a second nearby location (2), namely , $\Delta\epsilon = \epsilon_2 - \epsilon_1$, where the energy $\epsilon \equiv mv^2/2$. The amount of current integrated over time that results from this change in the initial coordinate of the electron is $\Delta j_\parallel = j_\parallel(\mathbf{v_2}) - j_\parallel(\mathbf{v_1})$. The efficiency which is a function of $\mathbf{v_2}$ and $\mathbf{v_1}$ is then given by

$$\frac{J}{P_d} = \Delta j_\parallel/\Delta\epsilon. \tag{4}$$

The current-drive efficiency is one of several important transport quantities that is a function only of where, in velocity space, the rf power is absorbed. A second transport quantity associated with the rf heated plasma is the enhanced conductivity due to the heating. This conductivity is similarly a function only of where in velocity space the rf heating takes place. The rf waves, however, because of the resonance condition, might interact with electrons in only a small region of velocity space. This forces a significant change in emphasis here, something of a "kinetic generalization," in which conductivity is thought of as a quantity with a velocity dependency, rather than as an integrated quantity, as it is usually defined.

The electron runaway effect can also be approached as a single particle, or kinetic, effect. We replace the binary description of electrons running away or not with an electron probability of running away, based on its initial velocity coordinates. To be specific, the runaway probability is the probability that under the influence of collisions and a dc electric field, an electron initially with velocity $\mathbf{v}$ will be accelerated to infinite speeds before slowing down to a thermal speed. Using the kinetic description, it is possible to calculate the increase in the number of ruaway electrons in an rf heated tokamak. In analogy to the treatment of the current drive efficiency, consider the change in the runaway probability $R(\mathbf{v})$ when an electron is accelerated from $\mathbf{v_1}$ to $\mathbf{v_2}$. Since it takes energy $\epsilon_2 - \epsilon_1$ to accomplish this acceleration, the *incremental* runaway production, due to the rf power, P_d, can be written as

$$\dot{N}_R = P_d\frac{R(\mathbf{v_2}) - R(\mathbf{v_1})}{\epsilon_2 - \epsilon_1}. \tag{5}$$

In other words, the absorption of rf power is accompanied by an increase in the runaway probability, so that the number of additional runaways is linear in the rf power.

The actual caculation of the rf-induced conductivity, the runaway probability function, or the current-drive efficiency is facilitated by the high-velocity approximation to the collision operator.

EVIDENCE FOR THE CLASSICAL DESCRIPTION

There is ample evidence that in an ideal (weakly coupled) plasma, particles interact by means of Coulomb collisions; what is not obvious is whether there are not collisionless mechanisms that might dominate the slowing down and scattering of an enhanced tail of superthermal electrons in a tokamak. Such collisionless mechanisms, in principle, might involve instabilities associated with an asymmetric velocity distribution function. More fundamentally, it might be imagined that Cerenkov or other collective emission could play a larger than expected role. However, there have now been a host of experiments that rule out any such surprises through the documentation of the theoretical predictions concerning the current-drive effect, in which as much as 2 MA of current have been driven by waves. Since the effect itself relies upon a detailed description of the electron collisions, the evidence that documents the current drive effect also shows that the dynamics of the fast electrons is indeed dominated by the classical collision processes.

The most detailed evidence comes from the PLT (Princeton Large Torus) current-drive and ramp-up experiments.[2] In the ramp-up experiments, in which the current is rising due to the rf waves, as much as 40% of the rf power was converted into poloidal field energy. The fact that the energy conversion can be so efficient, something that is consistent with the theory, is strong evidence for the model. These experiments spanned several parameter regimes, leading to different physics regimes too, including that of steady-state current drive, ramp-up of the current, and even the unsuccessful sustainment of the current.

In the interpretation of PLT experiments, an attempt was made to check the theory of the electron dynamics without making many assumptions concerning the details of the theory of wave propagation and damping.[3-4] This was accomplished by comparing dimensionless quantities each of which depended upon the wave being absorbed. Over 250 shots were tabulated, and with essentially only one free adjustable parameter, namely how much rf power could be absorbed, the fit to theoretical prediction was remarkable. What were the implications of these experiments? The immediate interpretation concerns the parallel velocity space dynamics of fast electrons. The dependencies of the electron dynamics upon density, electric field, and parallel phase velocity must, to make such a fit, be as predicted classically, leaving little room for any anomalous effects. Note that this is a far deeper statement on the nature of the electron dynamics than could be offered merely through a detailed study of Spitzer resistivity, which is an integrated quantity.

The second implication concerns confinement of the fast electrons. The fact that the current drive and current ramp-up effects were observed indicates immediately that the fast electrons must be reasonably well confined, at least on the order of their collision time. What has not been appreciated is that these experiments also give an upper bound to the confinement time of these electrons.

The upper bound arises from the efficiency of the ramp-up phenomenon. The conversion of wave energy to poloidal field energy is efficient only if the rf works primarily against an opposing electric field rather than against collisions. This implies that the electric field must be strong enough to produce a number of runaway electrons. Furthermore, the rf waves convert wave energy to field energy most efficiently precisely when they have parallel phase velocities that most enhance this production of runaway electrons. The problem with the runaway electrons is that, in undergoing constant acceleration by the dc electric field, they convert the field energy back to electron kinetic energy. Even a small number of runaways ruins the ramp-up effect. Therefore, a necessary assumption in explaining the observed data was that these runaway electrons could not be long confined.

Thus, the PLT data gives the following rough bounds on the confinement time $\tau_{\text{conf}}(\mathbf{v})$ of the fast electrons

$$\frac{\tau_c}{R(\mathbf{v})} > \tau_{\text{conf}} > \tau_c, \tag{7}$$

where $\tau_c(\mathbf{v})$ is the slowing down time of an electron initially with velocity $\mathbf{v}$. The second inequality states merely that electrons are confined long enough to be slowed down classically, something we believe to occur because of the then remarkable fit of the theory to the data. The first inequality states that the effect of runaway electrons is less important than the effect of resonant electrons. To arrive at this inequality, consider that every resonant electron supports the current ramp-up for a slowing down time, but works against it, with about the same efficacy, for a confinement time in the event that it becomes a backwards runaway. The first inequality follows, since that event occurs with probability $R(\mathbf{v})$. Typical slowing down times in these experiments were on the order of ten milliseconds and typical runaway probabilities in the resonant region were on the order of 0.1.

Diagnostic Possibilities

There are a number of ways in which tokamak emissions, which are dominated by superthermal electrons, can be used to inform on tokamak parameters. A standard measurement is to view the synchrotron radiation horizontally in order to find the electron temperature. This measurement relies on the distribution function being near maxwellian and, near a cyclotron resonance, behaving as a blackbody emitter. Comparing the synchrotron[5] or bremmstrahlung[6] radiation with numerically generated radiation from a presumed steady state nonthermal electron distribution has also been used.

Given the solid evidence for the classical description, there has recently been the suggestion to use the physical constraints imposed by the electron dynamics in the deduction of plasma parameters from synchrotron emission.[7-10] Here, it is contemplated that brief heating of the superthermal tokamak electrons will produce a transient, synchrotron radiation signal, in frequency-time space. It turns out that the two-dimensional radiation pattern is sensitive in different ways to different parameters, so that important parameters parameters are roughly *orthogonal*; *i.e.*, remarkably, they can be inferred almost independently. It may be possible to infer, for example, the dc toroidal electric field in the tokamak interior, which is not available by other means, and this inference is not significantly affected by uncertainty regarding, say, the density, n, or the ion charge state, Z_{eff}.

This suggestion exploits several fortuitous circumstances: First exploited is a separation of time-scales: $1/\omega \ll \tau_{\text{det}} \ll \tau_c \ll \tau_{\text{par}} \ll \tau_{\text{exp}}$. That there are many detector observations, τ_{det}, in a slowing down time, τ_c, can multiply by $10^2 - 10^3$ the amount of processible information in a non-transient analysis. Tokamak parameters are changing only on the time scale τ_{par}, which is on the order of seconds. Also exploited is that the dynamics of the perturbed high-velocity, weakly-relativistic electrons is an entirely linear problem, dominated by Coulomb collisions and the dc electric field, with energy diffusion ignorable. It turns out that the Green's function for the radiation response can be defined, independent of plasma temperature; and, very luckily, a fully relativistic analytic solution is obtainable in the limit of small dc electric field. The Green's function is used to consider efficiently many different initial conditions for the incremental distribution function.

As a diagnostic tool, the radiation data must be efficiently compared to the theoretical expectation for many different plasma parameters. This task would be unthinkable but for the Green's function response. Furthermore, the task is bolstered by several scale-invariant transformations of the radiation response $\eta(\omega, t)$. In other words, having solved for $\eta(\omega, t; n, B, E, Q, Z_{\text{eff}}, \theta) = \eta_0 + E\,\eta_1$, where Q measures the perturbation amplitude and θ is the observation angle, further radiation responses are immediately available through the transformations

$$\eta(\omega, t; \alpha B) = \alpha \eta(\omega/\alpha, t; B),$$
$$\eta(\omega, t; \beta Q) = \beta \eta(\omega, t; Q),$$
$$\eta(\omega, t; \gamma n) = \eta_0(\omega, t/\gamma; n) + \gamma^{-1} E\,\eta_1(\omega, t/\gamma; n). \tag{8}$$

The fact that the electron dynamics are sensitive to only relatively few plasma parameters usefully *constrains* the solution space; a great deal of data informs on but a few choice parameters. Essentially all competing sets of plasma parameters that might possibly explain the data can be considered, and the powerful analytic tools make feasible a numerical analysis of data that would otherwise be unthinkable. Using a model for noise, in order to account for either experimental or theoretical uncertainties, data can be simulated numerically, so that rigorous estimates can be made of the informative worth of the data prior to obtaining it.

This diagnostic system includes both the probing rf signal that leads to the incremental synchrotron signal, and an array of frequency detectors, oriented to view the plasma vertically, with submillisecond time resolution. Simulations of the system indicate the possibility of measuring accurately certain plasma parameters, including the dc toroidal field. The toroidal field, only about 1 V/m, would be extremely difficult to measure otherwise in the plasma center.

CONCLUSIONS

The trend towards hotter tokamak confinement devices, especially rf-driven steady state devices, promotes to greater importance the study of mildly relativistic electrons. These electrons play a role in a number of interesting processes. Our understanding of the transport properties of these electrons enjoys relatively detailed experimental evidence.

ACKNOWLEDGEMENTS

This work was supported by United States Department of Energy under contract numbers DE–AC02–76–CHO3073 and DE–FG02–84–ER53187.

REFERENCES

[17] N. J. Fisch, Rev. Mod. Phys. **59**, 175 (1987).

[2]F. C. Jobes, *et al.*, *Phys. Rev. Lett.* **55**, 1295 (1985).

[3]N. J. Fisch, and C. F. F. Karney *Phys. Rev. Lett.* **54**, 897 (1985) .

[4]C. F. F. Karney, N. J. Fisch, and F. C. Jobes, *Phys. Rev. A* **32**, 2554 (1985).

[5]T. Luce, P. Efthimion and N. J. Fisch, *Rev. Sci. Instr.* (1988).

[6]J. E. Stevens, *et al.*, *Nucl. Fusion* **25**, 1529 (1985).

[7]N. J. Fisch, *Plasma Phys. Control. Nucl. Fusion.* **30**, 1059 (1988).

[8]N. J. Fisch, and A. H. Kritz, *Phys. Rev. Lett.* (1989).

[9]N. J. Fisch, and A. H. Kritz, *Plasma Phys. Control. Nucl. Fusion* **31**, 1407 (1989).

[10]N. J. Fisch, and A. H. Kritz, to appear in *Plasma Phys. Control. Nucl. Fusion* (1990).

STATUS OF MAGNETIC FUSION *

Dale M. Meade

Princeton University, Princeton Plasma Physics Laboratory,
P.O. Box 451, Princeton, NJ 08543

Research in magnetic fusion for the past 40 years has concentrated on the issue of scientific feasibility. Recent results from large tokamak devices throughout the world have produced the plasma parameters in the range required for fusion reactors. Ion temperatures of 30 keV have been produced in the Tokamak Fusion Test Reactor (TFTR), densities of 10^{21} m^{-3} have been obtained in the ALCATOR-C tokamak and confinement times of 1.8 sec were achieved in the Joint European Torus (JET). In TFTR and JET, deuterium plasma parameters have been attained that would produce near breakeven conditions in a D-T plasma. Experiments with D-T plasmas planned for 1993-1994 on TFTR and 1995-1996 on JET are expected to produce $\sim$ 10-20 MW of fusion power.

In addition to achieving the plasma parameter range required for D-T fusion, significant progress has been made in increasing the efficiency of a magnetic fusion reactor. The Doublet III tokamak has achieved β = plasma pressure/magnetic pressure of $\sim$ 10%, which exceeds the values needed in tokamak fusion reactor designs. Present tokamak experiments treat the plasma as the secondary of a transformer, thereby producing pulsed discharges of $\sim$ 6 MA for several seconds. A steady-state plasma is highly desirable since it would eliminate thermal and mechanical fatigue and thereby enhance the reliability of a reactor. The stellarator is inherently steady-state, and is now under investigation in the U. S. (ATF), Japan (Hcliotron E) and Germany (Wendelstein). A steady-state current can be driven in a tokamak using RF waves (1.5 MA in JT-60) or neutral beams (1.0 MA in TFTR). These driven currents can be enhanced using the tokamak bootstrap current discovered on TFTR, which can increase the current drive efficiency by up to 4 times.

The progress in extending plasma parameters has been made possible by the development of plasma diagnostics to measure all plasma parameters, and sophisticated theoretical models for plasma behavior. The macroscopic magnetohydrodynamic (MHD) models accurately describe the equilibrium and stability of plasmas with complex shapes. MHD instabilities (kinks) are well understood and can be controlled experimentally.

The transport of particles along magnetic field lines in tokamaks is reasonably well understood. The measured parallel electrical resistivity, currents driven by RF and neutral beams and bootstrap current are all in good agreement with the neoclassical theory.

However, cross-field energy and particle transport are not understood in terms of basic theory. Nonetheless, global scaling models have been developed that quite accurately describe the plasma energy confinement under a variety of conditions in many tokamak devices. Detailed measurements of local cross-field transport coefficients show that the electron thermal diffusivity χ_e is comparable to the ion thermal diffusivity and both are 10 - 100 times the neoclassical cross-field diffusivities. These results are consistent with transport that is produced by electrostatic microturbulence in the plasma. Detailed measurements are now being made to determine the connection between microturbulence and plasma transport. This knowledge would help develop plasma regimes with improved confinement.

The success of the present generation of tokamak experiments has provided the technical basis for embarking on the next step of magnetic fusion - the study of burning plasmas and the development of fusion engineering. In the U.S., a Compact Ignition Tokamak (CIT) has been proposed to study the physics of burning plasmas. The CIT would produce over 100 MW of fusion power and could start operation about the year 2000. An International Thermonuclear Experimental Reactor (ITER) with world-wide participation is proposed as an engineering test reactor, which would begin operation around 2005.

* This work supported by the U.S. Department of Energy under contract DE-AC02-76-CHO-3073.

B. Alternative Magnetic Confinement

A D-He^3 Fusion Reactor Based on a Dipole Magnetic Field

Akira Hasegawa

AT&T Bell Laboratories
Murray Hill, New Jersey 07974

ABSTRACT

An innovative fusion reactor suitable for D-He^3 fuel is proposed, based on a dipole magnetic field produced by a levitated superconducting coil. The equilibrium plasma, whose phase-space density satisfies $\partial f_0 \, (\mu, \, J, \, \psi)/\partial\psi = 0$, where ψ is the flux function, has a steep enough pressure profile for an efficient fusion reaction yet is stable for low frequency instabilities for local beta exceeding unity. At the outerwall, the plasma is stabilized by line-tying or localized magnetic cusps which can be used for direct conversion. The fusion product confinement time can be controlled for ash removal by breaking the axisymmetry of the dipole magnetic field.

LARGE ORBIT CONFINEMENT FOR ANEUTRONIC SYSTEMS

N. Rostoker
University of California, Irvine, CA 92717

H. Rahman
University of California, Riverside, CA 92521

ABSTRACT

A Field Reversed Configuration (FRC) is considered in which
almost all of the ions have large orbit radius and are therefore
non-adiabatic. Electrons with much smaller energy and mass are
adiabatic. Self-consistent equilibria are obtained for infinitely
long FRC's and the finite problem is formulated. Formation of such
FRC's with the technology of intense ion beams is described. The
stability problem differs substantially from the usual case where
almost all ions are adiabatic. Some of the essential physical
differences are described and some results are given for infinitely
long systems.

INTRODUCTION

Aneutronic reactions require higher energy particles. For
example, D-He^3 reactions require average energy of 100's of keV
compared to 10's of keV for D-T neutronic reactions. The reaction
products of a D-He^3 reactor are 3.6 MeV-He^4 and a 14.7 MeV proton.
To contain the reaction products for ignition requires a very large
volume or a very large magnetic field if the confinement is conventional as in a tokamak. In this case the ion-gyroradii are small
compared to characteristic dimensions such as the minor radius and
the particle dynamics is adiabatic. This, in turn, leads to the
Kruskal-Shafranov limit. The safety factor must be greater than
unity so that a large toroidal magnetic field must be maintained
for no reason other than to guarantee gross stability. We consider
departing from the constraints of a conventional tokamak by confining only high energy ions so that the gyroradius is as large as the
characteristic size of the confined plasma. The ion dynamics are
similar to the particle dynamics of a conventional accelerator; the
electrons remain adiabatic. It is well known from accelerators that
non-adiabatic ions can be magnetically confined. Indeed the confinement is much better than the confinement of adiabatic ions in a
plasma. It has been assumed that it obtains only for a low density.
It has, however, been shown experimentally that large orbit non-
adiabatic electrons of high density can be confined for long times[1]

in reverse field configurations. Also, large orbit ions[2] have been
confined at high density in a similar configurations.

If only the primary particles (e.g. D and He^3) are confined,
the reactor can be of a modest size. Ignition is not then possible,

and the fusion reactor is only an energy amplifier. We consider a reverse field configuration where D-ions have an average energy of .8 MeV and He^3 ions have an average energy of 1.2 MeV. The two species have the same average azimuthal velocity and for both ions the distribution function is approximately a Maxwell distribution with a temperature of about 100 keV in a frame of reference moving with the velocity corresponding to .8 MeV or 1.2 MeV He^3. Such ion distributions would not be changed by ion-ion collisions. Confinement could thus last long enough for most of the ions to undergo fusion reactions, in which case the energy gain would be 18.3/2. Considering the fact that the reaction products are charged particles, direct conversion of their energy to electric power is possible without a heat cycle so that a practical compact reactor without ignition and with very few neutrons is possible.

The technology of intense ion beams has already been developed so that it is adequate for a reactor. Beams of about 1 MeV ions are routinely produced at currents of 100 kA, current densities of 1 kA/cm^2 and pulse duration up to 1 μsec. This is sufficient for maintaining a reverse field configuration of D-He^3 for a compact reactor with a pulse repetition rate of about 1 Hz. The beams can be directly injected into a reverse field configuration and trapped by a collective process.[3] The beam carries neutralizing electrons with it. It crosses a magnetic field and vacuum (or a low density plasma) by means of self-polarization and ExB drift. When it reaches a dense plasma the polarization is shielded mainly by electrons; the beam curves in the magnetic field like single particles and is therefore trapped.

The physical principles and the required technology for a compact D-He^3 reactor are available now; the stability picture is not complete and never will be unless some experiments are done. There are questions regarding the availability of He^3. It can be obtained by radioactive decay of Tritium which has a 12.6 year half life. It can be mined on the moon where it is known to be abundant, being carried to the moon continuously by the solar wind.

SELF-CONSISTENT F.R.C. EQUILIBRIA

The constants of the motion are the canonical angular momentum and the energy

$$L_j = m_j(xv_y - yv_x) + (e_j/c)r \, A_\theta(r) \tag{1}$$

$$\epsilon_j = m_j v^2/2 + e_j \, \Phi(r) \tag{2}$$

The index j refers to electrons and each species of ion. $\Phi(r)$ and $A_\theta(r)$ are scalar and vector potentials. A self-consistent solution of the Vlasov equation is

$$f_j(\underline{x},\underline{v}) = A_j \exp - (\epsilon_j + \omega_j L_j)/T_j \tag{3}$$

$$= \left(\frac{m_i}{2\pi T_j}\right)^{3/2} n_j(r) \exp - \left\{\frac{m_i}{2T_j}\left[(v_x - \omega_j y)^2 + (v_y + \omega_j x)^2\right]\right\}$$

T_j is temperature and $n_j(r)$ is the density which can be determined with Maxwell's equations. The mean velocity is

$$\langle v_\theta \rangle = -\omega_j r \tag{4}$$

We assume quasi-neutrality

$$n_e = \sum_i n_i Z_i \tag{5}$$

where n_e is electron density and the sum is over ion species of charge Z_i. The current density is

$$J_\theta = \sum_j n_j e_j \langle v_\theta \rangle_j = e\, n_e r(\omega - \omega_e) \tag{6}$$

It has been assumed that $\omega_i = \omega$ is the same for all ions and $\omega_j = \omega_e$ for electrons. The electric and magnetic fields are $E_r = -\partial\Phi/\partial r$ and $B_z = (1/r)\,\partial(rA_\phi)/\partial r$. The particle moment equations for distributions like Eq. (3) are

$$- n_j m_j r \omega_j^2 = n_j e_j\left(E_r - \frac{r\omega_i}{c} B_z\right) - T_j \frac{dn_j}{dr} \tag{7}$$

Combining these equations with Maxwell's equation

$$\frac{dB_z}{dr} = \frac{4\pi}{c} n_e\, e(\omega - \omega_e)\, r \tag{8}$$

leads to equations for the particle densities. Equation (7) for electrons can be solved for

$$E_r = - (T_e/en_e)(dn_e/dr) + (r\omega_e/c)B_z - (m\, r\omega_e^2/e)$$

and thus E_r can be eliminated from the ion moment equations

$$\frac{1}{r}\frac{d\log n_i}{dr} = -\frac{Z_i e}{c}\left(\frac{\omega-\omega_e}{T}\right) B_z - \left(\frac{T_e}{T}\right)\left(\frac{Z_i}{r}\right)\frac{d\ln n_e}{dr} + \frac{m_i \omega_i^2}{T} \tag{9}$$

It has been assumed that $T_i = T$ for all ions and $m\omega_e^2$ has been neglected compared with $m_i \omega_i^2$. By differentiating Eq. (9) and substituting Eq. (8), B_z can be eliminated

$$\frac{d}{dr}\frac{1}{r}\frac{d\log n_i}{dr} = -\frac{4\pi n_e Z_i e^2}{T}\frac{(\omega-\omega_e)^2}{c^2} r - \frac{Z_i T_e}{T}\frac{d}{dr}\frac{1}{r}\frac{d\ln n_e}{dr} \tag{10}$$

Equations (5) and (10) determine the densities. To obtain a dimensionless form, let $N_j = n_j/n_o$ where n_o is the peak electron density at $r = r_o$, $\xi = r^2/2r_o^2$ and

$$\Lambda^2 = \frac{8\pi n_o e^2}{c^2}\frac{(\omega-\omega_e)^2}{(T+T_e)} r_o^4 \tag{11}$$

$$\frac{d^2\log N_i}{d\xi^2} = -\frac{\Lambda^2}{2} Z_i\left(1 + \frac{T_e}{T}\right)N_e - (Z_i T_e/T)\frac{d^2\ln N_e}{d\xi^2} \tag{12}$$

$$N_e = \sum_i Z_i N_i \tag{13}$$

The simplest case is for one hydrogen-like ion species in which case $N_e = N_i$, $Z_i = 1$

$$\frac{d^2}{d\xi^2}\log N = -\Lambda^2 N/2 \tag{14}$$

This equation has a simple solution

$$N(\xi) = \operatorname{sech}^2\left[\frac{\Lambda}{2}(\xi-\xi_o)\right] \tag{15}$$

Equation (8) can now be solved for the magnetic field.

$$B_z(r) = B_o + \left\{B_1 - B_o\right\} \left\{\frac{\tanh(\Lambda/4) + \tanh(\Lambda/4)[(r/r_o)^2 - 1]}{1 + \tanh(\Lambda/4)}\right\} \quad (16)$$

B_o is the magnetic field at r=o and $B_1 = B_z(\infty)$. They can be expressed as

$$B_1 = \frac{M\omega^2 c}{e(\omega - \omega_e)} \left[1 + \frac{\Lambda}{W_o} (T_e + T)\right] \quad (17)$$

$$B_o = \frac{M\omega^2 c}{e(\omega - \omega_e)} \left[1 - \frac{\Lambda}{W_o} (T_e + T)\right] \quad (18)$$

where

$$W_o = \frac{M}{2} (r_o \omega)^2$$

is the average ion kinetic energy at $r = r_o$. Field reversed equilibria obtain if, and only if,

$$\frac{\Lambda}{W_o} (T_e + T) > 1$$

As a numerical example consider D ions with T = 100 keV, T_e= 20 keV and W_o = 1 MeV. The line density is

$$N_L = \int_o^\infty n_e\ 2\pi r\ dr = \frac{4\pi n_o r_o^2}{\Lambda} \left[1 + \tanh \frac{\Lambda}{4}\right] \quad (19)$$

Assume that $N_L = 2\pi r_o \Delta r n_o$; $r_o = 25$ cm, $\Delta r = 5$ cm and $n_o = 2 \times 10^{14}$ cm^{-3}. Then

$$\Lambda/[1 + \tanh(\Lambda/4)] = 4\pi n_o r_o^2/N_L = 2\ r_o/\Delta r$$

This identifies $\Lambda \simeq 20$ and $(\Lambda/W_o)(T_e + T) = 2.4$. The magnetic fields of Eq. (17) are $B_1 = 36$ kG and $B_o = -15$ kG. For multiple ion species Eqs. (12) and (13) have been solved numerically for the following data.

D-T

$$r_o = 30 \text{ cm}$$

$$n_D(r_o) = 10^{14} \text{ cm}^{-3}$$

$$W_o(D) = 500 \text{ keV} \qquad W_o(T) = 750 \text{ keV}$$

$$T = 50 \text{ keV}$$

$$T_e = 10 \text{ keV}$$

$$\Lambda = 15$$

D-He^3

$$W_o(D) = 800 \text{ keV} \qquad W_o(He^3) = 1200 \text{ keV}$$

$$T = 100 \text{ keV}$$

$$T_e \ 20 \text{ keV}$$

$$\Lambda = 12$$

Particle densities and the magnetic field B_z are plotted in Figs. 1 and 2. It should be noted that the peak densities for each ion species is at a different value of r, but the difference is small for a field reversed configuration.

The generalization for an FRC that is finite in the z-direction involves $\Phi(r,z)$ and $A_\theta(r,z)$. Equation (14) is replaced by

$$\frac{\partial^2}{\partial \xi^2} \log N + \frac{2\epsilon^2}{\xi} \frac{\partial^2 \log N}{\partial \eta^2} = -\frac{\Lambda^2}{2} N(\xi,\eta) \quad . \qquad (20)$$

In this equation $\epsilon = r_o/L$ where L is a scale length in the z-direction. $\xi = r^2/2r_o^2$ and $\eta = z/L$. Electric and magnetic fields can be determined from $n_e(r,z)$. Equations (12) and (13) for multiple ion species can be similarly generalized. Numerical solutions are being developed. Previous work on this problem has been restricted to an MHD model and solving the Grad-Shafranov equation with a scalar pressure. 2-D equilibria with arbitrary plasma rotation have been elaborated. However, the present parameters involve large orbit radius for most of the ions in which case the MHD models are inadequate.

The distribution functions of Eq. (3) have interesting properties. If we consider classical particle collisions

$$\left. \frac{\partial f_i}{\partial t} \right)_c = - \frac{2e^4}{m_i} \sum_j \int \frac{d\underset{\sim}{k}}{k^4} \underset{\sim}{k} \cdot \frac{\partial}{\partial \underset{\sim}{v}} \int d\underset{\sim}{v}' \; \delta\left[\underset{\sim}{k} \cdot (\underset{\sim}{v}' - \underset{\sim}{v}) \right] \underset{\sim}{k}$$

$$\cdot \left[\frac{f_i(v)}{m_j} \frac{\partial f_j}{\partial \underset{\sim}{v}'} - \frac{f_j(v')}{m_i} \frac{\partial f_i}{\partial \underset{\sim}{v}} \right] \tag{21}$$

For ion-ion collisions $\dfrac{\partial f_i}{\partial \underset{\sim}{v}} = - \dfrac{m_i}{T_i} (\underset{\sim}{v} - \underset{\sim}{V}_i) f_i$ where $\underset{\sim}{V}_i =$

$(\omega_i y, \, - \omega_i x, o)$ from Eq. (3).   For ions $\omega_i = \omega$ independent of
species and $T_i = T$.   In this case ion-ion collisions make no change
in the distribution function.   Only ion-electron collisions can
change the distribution function and they will do so very slowly
because of the large mass difference and because electrons are hot.
The collision time for electrons to make a large angle scattering is
$\tau_{ei} \cong 10^9 \, T_e^{3/2}/n \cong 1$ millisec. assuming an average $n \sim 10^{14}$ cm^{-3}
and $T_e = 20$ keV.   The ion-ion collision time is longer by a factor
of $(m_i/m)^{1/2}$ and the time for ions to lose energy to electrons is
longer by a factor of m_i/m.   The slowing down time due to ion-electron collisions is $\tau_s \cong 3.6 - 4.5$ sec. $n\tau_s \cong 10^9 \, T_e^{3/2} (m_i/m)$
$\sim 4 \times 10^{14}$ depends only on electron temperature. T_e will be determined by the balance between radiation losses and heating by ions.

STABILITY

In the present treatment the physical effects on stability of
large orbit particles will be compared with small orbit (adiabatic)
particles. The comparison will be based on simple calculations. The
main difference is that adiabatic particles follow magnetic field
lines to a first approximation.   Large orbit particles do not. This
is illustrated in Fig. 3 with a typical orbit in a tokamak and a
betatron.   A consequence for tokamak orbits is that the safety
factor $q - 2\pi/\Delta\theta > 1$ is required for stability. $q - 1$ is called the
Kruskal-Shafranov limit.   A large toroidal magnetic field $B_t =$
$B_p(R/r)q$ must be maintained for stability. Since large orbit particles do not follow field lines there is no such requirement. Stable
confinement of large orbit electrons[1] has been observed with $q < 1$
and large orbit[2] ions with $q = 0$.

MHD calculations predict that the tilt mode should be unstable.

That it is not observed in FRC experiments is usually attributed to finite Larmor radius effects.[6] Computer simulations[7] show that plasma rotation with $V_\theta \gtrsim 1.5\ C_s$ (where $C_s = \sqrt{\gamma P/\rho}$ is sound velocity) stabilizes the tilt mode. More recently numerical calculations[8] show that some large orbit ions can stabilize the tilt mode in an FRC with mainly small orbit ions and electrons; the large orbit ions require less than 20% of the total plasma energy.

In order to understand non-adiabatic effects on the stability problem, we consider the simplest problem of a rotating plasma in a constant magnetic field. In the laboratory frame $\underset{\sim}{B} = \underset{\sim}{e}_z B_o$, $V_\theta = r\Omega$ $E_r = -\Omega B_o r/c$, $n = n(r)$ describes the equilibrium. Consider kink modes where all perturbed quantities are proportional to $\exp -i(\omega t - \ell\theta)$ where ℓ is an integer. For adiabatic particles the equation for the guiding center motion is

$$\frac{d\underset{\sim}{x}_c}{dt} = v_\parallel \underset{\sim}{e}_z + c(\underset{\sim}{E} \times \underset{\sim}{B}/B^2) + cm_j(\dot{\underset{\sim}{v}} \times \underset{\sim}{B})/e_j B^2 \tag{22}$$

The equilibrium drift is

$$\underset{\sim}{V}_c = r\Omega\ \underset{\sim}{e}_\theta + (\Omega^2 r/\Omega_j)\ \underset{\sim}{e}_\theta \tag{23}$$

where the first term comes from E_r and the second term is inertial drift from the centipetal acceleration. The perturbed drift is

$$\delta\underset{\sim}{V}_c = \frac{c}{B_o}\left(\delta E_\theta \underset{\sim}{e}_r - \delta E_r \underset{\sim}{e}_\theta\right) - \frac{c}{B_o}\frac{i(\omega-\ell\Omega)}{\Omega_j}\delta\underset{\sim}{E}_\perp - \frac{2\Omega}{\Omega_j}\frac{c}{B_o}\left(\delta E_\theta \underset{\sim}{e}_r - \delta E_r \underset{\sim}{e}_\theta\right) \tag{24}$$

The first term is $\delta E \times B$-drift. The second term is polarization-inertial drift. The last term comes from $\underset{\sim}{\Omega} \times \delta\underset{\sim}{v}$ acceleration and part of the polarization drift $(m_j c^2/e_j B^2)\ d\ \delta\underset{\sim}{E}_\perp/dt = (c/\Omega_j B_o) - [i\omega\delta\underset{\sim}{E}_\perp + (\underset{\sim}{v} \cdot \nabla\delta\underset{\sim}{E}_\perp)]$. The last two terms of $\delta\underset{\sim}{V}_c$ are significant only for ions since $\Omega_j = e_j B_o/m_j c$ is much larger for electrons. For electrons the continuity equation

$$\frac{\partial\delta n_e}{\partial t} + \nabla \cdot \left(\delta n_e \underset{\sim}{V}_e + n_e \delta\underset{\sim}{V}_e\right) = 0$$

gives
$$\delta n_e = \left(c/B_o\right)\left[\delta E_\theta/i\,(\omega-\ell\Omega)\right]\left(dn/dr\right)$$

Quasi-neutrality, or $\delta n_e \simeq \delta n_i$ is assumed. The perturbed charge density is calculated from $\delta\rho = \nabla\cdot\delta\underline{j}/i\omega$ where

$$\delta\underline{j} = \sum_j \left(\delta n_j \underline{V}_j + n_j\,\delta\underline{V}_j\right)e_j \tag{25}$$

$\delta\rho = \delta\rho^{(1)} + \delta\rho^{(2)}$ where

$$\delta\rho^{(1)} = \frac{1}{i\omega}\,\nabla\cdot\sum_j \delta n_j \underline{V}_j e_j = -\,(nec/B_o)(\Omega^2/\Omega_i)(d\log n/dr)\,\cdot$$

$$(i\ell/r\omega)\left[\delta E_\theta/(\omega-\ell\Omega)\right]$$

and this comes from the centripetal force drift.

$$\delta\rho^{(2)} = \frac{1}{i\omega}\,\nabla\cdot\sum_j n_j\,\delta\underline{V}_j e_j = \frac{nm_i c^2}{B_o^{\,2}}\left[(\omega-\ell\Omega)\nabla\cdot(\delta\underline{E}_\perp/\omega) + (d\log n/dr)\delta E_r\right]$$

$$-\,(2m_i c^2/B_o^{\,2})(dn/dr)(\Omega/i\omega)\delta E_\theta$$

The first term is from the polarization drift of ions and the rest of the terms come from polarization and inertial drifts. Poisson's equation reduces to $\delta\rho \simeq 0$ because $nm_i c^2/B_o^{\,2} \gg 1$. In terms of the potential defined by $\delta\underline{E} = -\,\nabla\delta\Phi$ where $\delta\Phi = f(r)\,\exp(-i\omega t + i\ell\theta)$ $\delta\rho^{(1)} + \delta\rho^{(2)} \simeq 0$ is the following differential equation

$$\frac{1}{r}\frac{d}{dr}\left(nr\frac{df}{dr}\right) - \frac{\ell^2}{r^2}\left[n + \frac{r}{\ell^2}\frac{dn}{dr}\,\nu\right]f = 0 \tag{26}$$

where $\nu = \left[(\ell\Omega)^2 + 2\bar\omega\ell\Omega\right]/\bar\omega^2$ and $\bar\omega = \omega - \ell\Omega$

This equation can be solved for

$$\frac{dn}{dr} = -\,n_o\delta(r-R)$$

or
$$n(r) = n_o e^{-(r/r_o)^2}$$

For both cases the eigenvalue parameter is $\nu = \ell$ so that

$$\bar{\omega} = \Omega\left[1 \pm \sqrt{1 - \ell}\right] \tag{27}$$

This is a well-known rotational instability of MHD fluid theory. Finite Larmor radius introduces a stabilization because the electric field $\delta\underline{E}$ is different for electrons and ions ie.

$$\delta\underline{E}(i) - \delta\underline{E}(e) \cong a_i^2 \nabla^2 \delta\underline{E} \tag{28}$$

In addition the rotation velocity of the fluid is modified by the ion "diamagnetic drift" to the same order.

$$\Omega' = \Omega - 2\left(a_i/r_o\right)^2\Omega_i$$

a_i is the mean gyroradius for ions. If the rotation produced by a radial electric field does not exceed $2\,(a_i/r_o)^2\Omega_i$ by an appreciable factor the instability[9] will be stabilized.

If the ions have large betatron-like orbits the drift equations are not applicable and the basic equations of motion must be reconsidered.

In cylindrical coordinates they are

$$\frac{dv_r}{dt} - \frac{v_\theta^2}{r} = \frac{e}{m_i} \frac{v_\theta}{c} B_z + \frac{e}{m_i} E_r \tag{29}$$

$$\frac{d}{dt}\left(rv_\theta + \frac{e}{m_i c} rA_\theta\right) = \frac{e}{m_i} rE_\theta \tag{30}$$

Assume the unperturbed state is $v_\theta = -\,r\Omega_i$, $v_r = 0$, $E_\theta = E_r = 0$, $B_z = $ const. and $A_\theta = B_o r/2$. The perturbed equations are

$$\frac{d^2}{dt^2} \delta v_r + \Omega_i^2 \, \delta v_r = \frac{e}{m_i}\left(\frac{d}{dt} \delta E_r - \Omega_i \delta E_\theta\right)$$

$$\frac{d}{dt} \delta v_\theta = \frac{e}{m_i} \delta E_\theta$$

Assuming all perturbations are proportional to $\exp(-i\omega t + i\ell\theta)$ where $\omega \ll \Omega_e$, the perturbed velocity for ions is

$$\delta \underset{\sim}{V}_i = - \frac{e}{m_i}\left\{\left[\frac{i\bar{\omega}\,\delta E_r + \Omega_i \delta E_\theta}{\Omega_i^2 - \bar{\omega}^2}\right]\underset{\sim}{e}_r + \frac{\delta E_\theta}{i\bar{\omega}}\,\underset{\sim}{e}_\theta\right\} \tag{31}$$

where $\bar{\omega} = \omega + \ell\Omega_i$.

For the electrons $\underset{\sim}{V}_e \cong 0$ and

$$\delta \underset{\sim}{V}_e \cong \frac{c}{B_o}\left(\delta E_\theta \underset{\sim}{e}_r - \delta E_r \underset{\sim}{e}_\theta\right) \tag{32}$$

$$\delta n_e \cong \left(c/B_o\right)\left[\delta E_\theta/i\omega\right]\,dn/dr \cong \delta n_i$$

$\delta\rho$ can be calculated as in the previous case and $\delta\rho \cong 0$ gives the following differential equation

$$\frac{1}{r^s}\frac{d}{dr}(nr^s)\frac{df}{dr} + \left[\left(\frac{\Omega_i^2 - \bar{\omega}^2}{\bar{\omega}^2}\right)\frac{\ell^2}{r^2}\left(n + \frac{\bar{\omega}}{\omega} r \frac{dn}{dr}\right) + \frac{\ell\Omega_i}{\bar{\omega}}\frac{1}{r}\frac{dn}{dr}\right]f = 0 \tag{33}$$

$$s = 1 + \frac{\Omega_i}{\bar{\omega}}\ell$$

Consider the case of constant density $n = n_o$ for $r \leq R$ and $n = 0$ for $r > R$.

$$\frac{1}{r^s}\frac{d}{dr} r^s \frac{df}{dr} + \left(\frac{\Omega_i^2 - \bar{\omega}^2}{\bar{\omega}^2}\right)\frac{\ell^2}{r^2} f = 0 \tag{34}$$

$$f = r^{\alpha_\pm}$$

where $2\,\alpha_\pm = -(s-1) \pm \sqrt{(s-1)^2 - 4\left(\frac{\Omega_i^2 - \bar{\omega}^2}{\bar{\omega}^2}\right)\ell^2}$

$$f(r) = A(r/R)^{\alpha_+} \qquad r \leq a$$

$$= A(r/R)^{\alpha_-} \qquad r \geq a$$

Since $\dfrac{dn}{dr} = - n_o \, \delta(r-R)$

$$\left.\frac{df}{dr}\right)_{R-0} + \frac{f(R)}{R}\left[\ell^2\,\frac{\bar{\omega}}{\omega}\left(\frac{\Omega_i^2 - \bar{\omega}^2}{\bar{\omega}^2}\right) + \ell\frac{\Omega_i}{\bar{\omega}}\right] = 0 \tag{35}$$

or
$$\bar{\omega} = \ell\Omega_i(\alpha^+ - 1)/(\alpha^+ - \ell^2) \tag{36}$$

To compare this result with Eq. (27) consider $\bar{\omega} = -\Omega_i$ since for the present case $\Omega = -\Omega_i$ is the only rotation frequency for ions. For this case $s = 1-\ell$ and $\alpha_+ = \ell$. Equation (36) then confirms the assumed result. This mode which was unstable for small gyroradius ions is now stable. A similar result has been previously obtained as a finite Larmor radius correction.[9] However, stabilization took place for this mode only for low frequencies of rotation $|\Omega| \ll |\Omega_i|$. For large gyroradius the mode is stabilized at $|\Omega| = |\Omega_i|$. Equation (36) contains three other modes

$$\bar{\omega} = \Omega_i$$

$$\bar{\omega} = \left\{\Omega_i/2(\ell^2 - 1)\right\}\left\{-1 \pm \left[1 + 4(\ell^2 - 1)\right]^{1/2}\right\} \tag{37}$$

which are all stable.

The previous results for finite Larmor radius stabilization[9] were obtained by integrating the Vlasov equation along unperturbed orbits and then expanding the perturbed distribution function in the parameter a_i/r_o. The unperturbed orbits for small gyroradius particles look the same as for large gyroradius in this problem. ie., they are simply circles. The difference lies in the perturbed orbits which were obscured by the previous analysis and are explicit in the present treatment. The physical reason for the stabilization is that electrons drift in the radial direction and ions have a different dynamics so that there must be a radial current density that is stabilizing. A similar explanation has been given for the stability of kink modes for a pure electron plasma.[10]

The present methods have not yet been applied to the stability of the self-consistent F.R.C. equilibria described in this paper.

Very thin infinitely long FRC equilibria[11] have been investigated for low frequency kink instabilities ($\ell \geq 2$). The equilibria involve a beam and a background plasma and the kink modes are stable if the ratio of ion plasma density to ion beam density is below the critical value of $\ell^2 - 1$.

INTENSE NEUTRALIZED ION BEAMS

Intense ion beams are produced with a magnetically insulated diode as illustrated in Fig. 4. A current density of ions of 150 A/cm^2 can be produced[12] with a diode voltage of 200 kV. It has been documented experimentally[13] that an intense ion beam will cross a magnetic field as illustrated in Fig. 4 without deflection when[14]

$$\epsilon_\perp = \frac{4\pi nMc^2}{B^2} = \frac{2.9 \times 10^3}{B^2} \frac{J}{W^{1/2}} > \left(\frac{m_i}{m}\right)^{1/2} \tag{38}$$

For example, if $J = 150$ A/cm^2, $W = 200$ kV and $B = 6$ kG, $\epsilon_\perp = 856 >> (m_i/m)^{1/2}$. For a cylindrical beam as illustrated in Fig. 5, the beam polarizes so that there is a uniform electric field $E_y = -VBx/c$ where V is the beam velocity. The surface charge density is $\sigma = \left[|E_y|/2\pi\right]\sin\theta$ The thickness of the charge layer is $\Delta r \sim \left(|E_y|/2\pi ne\right)\sin\theta$. The polarization charge produces a potential

$$\Phi = \left[\pi a^2 |E_y|/r\right]\sin\theta \qquad r > a \tag{39}$$

$$= \pi r |E_y|\sin\theta \qquad r < a$$

where a is the beam radius and θ is an angle measured clockwise from the x-axis. If the beam encounters a plasma the process of shielding takes place mainly by plasma electrons moving along the field lines as in Fig. 5. On a time scale of ω_{pe}^{-1} the potential becomes

$$\Phi = 2\pi a |E_y| \, I_1(a/L_D) \, K_1(r/L_D) \, \sin\theta \qquad r > a$$

$$= 2\pi a |E_y| \, I_1(r/L_D) \, K_1(a/L_D)\sin\theta \qquad r < a \tag{40}$$

ie., E_y is reduced by the factor $(L_D/a)(r/a)^{1/2}\exp - \left[(a-r)/L_D\right]$ on a time scale of the plasma period. L_D is the Debye length corresponding to the plasma electrons.

While electrons from the plasma shield the polarization, the polarization is recreated by the magnetic field on a time scale $\Delta r/V$ where $\Delta r \cong |E_y|/2\pi ne = VB_x/2\pi nec = 2a_i/\epsilon_\perp$. $\epsilon_\perp$ is given by Eq. (38) and $a_i = V/\Omega_i$ is the ion gyroradius. The beam should lose its polarization and be trapped as illustrated in Fig. 5 if

$$\Delta r/V > 2\pi/\omega_{pe} \qquad \text{or}$$

$$n_p > 4\pi^3 n^2 (mc^2/B^2) \qquad (41)$$

n_p is the plasma electron density and n is the beam density. The scaling law of Eq. (41) has been verified experimentally.[15] Typical experimental results are $n = 10^{11}$ cm^{-3}, $B = 200$ gauss for which $n_p \gtrsim 10^{13}$ cm^{-3} is required. Some simple experiments have been carried out which show trapping in a mirror[16] and a torus.[17]

ACKNOWLEDGEMENT

This work was supported by a University of California Energy Research Grant.

REFERENCES

1. H. H. Fleischmann et al., Phys. Fluids <u>17</u>, 2226 (1974); <u>19</u>, 728 (1976).
2. R. E. Siemon et al., Fusion Technology <u>9</u>, 13 (1986).
3. F. J. Wessel et al., Phys. Fluids <u>31</u>, 3778 (1988).
4. E. K. Maschke and H. Perrin, Plasma Phys. <u>22</u>, 579 (1980).
5. R. A. Clemente and R. Farengo, Phys. Fluids <u>27</u>, 776 (1984).
6. D. C. Barnes et al., Phys. Fluids <u>29</u>, 2616 (1986).
7. R. Horiuchi and T. Sato, Phys. Fluids B <u>1</u>, 581 (1989).
8. D. C. Barnes and Z. Mikie, Workshop on Compact Torus Research, Los Alamos, Nov. 6-8 (1989).
9. M. N. Rosenbluth, N. A. Krall and N. Rostoker, Nuclear Fusion 1962 Supplement, Part I, 143 (1962).
10. N. Rostoker and A. Fisher, Physica Scripta T2/2, 293 (1982).
11. R. Lovelace, Phys. Fluids <u>22</u>, 708 (1979); R. Lovelace, H. L. Berk, N. Rostoker and H. V. Wong, Sherwood Theory Meeting (1988); Bull. Amer. Phys. Soc., Oct. (1988).

12. S. Humphries, Jr., and G. W. Kuswa, Appl. Phys. Lett. $\underline{35}$, 13 (1979).
13. S. Robertson, H. Ishizuka, W. Peter and N. Rostoker, Phys. Rev. Lett. $\underline{17}$, 508 (1981).
14. W. Peter and N. Rostoker, Phys. Fluids $\underline{25}$, 730 (1982).
15. R. Hong, F. J. Wessel, J. Song, A. Fisher and N. Rostoker, J. Appl. Phys. $\underline{64}$, 73 (1988).
16. M. Wickham and S. Robertson, Plasma Physics $\underline{25}$, 103 (1983).
17. J. Katzenstein and S. Robertson, Proc. Compact Torus Symposium, Bellevue, Washington, Nov. 16-18 (1982).

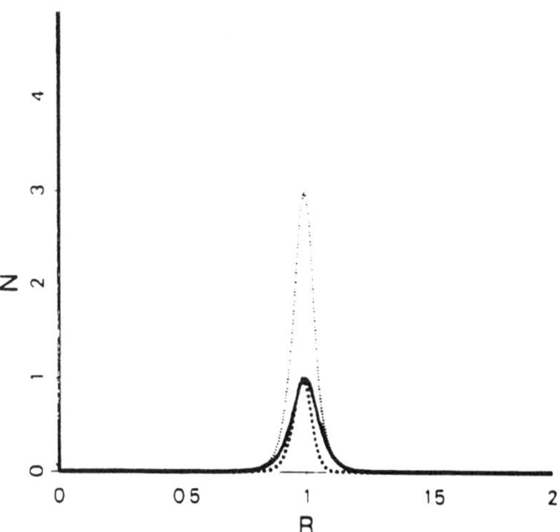

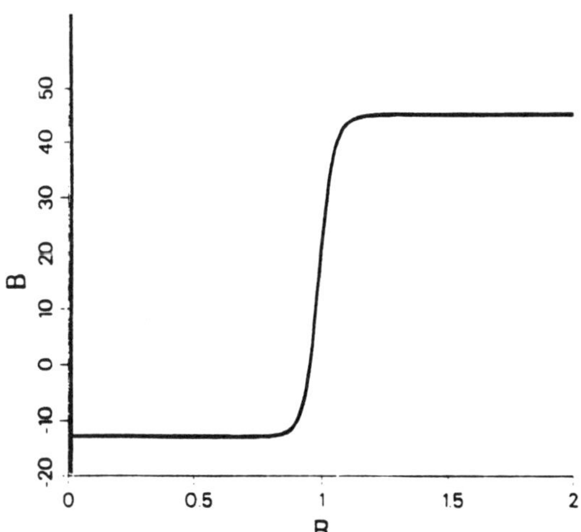

Fig. 1.   Self-consistent FRC equilibrium
for D-He^3

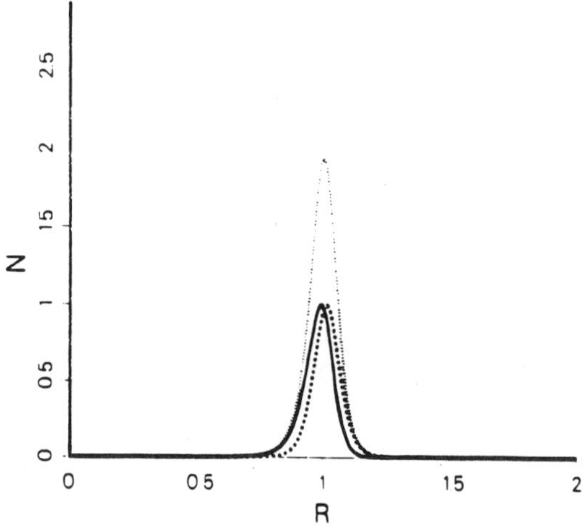

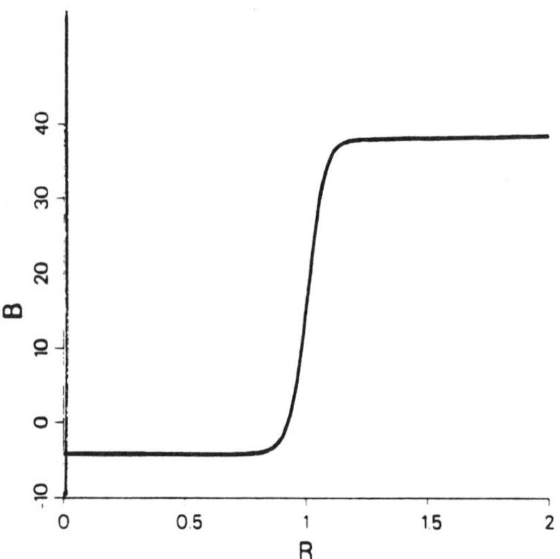

Fig. 2.   Self-consistent FRC equilibrium
          for D-T.

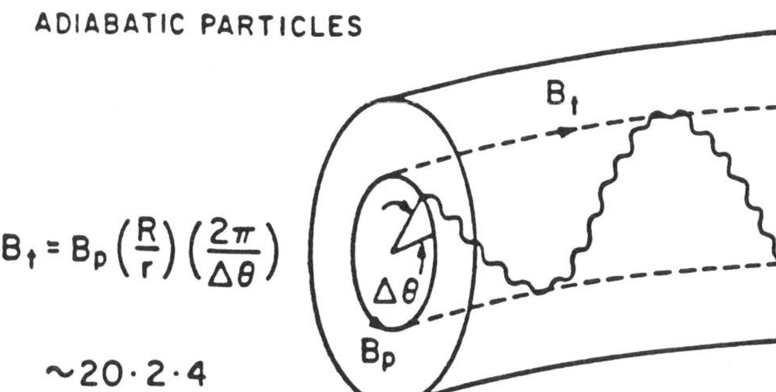

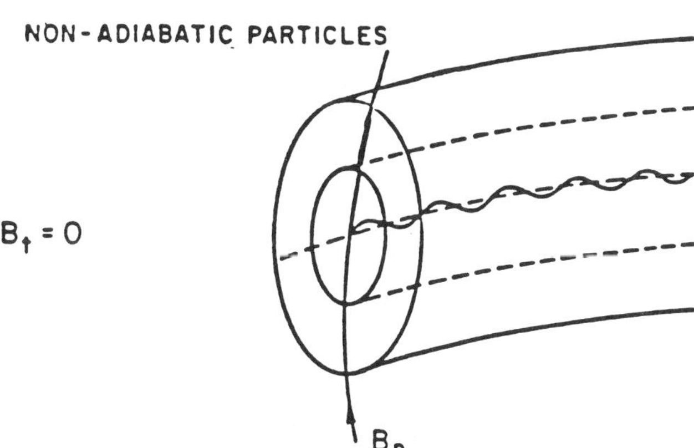

Fig. 3. Differences between adiabatic and
non-adiabatic particle orbits.

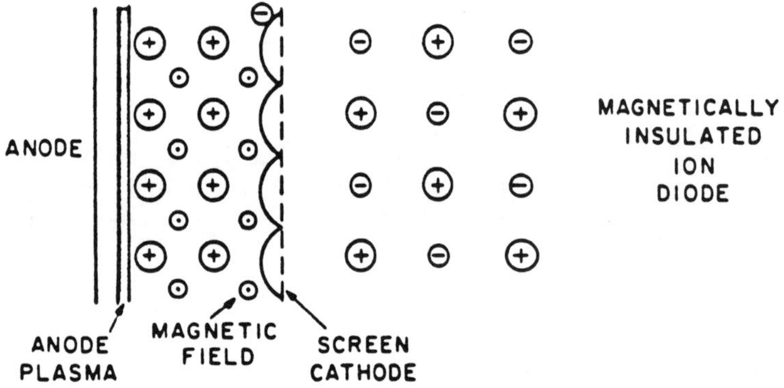

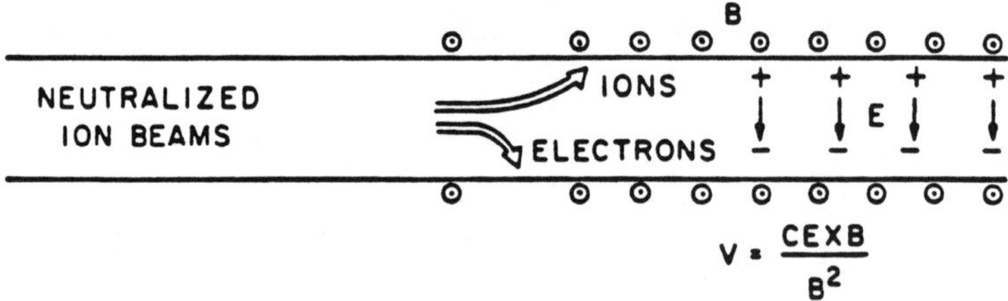

$$V = \frac{C E X B}{B^2}$$

Fig. 4.   Production and cross-field propagation
of a neutralized ion beam.

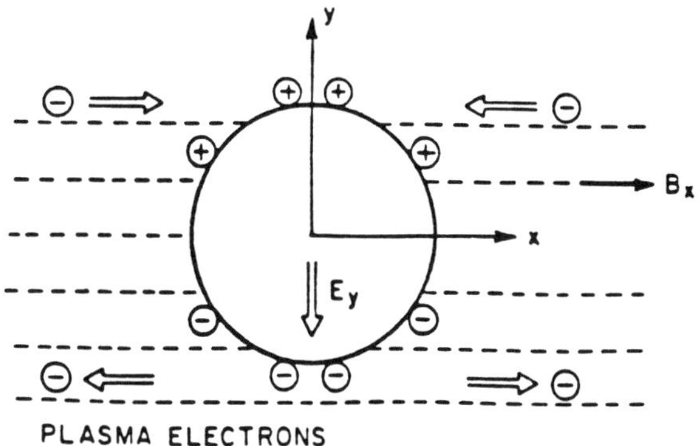

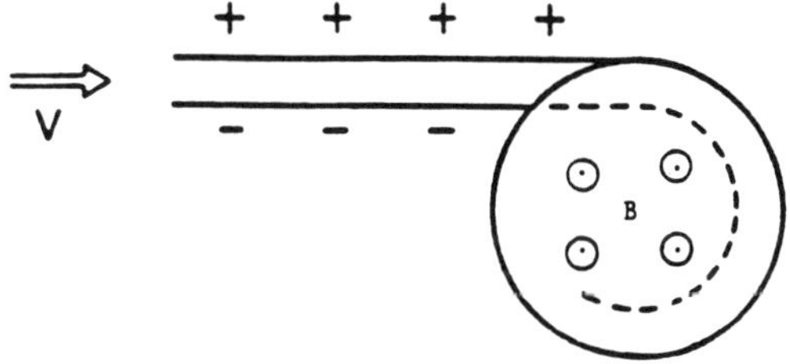

Fig. 5.   Trapping a neutralized beam

C. MHD Turbulence and Transport

EFFECT OF THE EQUILIBRIUM MAGNETIC FIELD ON PLASMA EDGE FLUCTUATIONS*

B. A. Carreras, J. N. Leboeuf, D. K. Lee,[†] J. A. Holmes,[†] and V. E. Lynch[†]

Oak Ridge National Laboratory, Oak Ridge, Tennessee 37831-8071

ABSTRACT

The magnetic field structure reflected in the safety factor (q) profile can play an important role in defining some of the basic structures of the magnetohydrodynamic (MHD) turbulence: it plays a role in the generation of spatial structures in the initial nonlinear phase, and it gives a radial dependence to the spectral average of the square of the poloidal mode number, $\langle m^2 \rangle^{1/2}$, that can modify the explicit dependences of an analytically derived expression for turbulence-induced anomalous losses.

* Research sponsored by the Office of Fusion Energy, U.S. Department of Energy, under contract DE-AC05-84OR21400 with Martin Marietta Energy Systems, Inc.

† Computing and Telecommunications Division, Martin Marietta Energy Systems, Inc.

I.   INTRODUCTION

A great deal of attention has been focused on the study of fluctuations and anomalous transport at the edge of toroidally confined plasmas. Understanding the transport properties of edge plasmas is essential for any magnetic confinement device. In recent years, abundant data on edge fluctuations have been gathered, mostly from the Texas Experimental Tokamak (TEXT)[1] but also from other devices such as the University of Wisconsin Tokapole[2] and the Advanced Toroidal Facility (ATF).[3] Theoretical models have also been developed for edge plasmas. It is now possible to undertake a detailed modeling of the plasma edge turbulence and compare these models with the experimental results. The use of fluid-type equations to study these plasmas facilitates the computational tasks.[4] Many of the available theoretical models are based on magnetohydrodynamic (MHD) turbulence. An example considered in this paper is the model proposed by Thayer and Diamond,[5] which is based on the coupling of convective cells driven by line radiation and the resistivity-gradient-driven turbulence. Many features of the resistivity-gradient-driven turbulence model with thermal instability drive agree with experimental tokamak edge fluctuation measurements. In particular, comparisons with TEXT experimental results show that the model can explain features of the data such as $\langle e\tilde{\Phi}/T_e \rangle > 50\%$, $\langle e\tilde{\Phi}/T_e \rangle \geq \langle \tilde{n}/n_0 \rangle$, and $\langle m \rangle_{\mathrm{rms}} \approx 15$ to 40.

A number of properties of the radial dependence of the average poloidal mode number and the real-space turbulence structures observed in the numerical calculations are common to most of the MHD turbulence modes. They depend only on the underlying magnetic field structure, and they have a general nature. These features are the ones we discuss in this paper.

In Sec. II, after a short discussion of the model, the radially integrated (over the global spectral structures) spectrum of the turbulence is described, showing the localization of the energy spectrum in (m, n) space. In real space, mode coupling structures are observed in the nonlinear calculations. They are related to the underlying magnetic field structure and described in Sec. III. The local spectrum depends on the local value of the safety factor q, which induces a radial dependence of the local spectrum that is discussed in Sec. IV. Finally, in Sec. V, the conclusions are stated.

II.   EQUATIONS AND GLOBAL SPECTRAL STRUCTURE

Although the results discussed in this paper are general to many MHD turbulence models, to focus the discussion we use the model of Thayer and Diamond.[5] This model couples the resistivity-gradient-driven turbulence, for which the underlying instability is the rippling mode,[6] with the thermal instability drive. The basic equations are Ohm's law,

$$\frac{\partial \Psi}{\partial t} = -R_0 \nabla_\parallel \tilde{\Phi} + R_0 \frac{d\eta}{dT} \tilde{T} \frac{E_{\parallel 0}}{\eta_0} + R_0 \eta_0 \tilde{J}_\parallel + R_0 \frac{d\eta}{dT} \tilde{T} \tilde{J}_\parallel \ , \tag{1}$$

the momentum balance equation,

$$\frac{\partial \left(\nabla_\perp^2 \tilde{\Phi} \right)}{\partial t} + \vec{V} \cdot \nabla \left(\nabla_\perp^2 \tilde{\Phi} \right) = \frac{B_0^2}{\rho_m} \nabla_\parallel \tilde{J}_\parallel + \mu \nabla_\perp^4 \tilde{\Phi} \ , \tag{2}$$

and the electron temperature equation,

$$\frac{\partial \tilde{T}}{\partial t} + \vec{V} \cdot \nabla \tilde{T} = \chi_{\parallel} \nabla_{\parallel}^2 \tilde{T} - \tilde{V}_r \frac{dT_0}{dr} - I_Z(T) + \chi_{\perp} \nabla_{\perp}^2 \tilde{T} \ . \tag{3}$$

Here Ψ is the poloidal magnetic flux, which allows one to write the magnetic field as $\vec{B} = \frac{1}{R} \nabla \Psi \times \hat{\zeta} + B_0 \hat{\zeta}$; Φ is the electrostatic potential; T is the electron temperature; $\vec{V} \equiv \nabla \frac{\tilde{\Phi}}{B_0} \times \hat{\zeta}$ is the flow velocity; the resistivity is η; the cross-field transport coefficient is $\chi_{\perp}$; the parallel conductivity is $\chi_{\parallel}$; the viscosity is μ; and the mass density is ρ_m. The intensity of the cooling caused by line radiation is given by I_Z, which is a function of the temperature. In most of our calculations, we have used the functional dependence on T given by the coronal radiation model, but the level has been normalized to experimental measurements. The parallel current $J_{\parallel}$ is related to the poloidal magnetic flux by

$$\tilde{J}_{\parallel} = \frac{1}{\mu_0 R_0} \nabla_{\perp}^2 \tilde{\Psi} \ . \tag{4}$$

The main nonlinearities are $E \times B$ nonlinearities in the convective terms and the magnetic nonlinearity in the parallel derivative operator

$$\nabla_{\parallel} \tilde{\Phi} = \frac{1}{|\vec{B}|} \vec{B} \cdot \nabla \tilde{\Phi} \ . \tag{5}$$

In these equations, the perturbed quantities are indicated by a tilde and the equilibrium quantities by the subindex 0.

These equations lead to a total energy conservation condition

$$\frac{d}{dt} (E_M + E_K + E_T) = - \int dV \left(\eta J_{\parallel}^2 + \chi_{\perp} \left| \vec{\nabla}_{\perp} \tilde{T} \right|^2 + \mu \left| \nabla_{\perp}^2 \tilde{\Phi} \right|^2 \right)$$
$$- \int dV \left[I_Z(T) + \tilde{V}_r \frac{dT_0}{dr} \right] \tilde{T} - \int dV \chi_{\parallel} \left| \nabla_{\parallel} \tilde{T} \right|^2 \ . \tag{6}$$

The first three terms on the right-hand side of the equation are the collisional dissipation terms, which in general are small in comparison with the other terms. The dominant terms are the next two terms, which can provide the drive through the temperature gradient and the line radiation, and the last term, which provides the main stabilization mechanism, through parallel conduction. By expanding the function I_Z around the equilibrium temperature value, we have $I_Z(T) \approx (dI_Z/dT) \tilde{T}$. Therefore, in the region where $dI_Z/dT < 0$, line radiation drives the instability. The corresponding energies are the magnetic energy

$$E_M = \int dV \frac{\vec{B}^2}{2\mu_0} = \int dV \frac{\left| \nabla_{\perp} \tilde{\Psi} \right|^2}{2R_0^2 \mu_0} \ , \tag{7}$$

the kinetic energy

$$E_K = \int dV \frac{\vec{V}_{\perp}^2}{2} = \int dV \frac{\left| \nabla_{\perp} \tilde{\Phi} \right|^2}{2B_0^2} \ . \tag{8}$$

and a measure of the thermal energy

$$E_T = \int dV \left| \tilde{T} \right|^2 \tag{9}$$

These nonlinear equations are solved numerically as an initial value problem. All fields are Fourier expanded in poloidal, θ, and toroidal, ζ, angles. A finite difference representation is used for the radial coordinate r. The numerical scheme used in these calculations treats all linear terms as fully implicit. The nonlinear terms are always treated explicitly. Details on the numerical scheme and its implementation are given in Ref. 7.

The analytical theory[8,9] shows that the turbulence saturation condition is the result of the enhanced parallel diffusion by the radial turbulent transport balancing the gradient and thermal instability drives. These analytical results have been confirmed by the numerical calculations.

The calculations discussed here are for a driven turbulence situation. Since the driving term is the radiation cooling, they were done with the radiation profile frozen in time. Therefore, in these calculations, the quasilinear modification of the line radiation profile is cancelled. This hypothesis is equivalent to the assumption that the heating is instantaneous and the impurity density of the profile is maintained. The reason for such an assumption is to have a steady-state turbulence with a prescribed radiation profile. This implies that the source of free energy for the instabilities is constant during the calculation and feeds the long-scale instabilities. The sink of energy is essentially the parallel conduction term. To reach a steady state, the turbulence level must adjust itself in such a way that the nonlinear broadening of the modes is balanced by the parallel conduction that restricts its radial extent. In this way, the dissipation balances the input energy that feeds the instability. Many convergence tests of the saturation level of turbulence as a function of the number of modes and radial grid size have been done.

Ψ and $\tilde{\Phi}$ can be written as a Fourier expansion on the poloidal and toroidal angles

$$\Psi = \sum_{m,n} \Psi^c_{mn} \cos(m\theta + n\zeta) \ , \tag{10}$$

$$\tilde{\Phi} = \sum_{m,n} \tilde{\Phi}^s_{mn} \sin(m\theta + n\zeta) \ . \tag{11}$$

Here, m and n are the poloidal and toroidal mode numbers, respectively. All quantities can be written in terms of the Fourier expansion. It is interesting to consider first the expansion of the energies and their spectral distribution in the (m,n) plane.

Because the driving term of the instability, the radiation cooling, is radially localized in an interval (r_1, r_2) where $dI_Z/dT < 0$, the relevant resonance surfaces are bounded by the values $q_1 = q(r_1)$ and $q_2 = q(r_2)$. The resonant components dominate the spectrum; hence the (m,n) components most relevant for these calculations fall into a wedge in the (m,n) plane delimited by the values of q_1 and q_2. Therefore, the corresponding energy spectrum is localized in the (m,n) plane. This localization of the global energy spectrum has been tested numerically. The kinetic energy spectrum at constant n shows a sharp falloff, indicating the localization of the spectrum in a wedge in the (m,n) space. The falloff of the energy spectrum with n along a fixed helicity is less sharp than at constant m.

This localization of the energy spectrum in the (m, n) plane is one of the consequences of the underlying magnetic structure effects on the characterization of the MHD turbulence. It has important consequences in facilitating the plasma turbulence calculations, as discussed in Ref. 4.

III. MODE COUPLING IN REAL SPACE

In these turbulence calculations, apart from the statistical analysis of the numerical results to determine average mode number and spectral decay index, the analysis is essentially visual. Sometimes this is the only way to understand some of the basic physics involved in the nonlinear processes. High-resolution contour plots of the fields (Fig. 1) are useful for this analysis. In this figure, the contours of constant electrostatic potential fluctuations are plotted in a rectangular box, the horizontal axis being the r range $(0.80 < r/a < 0.92)$ in which $dI_Z/dT < 0$ and the vertical axis being the poloidal angle between $0°$ and $180°$. The structures observed in the electrostatic potential fluctuations are very similar to the structures in the temperature fluctuations.

At the beginning of the nonlinear phase in some of the turbulence calculations, well-defined radial patterns are apparent for the electrostatic potential fluctuations (Fig. 1a). Deep into the nonlinear regime, a memory of the radial patterns seems to remain. It is, therefore, important to understand the cause of such structures. As we will discuss, these patterns reflect the structure of the lowest m numbers associated with the rational surfaces.

To calculate the lowest m number for each rational surface, we can generate the rationals between two given surfaces in a way similar to a Fibonacci sequence. Let us consider the case for which the unstable region is between the $q_1 = 2$ and $q_2 = 3$ surfaces. The lowest m modes at these two surfaces are $(m_1 = 2, n_1 = 1)$ and $(m_2 = 3, n_2 = 1)$, respectively. The next lowest m mode in this q interval is obtained by the beating of these two modes, that is, $(m_1 + m_2, n_1 + n_2) = (5, 2)$. This mode is associated with $q = (m_1 + m_2)/(n_1 + n_2) = 5/2$. We can proceed in this way and generate sequences of rational numbers such that the numerators and denominators are obtained by adding the numerators and denominators, respectively, of the two previous terms in the sequence. For the case considered here, an example of a three-step sequence is

3/1				5/2				2/1
3/1		8/3		5/2		7/3		2/1
3/1	11/4	8/3	13/5	5/2	12/5	7/3	9/4	2/1

Here, the rationals are ordered in a descending order which would correspond to a radial distribution of the rational flux surfaces. These sequences give the lowest relevant m at each radial position. The lowest m as a function of r, $m(r)$, is a Cantor set, and the correlation between r and m is rather complicated.

The set of m numbers for the fifth step in this sequence has been represented graphically in Fig. 1b. This has been done by assigning at each radial position a number of cells corresponding to the m value. These cells schematically represent the structure of a given m mode localized at the resonant surface. For neighboring modes, the nonlinear interaction is maximized at a few θ values. The phase-reinforcing lines form a well-defined pattern in the radial direction. These patterns are very similar to those obtained in the numerical calculation for the electrostatic potential contours (Fig. 1a). This indicates that the observed

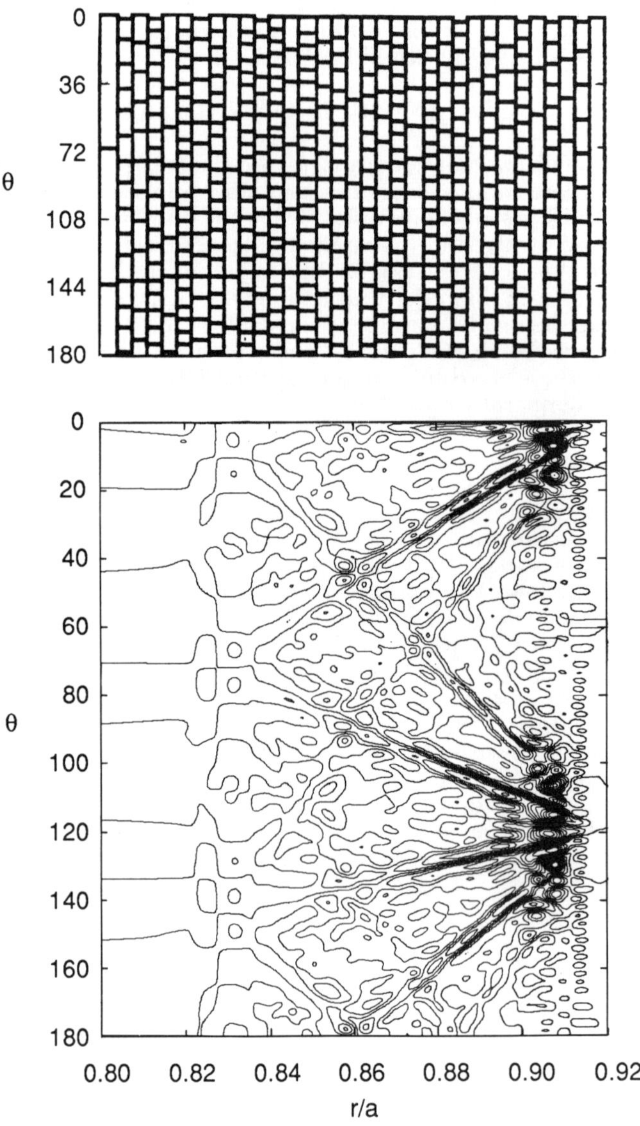

Fig. 1.　Comparison of (a) lowest m values at the resonance surfaces between $q = 2$ and $q = 3$ and (b) observed structures in the numerical calculations using Eqs. (1)–(3).

patterns in real space correspond to directions of maximum overlap between neighboring modes.

These structures are less likely to occur in more realistic models. The existence of a differential rotation between neighboring surfaces, because of diamagnetic effects or sheared electric fields, will change these patterns and can substantially change the nonlinear interactions. Their relevance for MHD turbulence remains when the growth of the instability is larger than the differential rotation.

IV. RADIAL DEPENDENCE OF THE SPECTRUM

The important role that the lowest m mode in each magnetic surface plays in the nonlinear interaction can be expected to affect the local mode spectrum. Numerical results show that the m spectrum at a fixed radius depends on the lowest m value associated with the nearest resonant surface. In Fig. 2, the m-spectra at two different radial positions have been plotted. These two positions correspond to q values of 8/3 and 20/7, respectively. The peak of the spectrum shifts as the lowest dominant m value changes, being lower at the lowest m value. The value of the spectral average of m^2 also changes with the radial position. In general, the spectrum-averaged value of m is proportional to the lowest local m value. The proportionality factor is between 1.25 and 1.5, depending on the particular problem. Therefore, determination of the lowest relevant m at a given radial position is important in determining the expected properties of the spectrum. To estimate the radial dependence of the spectral average of m^2, $\langle m^2 \rangle^{1/2}$,

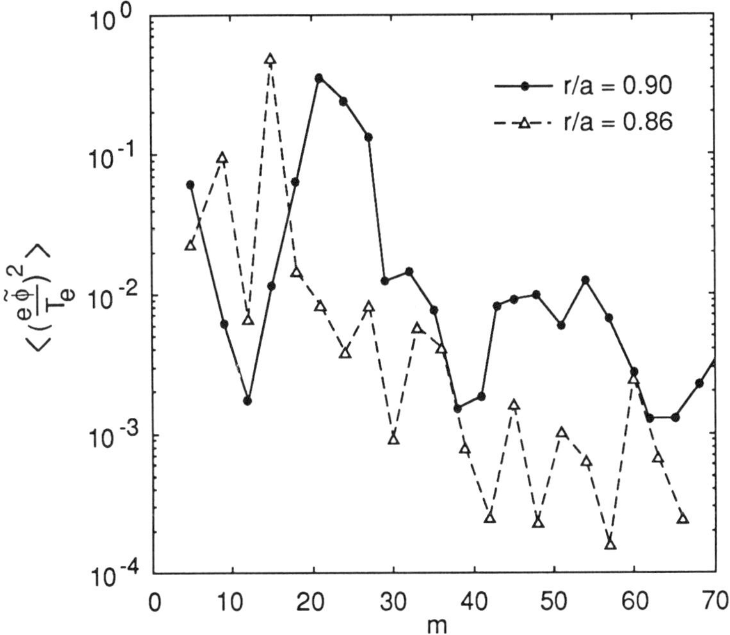

Fig. 2.   Electrostatic fluctuation m spectrum at two different radial positions corresponding to $q = 8/3$ and $q = 20/7$.

we calculate the function $m = M(r)$ as follows: (1) first, we calculate the lowest m at each radial position as described in Sec. III; (2) each rational surface is associated with radial width, $W(m)$, given by the nonlinear theory (in the case of the resistivity-gradient-driven turbulence this nonlinear width scales as $m^{-1/3}$); and (3) in radial regions where two widths overlap, the lowest m value is taken to be the dominant local m.

The resulting function $M(r)$ has a complicated structure (Fig. 3). If the nonlinear width $W(m)$ is increased for each mode, many of the details of the radial structure of $M(r)$ are washed out. For a given $W(m)$, $\langle m^2 \rangle^{1/2}$ scales as q in a radially averaged sense. Since $q(a) \approx 3$ in TEXT and $q(a) \approx 1$ in ATF, we expect $\langle m^2 \rangle^{1/2}$ for ATF to be about a factor of three lower than for TEXT. Because of the complicated dependence on r, the factor of three is only relevant in a radial averaged sense. Recent results from ATF[3] seem to verify this scaling. However, at present, low-temperature results from TEXT are difficult to compare with those from ATF because the TEXT results correspond to plasmas in the shadow of the limiter. But it will be interesting to continue the experimental evaluations of the radial dependence of $\langle m^2 \rangle^{1/2}$.

The numerical results for $\langle m^2 \rangle^{1/2}$ seem to follow the structure of the calculated $M(r)$ function (Fig. 4). The results plotted in Fig. 4 correspond to a

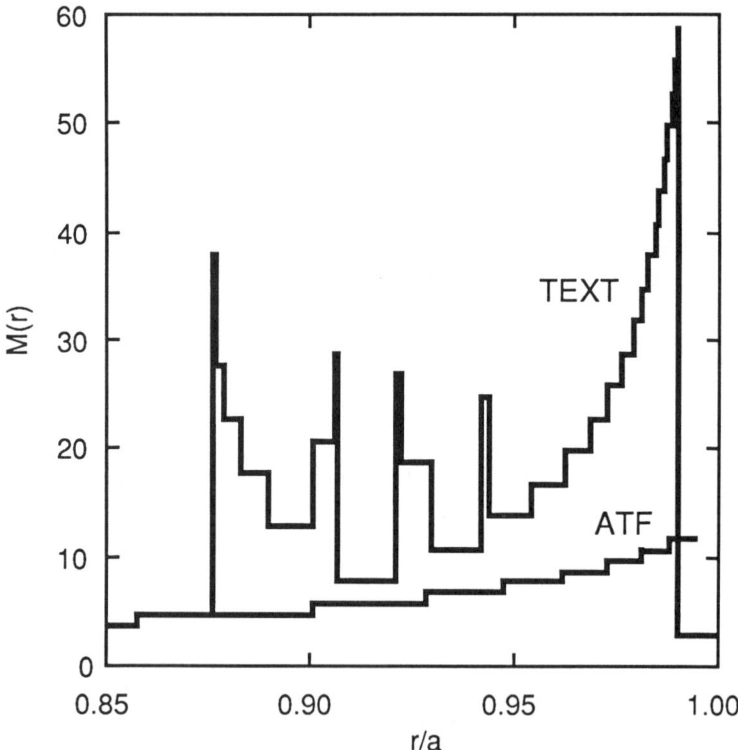

Fig. 3.   Lowest m value resonant at a given q surface as a function of radius for q profiles relevant to TEXT and ATF.

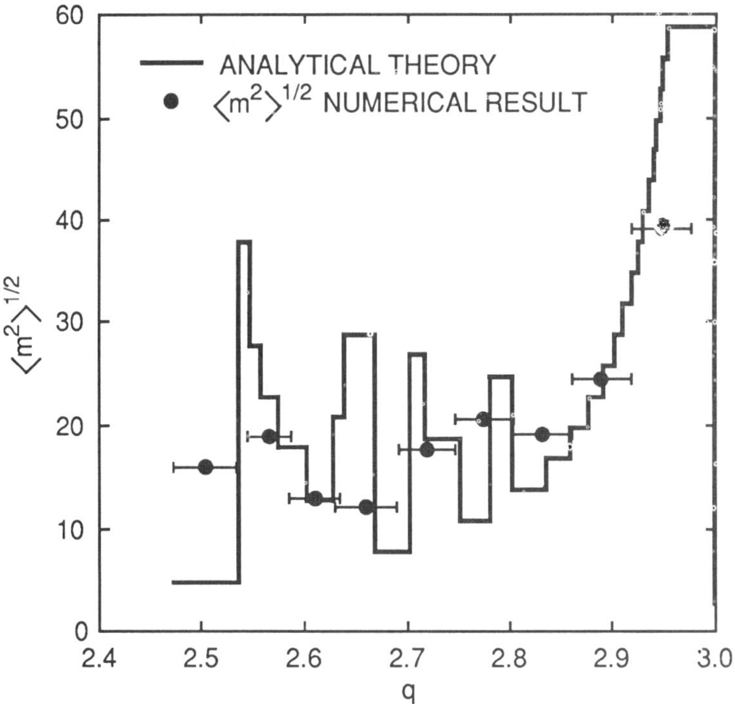

Fig. 4.   Comparison of the spectrum-averaged $\langle m^2 \rangle^{1/2}$ and the local lowest m value calculated as described in the text.

nonlinear calculation using the model described in Sec. II with parameters corresponding to TEXT edge plasmas. The numerical results have been averaged over the steady-state turbulence regime and over a small radial interval $\Delta r = 0.05a$.

For MHD turbulence, the induced thermal and particle losses are, in general, a function of $\langle m^2 \rangle^{1/2}$. In comparing the analytical results with experimental data in transport simulations, the term $\langle m^2 \rangle^{1/2}$ is normally taken to be a given constant. Therefore, the radial dependence discussed here is never included. Its inclusion will have obvious implications, such that low values of $\langle m^2 \rangle^{1/2}$ imply high levels of losses, and high values of $\langle m^2 \rangle^{1/2}$ imply low levels of losses. This hidden q dependence in $\langle m^2 \rangle^{1/2}$ could explain the step-like structures in the measured electron temperature profiles in the Tokamak Fusion Test Reactor (TFTR).[10] It also could explain the increased losses near the low rational values of q at the plasma edge. It will have an effect similar to the plasma edge mechanism proposed by R. Waltz et al.[11] to degrade confinement at high q.

In stellarators, a dependence of confinement on $q(a)$ has been experimentally established by the Wendelstein VIIA group.[12] This dependence is attributed to the existence of resonant magnetic islands, but could also come through electrostatic fluctuation-induced transport in which the basic scale length verifies the pattern shown in Fig. 3.

V.   CONCLUSIONS

These results indicate that even for cases in which the anomalous particle and thermal losses are dominated by electrostatic turbulence, the confinement can depend on the structure of the q profile. The magnetic field structure reflected in the q profile can play an important role in defining some of the basic structures of the MHD turbulence. It plays a role in the generation of spatial structures in the initial nonlinear phase that can modify the character of the turbulence in the steady state. It also plays a role in giving a complicated radial dependence to $\langle m^2 \rangle^{1/2}$. That is, the basic scales of the turbulence can change significantly near low-m resonant surfaces, enhancing the loses at these q values. The dependence of $\langle m^2 \rangle^{1/2}$ on the q value, and therefore on the radial position, can modify the explicit dependences of analytically derived anomalous transport coefficients.

REFERENCES

1. Ch. P. Ritz, R. V. Bravenec, R. V. Schoch et al., Phys. Rev. Lett. **62**, 1844 (1989).
2. D. E. Graessle, C. S. Prager, and R. N. Dexter, Phys. Rev. Lett. **62**, 535 (1989).
3. C. Hidalgo et al., paper presented at the 17th European Conference on Controlled Fusion and Plasma Physics, Amsterdam, The Netherlands, June 25–29, 1990.
4. B. A. Carreras, L. A. Charlton, N. Dominguez, J. B. Drake, L. Garcia, J. A. Holmes, J. N. Leboeuf, D. K. Lee, and V. E. Lynch, "Plasma Turbulence Calculations on Supercomputers," to be published in the International Journal of Supercomputer Applications (1990).
5. D. R. Thayer and P. H. Diamond, Phys. Fluids **30**, 3624 (1987).
6. H. P. Furth, J. Killeen, and M. N. Rosenbluth, Phys. Fluids **6**, 77 (1963).
7. L. Garcia, H. R. Hicks, B. A. Carreras, L. A. Charlton, and J. A. Holmes, J. Comput. Phys. **65**, 253 (1986).
8. L. Garcia, P. H. Diamond, B. A. Carreras, and J. D. Callen, Phys. Fluids **28**, 2147 (1985).
9. P. W. Terry, P. H. Diamond, K. C. Shaing, L. Garcia, and B. A. Carreras, Phys. Fluids **29**, 2501 (1986).
10. B. Grek, D. Johnson, W. Park, G. Taylor, K. McGuire, D. Mansfield, and H. Park, Bull. Am. Phys. Soc. **34**, 1965 (1989).
11. R. E. Waltz, R. R. Dominguez, S. K. Wong, P. H. Diamond, G. S. Lee, T. S. Hahm, and N. Mattor, in Plasma Physics and Controlled Nuclear Fusion Research (Proc. 11th Int. Conf. Kyoto, 1986), Vol. 1, IAEA, Vienna, p. 345 (1987).
12. W VII-A Team, NI Group, Nucl. Fusion **25**, 1593 (1985).

GYROKINETIC SIMULATION STUDIES OF NONLINEAR DRIFT WAVE BEHAVIOR

A. M. Dimits
Laboratory for Plasma Research
University of Maryland, College Park, MD 20742

and

W. W. Lee
Plasma Physics Laboratory, Princeton University
Princeton, NJ 08543

INTRODUCTION

Drift instabilities driven by electron dissipation have been suggested as an explanations of small-scale, low-frequency fluctuations that have been observed in laser scattering experiments,[1] and anomalous electron thermal transport, both in the bulk plasma,[2] and at the edge[3,4] in tokamak discharges. The density gradients are generally considered to be quasistatic, thus precluding profile relaxation as a saturation mechanism.

Drift waves arise because of the balance between the near-adiabatic electron response $n^e \simeq n_0 e\phi/T_e$ and the ion density perturbations due to the $E \times B$ advection of the ions from an ion density profile with a nonzero density gradient. A variety of collisional and collisionless dissipation mechanisms,[5,6] associated with different regimes in the parameter space spanned by magnetically confined fusion plasmas, can drive drift-type modes unstable.

While the linear theory of drift instabilities[5,6] is well understood, the situation regarding the understanding of their nonlinear behavior, even in simple geometries, is not so satisfactory. This situation motivated the development of the gyrokinetic particle simulation method,[7] which enabled the first direct simulations of steady-state drift wave turbulence taking nonlinear kinetic effects into account. By using conventional particle simulation methods to solve a gyro-averaged Vlasov-Poisson (or Vlasov-Maxwell) system of equations rather than the basic dynamical particle equations, timesteps longer than the gyroperiod Ω_i^{-1} and the plasma period ω_{pe}^{-1}, and grid sizes longer than the Debye length can be used.[8] This represents a significant relaxation of the timestep and grid size requirements compared with simulations which follow the full ion dynamics. Also, to obtain a given noise level, $\lambda_{\mathrm{De}}/\rho_{\mathrm{s}}$ fewer particles are needed in a (two-dimensional) gyrokinetic simulation, where λ_{De} is the Debye length and $\rho_{\mathrm{s}} = c_{\mathrm{s}}/\Omega_{\mathrm{i}}$, where c_{s} is the ion sound speed. This ratio is a small number for plasmas of fusion interest.

In this paper, we report the results of investigations of the basic mechanisms which occur in gyrokinetic particle simulations of drift waves. We first review the basic gyrokinetic algorithm and then discuss some important mathematical properties of the model. We then review the phenomenology observed in gyrokinetic simulations of drift waves and go on to develop a nonlinear model which fits this phenomenology. Finally, the results of the paper are summarized along with a discussion of unresolved issues immediately suggested by this work.

GYROKINETIC PARTICLE SIMULATION

The gyrokinetic particle simulation method[8] uses particle simulation methods to solve a gyro-averaged Vlasov-Poisson (or Vlasov-Maxwell) system of equations[7,9] valid in the gyrokinetic ordering,

$$\frac{q_\alpha \phi}{T_\alpha} \sim \frac{\omega}{\Omega_\alpha} \sim \frac{\rho_\alpha}{L} \sim \epsilon \ll 1, \quad L \sim L_\parallel,$$

where $\rho_\alpha \equiv v_{t\alpha}/\Omega_\alpha$, $\Omega_\alpha \equiv q_\alpha B/m_\alpha c$, $v_{t\alpha} \equiv \sqrt{T_\alpha/m_\alpha}$, q_α, m_α, and T_α are respectively the charge, mass and temperature for species α, c is the speed of light, B is the magnetic field strength, ϕ is the electrostatic potential, ω is the frequency of the perturbation, L is a characteristic perpendicular equilibrium scale length of the system, and $L_\parallel$ is the characteristic parallel wavelength of the perturbation. We will drop the species index α when we are referring to either species and there is no ambiguity.

The electrostatic gyrokinetic Vlasov equation for a plasma in a uniform magnetic field is

$$\frac{\partial F}{\partial t} + v_\parallel \hat{b} \cdot \frac{\partial F}{\partial \boldsymbol{R}} - \frac{c}{B}\frac{\partial}{\partial \boldsymbol{R}} \cdot \left[\left(\frac{\partial \Psi}{\partial \boldsymbol{R}} \times \hat{b} \right) F \right] - \frac{q}{m}\frac{\partial \Psi}{\partial \boldsymbol{R}} \cdot \hat{b}\frac{\partial F}{\partial v_\parallel} = C(F), \tag{1a}$$

where

$$\Psi(\boldsymbol{R}) \equiv \bar{\phi} - \frac{q}{2T}\left(\frac{v_t}{\Omega}\right)^2 \left|\frac{\partial \phi}{\partial \boldsymbol{R}}\right|^2, \tag{1b}$$

$$\bar{\phi} \equiv \sum_{\boldsymbol{k}} \phi(\boldsymbol{k}) J_0\left(\frac{k_\perp v_\perp}{\Omega}\right) \exp(i\boldsymbol{k} \cdot \boldsymbol{R}), \tag{1c}$$

$$\boldsymbol{R} \equiv \boldsymbol{x} + \frac{v_\perp \times \hat{b}}{\Omega}, \tag{1d}$$

x is the particle position, $v_\perp$ is the perpendicular velocity, $\bar\phi(R)$ is the gyro-averaged electrostatic potential, $F(R,\mu,v_\parallel,t)$ is the full gyro-averaged distribution function, X is the x component of R, $\mu \equiv v_\perp^2/2$, and $C(F)$ is the collision operator. The electrostatic potential ϕ is given by the gyrokinetic Poisson equation which, for a single ion species i, is

$$\nabla^2\phi - \frac{\tau(\phi - \tilde\phi)}{\lambda_D^2} + \left(\frac{\rho_s}{\lambda_D}\right)^2 \nabla_\perp\cdot\left[\frac{(n^i - n_0)}{n_0}\nabla_\perp\phi\right] = -4\pi e(\bar n^i - n^e) \quad (1e)$$

where

$$\tilde\phi(x) \equiv \sum_k \phi(k)\Gamma_0(k_\perp^2\rho_i^2)\exp(ik\cdot x), \quad (1f)$$

$$\bar n(x) \equiv \sum_k \int F(k)J_0\left(\frac{k_\perp v_\perp}{\Omega}\right)\exp(ik\cdot R)\,d\mu\,dv_\parallel, \quad (1g)$$

and where $\tau \equiv T_e/T_i$, $\rho_s \equiv c_s/\Omega_i$, $c_s \equiv \sqrt{T_e/m_i}$, $k_\perp$ is the perpendicular wave number, $\lambda_D \equiv \sqrt{T_e/4\pi n_0 e^2}$ is the electron Debye length, n_0 is the background ion number density, n^α is the number density of species α, and $\Gamma_0(b) \equiv I_0(b)e^{-b}$. For ions, $C(F) = 0$. For the electrons, finite gyroradius effects are negligible so that $\Psi = \bar\phi = \tilde\phi = \phi$. Pitch-angle scattering is retained for the electrons, via a Monte-Carlo model for the Lorentz collision operator,[10]

$$C(F) = \frac{\nu_{ei}}{2\sin\xi}\frac{\partial}{\partial\xi}\left(\sin\xi\frac{\partial F_e}{\partial\xi}\right).$$

Here ν_{ei} is the electron pitch-angle scattering frequency, which in these studies is taken to be velocity-independent, and ξ is the pitch angle. Since the parallel phase velocity of the waves is smaller than the electron thermal velocity, the population of runaway electrons should be negligible. The intrinsic collision time due to the particle discreteness in the simulation is several orders of magnitude greater than the typical run time.[10]

The geometry used is a periodic 2-1/2-dimensional shear-free slab in which one position coordinate, which we take to be z, is ignorable. The magnetic field is uniform and inclined to the z axis, $B = B(\hat z + \theta\hat y)$, where $\hat y$ and $\hat z$ are unit vectors in the y and z directions. The parallel wavenumber is then $k_\parallel = \theta k_y$. The inclination angle θ is a small constant in the shear-free slab models. As defined above, the background gradient is in the x direction.

To model the driving due to the gradients in a periodic geometry, we use a simple particle conserving generalization of the standard density gradient driving source term commonly used in linear theory.[6] The perpendicular configuration-space velocity in Eq. (1a), is modified so that it becomes

$$v_{\mathrm{perp}} \equiv \frac{d\boldsymbol{R}}{dt} = -\frac{c}{B}\left[\frac{\partial\Psi}{\partial\boldsymbol{R}} + \kappa\Psi\right]\times\hat{\boldsymbol{b}}, \tag{2}$$

where $\boldsymbol{\kappa} \equiv -\nabla\ln F_{\mathrm{M}} = \kappa\hat{\boldsymbol{x}}$ and $\hat{\boldsymbol{b}} \equiv \hat{\boldsymbol{z}} + \theta\hat{\boldsymbol{y}}$. Here, F_{M} is the background Maxwellian distribution function with the associated density $n_0(x)$ and temperature $T_0(x)$. We can write κ as $\kappa = \kappa_n\left[1 + \eta\left(v_\parallel^2/v_{t\alpha} - \frac{1}{2}\right)\right]$, where $\eta \equiv (d\ln T_0/d\ln n_0)$. For $\eta = 0$, the subscript n will be dropped so that κ will appear instead of κ_n.

This velocity has the important property

$$\nabla\cdot v_{\mathrm{perp}} = -\kappa\dot{X}, \tag{3}$$

where $\dot{X} = -(c/B)(\partial\Psi/\partial Y)$ is the x component of v_{perp}.

HAMILTONIAN FORM OF EQUATIONS OF MOTION

The gyrocenter equations of motion used by the particle simulation, neglecting collisions, are given by Eq. (2), along with the parallel Newton's law

$$\dot{v}_\parallel = -\frac{q}{m}\hat{\boldsymbol{b}}\cdot\nabla\Psi.$$

In the $2\frac{1}{2}$-dimensional geometry, $\dot{y} = v_{\mathrm{py}} + \theta v_\parallel$ where v_{py} is the y component of v_{perp}. Upon taking this into account and using $\partial/\partial z = 0$ to express $E_\parallel$ in terms of $\dot{x}$, we obtain

$$\dot{x} = -\frac{c}{B}\frac{\partial\Psi}{\partial y}, \tag{4a}$$

$$\dot{y} = \frac{c}{B}\left(\frac{\partial\Psi}{\partial x} + \kappa\Psi\right) + \theta v_\parallel - v_{\mathrm{ph}}, \tag{4b}$$

$$\dot{v}_\parallel = \zeta\dot{x}, \tag{4c}$$

where

$$\zeta = \begin{cases} -\Omega_e\theta & \text{(electrons)}, \\ \Omega_i\theta & \text{(ions)}. \end{cases} \tag{4d}$$

A Galilean transformation to a frame moving with fixed velocity v_{ph} in the y direction has been made. By making a transformation to the wave frame, i.e., setting $v_{\rm ph}$ equal to the phase velocity of the waves in the y direction, if one exists, Ψ and Eqs. (4) can be made time independent.

These equations can be recast in Hamiltonian form as follows. First, Eq. (4c) is integrated with respect to time to yield the constant of motion

$$C_1 = v_\| - \zeta x. \tag{5}$$

This constant is then used to eliminate either x or $v_\|$ from Eqs. (4a) and (4b). If the parallel-trapping frequency is small or comparable to the $\boldsymbol{E} \times \boldsymbol{B}$-trapping frequency, then it is more physical to eliminate $v_\|$. An integrating factor $I(x)$ for Eq. (4b) can be found from the $(\partial\Psi/\partial x)$ term and the $\kappa\Psi$ terms. For $\eta = 0$, we have $I(x) = \exp(\kappa x)$. For nonzero η,

$$I(x) = \exp\left\{\kappa_n\left(x + \eta_{\rm i}\left[\frac{(\zeta x + C_1)^3}{(3\zeta v_{t\alpha}^2)} - \frac{1}{2}x\right]\right)\right\}.$$

Equation (4b) can then be rewritten as

$$\dot{y} = \frac{1}{I(x)}\frac{c}{B}\frac{\partial}{\partial x}[I(x)\Psi] + V(x),$$

where

$$V(x) = \zeta\theta x + C_1\theta - v_{\rm ph}.$$

Equations (4a) and (4b) can thus be written in the Hamiltonian form

$$\frac{dx}{ds} = -\frac{\partial}{\partial y}H,$$

$$\frac{dy}{ds} = \frac{\partial}{\partial x}H,$$

where

$$\frac{d}{ds} \equiv I(x)\frac{d}{dl}$$

and

$$H = \frac{c}{B}I(x)\Psi(x) + \int^x dx' I(x')V(x'). \tag{6}$$

A few comments are in order. This procedure takes the parallel acceleration into account for nonzero ζ. When $v_\|$ is eliminated using Eq. (5), the parallel velocity is expressed as a function of x. The terms representing the effects of the parallel acceleration are those containing nonzero powers of ζ.

The transformation to the new "time" variable s has the interpretation that along their trajectories, the particles move more slowly for higher values of $I(x)$. Thus, the x-y area element containing a group of gyrocenters is compressed by a factor $I(x_2)/I(x_1)$ as it travels from a position with x coordinate x_1 to one with x coordinate x_2. The rate of compression is given by Eq. (3). If the background density (of simulation particles) is flat, then this results in a density perturbation. The imposition of periodic boundary conditions, however, forces the mean large-scale simulation particle density to be flat. Thus, there is a permanently enforced separation between the mean particle density and the equilibrium particle density. This is an alternative mathematical representation of the physics of the density increase due to incompressible advection of particles from a nonuniform density profile. We note that an alternative way to cast Eqs. (4) into Hamiltonian form is to transform x instead of t, and use the new variable $X \equiv \exp(\kappa x)$.

For $\eta_i = 0$, it is straightforward to write the Hamiltonian of Eq. (6) in closed form,

$$H = e^{\kappa x}\left\{\frac{c}{B}\Psi + \frac{\zeta\theta}{\kappa}x + \frac{1}{\kappa^2}[(\theta C_1 - v_{\rm ph})\kappa - \zeta\theta]\right\}.$$

For the ions, particularly in the case of the collisionless drift mode in a shear-free slab, the limit of zero parallel velocity is of interest. The result is

$$H = e^{\kappa x}\left(\frac{c}{B}\Psi - \frac{v_{\rm ph}}{\kappa}\right).$$

The $\kappa = 0$ limit of this result has been either the focus of, or the starting point for, several analyses of the nonlinear effects of drift waves on gyrocenter orbits.[11-12] The summarize the $\kappa \to 0$ limits of the above results are easily obtained (after subracting off appropriate constants) as

$$H = \begin{cases} c\Psi/B - v_{\rm ph}x & (v_\| = 0), \\ c\Psi/B + \frac{1}{2}\theta\zeta x^2 + (\theta C_1 - v_{\rm ph})x & (\text{no shear}). \end{cases} \tag{7}$$

The first of these was given by Ching.[11] The limit of zero parallel acceleration can be obtained by setting $\zeta = 0$ and $C_1 \to v_\| =$const. in any of the above results. Thus, for example, the Hamiltonian for free-streaming electrons in a shear-free magnetic field of Smith *et al.*[13]

$$H = \frac{c}{B}\Psi + (\theta v_\| - v_{\rm ph})$$

follows from the second result of Eq. (7).

The existence of the above two-dimensional Hamiltonians for the gyrokinetic system has important implications. Firstly, if by transforming to the wave frame, Ψ can be made independent of time, then the gyrocenter trajectories are integrable. In order for there to be a steady state $E \times B$-diffusion, the potential Ψ must therefore be time dependent. Secondly, the analysis of the flow in time-dependent potentials is greatly simplified since the instantaneous streamlines of the flow field in phase space are contours of constant H and C_1.

It is straightforward to generalize the above results to include the effect of weak shear by taking form $\theta = x/L_s$ where L_s is the shear length.[14] We have, however, been unable to find Hamiltonians for general three-dimensional potentials with finite κ. Whether they exist, and the consequences if they do not need to be investigated if the multiple-scale model is to be used in three-dimensional simulations. At the very least, checks need to be made against simulations which do not use it.

EQUILIBRIA

It is important to understand the solutions of the gyrokinetic Vlasov for the particle distribution in a given steadily-propagating wave, since such solutions are the ones that the distribution tries to relax to, either by collisions or by mixing due to the sheared $E \times B$ flows. The gyrokinetic Vlasov, equation for solutions which are time-independent in the wave frame, can be written as

$$\frac{\partial}{\partial x}(\dot{x}f) + \frac{\partial}{\partial y}(\dot{y}f) + \frac{\partial}{\partial v_\parallel}(\dot{v}_\parallel f) = C(f), \tag{8}$$

where $\dot{x}$, $\dot{y}$, and $\dot{v}_\parallel$ are given by Eqs. (4).

For $\nu_{ei} = 0$ [so that $C(f) = 0$], but $\kappa \neq 0$, Eq. (8) can be solved by the method of characteristics to yield

$$f = I(x)\hat{f}(H, C_1),$$

where $\hat{f}$ is an arbitrary function of H and C_1.

For $\kappa = 0$, but $C(f) \neq 0$, the steady solution is a simple generalization of the standard adiabatic response.

$$f = F_M^e(v_\parallel)n(\boldsymbol{x}), \tag{9a}$$

where

$$n(x) = \exp\left(-\frac{\mathrm{sgn}(q)H_i}{\rho v_{\mathrm{t}}}\right), \tag{9b}$$

and

$$H_i \equiv \frac{c}{B}\phi - v_{\mathrm{ph}}x. \tag{9c}$$

Note that there is a density gradient in this "equilibrium" solution associated with the $v_{\mathrm{ph}}x$ term in H_i. This density gradient is the nonlinear manifestation of the ω term in the $\omega_* - \omega$ factor in the linear phase shift δ_{el} of Eq. (13), and results from the parallel acceleration that takes place simultaneously with the radial advection. Thus, waves with a real frequency ω tend to drive the density profile to one which has a diamagnetic frequency ω_* equal to ω.

When both $\kappa \neq 0$ and $\nu_{\mathrm{ei}} \neq 0$, exact solutions to Eq. (8) are more difficult to construct. For short wavelengths and small θ, such that the gradient driving term and the parallel acceleration can be linearized, the steadily-propagating solution consists of an adiabatic response plus a profile flattening

$$\frac{f}{F_{\mathrm{M}}} = -\frac{q\phi}{T} + \frac{x v_{\mathrm{ph}}}{\rho v_{\mathrm{t}}} + \kappa x. \tag{10}$$

If the potential is time independent and has lines across which there is no flow in the x direction, then the relaxation will be to a stepped profile in which the distribution is piecewise of this form. This can be used to show[14] that the levels of transport which are observed in the gyrokinetic simulations are much larger than could occur through relaxation if the potential were time independent.

DRIFT WAVES: INTRODUCTION

To understand the basic physics of drift waves, consider oscillations with wavenumber $\mathbf{k} = (k_{\mathrm{x}}, k_{\mathrm{y}}, 0)$ in a plasma in a magnetic field $\mathbf{B} = B(\hat{z} + \theta\hat{y})$, with a density gradient in the $-x$ direction, in the frequency-wavenumber regime $k_{\|}v_{\mathrm{ti}} \ll \omega \lesssim \omega_* \ll k_{\|}v_{\mathrm{te}}$, $k_{\|} \ll k_{\mathrm{y}} \sim \rho_{\mathrm{s}}^{-1}$. Here, $\omega_* \equiv k_{\mathrm{y}}\kappa_n c_{\mathrm{s}}\rho_{\mathrm{s}}$ is the diamagnetic frequency. Assume that the temperature ratio $T_{\mathrm{i}}/T_{\mathrm{e}} \ll 1$. In this regime, the ion dynamics are cold and linearly satisfy

$$\gamma n^{\mathrm{i}} \simeq -i\omega_*(1 - b)n_0 \left(\frac{e\phi}{T_{\mathrm{e}}}\right), \tag{11}$$

where $b \equiv k_\perp^2 \rho_i^2$ γ is the growth rate and the first term on the right hand side is the ion polarization density response. The electron dynamics are approximately adiabatic

$$n^e \simeq (1 - i\delta_e)n_0 \left(\frac{e\phi}{T_e}\right), \tag{12}$$

where δ_e is a small phase shift which, depending on the plasma parameter regime of interest, may be produced by collisional or collisionless dissipation associated with passing or magnetically trapped electrons. To be definite, the case of collisionless and weakly collisional dissipation due to passing electrons is treated here. For $\tilde{\omega}/(k_\parallel v_{te}) \ll 1$ and $\tilde{\omega}/(k_\parallel v_{ti}) \gg 1$, where $\tilde{\omega} = \omega + i\nu_{ei}$,

$$\delta_e \simeq \delta_{el} = \begin{cases} \sqrt{\pi/2}(\omega_* - \omega)/(k_\parallel v_{te}) & \nu_{ei} \ll k_\parallel v_\parallel, \\ (\nu_{ei}/k_\parallel v_\parallel)(\omega_* - \omega)/(k_\parallel v_{te}) & \nu_{ei} \gg k_\parallel v_\parallel. \end{cases} \tag{13}$$

For typical tokamak plasma parameters $\omega_*/\omega_{pe} \ll \lambda_{De}/\rho_s = v_a/c \ll 1$, $v_a \equiv B/\sqrt{4\pi n_0 m_i}$ is the Alfven speed, and $\omega_{pe} \equiv v_{te}/\lambda_{De}$ is the electron plasma frequency. Equation (1e) then reduces to the quasineutrality condition for the perturbed densities

$$k_\perp^2 \rho_s^2 \left(\frac{n_0 e\phi}{T_e}\right) = \bar{n}^i - n^e \tag{14}$$

Combining Eqs. (11), (12), and (14), gives the real frequency and growth rate as

$$\omega_r \simeq \frac{\omega_*}{1 + k_\perp^2 \rho_s^2},$$

$$\gamma \simeq \frac{\omega_r \delta_{el}}{1 + k_\perp^2 \rho_s^2}.$$

There are several nonlinear mechanisms that can cause drift instabilities to saturate. Ion $E \times B$ diffusion[16] and mode coupling through the ion polarization drift[17] and $E \times B$ nonlinearities[18,19,20] can cause saturation even if the electron dissipation which drives the drift waves unstable remains linear (the so-called "$i\delta$ model" assumption). A nonlinear treatment of the electrons is necessary, however, to account for the strong collisionality dependences of the particle flux and saturation levels in the gyrokinetic simulations since the linear electron phase shifts have a very weak dependence on the collisionality in the regimes studied here. Parallel trapping can result in steady nonlinear drift wave states,[21] although for typical parameters, the $E \times B$ nonlinearity is much larger for both species.[22,10] The coherent $E \times B$ advection of the resonant electrons, can completely switch off the electron phase shift.[22,13] This mechanism is important for cases cases where the flux is much less than quasilinear and in the initial stages of the saturation in other cases. When significant $E \times B$ particle fluxes are present, however, an electron phase shift remains and the ion nonlinearities become an essential part of the saturation mechanism.

PHENOMENA OBSERVED IN GYROKINETIC DRIFT WAVE SIMULATIONS

Many drift wave simulation runs using the above algorithm and geometry for drift waves have been made with parameters chosen so that the box size is not much longer than the wavelength of the unstable modes.[22,10,14]

Typical parameter values used for the simulations were $m_e/m_i = 1/1837$, $\tau \equiv T_e/T_i = 4$ or 100, $\rho_s/\Delta = 4.286$, $\lambda_D/\Delta = 1$, $\kappa\rho_s = 0.214$ or 0.107, $L_x = L_y = 32\Delta$, $N_p = 2^{14}$, $\theta = 0.002$ or 0.01, $\nu_{ei}/\Omega_i = 0$, 0.001, or 0.01, and $(k_x, k_y) = (0.842m, 0.842n)$, where m and n are integers. Here, Δ is the grid spacing (equal for x and y), L_x and L_y are the lengths of the simulation box in the x and y directions, and N_p is the number of particles per species. The typical timestep used was $dt \simeq \Omega_i^{-1} \simeq 0.18\omega_*^{-1}$.

Linearly, these parameters correspond to collisionless or weakly collisional drift waves, and satisfy $\omega/k_\parallel v_{te} \sim 0.5 - 1$. A numerical solution of the dispersion relation agrees fairly well with the analytical results given above and gives $(\omega_{lin} + i\gamma_{lin})/\Omega_i = \pm 0.06 + 0.011i$ for the lowest modes in the $\theta = 0.01$ cases and $\pm 0.034 + 0.027i$ for the $\theta = 0.002$ cases, with $\tau = 4$, and $\nu_{ei} = 0$. For the values of collisionality used, the collisions have a small effect on the linear behavior. The main phenomena observed can be summarized as follows.

The evolution has a pre-normal mode stage, a linear stage which fits the predictions of linear theory, a saturation stage in which the evolution is nonlinear but not statistically stationary in time, and steady-state stage. These stages can be seen in Fig. 1, which shows the time history of the $(m, n) = (1, 1)$ mode for a typical gyrokinetic run with $\theta = 0.002$, $\tau = 100$, and $\nu_{ei} = 0.001$.

In the steady-state stage, most of the energy is concentrated in the lowest available modes. These lowest modes are the $(m, n) = (1, 1)$, $(1, -1)$, $(2, 0)$ and $(0, 2)$ modes, where the available wavenumbers are $(k_x, k_y) = (2\pi m/L_x, 2\pi n/L_y)$ and L_x and L_y are the lengths of the sides of the simulation box. To make the notation more compact, these modes

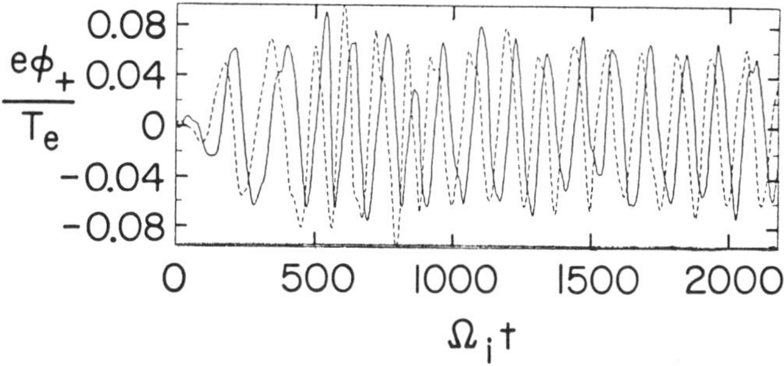

Figure 1: Time history of the $(m,n) = (1,0)$ mode of ϕ from a gyrokinetic code run with $\tau \equiv T_e/T_i = 100$, $\kappa\rho_s = 0.214$, $L_x = L_y = 32\Delta$, $\theta = 0.002$, and $\nu_{ei}/\Omega_i = 0.001$.

will be denoted by their n values respectively as the "+", "−", "0" and "2" modes in the remainder of this paper. The choice of parameters used was made deliberately, in order to produce relatively coherent states. A simulation run was made, however, with $L_y = 128\Delta$, $N_p = 2^{16}$, and $\nu_{ei} = 0.001$. In this run, the lowest available modes were still strongly dominant, even though they are no longer the most strongly unstable.

The + and − modes oscillate with nonlinearly modified real frequencies in the post-linear phases. The nonlinear modifications include a frequency shift, a frequency broadening, a frequency difference ($\omega_+ - \omega_- \simeq 0.007\Omega_i$ for $\theta = 0.002$ and $\nu_{ei} = 0.01$) implying a net x component of the phase velocity, and amplitude fluctuations implying an energy exchange between the + and − modes as well as a smaller fluctuation in the sum of the energies.

A steady-state particle flux due almost entirely to the + and − modes is observed. Its value is much less than expected from quasilinear theory, and increases strongly with ν_{ei}, and decreases with θ. Figure 2 shows a time history of the particle flux from the same run as in Fig. 1, as well as the time-averaged values of the flux at late time from runs with $\nu_{ei} = 0$

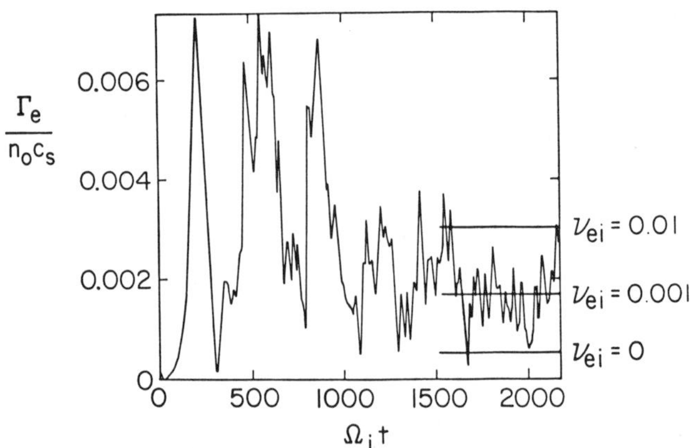

Figure 2: Time history of the particle flux from the same run as shown in Fig. 1, and time-averaged fluxes at late time for runs with $\nu_{ei}/\Omega_i = 0$, 0.001, and 0.01, with other parameters as for Fig. 1.

and 0.01, with all other parameters the same as for the curve.

"Globs" — regions of enhanced particle density are observed near the maxima of ϕ in the $\theta = 0.002$, $\nu_{ei} \neq 0$ cases. The observed x component of the phase velocity of the waves is consistent with the notion that the coherent advection of the globs, which are strongly localized near the potential maxima, contribute to the particle flux.

The nonlinear terms in Ψ and in the gyrokinetic Poisson equation and the ion parallel motion are unimportant for the steady-state phase.

MODEL FOR DRIFT WAVE SIMULATION RESULTS

In this paper we argue that the saturation levels and particle fluxes in the gyrokinetic simulations of drift waves can be understood as follows. The gyrokinetic Poisson equation imposes the constraint that the spatially averaged fluxes of the ions and electrons must exactly balance ("intrinsic ambipolarity"). As in the resonance-broadening picture,[16] we first calculate the fluxes of electrons and ions in given fluctuating steady-state electrostatic potential fields similar those observed in the gyrokinetic simulations. This calculation gives the dependences of these test-gyrocenter fluxes, which need not be the same, on the mean field amplitude and, for the electrons, the collisionality. At one or more values of the mean field amplitude, the fluxes of the electrons and ions intersect. We show that these values of the potential represent possible steady-state saturation levels. To understand the collisionality dependences, we consider the physics of the particle fluxes. These fluxes are due to the nonlinear $E \times B$ advection of those electrons and ions which are strongly $E \times B$ trapped. For $E \times B$ trapping to occur, the y component of the $E \times B$ drift must be somewhat larger than required to be able to cancel $\theta v_{\parallel} - v_{\mathrm{ph}}$ in Eq. (4b). This condition can be reexpressed as

$$|\omega_{\mathrm{r}} - k_{\parallel} v_{\parallel}| < \epsilon \Omega_{EB},$$

where ω_{r} is the real frequency, $v_{\parallel}$ is the parallel velocity, Ω_{EB} is the characteristic $E \times B$-trapping frequency, and $\epsilon \lesssim 1$. The drift waves studied here satisfy $k_{\parallel} v_{\mathrm{ti}} \ll \omega_{\mathrm{r}} \lesssim \omega_{*} \lesssim k_{\parallel} v_{\mathrm{te}}$. Thus, the ions are essentially all non-resonant for $\Omega_{EB} \lesssim |\omega_{\mathrm{r}}|$ and all become resonant for $\Omega_{EB} \gtrsim |\omega_{\mathrm{r}}|$. The test-gyrocenter ion $E \times B$ flux is therefore very small for small values of the mean field amplitude and grows very rapidly for mean field amplitudes above the trapping threshold. The electrons, on the other hand, have a significant population right at the resonant parallel velocity $v_{\mathrm{res}} \equiv \omega_{\mathrm{r}}/k_{\parallel}$. As the mean field amplitude is increased, therefore, the number of resonant electrons increases smoothly, roughly in linear proportionality with the amplitude. The test-electron flux therefore increases smoothly as a function of the mean field amplitude. Thus these test-gyrocenter fluxes, considered as functions of the field amplitude, have a clear intersection, which corresponds to a possible stable saturated nonlinear state of the system. So far, this picture is just an application of resonance-broadening applied to the specific spatially semi-coherent drift waves seen in the simulations. Because of the constancy of C_1 for an isolated resonance, however, electrons get accelerated out of the resonance as they $E \times B$ diffuse radially. This results in a significant reduction in the electron flux and therefore the saturation level. In the 2-1/2-dimensional model, collisions constitute the only mechanism present that can repopulate the resonance, resulting in the strong collisionality dependence of the flux in the saturated steady state. The time dependence of the mode amplitudes is primarily the result of the $E \times B$ advection of electrons and ions from nonuniform density distributions.

CALCULATION OF TEST ION AND ELECTRON FLUXES

A code was written to follow the motion of a large number of test gyrocenters and calculate the resulting fluxes in time-dependent electrostatic potentials resembling those observed in the gyrokinetic simulations. We use the equations of motion (4) along with the potential $\Psi = \bar{\phi} = \phi(1 - b)$ for the ions and $\Psi = \phi$ for the electrons. For the electrons, the equations of motion include the parallel acceleration and the pitch-angle scattering. Only the $+$ and $-$ modes of ϕ are kept. The amplitudes ϕ_+ and ϕ_- are taken to fluctuate with time with the same mean values and with a negative cross-correlation. For simplicity, the time dependence of the two modes is taken to be quasiperiodic with two frequency components. Although the actual time dependence of the mode amplitudes in the simulations is probably chaotic, the nonlinearity of the gyrocenter motion will reduce any spurious correlations between the time dependence of the mode amplitudes and the gyrocenter motion. The ϕ modes are taken to propagate in the y direction with phase velocity $v_{\rm ph}$ and in the x direction with velocity v_x. Thus, the resulting electrostatic potential is

$$
\begin{aligned}
\phi = &-2\phi_+(t)\cos[k_x(x - v_x t) + k_y(y - v_{\rm ph}t)] \\
&+ 2\phi_-(t)\cos[k_x(x - v_x t) - k_y(y - v_{\rm ph}t)],
\end{aligned}
\tag{15a}
$$

where

$$
\phi_\pm(t) \equiv \hat{\phi}(1 \pm \epsilon_1 \sin\omega_1 t \pm \epsilon_2 \sin\omega_2 t).
\tag{15b}
$$

The parameters k_x, k_y, v_x, $v_{\rm ph}$ $\hat{\phi}$, ϵ_1, ϵ_2, ω_1, and ω_2 are chosen to fit the data from particular runs of the gyrokinetic code. The gyrocenters can be loaded with a uniform Maxwellian distribution of selectively in phase space, for example at the resonant velocity or so as to give all particles the same value of C_1. Since there is no need to interpolate the particle density on a grid, this code is much quicker than the fully self-consistent code.

By suitably choosing the input parameters to make the potential resemble those observed in the full gyrokinetic simulations, the formation of globs, both for ions and electrons, has been observed, the dominant contribution of the resonant electrons and the effect of the collisional depopulation of the resonance in the electron flux has been verified, and good quantitative agreement has been obtained with the values of the flux given by the full gyrokinetic code.[14] The behavior of the test electron and ion gyrocenter fluxes is perhaps most clearly illustrated by plotting the fluxes as a function of the parameter $\hat{\phi}$. Figure 3 shows the flux as a function of $\hat{\phi}$ sequences with the parameter values $k_x\rho_s = k_y\rho_s = 0.842$, $k_y v_{\rm ph} = 0.042$, $v_x = 0$, $\epsilon_1 = 0.11$, $\epsilon_2 = 0.12$, $\omega_1 = 0.00237\sigma\Omega_{\rm i}$, $\omega_2 = 0.00625\sigma\Omega_{\rm i}$, $e\hat{\phi}/T_{\rm e} = 0.01\sigma$, and $\theta = 0.002$, where σ is a multiplier which ranges from 1 to 12. For the electrons sets of points for $\nu_{ei}/\Omega_i = 0.001$ and 0.01 are shown. In this sequence the frequencies of the amplitude fluctuations were increased in proportion to $\hat{\phi}$ so as to maintain a roughly constant relationship between these frequencies and the $E \times B$-trapping frequency. The primary effect of the presence of a nonzero radial propagation velocity v_x is to increase the values of the ion flux at moderately low values of the potential due to the formation and advection of globs. This effect could be taken into account by running sequences in which v_x is increased roughly in proportion with $\hat{\phi}$. Such a sequences would have the same qualitative character as those of Fig. 3. The contrast between the smooth dependence of the electron flux on $\hat{\phi}$ for moderate collisionality and the abrupt increase of the ion flux above a threshold value of $\hat{\phi}$ is clearly evident.

We now argue that it is plausible that that intersection of the curves of the test-gyrocenter flux, as a function of the mean saturation level, at which the ion flux rises more rapidly than the electron flux represents a possible steady state in the sense that it is stable at least to variations in the saturation level. It is intuitively clear that the ion nonlinearity has to be involved in the saturation process if, in the saturated state, there is a nonzero electron flux, since the linear ion gyrocenter fluid response $\bar{n}^i \simeq (\omega_*/\omega)(1 - b)\phi$ permits no flux without wave growth. In the linear regime, the electron density phase shift δ_{el} with respect to the potential has a very weak dependence on the growth rate, while that of the ions is directly proportional to the growth rate. If we write the ion flux as $\Gamma_i = \Gamma_0 + \gamma\Gamma_1$, where Γ_1 is positive, and assume that the electron flux Γ_e has a very weak explicit dependence on the growth rate, then the ambipolarity condition gives

$$\gamma = \frac{\Gamma_e - \Gamma_0}{\Gamma_1}.$$

If the test ion gyrocenter flux Γ_0 increases more rapidly as a function of $\hat{\phi}$ than the test electron flux Γ_e, at the intersection of the fluxes, then this intersection is stable. The properties of the fluxes need to be proved or established numerically in order to make this reasoning rigorous.

THREE-DIMENSIONAL EFFECTS

Three-dimensional effects can significantly alter the electron motion in drift waves. For parameters comparable to those used here, if the wave

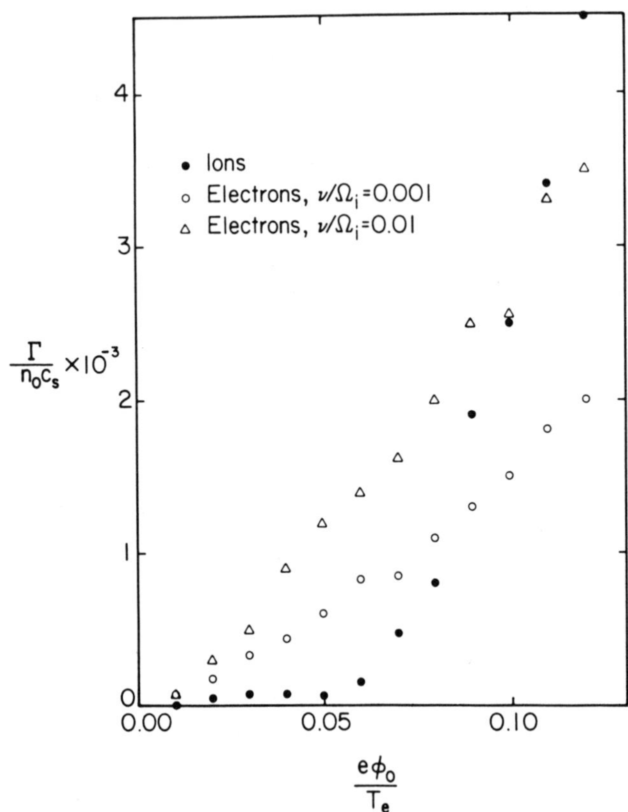

Figure 3: Test gyrocenter fluxes as a function of the mean saturation amplitude $\hat{\phi}$, for both electrons and ions. The uncertainties in the time-averaged values of the fluxes are of order 15%.

number spectrum of ϕ is dominated by the lower $k_\parallel$ modes, the ion behavior may still be adequately described by cold fluid equations.

The importance of three-dimensional effects can be shown by considering the following simple generalization of our picture of the two-dimensional electron behavior. We assume that the spectrum of ϕ is dominated by four modes of equal amplitudes with wave vectors (k_x, k_y, k_z), $(k_x, -k_y, k_z)$, $(k_x, k_y, -k_z)$, and $(k_x, -k_y, -k_z)$, where k_x, k_y, and k_z are fixed positive wave numbers. This generalizes the spectra observed in the $2\frac{1}{2}$-dimensional runs by symmetrically including modes with negative-θ (i.e., modes with $\theta = -|k_z/k_y|$). For the moment, we also neglect the parallel acceleration. The electrons at the positive $v_\parallel$ resonance see a stationary potential structure due to positive-θ waves and a moving one due to the negative-θ wave. In the x-y plane the negative-θ wave is moving past the positive-θ wave at twice the phase velocity in the $-y$ direction. This situation has been considered before in some detail.[23,12] Upon writing the stochasticity onset criterion of Horton,[12] for waves of comparable amplitude, one obtains $\Omega_{EB} > \omega_\mathrm{r}$, where $\Omega_{EB} = 4ck_xk_y|\phi_\pm|/B$ is the $\boldsymbol{E} \times \boldsymbol{B}$ trapping frequency due to waves of a single θ (positive, say). For a typical saturation levels, the electron orbits are stochastic by this criterion. If this stochasticity criterion is strongly satisfied, then the diffusion can still be small if the $\dot{x} = 0$ lines associated with the two sets of waves are approximately aligned.[23] A collisionless three-dimensional gyrokinetic simulation run for drift waves with the box size in the z direction $L_z = 500L_y$, $N_\mathrm{p} = 2^{16}$, and other parameters as for the the runs of Fig. 1 has been reported.[24] In this run, the particle flux is somewhat larger than in the corresponding two-dimensional run, but still considerably less than quasilinear. This strongly suggests that there is alignment of the bounding surfaces of waves for different θ, that is, a strong coherence of the wave fields. Parallel acceleration is no longer very effective in causing resonance depletion since when the positive and negative θ waves overlap, the particles are affected by both sets of waves. The constancy of C_1 is then broken. This reasoning has been verified using the test-gyrocenter code with positive and negative-θ waves included. For zero parallel acceleration, $\hat{\phi} = 0.08$, with the positive- and negative-θ waves out of phase by $\pi/4$ in the x direction, and other parameters as for the single-θ test-gyrocenter code, the flux $\Gamma_\mathrm{e} \simeq 0.07 n_0 \rho_\mathrm{s}$, which is of order the quasilinear expectation. The reduction in the flux due to the parallel acceleration with $\theta = \pm 0.002$ is is about than 30% irrespective of whether the diffusion is due to spatial or temporal stochasticity. The dependences of the fluxes and the saturation levels on the collisionality would therefore be expected to be weaker in three-dimensional simulations.

TURBULENT LIMIT

Finally, we show that if the drift wave fields are sufficiently turbulent, and resonance depopulation due to the isolation of resonances is unimportant, then $E \times B$ diffusion of electrons results in a phase shift between the electron density and the potential that is of the order of and has the same form as the linear result.

The diffusive electron flux can be written in the form

$$\Gamma = \kappa_{\text{eff}} D_{\text{ql}} \xi_{\text{r}}, \tag{16}$$

where the factors are as follows: The effective logarithmic density gradient κ_{eff} is the difference between the actual logarithmic density gradient and that to which drift waves with a real frequency ω drive the density profile. Its value, as found in the discussion immediately following Eq. (9) is

$$\kappa_{\text{eff}} = \kappa\left(1 - \frac{\omega}{\omega_*}\right). \tag{17}$$

The quasilinear diffusion coefficient D_{ql} is the diffusion coefficient valid for completely chaotic trajectories. From random walk arguments,

$$D_{\text{ql}} \simeq (\Delta x)^2/2\tau,$$

where $\Delta x \simeq \pi/k_x$ is the radial step size and $\tau \simeq \tau_{EB} \equiv 2\pi/(\Omega_{EB})$, is the decorrelation time, where $\overline{\phi}_0$ is a mean saturation amplitude of the $+$ or $-$ mode. Thus, the diffusion coefficient is given by

$$D_{\text{ql}} \simeq \pi\frac{k_y}{k_x}\overline{\phi}_0, \tag{18}$$

The factor ξ_{r} is the fraction of gyrocenters in resonance in the nonlinearly broadened resonance. Using Eq. 15 to estimate this width, we obtain

$$\xi_{\text{r}} \simeq \int_{v_-}^{v_+} dv_{\|}F_{\text{M}}(v_{\|}) \simeq \frac{2\epsilon\,\Omega_{EB}}{k_{\|}}F_{\text{M}}^{\text{e}}(v_{\text{res}}), \tag{19}$$

where $v_{\pm} \equiv v_{\text{res}} \pm \Omega_{EB}/k_{\|}$.

Upon combining Eqs. (16), (17), (18), and (19), and comparing this result with the relationship between the electron flux and the phase shift $\Gamma_{\text{e}} = 2k_y\delta_{\text{e}}(|\phi_+|^2 + |\phi_-|^2)$, we obtain

$$\delta_{\text{e}} \simeq \frac{2\epsilon\pi(\omega_* - \omega)}{k_{\|}}F_{\text{M}}^{\text{e}}(v_{\text{res}}) \simeq 2\epsilon\delta_{\text{el}},$$

where δ_{el} is the linear phase shift. The parameter ϵ characterizes the level of incoherence of the fields. For $\epsilon = 1$ the potential contours are straight and diagonal, while the $\epsilon = 0.2$ is representative of the semi-coherent states seen in the gyrokinetic simulations. Fully developed turbulence would be characterized by a value somewhere in between 0.2 and 1. Taking $\epsilon \simeq 0.5$, which gives the linear result.

SUMMARY AND DISCUSSION

We have discussed some of the mathematical properties for the gyrokinetic particle simulation model for a 2-1/2-dimensional geometry and given a theoretical model of the dynamics observed in gyrokinetic simulations of drift waves. The model was shown to predict the qualitative dependences observed in the gyrokinetic simulations, and the transition to less coherent states was examined. The model is not fully self-consistent in that it cannot predict the degree of coherence of the nonlinear states. The interesting questions regarding the degree of spatial coherence in more realistic siturations motivate further studies using well-diagnosed simulation runs with much larger box sizes and three-dimensional codes.

ACKNOWLEDGEMENTS

The authors acknowledge stimulating discussions with J. A. Krommes.

This work was supported by the U.S. Department of Energy, partly through Contract Nos. DE-ACO2-CHO-3073 and DEFG05-86ER53221.

References

[1] C. M. Surko and R. E. Slusher, Science **221**, 817 (1983).

[2] P. C. Liewer, Nucl. Fusion **25**, 543 (1985).

[3] A. Hasegawa and M. Wakatani, Phys. Rev. Lett. **50**, 682 (1983).

[4] P. W. Terry and P. H. Diamond, Phys. Fluids **28**, 1419 (1985).

[5] B. B. Kadomtsev and O. P. Pogutse, in Reviews of Plasma Physics, Vol. 5 [M. A. Leontovich, Ed., Consultants Bureau, New York, London (1970)], p249.

[6] W. M. Tang, Nucl. Fusion **18**, 1089 (1978).

[7] W. W. Lee, Phys. Fluids **26**, 556 (1983).

[8] W. W. Lee, J. Comp. Phys., **72**, 243 (1987).

[9] D. H. E. Dubin, J. A. Krommes, C. Oberman, and W. W. Lee, Phys. Fluids **26**, 3524 (1983).

[10] J. F. Federici, W. W. Lee, and W. M. Tang, Phys. Fluids **30**, 425 (1986).

[11] H. Ching, Phys. Fluids **16**, 130 (1973).

[12] W. Horton, Plasma Phys. **27**, 937 (1985).

[13] R. A. Smith, J. A. Krommes, and W. W. Lee, Phys. Fluids **28**, 1068 (1985).

[14] A. M. Dimits, Ph.D. thesis, Princeton University, 1988.

[15] L. I. Rudakov and R. Z. Sagdeev, Dokl. Akad. Nauk. SSSR **138**, 581 (1961) [Sov. Phys. Doklady, **6**, 415 (1961)].

[16] T. H. Dupree, Phys. Fluids **10**, 1049 (1967).

[17] A. Hasegawa and K. Mima, Phys. Fluids **21**, 87 (1978).

[18] P. W. Terry and W. Horton, Phys. Fluids **26**, 106 (1983).

[19] W. Horton, Phys. Fluids **29**, 1491 (1986).

[20] R. E. Waltz, Phys. Fluids **26**, 169 (1983).

[21] R. Z. Sagdeev and A. A. Galeev, *Nonlinear Plasma Theory* (W. A. Benjamin Inc., New York, Amsterdam, 1969).

[22] W. W. Lee, J. A. Krommes, C. Oberman, and R. A. Smith, Phys. Fluids **27**, 2652 (1984).

[23] R. G. Kleva and J. F. Drake, Phys. Fluids **27**, 1686 (1984).

[24] R. D. Sydora, W. W. Lee, T. S. Hahm, H. Natiou, J. M. Dawson, and V. K. Decyk, Bull. Am. Phys. Soc. **32**, 1889 (1987, paper 7T5).

IDEAL AND RESISTIVE MHD EQUILIBRIA AND
STATES OF MINIMUM DISSIPATION

X. Shan, H. Chen and D. Montgomery
Dartmouth College, Hanover, NH 03755

ABSTRACT

Ideal magnetohydrodynamic equilibria have been the organizing conceptual framework for interpreting fusion confinement experiments. In hydrodynamics, bounded, ideal equilibria are equally plentiful, but the only ones even approximately attainable in the laboratory are those which lie close to equilibria with small but non-zero dissipation coefficients. If the same criteria are demanded in MHD, the range of possibilities is greatly restricted. If the additional common requirements of axisymmetry and zero equilibrium flow velocity are demanded, we are left with a nearly empty set of solutions, and not very interesting ones at that. For example, in cylindrical periodic geometry with a dielectric-coated conducting liner, the sole possibility seems to be the uniform-current-density profile, if uniform resistivity is assumed. It has begun to seem likely that interesting MHD equilibria may *all* contain flows: and not just small-scale, turbulent, "dynamo" velocities, but organized flows that can affect the current distribution over its channel. These states seem to be approximatable as states of minimum energy dissipation rate. Experimental searches for flow velocities, on MHD plasmas cool enough to permit internal probes, seem to be indicated. Issues of MHD equilibrium, at a level more sophisticated than ideal, axisymmetric assumptions permit, need to be reopened. At present, the most important and revealing experiments that can be done seem to be numerical ones.

I.   INTRODUCTION

Several features observed in numerical solutions to the magnetohydrodynamic (MHD) equations, in straight-cylinder approximations to tokamak and reversed-field pinch geometries, can be qualitatively accounted for by assuming that the magnetofluid will seek a state of minimum energy dissipation rate.[1-4] A turbulent relaxation phase is followed by a much more smooth and quiescent state with only small and unsystematic time variations.

Minimum dissipation rate principles are: (1) rigorously correct as applied to current distributions in rigid, non-uniform electrical conductors;[5,6] (2) accurate but inadequately justified as predictors of hydrodynamic shear flows at low to moderate Reynolds numbers;[7,8] and (3) plausible but speculative for electrically driven magneto-fluids (again, at low to moderate Reynolds numbers).[1-4] Their intuitive physical justification lies in their closeness to the (gradient-free) thermodynamic equilibrium state to which continuous media would prefer to relax if their relaxation were not prohibited by external driving mechanisms. These driving mechanisms, usually manifested as boundary conditions, mandate the existence of non-zero spatial gradients, and minimum dissipation states minimize these gradients in a mean square sense.

Some consequences of the minimum dissipation rate assumption,[1-4] together with a likely connection to "maximum entropy" methods[9] (in the information theory sense) have previously been presented. Our purpose here is further exploration of their implications for the steady-state field distributions in periodic cylinders of voltage-driven, incompressible, dissipative magnetofluids. We believe this problem to contain the essential MHD involved in tokamaks and reversed-field pinches (RFPs), even though many important features of real plasmas in these devices (such as temperature gradients, compressibility, and kinetic effects) are left out of this MHD description.

In a recent paper it was shown[3] how to construct an explicit MHD solution for this geometry that, below certain thresholds in the dissipation coefficients, would bifurcate into an axisymmetric and a helically-deformed solution for the same parameters and boundary conditions. The helically-deformed state had the lower dissipation rate. The resulting helical current channels and vorticity distributions were seen, by visual inspection of contour plots and three-dimensional perspective plots, to be similar to those that had emerged previously from a series of time-dependent numerical solutions of MHD.[10-15] However, this explicit solution did not contain a prediction for the overall amplitudes of the helical components, which stood as a free parameter in the theory. One purpose here is to obtain an analytic estimate for these amplitudes rather than using them to fit available numerical data. Another is to predict the dominant toroidal and poloidal mode numbers.

In Section II the problem is posed in a sufficiently simplified analytical framework that an estimate of the helical amplitudes can emerge. The biggest single difficulty, by far, in the problem is the inclusion of the MHD Ohm's law in the variational calculation. The steady-state Ohm's law acts as a *pointwise* constraint relating the magnetic field locally to the velocity field. As will be discussed at greater length later, the non-standard nature of the constraint is what moves the problem out of reach of the usual methods of the calculus of variations and requires innovative, less well justified techniques. In order to include the Ohm's law, an important approximation is made: we assume that a single pair of poloidal and toroidal mode numbers (m, n) dominate the non-axisymmetric components of the fields. Even then, there remains some ambiguity about how many of the consequences of the Ohm's law to include; at one point, it will be seen that it is possible that past a certain point, inclusion of additional information might cause the quality of the approximations to deteriorate.

It is convenient to classify the possible regimes by the dimensionless "pinch ratio" Θ, which is the wall-averaged poloidal magnetic field divided by the volume-averaged toroidal magnetic field. We call $\Theta < 1$ the "tokamak" regime and $\Theta > 1$ the "RFP" regime, since toroidal field reversal at the wall typically occurs between $\Theta = 1$ and $\Theta = 2$. A very strongly driven RFP might have $\Theta \gtrsim 4$. Our emphasis here will be on the tokamak regime, $\Theta < 1$, since it is analytically simpler. We shall assume the magnetic field to be only a fractionally small distortion of the (uniform) vacuum magnetic field.

In Section III, some elementary consequences of the theory are developed.

Despite the rather drastic truncation that the assumption of a single pair of non-zero mode numbers represents, these consequences are surprisingly difficult to extract and also surprisingly rich. They include repeated discontinuous transitions in the dominant mode numbers from one pair of values to another as critical thresholds in the toroidal current are crossed. An inclusive catalogue of these transitions has not so far proved possible. We have concentrated on obtaining a few predictions that seem simple enough that they might in the future be testable with manageable numerical codes. One such test is presented.

True experimental tests would, of course, be far more desirable; but existing tokamak diagnostics do not appear to permit the internal, pointwise measurements of velocities, magnetic fields, current densities, and vorticities that would be required for a test of the predicted helical field distributions and flows in the current generation of very hot tokamaks.

Section IV summarizes and discusses desirable future extensions of the theory. In Appendix A the dimensionless units used throughout are derived. In Appendix B some properties of the Chandrasekhar-Kendall functions[16−18] (used as an expansion basis) are given. Neither appendix may be necessary for the thoroughly knowledgeable reader, but both are included for completeness.

What the present investigations imply, if correct, is a zeroth-order revision of what the MHD steady states are that can be considered relevant for confinement devices. Heretofore, these have been a multiply infinite class of *ideal*, axisymmetric, zero-flow equilibria. Their stability properties, which depend sensitively on current profiles, have been extensively investigated. Now it appears that there is a much narrower class of steady states, involving helical distortions and some mass flow in the interesting regimes, which are inherently *resistive* in nature. It is the non-ideal stability of these non-ideal partially-helical states which seems to be a candidate for stability analysis if anything is.

II.　FORMULATION AND METHODOLOGY

Our principal task is that of finding MHD fields, subject to boundary conditions, that will minimize the total energy dissipation rate, which may be taken to be:

$$R \equiv \int_V d^3x (\eta \mathbf{j}^2 + \nu \boldsymbol{\omega}^2). \tag{1}$$

We work throughout the dimensionless variables introduced in Appendix A. Here, $\mathbf{j} = \nabla \times \mathbf{B}$ is the electric current density, where $\mathbf{B}$ is the magnetic field. The fluid vorticity is $\boldsymbol{\omega} = \nabla \times \mathbf{v}$, where $\mathbf{v}$ is the velocity field, assumed solenoidal. The dimensionless resistivity and viscosity, to be assumed constant and spatially uniform, are η and ν, respectively. In the dimensionless variables they are equivalent to the reciprocals of Reynolds-like numbers and may be thought of as $<< 1$. The integration of Eq. (1) runs over the volume V of the magnetofluid. This volume is taken to be the interior of a right circular cylinder of radius $r = a$, axial length L_z. The cylinder is assumed to have periodically identified ends, the only concession to the toroidal nature of the laboratory problem. All field variables are assumed periodic in z, the axial coordinate, with period L_z; they are also periodic in φ, the azimuthal coordinate, with period 2π. We use cylindrical polar coordinates (r, φ, z).

In addition to the periodicities, there are four radial boundary conditions imposed on $\mathbf{B}$, $\mathbf{v}$, $\mathbf{j}$, and $\boldsymbol{\omega}$ at $r = a$. They are $\mathbf{B} \cdot \hat{n} = 0$, $\mathbf{v} \cdot \hat{n} = 0$, $\mathbf{j} \cdot \hat{n} = 0$, and $\boldsymbol{\omega} \cdot \hat{n} = 0$, where $\hat{n} = \hat{e}_r$ is the unit outward normal to the bounding cylinder at $r = a$. We imagine the cylinder wall as a rigid perfect conductor ($\mathbf{B} \cdot \hat{n} = 0$) coated with an infinitesimally thin layer of insulating dielectric ($\mathbf{j} \cdot \hat{n} = 0$). The wall is assumed mechanically impenetrable ($\mathbf{v} \cdot \hat{n} = 0$). For the fourth boundary condition ($\boldsymbol{\omega} \cdot \hat{n} = 0$), we would have preferred to have had either no-slip walls ($\mathbf{v} \times \hat{n} = 0$) or smooth, stress-free walls ($\hat{n} \times (\underset{\approx}{g} \cdot \hat{n}) = 0$, where $\underset{\approx}{g}$ is the viscous stress tensor). However, employing either of these boundary conditions seems to move the problem outside the realm of present tractability. Their employment would require a more subtle boundary-layer analysis than we have so far been able to give. Some comfort can be derived from the fact that, for the MHD states to be identified, $\boldsymbol{\omega} \times \mathbf{v} = 0$ at $r = a$, so that no mechanical work is being done at the boundary and $\nu \omega^2$ still represents a viscous energy dissipation density throughout the fluid. The replacement of our boundary condition $\boldsymbol{\omega} \cdot \hat{n} = 0$ (the condition $\boldsymbol{\omega} \times \hat{n} = 0$ *is* stress-free for plane walls[10−15] without curvature) by something more satisfactory must be deferred to the future; but hydrodynamic experience does not suggest that it will be a simple development. It should be emphasized that these boundary conditions, while being much the simplest analytically, are not the ones we have used in the simulations which are the stimulus for this study. It is one of its major limitations and not an easy one to remove.

An additional boundary condition that leads to a constraint which gives the problem most of its content, is that there will be assumed to be imposed an externally-maintained electric field, $\mathbf{E}_0 = E_0 \hat{e}_z$, at $r = a$. Without it the minimum-dissipation state would simply be $R = 0$ with the trivial values $\mathbf{v} = 0$, $\mathbf{j} = 0$, $\mathbf{B} = B_0 \hat{e}_z$ (where $B_0 \pi a^2$ is the initial toroidal magnetic flux, a constant of the motion). The product $E_0 L_z$ is the familiar "loop voltage."

The MHD Ohm's law in the steady state appears as

$$\mathbf{E} = \mathbf{E}_0 + \nabla \tilde{\Phi} = -\mathbf{v} \times \mathbf{B} + \eta \mathbf{j}, \tag{2}$$

where $\mathbf{E}$ is the electric field. $\mathbf{E}_0$ is its spatial average, which we assume to be maintained at a fixed value by the external circuit. $\tilde{\Phi}$ is a scalar potential which must obey

$$\nabla^2 \tilde{\Phi} = -\nabla \cdot (\mathbf{v} \times \mathbf{B}) \tag{3}$$

and the boundary condition $\hat{n} \times \nabla \tilde{\Phi} = 0$ at $r = a$. In a time-fluctuating steady state, Eq. (2) may also be interpreted as a time average, provided that the time-variable parts of $\mathbf{v}$ and $\mathbf{B}$ are small enough that their cross product may be neglected compared to the cross product of their steady-state, time-averaged parts.

Eq. (2) implies virtually all the interesting complications of the present problem. The solution of Eq. (3) provides $\tilde{\Phi}$ as a functional of $\mathbf{v}$ and $\mathbf{B}$, and when that is substituted back into Eq. (2), it becomes a *pointwise constraint*, relating $\mathbf{v}$ to $\mathbf{B}$ and $\nabla\times\mathbf{B}$, at every interior point of magnetofluid. This is a constraint of a non-standard kind, as far as the calculus of variations is concerned. Systematic methods exist for dealing with global, integral constraints by the method of Lagrange multipliers. Many such constraints are implied by Eq. (2). For example, dotting Eq. (2) with $\mathbf{B}$ and integrating over the volume gives the previously-used[1,2] constraint of constant rate of supply and dissipation of magnetic helicity. But squaring (2), or raising it to higher powers, yields upon integration over V, an *infinite* number of global, integral constraints which could be included individually by standard methods, but which become a qualitatively different phenomenon when taken *in toto*.

If Eq. (2) could be solved explicitly for either $\mathbf{v}$ or $\mathbf{B}$ in terms of the other and its derivatives, one field or the other could be eliminated from R in favor of the other. The remaining field could be varied independently and R could be minimized in terms of it. However, the connection between $\mathbf{v}$ and $\mathbf{B}$ implied by Eqs. (2) and (3) is quite implicit, and an explicit expression for one field in terms of the other seems quite unlikely. The reason for this is that $\tilde{\Phi}$ must be obtained for a general $\mathbf{v}$ and $\mathbf{B}$.

What can be done[4] is to consider Eqs. (2) and (3) as defining an *infinite sequence* of global, integral constraints of the following kind. If inner products of Eq. (2) are taken with the members of a sequence of orthogonal, solenoidal functions which are periodic in z and φ and whose radial components vanish at $r = a$, the $\nabla\tilde{\Phi}$ contributions drop out each time, and we are left with an infinite sequence of integral constraints involving $\mathbf{v}$, $\mathbf{B}$, and $\nabla\times\mathbf{B}$. These are standard constraints which can be dealt with by Lagrange multipliers. As previously remarked, dealing with an infinite number is out of the question. But there is reason to believe that a sequence of orthogonal functions might be found for which the convergence would be rapid enough that a low-order truncation of the sequence of constraints might be sufficient, at least for small enough values of the pinch ratio Θ.

In particular, the numerical solutions have shown a proclivity to evolve dynamically into nearly force-free states, dominated by only a very few pairs of mode numbers (m, n), with one pair clearly dominant, with magnetic energies which are large compared to kinetic energies, and axisymmetric magnetic field contributions which are large compared to the helical ones. This suggests, for a set of solenoidal, orthogonal functions, the Chandrasekhar-Kendall[16-18] normalized eigenfunctions of the curl. It has been noted previously[3] that the computed states were not badly approximated, for sufficiently low E_0, by adding to the axisymmetric, zero-flow state of uniform $\mathbf{j} = \mathbf{E}_0/\eta$ a single Chandrasekhar-Kendall helical function for $\mathbf{B}$ and $\mathbf{j}$ and a second, paired one for $\mathbf{v}$ and $\boldsymbol{\omega}$. The helical combination amounted to a threshold-unstable normal mode of the uniform-current-density, zero-flow equilibrium. (Only the overall multiplicative amplitude was undetermined, and was chosen as a free parameter to fit the simulation results.)

The Chandrasekhar-Kendall functions[16-18] are very likely a complete set for $\mathbf{v}$ and $\mathbf{B}$ in the presence of these boundary conditions. We proceed here by taking the simplest non-trivial approximation[4] in which the magnetic field is approximated by just two Chandrasekhar-Kendall functions, one axisymmetric and one helical, and the velocity field is approximated by a helical one alone. This is not something one naturally would have thought of had not the numerical solutions suggested it. It seems likely that such a severe truncation will work best in the "tokamak regime," $\Theta < 1$, but there is no harm in seeing how well it fits the $\Theta > 1$ results.

The fields will be written as spatial averages (indicated by angle brackets $<>$) plus spatially-variable parts with zero spatial averages. We write, specifically

$$\mathbf{j} = <\mathbf{j}> + \delta\mathbf{j} \tag{4a}$$

$$\mathbf{B} = <\mathbf{B}> + r <j_z> \hat{e}_\varphi/2 + \delta\mathbf{B} \tag{4b}$$

$$\mathbf{v} = \delta\mathbf{v} \tag{4c}$$

$$\omega = \delta\omega, \tag{4d}$$

where the spatial averages of $\delta\mathbf{j}$, $r <j_z> \hat{e}_\varphi/2$, $\delta\mathbf{B}$, $\delta\mathbf{v}$, and $\delta\omega$ are all zero. The term $r <j_z> \hat{e}_\varphi/2$ is necessary in order that the spatially averaged current density $<\mathbf{j}> = <j_z> \hat{e}_z$ be permitted to contribute to the axisymmetric poloidal magnetic field. For $<\mathbf{B}>$ we will get $<\mathbf{B}> = B_0\hat{e}_z$. Both B_0 and $<j_z>$ are spatially-uniform constants. B_0 is externally determined by fixing the toroidal magnetic flux, but $<j_z>$ is to be determined as part of the solution to the problem.

Eqs. (4) are general, but we now make a severe approximation on $\delta\mathbf{B}$ and $\delta\mathbf{v}$. In a highly abbreviated notation we truncate their expansions as

$$\delta\mathbf{B} = \xi_b\mathbf{A}^c, \quad \delta\mathbf{v} = \xi_v\mathbf{A}^s$$

$$\delta\mathbf{j} = \nabla\times\delta\mathbf{B} = \lambda\xi_b\mathbf{A}^c$$

$$\delta\omega = \nabla\times\delta\mathbf{v} = \lambda\xi_v\mathbf{A}^s. \tag{5}$$

The notation has the following meanings. $\mathbf{A}^c$ and $\mathbf{A}^s$ are "cosine" and "sine" Chandrasekhar-Kendall functions defined by (see Appendix B)

$$\mathbf{A}^{c,s} \equiv I_{nmq}^{-1/2} \left[\lambda_{nmq}\nabla\times\hat{e}_z\psi_{nmq}^{c,s} + \nabla\times(\nabla\times\hat{e}_z\psi_{nmq}^{c,s})\right], \tag{6}$$

with a fixed, but so far arbitrary, set of mode numbers n, m, q. The scalar functions $\psi_{nmq}^{c,s}$ are

$$\psi_{nmq}^{c,s} = J_m(\gamma_{nmq}r) \begin{bmatrix} \cos(m\varphi - k_n z) \\ \sin(m\varphi - k_n z) \end{bmatrix}, \tag{7}$$

where J_m is the mth order Bessel function of the first kind, $k_n = 2\pi n/L_z$, and m, n and q are integers. $\gamma_{nmq} > 0$ is determined by the requirement that $\hat{e}_r \cdot \mathbf{A}_{nmq} = 0$ at $r = a$, or by

$$m\lambda_{nmq}J_m(\gamma_{nmq}a) = ak_n dJ_m(\gamma_{nmq}a)/da. \tag{8}$$

Finally, $\lambda_{nmq} = \pm\sqrt{\gamma_{nmq}^2 + k_n^2}$, with both signs of $\lambda = \lambda_{nmq}$ allowed, but with a fixed sign maintained throughout a given possible solution. I_{nmq} is a normalizing constant to be chosen so that

$$\int_V (\mathbf{A}^c)^2 d^3x = 1 = \int_V (\mathbf{A}^s)^2 d^3x, \tag{9a}$$

while

$$\int_V \mathbf{A}^c \cdot \mathbf{A}^s d^3x = 0 = \int_V \mathbf{A}^c d^3x = \int_V \mathbf{A}^s d^3x. \tag{9b}$$

For each n, m, q and $\lambda = \lambda_{nmq}$ there are linearly independent "sine" and "cosine" modes, and the important thing is that $\delta\mathbf{v}$ and $\delta\mathbf{B}$ differ in the c, s superscripts. Since in our low-order truncation there is only one set of indices m, n and q, they may be omitted, hereafter.

After this drastic truncation, only three undetermined numbers are left in the problem: the real helical amplitudes ξ_v and ξ_b and the volume-averaged current density $< j_z >$, which can be thought of as determining an amplitude for the axisymmetric ($m = n = 0$) Chandrasekhar-Kendall function in the limit of small γ_{001} (tokamak limit). A thoroughly general formulation of the problem would also look like Eqs. (4) and (5) except that n, m, and q would be summed over.

The basic physical effect which leads to the helical components, as will be seen below, derives from the Ohm's law. The $< \mathbf{v} \times \mathbf{B} >$ emf that is obtained from non-zero ξ_v and ξ_b *opposes* the applied field $\mathbf{E}_0 = E_0\hat{e}_z$ and depresses $< j_z >$ below its $\mathbf{v} = 0$ value (E_0/η), sufficiently that the overall dissipation R drops below $(E_0^2/\eta)(\pi a^2 L_z)$, even though there are new positive-definite contributions to the dissipation that result from non-zero ξ_v and ξ_b. The integer mode numbers n, m, q are those of the helical mode that dominates the departure from axisymmetry, either because that mode represents the one which minimizes R, or sometimes at the threshold, is the only one that is dynamically accessible.

The Ohm's law now appears, when projected into this highly restricted subspace, as

$$\mathbf{E}_0 + \nabla\tilde{\Phi} \cong -\delta\mathbf{v}\times\left[< \mathbf{B} > +r < j_z > \hat{e}_\varphi/2 + \delta\mathbf{B}\right] + \eta < \mathbf{j} > +\eta\nabla \times \delta\mathbf{B}. \tag{10}$$

(The equality of Eq. (10) is now only approximate.)

We may now dot Eq. (10) with $\delta\mathbf{B}$, $\delta\mathbf{v}$, and $1\hat{e}_z$, and integrate over V ($1\hat{e}_z$ may be thought of as the unnormalized, axisymmetric, Chandrasekhar-Kendall function in the extreme tokamak limit, $\gamma a << 1$). First, $1\hat{e}_z$ gives

$$E_0 = - < \delta\mathbf{v}\times\delta\mathbf{B} > \cdot\hat{e}_z + \eta < j_z > . \tag{11}$$

Dotting with $\delta\mathbf{v}$ and integrating gives zero on both sides; and repeating the procedure with $\delta\mathbf{B}$ gives

$$0 = \mathbf{B}_0 \cdot \int \delta\mathbf{v}\times\delta\mathbf{B}d^3x + (< j_z > /2) \int \delta\mathbf{v}\times\delta\mathbf{B} \cdot r\hat{e}_\varphi d^3x+$$

$$\eta \int \delta\mathbf{B} \cdot \nabla \times \delta\mathbf{B} d^3 x. \tag{12}$$

Upon dividing Eq. (12) by V, it may be written as

$$-\eta\lambda\tilde{\xi}_b^2 = \beta\tilde{\xi}_v\tilde{\xi}_b, \tag{13}$$

where $\tilde{\xi}_v \equiv \xi_v/\sqrt{V}$, $\tilde{\xi}_b \equiv \tilde{\xi}_b/\sqrt{V}$, $V = \pi a^2 L_z$, and β is given by

$$\beta \equiv B_0 \hat{e}_z \cdot \int \mathbf{A}_1^s \times \mathbf{A}_1^c d^3 x + \frac{< j_z >}{2} \int \hat{e}_\varphi r \cdot (\mathbf{A}_1^s \times \mathbf{A}_1^c) d^3 x. \tag{14}$$

Eliminating $< j_z >$ from Eqs. (12)-(14) leads to

$$\tilde{\xi}_b = -(\eta\lambda)^{-1} \left[\beta_0 + \beta_1\tilde{\xi}_v\tilde{\xi}_b/2\right] \tilde{\xi}_v, \tag{15}$$

with β_0 and β_1 given by

$$\beta_0 \equiv B_0 \hat{e}_z \cdot \int \mathbf{A}_1^s \times \mathbf{A}_1^c d^3 x + \frac{E_0}{2\eta} \int \hat{e}_\varphi r \cdot (\mathbf{A}_1^s \times \mathbf{A}_1^c) d^3 x \tag{16}$$

$$\beta_1 \equiv \frac{1}{\eta} \left[\int \mathbf{A}_1^s \times \mathbf{A}_1^c \cdot \hat{e}_z d^3 x\right] \left[\int \mathbf{A}_1^s \times \mathbf{A}_1^c \cdot \hat{e}_\varphi r d^3 x\right], \tag{17}$$

and both independent of the unknowns $\tilde{\xi}_v$, $\tilde{\xi}_b$ and $< j_z >$.

For verbal simplicity, Eq. (15) will be called, not altogether accurately, "the" constraint. It is the sole surviving consequence of the Ohm's law which is other than straightforward, as contrasted with Eq. (11), another constraint whose content is transparent.

With the same truncation R now becomes

$$\frac{R}{V} = \frac{1}{\eta} \left[E_0 + \frac{k}{\lambda}\tilde{\xi}_v\tilde{\xi}_b\right]^2 + \nu\lambda^2\tilde{\xi}_v^2 + \eta\lambda^2\tilde{\xi}_b^2, \tag{18}$$

where we have used Eq. (11) and the result

$$\int \mathbf{A}_1^s \times \mathbf{A}_1^c d^3 x = (k/\lambda)\hat{e}_z. \tag{19}$$

Upon performing a few integrals, it may also be demonstrated that

$$\int \mathbf{A}_1^s \times \mathbf{A}_1^c \cdot r\hat{e}_\varphi d^3 x = -\frac{m}{\lambda}. \tag{20}$$

From these two relations it follows that

$$\beta_0 = \frac{k_n B_0}{\lambda} - \frac{E_0}{2\eta}\frac{m}{\lambda} \tag{21}$$

and

$$\beta_1 = -\frac{mk_n}{\eta\lambda^2}. \tag{22}$$

The problem is now to find the minimum of R/V as given by Eq. (18), subject to the constraint (15), by varying the amplitudes $\tilde{\xi}_v$ and $\tilde{\xi}_b$. What had been a problem with an infinite number of degrees of freedom has been truncated to one with only two remaining, but it is still more complicated than it might appear because all possible sets of m, n and λ must still be considered. After Eq. (18) is minimized with respect to $\tilde{\xi}_v$ and $\tilde{\xi}_b$ for fixed m, n and λ, the minimum over possible sets of m, n, and λ must be sought in a subsequent minimization.

We should note the option of minimizing the expression (18) alone, without taking the constraint (15) into account. As will be seen in the next section, not only is this a simpler problem, but it appears that for some purposes the capacity of this "unconstrained minimum" to predict the simulation results may be of some significance. The reasons for this are not altogether clear but are speculated upon in Section IV.

III. SINGLE MODE-NUMBER STATES, $\Theta < 1$

It is possible to minimize R/V as given by Eq. (18) with or without including the constraint (15). An easier problem results from omitting it and somewhat surprisingly, the answers are not vastly inferior predictors of the simulation results available so far. We approach the problem from both points of view, taking the simpler case first.

For $|E_0 k/\lambda|$, which is always $< E_0$, sufficiently small, the unconstrained minimum of R/V obtained from Eq. (18) lies at the origin, $\tilde{\xi}_v = 0 = \tilde{\xi}_b$. This is the familiar, zero-flow, axisymmetric, uniform-current-density state: the same one to be expected for a piece of copper wire in response to a uniform electrostatic potential drop along it.

As E_0 is increased, a threshold is reached at which the unconstrained minimum moves off to the point $(\tilde{\xi}_{v0}, \tilde{\xi}_{b0})$, with

$$\tilde{\xi}_{v0}^2 = (\eta/\nu)\tilde{\xi}_{b0}^2 \tag{23a}$$

and

$$\tilde{\xi}_{b0}^2 = \mp\sqrt{\nu/\eta}\left(\frac{\lambda}{k}\right)\left[E_0 \pm \frac{\eta\sqrt{\eta\nu}\lambda^3}{k}\right], \tag{23b}$$

and where the upper (lower) sign in Eq. (23b) refers to negative (positive) λ/k. The appearance of the minimum can be expressed in terms of the pinch ratio $\Theta_0 = B_\varphi(a)/B_0$ associated with the axisymmetric, uniform-current, zero-flow solution for the same E_0:

$$\Theta_0 = (E_0 a)/(2\eta B_0). \tag{24}$$

(Note that Θ_0 will *not* be the computed Θ for the helical states.) The unconstrained minimum moves to the location (23) whenever

$$\Theta_0 \geq \frac{\sqrt{\eta\nu}}{2aB_0}(\lambda a)^2\left|\frac{\lambda a}{ka}\right|. \tag{25}$$

Notice that the threshold (25) and the location (23) depend upon the m, n and λ, and so differ from mode to mode. The overall minimum must be found by searching through the m, n, and λ, and requires a stored numerical table of λ's obtained from solving Eq. (8) numerically. Evaluated at the location (23), the unconstrained minima are

$$\left(\frac{R}{V}\right)_{min} \quad \text{(unconstrained)} \quad = 2\left|\frac{\lambda^3}{k}\right|\sqrt{\eta\nu}\left[E_0 - \frac{\eta\sqrt{\eta\nu}}{2}\left|\frac{\lambda^3}{k}\right|\right], \qquad (26)$$

which is less than E_0^2/η any time the inequality (25) is satisfied. When E_0 is $>>$ the latter term in the square brackets in Eq. (26), as it has often been in the simulations, we are above the threshold and it is apparent that the minimization over m, n and λ is achieved by seeking the numerically smallest $|\lambda^3/k|$. We take two sets of linear dimensions which have been typical of the (square-boundary) computations that have been done. For $L_z = 2\pi$ and $a = \pi/2$, the minimum $|\lambda^3/k|$ is achieved for $m = 1$, $n = 1$, $\lambda \cong -2.01$, and is $\cong 8.11$; for $L_z = 4\pi$ and $a = \pi/2$ the minimum $|\lambda^3/k|$ is achieved for $m = 1$, $n = 3$, $\lambda \cong -2.23$, and is $\cong 7.35$. For the former aspect ratio, RFP computations[11] have typically shown the dominant $(m, n) = (1, 1)$ or $(1, 2)$, and for the latter,[14,15] $(m, n) = (1, 3)$ or $(1, 4)$. It has been observed that $< \boldsymbol{\omega} \cdot \mathbf{v} > / < \mathbf{v}^2 >$ was negative, consistent with $\lambda < 0$. Unfortunately, closer comparisons with previous cases seem difficult, for the computations typically have had: only numerical viscosity, if any; shaped resistivity profiles which depend on space; bare, perfectly-conducting boundary conditions; and square boundaries. All these render the computational and analytical situations sufficiently different that a quantitative comparison of amplitudes would seem to be of uncertain utility. The tokamak computations ($\Theta < 1$) have also been done in the Strauss or "reduced MHD" approximation[10,12,13] rather than with the full set of 3D MHD equations, which renders the comparisons even more between disparate entities. Finally, neither set of simulations has really satisfied the weak perturbation (of the uniform current profile) criterion that has motivated treatment of the axisymmetric current contribution as independent of r. It is interesting, but far from conclusive, that the dominant m and n numbers agree as well as they do. It is also perhaps fortuitous, but still interesting, that the Strauss computations as well have usually revealed $(m, n) = (1, 1)$ for these aspect ratios. At least we can say that the "minimum $|\lambda^3/k|$" criterion has not in the past given wildly incorrect results, as given by the simulations.

Parenthetically, we remark that in Eq. (25) the dimensionless number $B_0 a/\sqrt{\eta\nu}$ has the significance of being the geometric mean of the poloidal Lundquist number $S \equiv B_0 a/\eta$ and the viscous analogue of it, $M \equiv B_0 a/\nu$. In another context, this mean is called the "Hartmann number," H. All the other terms in Eq. (25) are determined entirely in terms of the geometry of the cylinder through Eq. (8).

The critical Θ_0 defined by Eq. (25) is not to be confused with the threshold Θ_i for the onset of linear instability, since $\Theta_i >$ this Θ_0, always. Thus, it is imaginable that by ramping up the current slowly and carefully, the threshold (25) might be passed through before instabilities made the helical states dynamically accessible. The instability threshold Θ_i for the uniform current density state is analytically calculable and the result may be written as

$$\Theta_i = \frac{\alpha n \left(m + \frac{\alpha n}{x}\right) - \left[\frac{\alpha^4 n^4}{x^2} - \frac{m \eta \nu}{a^2 B_0^2}\left(m + \frac{2\alpha n}{x}\right) x^4\right]^{1/2}}{m\left(m + (2\alpha n/x)\right)} \qquad (27)$$

where $x \equiv \lambda a$, $\alpha \equiv 2\pi a/L_z$ is the inverse aspect ratio, and $k_n a = \alpha n$. On the right-hand side of Eq. (27), m and n are both to be regarded as non-negative, but λ and x can have either sign.

We may now consider the more demanding problem defined by adding the constraint (15), which not only considerably increases the difficulty, but sometimes may not always improve the capacity to predict the simulation results (we defer to Section IV a discussion of why this might be so). The situation may best be understood with reference to Figure 1, for a typical case. Figure 1, for which the parameters are given in the figure caption, shows contours of constant R/V as given by Eq. (18) with the two minima clearly visible. Superimposed upon these is the constraint curve defined by Eq. (15) for the same parameters. The achievable minimum dissipation state, indicated by a cross on the constraint curve, is that point on it which reaches the lowest value of R/V. As far as we can tell, this constrained minimum is only obtainable numerically. Once it has been obtained, a second numerical search, over the table of m, n, and λ, is required to fix the constrained minimum dissipation state.

One analytical calculation involving the constraint (15) is possible. This involves finding the minimum Θ (call it Θ_c) such that there *exists* a constrained minimum away from the origin. The expression for Θ_c, which may be obtained by eliminating $\tilde{\xi}_b$ between Eqs. (15) and (18) and asking for the condition that a minimum with $\tilde{\xi}_v \neq 0$ exist, is remarkably similar to the expression (27) and can be written as

$$\Theta_c = \frac{\alpha n \left(m + \frac{2\alpha n}{x}\right) - \left[\frac{4\alpha^4 n^4}{x^2} - \frac{m \eta \nu}{a^2 B_0^2}\left(m + \frac{4\alpha n}{x}\right) x^4\right]^{1/2}}{m\left(m + (4\alpha n/x)\right)} \qquad (28)$$

The symbols in Eqs. (28) and (27) are the same. The inequality $\Theta_i > \Theta_c > \Theta_0$ as given by Eq. (25) always holds. A plot of Θ_c and Θ_i vs. $\eta\nu(a^2 B_0^2)^{-1} = (SM)^{-1} = H^{-2}$ is given in Figure 2a for $L_z = 4\pi$ and $a = \pi/2$. In Figure 2b, the corresponding m, n and λa for the minimum-dissipation threshold Θ_c are displayed as functions of the same variable, for the same case. This curve also has been obtained by a numerical search using the expression (28) and the λ table.

We find the abrupt jumps in the threshold m, n and λ values seen in Figure 2b to be rather remarkable. Physical intuition has proved an inadequate guide to understanding them. If $\sqrt{\eta\nu}$ is thought of as a function of a uniform temperature, as the magnetofluid becomes hotter, one moves to the left on the graph. If Θ is fixed at a definite value above threshold, a numerical search among the allowed constrained minima is required to ascertain the minimum dissipation state. No

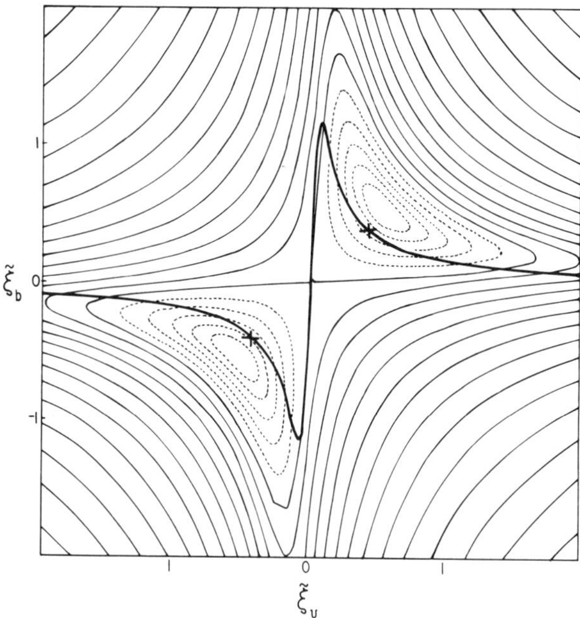

Figure 1. The $\tilde{\xi}_v$ $\tilde{\xi}_b$ plane for a typical case, with contours of constant R/V, as given by Eq. (18) shown with a logarithmic contour interval. Superposed is the constraint curve, Eq. (15); the cross on it indicates the location of the location of the constrained minimum ($B_0 = 4.5$, $\eta = \nu = 10^{-2}$, $L_z = 4\pi$, $a = \pi/2$, $E_0 = 0.051$, $m = 2$, $n = 1$, $\lambda a = -4.81$.)

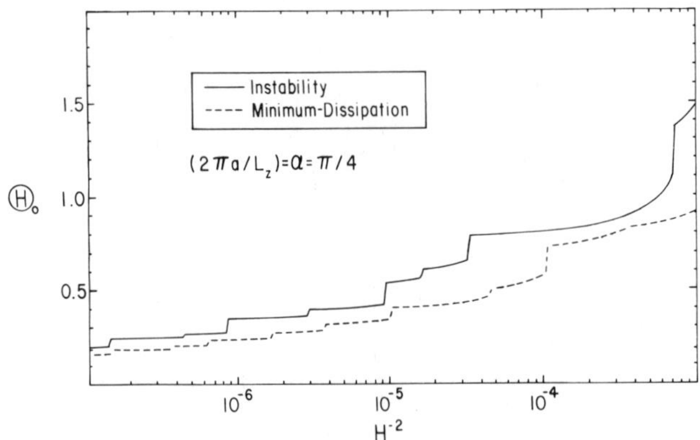

Figure 2 a. Thresholds in pinch ratio ($L_z = 4\pi$, $a = \pi/2$) for linear instability of the axisymmetric state (solid curve), and for the existence of a helically-distorted minimum dissipation state (dashed curve), as functions of the variable $H^{-2} \equiv \eta\nu(a^2 B_0^2)^{-1}$.

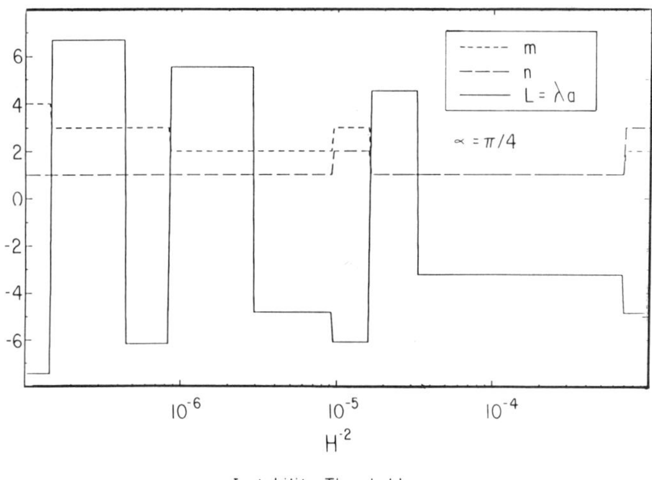

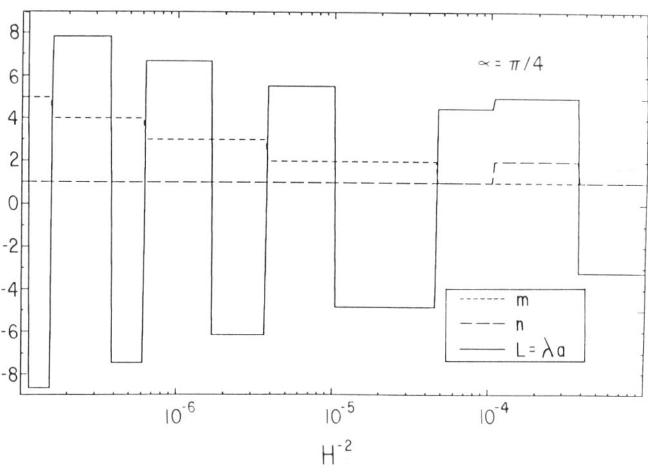

Figure 2 b. Threshold values of m, n and λ as functions of H^{-2}: above, for the first linearly unstable mode; below, for the first helically-distorted minimum-dissipation state that appears as Θ_0 is raised.

simple transparent criterion for the dominant m, n, and λ has revealed itself, for the constrained minimum case, in contrast with the rather simple "minimum $|\lambda^3/k|$" criterion for the unconstrained minimum, which depends only on the dimensions of the container.

Figure 3 displays the results of the numerical search, for the parameters given in the figure caption, for the minimum dissipation state. Figure 3a plots the parabola E_0^2/η vs. Θ_0, the axisymmetric, uniform-current dissipation curve. It also shows, below it, the numerically-ascertained minimum-dissipation curve, also as a function of Θ_0. The two part company at the value of Θ_0 obtained from Figure 2b. Figure 3b shows the dominant m, n and λ between the transitions. Figure 3c shows the computed amplitudes $\tilde{\xi}_v$ and $\tilde{\xi}_b$ associated with the helical deformations.

The previous set of driven RFP computations were done[15] with $\nu = 0$ and a shaped resistivity profile $\eta(x, y)$ inside of a square poloidal boundary. The volume average of η can be estimated as 0.01. A numerical exploration of the *constrained* minimum can be carried out for $\nu = 0$. The results of doing so are shown in Figure 4. The values of Θ_0 are so large that there is no reason to expect the theory to be accurate. The following list shows the parameters for the three principal runs reported,[15] for $L_z = 4\pi$, $a = \pi/2$, $< \eta > \cong 0.01$.

ν	$<\eta>$	Θ_0	Θ	E_0	(m,n) (predicted)	(m,n) (computed)
0	0.01	3.92	1.3	0.025	(1.5)	(1,4)
0	0.01	7.85	2.2	0.05	(1,4)	(1,3) or (1,4)
0	0.01	31.4	4.5	0.2	(1,9)	(1,4)

Allowing for the fact that the theoretically predicted values follow from a truncation that is supposed to be justified by $\Theta_0 < 1$, the fit is not totally discouraging, even though the sign of λa is positive. Recall also the computational boundary is a square while the theory is for a circle, while the theory has a dielectric coated conducting boundary in a contrast to the bare conductor of the computations.

A serious potential defect of the $\nu = 0$ theory is also illustrated by Figure 4(b). The cusps at the transitions actually do correspond to $\tilde{\xi}_v^2$ becoming unbounded. The assumption of low-order truncation would undoubtedly break down as that occurred. Moreover, any viscosity, however small, would prohibit it. All we can say is, at best, that immediately below each transition between one mode and the next, there is a neighborhood of undetermined width where the theory fails.

Though the full variational problem defined by Eqs. (15) and (18) seems to require numerical treatment, it becomes algebraically tractable away from the threshold in a "strong field" limit. Solving Eq. (15) for $\tilde{\xi}_b$ gives

$$\tilde{\xi}_b = -\frac{(\beta_0/\eta\lambda)\tilde{\xi}_v}{1 + (\beta_1\tilde{\xi}_v^2/2\eta\lambda)}. \tag{29}$$

When $|\tilde{\xi}_v^2\beta_1/2\eta\lambda| >> 1$, a condition whose requirements will presently be in-

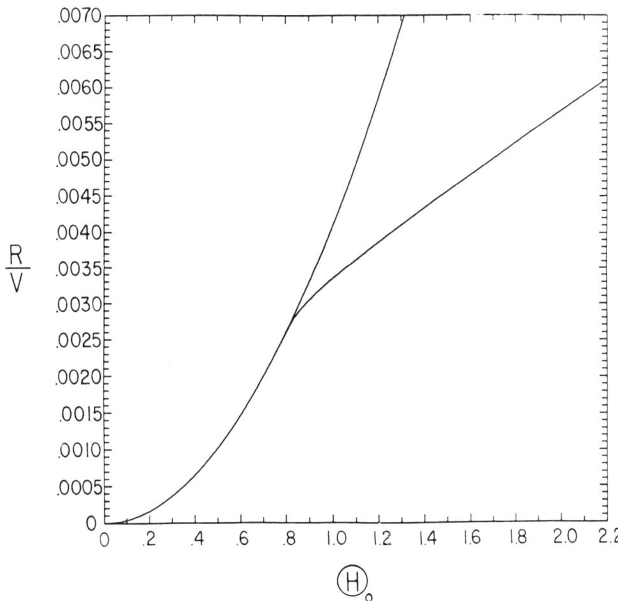

Figure 3.a Volume averaged Ohmic dissipation rate vs. Θ_0. Upper curve is the axisymmetric state, lower curve is the constrained helically distorted minimum. (Throughout Figure 3, $B_0 = 0.5$, $\nu = \eta = 0.01$, $L_z = 4\pi$, and $a = \pi/2$.)

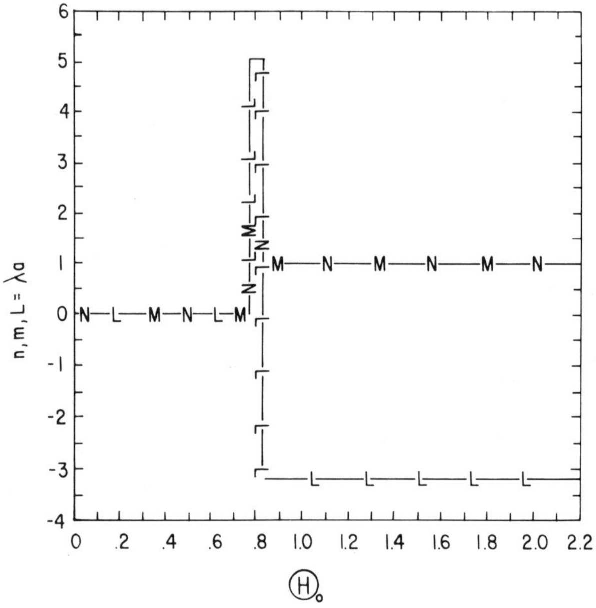

Figure 3.b Associated m, n, and $L = \lambda a$ for the interval of Θ_0 shown in (a); $m = n = 1$ above $\Theta_0 \gtrsim 0.8$.

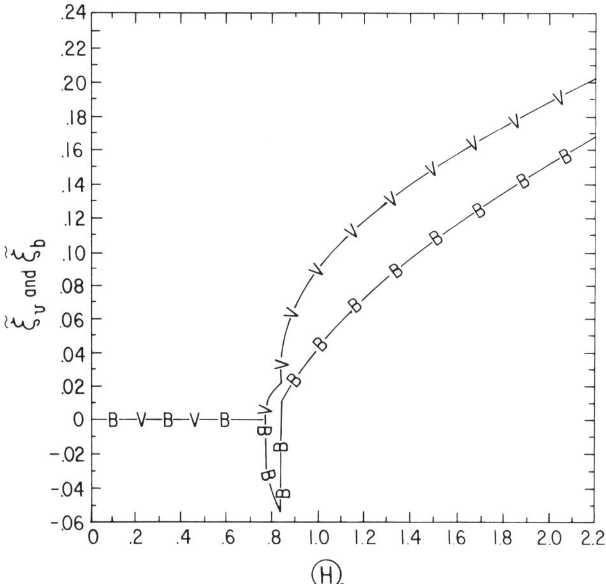

Figure 3.c Helical amplitudes $\tilde{\xi}_v$, $\tilde{\xi}_b$ for the situation shown in (a) and (b). $\tilde{\xi}_v$ is in units of the toroidal Alfvén speed, and the perturbation theory is probably only accurate near the threshold.

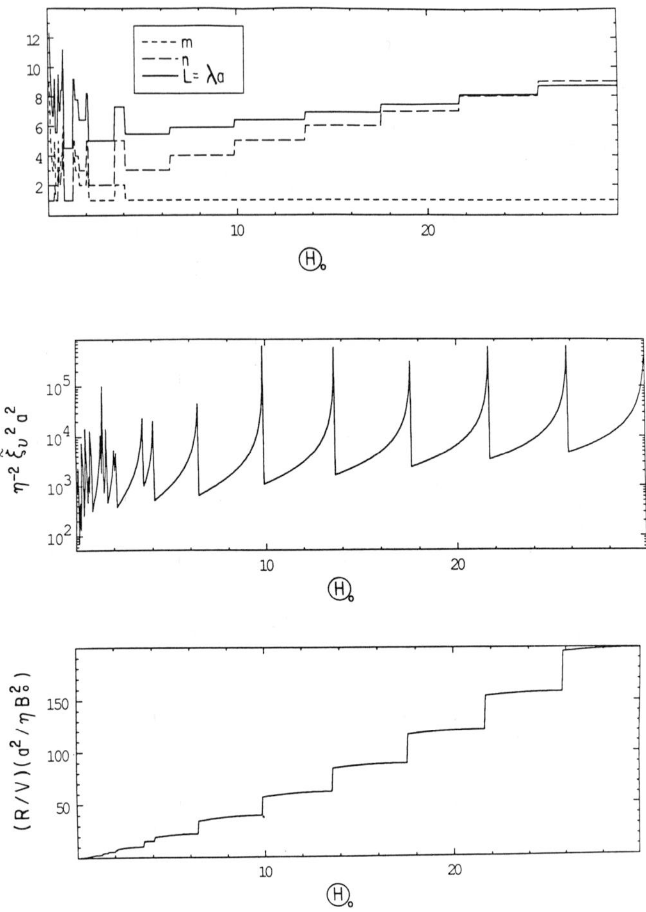

Figure 4. Characteristics of the minimum dissipation state for zero viscosity as a function of Θ_0. The parameters chosen are $B_0 = 0.5$, $\eta = 0.01$, $L_z = 4\pi$, $a = \pi/2$, though the ordinates all depend only upon Θ_0 and the inverse aspect ratio $\alpha = 2\pi a/L_z = \pi/4$. (a) dominant m, n, and $L = \lambda a$ vs. Θ_0; (b) $\eta^{-2}\tilde{\xi}_v^2 a^2$, measuring the helical kinetic energy; (c) $(R/V)(a^2/\eta B_0^2)$, measuring the minimum dissipation, which is much less than that associated with the parabolic curve of the axisymmetric configuration.

quired into, Eq. (29) collapses into a simple hyperbolic constraint

$$\tilde{\xi}_b \tilde{\xi}_v \cong -2\beta_0/\beta_1, \tag{30}$$

and reduces Eq. (18) to ($\nu \neq 0$):

$$\frac{R}{V} = \nu\lambda^2 \tilde{\xi}_v^2 + \eta\lambda^2 \left(\frac{4\beta_0^2}{\beta_1^2 \tilde{\xi}_v^2}\right) + \quad \text{constant.} \tag{31}$$

The minimum R/V now occurs at

$$\tilde{\xi}_v^2 = \left(\frac{\eta}{\nu}\frac{4\beta_0^2}{\beta_1^2}\right)^{1/2}, \quad \tilde{\xi}_b^2 = \left(\frac{\nu}{\eta}\frac{4\beta_0^2}{\beta_1^2}\right)^{1/2}, \tag{32}$$

and is

$$\left(\frac{R}{V}\right)_{min} \quad \text{(strong field)} \quad = \frac{1}{\eta}\left[E_0 - \frac{2k\beta_0}{\beta_1}\right]^2 + 4\lambda^2\sqrt{\eta\nu}\left|\frac{\beta_0}{\beta_1}\right|. \tag{33}$$

Using the expressions for β_0 and β_1 in Eq. (33),

$$\left(\frac{R}{V}\right)_{min} \quad \text{(strong field)} \quad = \frac{4k_n^2 B_0^2}{m^2}\eta + 2\sqrt{\eta\nu}\left|\frac{\lambda^3}{k}\right| \cdot \left|E_0 - \frac{2\eta k B_0}{m}\right|. \tag{34}$$

The condition that $|\tilde{\xi}_v^2\beta_1/2\eta\lambda| \gg 1$ can now be seen to be

$$\left(\frac{mE_0}{2\eta} - kB_0\right)^2 \gg \eta\nu\lambda^4 \tag{35a}$$

or

$$\left|\Theta_0 - \frac{ka}{m}\right| \gg \frac{\sqrt{\eta\nu}(\lambda a)^2}{|m|(B_0 a)}. \tag{35b}$$

For sufficiently low values of the transport coefficients, the inequalities (35) are not difficult to satisfy; expressed in terms of the tokamak "safety factor" $q(a)$, the implication of Eq. (35b) is:

$$\left(\frac{1}{q(a)} - \frac{m}{n}\right)^2 \gg \frac{\eta\nu a^2}{4\pi B_0^2}\frac{(\lambda a)^4}{m^2}\left(\frac{L_z}{a}\right)^2. \tag{36}$$

Eq. (36) expresses a requirement that $q(a)$ for the minimum dissipation mode, as determined from Eq. (34), be sufficiently less than m/n that Eq. (36) is satisfied. It is interesting, however, that for sufficiently large E_0, the crucial combination for determining m and n once again turns out to be the minimum value of $|\lambda^3/k|$. The repeated appearance of this combination in such disparate cases suggests a deeper significance for it, but we have been unable to perceive what it is. The linear dependence of R_{min} on $|E_0 - 2\eta k B_0/m|$, both at the threshold and far above it, may also be significant; surely it makes clear the ultimate victory of the helical state over the axisymmetric (E_0^2/η) one at minimizing the overall dissipation rate as the driving strength increases.

A driven MHD run has been very recently performed with finite, spatially-independent $\nu = \eta = 0.01$, $B_0 = 4.5$, $L_z = 4\pi$, $a = \pi/2$, $E_0 = 0.51$, $\Theta_0 \cong 0.89$. An initially uniform current profile with a low level of random noise was used to minimize the time necessary to develop the helically distorted state. The time histories of some global quantities are shown in Figure 5. Θ_0 was unfortunately too high to be considered close to the instability threshold in the sense of Figure 2, but it is interesting to report the run in any case, since $m = 3$ distortions characterize the fully developed discharge. These are clearly visible in the 3D perspective plot of the surface $v_z = 0.1$ at $t = 190$ as shown in Figure 6. Figure 7 reveals the results of the theory for the single-mode minimum dissipation state. It is seen that immediately above the threshold, the prediction for the dominant mode is $m = 3$, $n = 1$ (and $\lambda < 0$). However, for $\Theta_0 \cong 0.89$, the dominant mode has changed to one which is not in agreement with the simulation results ($m = 2$, $n = 1$ and $\lambda < 0$). This discrepancy is presently unexplained.

IV.   DISCUSSION

We have   given the   simplest possible   non-trivial   framework in which a minimum-dissipation, driven MHD state could be calculated employing the Ohm's law. The Ohm's law is equivalent to an infinite number of integral constraints to be retained, at any finite level of truncation. In our case, there were three unknowns, $\tilde{\xi}_v$, $\tilde{\xi}_b$, and $< j_z >$, and three constraints: the results of taking the inner product, with the Ohm's law, of $1\hat{e}_z$, $\delta\mathbf{B}$, and $\delta\mathbf{v}$. If the third had not trivially reduced to zero equals zero, there would have been no minimization left to perform. We have been unable so far to formulate rules of any generality that seem convincing for the optimum number of constraints to retain for a given level of truncation. Much experience with this non-standard class of variational problem remains to be acquired. It is noteworthy that even with these drastic simplifying assumptions an extraordinarily rich class of results emerges in such figures as Figures 2 or 4.

Unfortunately, none of the simulations (including the new one reported here) corresponds to a low enough value of Θ_0 to be a sharp test of the $\Theta_0 < 1$ theory as presented here. It is a high priority item to repeat the computations with values of Θ_0 only slightly above Θ_c; (say $\Theta_0 \cong 0.4$, for the situation shown in Figure 6). Even then, disparities in boundary conditions between the codes available and the theory imply the need for tentative and cautious interpretation of any results.

We close with an appeal to resurrect cold-tokamak internal probe diagnostics. The numerical results strongly suggest the presence of helical distortions and vortex cells as a feature of tokamaks in interesting regimes. Proof or disproof of their physical significance can only come from direct experimental test, however imaginative use of computer simulation may be made, or however zealously dissipation rates may be minimized. We should also remark that the variational

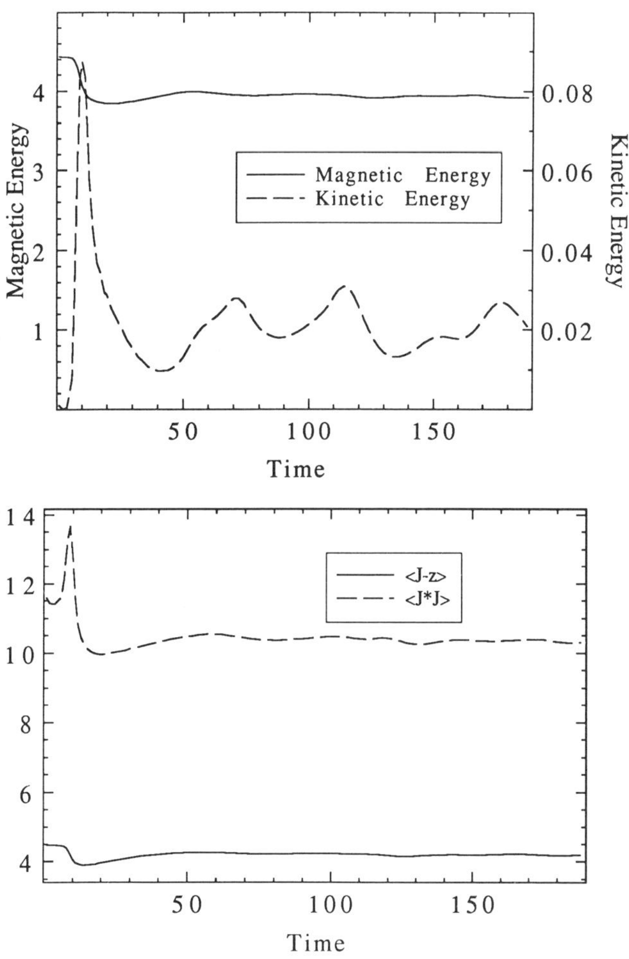

Figure 5. Global variables vs time (in poloidal Alfvén transit times) for a new run, not previously published. Parameters are the same as in Figure 7. (a) Magnetic and kinetic energy vs time. (b) $< \mathbf{j}^2 >$ and $< j_z >$ vs. time.

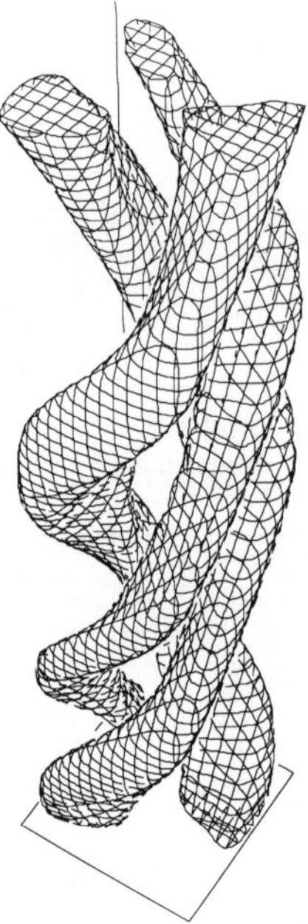

Figure 6. Isosurface $v_z =$ constant at $t = 190$ for new run shown in Figure 5. The persistent $m = 3$, $n = 3$ structure is clearly visible. Figure 7 predicts the dominant mode should be $m = 3$, $n = 1$ just above the threshold, but $m = 2$, $n = 1$ at the computed value of $\Theta_0 \cong 0.89$.

.26
.24
.22
.20
.18
.16
.14
.12
.10
.08
.06
.04
.02
0
R/V
0 .2 .4 .6 .8 1.0 1.2 1.4 1.6 1.8 2.0 2.2
H₀

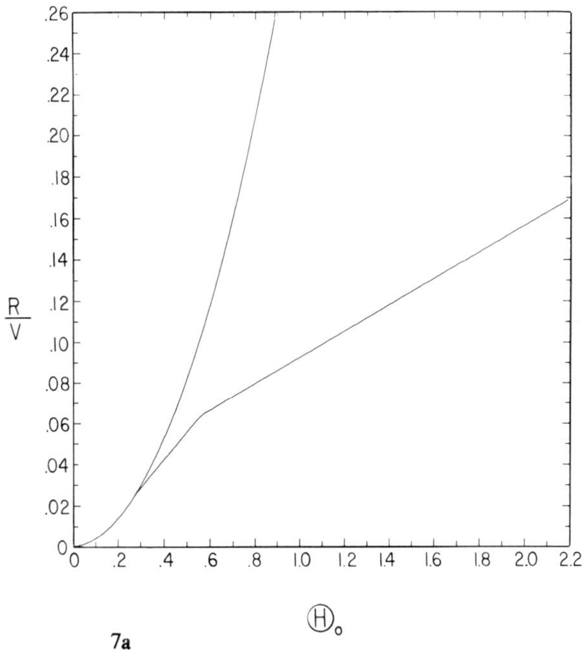

7a

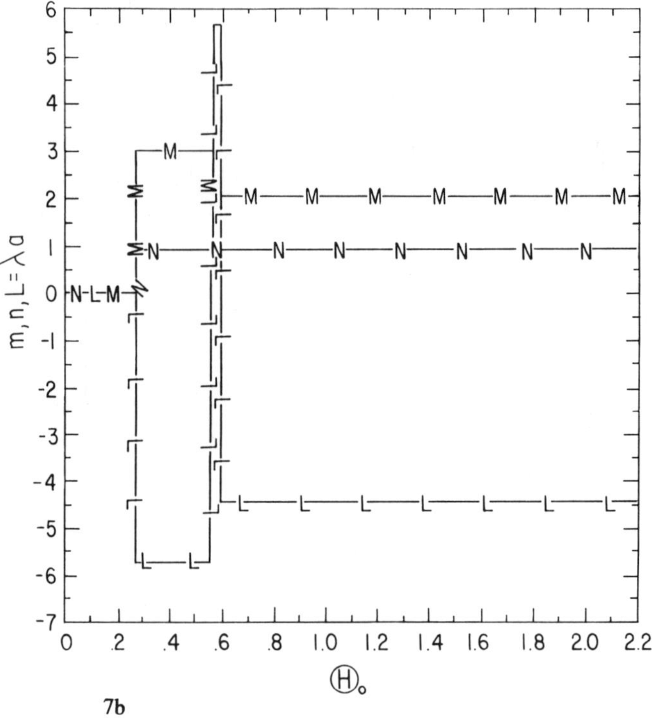

7b

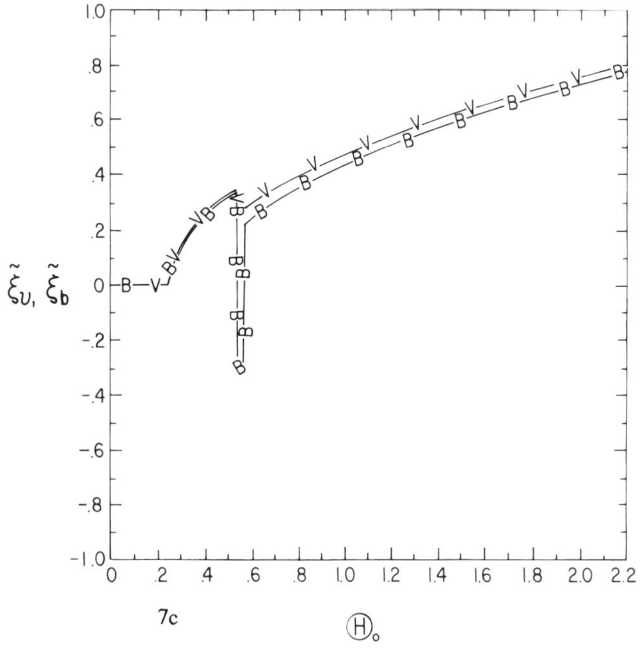

Figure 7. Results of the search for the constrained minimum for $B_0 = 4.5$, $\nu = \eta = 0.01$, $a = \pi/2$, $L_z = 4\pi$ as a function of Θ_0. (a) Dissipation curves vs. Θ_0. (Upper curve is the parabola of the axisymmetric state.) (b) Dominant m, n, and $L = \lambda a$ as a function of Θ_0. (c) $\tilde{\xi}_v$ and $\tilde{\xi}_b$ as a function of Θ_0.

problems which have been so troublesome here will also arise in virtually identical form in "minimum energy" calculations of the Taylor type,[20] if it is attempted to include the Ohm's law in the formalism.

ACKNOWLEDGEMENTS

We are indebted to Dr. Lee Phillips for useful discussions and to Dr. Jill Dahlburg for the numerical background and the gift of her 3D RFP code for further development of it.

This work was supported in part by U.S. Department of Energy Grants DE-FG02-89ER53298 and DE-FG02-85ER53194, and NASA Grant NAG-W-710.

APPENDIX A: DIMENSIONLESS VARIABLES

In Gaussian units, the MHD equations we work with are

$$\rho_0 \left(\frac{\partial \mathbf{v}}{\partial t} + \mathbf{v} \cdot \nabla \mathbf{v} \right) = -\nabla p + \frac{\mathbf{j}}{c} \times \mathbf{B} + \rho \nu \nabla^2 \mathbf{v} \qquad (A1)$$

$$\frac{\partial \mathbf{B}}{\partial t} = \nabla \times (\mathbf{v} \times \mathbf{B}) - \frac{c}{\sigma} \nabla \times \mathbf{j} \qquad (A2)$$

$$\mathbf{E} + \frac{\mathbf{v}}{c} \times \mathbf{B} = \frac{\mathbf{j}}{\sigma} \qquad (A3)$$

$$\nabla \times \mathbf{B} = \frac{4\pi}{c} \mathbf{j}, \qquad (A4)$$

supplemented by $\nabla \cdot \mathbf{B} = 0 = \nabla \cdot \mathbf{v}$. p is the mechanical pressure, σ is the electrical conductivity, c is the speed of light, $\mathbf{E}$ is the electric field, ρ_0 is the (uniform) mass density, and the other fields have been previously defined.

We introduce dimensionless units, indicated by tildes $(\tilde{\ })$, according to $t = \tilde{t} L_0/U_0$, $\mathbf{x} = \tilde{\mathbf{x}} L_0$, $\mathbf{v} = \tilde{\mathbf{v}} U_0$, $\mathbf{B} = \tilde{\mathbf{B}} B_0$, $\mathbf{j} = \tilde{\mathbf{j}} j_0$, etc., where T_0, U_0, B_0, j_0, ... are units in which we measure the fields and independent variables. If we choose $j_0 = cB_0/4\pi L_0$, $p_0 = \rho_0 U_0^2$, and $U_0^2 = C_A^2 \equiv B_0^2/4\pi\rho_0$, $\tilde{\nu} = \nu(L_0 U_0)^{-1}$, and $\tilde{\eta} \equiv c^2/(4\pi\sigma U_0 L_0)$, Eqs. (A1)-(A4) can be combined to give

$$\frac{\partial \tilde{\mathbf{v}}}{\partial \tilde{t}} + \tilde{\mathbf{v}} \cdot \tilde{\nabla} \tilde{\mathbf{v}} = -\tilde{\nabla} \tilde{p} + (\tilde{\nabla} \times \tilde{\mathbf{B}}) \times \tilde{\mathbf{B}} + \tilde{\nu} \tilde{\nabla}^2 \tilde{\mathbf{v}} \qquad (A5)$$

and

$$\frac{\partial \tilde{\mathbf{B}}}{\partial \tilde{t}} = \tilde{\nabla} \times (\tilde{\mathbf{v}} \times \tilde{\mathbf{B}}) + \tilde{\eta} \tilde{\nabla}^2 \tilde{\mathbf{B}}, \qquad (A6)$$

while (A3) becomes

$$\tilde{\mathbf{E}} + \tilde{\mathbf{v}} \times \tilde{\mathbf{B}} = \tilde{\eta} \tilde{\mathbf{j}}. \qquad (A7)$$

Eqs. (A5)-(A7) are the dimensionless set of MHD equations which we have typically used. Dropping the tildes, the integral which measures the energy dissipation rate for stationary boundaries predicted by Eqs. (A5), (A6) is Eq. (1) of the text. The following table displays explicitly the units used:

$$\mathbf{B} \quad : \quad B_0$$

$$\mathbf{v} \quad : \quad U_0 = C_A = B_0(4\pi\rho_0)^{-1/2}$$

$$\mathbf{j} \quad : \quad j_0 = cB_0(4\pi L_0)^{-1}$$

$$\tilde{\nu} = \nu(U_0 L_0)^{-1} = \nu(C_A L_0)^{-1}$$

$$\mathbf{E} \quad : \quad E_0 = C_A B_0/c$$

$$\tilde{\eta} = c^2(4\pi\sigma C_A L_0)^{-1}$$

$$L_0 = a.$$

The purpose of the non-dimensionalization, in addition to rendering the equations of motion visually simple, is to bring the most basic length/time scales and field variables, such as $\mathbf{B}$, $\mathbf{v}$, L_0, U_0, into the range 0.1 to 10, approximately. To represent a laboratory experiment, for example, the basic magnetic field might be measured in units of 10^3 gauss. When, in the text, we write B_0 for the axial field strength (as in Eq. (25)), the numerical value of a 4.5 kG axial field would then be: $B_0 = 4.5$.

If we wish to write the critical laboratory electric field, in stat-Volts/cm, from an equation like (25) for example, the easiest way is to write everything first in dimensionless units. Thus, from (23b), in cgs laboratory units, the threshold electric field that will make $\tilde{\xi}_{b0}^2 \geq 0$ is

$$\frac{E_{critical}}{B_0} = \sqrt{\frac{\nu}{\eta}} \left(\frac{c^2}{4\pi\sigma a C_A}\right)^2 \left|\frac{(\lambda a)^3}{(ka)}\right| \left(\frac{C_A}{c}\right), \tag{A8}$$

where all variables are in cgs units on the right-hand side; and the corresponding expression in the dimensionless units used in the text is

$$\frac{E_{critical}}{B_0} = \frac{\eta\sqrt{\eta\nu}}{B_0} \left|\frac{(\lambda a)^3}{ka}\right|. \tag{A9}$$

APPENDIX B: CHANDRASEKHAR-KENDALL FUNCTIONS[16−18]

If one starts with any solution of the scalar Helmholtz equation

$$(\nabla^2 + \lambda^2)\psi = 0, \tag{B1}$$

and forms the combination

$$\mathbf{A} = \lambda\nabla\times\hat{e}\psi + \nabla\times(\nabla\times\hat{e}\psi), \tag{B2}$$

where $\hat{e}$ is any constant unit vector, it is elementary to show that

$$\nabla \times \mathbf{A} = \lambda \mathbf{A}. \tag{B3}$$

The property (B3) is useful in expanding solutions to vector field equations like those of MHD, where all fields are solenoidal and curls are taken repeatedly. For cylindrical geometry, the appropriate choices are $\hat{e} = \hat{e}_z$, and ψ is the real or imaginary part of $J_m(\gamma r)\exp(im\varphi - ik_n z)$, with $\gamma^2 + k_n^2 = \lambda^2$. These lead to Eqs. (6) of the text, with I_{nmq} given by

$$I_{nmq} = \pi L_z J_m^2(\gamma_{nmq}a)\left[\frac{m\lambda_{nmq}\gamma_{nmq}^2}{k_n} + \right.$$

$$\left. \lambda_{nmq}^2(\gamma_{nmq}a)^2\left(1 + \frac{m^2}{k_n^2 a^2}\right)\right]. \tag{B4}$$

The choice (B4) makes the eigenfunctions (6) orthonormal.

The expressions (6) and (B4) are appropriate for $m^2 + n^2 \neq 0$. The axisymmetric $m = 0 = n$ solutions require special treatment since the boundary condition (8) does not determine γ_{00q}. A convenient choice is to put all the (conserved) toroidal magnetic flux into the $\mathbf{A}_{001}$ eigenfunction and to determine the $\mathbf{A}_{00q}$ for $q \geq 2$ by the requirement that all the $\mathbf{A}_{00q}$ also be orthonormal. In the present "tokamak limit," $\gamma_{00q}^2 a^2 << 1$, and this is not necessary. The axisymmetric 001 expansion function is just taken to be $< j_z > r\hat{e}_\varphi/2 + \mathbf{B}_0$, normalized so that its volume integral is unity, which to the accuracy we are going is equivalent to dividing it by $B_0\sqrt{\pi a^2 L_z}$. The ξ_{001} amplitude is then just $B_0\sqrt{\pi a^2 L_z}$, and the field resulting, $< \mathbf{B} > + r < j_z > \hat{e}_\varphi/2$, is just that appearing in Eq. (4b) as the axisymmetric contribution. Considerably more detailed attention would have to be paid to the axisymmetric $\mathbf{A}_{00q}$ if we wanted to work in the RFP limit.

REFERENCES

1. D. Montgomery and L. Phillips, *Phys. Rev. A*, **38**, 2953 (1988).
2. D. Montgomery and L. Phillips, *Physica*, **D37**, 215 (1989).
3. D. Montgomery, L. Phillips, and M.L. Theobald, *Phys. Rev. A*, **40**, 1515 (1989).
4. D. Montgomery, "Relaxed States in Driven, Dissipative Magnetohydrodynamics: Helical Distortions and Vortex Pairs," to appear in *Trends in Theoretical Physics*, **Vol. I**, ed. by P.J. Ellis and Y.C. Tang, Addison-Wesley, New York (1990).
5. G.D. Kirchoff, *Ann. Phys.*, **75**, 189 (1848).
6. E.T. Jaynes, *Ann. Rev. Phys. Chem.*, **31**, 579 (1980).
7. H. Lamb, *Hydrodynamics*, 6th ed., Dover, New York (1945), pp. 617-619,.
8. G.K. Batchelor, *An Introduction to Fluid Dynamics*, Cambridge University Press, Cambridge, UK, (1967), 228.

9.   D. Montgomery and L. Phillips, "Minimum Dissipation and Maximum Entropy," to appear in Proc. MAXENT '89 Workshop, P.F. Fougere, editor; Kluwer Academic Publishers (1990).

10.   M.L. Theobald, D. Montgomery, G.D. Doolen, and J.P. Dahlburg, *Phys. Fluids* **B1**, 766 (1989).

11.   J.P. Dahlburg, D. Montgomery, G.D. Doolen and L. Turner, *Phys. Rev. Lett.*, **57**, 428 (1986).

12.   J.P. Dahlburg, D. Montgomery, and W.H. Matthaeus, *J. Plasma Phys.*, **34**. 1 (1985).

13.   J.P. Dahlburg, D. Montgomery, G.D. Doolen, and W.H. Matthaeus, *J. Plasma Phys.*, **35**, 1 (1986).

14.   J.P. Dahlburg, D. Montgomery, G.D. Doolen, and L. Turner, *J. Plasma Phys.*, **37**, 299 (1987).

15.   J.P. Dahlburg, D. Montgomery, G.D. Doolen, and L. Turner, *J. Plasma Phys.*, **40**, 39 (1988).

16.   S. Chandrasekhar and P.C. Kendall, *Astrophys. J.*, **126**, 457 (1957).

17.   L. Turner, *Ann. Phys. (NY)*, **149**, 58 (1983).

18.   D. Montgomery, L. Turner, and G. Vahala, *Phys. Fluids,* **21**, 757 (1978).

19.   e.g., G. Bateman, *MHD Instabilities*, Cambridge, MA, MIT Press (1978).

20.   J.B. Taylor, *Phys. Rev. Lett.*, **33**, 1139 (1974) and *Revs. Mod. Phys.*, **58**, 741 (1986).

THREE-DIMENSIONAL SIMULATIONS
OF SIMPLE ELECTRON DRIFT WAVE TURBULENCE
IN TOKAMAK GEOMETRY*

R.E. Waltz

General Atomics, San Diego, California 92138-5608

ABSTRACT

A three-dimensional code is used to study the scaling of an electrostatic electron drift wave turbulence with respect to shear and curvature of the tokamak magnetic field as well as the driving strength. The electrons are modeled with a simple formula for the deviation from a near Boltzmann response and the ions are modeled as a nonlinear fluid. For sheared slab geometry suppressing curvature (or toroidal) effects, the turbulence level and transport scale according to the mixing length rules with the mixing length given by the linear normal mode widths and the poloidal wavelengths scaled to the ion gyroradius. Toroidal coupling results in a more complex behavior.

1. INTRODUCTION

There is now a considerable body of experience with two-dimensional simulation of homogeneous drift wave turbulence applicable to transport in tokamaks. These studies range in complexity from simple driven-damped Hasegawa-Mima models[1] to electrostatic and electromagnetic isothermal collisional two-fluid and trapped electron three-fluid[2] simulations of plasma flow. Heat flows and thermal effects on plasma flows for both electron and ion drift modes have also been treated.[3] The simulations are performed on a (k_x, k_y) representation for the radial and poloidal wave numbers. Each mode is assigned the same average parallel wave number $(k_\parallel = 1/Rq)$, corresponding to the inverse connection length between regions of bad and good curvature, and an average curvature drift $(gv_*$ where v_* is the diamagnetic drift and $g = L_p/R$ with L_p the pressure length). Simulations based on these local two-dimensional models are reviewed in Ref. 4. There is no attempt to represent the shear or the variation in the curvature of the magnetic field.

This paper attempts to deal explicitly with the three-dimensional aspects of the shear and curvature of the magnetic field, although the electron dynamics is considerably simplified. A series of numerical experiments or cases is presented with a heuristic description of the results. It primarily on electrostatic drift wave turbulence in a finite periodic sheared slab with cold ions. A radial grid covers the slab spanned by modes with poloidal or toroidal harmonic numbers (m, n) having helicity values m/n between $q(r_1)$ and $q(r_2)$. The $q(r)$ profile is arranged to have a constant shear length $L_s = Rq/\hat{s}$ where $\hat{s} = d\ell n q/dr$. The electron dynamics is artificially modeled as a linear deviation from a Boltzmann response supported only near the singular surfaces where $q(r) = m/n$ and the parallel wave number $k_\parallel = (1/Rq)(qn - m)$ is small. This is the so-called $i\delta_k$ model where $\tilde{n}_k/n_0 = (e\tilde{\Phi}_k/T_e)(1 - i\delta_k)$ linearly relates the density and potential $(\tilde{\Phi})$ perturbations for a given wave number k. The potential is advanced by the ion continuity equation which includes polarization and $\tilde{E} \times B$ cross-field motion as

---

*This is a report of work sponsored by the U.S. Department of Energy under Contract No. DE-AC03-89ER53277.

well as parallel ion motion. The latter is advanced by the ion parallel momentum equation. We also treat toroidicity or field curvature by retaining the divergence of $\tilde{E} \times B$ motion which linearly couples modes with the same n but differing m. We restrict our discussion here to cold ions. Hot ion effects including finite Larmor radius (FLR) and coupling to an ion pressure equation have no important effect on electron drift turbulence. The simulation code with hot ions and an ion pressure equation was previously employed for limited three-dimensional (m, n, r) global simulations of ion temperature gradient mode turbulence in sheared and curved magnetic geometry.[5]

Previous simulations of electron drift wave simulations in a sheared magnetic field have been restricted to single helicity (essentially two-dimensional studies with one m/n value) and severely limited parametric variation. These have included a simulation of the $i\delta$ model[6] and a simulation with physical electron dynamics corresponding to collisional drift waves.[7] The former emphasized the possibility of subcritical turbulence, i.e., turbulence below the linear threshold for shear stabilization. The latter emphasized the importance of nonlinear electron dynamics in regions with strong deviation from the Boltzmann response as well as the need for special numerical algorithms to treat the stiff nature of the electron dynamics. In contrast to these earlier single helicity studies which are restricted to sheared slab geometry, the present multiple helicity addresses a computational demanding question: how does electron drift turbulence level and transport scale with shear and with toroidicity?

2. MODEL EQUATIONS

In the electrostatic limit, a suitable fluid model for the drift mode turbulence with cold ions is given by the ion continuity equation, the ion parallel momentum equation. The electrons are assumed to have a nearly Boltzmann response so that $\tilde{n}/n = (e\tilde{\Phi}/T_e) \cdot (1 - i\delta_k)$. In normalized units, these equations are

$$\frac{\tilde{d}}{dt} \left[\tilde{\phi}_k \left(1 - i\delta_k \right) + k_\perp^2 \rho_{s0}^2 \tilde{\phi}_k \right] = - i\hat{\omega}_* \left(\hat{1} - \tilde{g} \right) \tilde{\phi}_k - ik_\parallel \tilde{U}_k \quad , \tag{1}$$

$$\frac{\tilde{d}}{dt} \tilde{U}_k = - ik_\parallel \tilde{\phi}_k - \mu_\parallel k_\parallel^2 \tilde{U}_k \quad , \tag{2}$$

where

$$i\hat{\omega}_* \, \tilde{g} \, \tilde{f}mn = \frac{a}{R} \, i\rho_{s0} \left[\left(\frac{m}{r_*} \right) \left(\tilde{f}_{m+1,n} + \tilde{f}_{m-1,n} \right) \right.$$

$$\left. + \frac{\partial}{\partial r} \, \tilde{f}_{m+1,n} - \frac{\partial}{\partial r} \, \tilde{f}_{m-1,n} \right] \quad , \tag{3}$$

$$\frac{\tilde{d}}{dt} \, \tilde{f}mn = \frac{\partial}{\partial t} \, \tilde{f}mn + \mu_\perp k_\perp^2 \rho_{s0}^2 \, fmn + \rho_{s0}^2 \sum_{m_1 n_1} \left[\frac{\partial}{\partial r} \, \tilde{\phi}_{m_1 n_1} \frac{im_2}{r_*} \, \tilde{f}_{m_2 n_2} \right.$$

$$\left. - \frac{im_1}{r_*} \, \tilde{\phi}_{m_1 n_1} \frac{\partial}{\partial r} \, \tilde{f}_{m_2 n_2} \right] \quad , \tag{4}$$

where $m_2 = m - m_1$ and $n_2 = n - n_1$. The mode labels are given are $\tilde{f}_k = \tilde{f}_{mn}(r)$ $= \tilde{f}^*_{-m-n}(r)$ where m and n are the poloidal and toroidal harmonic numbers. The units for time and length are $(c_{s0}/a)^{-1}$ and $a = 1$ where $c_{s0} = (T_0/M_i)^{1/2}$.

The units for the fields are (ρ_{s0}/a) where $\rho_{s0} = c_{s0}/\Omega_i$, $\Omega_i = eB/M_i c$, $\tilde{\phi} = (e\tilde{\Phi}/T_0)/(\rho_{s0}/a)$, $\tilde{U} = (\tilde{u}_\parallel/c_s)/(\rho_{s0}/a)$, and $\hat{1} = a/L_n(r)$. In addition

$$k_\perp^2 = (m/r_*)^2 - (1/r)\,(\partial/\partial r_*)\,(r_*\partial/\partial r) \quad ,$$
$$\hat{\omega}_* = k_\theta \rho_{s0} = (m/r_*)\rho_{s0} \quad ,$$
$$k_\parallel = [a/R\,q(r)]\,[nq(r) - m] \quad .$$

The model for the Boltzmann deviation is

$$\delta_k^{(r)} = k_\theta \rho_{s0}\, \delta_0 \; exp\left[- \left(r - r_k\right)^2 \Big/ \rho_{s0}^2 \,\delta_1^2\right] \quad , \tag{5}$$

where r_k is the location of the mode rational surface where $k_\parallel = 0$.

For a "uniform slab" geometry avoiding cylindrical effects, we take $r_* = \frac{1}{2}$, otherwise $r_* = r$. To obtain a slab of uniform shear length L_s, we can use the q profile $1/q(r) = (1/q_1 - 1/q_0)r + 1/q_0$. $k_\parallel$ can then be written $k_\parallel = +k_\theta(r - r_{mn})/L_s$ where $L_s = -(1/q_1 - 1/q_0)r_*/R = \hat{s}/Rq$ with $\hat{s} = r^* d\ell n q/dr$. The location of the singular surfaces is $r_{mn} = r_*(1/q_0 - n/m)L_s/R$. The distance between surfaces at constant m is $\Delta_m = (L_s/R)(1/k_\theta)$ and at constant n it is $\Delta_n = (L_s/R)(1/k_\theta)n/(m+1) \approx 1/\hat{s}k_\theta$ with $q \simeq m/n \simeq (m+1)/n$. While L_s is uniform, $\hat{s} = Rq/L_s$ clearly is not. In the slab geometry, $r = 0$ and 1 have no particular significance and we can extend above or below these points.

The terms in the ion continuity equation [Eq. (1)] are as follows: the first and third terms from left to right are from the $\tilde{E} \times B$ convection of density; the second term is the divergence of polarization drift; the fifth term is the divergence of parallel flow. The fourth term corresponds to the divergence of $\tilde{E} \times B$ motion have the tokamak curvature operator $\vec{g}$ given by Eq. (3). The tokamak magnetic field $B/(1 + r/R\cos\theta)$ couples the mth to the $m\pm 1$ harmonic. The parallel flow is advanced by Eq. (2). Only nonlinear $\tilde{E} \times B$ convective motion is retained via the time derivative [Eq. (4)].

The parallel momentum equation contains parallel dissipation corresponding to ion Landau damping:

$$\mu_\parallel = (v_{ith}^2/|\omega_{0k}|) \cdot \left\{|\omega_{0k}|/|k_\parallel|v_{ith}\right\}_{min} \quad , \tag{6}$$

where $-i\omega_{0k} + \gamma_{0k} = \partial/\partial t(\ell n \tilde{\phi}_k)$ is the instantaneous frequency, and $v_{ith} = 2^{1/2}$ is the ion thermal velocity in c_{s0} units. (Here we retain hot ions.) In effect, the ion collision frequency ν_{ii} is replaced with $|\omega_{0k}|$ in parallel viscosity v_{ith}^2/ν_{ii} which is valid only $\nu_{ii} > |\omega_{0k}|$. For $k_\parallel$ larger than the ion Landau resonance value ω_{0k}/v_{ith}, momentum is damped at the flux limiter flow value.

The cross-field dissipation $\mu_\perp$ in Eq. (4) physically results from collisional magnetized ion viscosity at very short scales. While made artificially large, it is believed not to affect the scaling beyond logarithmic factors. This should be the case so long as the artificial dissipation does not truncate the production range of unstable modes. This is a problem for cold ion electron drift modes since the truncation of the driving spectrum at high k_θ is lost with the neglect of FLR. We have avoided the artificial truncation from the viscous damping $\mu_\perp k_\perp^2 \rho_{s0}^2$ by replacing this isotropic damping with $\mu_x k_x^2 \rho_{s0}^2$. This viscous damping has little effect on the leading electron instability but damps the large k_x perturbations as required for numerical stability.

The $\tilde{E} \times B$ plasma diffusivity is our primary interest. This is simply the flux surface average of the density perturbation $(\tilde{n})$ correlated with the radial $\tilde{E} \times B$ velocity perturbations $\tilde{V}_E$ divided by the density gradient

$$\hat{D}(r) = \sum_k \left[-\frac{im}{r*}\,\rho_{s0}\,\tilde{\phi}_k(r)\right]^* \left[-i\delta_k(r)\,\tilde{\phi}_k(r)\right] \Big/ \hat{1} \quad , \tag{7a}$$

in units of $\rho_{s0}^2\,c_s/a$. We also quote the turbulence level in terms of the rms density perturbations

$$\hat{N}(r) = \left[\sum_k \left|1 - i\delta_k(r)\right|^2 \tilde{\phi}_k^*(r)\tilde{\phi}_k(r)\right]^{1/2} \quad . \tag{7b}$$

3. LINEAR DISPERSION RELATION

Following the early work of Pearlstein and Berk,[8] it is straightforward to obtain the linear dispersion relation for Eqs. (1) and (2) in the case of a uniform shear slab without the curvature effects of $\vec{g}$. The linear normal modes are labeled by m and localized by around each singular surface r_{mn} as a function of $x = r - r_{mn}$. n is a redundant label giving only the location of the singular surfaces. It is convenient to use ρ_{s0} as the cross-field normalization length so that $k_\theta \rho_{s0} = (m/r_*)\rho_{s0} = \hat{k}_\theta$ and $x/\rho_{s0} = \hat{x}$. The well known Weber equation is obtained

$$\frac{\partial^2}{\partial \hat{x}^2}\,\tilde{\phi}(\hat{x}) - \Lambda\tilde{\phi}(\hat{x}) + \mu^2\hat{x}^2\tilde{\phi}(\hat{x}) = 0 \quad . \tag{8}$$

The outgoing wave energy solutions[8] are $\tilde{\phi}_\ell(\hat{x}) = H_\ell(\sqrt{i\mu}\,\hat{x})e^{-i\mu\hat{x}^2/2}$ where H_ℓ is the Hermite polynomial. These require the dispersion relation $\Lambda + i(2\ell+1)\mu = 0$. In the cold ion case, $\mu = (\hat{\omega}_*\,\hat{1}/\omega)(L_n/L_s)$ and $\Lambda = 1 + \hat{k}_\theta^2 - i\hat{\delta} - \hat{\omega}_*\,\hat{1}/\omega$. $\hat{\delta}_k = \langle\tilde{\phi}(\hat{x})\delta_k(\hat{x})\tilde{\phi}(\hat{x})\rangle/\langle\tilde{\phi}(\hat{x})\tilde{\phi}(\hat{x})\rangle$ is the perturbation theory solution where δ is a function of x. The dispersion relation can be rewritten as

$$\omega = \hat{\omega}_*\,\hat{1}[1 - i(2\ell+1)(L_n/L_s)] \Big/ \left[1 + \hat{k}_\theta^2 - i\hat{\delta}_k\right] \quad . \tag{9a}$$

For the least stable $\ell=0$ mode, $\hat{\delta}_k = \hat{k}_\theta\delta_0/[1 - i(L_s/L_n)/\delta_1^2]$ and the distance to the first zero (or longest quarter wavelength) is

$$\Delta x = \rho_s\,(\pi \cdot L_s/L_n \cdot |\omega|/\omega_*)^{1/2} \quad . \tag{9b}$$

In the case of toroidal coupling, n is the mode label, but we can pick a "reference" singular surface with $q = m/n$ and label the other surfaces with the same n as $m + j$ with $j = 0, \pm 1, \pm 2$

$$\tilde{\phi}(\hat{x},\theta) = \sum_{j=-\infty}^{\infty} \tilde{\phi}_j(\hat{x})\,exp\,[i(m+j)\theta - in\phi - i\omega t] \quad . \tag{10}$$

A system of coupled equations is obtained for $\tilde{\phi}_j(\hat{x})$ (see Ref. 9 and original references therein)

$$\frac{\partial^2}{\partial \hat{x}^2}\,\tilde{\phi}_j(\hat{x}) - \Lambda\tilde{\phi}_j(\hat{x}) + \mu^2\left(\hat{x} - j\hat{\Delta}_n\right)^2\tilde{\phi}_j(x) - \frac{M}{2}\left\{\left[\tilde{\phi}_{j+1}(\hat{x}) + \tilde{\phi}_{j-1}(\hat{x})\right]\right.$$

$$\left. + \hat{k}_\theta^{-1}\,\frac{\partial}{\partial \hat{x}}\left[\tilde{\phi}_{j+1}(\hat{x}) - \tilde{\phi}_{j-1}(\hat{x})\right]\right\} = 0 \quad . \tag{11}$$

$\hat{\Delta}_n = 1/\hat{k}_\theta \hat{s}$ is the ρ_{s0} normalized spacing between singular surfaces at fixed n. In the cold ion limit, $M = 2\,\epsilon_n\,\hat{\omega}_*\,\hat{1}/\omega$ where $\epsilon_n = L_n/R$. This term results from the divergence of $\tilde{E} \times B$ motion [second part of the third term in Eq. (1)]. If we ignore the only remaining nonuniformity from $\hat{s}$, then a system allowing all m and n is translationally invariant and we can write $\tilde{\phi}_j(\hat{x}) = e^{i\theta_0 j}\phi_0^{(\theta_0)}(\hat{x} - j\hat{\Delta}_n)$. When this is substituted into Eq. (11), a differential-difference equation is obtained for $\phi_0^{(\theta_0)}(\hat{x})$, i.e., there is coupling to $\phi_0^{(\theta_0)}(\hat{x} \pm \hat{\Delta}_n)$. It is convenient to Fourier transform this equation to $\hat{k}_x$ space. We obtain a form of the ballooning mode equation[10]

$$- \hat{k}_x^2 \phi^{(\theta_0)}\left(\hat{k}_x\right) - \Lambda\phi^{(\theta_0)}\left(\hat{k}_x\right) - \mu^2\,\frac{\partial^2}{\partial \hat{k}_x^2}\,\phi^{(\theta_0)}\left(\hat{k}_x\right)$$

$$- M\left[\cos\left(\theta_0 + \hat{k}_x\hat{\Delta}_n\right) + \hat{s}\,\hat{k}_x\hat{\Delta}_n\,\sin\left(\theta_0 + \hat{k}_x\hat{\Delta}_n\right)\right]\phi^{(\theta_0)}\left(\hat{k}_x\right) = 0 \quad. \tag{12}$$

θ_0 is the ballooning poloidal angle with $\theta_0 = 0(\pi)$ corresponding to outward (inward) ballooning. Equation (12) can be written as $-\partial^2/\partial\hat{k}_x^2\phi(\hat{k}_x) + V(\hat{k}_x)\phi(\hat{k}_x) = 0$ where $V(\hat{k}_x)$ is the "potential."

A Weber equation can be recovered from Eq. (12) by expanding around the extrema of the toroidal coupling term $\hat{k}_x = 0$. The least and most stable modes are near $\theta_0 = 0$ or π. These are closest to the "slab-like" branch. When the potential $V(\hat{k}_x)$ has an extrema at $\hat{k}_x \neq 0$, $\phi(\hat{k}_x)$ can be localized there and another "toroidicity induced" branch can be found.[11] We shall not treat that case here. The resulting Weber equation is

$$- \xi_\pm\hat{k}_x^2\phi_0^\pm\left(\hat{k}_x\right) - \Lambda'_\pm\phi_0^\pm\left(\hat{k}_x\right) - \mu^2\,\frac{\partial^2}{\partial\hat{k}_x^2}\,\phi^\pm\left(\hat{k}_x\right) = 0 \quad, \tag{13a}$$

where

$$\xi_\pm = \left[1 \pm 2\,\epsilon_n(\hat{\omega}_*\hat{1}/\omega)\left(\hat{s} - \tfrac{1}{2}\right)\hat{\Delta}_n^2\right] \quad, \tag{13b}$$

and

$$\Lambda'_\pm = 1 + \hat{k}_\theta^2 - i\hat{\delta}_k - \left(\hat{\omega}_*\hat{1}/\omega\right)\left(1 \mp 2\,\epsilon_n\right) \quad. \tag{13c}$$

The signs $\pm$ refer to $\theta_0 = 0(\pi)$. The normal mode solution is $\phi_0^\pm = H_\ell[(-i\xi_\pm^{1/2}/\mu)^{1/2}\hat{k}_x]\,exp(+i\hat{k}_x^2\xi_\pm^{1/2}/2\mu)$, provided the dispersion relation $\Lambda'_\pm + i(2\ell + 1)\mu\xi_\pm^{1/2} = 0$ is satisfied.

The "uniform" linear dispersion relation for the model including the effects of $\vec{g}$ is

$$\omega^\pm = \hat{\omega}_*\,\hat{1}\left[(1 \mp 2\,\epsilon_n) - i(2\ell + 1)(L_n/L_s)\,\xi_\pm^{1/2}\right]\Big/\left(1 + \hat{k}_\theta^2 - i\hat{\delta}_k\right) \quad, \tag{14a}$$

where

$$\xi_\pm = \left[1 \pm \frac{\hat{\omega}_*\hat{1}}{\omega}\,2\,\epsilon_n\left(\hat{s} - \tfrac{1}{2}\right)\hat{\Delta}_n^2\right] \quad. \tag{14b}$$

Transforming back to $\hat{x}$ space, the distance to the first zero of the poloidal harmonics (or reduced wave functions) for the $\ell=0$ mode is

$$\Delta_{\hat{x}}^\pm = \rho_s\left[\pi \cdot L_s/L_n \cdot |\omega|/\left(\hat{\omega}_*\hat{1}\right)\cdot|\xi_\pm|^{1/2}\right]^{1/2} \quad. \tag{14c}$$

Clearly, the toroidicity ϵ_n has two effects: first, to shift the frequency (and growth rate) up or down by $1 \mp 2\,\epsilon_n$, and second, the shear stabilization can be much smaller if $\xi_\pm < 1$ or even removed if $\xi_\pm < 0$. The outward ballooning modes are downshifted in frequency and have less shear damping where $\hat{s} < \frac{1}{2}$ and the inward ballooning modes have less shear damping where $\hat{s} > \frac{1}{2}$ and are upshifted in frequency.

4. NUMERICAL ILLUSTRATION OF A STANDARD CASE

We must first describe the standard case numerical parameters for the multiple helicity uniform sheared slab. The "radius" of the slab is spanned by 128 grids from $-0.25 < r < 1.25$ giving a spacing $\Delta r = 0.0117$. The $1/q$ profile is linear as noted in Section 2 with $q_0 = 0.8$ and $q_1 = 3.4$ at $r = 0$ and 1. (m, n) poloidal and toroidal harmonic modes with helicities m/n between $q = 1$ and 2 are retained. This gives an effective thickness for transport in the slab of 0.526 between the $q=1$ and 2 surfaces at $r = 0.262$ and 0.788. The slab is centered near $r_* = 0.5$. The maximum m is 23, giving 155 complex (m, n) harmonics. The value $\rho_{s0} = 0.032$ is chosen so maximum $k_\theta \rho_s 0 = \hat{k}_\theta = 1.5$ (recall $k_\theta = m/r_*$). $R = 3$ so that $L_s^{-1} = 0.159$. $\hat{s} = Rq/L_s$ varies from 0.48 to 0.95 between $q = 1$ and 2. $\hat{1} = 1/L_n = 1.0$. The Boltzmann derivation parameters are $\delta_0 = 1.2$ and $\delta_1 \rho_{s0} = 0.1$. For these parameters, Eq. (12) shows that modes with $k_\theta \rho_{s0} < 0.182\,(2\ell + 1)/\delta_0$ are stable. $\mu_x = 0.2$ is the standard case artificial damping coefficient. Without toroidal coupling, the distance to the first zero of the leading normal mode is $\Delta_x = 0.142(|\omega|/\omega_*)^{1/2}$. The distance between singular surfaces at fixed m is $\Delta_m = \rho_{s0}(L_s/R)/(k_\theta \rho_{s0}) = 0.067/k_\theta \rho_s$. The distance at fixed n is $\Delta_n = \rho_{s0}/(\hat{s}\,k_\theta \rho_{s0}) = (0.067 \text{ to } 0.03)/(k_\theta \rho_{s0})$ from $q = 1$ to 2. It should be made clear that since $k_\theta \rho_{s0}|max$ and (as we will show) the spectral average $k_\theta \rho_s$ are invariant, the mode widths and singular surface spacings both scale with ρ_s. Thus, going to larger maximum m, smaller ρ_{s0}, and necessarily smaller Δr does not change these length ratios. As long as ρ_{s0}/a is small there is no advantage to making m_{max} larger except to go to larger $k_\theta \rho_{s0}|max$.

We now proceed to illustrate the standard case simulation without toroidal coupling ($\hat{g} \equiv 0$). Figure 1 shows the time average radial profile and $r = r_*$ time history for the turbulence level $(\hat{n}/n)_{rms}$ normal to ρ_{s0}/a [$\hat{N}$ Eq. (7b)] and diffusion coefficient D normed to $\rho_s^2 c_s/a$ [$\hat{D}$ Eq. (7b)]. Time is in a/c_s units. The time steps are always smaller than 0.05 as required to satisfy the Courant stability limit $\tilde{V}_E \cdot \Delta t/\Delta r \ll 1$. $\tilde{V}_E$ is the radially directed $\tilde{E} \times B$ velocity. The fastest linear wave period is 4.8 and the slowest is 96.3. The simulations are run over periods long enough to obtain "good statistical averaging" (in this case, $t = 150$ to 700). We shall generally refer to space average of $\hat{N}$ and $\hat{D}$ between the $q=1$ and 2 surfaces, i.e., the effective thinness of the slab, and give some indication of the statistical uncertainty. In addition to the standard case $\delta_0 = 1.2$, runs at larger $\delta_0 = 1.6$ and smaller during $\delta_0 = 0.8$ are shown. The sheared slab result should be invariant under $q \to 2 \times q$ and $R \to \frac{1}{2} \times R$. This is verified by the dotted line. Note this variation is numerically identical to averaging the mode spacing from $\Delta n = 1$ to $\Delta n = 2$. These should then be the same if the turbulence in the 0 to πR sector of the doubly periodic slab is the same on time average as in the πR to $2\pi R$ sector. The dashed curves show a simple helicity simulation about the $q=1$ surface for the $\delta_0=0.8$ case. Surprisingly there is little numerical advantage to single helicity. There are fewer modes (in this case 23 as opposed to 155) so that the computational cost per time step is much reduced [$\sim (23/155)^2$ smaller]. However, this makes for much larger statistical fluctuations, particularly as the

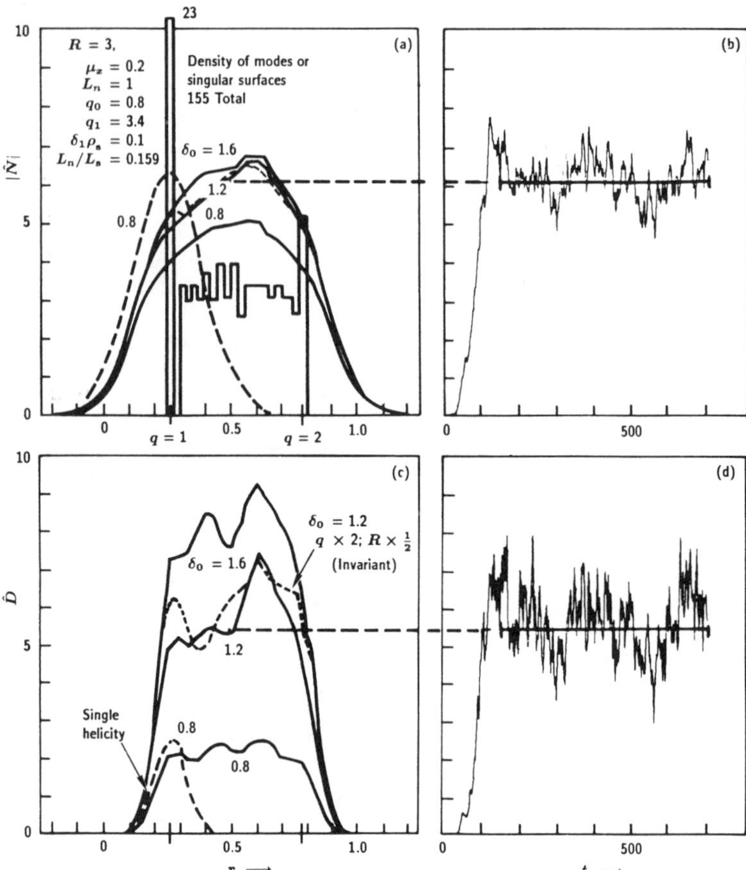

Fig. 1. Standard case turbulence and transport rms radial profiles and time histories. Density fluctuations normed to ρ_s/a in (a) and (b). Diffusion coefficient normed to $(c_s/a)\cdot\rho_s^2$ in (c) and (d). Density of 155 singular surfaces shown in (a).

turbulence intermittently settles into the low m state with very large amplitude. This means Δt must be smaller to prevent excursions beyond the Courant limit and the case must be run much longer to obtain good statistics.

Figure 2 shows the spectral properties for the standard case. Figure 2 (a) and (b) show the time average m-spectral contributions to $\hat{N}$ and $\hat{D}$, respectively [see Eq. (7)] at radii corresponding to $q = 1$, $q^* = 1.29$ at $r = r_*$ and $q = 2$. We shall take the peak value of $m = 6$ as the characteristic $k_\theta\rho_s = 0.4$. We can also define rms poloidal wavenumber values:

$$k_\theta^N \rho_s(r) = \left[\sum_m \left(k_\theta\rho_s\right)^2 \hat{N}_m^2(r) \bigg/ \sum_m \hat{N}_m^2(r)\right]^{1/2} , \qquad (15a)$$

$$k_\theta^D \rho_s(r) = \left[\sum_m \left(k_\theta \rho_s\right)^2 \hat{D}_m(r) \Big/ \sum_m \hat{D}_m^2(r) \right]^{1/2} \quad . \qquad (15b)$$

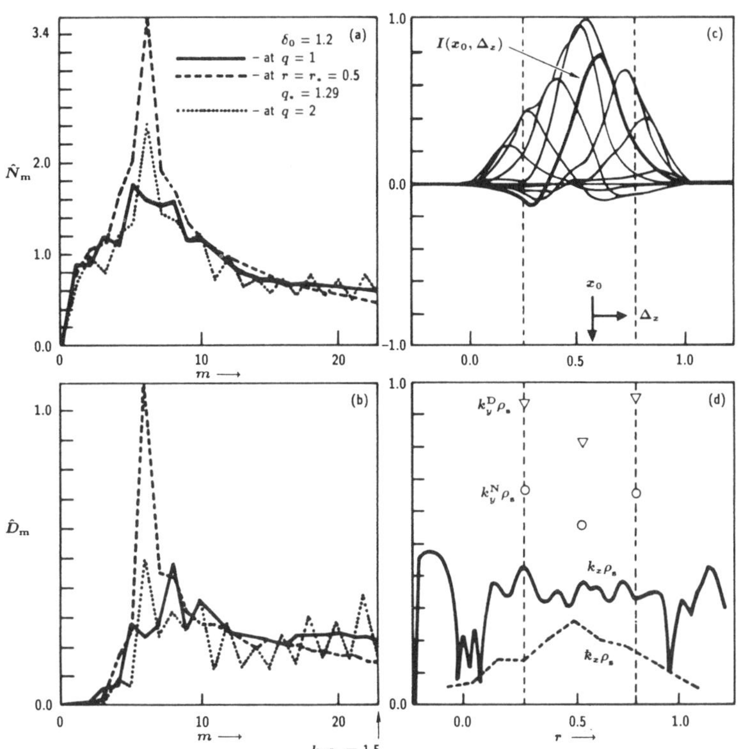

Fig. 2. Spectra and correlation functions for the standard case. Normed m-spectra of density perturbations in (a) and contributions to transport in (b). The density correlation function in (c) and the spectral average wave numbers defined in Eqs. (15) and (16) shown in (d).

From Eq. (7b) and the fact that the $i\delta$ model weights larger values of $k_\theta \rho_{s0}$, it is not surprising that $k_\theta^D \rho_s$ is larger than $k^N \rho_s$. We can also define a two point radial correlation function $I(x, \Delta x) = \langle \hat{N}(x_0 + \Delta x) N(x) \rangle$ where $\langle \ \rangle$ denotes time average, x_0 is the reference point, and $x_0 + \Delta x$ is the separation point. $I(x_0, \Delta x)$ is shown for several values of x_0 in Fig. 2 (c). There are at least two ways to define an inverse radial correlation length:

$$k_x \rho_s(x_0) = \left[-\rho_s^2 \frac{\partial^2}{\partial \Delta x^2} \ I(x_0, \Delta x) \Big/ I(x_0, \Delta x) \right]^{1/2}_{\Delta x \to 0} \quad , \qquad (16a)$$

$$\tilde{k}_x \rho_s\left(x_0\right) = \left[\rho_s^2 \int d\Delta x \, I\left(x_0, \Delta x\right) \bigg/ \int d\Delta x \, \Delta x^2 \, I\left(x_0, \Delta x\right)\right]^{1/2} \quad , \quad (16b)$$

These are identical for a Gaussian weighting in Δx. However, $\tilde{k}_x$ is smaller than k_x because it picks up the small but important long distance correlation apparent in Fig. 2 (c). Finally, Fig. 2 (d) shows a comparison of k^D, k^N, k_x, and $\tilde{k}_x$ at various radii.

5. MIXING LENGTH DESCRIPTION OF NUMERICAL SIMULATIONS

The conventional heuristic guide to describing the nonlinear saturation of the density fluctuations $(\tilde{n}/n)$ and turbulent transport coefficient (D) is given by the mixing length rules. For electrostatic drift wave turbulence driven by a density gradient $(1/L_n = d\ell n\, n/dr)$, these are $\tilde{n}/n = \Theta^{1/2}/k_x L_n$ and $D = \Theta \gamma^d/k_x^2$ where $\Theta = \{\gamma/\varpi, 1\}_{min}$.[12] The characteristic inverse wave number $1/k_x$ is the mixing length Δ_x^{ML} and γ^d is the driving rate. For the $i\delta$ model, $\gamma_k^d = \omega_* \delta_k/(1+\delta_k^2)$. $\omega_* = k_\theta \rho_s \cdot c_s/L_n$ is the drift wave frequency and k_θ is the characteristic poloidal wave number $[\rho_s = c_s/(eB/M_i c), \, c_s = (T_e/M_i)^{1/2}]$. From the definition of $\check{E} \times B$ plasma flow and the assumed linear relation between $\tilde{\phi}$ and $\tilde{n}$ in the $i\delta$ model, it follows directly and without approximation that $D = \gamma^d L_n^2 (\tilde{n}/n)^2$. In the limit of strong turbulence $\gamma/\varpi \geqslant 1$, where γ is the typical growth rate and ϖ is a characteristic frequency $(\omega - \omega_*)$, $\tilde{n}/n = 1/k_x L_n$ is heuristically justified as the maximum fluctuation required to flatten the density gradient over a mixing length or equivalently for the fluctuating $\tilde{E} \times B$ velocity to cancel the diamagnetic drift. The dependence of the saturation level on γ and the transition to weak turbulence $\gamma/\varpi \leqslant 1$ is a nontrivial feature of the mixing length rule. We find clear evidence of the weak turbulence factors (Θ) provided we properly interpret the mixing length Δ_x^{ML} (see Fig. 3).

The mixing length rules are of little use for estimating the transport unless we know the mixing length $1/k_x = \Delta_x^{ML}$ and the characteristic wave number k_θ to determine the driving rate γ^d. In the case of a sheared slab, it is well known that the shear localizes the normal modes near the singular surfaces where $k_\parallel \sim 0$.[8] It is natural to assume the mixing length is a characteristic length of the normal mode, but should it be the coherence length or the longest wavelength associated with the mode? The coherence length is the mode envelope length. This is the same as the rms derivation from the singular surface (or the inverse rms wave number) given by $\Delta_x' = \rho_s(L_s/L_n \cdot |\omega|/\omega_* \cdot |\omega|/\gamma)^{1/2}$. When $\gamma/|\omega|$ becomes very small, Δ_x' gets large but it should not exceed the distance to the ion Landau point $\rho_s \cdot L_s/L_n \cdot (|\omega|/\omega_*)$ where $k_\parallel v_{ith} = |\omega|$. The modes oscillate faster with distance from the singular surface and thus contain all wavelengths. The longest quarter wavelength or distance to the first node is $\Delta_x = \rho_s(L_s/L_n \cdot |\omega|/\omega_* \cdot \pi)^{1/2}$. Δ_x' and Δ_x are not easily distinguished except near marginality. The numerical simulations suggest that Δ_x, the longest wavelength is the best description of the mixing length. There is no evidence of the $|\omega|/\gamma$ factor and the $\rho_s(L_s/L_n)^{1/2}$ scaling is clearly evident (see Fig. 4). It should be clear that the mixing length comes in combination with the weak turbulence factor $\Theta^{1/2}\Delta_x^{ML} = \tilde{\Delta}_x^{ML}$. We could also identify $\tilde{\Delta}_x^{ML}$ with $[Re\langle\varphi x^2 \varphi\rangle/\langle\varphi\varphi\rangle]^{-1/2}$ or $[Re\langle\partial\phi/\partial x \, \partial\phi/\partial x/\rangle/\langle\varphi\varphi\rangle]^{-1/2} \propto (\gamma/\omega_*)^{1/2}\rho_s(L_s/L_n)^{1/2}$ [$\langle\,\rangle$ represents spatial average and φ is the normal mode wave function].

While it may suffice to know the mixing length $1/k_x$ to determine the scaling of the turbulence level $\tilde{n}/n$, we must also know the characteristic (or spectral

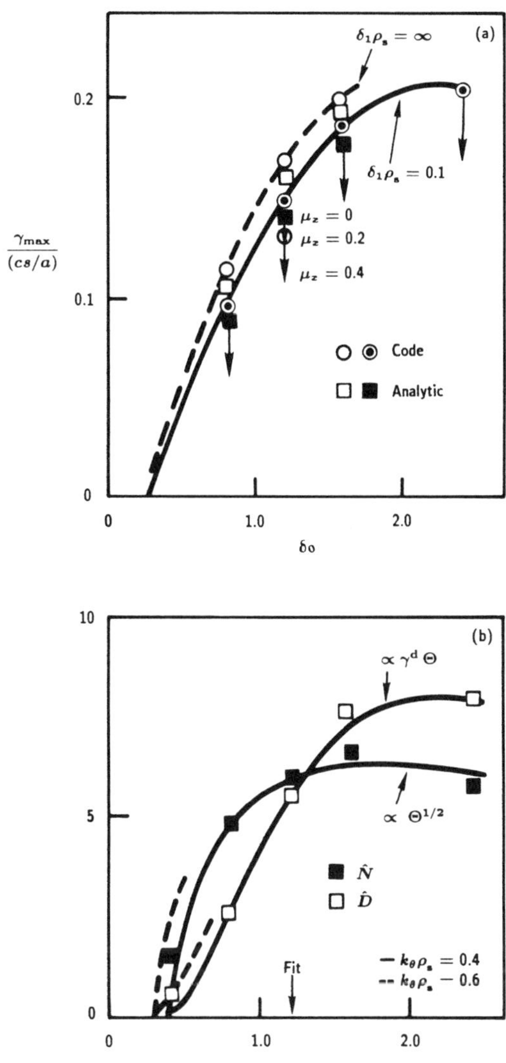

Fig. 3. Maximum growth rate (a) and normed density fluctuation
and transport levels (b) versus driving strength.

average) poloidal wave number k_θ to determine the scaling of the transport co-
efficient D. For our model in particular $\delta_k = k_\theta \rho_s \delta_0 \, exp[-x^2/(\delta_1 \rho_s)^2]$ with x the
distance from the singular surface, we expect $D \propto (c_s/L_n) \cdot \rho_s^2 (k_\theta \rho_s)^2 \delta_0/(k_x \rho_s)^2$.
Since wave-wave coupling tends to isotropize turbulence in the direction perpen-
dicular to the magnetic field, we might conjecture that k_θ (or k_y) scales like
k_x. Indeed, two-dimensional (k_x, k_y) representations even with driving on the
k_y axis show isotropization for strong turbulence[2] and in both simulations and
experiments,[13] k_θ and k_x are numerically comparable. In this case, D (for our

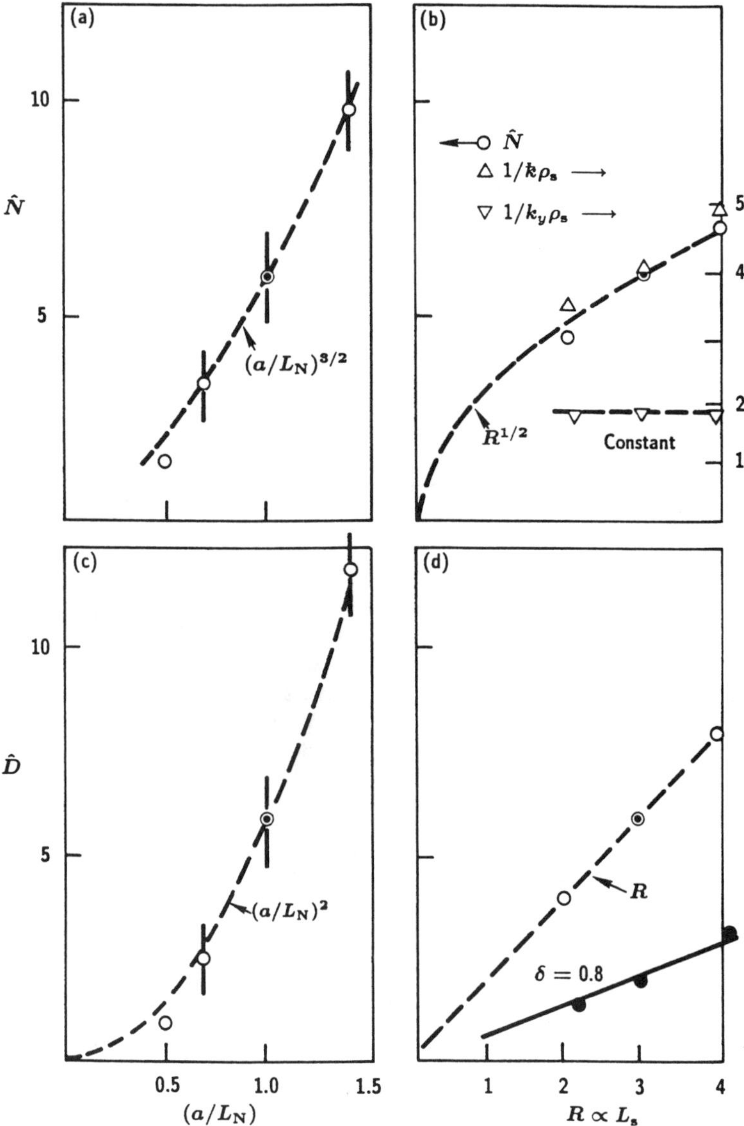

Fig. 4. Scaling of normed fluctuation levels [(a) and (b)] and transport [(c) and (d)] versus density gradient and shear lengths.

model) would be largely independent of shear by the cancellation of k_θ and k_x, whereas in fact the sheared slab simulations clearly show $D \propto L_s$. The simulations show the average $k_\theta\rho_s$ to be constant (0.4) nearly independent of all parametric variations. Figure 4 (b) shows constancy with respect to L_s and Table I with respect to ρ_s.

Table I: ρ_s Scaling

$\rho_s/0.032$	$k_y\rho_s\vert_{max}$	m_{max}	$\delta_1\rho_s/0.1$	N	D	m_{peak}	k_y^N/k_y^D	$k_y\rho_s$	$\overline{k}_x\rho_s$
1	1.5	23	1	6.0	5.5	6	0.60/0.89	0.35	0.25
1	$\sqrt{2}\times 1.5$	43	1	7.5	5.0	4	0.39/0.63	0.25	0.18
$\sqrt{2}$	$\sqrt{2}\times 1.5$	23	$\sqrt{2}$	7.3	5.0	3	0.37/0.70	0.25	0.20

It is worth noting that the scaling with gyroradius is not entirely trivial. In the large m "continuum limit," the nonlinear drift wave equations are functions only of the combination $k_\theta\rho_s$ (not k_θ and ρ_s separately). However, the simulations (and physical reality) are represented on a "quantized" m,n grid. Magnetized plasma turbulence is nearly two-dimensional and cascades energy to the longest wavelengths. Indeed, for some resistive MHD simulations,[14] the peak mode number is the lowest possible mode number, $n = 1$ with $m = q\cdot n \simeq q$ as the apparent scaling invariant rather than the parametric length invariant predicted by the scale transformation of the "continuum" equations.[15]

To summarize the scaling for the multiple helicity finite periodic sheared slab geometry, we find $\tilde{n}/n \propto \Theta^{1/2}(\rho_s/L_n)(L_s/L_n)^{1/2}$ and $D \propto \Theta\delta_0(c_s/L_n)$ $\cdot \rho_s^2(L_s/L_n)$ with D additionally proportional to δ_1, the support width of the $i\delta$ model. There is no other parametric dependence. In particular, there is no dependence on the relative spacings between singular surfaces $\Delta n/\rho_s$ provided these variations are made at constant slab radial thickness which physically should be invariant. As shown in Fig. 5, packing more or fewer surfaces into the same length at constant L_s does not change the diffusion. Thus, we obtain the scaling corresponding to the infinite sheared slab for which $\Delta n/\rho_s = 0$.

Our results are consistent with the analysis by Tange, Nishikawa, and Sen.[16] These authors developed an approximate renormalization or statistical nonlinear theory of a single helicity hot ion model with $i\delta = 0$. While nonlinear $\tilde{E}\times B$ convection of the parallel ion motion was retained, only nonlinear polarization drift was retained in the ion continuity equation. Since nonlinear $\tilde{E}\times B$ convection of density was neglected by taking $i\delta = 0$, the analysis is restricted to the weak driving (weak turbulence) regime. Nevertheless, the authors were able to make progress by retaining only the large driven shear harmonic normal modes even about the singular surfaces and a small admixture of the least shear damped odd harmonics nonlinearly driven by the large even harmonics. Since the large even harmonics are nonlinearly coupled only to a combination of even and odd harmonics, a tractable statistical theory of the large even modes could be obtained by eliminating the odd modes in terms of the even. They obtain a self-consistent nonlinear spectrum with $k_\theta\rho_s$ almost parametrically constant and a saturation which scales as $\tilde{n}/n \propto \Theta^{1/2}(\rho_s/L_n)\cdot(L_s/L_n)^{1/2}$.

For toroidal geometry, our heuristic notions about the mixing length are much less clear. The effect of toroidicity is parametrically complex. Assuming that the mixing length scales to ρ_s, it can depend on $L_n/L_s = L_n/R\cdot\hat{s}/q$ as in the sheared slab case but additionally on L_n/R and $\hat{s}$ separately. Note $\xi_\pm$ is not a simple multiplicative function of the latter variables. Furthermore, although we can keep L_s and L_n uniform over the simulation region, $\hat{s}$ is not uniform since q is not. In addition, the degree of toroidal coupling, or number of m coupled at each n, varies with n. These are, of course, real features of a physical plasma. Nevertheless, for rather strong toroidicity $\epsilon_n = \frac{1}{3}$ to $\frac{2}{3}$ and strong shear parameter $\frac{1}{2} < \hat{s} < 1$ to $1 < \hat{s} < 2$ where the $i\delta$ electron drift modes are inward ballooning, both γ and ω are upshifted by $(1+2\,\epsilon_n)$ and some shear stabilization

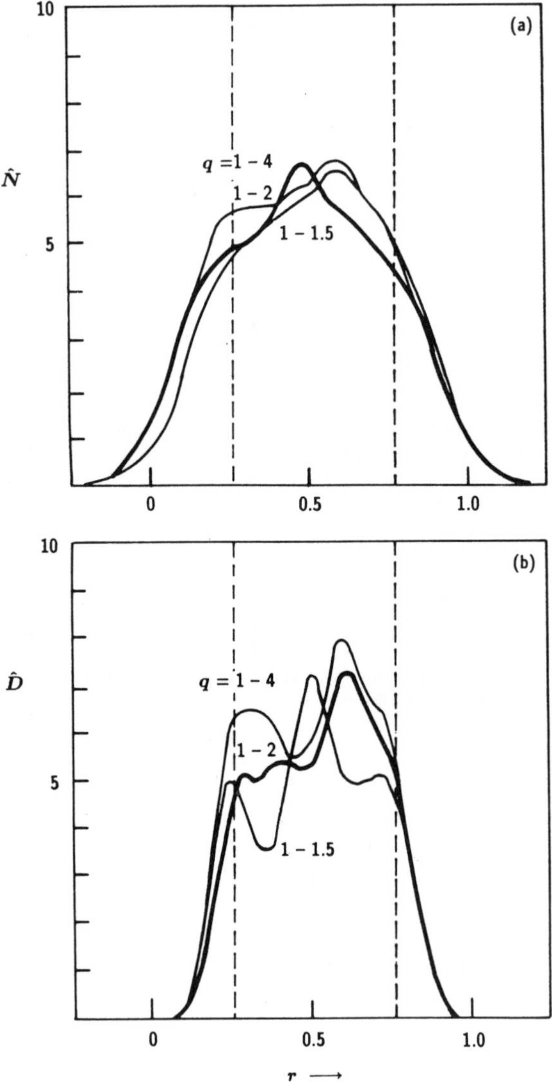

Fig. 5. Dependence of normed density fluctuations (a) and transport (b) on spacing between singular surfaces with effective slab thickness fixed.

remains, we can summarize our results by first examining several possible mixing lengths which do not describe the simulations.

The first possibility is that the turbulence is so strong that it breaks apart the ballooning mode completely so that the poloidal harmonic at a given n cannot linearly couple (line up in phase). There is neither more nor less stability

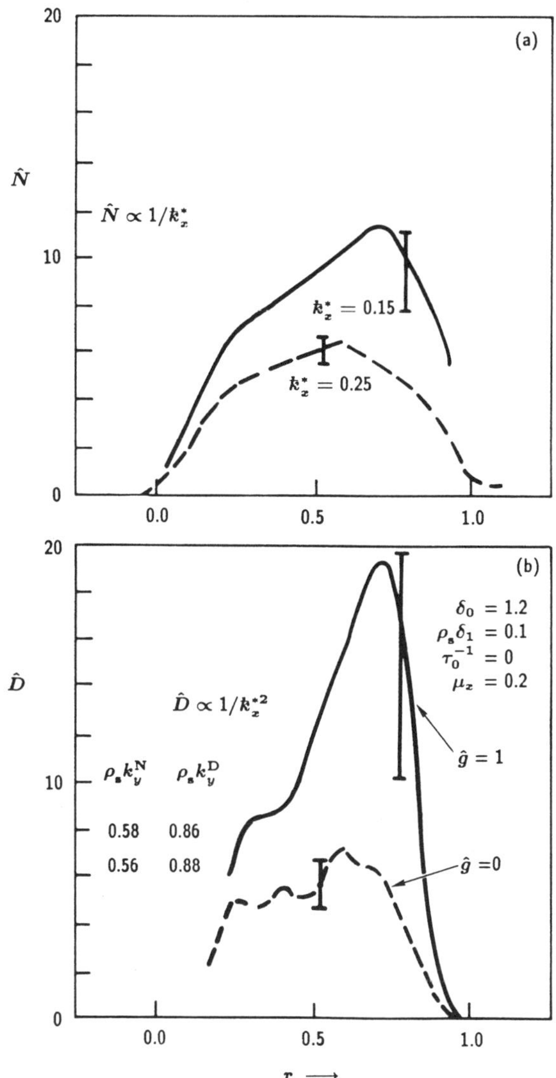

Fig. 6. Normed density fluctuations (a) and transport (b) versus
radii with (solid line) and without (dashed line) toroidal coupling.

than the slab and the mixing length is the same as the slab Δ_x. This seems
not to be the case, as shown in Fig. 6 the simulation shows larger turbulence
and transport as the toroidicity is turned "on" (at fixed L_n, L_s, $\hat{s}$, and ρ_s).
Figure 7 shows the instantaneous potential contours of the turbulence for the
slab [$g = 0$, Fig. 7 (a)] and toroidal [$g = 1$, Fig. 7 (b)] standard cases of Fig. 6.
There is clear evidence of strong turbulence where the modes balloon to the

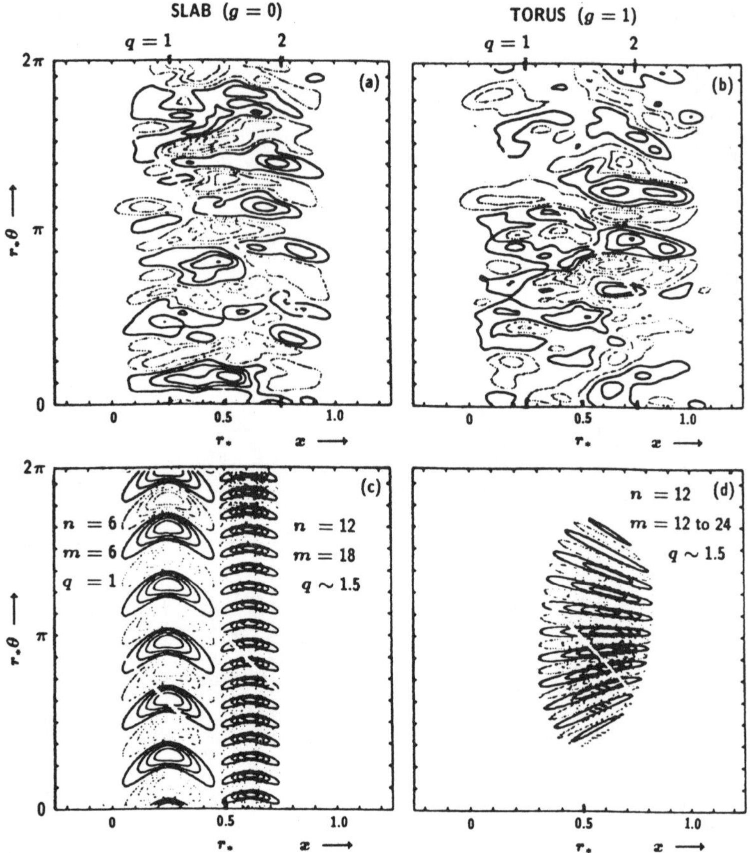

Fig. 7. Turbulent and normal mode potential contours. $\delta_0 = 1.2$, $L_n/L_s = 0.16$, $L_n/R = 1/3$, $a/L_n = 1$, and $\tau^{-1} = 0$.

inside ($\theta = \pi$). Figures 6 (c) and 6 (d) illustrate contours for a few selected linear normal modes to contrast the ballooning character of the torus from the slab. A second possibility is that the mixing length is the characteristic width of the poloidal subharmonics of the ballooning modes $\Delta_x^{\pm}$ as modified by the toroidicity $\xi_{\pm}$. This would seem the most likely *a priori* possibility from the sheared slab result. The frequency upshift $\omega/\omega_* \simeq (1 + 2\epsilon_n)$ would explain the increase in mixing length. However, $\Delta_x^{\pm}$ is proportional to $L_s^{1/2}$ and doubling of L_s at fixed ϵ_n and $\hat{s}$ does not double the transport [see Fig. 8 (a)]. Further, there is no significant change in the average $k_\theta \rho_s$. The third possibility is that only a fraction of the ballooning modes must hang together so that there is coherence over a minimum number of surfaces, roughly 2 or 3. In this case, the mixing length is (2 to 3) Δ_n ($\Delta_n = \rho_s/(k_\theta \rho_s)\hat{s}$). However, doubling $\hat{s}$ at fixed L_s and ϵ_n does not diminish the transport by four-fold and in fact not at all [see Fig. 8 (b)]. Again, there is no apparent change in the average $k_\theta \rho_s$.

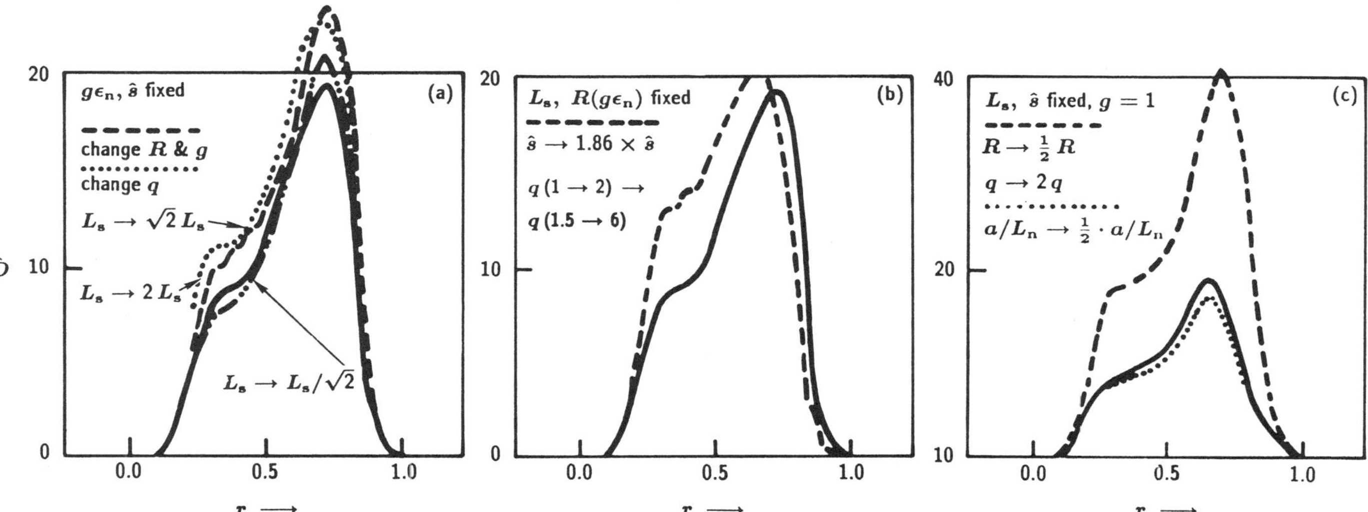

Fig. 8. Scaling of transport with toroidal couplings. Dependence on shear lengths L_s in (a), shear parameter $\hat{s}$ in (b), and toroidicity (c) compared to the standard toroidal case (solid line).

For strong toroidicity and well above linear thresholds, the simulations appear to be described by $\Delta_x^{\mathrm{ML}} \propto \rho_s (L_n/R)^{1/2}$. Consistent with a constant $k_\theta \rho_s$ spectrum, we find $D \propto \delta_1 \delta_0 \, c_s / L_n \cdot (\Delta_x^{\mathrm{ML}})^2$. There appears to be little dependence of Δ_x^{ML} on $\hat{s}$ or q. Figure 8 (c) illustrates the dependence on R and L_n at fixed q and $\hat{s}$. In the weak toroidicity limit $\epsilon_n \to 0$ or $R/L_n \to \infty$, we expect to recover the sheared slab result $\Delta_x^{\mathrm{ML}} \propto \rho_s (L_s/L_n)^{1/2}$. It appears that $L_n D \propto (\Delta_x^{\mathrm{ML}})^2 \propto \rho_s^2 (C \cdot L_n/R + R/L_n \cdot q/\hat{s})$ is a good summary if C is adjusted so that contributions to Δ_x^{ML} are equal at $R/L_n \sim 4$, $q \sim 2$, and $\hat{s} \sim 1$ (see Fig. 9). Thus, apart from any dependence through δ_0, the aspect ratio (R/a) scaling of electron drift turbulence is mixed. The exterior plasma where a/L_n is large contributes to unfavorable R/a scaling, whereas the toroidal region in the interior where L_n/a is large has favorable R/a scaling.

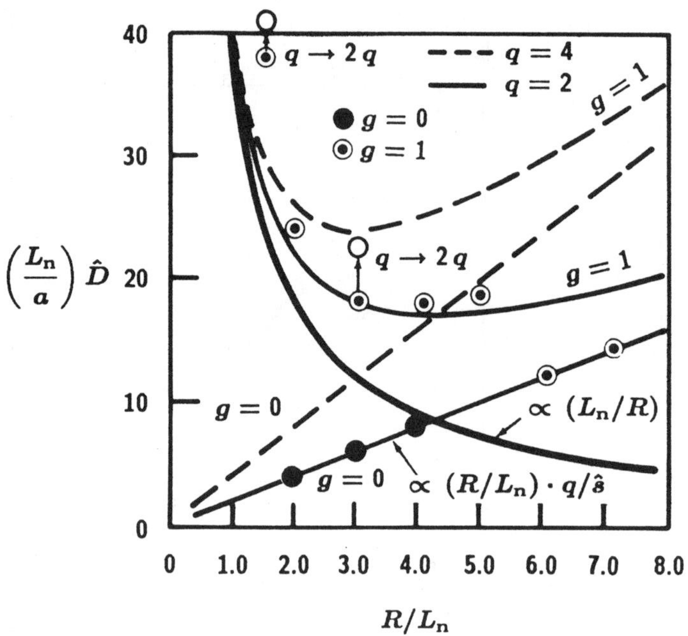

Fig. 9. Normalized diffusion versus local aspect ratio. $g = 0$ corresponds to sheared slab and $g = 1$ to toroidal coupling.

The mixing length scaling in the strong toroidicity limit is not understood quantitatively, particularly its weak dependence on $\hat{s}$ and q. However, the following heuristic and dimensional description may be useful to understand its dependence on toroidicity ϵ_n. We can imagine the ballooning mode as a "link chain" of poloidal harmonics locked in phase. Starting with a single reference harmonic m and n, we obtain a linear growth rate γ_0 corresponding to the shear slab normal mode. If we allow only one link in phase, say m to $m+1$ or m to $m-1$, we obtain the growth rate $\gamma_1 > \gamma_0$. Two links $m-1$, m, $m+1$ gives $\gamma_2 > \gamma_1$; three links $m-1$, m, $m+1$, $m+2$ or $m-2$, $m-1$, m, $m+1$ gives $\gamma_3 > \gamma_2$ and so on. An infinite number of links gives the canonical ballooning mode rate γ_∞.

In fact, the sequence typically saturates after roughly a half-dozen lengths. The difference $\gamma_\infty - \gamma_0$ increases with ϵ_n and represents the coupling strength of the chain. When the chain is imbedded in a turbulent bath with phase decorrelation rate $\Delta\omega$, the chain is broken into L links such that $(\gamma_L - \gamma_0) = (\gamma_\infty - \gamma_0) - \Delta\omega$. The larger ϵ_n, the stronger the chain and more links can survive as a phase correlated unit. Since the mixing length can be associated with the number of links and it will increase with ϵ_n. This is a variant on the third possible mixing length described above. Rather than only the minimum one link ($\Delta_x \propto \Delta_n$) surviving the turbulence, many links may survive depending on the toroidicity.

To understand why the mixing length is specifically proportional to $(L_n/R)^{1/2}$ for large L_n/R, it is useful to note that the strong toroidal limit can be approached by small R or large L_n. For vanishing density gradient $1/L_n$, the ballooning mode growth rates and all its characteristic lengths remain finite and independent of L_n. R is the only macrolength remaining. Thus, we should expect D to remain finite and independent of L_n. Since $D \propto 1/L_n \cdot (\Delta_x^{\mathrm{ML}})^2$, Δ_x^{ML} must be proportional to $L_n^{1/2}$ and hence to $R^{-1/2}$. This limit has the strange feature that the diffusivity D remains finite while the flow $\langle \tilde{v}_{E_x} \tilde{n} \rangle \equiv n_0 D/L_n$ and the turbulence vanishes. This is, however, perfectly consistent with $\tilde{n}/n_0 \propto \Delta_x^{\mathrm{ML}}/L_n \propto 1/\sqrt{L_n R}$. [Recall that we take the vanishing $1/L_n$ limit with δ fixed, whereas physically may δ vanishes and, therefore, D vanishes.]

ACKNOWLEDGMENTS

The author wishes to thank Drs. R.R. Dominguez, H. Biglari, and W.M. Tang for numerous discussions on this work. He is particularly grateful to Dr. M.N. Rosenbluth for emphasizing the need to recover a finite result for the infinite sheared slab.

This is a report of work sponsored by the U.S. Department of Energy under Contract No. DE-AC03-89ER53277.

REFERENCES

1. R.E. Waltz, Phys. Fluids **26**, 469 (1983).
2. R.E. Waltz, Phys. Fluids **28**, 577 (1985).
3. R.E. Waltz, Phys. Fluids **29**, 3684 (1986).
4. R.E. Waltz, *Nagoya Lectures in Plasma Physics and Controlled Fusion* (Y.H. Ishikawa and T. Kamimura, Eds.) (Tokai University Press, Tokyo, 1989).
5. R.E. Waltz, Phys. Fluids **31**, 1962 (1988).
6. D. Biskamp and M. Walter, Phys. Lett. **109A**, 34 (1985).
7. B.D. Scott, J. Comp. Phys. **78**, 114 (1988).
8. L.D. Pearlstein and H.L. Berk, Phys. Rev. Lett. **23**, 220 (1969).
9. R.E. Waltz, W. Pfeiffer, and R.R. Dominguez, Phys. Fluids **23**, 605 (1980).
10. J.B. Taylor, in *Plasma Physics and Controlled Nuclear Fusion Research, Bertchesgaden, 1976* (International Atomic Energy Agency, Vienna, 1977), Vol. 2, p. 323.
11. C.Z. Chang and L. Chen, Nucl. Fusion **21**, 403 (1981).
12. B.B. Kadomtsev, *Plasma Turbulence* (Academic Press, London, 1965).
13. C.M. Surko and R.E. Slusher, Phys. Rev. Lett. **37**, 1747 (1976).
14. L. Garcia, P.H. Diamond, B.A. Carreras, and J.D. Callen, Phys. Fluids **28**, 2147 (1985).
15. J.W. Connor, Nucl. Fusion **26**, 517 (1986).
16. T. Tange, K. Nishikawa, and A.K. Sen, Phys. Fluids **25**, 1592 (1982).

Part III

RELATIVISTIC PARTICLE BEAMS

PLASMA FOCUSING AND DIAGNOSIS OF
HIGH ENERGY PARTICLE BEAMS*

Pisin Chen

Stanford Linear Accelerator Center
Stanford University, Stanford, California 94309

ABSTRACT

Various novel concepts of focusing and diagnosis of high energy charged particle beams, based on the interaction between the relativistic particle beam and the plasma, are reviewed. This includes overdense and underdense thin plasma lenses, an (underdense) adiabatic plasma lens, and two beam size monitor concepts. In addition, we introduce another mechanism for measuring flat beams based on the impulse received by heavy ions in an underdense plasma. Theoretical investigations show promise of focusing and diagnosing beams down to sizes where conventional methods are not possible to provide.

1. INTRODUCTION

As the energy of circular colliding beam accelerators becomes higher, one must face the limitation imposed by the energy loss to synchrotron radiation. For this reason, it is likely that future lepton colliders will be linear machines. The disadvantage of the linear scheme is that the beams are used only once, and then discarded. In order to achieve desirable luminosity, $\mathcal{L} = f_{rep} N^2 H_D / 4\pi \sigma_0^{*2}$, where N is the number of particles per bunch, f_{rep} the collider repetition rate, H_D the beam–beam disruption enhancement factor, and σ_0^* the rms beam radius at collision, one must either increase the beam current $f_{rep} N$ or decrease the spot size. The current is constrained by many factors, e.g., power limitations or wakefield effects in an accelerator. On the other hand, the minimum spot sizes are presently limited by the strength of conventional focusing quadrupoles. Eventually, however, one would reach the so-called *Oide limit*,[1] where the stochastic nature of the synchrotron radiation triggered at the final focusing lens imposes a strong limitation on the minimal possible beam size.

The plasma lens, which uses the self-focusing wakefields of a bunched relativistic charged particle beam in a plasma, and promises a very strong focusing, was first proposed by Chen.[2] Subsequent works[3–5] pushed the idea further. In the case of the overdense plasma lens, the beam peak density n_b is much less than the ambient plasma density n_0 it encounters as it traverses the lens. In this case, assuming that the beam length σ_z is large compared to the plasma wavelength $\lambda_p = \sqrt{\pi r_e / n_0}$ (the response of the plasma electrons to the beam is adiabatic and not oscillatory), the beam width σ is small compared to the plasma wavelength (plasma response is radial), and the ions are stationary, then the plasma electrons move to approximately neutralize the beam charge, leaving the beam current self-pinching forces unbalanced. The focusing wakefields reduce, to a good approximation, to the magnetic self-fields of the beam.

---

*Work supported by Department of Energy contract DE–AC03–76SF00515.

These self-fields are quite strong, but as they are dependent on the configuration of the beam density, the resulting focusing is nonlinear and prone to aberration. This requires that the lens be placed very close to the interaction point to minimize aberration effects, which in turn means that, for parameters typical of the Stanford Linear Collider (SLC) design, the plasma lens must be very dense. This dense plasma is a source of a very large background event rate, primarily due to the inelastic e-p scatterings, and is undesirable for particle physics purposes.

The background and aberration problems motivate the investigation of the underdense plasma lens.[6,7] An underdense plasma reacts to an electron beam by total rarefaction of the plasma electrons inside the beam volume, producing a uniformly charged ion column of charge density en_0. This uniform column produces linear, nearly aberration-free focusing. For positron beams, however, plasma electrons do not behave simply, and the focusing is not linear. The luminosity enhancement is thus achieved by the disruption of the larger positron beam by the smaller electron beam. This process has been termed "bootstrap disruption,"[6] as it involves a cascade of beam-dependent focusing effects; the pre-focusing of the electron beam by its own self-fields and the subsequent strengthened disruption of the positron beam by the electron beam. Simulations[6] indicate that luminosity enhancement by a factor 3-5 may be possible above conventional focusing schemes for the SLC design parameters, with large reductions in background event rates from similar overdense plasma lens schemes.

As mentioned earlier, ultimately the attainable beam spot size reaches a limit due to synchrotron radiation at the lens set by Oide. This limit is generic and applies also to plasma lenses. It therefore appears that the primary motivation for plasma lenses, i.e., the very strong focusing field attainable, becomes obsolete. Ironically it occurs that plasmas provide a solution to overcome this limit. The idea of *adiabatic focusing*, proposed by Chen, Oide, Sessler, and Yu,[8] using for example a plasma column, promises to evade the synchrotron radiation limit set by Oide. This is achieved by implementing a beam optics system where the focusing gradient is continuously and slowly increased along the direction of beam propagation, such that the β-function decreases linearly along the lens. In such a focusing system, beam particles with different energies would always oscillate within a definite envelope and eventually be focused down to within the designated size. The problem of chromatic aberration associated with conventional discrete focusing lenses, including the plasma lenses, can thus be alleviated.

In addition to its potential as focusing devices, plasmas could also provide useful diagnosis on the beam spot size through beam-plasma interaction. The information of the beam size at its final focus is indispensable for the tuning of linear colliders. The conventional method of wire scanning detects the bremsstrahlung signals from the beam intercepted by metal wires. Experience at SLC shows that such wire (with diameter $\sim$ 2-3 μm) would not sustain bunches of size smaller than $\sigma \lesssim$ 1-2 μm with $N \gtrsim 1 \times 10^{10}$ particles. Since it is expected that the beam size in the next generation linear colliders will be as small as nanometers in the vertical dimension, the conventional method would become obsolete and novel approaches are required.

It has been suggested earlier[9,10] that ions, by interacting with the collective fields of the high energy beam, can provide useful information on the size of *round* $(\sigma_x = \sigma_y)$ beams. The idea is revived more recently with the attention to *flat* $(\sigma_x \gg \sigma_y)$ beams conceived for the next generation colliders.[11,12]

In this paper, we review various novel concepts on plasma focusing and diagnosis of high energy beams, and introduce another mechanism of flat beam measurement, which is an extension from Refs. 9 and 10, based on the responses of heavy ions in an underdense plasma intercepted by the focused beam. Materials in Secs. 2-4 are largely extracted from the published works in Refs. 2, 4, 6, and 8. This is merely due to convenience in presentation, and should not imply that these are the only contributions to the subject. Other than the review of Refs. 11 and 12, the investigation on the beam size monitor in Sec. 5 has not been published elsewhere.

2. THE OVERDENSE PLASMA LENS

Assuming that the unperturbed plasma velocity v_0 is zero and the perturbed plasma density n_1 is much smaller than its unperturbed density n_0, the equation of motion and the equation of continuity can be linearized as:

$$\begin{cases} -\partial_\zeta \vec{v}_1 = -\dfrac{e}{mc}\vec{E}_1 \ , \\[2mm] -c\partial_\zeta n_1 + n_0 \nabla \cdot \vec{v}_1 = 0 \ , \end{cases} \qquad (2.1)$$

where $\partial_z = \partial_\zeta$ and $\partial_t = -c\partial_\zeta$ have been used.

Combining the fluid equation with the Maxwell's equations in the Coulomb gauge, and assuming that the transverse motions of the bunch particles are negligibly small during the beam-plasma interaction, we obtain

$$\partial_\zeta^2 n_1 + k_p^2 n_1 = \mp k_p^2 \sigma(\vec{x}) \ , \qquad (2.2)$$

where the plasma wave number $k_p \equiv \omega_p/c = (4\pi e^2 n_0/mc^2)^{1/2}$, and $\mp e\sigma(\vec{x})$ is the density for electron $(-)$ and positron $(+)$ bunches, respectively. It can be further shown that the combination of the perturbed plasma potentials, $A_{1z} - \phi_1$, satisfies the following equation of motion:

$$(\nabla_\perp^2 - k_p^2)(A_{1z} - \phi_1) = -4\pi e n_1 \ , \qquad (2.3)$$

where $\nabla_\perp^2 = \nabla^2 - \partial_\zeta^2$.

High energy $e^+ e^-$ beam particles are normally in Gaussian distributions in the three dimensional space. For the ease of calculation without sacrificing the essential physics, we invoke a parabolic bunch distribution:

$$\sigma(\vec{x}) = \rho_b \left(1 - \frac{r^2}{a^2}\right)\left(1 - \frac{(\zeta + b)^2}{b^2}\right) \ , \qquad (2.4)$$

where $0 \leq r \leq a$ and $-2b \leq \zeta \leq 0$. The parabolic profiles in both r and ζ directions are introduced to approximate the Gaussian profiles. The constant, ρ, can be related to the total number of particles N in the bunch:

$$\rho_b = \frac{3N}{2\pi a^2 b} \ . \qquad (2.5)$$

With this density distribution, it is straightforward to find that within the bunch,

$$A_{1z} - \phi_1 = \mp \frac{8\pi e\rho_b}{k_p^2} \left[I_0(k_p r) K_2(k_p a) + \frac{1}{2}\left(1 - \frac{r^2}{a^2}\right) - \frac{2}{k_p^2 a^2} \right]$$

$$\times \left[\left(1 - \frac{(\zeta+b)^2}{b^2}\right) + \frac{2}{k_p b}\,\sin k_p\zeta + \frac{2}{k_p^2 b^2}\,(1 - \cos k_p\zeta) \right] . \tag{2.6}$$

The transverse force exerting on the beam is simply the transverse derivative of $A_{1z} - \phi_1$. Thus

$$\mathcal{F}_\perp = \frac{8\pi e^2 \rho_b}{k_p} \left[I_1(k_p r) K_2(k_p a) - \frac{r}{k_p a^2} \right]$$

$$\times \left[\left(1 - \frac{(\zeta+b)^2}{b^2}\right) + \frac{2}{k_p b}\,\sin k_p\zeta + \frac{2}{k_p^2 b^2}\,(1 - \cos k_p\zeta) \right] . \tag{2.7}$$

Notice that the transverse force is exerting on the like particles in the same bunch, thus $\mathcal{F}_\perp$ has the same sign for both electron and positron bunches and can be verified to be always focusing. In the case where $k_p r \le k_p a \ll 1$,

$$I_1(k_p r) K_2(k_p a) - \frac{r}{k_p a^2} \simeq -\frac{1}{4} k_p r \ , \tag{2.8}$$

and we have a focusing force that is linear in r. The requirement that the focusing force be linear in r, i.e., that $k_p a \ll 1$, can be rewritten as $n_0 \ll 1/4\pi r_e a^2$, where r_e is the classical electron radius. On the other hand, self-consistency in the linearized fluid theory that we employed requires that $n_0 \gg n_1$. Combining these two conditions we arrive at a chain inequality which the system must satisfy:

$$\frac{1}{4\pi r_e a^2} \gg n_0 \gg \frac{3N}{2\pi a^2 b} \left[\left(1 - \frac{(\zeta+b)^2}{b^2}\right) + \frac{2}{k_p b}\,\sin k_p b + \frac{2}{k_p^2 b^2}(1 - \cos k_p b) \right] . \tag{2.9}$$

An interesting situation is when $k_p b \gg 1$. Therefore $b \gg \lambda_p/2\pi \gg a$, and we have a long bunch where the longitudinal extent is much larger than the transverse extent. When this is satisfied, the focusing force is

$$\mathcal{F}_\perp(r,\zeta) \simeq \frac{3e^2 N}{a^2 b}\left(1 - \frac{(\zeta+b)^2}{b^2}\right) r \ , \qquad k_p b \gg 1 . \tag{2.10}$$

We see that in this limit the focusing force varies as the longitudinal profile of the beam and the maximum is at the midpoint along the bunch. This focusing force is very strong. For comparison, consider $N = 5 \times 10^9$, and $a = 100$ μm, $b = 1$ mm. We find the corresponding $G \sim 720$ KG/cm. In contrast, typical iron magnets ($G \sim 5$ KG/cm) and superconducting magnets ($G \sim 10$ KG/cm) are about $1 \sim 2$ orders of magnitude weaker.

Physically, this self-focusing effect arises because the electrons in the plasma are either expelled (for the case of interacting with an electron bunch) or pulled (for the case of interacting with a positron bunch) by the leading particles in the bunch, while on this time scale the ions in the plasma are essentially stationary. As a result, the trailing particles in the same bunch experience an attractive force due to the access charges in plasma within the volume of the bunch. Large self-pinching of the beam is thus induced. This effect has been observed in computer simulations,[13,3] and in experiments.[14,15]

3. THE UNDERDENSE PLASMA LENS

In the underdense regime, the electron beam experiences a linear, nearly aberration-free focusing. Simulations have shown that one needs to have $n_b \geq 2n_0$ to produce linear focusing over most of the bunch.[7] When this condition is satisfied, the beam β-function as a function of the distance down the beam-line s can be described by the third order linear differential equation:

$$\beta''' + 4K\beta' + 2K'\beta = 0 \quad , \tag{3.1}$$

where $\beta = \sigma^2/\epsilon_0$, ϵ_0 is the unnormalized transverse emittance and $K = 2\pi r_e n_0/\gamma$ is the focusing strength of the lens. To solve Eq. (3.1) we must first integrate through the δ-function in K' at the start of the lens to obtain $\Delta\beta_0'' = -2K\beta_0$. The other two initial conditions are just continuity requirements $\beta' = \beta_0'$, and $\beta = \beta_0$. Assuming the electron bunch to have a cylindrically symmetric bi-Gaussian distribution of rms length σ_z, then we can define the phase space density parameter $\zeta = Nr_e/\sqrt{8\pi}\epsilon_0\gamma\sigma_z$, and the focusing strength of an underdense plasma lens is, with $\beta = \beta_0$ and $n_0/n_b = 1/2$ at the start of the lens, $K = \zeta/\beta_0$. Using the initial conditions, we integrate Eq. (3.1) once to obtain

$$\beta'' + 4K\beta = 2/\beta_0^* + 2\zeta \quad , \tag{3.2}$$

where β_0^* is the minimum β-function achieved in the absence of the plasma lens. The solution for the β-function inside the lens is easily found from Eq. (3.2) to be

$$\beta = \frac{\beta_0}{2} + \frac{1}{2K\beta_0^*} + \left(\frac{\beta_0}{2} - \frac{1}{2K\beta_0^*}\right)\cos\nu(s-s_0) + \frac{2s_0}{\nu\beta_0^*}\sin\nu(s-s_0) \quad , \tag{3.3}$$

where $\nu^2 = 4K$.

It is straightforward to show from the above considerations that the maximum reduction in β^* that one can achieve with this lens occurs when one places the entrance of the plasma at a position $-s_0 \gg \beta_0^*$. This reduction is given by

$$\frac{\beta^*}{\beta_0^*} = \frac{1}{1 + K\beta_0^*(\beta_0 - \beta_1)} \simeq \frac{1}{1 + \zeta\beta_0^*} \tag{3.4}$$

where β_1 is the β–function at the exit of the plasma lens at $s = s_1$. For SLC design parameters ($\epsilon_n = 3 \times 10^{-5}$m-rad, $\sigma_z = 1$mm, $\beta_0^* = 7$mm, $\gamma = 10^5$, and $N = 5 \times 10^{10}$) we have $\zeta = 9.4 \times 10^2$m^{-1}, and a possible reduction in β of $1/7.5$. It is also interesting to note that according to this formula, one should never back off of the focus, i.e., make β_0^* larger, as the ultimate β^* attainable is inversely proportional to $\zeta + (1/\beta_0^*)$. This implies one should minimize β_0^*. It also says that if $\zeta\beta_0^* < 1$ then plasma lens is irrelevant, as it is not strong enough to overcome the inherent divergence in the beam. If one only reduces the spot size σ_-^* of the electron beam in the collisions and leaves the positron beam spot size σ_0^* unchanged, then the possible luminosity enhancement due to the lens H_L (excluding depth of focus and disruption effects) is easily shown to be

$$H_L = \frac{2(\sigma_0^*)^2}{(\sigma_-^*)^2 + (\sigma_0^*)^2} = \frac{2\beta_0^*}{\beta_-^* + \beta_0^*} \quad , \tag{3.5}$$

which is strictly less than two. For example, an electron spot size reduction of $\sigma_-^*/\sigma_0^* = 0.4$ gives a luminosity enhancement of 1.73. This is a very modest number; it is boosted, however, by the bootstrap disruption enhancement.

Previous calculations of the luminosity enhancement due to beam–beam disruption have treated symmetric beams. It has been found[16] that the disruption luminosity enhancement is influenced by two factors: the strength of the pinch, represented by the disruption parameter D,

$$D = \frac{N r_e \sigma_z}{\gamma \sigma_0^{*2}} = \frac{N r_e \sigma_z}{\gamma \beta_0^* \epsilon_0} \quad , \tag{3.6}$$

and the effects of the inherent divergence of the beam, represented by the divergence parameter $A = \sigma_z/\beta_0^*$. The disruption enhancement H_D is a strongly decreasing function of A when $A > 1$, due to the effects of depth of focus and inherent beam divergence, and a monotonically increasing function of D. Since both D and A are inversely dependent on β_0^*, there exists a maximum luminosity for some value of β_0^*. This is also true for bootstrap disruption enhancement, H_B.

Since simulations have shown that the underdense plasma lens can focus positrons,[7] albeit with strong aberrations, it is interesting to see what sort of luminosity enhancements are ultimately possible using two underdense lenses. A theory of aberration-prone focusing has been developed in Ref. 4. In terms of the quantity called the aberration power P, the transformations of the initial transverse phase space parameters $(\alpha_0, \beta_0, \epsilon_0)$ by an aberration-prone thin lens are

$$\alpha = (\alpha_0 + \beta_0/f)/P, \quad \beta = \beta_0/P, \quad \epsilon = \epsilon_0 P \quad , \tag{3.7}$$

where f is the lens focal length, $\alpha = -2\beta'$, and $P = \sqrt{1 + (\beta_0 \eta/f)^2}$. The parameter η corresponds to the 'rms variation of the focusing strength K in the lens. Simulations have shown that for a mildly underdense lens, $\eta \simeq 0.28$ for positron focusing. Note that in this model the aberration results in an emittance blowup which is dependent on the strength of the lens. The total reduction in spot size is thus

$$\frac{\sigma^*}{\sigma_0^*} = \left[\frac{\beta^* \epsilon}{\beta_0 \epsilon_0}\right]^{1/2} = \frac{P}{\sqrt{P^2 + (\alpha_0 + \beta_0/f)^2}} \quad . \tag{3.8}$$

Using this model and the computational results from Ref. 8, Chen et al.[6] simulate the collision of an electron beam focused by an underdense plasma lens to 0.4 of its original spot size with a positron beam focused, with aberrations, by a mildly underdense plasma lens, to 0.6 of the conventionally achieved spot size. All other parameters are taken from the SLC design. The luminosity is found to be 1.5×10^{31} cm^{-2} sec^{-1} and the total enhancement is approximately five. The physical parameters involved in this configuration are shown in Table I.

Table I A set of plasma lens parameters for SLC

Plasma Lens Parameters	Electrons	Positrons
n_0 [cm^{-3}]	1.5×10^{15}	4.8×10^{16}
l[cm]	5.0	0.33
Beam Parameters		
N	5×10^{10}	5×10^{10}
$\mathcal{E}$ [GeV]	50	50
ϵ_0 [m-rad]	3×10^{-10}	3×10^{-10}
σ_z [mm]	1.0	1.0
Beam Optics Parameters		
s_0 [cm]	20.0	1.3
β_0^* [mm]	7.0	7.0
ϵ[m-rad]	3×10^{-10}	4.2×10^{-10}
β^* [mm]	1.12	1.84
η	0	0.28
P	1.0	1.39
f [cm]	7.5	1.1
Luminosity Enhancement		
$\mathcal{L}_{00}$ [10^{30}cm^{-2}]	1.76	1.76
H_D	1.73	1.73
$\mathcal{L}_0(=H_D\mathcal{L}_{00})$ [10^{30}cm^{-2}]	3.0	3.0
$\sigma_\pm^*/\sigma_0^*$	0.4	0.6
H_B	5.0	5.0
$\mathcal{L}(=H_B\mathcal{L}_0)$ [10^{30}cm^{-2}]	15.0	15.0

This treatment of the positron focusing is approximate, but gives qualitative insight into the role aberrations play in this scheme.

4. ADIABATIC FOCUSING

In general, in a focusing (or defocusing) environment, a particle with coordinate y satisfies the equation of motion

$$\frac{d^2y}{ds^2} + K(s)y = 0 \ ,$$
(4.1)

and the well-known solution is [17]

$$y(s) = \sqrt{\epsilon\beta(s)} \, \cos[\psi(s) + \phi] \quad , \tag{4.2}$$

where ϵ is the emittance and

$$\frac{d\beta}{ds} = -2\alpha(s) \quad ; \qquad \psi(s) = \int^{s} \frac{ds'}{\beta(s')} \quad .$$

In an adiabatic focusing, we demand that the change in β, occurring in a length given by β, is small compared to β. For the sake of simplicity, we shall assume that

$$\frac{d\beta}{ds} = \text{constant} \quad . \tag{4.3}$$

Hence we take

$$\beta(s) = \beta_0 - 2\alpha_0 s \quad , \tag{4.4}$$

where α_0 is the initial condition and a constant of the system that characterizes the amount of adiabaticity.

Since $\alpha(s) = \alpha_0 = \text{constant}$, we have $d\alpha/ds = 0$, and the focusing strength along the channel varies as

$$K(s) = \frac{1 + \alpha_0^2}{\beta^2} = \frac{1 + \alpha_0^2}{(\beta_0 - 2\alpha_0 s)^2} \quad . \tag{4.5}$$

Notice that the focusing strength scales inverse quadratically with $\beta(s)$. The phase advance, on the other hand, varies as

$$\psi(s) = \frac{1}{2\alpha_0} \ln \frac{\beta_0}{\beta_0 - 2\alpha_0 s} \quad . \tag{4.6}$$

For a particle with less energy than the design energy E_0, i.e., $E = (1-\delta)E_0$, where $\delta \ll 1$, the focusing force K is larger by an amount $1/(1+\delta)$. According to Eq. (4.1), the matched β-function for the lower-energy particle becomes $\bar{\beta}(s) = \sqrt{1 - \delta}\beta(s)$, and the α-function is also reduced to $\bar{\alpha}(s) = \sqrt{1 - \delta}\alpha(s)$. The mismatched β-function can be shown to be

$$\widetilde{\beta}(s) = \beta(s)[1 - \delta \sin^2 \bar{\psi}(s)] \leq \beta(s) \quad , \tag{4.7}$$

where $\bar{\psi}(s) \equiv \psi(s)/\sqrt{1 - \delta}$.

Thus, the amplitude of the lower-energy particle never exceeds that of the reference particle. If one chooses the design energy of the focuser at the maximum energy of the incoming beam, the entire beam is expected to be focused. This *achromatic* nature of the focuser will hold true for a particle, which emits radiation while traversing the focuser, and is the very basis of the adiabatic focuser concept.

Insensitive to the energy variation as it is, an adiabatic focuser would render useless if a large fraction of the beam energy is lost during the process. The rate of energy loss of a relativistic electron due to synchrotron radiation is well-known.[18]

In order to perform simple analytic calculations, it is convenient to approximate the exact formula by the following expressions in the *classical,* the *transition,* and the *quantum* regimes: [19]

$$\frac{d\gamma}{ds} = -\frac{2}{3}\frac{\alpha}{\lambda_c}\begin{cases} \Upsilon^2 \ , & \Upsilon \lesssim 0.2 \ , \\ 0.2\Upsilon \ , & 0.2 \lesssim \Upsilon \lesssim 22 \ , \\ 0.556\Upsilon^{2/3} \ , & 22 \lesssim \Upsilon \ , \end{cases} \qquad (4.8)$$

where γ is the Lorentz factor of the electron, α the fine structure constant, and $2\pi\lambda_c$ the Compton wavelength. We see that the energy loss is uniquely determined by the parameter Υ, which is Lorentz invariant and defined as

$$\Upsilon \equiv \gamma\frac{B}{B_c} = \frac{\gamma^2\lambda_c}{\rho} \ . \qquad (4.9)$$

Here B is the local field strength, $B_c = m^2c^3/e\hbar \simeq 4.4 \times 10^{13}$ Gauss is the Schwinger critical field, and ρ is the instantaneous radius of curvature of the particle.

Since $1/\rho = K(s)y$, with the help of the relation $\sigma = \langle y^2 \rangle = \beta\epsilon$, where ϵ is the emittance of the beam, one could calculate the energy loss in terms of α_0, β_0, and the resulting β, in the three regimes. It is found[9] that there exists an optimal value of α_0 for attaining a desired β-function with minimum energy loss:

$$\alpha_0 = \begin{cases} \dfrac{1}{\sqrt{3}} \ , & \text{(classical)} \ , \\ 1 \ , & \text{(transition)} \ , \\ \sqrt{3} \ , & \text{(quantum)} \ . \end{cases} \qquad (4.10)$$

We see that the optimal situation does not rely on the condition $\alpha_0 \ll 1$, in the strict sense of the word *adiabatic.*

It should, in principle, be possible to set up an adiabatic focuser where the increase of its focusing strength varies in accordance with the three different optimum values given above. But the focuser may be experimentally more convenient if α_0 is fixed throughout the system. If a focuser covers all three regimes of radiation, an obvious compromise would be $\alpha_0 = 1$. With this choice there will be about 15% additional radiation in the classical regime and about 30% more in the quantum regime. Alternatively, since the radiation loss occurs primarily near the end of an adiabatic focuser, a choice of α_0 according to the final regime is most advisable.

With the choice of $\alpha_0 = \sqrt{3}$, we find that there exists a *critical emittance* above which a beam can never be focused so tight as to enter the quantum regime of radiation:

$$\epsilon_c \equiv \frac{3^{3/2} \cdot 15^3}{2^3 \cdot 4^2 \cdot 22}\frac{\lambda_c}{\alpha^3} = 6.17 \times 10^{-6} \text{ m} \ , \qquad (4.11)$$

which depends only on fundamental physical parameters.

If the actual normalized emittance ϵ_n in the system can indeed be lower than the above critical value, then the requirement that the fractional energy loss be much less than unity translates into a limit on the final beam size,

$$\sigma_{ad} \gg 1.39 \times 10^{-8} \left(\frac{\epsilon_n}{\epsilon_c}\right)^{2/3} \exp\left\{-1.12\left(\frac{\epsilon_n}{\epsilon_c}\right)^{-1/3}\right\} \text{ m} \quad . \qquad (4.12)$$

This is to be compared with the Oide limit on beam size at the focus which can be expressed as

$$\sigma \gtrsim 3.4 \times 10^{-4} \epsilon_n^{5/7} \quad , \qquad (4.13)$$

in the vertical dimension for flat beams. For an emittance $\epsilon_n = \epsilon_c/10$, we find that $\sigma_{ad} \gg 2.68 \times 10^{-9}$ m, whereas the Oide limit for the same emittance gives $\sigma \gtrsim 1.25 \times 10^{-8}$ m.

Numerical examples have been studied by Chen et al.[8] The first is a proof-of-principle case using the beam in the SLAC End Station. The second involves the use of a focuser on the SLC, and the third is a focuser on a TeV Linear Collider (TLC). Parameters of the beam, the focuser, and the expected performance are displayed in Table II. In the first two cases, round beams, i.e., $\sigma_y = \sigma_x$, are assumed, whereas in the third case for the TLC, the beam is assumed to be flat ($\sigma_y \ll \sigma_x$). In the End Station Focuser, the device is rather long and one may employ differential pumping to form the variation in plasma density, ramping from the initial value, n_0, to the final value, n^*, over a length L. The other two focusers require higher densities (ranging up to solid density), with variation over shorter distances.

Table II   Three examples of the adiabatic focuser

	SLAC End Station	SLC	TLC
Initial Beam Properties			
$\mathcal{E}$ [GeV]	15	50	500
ϵ_n [m]	1×10^{-4}	3×10^{-5}	1×10^{-8}
σ_0 [μm]	20	3	5×10^{-3}
β_0 [cm]	12	3	0.25
Focuser Properties			
α_0	5×10^{-2}	$1/\sqrt{3}$	$\sqrt{3}$
L [cm]	119	2.6	0.07
n_0 [cm^{-3}]	1.2×10^{14}	8.4×10^{15}	1.8×10^{19}
n^* [cm^{-3}]	1.2×10^{18}	8.4×10^{19}	1.8×10^{23}
Final Beam Properties			
δ	Negligible	3%	1%
σ^* [μm]	2	0.3	0.5×10^{-3}

5. BEAM SIZE MONITOR

For the purpose of monitoring the beam size, one again envisions an underdense plasma intercepting the focused beam. In principle, both the beam and the ions in the plasma carry useful signals on the beam size due to their mutual interaction. The plasma electrons, however, are too severely perturbed to provide any information. Regarding the plasma formation, either preformed plasma by external means, or self-ionization by the incoming beam can be conceived, depending on practical considerations on the device.

In the concept suggested by Chen et al.,[11] a heavy element gas jet with density n_g is ionized by a high energy beam of N particles. The number of ions created in the beam channel is

$$N_i = \frac{N\sigma_i}{4\pi\sigma_x\sigma_y}N_g \quad , \tag{5.1}$$

where σ_i is the ionization cross section and N_g is the total number of gas atoms in the channel:

$$N_g = 4\pi\sigma_x\sigma_y l n_g \quad . \tag{5.2}$$

Here l is the thickness of the gas target. The ionization is considered to be *non-saturated* if $N_i < N_g$, or $N\sigma_i/4\pi\sigma_x\sigma_y < 1$, from Eq. (5.1). In this regime one can easily see that N_i is independent of the beam size. However, if the ionization cross section is so large that $N\sigma_i/4\pi\sigma_x\sigma_y > 1$, then the ionization will be *saturated*, as there cannot be more ions than the available number of atoms in the channel. In this regime $N_i = N_g$, and N_i is now a function of the beam size as in Eq. (5.2).

As we have seen in Sec. 3, the focusing strength due to an ion column is proportional to the number of ions encompassed in the beam volume. The synchrotron radiation triggered by the ion focusing can again be described by the parameter Υ, in this case

$$\Upsilon = \frac{\sqrt{3}}{2}\frac{r_e\lambda_c\gamma N_i}{l\sigma_y(1+R)} \quad . \tag{5.3}$$

Inserting the expression in Eq. (5.2) for N_i, we find, to the accuracy of the order $1/R$, that

$$\Upsilon = 2\sqrt{3}\pi r_e\lambda_c\gamma n_g\sigma_y \quad . \tag{5.4}$$

Since the critical energy of synchrotron radiation is $\omega_c = (3/2)\Upsilon\mathcal{E}$, where $\mathcal{E}$ is the beam particle energy, one should in principle be able to measure σ_y by analyzing the spectrum of such a radiation.

Note that other than the gas density, the signal is a function of σ_y only, and is independent of σ_x, σ_z, and N of the beam, and the thickness l of the target. This is important because in the case of focusing flat beams the challenge is generally in the vertical dimension, where σ_y in the range of nanometers is conceived for the next generation linear colliders. Therefore, one expects that in the final focusing process the variation in σ_x is relatively mild, and the major uncertainty comes from the minor dimension, which is what we measure. We should also emphasize that in case saturation of self-ionization turns out to be unattainable, a preformed, fully ionized underdense plasma by external means would suffice the same purpose for radiation signals.

Buon[12] considers the case of light ions. For electron beams, the ions would experience an attractive potential well. So when the ions are light enough, they will oscillate during the beam passage. The nature of the beam collective field is such that the maximum field strength along the vertical direction is very close to that in the horizontal direction. (The actual maximum values in the two dimensions are slightly different, which will be discussed below.) In addition, the locations of these maximum fields are roughly at σ_x and σ_y, respectively. Therefore, the focusing strengths in x and y directions differ by a factor R : $K_y = RK_x$. From conservation of total energy, we have

$$\frac{1}{2}K_x\sigma_x^2 = \frac{1}{2}Mv_{x_m}^2 \quad ,$$
$$\frac{1}{2}K_y\sigma_y^2 = \frac{1}{2}Mv_{y_m}^2 \quad , \tag{5.5}$$

respectively. It is easy to see that the maximum horizontal and vertical velocities differ by a factor $\sqrt{R}$:

$$v_{y_m} = \sqrt{R}v_{x_m} \quad . \tag{5.6}$$

By measuring the angular distribution of the out-coming ions, one can then infer the beam aspect ratio.

Let us now introduce yet one more mechanism, using heavy ions, for monitoring flat beam size. Ionization saturation is not necessary in this approach. To a large extent, this is an extension from the ideas of Rees[9] and Prescott.[10] Consider a plasma density which is orders of magnitude lower than that of the beam. Under this situation the ion focusing on the beam, and therefore the change of the field strength in the beam, can be ignored. For heavy ions which are initially quiescent and immobile within the time scale of the transit of the beam, the impulse received by the ions are predominantly from the local electric component of the collective beam field. Since the beam field reaches a maximum on the "boundary" of the beam and decreases outside the beam, the ion momenta will be, in general, double-valued. This makes reconstruction of the entire transverse beam field ambiguous. The maximum fields, however, are single-valued. More importantly, the maximum fields in x and y directions do not scale the same in R. As we shall see below, by comparing the maximum momenta in the x and y directions one can in principle determine both σ_x and σ_y.

The strength of the electric field in a flat beam can be described by the Basseti-Erskine formula.[20] To the accuracy of the order $1/R$, we find

$$E_x(x,y,z) \simeq \frac{eN}{\sigma_x\sigma_z}\, e^{-z^2/2\sigma_z^2}\, \mathcal{I}m\, F(x,y) \quad ,$$
$$E_y(x,y,z) \simeq \frac{eN}{\sigma_x\sigma_z}\, e^{-z^2/2\sigma_z^2}\, \mathcal{R}e\, F(x,y) \quad , \tag{5.7}$$

where

$$F(x,y) = \mathcal{W}\!\left(\frac{x+iy}{\sqrt{2}\sigma_x}\right) - e^{-(x^2/2\sigma_x^2 + y^2/2\sigma_y^2)}\, \mathcal{W}\!\left(\frac{x/R+iRy}{\sqrt{2}\sigma_x}\right) \quad , \quad R \gg 1 \quad .$$

Here $\mathcal{W}$ is the complex error function. The maxima of E_x along x-axis and of E_y along y-axis can be found by solving the equations $\partial\,\mathcal{I}m\,F(x,0)/\partial x = 0$ and $\partial\,\mathcal{R}e\,F(0,y)/\partial y = 0$, respectively. We find that the maximum E_x and E_y locates at

$$\frac{x_m}{\sigma_x} = 1.307 + \frac{0.393}{R} \quad , \tag{5.8}$$

and

$$\frac{y_m}{\sigma_y} = \sqrt{2\ln R} + \frac{1}{R}\sqrt{\frac{\pi}{2}} \quad . \tag{5.9}$$

The corresponding field strengths at these locations are

$$E_{x_m} = \frac{eN}{\sigma_x\sigma_z}\, e^{-z^2/2\sigma_z^2}\,\sqrt{\frac{2}{\pi}}\frac{\sigma_x}{x_m}\left[1 - \frac{1}{R}\,e^{-x_m^2/2\sigma_x^2}\right] \quad ,$$

$$E_{y_m} = \frac{eN}{\sigma_x\sigma_z}\, e^{-z^2/2\sigma_z^2}\,\sqrt{\frac{2}{\pi}}\frac{\sigma_x}{y_m}\left[1 - \left(R + \frac{1}{R}\right)e^{-y_m^2/2\sigma_y^2}\right] \quad . \tag{5.10}$$

Inserting the explicit forms of x_m and y_m in Eq. (5.8) and Eq. (5.9), we obtain

$$E_{x_m} = \frac{eN}{\sigma_x\sigma_z}\, e^{-z^2/2\sigma_z^2}\,\sqrt{\frac{2}{\pi}}\frac{1}{1.307}\left[1 - \frac{0.726}{R}\right] \quad ,$$

$$E_{y_m} = \frac{eN}{\sigma_x\sigma_z}\, e^{-z^2/2\sigma_z^2}\left\{1 - \frac{\sqrt{\pi}}{2R}\left[\sqrt{\ln R} + \frac{1}{\sqrt{\ln R}}\left(\frac{1}{2} + \frac{2}{\pi}\right)\right]\right\} \quad . \tag{5.11}$$

We see that E_{x_m} and E_{y_m} are asymptotically independent of R, but the asymptotic values differ by a factor 0.610.

With the assumption that the ions are immobile during the beam transit time, the impulse received by the ions is a trivial function of the local field strength. This is obtained by integrating the fields in Eq. (5.11) over time ($t = z/c$). For singly charged ions we find

$$\Delta p_{x_m} = \frac{2}{1.307}\frac{e^2 N}{c\sigma_x}\left[1 - \frac{0.726}{R}\right] \quad ,$$

$$\Delta p_{y_m} = \sqrt{2\pi}\,\frac{e^2 N}{c\sigma_x}\left\{1 - \frac{\sqrt{\pi}}{2R}\left[\sqrt{\ln R} + \frac{1}{\sqrt{\ln R}}\left(\frac{1}{2} + \frac{2}{\pi}\right)\right]\right\} \quad . \tag{5.12}$$

As long as the total bunch population N, or the beam current, is known, measurements of Δp_{x_m} and Δp_{y_m} should in principle determine uniquely the values of σ_x and R, or in turn σ_x and σ_y. In practice, however, this scheme may suffer from low statistics. From Eq. (5.12) we see that the relative accuracy of R diminishes roughly as $1/R$:

$$\frac{\delta\Delta p_{x_m}}{\Delta p_{x_m}} = \frac{0.726}{R}\frac{\delta R}{R} \quad . \tag{5.13}$$

To have a 10% error in R, one needs to have an accuracy of 1% in Δp_{x_m} when R is around 10, and of 0.1% when $R \sim 100$. Evidently, a large collection of data is necessary to achieve this goal. This could be realized by increasing the gas jet density and the repetition of measurements. One other concern is that ions can never be infinitely heavy. This would result in a certain degree of smearing of the signal due to ion motions across the peak fields in the beam. The best way to assess this degradation is through computer simulations, which are still in progress.

It is also of interest to note that the angular distribution of the out-coming heavy ions should also provide useful information on the beam size. This can be appreciated by realizing first that the maximum field lies on the "boundary" of the beam. Secondly, the transverse field tends to "flip" from the horizontal direction towards the vertical direction as soon as one departs from x_m along the boundary. This asymmetry of the field pattern should also help to diagnose the aspect ratio.[21]

6. DISCUSSION

Confirmation of the existence of strong self-focusing in plasma wakefields has been experimentally verified in tests performed at Argonne Advanced Accelerator Test Facility.[14] More recently, a dedicated plasma lens experiment has been performed in Japan,[15] which covers both the overdense and the underdense regimes. These experiments use electron beams of energies of the order 20 MeV. Although the physics of self-focusing should not be different as long as the beams are relativistic, it is of interest to test the plasma lens with beam characteristics pertinent to that in a true linear collider. To this end, such a test at the Final Focus Test Beam (FFTB) now under construction at SLAC, with the possibility of testing both round and flat 50 GeV electron and positron beams, should be meaningful.

Examples of adiabatic focuser have been shown in Table II in Section 4. When applied to the beam parameters similar to those of the SLAC End Station, where the beam energy is 15 GeV, and those of SLC, the necessary parameters for the plasma adiabatic focuser are shown to be very reasonable; and, in principle, to yield a significant increase in the luminosity for the SLC. To apply the scheme to TeV-range linear colliders, it is found necessary to invoke liquid or even solid-state materials. Although the necessary technology for the focuser is yet to be developed, such a focuser should in principle be more compact than the conventional focusing system. In particular, for a focuser relevant to the SLAC End Station-type parameters, the requirements for the system seems to be immediately realizable.

The several ideas of beam size monitoring using plasmas were reviewed and introduced. These concepts have been intensely pursued recently. Experimental effort on such a monitor is now under way through an Orsay-SLAC collaboration. The device is expected to be tested at the FFTB at SLAC.

To conclude, we have reviewed various novel plasma focusing and diagnosis concepts, and have introduced a new scheme of beam size measurement. The potential applications of plasma physics to high energy particle beam instrumentation is seen to be very rich. Most importantly, many of these concepts have been followed by experimental efforts, and are in a very healthy state of progress. We expect that in a few years, more experimental verifications will be available.

ACKNOWLEDGEMENTS

I would like to thank my collaborators in Refs. 4, 6, 8, and 11 for using the materials directly from the papers. I would also like to thank J. Buon of Orsay, France, and D. L. Burke and R. D. Ruth of SLAC, Stanford, for many helpful discussions on beam size measurements.

REFERENCES

[1] K. Oide, Phys. Rev. Lett. **61**, 1713 (1988).

[2] P. Chen, Part. Accel. **20**, 171 (1987).

[3] P. Chen, J. J. Su, T. Katsouleas, S. Wilks, and J. M. Dawson, IEEE Trans. Plasma Sci. **PS-15**, 218 (1987).

[4] J. B. Rosenzweig and P. Chen, Phys. Rev. D**39**, 2039 (1989).

[5] D. B. Cline, B. Cole, J. B. Rosenzweig and J. Norem, *Proceedings of the 1987 Washington Accelerator Conference* (IEEE, Washington, 1987) p. 241.

[6] P. Chen, S. Rajagopalan, and J. B. Rosenzweig, Phys. Rev. D**40**, 923 (1989).

[7] J. J. Su, T. Katsouleas, J. Dawson, and R. Fedele, UCLA Report No. PPG–1177, 1988; to appear in Phys. Rev. A.

[8] P. Chen, K. Oide, A. Sessler, and S. Yu, Phys. Rev. Lett. **64**, 1231 (1989).

[9] J. Rees, SLAC Report SLAC–AP–24, 1984.

[10] C. Prescott, SLAC internal memorandum, 1984, unpublished.

[11] P. Chen, D. L. Burke, M. D. Hildreth, and R. D. Ruth, SLAC Report SLAC–AAS–41, 1988, presented at the International Workshop on the Next Generation of Linear Colliders, Stanford, 1988.

[12] J. Buon, Part. Accel. **31**, 39 (1990).

[13] T. Katsouleas, Phys. Rev. A**33**, 2056 (1986).

[14] J. Rosenzweig, P. Schoessow, B. Cole, C. Ho, W. Gai, R. Konecny, S. Mtingwa, J. Norem, M. Rosing, and J. Simpson, Fermilab Report FERMILAB–PUB–89/213, 1989; to appear in Phys. Fluids B.

[15] A. Ogata *et al.*, KEK Report, in preparation.

[16] P. Chen and K. Yokoya, Phys. Rev. D**38**, 987 (1988).

[17] E. D. Courant and H. S. Snyder, Ann. Phys. **3**, 1 (1958).

[18] A. A. Sokolov and I. M. Ternov, *Synchrotron Radiation* (Pergamon, Berlin, 1968).

[19] This approximation was first suggested by P. B. Wilson, *Proceedings of the IEEE Particle Accelerator Conference*, IEEE Catalog No. 87CH2387–9, 53 (1987). Our expression, however, has different boundary values.

[20] M. Basseti and Erskine, CERN Report CERN–ISR–TH/80–06, 1980.

[21] P. Chen and R. D. Ruth, in preparation.

THE PHYSICS AND APPLICATIONS OF MODULATED INTENSE RELATIVISTIC ELECTRON BEAMS

M. Friedman, V. Serlin, Y.Y. Lau and J. Krall
Plasma Physics Division
Naval Research Laboratory, Washington, DC 20375-5000

ABSTRACT

This paper summarizes ten years of research in the generation, physics and application of modulated intense relativistic electron beams. With the knowledge accumulated during this period of time we can now project fully modulated electron beam sources to very high energies. These sources can drive high power RF sources, power directly high voltage gradient particle accelerators and be used to heat plasmas to thermonuclear temperature. The modulated intense relativistic electron beam sources, together with the repetitive HV sources that are now being developed, can revolutionize RF sources, accelerators and fusion devices.

BACKGROUND

The physics of intense relativistic electron beams (IREBs) is governed by the beam electric and magnetic self fields. There are many applications of IREBs that exploit these self fields. Examples can be found in research areas on high-power microwave generation [1], collective acceleration of charged particles [2], and plasma heating to thermonuclear temperatures [3]. NRL has an interest in these areas of research, and for the last 10 years an investigation was undertaken to understand the role of self-fields in the various applications. A new mechanism emerged from this research that makes it possible to efficiently modulate IREBs at power levels exceeding 10^9 to 10^{10}W [4-7]. This kind of power makes such devices interesting for heating plasmas to thermonuclear temperatures and for powering large RF accelerators.

PHENOMENOLOGY[5]

During the experimental investigation of IREB transport through drift space, we found that high levels of coherent oscillation appeared on the IREB current. The drift space consisted of a smooth metallic tube in which two or more coaxial cavities were inserted (Figure 1).

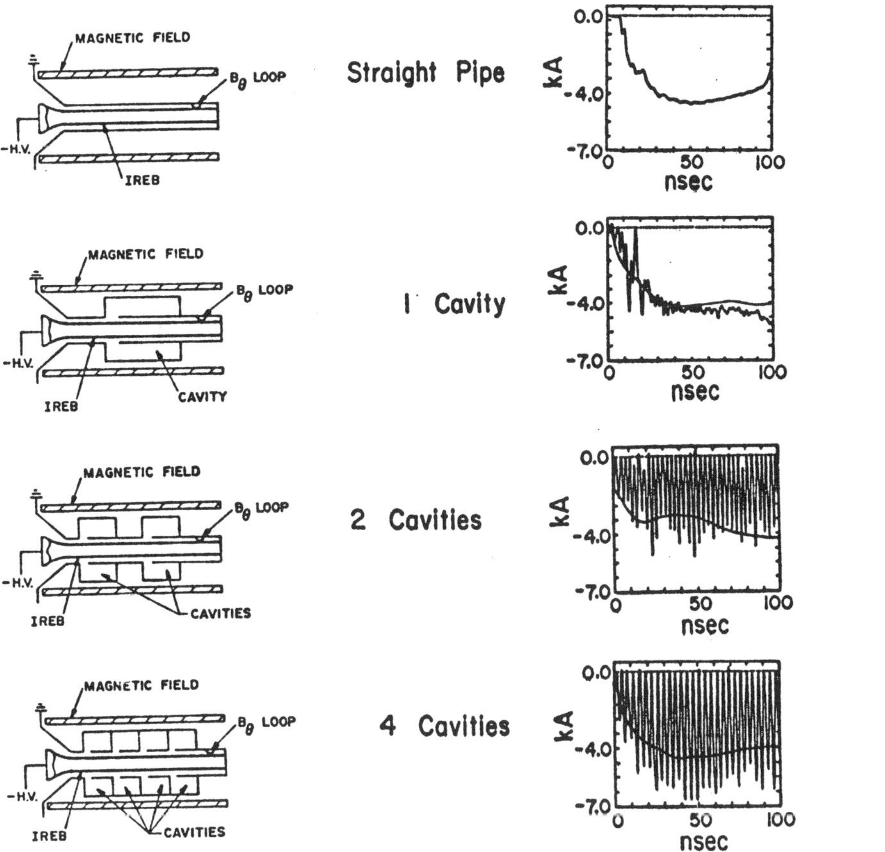

Fig.1   IREB Propagation through structures. (*Top*) smooth drift tube. (*2nd from top*) smooth drift tube with one cavity. (*3rd from top*) smooth drift tube with two cavities. (*Botton*) four cavities. The right side of the figure displays the output I.R.E.B. current.

The following important characteristics were observed:

* The frequency of oscillation depended strongly on the geometry and weakly on IREB current and voltage.

* The frequency of modulation was monochromatic and could be tuned between 60 MHz and 3 GHz.

* Electron beams at various accelerating potentials and currents were fully modulated with an efficiency of nearly 100%.

THEORY[5]

A simple theoretical model based on space-charge waves on IREBs gave excellent agreement with the experimental observation. Numerical simulations agreed with the theory and extended it into the nonlinear region. Both theory and simulation showed that the electric and magnetic self-fields of the IREB and the induced electric field that originated from IREB propagation through cavities caused redistribution of energy within the IREB in such a way that coherent bunches of electrons were formed.

RF GENERATION[7,8]

It is well known that RF power can be extracted from modulated electron beams (e.g. klystron). Thus a device was built that converted the kinetic energy of the modulated IREBs into either trains of high-voltage pulses or an RF pulse. Nearly 50% efficiency was achieved.

AMPLIFIER

During the investigation of self-modulation of IREBs described earlier, we were able to show theoretically the existence of a mechanism that can be used to construct an RF amplifier of power over 10^9 W. This theoretical prediction was verified experimentally, and an RF amplifier[7,9] using an IREB was built at NRL. Figure 2 shows the schematics of the high-power RF amplifier. The beam parameters used in the amplifier experiments were: current $\cong$ 16 kA, energy $\cong$ 500 kV, single pulse of duration $\cong$ 100 ns.
the most significant results from the amplifier work are:

* A low-power RF source completely modulates an IREB that has 10^4 to 10^5 times higher power. (Figure 3).

* Unlike a conventional klystron, a long drift length for beam bunching is not necessary.

* The frequency of modulation is 1.3 GHz, and phase variation is less than 4°.

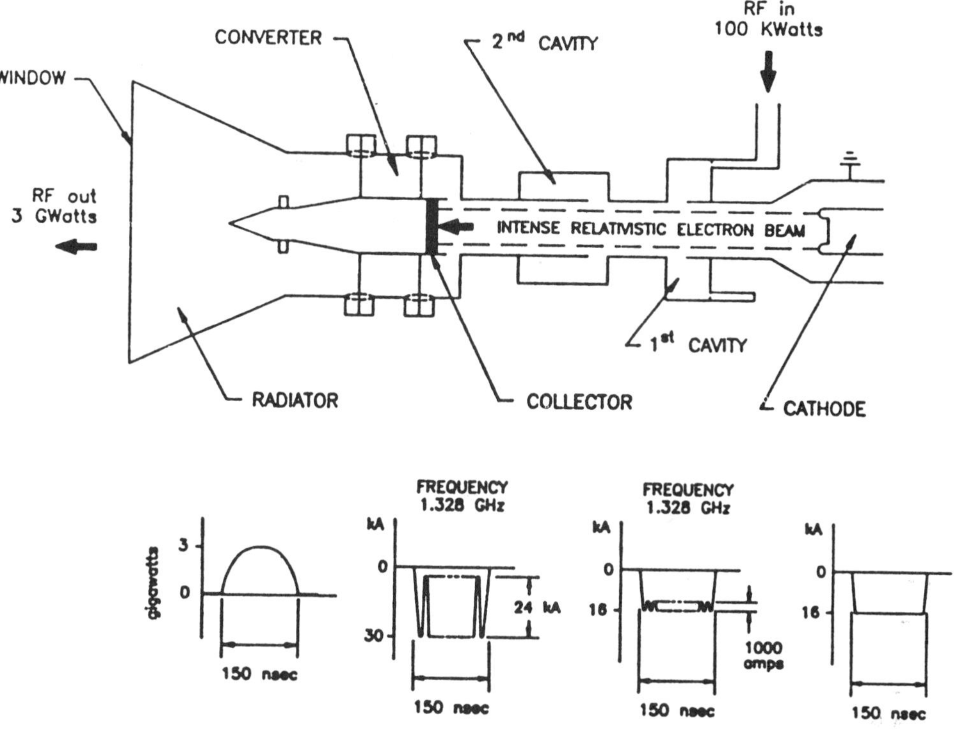

Fig. 2 High Power Relativistic Klystron Amplifier

* RF Power extraction with an efficiency of 40% has been achieved at power of 3 gigawatts. (Figure 4).

* The DC space charge of the annular beam provides significant electrostatic insulation [10] against RF breakdown at the gap.

* Theoretical analysis predicts that beam modulation and RF power extraction at higher frequencies (probably as high as 10 GHz) is possible with similar gain and power level.

$$2I_1 = 17.5 \ kA$$

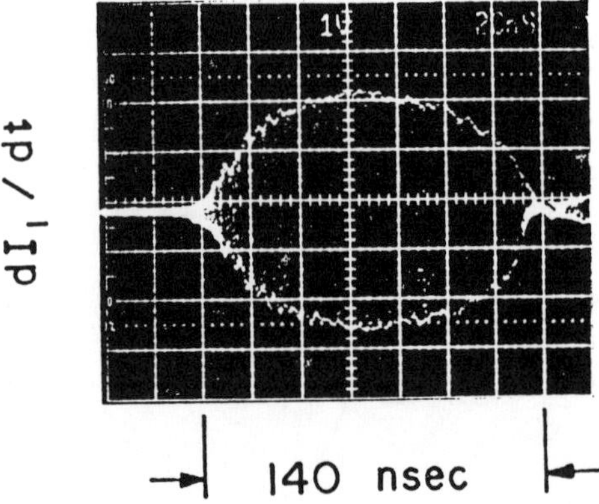

Fig. 3. Modulated IREB current measured at an axial position between the 2nd cavity and the converter.

APPLICATIONS OF MODULATED INTENSE RELATIVISTIC ELECTRON BEAMS TO CONTROL THERMONUCLEAR FUSION.

The interaction of modulated IREBs with plasmas can be used in thermonuclear fusion research. We suggested[12] that this interaction:

(1) Is a very efficient resonant-heating process that can be tuned to deliver energy on a controlled surface within the plasma.

(2) Allows greater RF penetration since the RF field is carried into the plasma via modulated IREBs.

(3) Does not need antennas to be placed near the plasma boundary.

(4) Does not have problems associated with high density cutoff and RF penetration into the plasma.

(The interested reader should consult references [13-16].

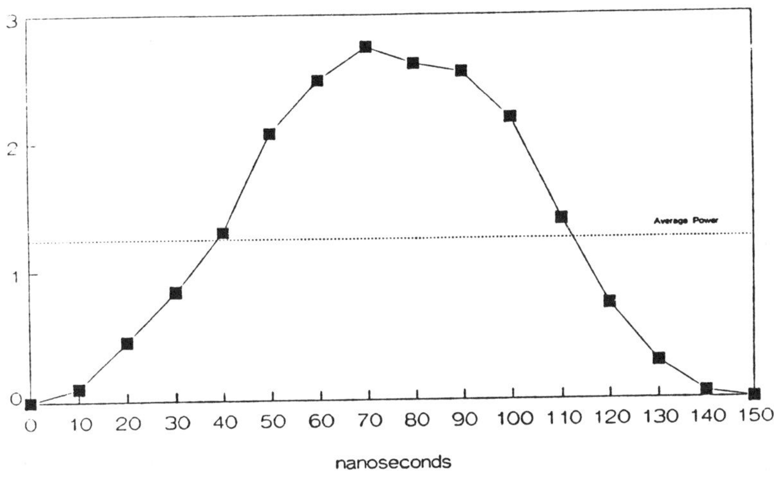

Fig. 4. Output RF Power at Frequency 1.328 GHz

ELECTRON ACCELERATORS DRIVEN BY MODULATED INTENSE RELATIVISTIC ELECTRON BEAMS

A modulated intense relativistic electron beam was used to power an accelerator structure directly[17]. The new mechanism of acceleration employs two beams of charged particles which interact via a metallic structure. The first beam, the MIREB, generated a standing wave electromagnetic field in a slow wave structure which in turn accelerated a low current secondary beam to high energy (Figure 5). Electrons of peak energy of 60 MeV were detected with a peak current of 200 Amp. An average voltage gradient of 60 MV/M and peak voltage gradient of 120 MV/M were established in the 1 meter long accelerator structure. We believe that in a future experiment an average voltage gradient >200 MV/M can be established along a similar structure. The reasons for that are:

(1) By using a traveling wave accelerating structure the average and peak electric fields are equal. Hence, an average voltage gradient of 100 MV/M can be achieved.

(2) The MIREB power can be increased by a factor of 10 from the present experiment resulting in an increase in the average electric field by a factor of 3.

(3) Using a MIREB of high frequency can increase the electric field intensity (e.g. a factor of 2 for a 5 GHz modulation).

One can look at this device as a single stage of a multistage high energy accelerator.

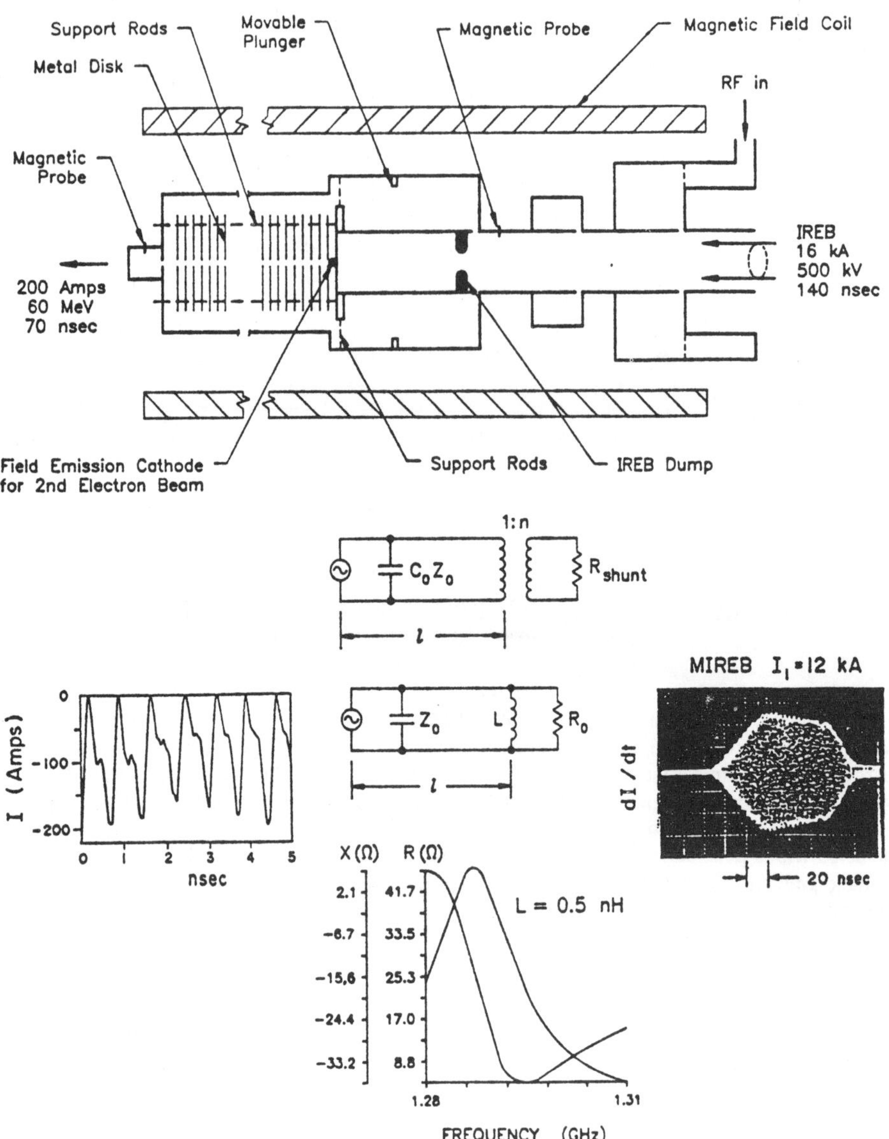

Fig. 5. *(Top)* accelerating structure powered by a modulated intense relativistic electron beam, *(Middle Right)* MIREB current, *(Middle Left)* accelerating current, *(Middle Center)* electric model of the structure, *(Bottom)* input impedance seen by the MIREB.

REFERENCES

1. N.F. Kovalev, M.I. Petelin, M.D. Raizer, A.V. Smogonskii, and L.E. Tsopp, Pisma Za. Eksp. Theor, Fiz $\underline{18}$, 232 (1973) , JETP Lett. $\underline{18}$, 138 (1973).

2. M. Friedman, IEEF Trans. Nucl. Sci, $\underline{\text{NS-26}}$, 4186 (1979).

3. J.P. VanDevender (Private Communication, 1982).

4. M. Friedman, V. Serlin, A. Drobot and L. Seftor, Phys. Rev. Lett $\underline{50}$, 1922 (1983).

5. M. Friedman, V. Serlin, A. Drobot and L. Seftor, J. App. Phys. $\underline{56}$, 2459, (1984).

6. M. Friedman, V. Serlin, Phys. Rev. Lett. $\underline{55}$, 2860 (1985).

7. M. Friedman, J. Krall, Y.Y. Lau, and V. Serlin, J. App. Phys. $\underline{69}$, 3353 ·1988).

8. M. Friedman, V. Serlin, Rev. Sci. Inst. $\underline{54}$, 1764 (1983).

9. M. Friedman, V. Serlin, Y.Y. Lau, J. Krall, to be published in the S.P.I.E. Proceedings (1989).

10. M. Friedman and V. Serlin, IEEE Trans. Elect. Insul. $\underline{\text{EI-23}}$, 51 (1988).

11. M. Friedman and V. Serlin, Rev. Sci. Instrum $\underline{54}$, 1764 (1983).

12. M. Friedman, V. Serlin, A. Drobot and A. Mondelli, IEEE Trans. Plasma Physics Science $\underline{\text{PS-14}}$, 201 (1982).

13. A.A. Mondelli and P.L. Auer, Plasma Phys. $\underline{17}$, 213 (1975).

14. Ya. B. Fainberg, Czech J. Phys. $\underline{\text{B18}}$, 652 (1968).

15. G.R. Allen et al., Phys. Rev. Lett. $\underline{41}$, 1045 (1978).

16. K. Yatsui et al., Phys. Rev. Lett. $\underline{73}$, 321 (1979).

17. M. Friedman, V. Serlin, Y.Y. Lau and V. Serlin, Phys. Rev. Lett $\underline{63}$, 2468 (1989).

This work is supported by the U.S. Office of Naval Research

TEST-PARTICLE MOTION IN A RELATIVISTIC PLASMA

Ph. de GOTTAL

Faculté des Sciences, Université de Mons, B-7000 MONS, Belgique

J. GARIEL

Université Paris VI, Institut Henri Poincaré, 11, rue P. et M. Curie,
F - 75231 Paris Cedex 05, France

ABSTRACT

Starting from the covariant relativistic Balescu-Lenard (or Silin-Klimontovich) equation we derive a classical relativistic Fokker-Planck equation for a test-particle in a plasma with collective effects. A H-theorem is proven and the geometrical form of the diffusion tensor and friction 4-vector is given. Longitudinal oscillations of the plasma appear, but transversal ones (Cerenkov effect) do not. In the case of the weak coupling (or Landau) approximation, the behaviour of the diffusion tensor and the friction 4-vector is analyzed for various dynamical conditions. It is shown that no phenomenological Langevin-type theory can cover all cases. Finally we study the perfect Lorentz-gas : we derive the relativistic Fokker-Planck equation for a light test-particle in a plasma of infinitely heavy ions. A partial H-theorem results and the solution to this equation may be explicitly computed.

INTRODUCTION

One knows that for a classical relativistic plasma the appropriate kinetic equation is the Silin-Klimontovich (or relativistic Balescu-Lenard) equation, which includes the plasma collective behaviour. We start from a covariant form of this equation, transform it into the Fokker-Planck form and present the general geometrical form of the diffusion tensor and friction four-vector, as well as a relation between them. We calculate the scalar coefficients of the diffusion tensor and express them in function of the longitudinal ε_l and transversal ε_t components of the isotropic dielectric tensor in the Lindhard representation. We display a discontinuity in the longitudinal invariant coefficient, which corresponds to longitudinal oscillations of plasma, and show that there is no similar discontinuity for the transverse part

"

(no Cerenkov radiation). We then demonstrate a H-theorem. In the case of a very dilute gas, we can use the weakly coupled (or Landau) approximation, which leads to the recovery of earlier results. In the case of an interesting limiting case, namely that of the perfect Lorentz gas, we present the exact solution to the kinetic equation.

We work in a Minkowski space with metric $g_{\mu\nu}$ (μ, ν = 0,1,2,3, : +,-,-,-). $\underline{a}$ is the euclidean 3-vector associated with any 4-vector a^μ. ∇_μ is the 4-space gradient, while ∂_μ is the derivative with respect to the 4-velocity u^μ. We use units such that the Boltzmann constant k = 1 and the velocity of light in vacuo c = 1. k^μ will designate the unit space-like

$$\text{wave 4-vector}: \quad \hat{k}^\mu = \frac{k^\mu}{(-k^2)^{1/2}} \quad , \hat{k}^\mu \hat{k}_\mu = -1 \ .$$

RELATIVISTIC COVARIANT BALESCU-LENARD EQUATION

This classical equation for the one-particle distribution function (d.f.) is the relativistic generalisation by Y. Klimontovich[3] of the Balescu-Lenard[1,2] non relativistic equation. We present it in a covariant form :

$$u_1.\nabla_1.f(1) = \frac{2e^4}{m^2} \int d^4x_2 \, \delta^4 \left(x_2 - x_1\right) \int d^4u_2 \int d^4k \, \partial_{1.k} \, |D : u_1 u_2|^2$$
$$\delta\left(k.u_1\right) \delta\left(k.u_2\right) k.\left(\partial_1 - \partial_2\right) f(1) f(2) \tag{1.1}$$

where $f(i) = f(x_i, u_i)$ (i = 1,2) is the d.f. of the particle i; (e, m) are the (charge, mass) of a medium particle (the medium is completely ionized). The D tensor appears because we take into account the collective effects

$$D_{\rho\beta} = L^{-1}\left[\mu \, \bar{\Delta}_{\rho\beta} + \varepsilon^{-1} \, \bar{u}_\rho \bar{u}_\beta \right] \tag{1.2}$$

with

$$L = -\epsilon] \mu \left(k.\bar{u}\right)^2 - \left(\bar{\Delta} : kk\right) , \tag{1.3}$$

$$\bar{\Delta}^{\rho\sigma} = g^{\rho\sigma} - \frac{\bar{u}^{\rho}\bar{u}^{\sigma}}{\bar{u}^2} \quad \text{(local barycentric projector)}$$

Let us mention that W.B. Thompson[4] uses a different interesting approach, the so-called "dressed-particle" approach, based on the relativistic Vlasov equation (self-consistent field).

FOKKER-PLANCK EQUATION

We now distinguish the test-particle (we suppress the "1"-index) with mass M, charge q, and suppose the medium at local equilibrium with a temperature T, density n and barycentric 4-velocity $\bar{u}^{\mu}$, so that

$$f(2) = f^{eq}(2) = \frac{n\alpha}{4\pi \, K_2(\alpha)} \, e^{\alpha \bar{u} u_2} \tag{2.1}$$

where $\alpha = \dfrac{m}{T}$ and $K_2(\alpha)$ is the Kelvin function of order 2. Eq. (2.1) is the Maxwell-Jüttner d.f.

(1.1) may easily be brought to the usual Fokker-Planck form

$$u. \, \nabla f = -\partial_{\mu} \left[F^{\mu} \, f \right] + \frac{T}{2M} \, \partial^2_{\mu\nu} \left[T^{\mu\nu} \, f \right] \tag{2.2}$$

with the diffusion tensor and the friction four-vector given by

$$T^{\mu\nu} = A \int d^4k \; k^{\mu}k^{\nu} \, \delta\,(k.u) \, \Phi_B \tag{2.3}$$

$$F^{\mu} = \frac{1}{2}\left(\frac{T}{M} \, \partial_{\nu} - \bar{u}_{\nu}\right) T^{\mu\nu} \tag{2.4}$$

where

$$A = \frac{2q^2 e^2 n\alpha}{MT\,K_2(\alpha)} \quad , \tag{2.5}$$

and Φ_B is the relativistic Balescu-Lenard kernel that we can calculate in the barycentric frame using the Lindhard[6] formalism :

$$\Phi_B = \frac{2}{k^4}\left\{ \frac{X^2}{|\varepsilon|^2}\left[2\alpha^2\phi'' + (1-y^2)\phi' \right] + \frac{(1-y^2)^2\left[1 - X^2(1-y^2)\right]}{\left| 1 - \varepsilon_t y^2 \right|^2}\,\phi' \right\} \tag{2.6}$$

where $\varepsilon = \varepsilon_l$ and ε_t are the longitudinal and transversal components of the isotropic dielectric tensor, and

$$\phi = \frac{1}{\left(-k^2\right)^{1/2}}\cdot\frac{e^{-Y^{1/2}}}{Y^{1/2}} \quad , \qquad \phi' = \frac{d\phi}{dY} \quad , \qquad \phi'' = \frac{d^2\phi}{dY^2} \quad , \qquad Y = -\alpha^2 u^2 \Delta^{-\mu\nu}\hat{k}_\mu\hat{k}_\nu \cdot \tag{2.7}$$

The most general form of the diffusion tensor $T^{\mu\nu}$ given by (2.3) is

$$T^{\mu\nu} = -\Psi_{B1}\,Z^\mu Z^\nu + \Psi_{B2}(\Delta^{\mu\nu} + Z^\mu Z^\nu) \tag{2.8}$$

where

$$Z^\mu = \frac{\Delta^{\mu\rho}\bar{u}_\rho}{\sqrt{\dfrac{X^2}{u^2} - \bar{u}^2}} \quad , \quad Z^2 = -1 \ , \ Z.u = 0 \tag{2.9}$$

From (2.4) and (2.8) we deduce the general geometrical form of the friction 4-vector :

$$2\,F^{\mu} = \left\{ \frac{T}{M}\left[\frac{\partial \Psi_{B1}}{\partial X} + \frac{2X}{X^2 - u^2}\left(\Psi_{B1} - \Psi_{B2} \right) \right] - \Psi_{B1} \right\} \sqrt{\frac{X^2}{u^2} - 1}\; Z^{\mu}$$

$$- \frac{T}{M}\left(\Psi_{B1} + 2\,\Psi_{B2} \right)\frac{u^{\mu}}{u^2} \tag{2.10}$$

DIFFUSION TENSOR

Ψ_{B1} and Ψ_{B2} are related in a simple way to the two fundamental invariants of $T_{\mu\nu}$:

$$T_1 = T^{\mu}{}_{\mu} \quad , \quad T_2 = T^{\mu\nu}\,\bar{u}_{\mu}\,\bar{u}_{\nu} \tag{3.1}$$

$$\Psi_{B1} = \frac{-\,T_2}{\dfrac{X^2}{u^2} - 1} \quad , \quad \Psi_{B2} = \frac{1}{2}\left(T_1 + \frac{T_2}{\dfrac{X^2}{u^2} - 1} \right) \tag{3.2}$$

Introducing (2.3), (2.6) and (2.7) in (3.1), and using spherical coordinates, we are led to :

$$\left. \begin{matrix} T_1 \\ T_2 \end{matrix} \right\} = \frac{-\,4q^2}{u_o\,v\,\pi\,M}\int_0^{\infty} d|k|\,|k|$$

$$\int_0^{\infty} dy\,\frac{\left(1 - y^2\right)}{y}\left\{ X^2\,\frac{\varepsilon_{12}}{|\varepsilon|^2} - \frac{\left[1 - X^2\left(1 - y^2\right) \right]}{\left| 1 - \varepsilon_t\,y^2 \right|^2}\,y^2\,\varepsilon_{t2} \right\}\left\{ \begin{matrix} 1 \\[4pt] \dfrac{-\,y^2}{1 - y^2} \end{matrix} \right. \tag{3.3}$$

where we have used some results of the relativistic theory of the dielectric "constant" for a plasma (Sitenko[7], Silin[8]). ε_{12} and ε_{t2} are the imaginary parts of ε and ε_t respectively; v is the particle velocity in the barycentric frame. Using the reduced variable $\left|\lambda\right| = \dfrac{\left|k\right|}{\kappa}$ (κ: inverse Debye radius) and separating the longitudinal and tranverse contributions in (3.3), leads to :

$$\left.\frac{T_1}{T_2}\right\} = \left.\frac{T_1}{T_2}\right\}_l + \left.\frac{T_1}{T_2}\right\}_t \tag{3.4}$$

longitudinal term

$$\left.\frac{T_1}{T_2}\right\}_l = \frac{-2q^2\kappa^2}{u_0 v\pi M} \int_0^v dy \frac{(1-y^2)}{y} X^2 \, f_2 \, J_1 \left\{\begin{array}{l} 1 \\[2ex] \dfrac{-y^2}{1-y^2} \end{array}\right. \tag{3.5}$$

$$J_1 = \int_0^\infty d\mu \frac{\mu}{\left(\mu + f_1\right)^2 + f_2^2} \qquad \left(\mu = \left|\lambda\right|^2\right) \tag{3.6}$$

$$f_1 = \frac{1}{\alpha K_2(\alpha)} \int_{-1}^{+1} \frac{z\,dz}{(z-y)} \left[\frac{1}{\alpha} + \frac{1}{\left(1-z^2\right)^{1/2}} + \frac{\alpha}{2\left(1-z^2\right)}\right] e^{-\alpha/\left(1-z^2\right)^{1/2}} \tag{3.6a}$$

$$f_2 = \frac{\pi}{\alpha K_2(\alpha)} \frac{y}{\left(1-y^2\right)} \left[\frac{(1-y^2)}{\alpha} + \left(1-y^2\right)^{1/2} + \frac{\alpha}{2}\right] e^{-\alpha/\left(1-y^2\right)^{1/2}} \tag{3.6b}$$

transversal term :

$$\left.\begin{matrix} T_1 \\ T_2 \end{matrix}\right\}_t = \frac{-2q^2\kappa^2}{u_o v \pi M} \int_0^v \frac{dy}{y} \left[1 - X^2(1 - y^2)\right] g_2 \, J_t \begin{cases} 1 \\ \\ \dfrac{-y^2}{1-y^2} \end{cases} \tag{3.7}$$

$$J_t = \int_0^\infty d\mu \, \frac{\mu}{\left(\mu - g_1\right)^2 + g_2^2} \tag{3.8}$$

$$g_1 = \frac{y}{2(1-y^2)} \frac{1}{\alpha K_2(\alpha)} \int_{-1}^{+1} \frac{dz}{(z-y)} \left[(1-z^2)^{1/2} + \frac{(1-z^2)}{\alpha}\right] e^{-\alpha/(1-z^2)^{1/2}} \tag{3.8a}$$

$$g_2 = \frac{y}{2(1-y^2)} \frac{\pi}{\alpha K_2(\alpha)} \left[1 + \frac{(1-y^2)^{1/2}}{\alpha}\right] e^{-\alpha/(1-y^2)^{1/2}} \tag{3.8b}$$

OSCILLATIONS OF PLASMA

We follow a line of reasoning similar to the one used by Balescu[5] in the N.R. case. For the longitudinal term, we calculate the integral J_1 (eq. (3.6)), in which we use the cut-off

$$\Lambda^2 = \left(\frac{k_M}{\kappa}\right)^2 \gg 1 \tag{4.1}$$

where k_M is the inverse critical impact parameter. We find, in the approximation where $\Lambda^2 \gg f_1, f_2$:

$$J_1 \approx 2 \ln \Lambda - \frac{f_1}{f_2}\left(\text{Arctg}\,\frac{f_2}{f_1} + n\,\pi\right) \tag{4.2}$$

$$n = 0 \qquad \text{for} \qquad \frac{f_1}{f_2} > 0$$

$$n = 1 \qquad \text{for} \qquad \frac{f_1}{f_2} < 0$$

so that (3.5) gives two contributions : the first one, with the Coulomb logarithm $\ln \Lambda$, is part of the Landau T_1 which we found[10] (the other part will stem from the transversal term, see below). The second contribution comes from collective effects, and may be written as :

$$\frac{q^2 \kappa^2}{v\,M} \frac{\chi^2}{u_o} \int_{v_c}^{v} dy\,\frac{\left(1 - y^2\right)}{y}\,f_1 \qquad \left(v \geq v_c\right) \tag{4.3}$$

where the critical velocity v_c of the test-particle is given by : $f_1(v_c) = 0$ (or $\varepsilon_{11} - 1 = 0$). Eq. (4.3) clearly represents the loss of energy through plasma oscillations which start when $v \geq v_c$, ("polarization losses").

If we again follow the same line of reasoning for the transversal term (3.7) with a similar type of approximation, namely $\Lambda^2 >> (g_1, g_2)$, we find results similar to (4.2) with f_1 replaced by $(-g_1)$. But now g_1 (or $\varepsilon_{t1} - 1$) never vanishes. This means there are no transversal plasma oscillations. In other words, there is no Cerenkov emission, in accordance with refs. 7,9.

Reverting to the longitudinal plasma oscillation term (4.3), we can evaluate it when we use the limit of f_1 (3.6a) for great values of wave length :

$$\lim_{k \to 0} f_1 = \lim_{k \to 0} (\varepsilon - 1)|\lambda|^2 \equiv \lim_{k \to 0} (\varepsilon_t - 1)|\lambda|^2 = -\frac{|\lambda|^2}{\omega^2}\,\omega_p^2 \tag{4.4}$$

where ω_p is the relativistic plasma frequency

$$\tilde{\omega}_p^2 = \frac{\omega_p^2}{K_2(\alpha)} \int_o^1 \exp\left[\frac{-\alpha}{\left(1-y^2\right)^{1/2}}\right]\left[\left(1-y^2\right)^{1/2} + \frac{\left(1-y^2\right)}{\alpha}\right] dy .$$

(4.5)

The "plasma oscillation" contribution to $T^{\mu\nu}$ is :

$$T^{pl}_{\mu\nu\,1} = \frac{-q^2\,\tilde{\omega}_p^2}{2\,v^3\,M}\,x$$

$$\left\{ -\ln\frac{v}{v_c}\left[\Delta_{\mu\nu} + \left(3 - 2v^2\right)Z_\mu Z_\nu\right] + \frac{1}{2}\left[\left(\frac{v}{v_c}\right)^2 - 1\right]\left[\Delta_{\mu\nu} + Z_\mu Z_\nu\right]\right\}$$

(4.6)

In the double non relativistic limit ($\alpha \gg 1$ and $v \ll 1$), we recover the results of Balescu[5].

H-THEOREM

The Fokker-Planck equation (2.2) may be written as

$$u \cdot \nabla f = \frac{T}{2M}\,\partial_\mu\left[T^{\mu\nu} f^{eq}\left(\partial_v \frac{f}{f^{eq}}\right)\right]$$

(5.1)

where f^{eq} is the maxwell-Jüttner d.f. (2.1) for the particle with mass M.

The neguentropy 4-vector

$$H^{\mu} = \int_{+} d^4u\, f\, u^{\mu}\, \ln \frac{f}{f^{eq}} \tag{5.2}$$

has a semi-negative 4-divergence

$$\nabla_{\mu} H^{\mu} = -\frac{T}{2M} \int_{+} d^4u\, f\, T^{\mu\nu} \left(\partial_{\mu} \ln \frac{f}{f^{eq}} \right) \left(\partial_{\nu} \ln \frac{f}{f^{eq}} \right) \leq 0 \tag{5.3}$$

if and only if the quadratic form under the integral sign is definite positive. It depends on the signs of the eigenvalues of $T^{\mu\nu}$ which are : $(0,\ \Psi_{B1},\ \Psi_{B2})$ with respective eigenvectors $(u^{\mu}, Z^{\mu}, \Delta^{\mu\nu} + Z^{\mu} Z^{\nu})$. It is easy to show from (3.3) and (3.2) with the condition $y \leq v$, that Ψ_{B1} and Ψ_{B2} are always negative. The evolution of the test-particle brings it to the Maxwell-Jüttner (2.1) form (with m replaced by M).

THE LANDAU APPROXIMATION

For a very dilute gas the medium is very weakly coupled and we can use the Landau equation in the place of the Silin-Klimontovich equation. The former is obtained from the latter by putting $\varepsilon = \mu = 1$ or $\varepsilon_l = \varepsilon_t = 1$. We recover all the results that we obtained in a earlier paper[10], particularly for the limiting cases :

Non relativistic medium (NR : $\alpha \gg 1$), non relativistic test-particle ($v \ll 1$) : we recover the results of Chandrashekar[11]. In this double limit, it is possible to introduce the brownian approximation ($m \ll M$), which leads to a degeneracy of the eigenvalues of $T^{\mu\nu}$ ($\Psi_1 = \Psi_2$) or to the isotropy of $T^{\mu\nu} - \Psi\Delta^{\mu\nu}$, as well as to a constant diffusion coefficient

$$\psi = -\frac{2}{3} \frac{\overset{NR}{v}}{\sqrt{\pi}} \tag{6.1}$$

where v^{NR} is the non relativistic collision frequency (or inverse of the relaxation time):

$$\nu^{NR} = \frac{8\pi q^2 e^2 n \ln \Lambda}{MT} \sqrt{\frac{\alpha}{2}} \tag{6.2}$$

Ultrarelativistic plasma (UR : $\alpha \ll 1$) : We find a collision frequency of the form

$$\nu^{UR} = \frac{8\pi q^2 e^2 n \ln \Lambda}{MT} \tag{6.3}$$

i.e. independent of the little mass m.

It does not seem generally possible to construct a phenomenological model for brownian motion from a Langevin type equation

$$m \frac{d u^\mu}{ds} = f^\mu \qquad (f.u. = 0) \tag{6.4}$$

This clearly follows from the fact that F^μ is generally not orthogonal to u^μ, as seen in eq. (2.10).

THE LORENTZ GAS

Another interesting limiting case which we have studied in another article[12] is that of a low density gas of electrons (mass m) interacting with infinitely heavy ions (mass M) at equilibrium. The Landau equation reduces to a Fokker-Planck equation which may be written as

$$u. \nabla f = \frac{T}{2m} \partial_\mu T^{\mu\nu} \partial_\nu f \tag{7.1}$$

with

$$T^{\mu\nu} = \Psi_2 \left(\overline{\Delta}^{\mu\nu} + \overline{Z}^\mu \overline{Z}^\nu \right) \tag{7.2}$$

$$F^\mu = \frac{T}{2m} \frac{\Psi_2}{\sqrt{X^2 - u^2 \bar{u}^2}} \bar{Z}^\mu \qquad (7.3)$$

$$\Psi_2 = \frac{-4\pi q^2 e^2 \ln \Lambda}{mT} \frac{X^2}{\sqrt{X^2 - u^2 \bar{u}^2}} \qquad (7.4)$$

$$\bar{Z}^\mu = \frac{\Delta^{-\mu\rho} \hat{u}_\rho}{\sqrt{\hat{X}^2 - 1}} \ , \quad \bar{Z}^2 = -1 \ , \quad \bar{Z}.\bar{u} = 0 \qquad \left(X = \hat{X}\, u\bar{u}\right) \qquad (7.5)$$

The Fokker-Planck operator is an orbital angular momentum operator in 4-velocity space, the geometrical part of which is

$$\bar{L}^2 f := \partial_\mu \left(\Delta^{-\mu\nu} + \bar{Z}^\mu \bar{Z}^\nu \right) f \qquad (7.6)$$

which may split as

$$\bar{L}^2 = \bar{L}^\nu \bar{L}_\nu \qquad (7.7)$$

with

$$\bar{L}^\nu := \left(\Delta^{-\nu\sigma} + \bar{Z}^\nu \bar{Z}^\sigma \right) \partial_\sigma$$

Let us mention that Anderson[13] introduced another slightly different Laplacian operator which still contains a radial part.

The solution of the Fokker-Planck equation (7.1) may be computed as an expansion in terms of the relativistic spherical harmonics

$$f\left(x^\alpha, u^\beta\right) = \sum_{l=0}^{\infty} C_{\mu 1 \ldots \mu l}\left(x^\alpha, \bar{z}, s\right) x^{\mu 1 \ldots \mu l} \tag{7.8}$$

$$X^{\mu_1 \ldots \mu_l} := \frac{1}{l!} \bar{z}^{-l+1} \frac{\partial}{\partial \bar{z}_{\mu 1} \ldots \partial \bar{z}_{\mu l}} \frac{1}{\bar{z}} \qquad \left(\bar{z}^\alpha = \bar{z}\bar{Z}^\alpha\right) \tag{7.9}$$

$$C_{\mu 1 \ldots \mu l}(s) = C_{\mu 1 \ldots \mu l}(0)\, e^{-\nu_l s} \tag{7.10}$$

with a spectrum of relaxation times

$$\nu_l^{-1} = \frac{m^2 \bar{u}\, \bar{z}^{-3}}{2\pi e^2 q^2 n \ln \Lambda\, X^2 l(l+1)} \qquad (l = 0, 1, \ldots, \infty) \tag{7.11}$$

Finally, we obtain a partial H-theorem in the sense that the initial d.f. will evolve to a final state in which it will be isotropic in the proper 3-space of the medium as previously noticed by D. Mosher[14]. Any d.f. depending on the norm z, and not only the Maxwell-Jüttner distribution, will be a stationary solution.

REFERENCES

[1] R. Balescu : Physics of Fluids $\underline{3}$, 52 (1960).

[2] A. Lenard : Ann. of Ph. $\underline{10}$, 390 (1960).

[3] Y.L. Klimontovich : The Statistical Theory of Non-Equilibrium Processes in a Plasma, MIT Press, Cambridge, MA, 1967.

[4] W.B. Thompson : Advanced Plasma Theory (Academic Press, 1964).

[5] R. Balescu : Statistical Mechanics of Charged Particles (Wiley-Interscience, London, 1963).

[6] J. Lindhard : Kgl. Danske Videnskab. Selskab, Mat.fys.Medd $\underline{28}$ n°8 (1954).

[7] A.G. Sitenko : Electromagnetic Fluctuations in Plasma (Academic Press, NY, 1967).

[8] V.P. Silin : JETP $\underline{13}$ n°2, 430 (1961).

[9] A.F. Alexandrov and al. : Principles of Plasma Electrodynamics (Springer, Berlin, 1984).

[10] Ph. de Gottal and J. Gariel : Physica A $\underline{157}$ (1989) 1059-1073.

[11] S. Chandrasehkar : Principles of Stellar Dynamics (University of Chicago Press, Chigago, IL, 1942).

[12] Ph. de Gottal and J. Gariel : Physica A (under press).

[13] J.L. Anderson : J. of Math.Phys. $\underline{15}$, 1116 (1974).

[14] D. Mosher : Phys. Fluids $\underline{18}$, 846 (1975).

REMARKS ON SHORT WAVELENGTH FREE ELECTRON LASERS

C. Pellegrini

Physics Department, University of California at Los Angeles
Los Angeles, CA 90024-1547

ABSTRACT

We review the main physics issues and the scaling laws for short wavelength Free Electron Lasers. We discuss the two main operating modes, oscillator and Self amplified Spontaneous Emission. We also compare different electron beam drivers, like RF linacs and storage rings. The favourable scaling laws and the recent progress in FELs physics and technology show that the FEL is a good candidate for a tunable Soft-X-Ray laser.

1. INTRODUCTION

The interest in Free electron Laser is due to: its large wavelength range, which at present extends from about one centimeter to 0.24 micrometer; its tunability; its high peak power[1,2]. There are two main regions of interest for FELs applications, where they can be far superior to other sources: one is in the IR and millimeter to centimeter region; the second is in the short wavelength region, below 0.1 micrometer. In this paper we will discuss the present status of the research to produce a FEL in the short wavelength region, and the characteristics and performance of such a system.

Because of its flexibility the FEL can find applications in many areas, such as particle acceleration, heating of fusion plasmas, material, biological, medical and solid state research. We just want to mention here some applications in the short wavelength region. The VUV region is of interest for chemistry. For pulse length of one picosecond or less, and large peak intensity, one can, for instance, perform fast spectroscopy and fast timing experiments in the region of 50 to 100 nm. The region around 1 to 5 nanometers would be particularly useful for X-ray microscopy and holography. X-ray microscopy of biological samples will produce images resolving cellular substructure in the natural state, without dehydration or staining, and on a picosecond time scale, before the hydrodynamic expansion of the sample becomes important. If one can produce even shorter wavelengths, below 1 nm , with the required intensity and spectral characteristics, one might be able to sequence DNA pairs by direct imaging.

The first operation of a FEL was obtained in 1976, by Madey and co-workers [3,4] at infrared wavelength. During the 13 years that have passed since many more FELS have been successfully built and operated at wavelength ranging from the centimeter to the UV, and power levels ranging from the GW to mW. A complete review of these experiments can be found in reference 1, and also in a recent paper by Roberson and Sprangle [5]. Here we will limit ourselves to mention some of the main results in the visible to UV region, which is the one more interesting for the extension to the soft X-Ray region.

FELs use different types of electron accelerators; in Table 1 we have shown some typical values for their main beam parameters. The two accelerators of interest for the short wavelength region are RF linacs and storage rings.

TABLE 1. Accelerators for FELs.

Accelerator	Energy	Peak Current	Pulse length	Wavelength
Pulsed diode	~1 MeV	1-1000 KA	~10-100 ns	cm to mm
Induction Linac	1-50 MeV	1-10 KA	~10-100 ns	cm to mm
Electrostatic	1-10 MeV	1-5 A	~10 mus	mm to 0.1 mm
RF Linac	1-1000 MeV	1-1000 A	1-10 ps	30 μm to nm
Storage Ring	0.1-10 GeV	1-1000 A	30 ps-1 ns	1 μm to nm

The Stanford superconducting RF linac was the accelerator used for the first FEL in 1976-77, and is still being used now by a Stanford-TRW group. This group has reported the operation of the first visible FEL, with a power of 21 KW at 0.52 μm [6]. Room temperature linacs have been used in the following years at Stanford[7,8,9], Los Alamos[10,11], Boeing[12] to drive FELs from 35 μm to the visible, with peak powers up to 40 MW, and pulse length as short as one picosecond. Both oscillators and master oscillator power amplifier [13] configurations have been used. Optical guiding, sidebands and harmonic generation have been observed. The RF linac can provide high quality beams of energies from a few MeV to Gev, to drive FELs in the infrared, visible or UV spectral regions. They

cation; effects like diffraction, beam energy spread and beam focusing are included; other effects, as for instance undulator
imperfections, are not and will have to be considered in a real
design. The notations we use are those of reference[20] and are: beam
energy (units mc^2), γ; Beam particle density, n_e; radiation wavelength,
λ; undulator period, λ_u; undulator field, B_u; undulator parameter,
$K = eB_u\lambda_u/2\pi mc^2$; undulator frequency, $\omega_0 = 2\pi c/\lambda_u$; beam plasma frequency $\Omega_p = (4\pi r_e c^2 n_e/\gamma)^{1/2}$.

   With these notations, and considering for simplicity a helical
undulator, we can write the FEL synchronism condition as

$$\lambda = \frac{\lambda_u}{2\gamma^2}(1 + K^2) \tag{1}$$

In the 1-D FEL theory , and for a cold beam, the radiation field in
the undulator grows exponentially until it saturates; the exponential
gain length, L_G, and the saturation power are determined by one parameter[20,22],

$$\rho = \left(\frac{K}{4\gamma}\frac{\Omega_p}{\omega_0}\right)^{2/3} \tag{2}$$

The gain length is

$$L_G = \frac{\lambda_u}{2(3)^{(1/2)}\pi\rho} \tag{3}$$

the laser power at saturation is related to the beam power, P_L, by

$$P_L \sim \rho P_{beam} \tag{4}$$

and the saturation length is

$$L_s \cong \frac{\lambda_u}{\rho} \tag{5}$$

The gain and the gain length will be changed, and usually reduced, by effects like energy spread, diffraction effects, and by how much we focus the beam through the undulator; (3) and (4) give a good approximation if some additional conditions are satisfied:

(a) limit on beam energy spread: $\sigma_\epsilon < \rho$;

(b) limit on beam emittance: $\epsilon < \frac{\lambda}{2\pi}$;

(c) condition for optical guiding: $\frac{Z_R}{L_G} > 1$;

where $Z_R = \pi \sigma_r^2 / \lambda$, is the Rayleigh range, σ_r is the beam radius, and σ_ϵ is its relative energy spread. The gain length is a very important quantity; it determines the scale length over which effects have to occur to influence the exponential growth rate; all effects which take place over a distance larger than the gain length will have very little effect on the FEL performance.

One way to increase ρ, and decrease the gain length, is to strongly focus the beam through the undulator, reducing the betatron oscillation wavelength. This, however, can produce a reduction of the gain, because during an oscillation a particle changes its position and velocity, and as a result can get out of synchronism[23]. There is one remarkable case when this reduction does not occur, and that is when the betatron oscillation wavelength is that given by the transverse focusing produced by the undulator field only

$$\lambda_{\beta 0} = \frac{2^{1/2} \gamma \lambda_u}{K} \tag{6}$$

For the typical parameters of a X-ray FEL, with a large value of gamma, this quantity tends to be larger than ten meters, limiting the value of ρ. It is then convenient to introduce extra focusing to make the FEL parameter large, accepting at the same time some reduction in gain. This reduction is small if we satisfy the condition

$$\lambda_\beta > L_G$$

so that the effect of betatron oscillations in a wavelength is negligible. The additional focusing can be obtained with external focusing elements, like quadrupoles[22], or, as proposed by Barletta and Sessler[24], with ion focusing.

It is important to notice that if conditions b, and c are satisfied than also the condition on the focusing is satisfied, since $\sigma_r^2 = \epsilon \lambda_\beta / 2\pi$. In a more complete theory of the FEL it is possible to write the gain as [25]

$$\frac{1}{L_G} = \frac{1}{L_{GO}} f\left(\frac{\epsilon}{2\pi\lambda}, \frac{\sigma_E}{\rho}, \frac{Z_R}{L_{GO}} \right) \tag{7}$$

where the function f is of the order of 1 if conditions a,b,c are satisfied; a value for f can be obtained either from a computer simulation of the FEL system, or using the results of ref. [25].

All the basic physics described in these formulae, like exponential growth from noise or optical guiding, has been proved experimentally in the near or far infrared. Assuming that the same physics remain valid at shorter wavelength, we can use these formulae to design a soft X-ray FEL.

3.   FEL SCALING LAWS

In the design of a FEL we want to maximize ρ for a given wavelength and beam characteristics. To this end we rewrite it using the beam invariants ϵ_N, ϵ_L, transverse and longitudinal normalized rms emittances (we assume for simplicity a cylindrically symmetric beam), and the longitudinal brilliance[22,26]

$$B_L = \frac{eNc}{(2\pi)^{1/2}\epsilon_L} \tag{8}$$

Using these quantities we obtain

$$\rho = \left\{ \frac{\lambda}{4\pi} \frac{K}{1+K^2} \left(\frac{\sigma_E}{E} \right)^{1/2} \left(\frac{4\pi B_L}{\lambda_\beta \epsilon_N I_A} \right)^{1/2} \right\}^{2/3} \gamma \tag{9}$$

where $I_A = ec/r_e$, and λ_β is the wavelength of the transverse (betatron) oscillation of the electrons in the undulator, produced by a focusing system. It is interesting to notice that the dependence of ρ on λ is not strong, so that a FEL at short wavelength seems feasible; in addition (9) shows that it is convenient to use a large beam energy. The cost of using a large beam energy is that, for the same wavelength, we have to increase the undulator period, and the undulator becomes longer. However, since $N_u \sim 1/\rho$ the undulator length increases only linearly with the beam energy.

The value of the FEL parameter depends now on very few quantities, the beam invariants, energy spread, and the betatron oscillation wavelength.

4. A SOFT X-RAY FEL

To design a FEL we have to maximize ρ, and at the same time satisfy the conditions a,b,c. In addition there is also another choice to be made for the FEL mode of operation. One possibility is to use an oscillator configuration, with an optical cavity[27], the other is to operate in the Self Amplified Spontaneous Emission (SASE) mode[28]. The oscillator requires a smaller gain and a shorter undulator than SASE, if the optical cavity has small losses, and this condition is difficult to satisfy at short wavelength because also the best multilayered mirrors have reflexivity of about 50%, and they are easily damaged. This damage can be enhanced by the small laser spot size, and the corresponding high power density of the radiation. To increase the mirror reflexivity Newnam has suggested to use multifaceted metal mirrors operating by total external reflection [29]. Another improvement can be made changing the optical cavity configuration from a two mirror design with near perpendicular angle of incidence, to a ring resonator with many mirrors and glazing angle of incidence[27].

The SASE avoids the mirrors and optical cavity problems, at the cost of using a longer undulator. In a SASE FEL one sends the electron beam through a long undulator; the spontaneous emission produced in the initial part of the undulator couples to the transverse particle velocity to modulate the electron energy on the scale of the radiation wavelength. This energy modulation leads to a bunching of the particles on the same scale length, and this bunching enhances the emission of radiation. It is clear that this process leads to an instability and to exponential growth of the radiation. This mechanism is similar to that of the negative mass instability in particle accelerators, although the radiation-beam coupling is with the transverse beam velocity instead of the longitudinal velocity. The radiation growth saturates when the electrons are trapped in the potential well formed by the combined action of the radiation and undulator field (ponderomotive potential well).

Using the model discussed above, one can design a Soft X-ray FEL based on SASE. Several designs have been made; an example of such a system is given in Table 2.

TABLE 2.   Example of Soft X-Ray FEL

Wavelength, nm	5
Normalized emittance (rms), mm mrad	1
Electron energy, GeV	1.0
Longitudinal Brilliance, A	5000
Energy spread, (percent)	0.1
Peak Current, A	600
λ_β, m	3.3
ρ	0.0024
Undulator Period, cm	2
Gain Length, m	1.8
Rayleigh Length, m	1.7
Beam Power, GW	600
Laser Power, GW	1.4

The most stringent requirements for the realization of a X-ray FEL are those on the electron beam. A high beam density is needed to obtain a value of ρ in the range of 10^{-3}, higher in fact that what has been achieved up to now. It is interesting to notice that a similarly high density is also needed for high energy linear colliders and high brightness synchrotron radiation sources. As we already mentioned two routes to high density electron beams are being followed, using storage rings at Duke University, Dortmund and Berkeley, or linacs at Los Alamos, Brookhaven and UCLA. These have

been reviewed at a Workshop held at Brookhaven in 1987[30], were it
was assumed that the beam parameters needed for a Soft X-ray Fel are
$B_L = 200A$, and $\epsilon_N = 1\,mmmrad$.

As an example of small emittance, high current storage ring we
can consider the SLAC damping ring[31], with an energy of 1.2 GeV, normalized emittance of 20 mm mrad, and a longitudinal brilliance of 120
A. When trying to further reduce the emittance we are faced with the
fact that also the ring momentum compaction decreases, and this tends
to lower the threshold for the microwave instability, which is also
proportional to the momentum compaction, thus limiting the longitudinal brilliance[32]. Some new ideas, like wiggler rings, were also
discussed at the Brookhaven workshop, and work is continuing in these
directions.

Progress has also been made in the production of small emittance, large brilliance beams from electron guns, making the linac a
possible option. Very good results have been obtained using large (30
to 100 MV/m) accelerating fields on the cathode, and laser driven
photocathodes for picosecond pulses, or longer pulses followed by
magnetic compression to reduce space charge effects at low beam energy[33,34,35]. As an example, the Los Alamos gun, operating at about 1.3
GHz, with a Cs_3Sb cathode, and a field on it of 100 MV/m, has
produced a beam with a normalized rms emittance of about 10 mm mrad,
and a longitudinal brilliance of 2000A.

In our examples in Table 2 we have assumed that it will be possible in the near future to reduce the normalized emittance to about
1 mm mrad. The case with an emittance of 3 mm mrad is given to show
the sensitivity of the FEL performance to this beam parameter. The
limitations on the beam emittance produced by a RF gun have been
analyzed[36]; the main effects producing an emittance blow up are:
space charge, non linear and time dependent RF fields, cathode temperature and current density. For present systems the main contribution to the emittance is produced by space charge effects near the
cathode, before the beam becomes relativistic. Particularly important
is the fact that the space charge force is proportional to the local
longitudinal charge density, and so it varies from the center to the
tails of the bunch. This effect, as well as similar other effects,
can be at least partly controlled by shaping the laser pulse, to
produce a step like density distribution, or by selecting the core of
the bunch out of a longer one, or by introducing time dependent
focusing elements. The core selection can be done by utilizing the
particle phase-energy correlation at the gun output, and by filtering

the beam through a momentum selecting slit. We expect that using
these techniques we will reduce the emittance value by one order of
magnitude in the near future.

The longitudinal brilliance produced by existing RF gun is
already adequate for a X-Ray FEL. However beam loading effects during
the beam acceleration usually blow it up. To reduce this effect and
achieve a large peak current and a small energy spread we can use
the bunch compression technique at several stages during the acceleration process. For the example of Table 4 we have assumed that
$\epsilon_L = 2.10^-5m$, at 5 MeV corresponding to a charge in the bunch of 1 nC,
a 0.3 % energy spread and a rms bunch length of 0.2 mm; at 100 MeV we
assume that the emittance is $2.4.10^{-4}$, for an energy spread of 0.2%,
determined by beam loading, and the same pulse length; at this energy
we can compress the bunch increasing the energy spread by 3 and
reducing the pulse length to 0.2 mm; subsequent acceleration to 1 GeV
would reducing the energy spread to less than 0.1% and for the same
bunch length; in effect the energy spread is determined again by beam
loading, and for this very short bunch length we expect this to
remain at 0.1%. Using this beam manipulation the final peak current
is 600 A, as shown in Table 2. This value is consistent with results
obtained in the SLC, or for the proposed Final Focus Test Facility at
SLAC, where the peak current is 400 A with an energy spread of 0.1%.

This recent progress in the production of high quality electron
beams, using either storage rings or electron guns and linacs, gives
us confidence that it will be possible , during the next few years,
to produce beams with characteristics as in Table 4, adequate to
drive an FEL in the few nm region.

Several studies and numerical simulations have also been done to
determine the effect of errors in the undulator. The effect of
wiggler field errors is estimated assuming that one can correct the
trajectory but not the phase errors; in this case field errors on the
order of 0.2% can be tolerated without any appreciable reduction in
output power. A soft X-ray FEL will thus require the fabrication of
long, high accuracy, high field undulators, with precision beam control. The considerable progress achieved in this area in the last few
years let us believe that this can be done. As an example an 80
period , two meters long undulator built by Rocketdyne for MarkIII
FEL[37], has demonstrated uncorrelated field errors which would satisfy our requirements.

To reduce the undulator length and make a soft X-Ray FEL smaller and less expensive, it is possible to split the undulator in two parts separated by a bunching section [38], as in an Optical Klystron. This can reduce the undulator length in half. For the example of Table 4, the undulator would then be about 10 m long.

5. CONCLUSIONS

As already discussed the key issues in the development of the Soft X-Ray FEL are: development of electron beams with the required six-dimensional phase space density; development of high precision undulators; verification that our understanding of the scaling laws can be extended to the shorter wavelength. Programs in these directions exist now at several laboratories, like Duke, Los Alamos, Brookhaven and UCLA. The UCLA program is starting now; its goals are:

1. Research on the electron source using the photocathode and RF gun approach, to reduce the beam emittance by shaping the laser pulse, and or selecting the beam core; development of bunch compression techniques that do not blow up the beam emittance;

2. Verification of the FEL physics and of the possibility of using the Optical Klystron in the high gain; regime in the wavelength region of 1 to 10 μm.

This experiment will use a 20 MeV high gradient S band linac with a built in photocathode and a 1.5 cm period undulator consisting of three sixty cm long parts. The expected initial beam characteristics are: normalized emittance smaller than 10 mm mrad, relative energy spread of 0.2%, charge per bunch of 1 nC, and peak current of 200 A. The corresponding gain length is 15 cm. The experiment will utilize a single electron bunch traversing the undulators. Combining the three undulator parts we can explore the optical klystron and the Self Amplified Spontaneous Emission regime, optical guiding and other effects like sidebands instabilities. After initial operation we will start a program to improve the beam characteristics using the radiation from the undulator as a diagnostic tool. As the beam quality improves one can also increase the beam energy and reduce the radiation wavelength.

The work now being done to improve and extend the FEL will lead to a better radiation source in the near future.

REFERENCES

1. W. Colson, C. Pellegrini, and A. Renieri, eds. *Free Electron Laser Handbook*, North Holland, Amsterdam, (in press).

2. T.C. Marshall, *Free Electron Laser*, McMillan, New York (1985).

3. L.R. Elias et al., Phys. Rev. Lett. 36, 717 (1976).

4. D.A.G. Deacon et al., Phys. Rev. Lett., 38, 892 (1977).

5. C.W. Roberson and P. Sprangle, Phys. Fluids, B1, 3 (1989).

6. J.A. Edighoffer et al., Appl. Phys. Lett. 52, 1569 (1988).

7. S.V. Benson and J.M.J. Madey, Nucl. Instrum. Methods, A32, 55 (1984).

8. S.V. Benson, et al. Nucl. Instrum. Methods, A272, 22 (1988).

9. J.E. La Sala, D.A.G. Deacon, and J.M.J. Madey, Phys. Rev. Lett. 59, 2047 (1988).

10. B.E. Newnam et al., IEEE J. Quantum Electron, QE-21, 867 (1985).

11. D.W. Feldman et al., IEEE J. Quantum Elec. QE-23, 1476 (1987).

12. T.W. Meyer et al., Nucl. Instrum. Methods, in course of publication.

13. L. Vintro et al., Phys. Rev. Lett.

14. M. Billardon et al., IEEE J. Quantum Electron. QE-21, 805 (1985).

15. M. Billardon et al., Nucl. Instrum. Methods, A259, 72 (1987).

16. N.A. Vinokurov and A.N. Skrinsky, preprint of the Institute of Nuclear Physics, 77-59, Novosibirsk (1977).

17. K. Robinson, in Proc. of Intern. Workshop on Coherent and Collective Properties in the Interaction of Relativistic Electrons and Electromagnetic Radiation, North Holland, Amsterdam, p. 111 (1985).

18. R. Prazeres et al., Nucl. Instrum. and Methods A272, 199 (1988).

19. I.B. Drobyazko et al., "Lasing in Visible and Ultraviolet Regions in Optical Klystron Installed on the Vepp-3 Storage Ring,'' to be published in Proc. of the International Congress on Optical Science and Engineering, Paris (1989).

20. R. Bonifacio, C. Pellegrini, and L. Narducci, Opt. Commun. 50 373 (1984).

21. P. Sprangle and R.A. Smith, Phys. Rev. A21, 293 (1980).

22. C. Pellegrini, Nucl. Instrum. Methods, $\underline{A272}$ 364 (1988).

23. W.M. Fawley, D. Prosnitz, and E.T. Scharlemann, Phys. Rev. $\underline{A30}$ 2472 (1984).

24. W.A. Barletta and A.M. Sessler,"Radiation from Fine Intense Self Focused Beams at High Energy," to be published in Nucl. Instrum. Methods, also LLNL Report UCRL-98767 (1989).

25. L.H. Yu et al., to be published in Phys. Rev. Lett.

26. C.W. Roberson, IEEE J. Quantum Electr. QE-21, 860 (1985).

27. J.C. Goldstein, B.D. McVey and C.J. Elliott, Nucl. Instrum. Methods, $\underline{A272}$, 177 (1988).

28. J.B. Murphy and C. Pellegrini, J. Optic. Soc. Am. $\underline{B2}$, 259 (1985).

29. B.E. Newman, in "Laser Induced Damage in Optical Materials," H.E. Bennett et al., eds. NBS (1985).

30. K. Wille, Nucl. Instrum Methods $\underline{A272}$, 59 (1988).

31. G.E. Fisher et al., "A 1.2 GeV Damping Ring Complex for the Stanford Linear Collider," Proc. 12th Intern. Conf. on High Energy Accelerators, Fermi National Accelerator Laboratory (1983).

32. See for instance the papers in Ref. 45 by J.B. Murphy, "Storage Ring Lattice Considerations for Short Wavelength Single pass FEL," p. 197, or A. van Steenbergen, "Collider Damping Ring: Theoretical Minimum Emittance Structure," p. 60, or A. Ruggiero, "Conceptual Design of the Damping Ring," p. 45.

33. R.L. Sheffield, E.R. Gray, and J.S. fraser, Nucl. Instrum. Meth-Methods, $\underline{A272}$, 222 (1988).

34. S.V. Benson et al., Nucl. Instrum. Methods, $\underline{A250}$, 39 (1986).

35. K. Batchelor et al., "Development of a High Brightness Electron Gun for the Accelerator Test Facility at Brookhaven," Proc. European Particle Accelerator Conf. p. 954, Rome (1988).

36. K.J. Kim "Coherent and Collective Properties in the Interactions of Relativistic Electron and Electromagnetic Radiation," R. R. Bonifacio, F. Casagrande, and C. Pellegrini eds., North Holland, Amsterdam (1985), also Nucl. Instrum. Methods, $\underline{A239}$, (1985).

37. M. Curtin et al., Nucl. Instrum. Methods $\underline{A272}$, 187 (1988).

38. J. Gallardo and C. Pellegrini, Optics Comm., $\underline{77}$, 45 (1990).

Intense Electron Beam and Free Electron Lasers[*]

D. Prosnitz
Lawrence Livermore National Laboratory
Livermore, California

ABSTRACT

The advent of high-quality high-current electron beams has not only permitted the operation of high-gain free-electron lasers but has also allowed other devices to operate in regimes not previously accessible. This talk will review the status of high-current induction-linac-produced electron beams, high-gain free-electron lasers, and the applications and challenges they present.

In particular, it will highlight the need for closed-loop control of accelerator power (modulator) systems and real-time micro-control of electron beam characteristics such as energy, current, and position. Progress in these areas will be discussed.

[*]Performed jointly under the auspices of the U.S. DOE by LLNL under W-7405-ENG-48 and for the DOD under SDIO/SDC-ATC MIPR No. W31RPD-9-D5007.

The Beam-Beam Interaction

Jonathan S. Wurtele
Department of Physics and Plasma Fusion Center
Massachusetts Institute of Technology
Cambridge, MA 02139

Abstract

The development of the next generation of high energy linear electron-positron colliders has motivated numerical and analytical investigations of the radiation and disruption of colliding electron-positron bunches. I will present results on beam-beam collisions using two different models. The first model simplifies the interaction to a set of coupled envelope equations and exhibits the qualitative behaviour of two dimensional plasma simulations at a fraction of the computational effort. In the second model the beams are broken into longitudinal slices and each slice is represented by a collection of macroparticles. As the bunches collide, particles in a given slice interact only with those in that slice of the opposing bunch which is at the same axial position. The interaction is modelled by the Coulomb force superposition. The results of three dimensional calculations will be presented.

Part IV

COHERENT RADIATION GENERATION AND PARTICLE ACCELERATORS

CHAOTIC ELECTRON MOTION INDUCED BY TRANSVERSE FIELD INHOMOGENEITIES IN FREE ELECTRON LASERS[†]

Chiping Chen and Ronald C. Davidson
Plasma Fusion Center
Massachusetts Institute of Technology
Cambridge, Massachusetts 02139

ABSTRACT

The motion of a relativistic test electron in a free electron lasers can be altered significantly by the equilibrium self-field effects produced by the beam space charge and current and by the transverse spatial inhomogeneities in a realizable magnetic wiggler field. In a field configuration consisting of an ideal (constant-amplitude) helical wiggler field and a uniform axial guide field, it is shown that the inclusion of self-field effects destroys the integrability of the motion, and consequently the Group-I orbits and the Group-II orbits become chaotic at sufficiently high beam density. An analytical estimate of the threshold value of the self-field parameter $\epsilon_s = \omega_{pb}^2/c^2 k_w^2$ for the onset of chaoticity is obtain. In addition, the effects of transverse spatial gradients in a realizable helical wiggler field with three-dimensional spatial variations are investigated in the absence of an axial guide field, but including self-field effects. For a thin electron beam ($k_w^2 r_b^2 \ll 1$) and small wiggler amplitude ($a_w^2 \ll \gamma_b^2$), it is shown that the motion is regular and confined radially provided that $\epsilon_s < \gamma_b a_w^2/(1+a_w^2)$. However, because of the intrinsic nonintegrability of the motion, the regular region in phase space diminishes in size as the wiggler amplitude is increased. Moreover, it is found in both cases that chaoticity developes on a time scale comparable with the beam transit time through one wiggler period.

I. INTRODUCTION

Hamiltonian chaos[1−3] has been an active area of research in physics and applied sciences. The classic work of Kolmogorov,[4] Arnol'd[5] and Moser[6] shows that the generic phase space of integrable classical Hamiltonian systems, subject to small perturbations, contains three types of orbits: stable periodic orbits, stable quasiperiodic orbits (KAM tori), and chaotic orbits. Chaotic orbits are sensitive to the initial conditions. In nonintegrable Hamiltonian systems with three-dimensional phase space, different chaotic regions are isolated by the KAM tori. As the perturbation increases in strength, the KAM tori destabilize and

---

[†]Research supported in part by the Department of Energy High Energy Physics Division, the Office of Naval Research, and the Naval Research Laboratory Plasma Physics Division.

become discrete *fractal* sets.[7] In wave-particle interactions, the breakdown of the last global KAM torus results in stochastic acceleration of particles. An example of such a phenomenon is the stochastic ion heating by a single electrostatic wave in a magnetized plasma.[8-10] The purpose of this paper is to examine chaotic behavior in particle orbits in free electron lasers. In contrast to the stochastic heating of ions, which is useful in controlled thermonuclear fusion research, the presence of chaoticity in the particle orbits in free electron lasers poses potential problems for laser operation in certain parameter regimes.

The free electron laser (FEL)[11-13] makes use of the unstable interaction of a relativistic electron beam with a transverse wiggler magnetic field to generate coherent electromagnetic waves. As demonstrated in various experiments,[14-18] free electron lasers have several remarkable properties, including frequency tunability, high efficiency, high power, and optical guiding by the electron beam. An important parameter characterizing free electron laser operation is the small-signal gain (growth rate). According to linear theory,[19,20] the gain increases as the beam density and the strength of the wiggler field are increased, whereas the gain decreases as the axial momentum spread of the electrons is increased. However, in the high-current (high-density) regime and the intense wiggler field (strong-pump) regime, the electron orbits can be modified significantly by the equilibrium self fields of the electron beam and the transverse spatial gradients in the applied wiggler field. This raises important questions regarding beam transport and the viability of the free electron laser interaction process in these regimes.

This paper examines the motion of a relativistic test electron in a helical-wiggler free electron laser in the absence of any electromagnetic signal wave. Of particular interest are the effects of transverse gradients in the beam-produced self fields and the realizable helical wiggler field on the dynamics of the test electron. In the high-current (high-density) regime, the self-electric and self-magnetic fields[21] of a nonneutral electron beam play an important role in altering the particle orbits, and an axial guide field is often used to provide transverse confinement of the beam electrons. The particle orbits in a helical wiggler field and a uniform axial guide field have been calculated,[22,23] including the effects of transverse spatial inhomogeneities in a realizable wiggler field.[24,25] However, the effects of equilibrium self fields are neglected arbitrarily in most treatment. In Sec. III, we analyze the particle motion in the combined field configuration consisting of an ideal (constant-amplitude) helical-wiggler field $\vec{B}_w^{(0)}(z) = -B_w[\vec{e}_x \cos(k_w z) + \vec{e}_y \sin(k_w z)]$ (with $B_w = $ const.), a uniform axial guide field $\vec{B}_0 = B_0 \vec{e}_z$, and the self-electric and self-magnetic fields produced by the space charge and current of a uniform-density electron beam.[26] It is shown that the inclusion of self-field effects destroys the integrability of the motion, and consequently the Group-I orbits and the Group-II orbits become fully chaotic when the self fields are sufficiently large (which requires sufficiently high beam density). An analytical estimate of the threshold value of the self-field parameter $\epsilon_s = \omega_{pb}^2/c^2 k_w^2$ for the onset of the chaoticity is obtained and found to be in good agreement with computer simulations. In addition, the characteristic time scale for self-field-induced changes in the particle orbits is shown to be of order the time require for a beam electron to transit one wiggler period.

In contrast to the high-current (high-density) regime, an intense (realizable) wiggler field provides a betatron focusing force so that, in the absence of a uniform axial guide field, the electron beam can be confined radially for the case of a helical-wiggler field configuration. For sufficiently small wiggler amplitude, the particle orbits are a superposition of well-defined helical motion and betatron oscillations. In Sec. IV, a condition for radial confinement of particle orbits is derived for a thin electron beam and small wiggler amplutide. In addition, it is shown for sufficiently large wiggler amplitude that the motion can become chaotic, particularly for off-axis particle orbits. For the special case where self-field effects are negligibly small, it is found that the onset of chaoticity occurs whenever the dimensionless parameter $\Delta = a_w/[2(\gamma_b^2 - 1 - a_w^2)]^{1/2}$ exceeds the critical value $\Delta_c = 0.28$, which corresponds to the maximum allowed wiggler amplitude $a_w^c \cong 0.37(\gamma_b^2 - 1)^{1/2}$ for the existence of regular helical orbits for given electron energy γ_b.[24] This suggests that there is an upper bound on the wiggler field strength for free electron laser operation. Similar results are also obtained for off-axis particle orbits in a planar-wiggler field configuration.

The organization of this paper is as follows. In Sec. II, a general formulation of the dynamical problem is given in canonical variables. In Sec. III, equilibrium self-field effects on particle orbits are examined in the applied field configuration consisting of a uniform axial guide field and an ideal helical wiggler field. In Sec. IV, the particle orbits are examined in a realizable helical wiggler field in the absence of an axial guide field, but including self-field effects. Finally, conclusions are given in Sec. V.

II. THEORETICAL MODEL AND ASSUMPTIONS

We consider a relativistic, cylindrical electron beam with radius r_b propagating in the z-direction through the externally applied magnetic field configuration

$$
\begin{aligned}
\vec{B}^{ext}(\vec{x}) &= B_0 \vec{e}_z + \vec{B}_w(\vec{x}) \\
&= B_0 \vec{e}_z - B_w\{[I_0(k_w r)\cos(k_w z) + I_2(k_w r)\cos(k_w z - 2\theta)]\vec{e}_x \\
&\quad + [I_0(k_w r)\sin(k_w z) - I_2(k_w r)\sin(k_w z - 2\theta)]\vec{e}_y \\
&\quad - 2I_1(k_w r)\sin(k_w z - \theta)\vec{e}_z\} \, .
\end{aligned}
\tag{1}
$$

Here, $B_0\vec{e}_z$ is the uniform axial guide field, and $\vec{B}_w(\vec{x})$ is the realizable helical

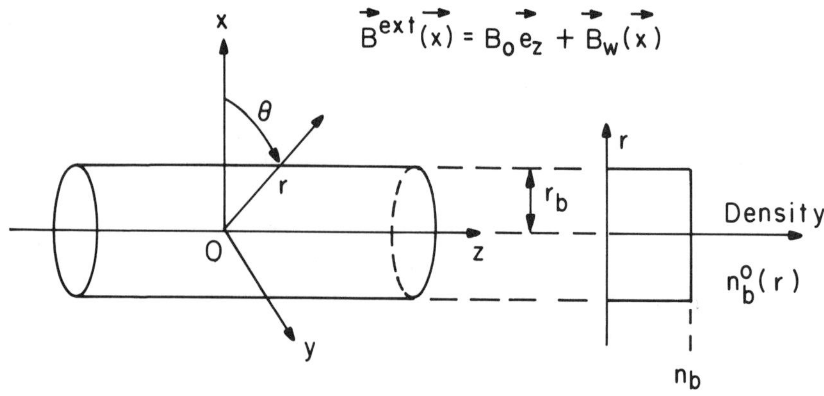

Fig. 1. Schematic of relativistic electron beam and coordinate system. Here, n_b, r_b, and V_{zb} are the density, radius, and average axial velocity of the electron beam, $B_0 \vec{e}_z$ is the axial guide field, and $\vec{B}_w(\vec{x})$ is the transverse wiggler field.

wiggler field with amplitude $B_w = $ const. and wiggler wavelength $\lambda_w = 2\pi/k_w = $ const. In Eq. (1), $I_n(x)$ is the modified Bessel functions of the first kind of order n, $r = (x^2 + y^2)^{1/2}$ is the radial distance from the axis of the helix, and (r, θ) are cylindrical polar coordinates with $x = r\cos\theta$ and $y = r\sin\theta$ (Fig. 1). It is readily shown that the wiggler field defined by Eq. (1) satisfies the vacuum Maxwell equation $\nabla \times \vec{B}_w(\vec{x}) = 0$.[24] In addition, the electron beam is assumed to have uniform density

$$n_b^0(r) = \begin{cases} n_b = \text{const.}, & 0 \leq r < r_b \ , \\ \\ 0 \ , & r > r_b \ , \end{cases} \qquad (2)$$

and uniform axial current density $J_{zb}^0(r) = -n_b^0(r)e\beta_{zb}c$ over the radial cross section of the electron beam. Here, $-e$ is the electron charge, c is the speed of light *in vacuo*, and $V_{zb} \equiv \beta_{zb}c = $ const. is the average axial velocity of the electron beam. It is readily shown from the steady-state Maxwell equations that the beam space charge and current generate the self-electric and self-magnetic fields[21]

$$\vec{E}_s = -\frac{m\omega_{pb}^2}{2e}(x\vec{e}_x + y\vec{e}_y) \ , \qquad (3)$$

and

$$\vec{B}_s = \frac{m\omega_{pb}^2\beta_{zb}}{2e}(y\vec{e}_x - x\vec{e}_y) \ , \qquad (4)$$

in the beam interior ($0 \leq r < r_b$). In Eqs. (3) and (4), m is the electron rest mass, and $\omega_{pb} = (4\pi n_b e^2/m)^{1/2}$ is the nonrelativistic plasma frequency of the beam electrons.

It is the primary purpose of this paper to examine the motion of an individual test electron in the combined applied field configuration and self fields described by Eqs. (1), (3) and (4). In this regard, it is convenient to represent the equilibrium fields as

$$\vec{E}^s(\vec{x}) = -\nabla\Phi_s(\vec{x}) , \tag{5}$$

and

$$\vec{B}^{ext}(\vec{x}) + \vec{B}^s(\vec{x}) = \nabla \times \vec{A}(\vec{x}) , \tag{6}$$

where

$$\Phi_s(\vec{x}) = \frac{m\omega_{pb}^2}{4e}(x^2 + y^2) \tag{7}$$

is the electrostatic potential for $0 \leq r < r_b$. In Eq. (6), the total vector potential $\vec{A}(\vec{x})$ can be expressed as

$$\vec{A} = B_0 x \vec{e}_y + \vec{A}_w(\vec{x}) + \beta_{zb}\Phi_s(\vec{x})\vec{e}_z , \tag{8}$$

where $\nabla \times (B_0 x \vec{e}_y) = B_0 \vec{e}_z$, $\nabla \times [\beta_{zb}\Phi_s(\vec{x})\vec{e}_z] = \vec{B}^s(\vec{x})$, and the vector potential for the helical wiggler field is defined by

$$\vec{A}_w(\vec{x}) = \frac{mc^2 a_w}{e}\{[I_0(k_w r)\cos(k_w z) - I_2(k_w r)\cos(k_w z - 2\theta)]\vec{e}_x$$

$$+[I_0(k_w r)\sin(k_w z) + I_2(k_w r)\sin(k_w z - 2\theta)]\vec{e}_y\} . \tag{9}$$

In Eq. (9), $a_w = eB_w/mc^2 k_w$ is the usual dimensionless measure of the wiggler field amplitude.

The equations of motion for a test electron within the beam ($0 \leq r < r_b$) can be derived from the Hamiltonian

$$H = [(c\vec{P} + e\vec{A})^2 + m^2 c^4]^{1/2} - e\Phi_s \equiv \gamma mc^2 - e\Phi_s . \tag{10}$$

In Eq. (10), $\vec{P}$ is the canonical momentum, $\gamma = [1 + (\vec{p}/mc)^2]^{1/2}$ is the relativistic mass factor, $\vec{p} = \vec{P} + e\vec{A}/c$ is the mechanical momentum, the electrostatic potential $\Phi_s(\vec{x})$ is defined in Eq. (7), and the vector potential $\vec{A}(\vec{x})$ is defined in Eqs. (8) and (9). Because H is independent of time, the Hamiltonian is a constant of the motion, i.e.,

$$H(x, y, z, P_x, P_y, P_z) = \gamma mc^2 - e\Phi_s = \text{const.} , \tag{11}$$

which corresponds to the conservation of total energy (kinetic plus potential energy) of the test electron.

For notational convenience, in the subsequent analysis we introduce the dimensionless potentials, $\hat{\vec{A}}(\vec{x})$, $\hat{\vec{A}}_w(\vec{x})$, and $\hat{\Phi}_s$, and Hamiltonian $\hat{H}$ defined by

$$\hat{\vec{A}} = \frac{e\vec{A}(\vec{x})}{mc^2} \; , \quad \hat{\vec{A}}_w = \frac{e\vec{A}_w(\vec{x})}{mc^2} \; , \quad \hat{\Phi}_s(\vec{x}) = \frac{e\Phi_s(\vec{x})}{mc^2} \; , \quad \hat{H} = \frac{H}{mc^2} \; . \tag{12}$$

In addition, the notation

$$a_0 = \frac{eB_0}{mc^2 k_w} \; , \quad a_w = \frac{eB_w}{mc^2 k_w} \; , \quad \epsilon_s = \frac{\omega_{pb}^2}{c^2 k_w^2} \; , \tag{13}$$

is introduced, where a_0 is a dimensionless measure of the axial guide field (B_0), a_w is a dimensionless measure of the wiggler field amplitude (B_w), and ϵ_s is a dimensionless measure of the strength of the equilibrium self fields. Combining Eqs. (12) and (13) with Eqs. (7), (8), and (10) then gives

$$\hat{H} = [(\vec{P}/mc + \hat{\vec{A}})^2 + 1]^{1/2} - \hat{\Phi}_s \; , \tag{14}$$

where

$$\hat{\Phi}_s(\vec{x}) = \frac{1}{4}\epsilon_s k_w^2 (x^2 + y^2) \; , \tag{15}$$

and

$$\hat{\vec{A}}(\vec{x}) = a_0 k_w x \vec{e}_y + \hat{\vec{A}}_w(\vec{x}) + \beta_{zb}\hat{\Phi}_s(\vec{x})\vec{e}_z \; . \tag{16}$$

From Eqs. (14)-(16), it is clear that there is a large region of the parameter space (a_0, a_w, ϵ_s) in which the motion of an individual test electron can be investigated. The remainder of this paper focuses on the following two cases: (a) in Sec. III, electron motion is investigated for a thin ($k_w^2 r_b^2 \ll 1$) electron beam propagating parallel to a strong axial guide field ($B_0 \neq 0$ and $a_0 > a_w$); (b) in Sec. IV, electron motion is investigated for a beam propagating through zero axial guide field ($B_0 = 0$ and $a_0 = 0$) and a strong-focusing wiggler field ($a_w \neq 0$ and $k_w^2 r_b^2 < 1$, but not necessarily $k_w^2 r_b^2 \ll 1$).

In case (a) (Sec. III), the assumption $k_w^2 r_b^2 \ll 1$ allows us to approximate the vector potential for the wiggler field by the ideal value

$$\hat{\vec{A}}_w = \hat{\vec{A}}_w^{(0)} = a_w[\vec{e}_x \cos(k_w z) + \vec{e}_y \sin(k_w z)] \; . \tag{17}$$

Because $a_0 > a_w$ is assumed, the axial magnetic field $B_0 \vec{e}_z$ plays an important role in providing radial confinement of the electron orbits in the presence of the (defocusing) space-charge field $\vec{E}^s(\vec{x})$. Indeed, for the special case where $a_w = 0$, an electron with axial velocity $v_z \cong V_{zb} = \beta_{zb}c$ and small transverse momentum $(p_x^2 + p_y^2)^{1/2} \ll \gamma_{zb}mc$ is radially confined provided[21]

$$2\gamma_{zb}\epsilon_s(1 - \beta_{zb}^2) < a_0^2 \; , \tag{18}$$

where $\gamma_{zb} = (1 - \beta_{zb}^2)^{-1/2}$, $\epsilon_s = \omega_{pb}^2/c^2 k_w^2$, and $a_0 = eB_0/mc^2 k_w \equiv \omega_{cz}/ck_w$. Equation (18) is equivalent to the familiar inequality $2\gamma_{zb}\omega_{pb}^2(1 - \beta_{zb}^2) < \omega_{cz}^2$, required for radial confinement of a nonneutral electron beam by an axial guide field $B_0\vec{e}_z$.

By contrast, in case (b) (Sec. IV), the axial guide field is zero ($B_0 = 0$ and $a_0 = 0$), and the assumption $k_w^2 r_b^2 < 1$ allows us to expand $\vec{\hat{A}}_w(\vec{x})$ correct to order $a_w k_w^2 r^2$, i.e.,

$$\vec{\hat{A}}_w(\vec{x}) = \vec{\hat{A}}_w^{(0)}(\vec{x}) + \vec{\hat{A}}_w^{(2)}(\vec{x}) + O(a_w k_w^4 r^4) \ . \tag{19}$$

In Eq. (19), $\vec{\hat{A}}_w^{(0)}$ is defined in Eq. (17), and $\vec{\hat{A}}_w^{(2)}(\vec{x})$ is defined by

$$\vec{\hat{A}}_w^{(2)} = \frac{a_w}{8}\left\{[(k_w^2 x^2 + 3k_w^2 y^2)\cos(k_w z) - 2k_w^2 xy \sin(k_w z)]\vec{e}_x\right.$$

$$\left. + [(k_w^2 y^2 + 3k_w^2 x^2)\sin(k_w z) - 2k_w^2 xy \cos(k_w z)]\vec{e}_y\right\} \ . \tag{20}$$

In this case, the (focusing) magnetic force associated with $\vec{\hat{A}}_w^{(2)}$ provides radial confinement of the electron orbits in the presence of the (defocusing) space-charge field $\vec{E}_s(\vec{x})$. For an electron with axial velocity $v_z \cong V_{zb} = \beta_{zb}c$, perpendicular momentum $(p_x^2 + p_y^2)^{1/2} \cong mca_w$, and total mechanical energy $\gamma \cong \gamma_b = [(1 + a_w^2)/(1 - \beta_{zb}^2)]^{1/2} = \gamma_{zb}(1 + a_w^2)^{1/2}$, it can be shown that the condition for radial confinement of the electron orbits is given by [Eq. (53)]

$$\gamma_b \epsilon_s (1 - \beta_{zb}^2) < a_w^2 \ . \tag{21}$$

Here, $\epsilon_s = \omega_{pb}^2/c^2 k_w^2$ and $a_w = eB_w/mc^2 k_w \equiv \omega_{cw}/ck_w$, and the inequality in Eq. (21) can be expressed in the equivalent form $\gamma_b \omega_{pb}^2(1 - \beta_{zb}^2) < \omega_{cw}^2$. For specified beam density n_b, note from Eq. (21) that sufficiently large wiggler amplitude a_w is required for confinement of the electron orbits.

III. PARTICLE ORBITS IN COMBINED AXIAL GUIDE FIELD AND IDEAL HELICAL WIGGLER FIELD

We first examine the motion of an individual test electron for the case where $B_0 \neq 0$ and the axial guide field is sufficiently strong that

$$a_0 > a_w \ . \tag{22}$$

For a thin electron beam with $k_w^2 r_b^2 \ll 1$, it follows from Eqs. (14)-(17) that the Hamiltonian $\hat{H} = H/mc^2$ can be approximated by (for $r < r_b$)

$$\hat{H} = [(\vec{P}/mc + \vec{\hat{A}})^2 + 1]^{1/2} - \frac{1}{4}\epsilon_s k_w^2 (x^2 + y^2) \,, \tag{23}$$

where

$$\vec{\hat{A}}(\vec{x}) = a_0 k_w x \vec{e}_y + a_w[\vec{e}_x \cos(k_w z) + \vec{e}_y \sin(k_w z)] + \frac{1}{4}\beta_{zb}\epsilon_s k_w^2 (x^2 + y^2) \,. \tag{24}$$

A. Hamiltonian in Guiding-Center Variables

As stated in Sec. II, because $\hat{H}$ does not depend explicitly on time t, the total energy $\hat{H} = \gamma - \epsilon_s k_w^2 (x^2 + y^2)/4$ is a constant of the motion. In order to find an additional constant of the motion and calculate the resonances, it is useful to perform the canonical transformation to the new variables $(\phi, \psi, k_w z', k_w P_\phi/mc,$ $k_w P_\psi/mc, P_{z'}/mc)$ defined by[26]

$$k_w x = \left(\frac{2}{a_0}\frac{k_w P_\phi}{mc}\right)^{1/2} \sin(\phi + k_w z') - \left(\frac{2}{a_0}\frac{k_w P_\psi}{mc}\right)^{1/2} \cos(\psi - k_w z') \,, \tag{25}$$

$$k_w y = \left(\frac{2}{a_0}\frac{k_w P_\psi}{mc}\right)^{1/2} \sin(\psi - k_w z') - \left(\frac{2}{a_0}\frac{k_w P_\phi}{mc}\right)^{1/2} \cos(\phi + k_w z') \,, \tag{26}$$

$$k_w z = k_w z' \,, \tag{27}$$

$$\frac{P_x}{mc} = \left(2a_0 \frac{k_w P_\phi}{mc}\right)^{1/2} \cos(\phi + k_w z') \,, \tag{28}$$

$$\frac{P_y}{mc} = \left(2a_0 \frac{k_w P_\psi}{mc}\right)^{1/2} \cos(\psi - k_w z') \,, \tag{29}$$

$$\frac{P_z}{mc} = \frac{P_{z'}}{mc} - \frac{k_w P_\phi}{mc} + \frac{k_w P_\psi}{mc} \,, \tag{30}$$

where $a_0 \equiv eB_0/mc^2 k_w$. It is shown later in Sec. III.B [see Eqs. (38)-(40)] that $k_w r_c = (2k_w P_\phi/a_0 mc)^{1/2}$ and $k_w r_g = (2k_w P_\psi/a_0 mc)^{1/2}$ are the normalized gyroradius and guiding-center radius, respectively, of the steady-state orbits. In Eqs. (25)-(30), we introduce the dimensionless variables

$$\hat{P}_\phi = \frac{k_w P_\phi}{mc} \,, \quad \hat{P}_\psi = \frac{k_w P_\psi}{mc} \,, \quad \hat{P}_{z'} = \frac{P_{z'}}{mc} \,. \tag{31}$$

Some straightforward algebra then shows that the Hamiltonian $\hat{H} = H/mc^2$ in the new variables can be expressed as

$$\begin{aligned}
&\hat{H}(\phi, \psi, \hat{P}_\phi, \hat{P}_\psi, \hat{P}_{z'} = \text{const.}) \\
&= [2a_0\hat{P}_\phi + 2a_w(2a_0\hat{P}_\phi)^{1/2}\cos\phi + (\hat{P}_{z'} - \hat{P}_\phi + \hat{P}_\psi + \beta_{zb}\hat{\Phi}_s)^2 + a_w^2 + 1]^{1/2} - \hat{\Phi}_s \\
&= \text{const.}
\end{aligned} \tag{32}$$

Here, $a_w = eB_w/mc^2 k_w$, and the normalized self-field potential $\hat{\Phi}_s = e\Phi_s/mc^2$ is defined by

$$\hat{\Phi}_s = \frac{\epsilon_s}{2a_0}[\hat{P}_\phi + \hat{P}_\psi - 2(\hat{P}_\phi \hat{P}_\psi)^{1/2}\cos(\phi + \psi)] . \tag{33}$$

Because $\hat{H}$ in Eq. (32) does not depend explicitly on z', it follows that $\hat{P}_{z'} = $ const.

B. Integrable Limit ($\epsilon_s = 0$)

In the limit where self-field effects are negligibly small ($\epsilon_s = 0$ and $\hat{\Phi}_s = 0$), the Hamiltonian in Eq. (32) reduces to

$$\hat{H}_0(\phi, \hat{P}_\phi, \hat{P}_\psi, \hat{P}_{z'})$$
$$= [2a_0\hat{P}_\phi + 2a_w(2a_0\hat{P}_\phi)^{1/2}\cos\phi + (\hat{P}_{z'} - \hat{P}_\phi + \hat{P}_\psi)^2 + a_w^2 + 1]^{1/2} \equiv \gamma_0 . \tag{34}$$

Equation (34) possesses three constants of the motion, namely, $\hat{P}_\psi$, $\hat{P}_{z'}$ and γ_0. The motion is integrable and has been analyzed by several authors.[22–29] It is readily shown from Eq. (34) that the steady-state orbits $(\phi_0, \hat{P}_{\phi 0})$ are given by

$$\cos\phi_0 = \pm 1 , \tag{35}$$

$$(2a_0\hat{P}_{\phi 0})^{1/2} = \pm\frac{a_w a_0}{\hat{p}_{z0} - a_0} > 0 , \tag{36}$$

where $\hat{p}_{z0} = \hat{P}_{z'} - \hat{P}_{\phi 0} + \hat{P}_\psi$ is the normalized axial mechanical momentum. Substituting Eqs. (35) and (36) into Eq. (34) yields

$$\hat{p}_{z0}^2\left[1 + \frac{a_w^2}{(\hat{p}_{z0} - a_0)^2}\right] + 1 = \gamma_0^2 , \tag{37}$$

which determines the values of $\hat{p}_{z0} = p_{z0}/mc$ in terms of the parameters a_w, a_0, and γ_0. Equation (37) is a fourth-order algebraic equation for $\hat{p}_{z0}$, which has at most four real roots. Figure 2 shows the dependence of $\hat{p}_{z0}$ on the strength of the axial guide field a_0 for the case $a_w = 0.2$ and $\gamma_0 = 3.0$. The stable orbit with $\hat{p}_{z0} < a_0$ is known as the Group-II orbit, whereas the stable orbit with $\hat{p}_{z0} > a_0$ is known as the Group-I orbit.[22,23] The Group-I orbit merges with an unstable orbit at $a_0 = a_0^{cr} \cong 2.1$. (In general, the value of a_0^{cr} for the merging of the Group-I orbit and the unstable orbit depends on γ_0 and a_w.) In free electron laser operation, the electron beam is injected typically into the Group-I orbit or the Group-II orbit. Substituting $\phi = \phi_0$, $\psi = \psi_0 + \beta_{z0}\tau$, $k_w z' = k_w z_0' + \beta_{z0}\tau$, $\hat{P}_\phi = \hat{P}_{\phi 0}$ and $\hat{P}_\psi = \hat{P}_{\psi 0}$ into Eqs. (25)-(27), it is readily shown that the steady-state trajectories can be expressed in Cartesian coordinates as

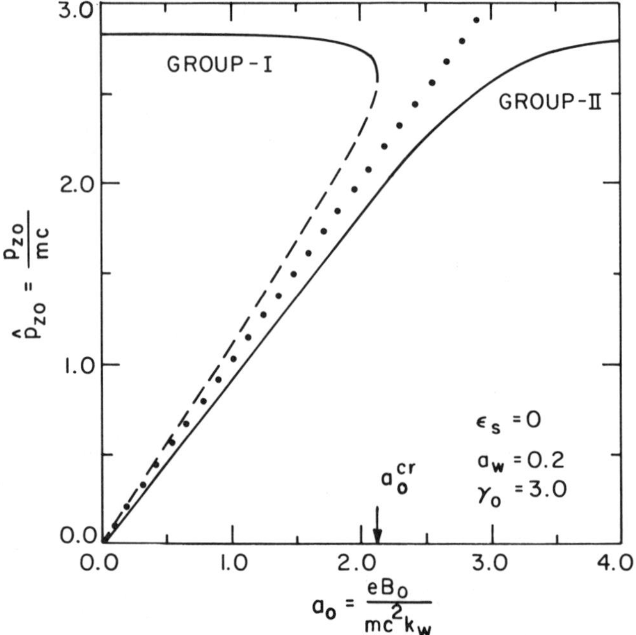

Fig. 2. Plot of the integrable steady-state orbits calculated from Eq. (32) for $\epsilon_s = 0$, $\gamma_0 = 3.0$, and $a_w = 0.2$. The solid (dashed) curves correspond to the stable (unstable) orbits, and the dotted straight line designates the magnetoresonance condition $\hat{p}_z = a_0$.

$$k_w x(\tau) = \pm(2\hat{P}_{\phi 0}/a_0)^{1/2} \sin[k_w z(\tau)] - (2\hat{P}_{\psi 0}/a_0)^{1/2} \cos\psi_0, \tag{38}$$

$$k_w y(\tau) = \mp(2\hat{P}_{\phi 0}/a_0)^{1/2} \cos[k_w z(\tau)] + (2\hat{P}_{\psi 0}/a_0)^{1/2} \sin\psi_0, \tag{39}$$

$$k_w z(\tau) = k_w z_0 + \beta_{z0}\tau, \tag{40}$$

for $\cos\phi_0 = \pm 1$. Here, $\beta_{z0} \equiv \hat{p}_{z0}/\gamma_0$, and $\tau \equiv ck_w t$ is the normalized time variable. Equations (38)-(40) describe helical trajectories with normalized gyro-radius $k_w r_c = (2\hat{P}_{\phi 0}/a_0)^{1/2}$ and guiding-center radius $k_w r_g = (2\hat{P}_{\psi 0}/a_0)^{1/2}$.

Figure 3 shows typical integrable phase-space structure for the two cases $0 < a_0 < a_0^{cr}$ and $a_0 > a_0^{cr}$. Here, the elliptic (hyperbolic) fixed points correspond to the stable (unstable) steady-state orbits. The Group-I orbit has greater axial momentum than the Group-II orbit in Fig. 3(a), whereas only the Group-II orbit is allowed in Fig. 3(b). An orbit which deviates slightly from the stable (Group-I or Group-II) orbit, i.e., $|\delta\phi| = |\phi - \phi_0| \ll 1$ and $|\delta\hat{P}_\phi| = |\hat{P}_\phi - \hat{P}_{\phi 0}| \ll 1$, exhibits

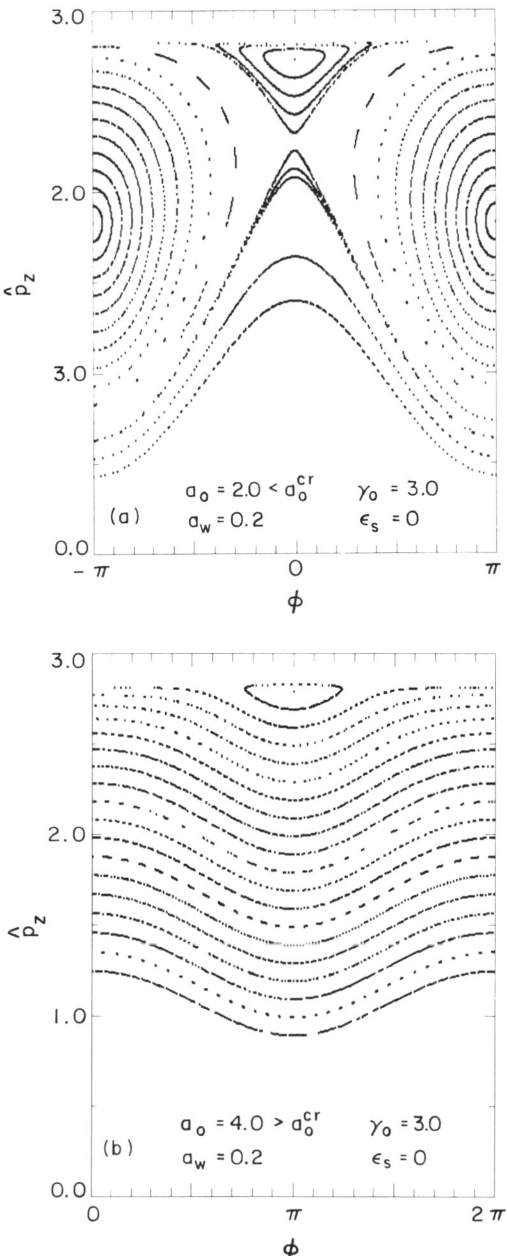

Fig. 3. Contour plots in the integrable phase plane (ϕ, p_z) calculated from Eq. (34) for $\epsilon_s = 0$, $\gamma_0 = 3.0$, and $a_w = 0.2$. The two cases correspond to (a) $a_0 = 2.0 < a_0^{cr} \cong 2.1$, and (b) $a_0 = 4.0 > a_0^{cr}$.

harmonic, guide-field-induced betatron oscillations. It is straightforward to show from the equations of motion for ϕ and $\hat{P}_\phi$ that the frequency of the betatron oscillations is given by[26]

$$\hat{\omega}_{\beta 0} \equiv \frac{\omega_{\beta 0}}{ck_w} = \frac{|\hat{p}_{z0} - a_0|}{\gamma_0}\left[1 - \frac{a_0}{a_w}\left(\frac{\hat{p}_{t0}}{\hat{p}_{z0}}\right)^3\right]^{1/2}, \tag{41}$$

where $\hat{p}_{t0} = a_w\hat{p}_{z0}/(\hat{p}_{z0} - a_0)$ is the normalized transverse mechanical momentum of the steady-state orbit.

C. Chaotic Motion ($\epsilon_s \neq 0$)

For $\epsilon_s \neq 0$, the self-field contribution $\hat{\Phi}_s \neq 0$ in Eq. (32) invalidates the constancy of $\hat{P}_\psi$. The motion described by the Hamiltonian in Eq. (32) occurs in the three-dimensional phase space $(\phi, \psi, \hat{P}_\phi)$, because $\hat{P}_\psi$ is determined from $\hat{H} = \text{const}$. The time scale T_s for self-field-induced changes in the particle orbit can be estimated from the rate of change of the phase $\phi + \psi$ in the electrostatic potential $\hat{\Phi}_s$ defined in Eq. (33). For an electron with $\phi \cong \phi_0$ and $\hat{p}_z \cong \hat{p}_{z0} \cong \beta_{zb}\gamma_b$, because $d(\phi + \psi)/d\tau \cong d\psi/d\tau = \partial\hat{H}/\partial\hat{P}_\psi = \hat{p}_z/\gamma + O(\epsilon_s) \cong \beta_{zb}$ or $d(\phi + \psi)/dt \cong k_wV_b$, the time required for the phase $\phi + \psi$ to advance by 2π is given by

$$T_s = \frac{2\pi}{k_wV_b} = \frac{\lambda_w}{V_b}. \tag{42}$$

In Eq. (42), $\lambda_w = 2\pi/k_w$ is the wiggler period, and T_s is the characteristic time scale for the electron to experience self-field-induced modifications as the electron undergoes the helical motion described by Eqs. (38)-(40). A detailed analysis of resonances between the guide-field-induced betatron oscillations and the helical motion has been carried out.[26] It is found that the resonance condition and the resonance width $\hat{w}_n$ of order n are given by

$$n\hat{\omega}_{\beta 0} + \frac{\hat{p}_{z0}}{\gamma_b} - \frac{\epsilon_s}{2a_0}\left(1 - \beta_{zb}\frac{\hat{p}_{z0}}{\gamma_b}\right) = 0, \tag{43}$$

and

$$\hat{w}_n = \left[\frac{8\epsilon_s\gamma_b}{a_0}|J_n(\delta\phi_0)|\right]^{1/2}(\hat{P}_{\phi 0}\hat{P}_{\psi 0})^{1/4} = 4\left[\frac{\gamma_b r_c r_g I_b|J_n(\delta\phi_0)|}{\beta_{zb}r_b^2 I_A}\right]^{\frac{1}{2}}. \tag{44}$$

Here, n is an integer, the normalized betatron oscillation frequency $\hat{\omega}_{\beta 0}$ is defined in Eq. (41), and $\hat{w}_n$ is the width of the separatrix of the resonance of order n projected along the $\hat{p}_z$-axis. In Eqs. (43) and (44), r_b, I_b and $\gamma_b mc^2$ are the radius, current and energy of the electron beam; $I_A = mc^3/e \cong 17$ kA is the Alfvén current; and $r_c = (2\hat{P}_{\phi 0}/k_w^2 a_0)^{1/2}$ and $r_g = (2\hat{P}_{\psi 0}/k_w^2 a_0)^{1/2}$ are the gyroradius and guiding-center radius, respectively. Figure 4 shows plots of the resonance curves corresponding to the solutions to Eq. (43).

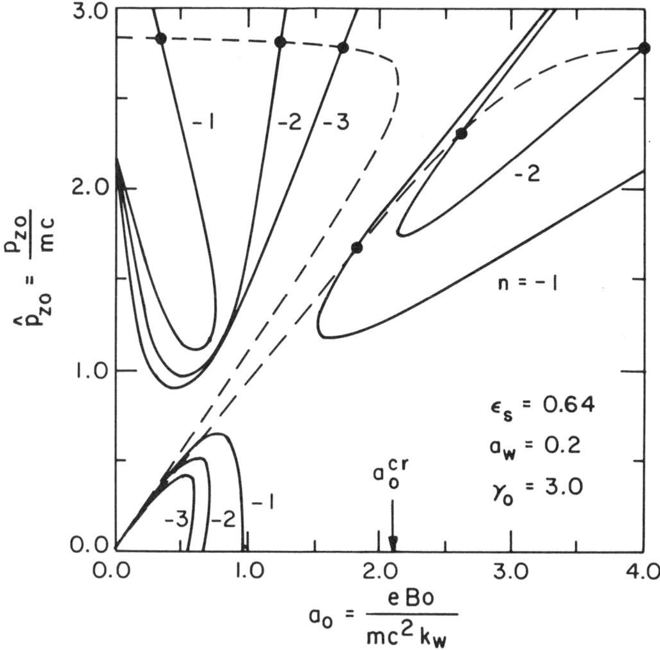

Fig. 4. The equilibrium self-field resonance curves (solid curves) correspond to the solutions to Eq. (43) for $\epsilon_s = 0.64$, $\gamma_0 = 3.0$, $a_w =$
0.2, and $n = -1, -2$, and -3. The dashed curves are the integrable steady-state orbits calculated in Fig. 2, and the dots
mark the intersections between the resonance curves and the
steady-state orbits.

In order to demonstrate that the particle motion is indeed chaotic, Poincaré
surface-of-section maps have been generated by numerically integrating the equations of motion derived from the Hamiltonian in Eq. (32). Figure 5 shows nonintegrable surface-of-section plots for $\hat{H} = 3.0$, $a_w = 0.2$ and the two cases: (a)
$0 < a_0 = 2.0 < a_0^{cr} \cong 2.1$ and $\epsilon_s = 0.16$, and (b) $a_0 = 4.0 > a_0^{cr}$ and $\epsilon_s = 0.64$.
The integrable limits corresponding to Figs. 5(a) and 5(b) are shown in Figs. 3(a)
and 3(b), respectively, for the case $\epsilon_s = 0$. In Fig. 5, the initial condition for $\hat{P}_\psi$
is fixed at the value $k_w r_g = (2\hat{P}_{\psi 0}/a_0)^{1/2} = 0.25$, whereas the initial condition
for $\hat{p}_z$ is allowed to vary. The second-order island appearing near the Group-II
orbit in Fig. 5(b) occurs near the intersection between the $n = -2$ resonance
curve and the Group-II orbit at $a_0 = 4.0$ in Fig. 4. It is evident from Fig. 5 that
the self fields are not intense enough [$\epsilon_s = 0.16$ in Fig. 5(a), and $\epsilon_s = 0.64$ in
Fig. 5(b)] to cause high-degree chaoticity in the vicinity of either the Group-I
orbit in Fig. 5(a) or the Group-II orbit in Fig. 5(b).

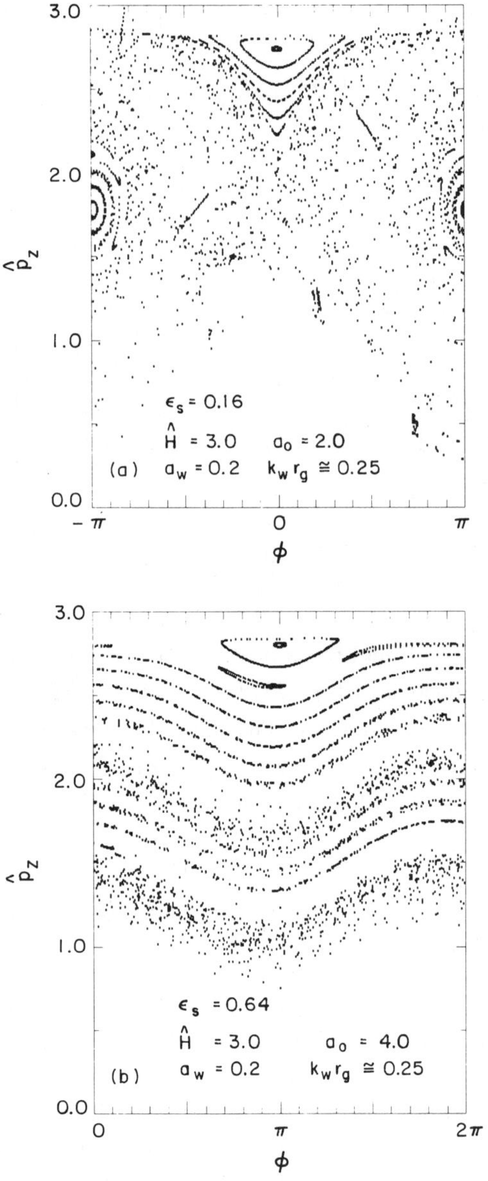

Fig. 5. Typical nonintegrable surface-of-section plots in the $(\phi, \hat{p}_z)$ plane at $\psi = 0 \pmod{2\pi}$ for the two cases: (a) $0 < a_0 = 2.0 < a_0^{cr} \cong 2.1$, and (b) $a_0 = 4.0 > a_0^{cr}$. Other system parameters are: (a) $\epsilon_s = 0.16$, $\hat{H} = 3.0$, $a_w = 0.2$, and $\beta_{zb} = 0.91$, and (b) $\epsilon_s = 0.64$, $\hat{H} = 3.0$, $a_w = 0.2$, and $\beta_{zb} = 0.93$.

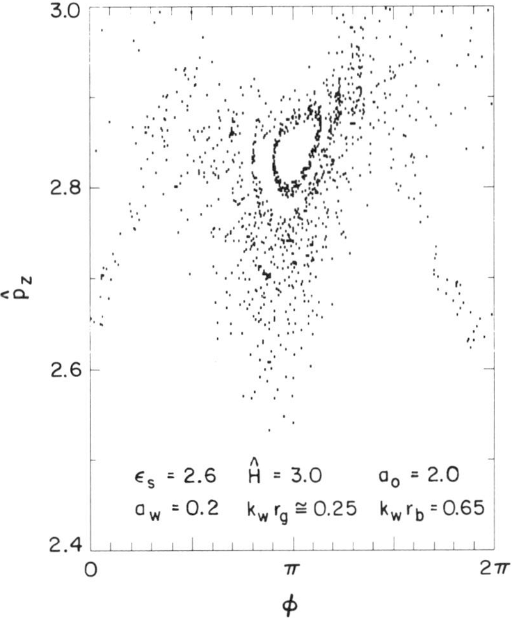

Fig. 6. Surface-of-section plot at the onset of chaoticity in the Group-II orbit for the choice of system parameters $\epsilon_s = 2.5$, $a_0 = 4.0$, $\hat{H} = 3.0$, $a_w = 0.2$, and $\beta_{zb} = 0.93$, corresponding to Fig. 5(b). Here, the normalized effective gyroradius $k_w r_c \cong (2\hat{P}_\phi/a_0)^{1/2}$ ranges from 0.17 to 0.35, the normalized guiding center radius is $k_w r_g \cong (2\hat{P}_\psi/a_0)^{1/2} \cong 0.25$, and the normalized beam radius is $k_w r_b = 0.65$.

The threshold values of the parameter ϵ_s for the onset of chaoticity in the Group-I or Group-II orbits can be estimated, using the scaling relation in Eq. (44). The criterion used here is that the onset of chaoticity occurs whenever the half width of the self-field-induced resonance is greater than the momentum separation between the resonance and the steady-state orbit. For example, the onset of chaoticity for the case corresponding to Fig. 5(b) can be estimated by making use of the secondary resonance at $\phi \cong 0.85\pi$ and $\hat{p}_z \cong 2.5$ in Fig. 5(b). Substituting the values $n = -2$, $\delta\phi_0 \cong 0.15\pi$, $\gamma_b = 3.0$, $a_0 = 4.0$, $k_w r_g = (2\hat{P}_{\psi 0}/a_0)^{1/2} = 0.25$, and $k_w r_c = (2\hat{P}_{\phi 0}/a_0)^{1/2} = [a_w/(a_0 - \hat{p}_{z0})]^{1/2} = [0.2/(4.0 - 2.8)]^{1/2} = 0.4$ into Eq. (44), we find that the width of the resonance scales as $\hat{w}_{-2} = 0.43\epsilon_s^{1/2}$. Note from Fig. 5(b) that the momentum separation between the Group-II orbit and the resonance is $\Delta\hat{p}_z \cong 0.3$. Therefore, it follows from

$\hat{w}_{-2}/2 = \Delta\hat{p}_z$ that the estimated value of ϵ_s for the onset of the chaoticity is given by $\epsilon_s \cong 1.2$. In reality, the actual onset of chaos for the Group-II orbit [corresponding to the Group-II orbit in Fig. 5(b)] occurs at $\epsilon_s \cong 2.5$ and is shown in Fig. 6, where $a_0 = 4.0$, $\hat{H} = 3.0$, $a_w = 0.2$, $\beta_{zb} = 0.93$, and $k_w r_b = 0.65$. As an example, for $\lambda_w = 3.0$ cm, the dimensionless parameters in Fig. 6 correspond to $r_b = 0.31$ cm, $I_b = 4.3$ kA, $B_w = 710$ G, $B_0 = 14.2$ kG, $\beta_{zb} = 0.93$, and $\gamma_b = 3.0$.

IV. PARTICLE ORBITS IN A REALIZABLE HELICAL WIGGLER FIELD

In this section, we examine the motion of an individual test electron for the case where the axial guide field is zero ($B_0 = 0$ and $a_0 = 0$) and the wiggler magnetic field is described by a realizable helical wiggler. For an electron beam with $k_w^2 r_b^2 < 1$, it follows from Eqs. (14)-(16) and (19) that the Hamiltonian $\hat{H} = H/mc^2$ can be approximated by (for $r < r_b$)

$$\hat{H} = [(\vec{\hat{P}} + \vec{\hat{A}})^2 + 1]^{1/2} - \frac{1}{4}\epsilon_s k_w^2(x^2 + y^2) \ . \tag{45}$$

Here, $\vec{\hat{P}} = \vec{P}/mc$ is the normalized canonical momentum, and the dimensionless vector potential is defined by

$$\vec{\hat{A}}(\vec{x}) = \vec{\hat{A}}_w^{(0)}(\vec{x}) + \vec{\hat{A}}_w^{(2)}(\vec{x}) + \frac{1}{4}\beta_{zb}\epsilon_s k_w^2(x^2 + y^2)\vec{e}_z \ . \tag{46}$$

In Eq. (46), the approximation $\vec{\hat{A}}_w(\vec{x}) \cong \vec{\hat{A}}_w^{(0)}(\vec{x}) + \vec{\hat{A}}_w^{(2)}(\vec{x})$ has been made for a realizable helical wiggler field, and $\vec{\hat{A}}_w^{(0)}(\vec{x})$ and $\vec{\hat{A}}_w^{(2)}(\vec{x})$ are defined in Eqs. (17) and (20).

A. Condition for Radial Orbit Confinement

For a thin electron beam with $k_w^2 r_b^2 \ll 1$ and small wiggler amplitude ($a_w^2 \ll \gamma_b^2$), because $\hat{P}_x^2 + \hat{P}_y^2 < k_w^2 r_b^2 a_w^2$ [see Eq. (57)], the Hamiltonian defined in Eqs. (45) and (46) can be expanded to order $k_w^2 r^2$. For $r < r_b$, this yields

$$\hat{H} \cong \hat{H}_0 + \hat{H}_1 \ , \tag{47}$$

where

$$\hat{H}_0(k_w z, \hat{P}_x, \hat{P}_y, \hat{P}_z) = \{\hat{P}_z^2 + 2a_w[\hat{P}_x \cos(k_w z) + \hat{P}_y \sin(k_w z)] + a_w^2 + 1\}^{1/2} \equiv \gamma_0 \ , \tag{48}$$

and

$$\hat{H}_1 = \frac{1}{2\gamma_0}\left[\hat{P}_x^2 + \hat{P}_y^2 + 2\vec{\hat{A}}_w^{(0)}\cdot\vec{\hat{A}}_w^{(2)}\right] - \frac{\epsilon_s}{4}\left(1 - \beta_{zb}\frac{\hat{P}_z}{\gamma_0}\right)k_w^2(x^2 + y^2)\ . \tag{49}$$

For the case of zero transverse canonical momentum with $\hat{P}_x = \hat{P}_y = 0$, it follows from Eq. (48) that the lowest-order (helical) particle orbit is described by

$$x_0(\tau) = r_c\sin[k_w z_0(\tau)] + x_g\ ,$$

$$y_0(\tau) = -r_c\cos[k_w z_0(\tau)] + y_g\ , \tag{50}$$

$$z_0(\tau) = (\beta_{z0}/k_w)\tau + z_0(0)\ .$$

In Eq. (50), $\tau = ck_w t$, $\beta_{z0} = [1 - (1 + a_w^2)/\gamma_0^2]^{1/2} = \text{const.}$ is the normalized axial velocity, $r_c = a_w/k_w\gamma_0\beta_{z0}$ is the radius of the helical orbit, and x_g and y_g are slow variables describing the guiding center of the helix.

To calculate the guiding-center trajectories, we substitute Eqs. (17), (20) and (50) into Eq. (49) and average over τ for one period $2\pi/\beta_{z0}$. For $\gamma_0 \cong \gamma_b$ and $\beta_{z0} \cong \beta_{zb} = [1 - (1 + a_w^2)/\gamma_b^2]^{1/2}$, some straightforward algebra shows that the average Hamiltonian can be expressed in the quadratic form

$$\langle\hat{H}_1\rangle = \frac{1}{2\gamma_b}\left\{\hat{P}_x^2 + \hat{P}_y^2 + \gamma_b^2\hat{\omega}_{\beta w}^2\left[1 - \frac{\gamma_b\epsilon_s}{a_w^2}(1 - \beta_{zb}^2)\right]k_w^2(x^2 + y^2)\right\}\ , \tag{51}$$

where

$$\hat{\omega}_{\beta w} \equiv \frac{\omega_{\beta w}}{ck_w} = \frac{a_w}{\sqrt{2}\gamma_b} \tag{52}$$

is the normalized frequency of the wiggler-field-induced betatron oscillations in the absence of self fields. It follows from Eq. (51) that the guiding center of the helical orbit oscillates harmonically about $r = 0$ when $\gamma_b\epsilon_s(1 - \beta_{zb}^2) < a_w^2$, and diverges radially when $\gamma_b\epsilon_s(1 - \beta_{zb}^2) > a_w^2$. Therefore, the condition for radial confinement of the particle orbits can be expressed as[29]

$$\gamma_b\epsilon_s(1 - \beta_{zb}^2) < a_w^2\ , \tag{53}$$

or equivalently,

$$\gamma_b\omega_{pb}^2(1 - \beta_{zb}^2) < \omega_{cw}^2\ . \tag{54}$$

Here, $\omega_{pb} = (4\pi e^2 n_b/m)^{1/2}$ is the nonrelativistic plasma frequency, and $\omega_{cw} = eB_w/mc = ck_w a_w$ is the nonrelativistic cyclotron frequency associated with the wiggler field amplitude B_w. Note that the condition in Eqs. (53) and (54) is analogous to the condition for radial confinement of particle orbits in a nonneutral electron beam by a uniform axial magnetic field.[21] Expressing $\epsilon_s = \omega_{pb}^2/c^2 k_w^2 = (4/\beta_{zb}k_w^2 r_b^2)(I_b/I_A)$, where I_b is the beam current and $I_A \equiv mc^3/e \cong 17$ kA is the Alfvén current, it readily follows that the condition in Eqs. (53) and (54) can be expressed in the equivalent form

$$I_b < I_b^{cr} \equiv \frac{\gamma_b \beta_{zb} k_w^2 r_b^2}{4} \frac{a_w^2}{1 + a_w^2} I_A \; .$$ (55)

As an example, for $a_w = 0.4$, $k_w r_b = 0.2$, $\gamma_b = 3.0$, and $\beta_{zb} = [1 - (1 + a_w^2)/\gamma_b^2]^{1/2} = 0.93$, the critical value of beam current defined in Eq. (55) is $I_b^{cr} = 70$ A.

Solving the equations of motion determined from $\langle \hat{H}_1 \rangle$ in Eq. (51) for radially confined orbits, we find that the guiding center trajectories are given by $x_g(\tau) = x_m \cos(\hat{\omega}_{\beta w}^s \tau + \alpha_x)$ and $y_g(\tau) = y_m \cos(\hat{\omega}_{\beta w}^s \tau + \alpha_y)$. Here, α_x and α_y are the phases of the betatron oscillations, x_m and y_m are the amplitudes, and

$$\hat{\omega}_{\beta w}^s = \hat{\omega}_{\beta w} \left[1 - \frac{\gamma_b \epsilon_s}{a_w^2}(1 - \beta_{zb}^2) \right]^{1/2}$$ (56)

is the normalized frequency of the wiggler-field-induced betatron oscillations including self-field effects. Because $\hat{P}_x = \gamma_b d(k_w x_g)/d\tau$, $\hat{P}_y = \gamma_b d(k_w y_g)/d\tau$, and $x_m^2 + y_m^2 < r_b^2$, it is readily shown that

$$\hat{P}_x^2 + \hat{P}_y^2 < k_w^2 r_b^2 a_w^2 \; ,$$ (57)

which assures the validity of the expansion in Eq. (47).

Figure 7 shows typical transverse trajectories for the two cases: (a) $\epsilon_s < \epsilon_s^{cr} \equiv a_w^2/\gamma_b(1 - \beta_{zb}^2) = \gamma_b a_w^2/(1 + a_w^2)$, and (b) $\epsilon_s > \epsilon_s^{cr}$. The orbits in Fig. 7 are obtained by integrating numerically the equations of the motion derived from the Hamiltonian defined in Eqs. (45) and (46). In Fig. 7(a), because the focusing force due to the wiggler and self-magnetic fields is greater than the defocusing force of the self-electric field ($\epsilon_s < \epsilon_s^{cr}$), the guiding center of the orbit oscillates about the axis of the wiggler helix, corresponding to a real value of $\hat{\omega}_{\beta w}^s$. In Fig. 7(b), because the defocusing force exceeds the focusing force ($\epsilon_s > \epsilon_s^{cr}$), the radius of the guiding center of the orbit oscillates between some minimum radius r_{min} and maximum radius r_{max}. The focusing force provided by higher-order terms in the vector potential expansion in Eq. (19), which become increasing large as r increases, prevents the particle orbits from diverging indefinitely in the radial direction in Fig. 7(b). Figure 8 shows the plots of the parameter ϵ_s^{cr}/γ_b versus a_w. Here, the solid curve corresponds to the analytical estimate $\epsilon_s^{cr}/\gamma_b = a_w^2/(1 + a_w^2)$, and the dashed curves are obtained from numerical integration of the equations of motion. In Fig. 8, the two dashed curves correspond to $\gamma_b = 4$ and $\gamma_b = 10$ used in the simulations. It is evident from Fig. 8 that the analytical and numerical results are in good agreement.

B. Chaotic Motion in the Strong-Pump Regime

We now examine the particle orbits in the regime where the wiggler field amplitude a_w is sufficiently large that

$$a_w \sim \gamma_b \beta_{zb} \; .$$ (58)

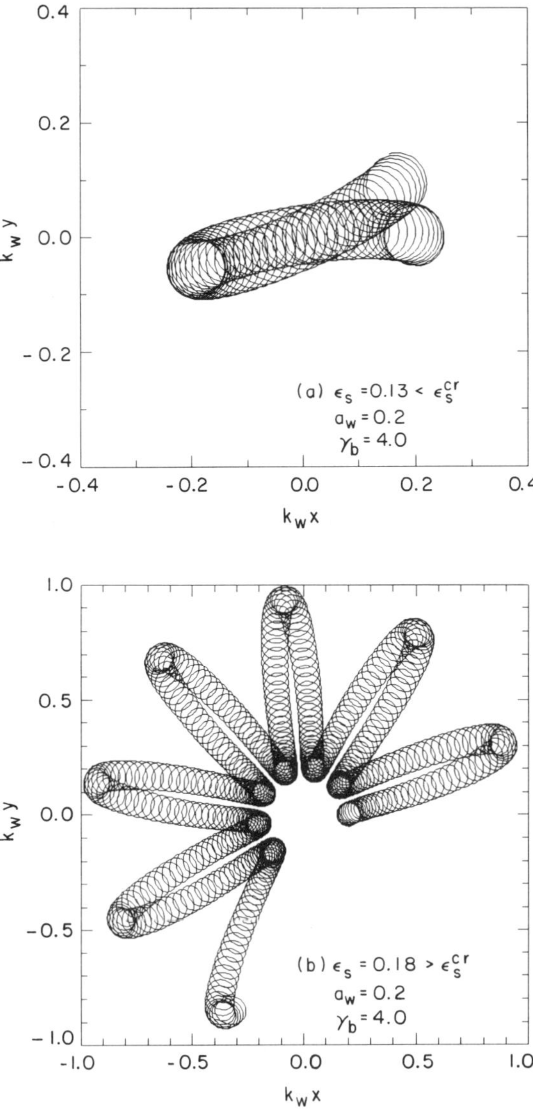

Fig. 7. Plots of typical transverse trajectories for the two cases:
(a) $\epsilon_s < \epsilon_s^{cr} \equiv \gamma_b a_w^2/(1 + a_w^2)$, and (b) $\epsilon_s > \epsilon_s^{cr}$. Here, the choices
of the system parameters for the two cases are: (a) $\epsilon_s = 0.13$,
$a_w = 0.2$, $\gamma_b = 4.0$, and $\epsilon_s^{cr} = 0.154$, and (b) $\epsilon_s = 0.18$, $a_w = 0.2$,
$\gamma_b = 4.0$, and $\epsilon_s^{cr} = 0.154$.

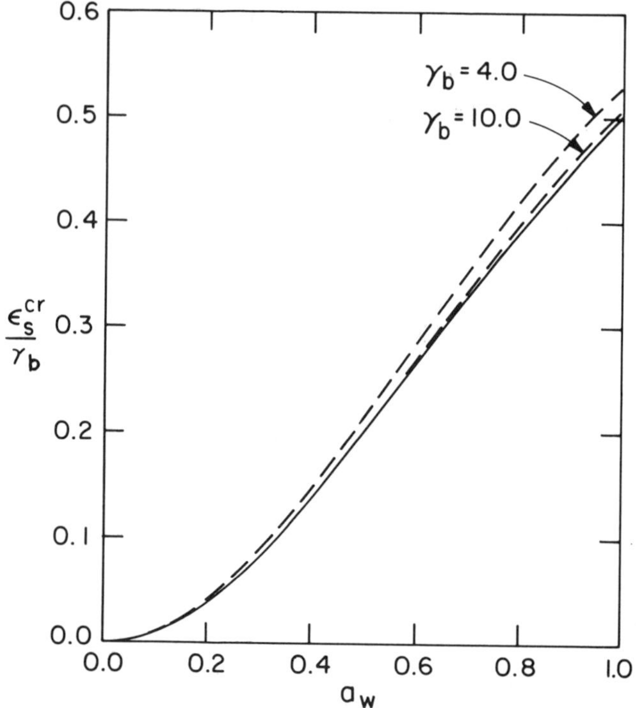

Fig. 8. Plots of ϵ_s^{cr}/γ_b versus a_w. Here, the solid curve corresponds to the analytical estimate $\epsilon_s^{cr}/\gamma_b = a_w^2/(1 + a_w^2)$, and the dashed curves are obtained by solving numerically the equations of motion for $\gamma_b = 4$ and $\gamma_b = 10$.

Because the normalized beam radius $k_w r_b$ and gyroradius $k_w r_c = a_w/\gamma_b \beta_{zb}$ are allowed to be of order unity, the analytical treatment in Sec. IV.A is no longer valid. For $k_w r \leq k_w r_b < 1$, however, the approximate Hamiltonian defined in Eqs. (45) and (46) still provides an adequate description of the particle motion. In the remainder of this section, we show that the motion is nonintegrable and exhibits chaotic behavior when a_w is sufficiently large.

To simplify the analysis, we assume that self-field effects are negligibly small ($\epsilon_s = 0$ and $\Phi_s = 0$), and focus on the region in phase space in the vicinity of helical orbits with guiding center on the axis of the wiggler helix ($r_g = 0$), electron energy $\gamma = \gamma_b$, and normalized axial velocity $\beta_z \cong \beta_{zb} = [1 - (1 + a_w^2)/\gamma_b^2]^{1/2}$. In addition, it is useful to introduce the dimensionless parameter

$$\Delta = \frac{\omega_{\beta w}}{c k_w \beta_{zb}} = \frac{a_w}{[2(\gamma_b^2 - 1 - a_w^2)]^{1/2}} \, , \tag{59}$$

which is a measure of the nonintegrability of the motion. Physically, λ_w/Δ is the axial distance through which an electron with energy $\gamma = \gamma_b$ and axial velocity $v_z = \beta_{zb}c$ travels in one betatron oscillation period $2\pi/\omega_{\beta w}$.

For present purposes, it is convenient to describe the particle motion in cylindrical polar coordinates (r, θ, z). The Hamiltonian defined in Eqs. (45) and (46) can be expressed as

$$\hat{H} = \left\{ \left[\hat{P}_r + a_w \left(1 + \frac{k_w^2 r^2}{8} \right) \cos(\theta - k_w z) \right]^2 + \right.$$

$$\left. \left[\frac{\hat{P}_\theta}{k_w r} - a_w \left(1 + \frac{3k_w^2 r^2}{8} \right) \sin(\theta - k_w z) \right]^2 + \hat{P}_z^2 + 1 \right\}^{1/2}, \tag{60}$$

where use has been made of Eqs. (17) and (20), and the dimensionless variables

$$\hat{P}_r = \frac{P_r}{mc} \quad \text{and} \quad \hat{P}_\theta = \frac{k_w P_\theta}{mc} \tag{61}$$

have been introduced. Because the combination $\theta - k_w z$ appears in $\hat{H}$, it is useful to perform the canonical transformation to the new variables $(k_w r, \chi, k_w z', \hat{P}_r, \hat{P}_\chi, \hat{P}_{z'})$ defined by

$$\chi = \theta - k_w z, \quad k_w z' = k_w z, \tag{62}$$

$$\hat{P}_\chi = \hat{P}_\theta, \quad \hat{P}_{z'} = \hat{P}_z + \hat{P}_\theta. \tag{63}$$

Here, the generating function is given by $F_2(k_w z, \theta; \hat{P}_{z'}, \hat{P}_\chi) = k_w z \hat{P}_{z'} + (\theta - k_w z)\hat{P}_\chi$. The Hamiltonian in the new variables can be expressed as

$$\hat{H}(k_w r, \chi, \hat{P}_r, \hat{P}_\chi, \hat{P}_{z'} = \text{const.})$$

$$= \left\{ \left[\hat{P}_r + a_w \left(1 + \frac{k_w^2 r^2}{8} \right) \cos \chi \right]^2 + \left[\frac{\hat{P}_\chi}{k_w r} - a_w \left(1 + \frac{3k_w^2 r^2}{8} \right) \sin \chi \right]^2 \right.$$

$$\left. + (\hat{P}_{z'} - \hat{P}_\chi)^2 + 1 \right\}^{1/2}$$

$$= \text{const.} \tag{64}$$

Equation (64) possesses two constants of the motion, namely, $\hat{H}$ and $\hat{P}_{z'}$. The motion occurs in the three-dimensional phase space $(\chi, \hat{P}_\chi, \hat{P}_r)$, because $k_w r$ can be determined from $\hat{H} = \text{const.}$

The (helical) steady-state orbits with guiding center on the z-axis are the solutions of the steady-state equations of motion derived from the Hamiltonian in Eq. (64). It can be shown for $0 \leq \Delta < \Delta_c \cong 0.28$ that the steady-state orbits are given by[24,29]

$$k_w r = k_w r_0, \quad \chi = \chi_0 = 3\pi/2, \quad \hat{P}_r = \hat{P}_{r0}, \quad \hat{P}_\chi = (3a_w/4)(k_w r_0)^3, \tag{65}$$

where the normalized gyroradii $k_w r_0 = k_w r_0^<$ and $k_w r_0 = k_w r_0^> > k_w r_0^<$ are the solutions of the algebraic equation

$$f(k_w r_0) \equiv \left[2\left(1 + \frac{1}{k_w^2 r_0^2}\right)\left(1 + \frac{9 k_w^2 r_0^2}{8}\right)^2 - 2 \right]^{-1/2} = \Delta \, . \tag{66}$$

Because the function $f(k_w r_0)$ satisfies $f(0) = f(\infty) = 0$ and has a (single) maximum $f_m = \Delta_c \cong 0.28$ at $k_w r_0 \cong 0.62$, it follows that Eq. (66) has two real solutions when Δ is in the interval $0 \le \Delta < \Delta_c$, and no real solution otherwise.

Poincaré surface-of-section maps are generated to demonstrate the chaoticity in the phase space in the vicinity of the steady-state orbit in Eq. (65) with $k_w r_0 = k_w r_0^<$, where $r_0^<$ is the smaller of the two solutions to Eq. (66). Figure 9 shows the Poincaré surface-of-section plots in the $(\chi, \hat{P}_\chi)$ plane at $\hat{P}_r = 0$ for $\hat{H} = \gamma_b = 6.0$ and $\epsilon_s = 0$, corresponding to the two cases: (a) $\Delta = 0.18 < \Delta_c \cong 0.28$ (or $a_w = 1.5$) and (b) $\Delta = 0.22 < \Delta_c$ (or $a_w = 1.8$). The orbits in Fig. 9 are calculated numerically from the equations of motion derived from the Hamiltonian in Eq. (64). It is evident that the phase space contains regular and chaotic orbits. In Fig. 9, the fixed point at $\chi = \chi_0 = 3\pi/2$ and $\hat{P}_\chi = \hat{P}_{\chi 0}$ corresponds to the steady-state orbit defined in Eq. (65). Each contour in Fig. 9 corresponds to an orbit with the guiding center oscillating about $r = 0$ approximately at the betatron frequency $\omega_{\beta w}$. As the contour increases in size, because the amplitude of the (betatron) oscillation increases, the coupling between the helical motion and the betatron oscillations is enhanced, which is the source of the chaoticity. As the value of the parameter Δ (or a_w) is increased, the area of the regular region in the phase plane decreases [compare Fig. 9(b) with Fig. 9(a)]. Indeed, the phase plane becomes fully chaotic if Δ exceeds the threshold value $\Delta = \Delta_c \cong 0.28$. Solving for $a_w = a_w^c$ from Eq. (59) with $\Delta = \Delta_c = 0.28$, it readily follows that the threshold value of the dimensionless wiggler amplitude for the onset of chaoticity is given approximately by

$$a_w^c = 0.37(\gamma_b^2 - 1)^{1/2} \, . \tag{67}$$

For given γ_b, the phase space is fully chaotic if $a_w > a_w^c(\gamma_b)$, whereas there is a regular region with some finite area in phase phase if $a_w < a_w^c(\gamma_b)$.

V. CONCLUSIONS

We have investigated the effects of equilibrium self fields and an inhomogeneous wiggler field on the dynamics of a test electron in a helical-wiggler free electron laser in the absence of electromagnetic signal wave. It was shown that the transverse spatial gradients in the self fields and a realizable helical wiggler field can cause chaoticity in the particle orbits. In addition, the characteristic time scale for radial-gradient-induced changes in the particle orbits is of order

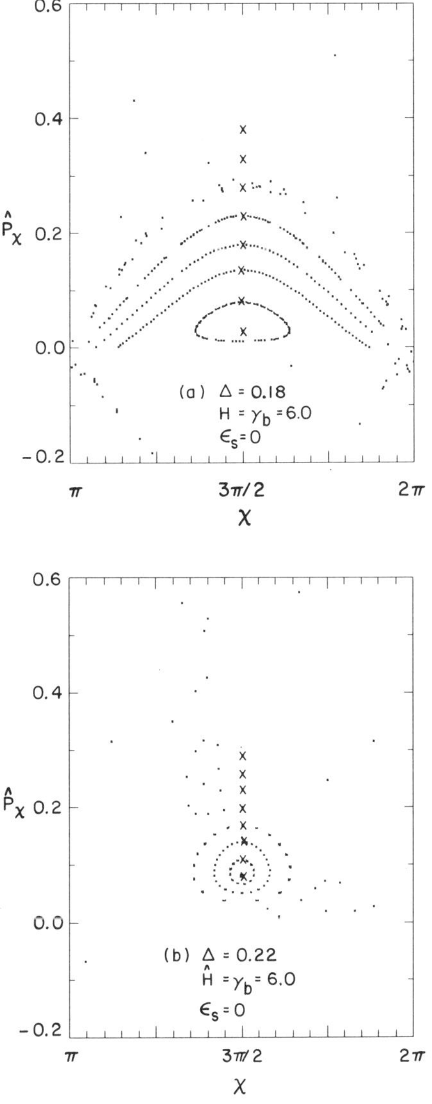

Fig. 9. Poincaré surface-of-section plots in the $(\chi, \hat{P}_\chi)$ plane at $\hat{P}_r = 0$. Here, 50 iterations are plotted for each orbit with the initial conditions marked by the crosses. The choices of system parameters for the two cases are: (a) $\Delta = 0.18$ $(a_w = 1.5)$, $\hat{H} = \gamma_b = 6.0$, and $\epsilon_s = 0$, and (b) $\Delta = 0.22$ $(a_w = 1.8)$, $\hat{H} = \gamma_b = 6.0$, and $\epsilon_s = 0$.

the beam transit time through one wiggler period. The following is a brief summary of the principal results and conclusions.

First, the influence of equilibrium self fields on the particle orbits was analyzed in the field configuration consisting of an ideal helical wiggler field and a uniform axial guide field. It was shown that the inclusion of equilibrium self-field effects destroys the integrability of the motion, and results in chaotic particle motion at sufficiently high beam density. In particular, the Group-I orbits and the Group-II orbits can become fully chaotic. The origin of this chaoticity is the coupling between the guide-field-induced betatron oscillations and the helical motion, modified by the radial gradient of the self fields. An analysis of the self-field-induced resonances was carried out, and scaling relations for the resonance widths were derived. Good agreement was found between the computer simulations and the analytical estimate of the threshold value of the self-field parameter for the onset of chaoticity.

Second, the effects of wiggler-induced betatron oscillations on the particle orbits were analyzed for a realizable helical-wiggler field configuration in the absence of an axial guide field, but including the influence of equilibrium self fields. A condition for radial confinement of the particle orbits was derived analytically and verified in computer simulations for a thin electron beam and small wiggler amplitude. Although the particle orbits consist of well-defined helical motion and betatron oscillations when the wiggler amplitude is small, it was shown that the particle trajectories become strongly chaotic when the wiggler amplitude is sufficiently large. As the wiggler amplitude is increased, the area of the regular region in phase space decreases in the Poincaré surface-of-section plots. For the special case where self-field effects are negligibly small, the threshold value of the wiggler amplitude for the onset of chaoticity was found to be $a_w^c \cong 0.37(\gamma_b^2 - 1)^{1/2}$, which corresponds to the maximum allowed value of the wiggler amplitude for the existence of regular helical orbits for given electron energy γ_b.

REFERENCES

1.   M.V. Berry, in *Topics of Nonlinear Dynamics*, ed., S. Jorna, AIP Conference Proceedings, No. 46, (American Institute of Physics, New York, 1978), p. 16.

2.   A.J. Lichtenberg and M.A. Lieberman, *Regular and Stochastic Motion*, Springer-Verlag, New York (1983).

3.   R.S. MacKay and J.D. Meiss, eds., *Hamiltonian Dynamical Systems*, Adam-Hilger, Philadelphia (1987).

4.   A.N. Kolmogorov, Dokl. Acad. Nauk, USSR **98**, 527 (1954).

5.   V.I. Arnol'd, Sov. Math. Dokl. **2**, 501 (1961); Russian Math. Surveys **18**, 5 (1963).

6. J. Moser, *On Invariant Curves of Area Preserving Mappings on an Annulus*, Nachr. Acad. Wiss. Göttingen. Math. Phys. K1., p. 1 (1962); Math. Ann. **169**, 163 (1967).

7. R.S. MacKay, J.D. Meiss, and I.C. Percival, Physica **13D**, 55 ((1983), and references therein.

8. G.R. Smith and A.N. Kaufman, Phys. Rev. Lett. **34**, 1613 (1975).

9. C.F.F. Karney and A. Bers, Phys. Rev. Lett. **59**, 550 (1977).

10. F. Skiff, F. Anderegg, and M.Q. Tran, Phys. Rev. Lett. **58** 1430 (1987).

11. W. Colson, C. Pellegrini, and Renieri, eds., *Free Electron Laser Handbook*, North-Holland, Amsterdam (1989).

12. C.W. Roberson and P. Sprangle, Phys. Fluids B**1**, 3 (1989).

13. T.C. Marshall, *Free Electron Lasers*, Macmillan, New York (1985).

14. D.A.G. Deacon, L.R. Ellis, J.M.J. Madey, G.J. Ramian, H.A. Schwettman, and T.I. Smith, Phys. Rev. Lett. **38**, 892 (1977).

15. R.W. Warren, B.E. Newman, and J.C. Goldstein, IEEE J. Quantum Electron. **QE-21**, 882 (1985).

16. J. Fajans, G. Bekefi, Y.Z. Yin, and B. Lax, Phys. Fluids **28**, 1995 (1985).

17. J. Fajans, J.S. Wurtele, G. Bekefi, D.S. Knowles, and K. Xu, Phys. Rev. Lett. **57**, 579 (1986).

18. T.J. Orzechowski, B.R. Anderson, J.C. Clark, W.M. Fawley, A.C. Paul, D. Prosnitz, E.T. Scharlemann, S.M. Yarema, D.B. Hopkins, A.M. Sessler, and J.S. Wurtele, Phys. Rev. Lett. **57**, 2172 (1986).

19. R.C. Davidson and H.S. Uhm, Phys. Fluids **23**, 2076 (1980).

20. P. Sprangle, R.A. Smith, and V.L. Granatstein, in *Infrared and Millimeter Waves*, Volume **1**, p. 297, ed., K.J. Button, Academic Press, New York (1979).

21. R.C. Davidson, *Physics of Nonneutral Plasmas*, Addison-Wesley, Reading, Massachusetts (1990).

22. L. Friedland, Phys. Fluids **23**, 2376 (1980).

23. H.P. Freund and A.T. Drobot, Phys. Fluids **25**, 736 (1982).

24. P. Diament, Phys. Rev. **A23**, 2537 (1981).

25. H.P. Freund and A.K. Ganguly, Phys. Rev. **A28**, 3438 (1983); Phys. Rev. **A34**, 1242 (1986); IEEE J. Quantum Electron. **QE-23**, 1657 (1987).

26. C. Chen and R.C. Davidson, Phys. Fluids **B2**, 171 (1990).

27. H.P. Freund, P. Sprangle, D. Dillenburg, E.H. da Jornada, R.S. Schneider, and B. Liberman, Phys. Rev. **A26**, 2004 (1982).

28. C. Chen and G. Schmidt, Comments in Plasma Physics and Controlled Fusion **12**, 83 (1988).

29. C. Chen and R.C. Davidson, submitted for publication (1990).

NONLINEAR WAKEFIELDS AND OPTICAL GUIDING OF INTENSE LASER PULSES IN PLASMAS

E. Esarey, P. Sprangle and A. Ting

Beam Physics Branch, Plasma Physics Division
Naval Research Laboratory, Washington, DC 20375-5000

Abstract

The generation of nonlinear plasma wakefields by an intense, short laser pulse and the relativistic optical guiding of intense laser pulses in plasmas are studied with a nonlinear, self–consistent model of laser–plasma interactions. Nonlinear steepening and period lengthening of the plasma waves are observed, and expressions are obtained for various nonlinear wakefield quantities. Relativistic focusing with the self–consistent plasma response shows that laser pulse fronts and laser pulses shorter than a plasma wavelength, $2\pi c/\omega_p$, are not relativistically guided and will continuously erode due to diffraction.

I. Introduction

The interaction of ultra–high power laser beams[1] with plasmas is rich in a variety of wave–particle phenomena. These phenomena become particularly interesting and involved when the laser power is high enough to cause the electron oscillation (quiver) velocity to become highly relativistic. Some of the interesting laser–plasma processes include: a) the generation of large amplitude plasma waves[2-12] (wakefields), b) relativistic optical guiding[13-23] of the laser beam, c) the excitation of coherent radiation at harmonics of the fundamental laser frequency,[8] d) frequency shifts induced in the laser pulse by plasma waves,[8,24,25] e) frequency amplification using an ionization front[8] and f) single particle acceleration in a laser pulse.[8] In the following, the development of a nonlinear, self–consistent model of intense laser–plasma interactions is discussed, and this model is then used to examine the first two of the above mentioned phenomena.

Recent advances in the generation of ultra–high power, short–pulse laser beams[1] ($P \gtrsim 10^{15}$ W, $\tau_L \sim 2\pi\omega_p^{-1} \sim 1$ psec) have introduced the possibility of producing large amplitude ($E \gtrsim 1$ GeV/m) wakefields by propagating such laser pulses in plasmas. A correctly placed trailing electron bunch can be accelerated by the longitudinal electric field and focused by the transverse electric field of the wake plasma waves, as recently proposed in the laser wakefield accelerator (LWFA) concept.[2-8] This mechanism is similar to the plasma wakefield accelerator (PWFA)[9,10] except the plasma responds to the ponderomotive forces of the laser pulse as opposed to the self–fields of the driving electron beam in the PWFA. However, in order to achieve efficient acceleration, it may be necessary to accurately tailor the axial profile of the driving beam in the PWFA, but no such tailoring is necessary in the LWFA.[4] The LWFA also has advantages over the plasma beat wave accelerator (PBWA)[2] where plasma waves are resonantly excited and, therefore, fine tuning is required of the plasma density and the frequencies of the two laser beams. Plasma wave generation in the LWFA is a nonresonant process and restrictions on the plasma uniformity and laser pulse length are far less stringent.[4]

A characteristic parameter of a high–power laser beam is the normalized vector potential $|a| \equiv |eA_\perp|/(mc^2)$, where $A_\perp$ is the transverse vector potential, e is the electronic charge, m is the mass of an electron and c is the speed of light. Conservation of transverse canonical momentum in one–dimensional (1D) systems shows that $a = \gamma\beta_\perp$, where γ is the relativistic mass factor and $\beta_\perp = v_\perp/c$ is the normalized transverse quiver velocity of the electron in the laser field. Therefore, when $|a| \gtrsim 1$, the electron quiver motion is highly relativistic. Relativistic effects at $|a| > 1$ introduce many difficulties in analyzing laser–plasma interactions. For example, previous analyses of the LWFA[3,4] utilize a simple expression of the ponderomotive force of a laser pulse for $|a| < 1$. However, in the regime of $|a| > 1$, it becomes difficult to extract an expression for the ponderomotive force and the LWFA has to be analyzed self–consistently. Previous analyses of the laser–plasma interactions (especially for linearly polarized lasers) have been limited to situations where $|a| < 1$.

A fully nonlinear 1D model has been developed which describes the self–consistent interaction of intense laser pulses with plasmas,[8] and is valid for arbitrary $|a|$, including $|a| > 1$. This model is briefly outlined in Section II. By assuming a "quasi–static" cold fluid plasma response, a set of coupled nonlinear equations is derived for the vector potential of the radiation field and for the electrostatic potential of the plasma. The quasi–static approximation assumes that for a short laser pulse, the plasma fluid sees a nearly steady–state radiation field as it transits through the laser pulse when the system is viewed in a window moving at the speed of light. The resulting nonlinear equations can be used to examine various laser–plasma interaction phenomena.

For an intense, short laser pulse with $|a| > 1$, nonlinear wakefields will be generated. This is analyzed with the nonlinear self–consistent model of laser–plasma interactions in Section III. Analytical and numerical results for various pulse shapes are presented. Nonlinear effects[5−8,10−12] of wave steepening and period lengthening are observed, and expressions are derived for the maximum wakefield amplitude and the nonlinear wakefield period. Nonlinear corrections to the pump depletion length and the phase detuning length are also obtained. The large acceleration gradients associated with the nonlinear wakefields may accelerate electrons to high energies for long interaction lengths. However, the interaction length is primarily limited by the diffraction of the laser pulse.[4] For intense laser pulses, relativistic focusing may provide the optical guiding necessary to extend the interaction length to beyond the diffraction limit.

Relativistic optical guiding[13-23] is a result of the relativistic quiver motion of the electrons by the laser field. Analysis[16-20] has shown that as the laser power exceeds a critical threshold, diffraction can be overcome, resulting in optical guiding of the laser pulse. Previous analyses of relativistic guiding have attributed this effect to only the transverse quiver motion of the electrons in the plasma response current.[13-23] Relativistic guiding was also believed to occur on a fast time scale (on order of the inverse laser frequency). However, the present analysis finds this not to be the case. In Section IV, the nonlinear analysis of relativistic optical guiding is presented, where the electron density response and the longitudinal electron motion are included self–consistently. It is shown that for short laser pulses (pulse lengths less than a plasma wavelength, $\lambda_p = 2\pi c/\omega_p$), the combined effects of the density response and the longitudinal motion of the electrons significantly reduce the relativistic guiding effect. It is found that relativistic guiding occurs only in the main portion of long pulses (greater than a plasma wavelength). However, the front portion of a long laser pulse behaves like a short pulse and will continuously diffract. The pulse front will therefore erode and the laser pulse will be continuously shortened as it propagates through the plasma. The possibility of guiding a laser pulse with the wakefield generated by a driving electron or laser beam is discussed and a conclusion is provided in Section V.

II. Nonlinear Self–Consistent Model

The 1D fields associated with the laser–plasma interaction can be described by the normalized transverse vector and scalar potentials, $a(z,t) = |e|\mathbf{A}_\perp/mc^2$ and $\phi(z,t) = |e|\Phi/mc^2$, respectively. The electrons are assumed to obey the relativistic cold fluid equations and the ions are assumed to be stationary. Thermal effects may be neglected provided i) the electron quiver velocity is much greater than the electron thermal velocity, and ii) the thermal energy spread is sufficiently small such that electron trapping in the plasma wave[26,27] is insignificant. It proves convenient to perform an algebraic transformation from the laboratory independent space and time coordinates (z,t) to the independent coordinates (ζ,τ) in a window moving at the speed of light where $\zeta = z - ct$ and $\tau = t$. This configuration is shown in the schematic diagram in Fig. 1.

The electron fluid response can be greatly simplified by noting that, in the speed of light window, under certain conditions a quasi–static state will exist[8] in the plasma fluid quantities n, β and γ, where n is the electron density (n_0 is the ambient density), $\beta = v_z/c$ is the normalized longitudinal electron fluid velocity and $\gamma = (1 + a^2)^{1/2}/(1 - \beta^2)^{1/2}$ is the relativistic factor associated with the electron fluid. That is, if the laser pulse is sufficiently short, the fields a and ϕ which drive the plasma are expected to change little during a transit time of the plasma through the laser pulse. The envelope of a can be shown to change on a characteristic time[8] $\tau_e \sim 2\gamma |n_0/n|(\omega/\omega_p)/\omega_p$, where ω is the laser frequency and $\omega_p = (4\pi e^2 n_0/m)^{1/2}$ is the ambient electron plasma frequency. Assuming $\omega >> \omega_p$ implies that τ_e is long compared to a plasma period. If the laser pulse duration τ_L is small compared to τ_e, then the quasi–static approximation is valid. In addition, the validity of the 1D model requires that the laser beam vacuum diffraction time, $\tau_d = \pi r_s^2/(\lambda c)$, be long compared to τ_e, where r_s is the laser spot size and λ is the laser wavelength. This is satisfied when $r_s^2 >> \lambda_p^2$.

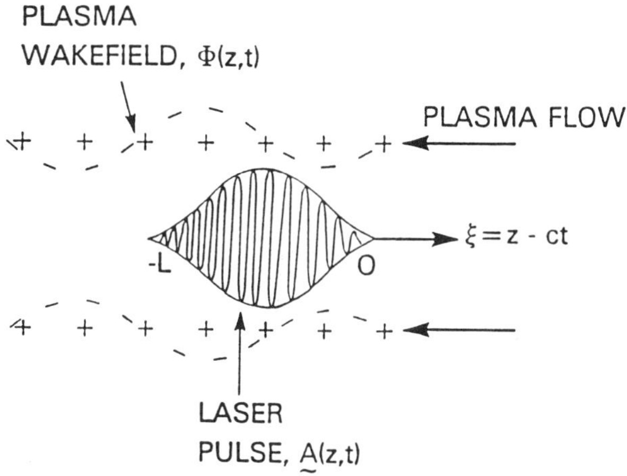

Fig. 1. Schematic showing the laser pulse in the speed of light frame (ζ, τ). The pulse extends from $-L \leq \zeta \leq 0$ and the front of the pulse is at $\zeta = 0$. In this frame the plasma flows from right to left and a quasi–static state exists.

Under the quasi-static approximation, the continuity equation and the axial momentum equation can be integrated[8] to give $n(1 - \beta) = n_0$ and $\gamma(1 - \beta) = 1 + \phi$. The transverse electron current is then given by $J_\perp \sim n\beta_\perp = na/\gamma = n_0 a/(1 + \phi)$. Using these relations, the wave equation and Poisson's equation may be written, respectively, as

$$\left(2\frac{\partial}{\partial \zeta} - \frac{1}{c}\frac{\partial}{\partial \tau}\right)\frac{1}{c}\frac{\partial a}{\partial \tau} = k_p^2 \frac{a}{1 + \varphi}, \tag{1}$$

$$\frac{\partial^2 \phi}{\partial \zeta^2} = -\frac{k_p^2}{2}\left[1 - \frac{(1 + a^2)}{(1 + \varphi)^2}\right]. \tag{2}$$

This coupled set of equations completely describes the 1D nonlinear laser–plasma interaction within the quasi-static approximation. This model is valid for laser pulses of arbitrary polarizations and arbitrary intensities (including $a^2 \geq 1$). Consistent with the quasi–static assumption, a variety of interesting phenomena can be studied concerning intense laser pulse propagation in plasmas.

III. Nonlinear Wakefield Generation

A. Analytic Theory

Aspects of nonlinear wakefield generation[5-8] may be examined analytically by solving Eq. (2) for a circularly polarized laser pulse with a square pulse profile for the laser envelope, $a_L = a_{L0}$ for $-L \leq \zeta \leq 0$ and $a_L = 0$ otherwise. Using the initial conditions $\phi = \partial\phi/\partial\zeta = 0$ at $\zeta = 0$, the formal solution for $\phi(\zeta)$ within the laser pulse, $-L \leq \zeta \leq 0$, is given by

$$k_p\zeta = -2\gamma_0 E(\alpha,\rho) + 2\left[(a_{L0}^2 - \phi)\phi/(1 + \phi)\right]^{1/2}, \tag{3}$$

where $\gamma_0^2 = 1 + a_{L0}^2$, $\rho = a_{L0}/\gamma_0$, $\alpha = \sin^{-1} y$ and $y^2 = \phi\gamma_0^2/\left[(1 + \phi)a_{L0}^2\right]$. Here, $E(\alpha,\rho)$ is the elliptic integral of the second kind. The above equation indicates that, inside the pulse, the potential ϕ lies in the range $a_{L0}^2 \geq \phi \geq 0$. The maximum value of $\phi = a_{L0}^2$ is reached at $\zeta = -L_0$, where L_0 is the optimal pulse length (corresponding to maximum wakefield generation) given by

$$L_0 = \frac{2}{k_p}\gamma_0 E(\rho) \simeq \frac{a_{L0}}{\pi}\lambda_p, \tag{4}$$

for $a_{L0}^2 >> 1$. Notice that when $L = 2L_{op}$, $\phi = 0$ and, hence, there is no wakefield behind the pulse ("wakeless" mode of operation).

For $L = L_{op}$, the wakefield behind the laser pulse ($\zeta < -L$) is given by

$$k_p\zeta = -k_p L_{op} - 2\gamma_0 E(\bar{\alpha},\bar{\rho}), \tag{5}$$

where $\bar{\rho}^2 = 1 - 1/\gamma_0^4$ and $\bar{\alpha} = \sin^{-1} \bar{y}$, with $\bar{y}^2 = \gamma_0^2(a_{L0}^2 - \phi)/(\gamma_0^4 - 1)$. Behind the pulse, the potential ϕ of the wakefield oscillates in the range $1/\gamma_0^2 \leq 1 + \phi \leq \gamma_0^2$. The distance over which $1 + \phi$ goes from γ_0^2 to $1/\gamma_0^2$ defines $\lambda_p^{NL}/2$, where λ_p^{NL} is the nonlinear plasma wavelength,

$$\lambda_p^{NL} = \frac{4}{k_p}\gamma_0 E(\bar{\rho}) \simeq \frac{2}{\pi}a_{L0}\lambda_p, \tag{6}$$

for $a_{L0}^2 >> 1$.

The normalized axial electric field of the wakefield $\hat{E}_z = E_z/E_0$ is related to ϕ by

$$\hat{E}_z^2 = a_{L0}^2 - \phi + \gamma_0^{-2} - (1 + \phi)^{-1}, \tag{7}$$

where $E_0 = k_p mc^2/|e| \simeq (n_0[\text{cm}^{-3}])^{1/2}$ V/cm is the nonrelativistic cold wavebreaking amplitude. The maximum axial electric field $\hat{E}_z^{max}$ occurs at $\phi = 0$ and is given by

$$\hat{E}_z^{max} = \frac{a_{L0}^2}{(1 + a_{L0}^2)^{1/2}}. \tag{8}$$

Notice that $\hat{E}_z^{max} \simeq a_{L0}$ for $a_{L0}^2 >> 1$ and $\hat{E}_z^{max} \simeq a_{L0}^2$ for $a_{L0}^2 << 1$.

The above expressions apply to nonlinear plasma waves with phase velocities near the speed of light, $v_{ph} \simeq c$. It is possible to show using cold fluid theory that, by transforming to a frame moving at an arbitrary phase velocity, wavebreaking of the nonlinear wakefield occurs when $\epsilon_{wb} \equiv (1 - \beta_{ph}^2)(1 + a^2)/(1 + \phi)^2 \geq 1$, where $\beta_{ph} = v_{ph}/c$. Equations (3)-(8) remain valid in the "sub–wavebreaking" limit of $\epsilon_{wb} << 1$. For the case of the nonlinear laser wakefield accelerator, in which the plasma wave has a phase velocity approximately equal to the group velocity of the laser pulse, $\beta_{ph} \simeq 1 - \omega_p^2/(2\gamma_0\omega^2)$, the condition for the wakefield to be well below wavebreaking implies $\gamma_0^3 \omega_p^2/\omega^2 << 1$. For a given value of ω_p^2/ω^2, this inequality gives an upper limit on a_{L0}^2 for which the fluid model remains valid.

The above expressions may be generalized quite readily to the case of a linearly polarized laser pulse with a square profile by noting that when $\omega_p^2/\omega^2 << 1$, Eq. (2) implies that $|\phi_f/\phi_s| << 1$, where ϕ_f is the fast part of the potential and ϕ_s is the slow part of the potential. Hence, for a linearly polarized square pulse, let $\phi \to \phi_s$ and $a_{L0}^2 \to a_{L0}^2/2$ in Eqs. (3)-(8).

B. *Numerical Example*

For more realistic laser pulse profiles, Eq. (2) may be solved numerically. Figures 2(a,b) show the plasma density variation $\delta n/n_0 = n/n_0 - 1$ and the corresponding axial electric field E_z for a laser pulse envelope given by $a_L = a_{L0}\sin(\pi\zeta/L)$ for $-L \leq \zeta \leq 0$. In this figure, $L = \lambda_p = 0.03$ cm, $\lambda = 10$ μm, and (a) $a_{L0} = 0.5$ and (b) $a_{L0} = 2$. The steepening of the electric field and the increase in the period of the wakefield[5−8,10−12] are apparent for the highly nonlinear situation shown in Fig. 2(b) ($a_{L0} = 2$) as compared to the slightly nonlinear case shown in Fig. 2(a) ($a_{L0} = 0.5$). Figure 3 shows that the electrostatic potential ϕ is predominantly slowly varying within the laser pulse even though $\delta n/n_0$ has rapidly varying components. In addition, the figure shows that ϕ is positive within the laser pulse and oscillates behind the laser pulse.

C. *Limitations on the Acceleration Distance*

The large amplitude axial electric fields associated with the plasma waves can be utilized to accelerate an injected beam of electrons to high energies (LWFA).[2−8] In the LWFA, the distance over which an electron may

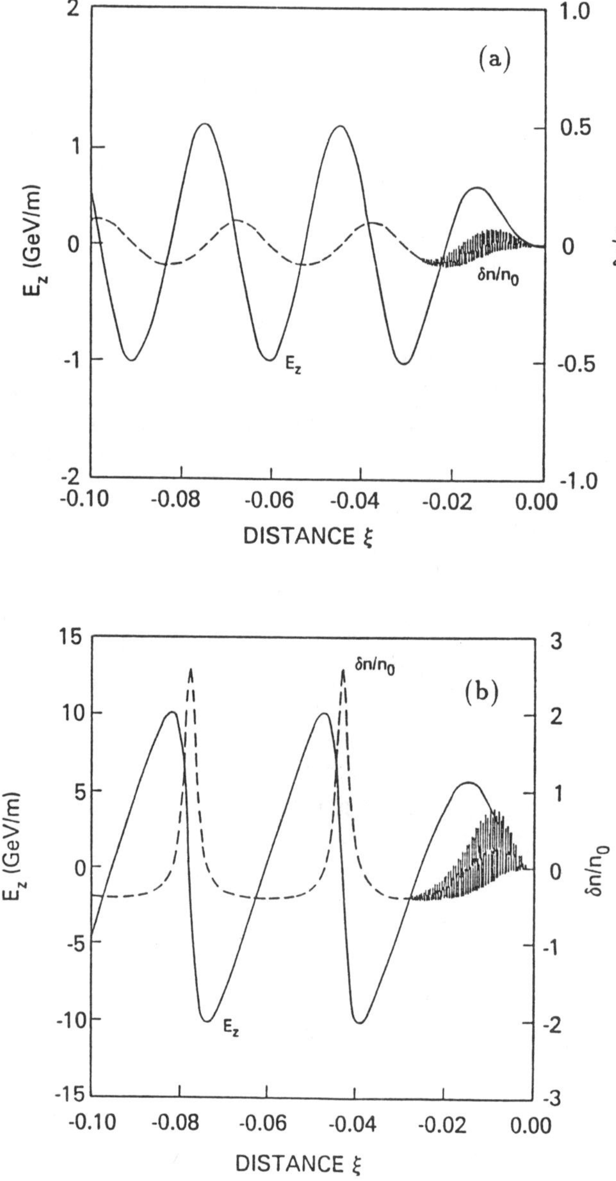

Fig. 2. Density variation $\delta n/n_0 = n/n_0 - 1$ and axial electric field E_z in GeV/m for a laser pulse located within the region $-L \leq \zeta \leq 0$, where $L = \lambda_p = 0.03$ cm, and (a) $a_{L0} = 0.5$ and (b) $a_{L0} = 2.0$.

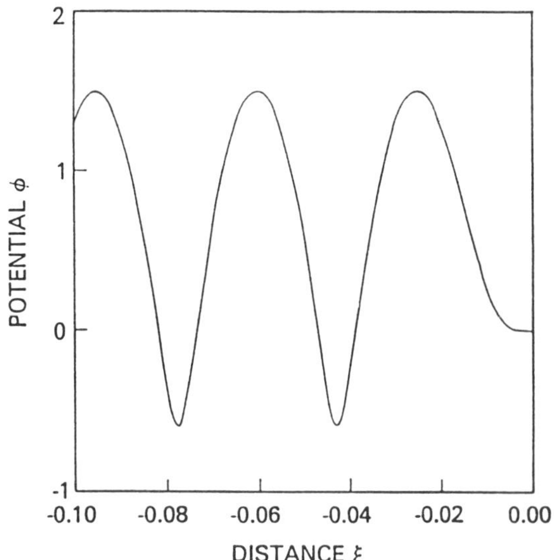

Fig. 3. Electrostatic potential ϕ for a laser pulse located within the region $-L \leq \zeta \leq 0$, where $L = \lambda_p = 0.03$ cm and $a_{L0} = 2.0$.

be accelerated in a single stage will be limited by (neglecting the effects of laser–plasma instabilities[28]) i) phase detuning between the plasma wave and the accelerated electron, ii) depletion of the energy in the driving (pump) laser pulse and iii) diffraction of the laser pulse. The phase detuning distance, L_t, in which an ultra–relativistic electron will outrun the accelerating region of a plasma wave, may be estimated by $L_t(1 - \beta_{ph}) = \lambda_p^{NL}/2$. Assuming $\beta_{ph} \simeq 1 - \omega_p^2/(2\gamma_0\omega^2)$, gives

$$L_t = \frac{4}{k_p}\gamma_0^2\frac{\omega^2}{\omega_p^2}E(\rho) \simeq \frac{2}{\pi}a_{L0}^2\frac{\omega^2}{\omega_p^2}\lambda_p, \tag{9}$$

for $a_{L0}^2 \gg 1$, where Eq. (6) has been used.

The pump depletion distance L_p may be estimated by equating the energy in the driving laser pulse with the energy left behind in the nonlinear wakefield, $L_p\langle \hat{E}_z^2 \rangle \equiv L_0 a_{L0}^2$, where $\langle \hat{E}_z^2 \rangle$ represents an averaging of $\hat{E}_z^2$ over a period of the nonlinear wakefield. Using Eqs. (4) and (7) gives

$$L_p = \frac{6(\omega^2/\omega_p^2)\gamma_0 a_{L0}^2 E(\rho)/k_p}{[\gamma_0^2 + \gamma_0^{-2} - 2\gamma_0^{-2}K(\bar{\rho})/E(\bar{\rho})]} \simeq \frac{a_{L0}}{3\pi}\frac{\omega^2}{\omega_p^2}\lambda_p, \tag{10}$$

for $a_{L0}^2 \gg 1$.

If the laser pulse in the LWFA is assumed to undergo vacuum diffraction, then the laser diffraction distance L_d will be approximately $L_d \simeq 2z_R$, where

$z_R = \pi r_s^2 / \lambda$ is the vacuum Rayleigh length. Assuming $r_s = 2\lambda_p$, $a_{L0} \geq 1$ and $\omega_p^2/\omega^2 << 1$, then $L_d \simeq 8\pi(\omega/\omega_p)\lambda_p << L_p < L_t$. Hence, it is clear that diffraction is the most severe limitation on the single stage acceleration distance. The plasma, however, may provide some form of optical guiding, thus preventing the laser from undergoing vacuum diffraction.

IV. Nonlinear Optical Guiding

A. *Nonlinear Index of Refraction*

The nonlinear index of refraction of the laser beam within the plasma determines, among other things, the optical guiding properties of the plasma. For the purpose of the present discussion, the laser field is assumed to be linearly polarized, $a = a_L \exp(ik\zeta)$, where a_L represents the laser envelope and k is the wavenumber. The characteristic spatial variation in the laser envelope is assumed to be of order L and is long compared to the laser wavelength, $\lambda = 2\pi/k$, i.e., $\partial|a_L|/\partial\zeta \simeq |a_L|/L << k|a_L|$.

Using this representation in Eq. (1), the refractive index, $\eta = ck/\omega$, is given by

$$\eta \simeq 1 - \frac{\lambda^2}{2\lambda_p^2} \frac{1}{(1 + \phi_s)}, \tag{11}$$

where ϕ_s is the slow part of the scalar potential and $\lambda << \lambda_p$. In obtaining this expression, it has been assumed that $|\phi_f| << |\phi_s|$, where ϕ_f is the rapidly varying part of ϕ, which is valid as long as $\lambda << \lambda_p$.

Although the present analysis is 1D, we expect that for a slowly varying transverse laser profile the index of refraction will depend on the transverse coordinates through the laser amplitude $|a_L|$. Since the actual laser beam amplitude falls off transversely, $\partial|a_L|/\partial r < 0$, so will the refractive index, i.e., $\partial\eta/\partial r < 0$. The negative transverse gradient of the refractive index can lead to optical guiding. Since the refractive index is a function of the laser amplitude, the condition for optical guiding places a lower limit on $|a_L|$. For the case of conventional relativistic focusing,[13-23] the refractive index (for a linearly polarized laser) is given by

$$\eta = 1 - \frac{\lambda^2}{2\lambda_p^2} \frac{1}{(1 + |a_L|^2/2)^{1/2}}. \tag{12}$$

With this refractive index, a critical laser power necessary for relativistic optical guiding can be obtained and is given by[16-20] $P_{crit} \simeq 17.4(\lambda_p^2/\lambda^2)$ [GW].

Two cases of interest may be considered, depending on the envelope scale length L compared to the plasma wavelength λ_p.

B. *Short Pulse Limit*

Consider the short laser pulse limit, $L \lesssim \lambda_p$. When $|\phi| << 1$, Eq. (2) can be solved for an arbitrary laser field $a(\zeta)$,

$$\phi \simeq \frac{k_p}{2} \int_{\zeta}^{0} d\zeta' a^2(\zeta') \sin k_p(\zeta' - \zeta), \tag{13}$$

where the boundary conditions $\phi(\zeta = 0) = \partial\phi(\zeta = 0)/\partial\zeta = 0$ have been used. If the pulse envelope is given by $a_L = a_{L0} \sin(\pi\zeta/L)$ for $-L \leq \zeta \leq 0$ and $a_L = 0$ otherwise, the scalar potential within the laser pulse is given by

$$\phi \simeq \frac{a_{L0}^2}{8} \left\{ 1 - \left(k_p^2 - \frac{4\pi^2}{L^2} \right)^{-1} \left[k_p^2 \cos\left(\frac{2\pi\zeta}{L} \right) - \left(\frac{4\pi^2}{L^2} \right) \cos(k_p\zeta) \right] \right\}, \tag{14}$$

where terms of order $\lambda^2/\lambda_p^2 << 1$ have been neglected. For $L << \lambda_p$, this gives $\phi_s \simeq (a_{L0}k_p/4)^2 g(\zeta)$, where $g(\zeta) = \zeta^2 - 2(L/2\pi)^2[1 - \cos(2\pi\zeta/L)]$. Notice that for $L << \lambda_p$, ϕ_s is maximum at $\zeta = L$ where $\phi_s \simeq (a_{L0}k_pL/4)^2$. Also, notice that even for $|a_{L0}| > 1$, the assumption that $\phi_s << 1$ is still valid as long as $L << \lambda_p$. The index of refraction in the short pulse limit is, therefore,

$$\eta \simeq 1 - \frac{\lambda^2}{2\lambda_p^2} \left[1 + \frac{\pi^2 a_{L0}^2}{4\lambda_p^2} g(\zeta) \right]^{-1}, \tag{15}$$

where $-L \leq \zeta \leq 0$.

In the short pulse limit, the fact that $\phi_s << 1$ implies that the optical guiding effect is reduced significantly, by more than the factor $(\pi^2/2)(L^2/\lambda_p^2)$ $<< 1$. The critical power, therefore, is increased by the inverse of this factor and, in addition, the degree of guiding varies along the pulse. Hence, it is unlikely that relativistic optical guiding can be effectively utilized in short, $L \lesssim \lambda_p$, laser pulses.

C. *Long Pulse Limit*

In the limit of a long pulse with a rise time and pulse duration longer than an inverse plasma period, $L >> \lambda_p = 2\pi c/\omega_p$, the left-hand side of Eq. (2) can be neglected and ϕ_s can be approximated by $1 + \phi_s \simeq (1 + |a_L|^2/2)^{1/2}$. Substituting this into Eq. (11) gives the refractive index of Eq. (12), which is the result of the conventional theory of relativistic focusing.[13-23] This result shows that the main portion of a long laser pulse, assuming the various laser–plasma instabilities can be controlled, may be optically guided by the relativistic focusing effect provided the laser power is greater the critical power for relativistic focusing. However, since the laser power in the head and tail of the laser pulse is necessarily less than the critical power, these regions will undergo diffraction. Initially, that portion of the head of a long rise time pulse in which the local power is less than P_{crit} will diffract. Once this portion has diffracted away, the pulse will exhibit "short pulse" diffractive behavior, i.e., the front region ($\sim \lambda_p$) will continue to diffract. The erosion of the front of the pulse due to diffraction will propagate back through the body of the pulse. The erosion velocity back through the body of the pulse (in the $\zeta = z - ct$ frame) may be estimated by $v_E \simeq (\lambda_p/z_R)c$, where $z_R = \pi r_s^2/\lambda$ is the vacuum Rayleigh length.

Equation (11) indicates that a large amplitude plasma wakefield may be capable of optically guiding[29] a properly phased trailing laser pulse. Guiding may occur if a short ($< \lambda_p/2$) laser pulse is positioned about the maximum in ϕ of the wakefield, such that $\partial\eta/\partial r < 0$. Assuming the amplitude of the trailing laser pulse is not large enough to perturb the plasma wave, the laser pulse may be optically guided by the wakefield.

V. Conclusions

Based on a 1D nonlinear quasi-static model, the phenomena of nonlinear wakefield generation and relativistic optical guiding have been analyzed. The generation of nonlinear plasma wakefields by intense, short pulses was examined both analytically and numerically. Wave steepening and period lengthening are observed for very intense laser pulses, and expressions were obtained for the maximum wakefield amplitude and the nonlinear wakefield period. In particular, the maximum wakefield amplitude is found to scale linearly with $|a|$, for $|a| > 1$. Nonlinear corrections to phase detuning length and pump depletion length of the laser pulse were also obtained. The plasma wakefields may be applied to accelerate a trailing electron bunch (laser wakefield acceleration), or to optically guide a trailing laser pulse. Relativistic optical guiding is found to depend strongly on the laser pulse duration. In the short pulse regime, relativistic guiding effects are greatly diminished by the density response and the longitudinal motion of the electrons. In the long pulse regime, optical guiding requires a minimum level of total laser power, i.e., the critical power. However, the leading portion of the pulse will experience diffraction and continuously erodes backward and thus shortens the laser pulse length.

Acknowledgments

This work was supported by the U.S. Department of Energy and the Office of Naval Research.

References

1. P. Maine, D. Strickland, P. Bado, M. Pessot and G. Mourou, IEEE J. Quantum Electron. **QE-24**, 398 (1988).
2. T. Tajima and J.M. Dawson, Phys. Rev. Lett. **43**, 267 (1979).
3. L.M. Gorbunov and V.I. Kirsanov, Zh. Eksp. Tear. Fiz. **93**, 509 (1987) [Sov. Phys. JETP 66, 290 (1987)].
4. P. Sprangle, E. Esarey, A. Ting and G. Joyce, Appl. Phys. Lett. **53**, 146 (1988); E. Esarey, A. Ting, P. Sprangle and G. Joyce, Comments Plasma Phys. Controlled Fusion **12**, 191 (1989).
5. V.N. Tsytovich, U. DeAngelis and R. Bingham, Comments Plasma Phys. Controlled Fusion **12**, 249 (1989).
6. T. Katsouleas, W.B. Mori and C.B. Darrow, in *Advanced Accelerator Concepts*, ed. by C. Joshi, AIP Conf. Proc. No. 193, (Amer. Inst. Phys., NY, 1989), p. 165.
7. V.I. Berezhiani and I.G. Murusidze, submitted to Phys. Lett. A (1989).
8. P. Sprangle, E. Esarey and A. Ting, accepted by Phys. Rev. Lett. (1990); Phys. Rev. A, Apr 15 (1990); NRL memo. report 6545 (1989).
9. P. Chen, J.M. Dawson, R.W. Huff and T. Katsouleas, Phys. Rev. Lett. **54**, 693 (1985).
10. J.B. Rosenzweig, Phys. Rev. Lett. **58**, 555 (1987).
11. A.I. Akhiezer and R.V. Polovin, Zh. Eksp. Teor. Fiz. **30**, 915 (1956) [Sov. Phys. JETP **3**, 696 (1956)].
12. A.C.L. Chian, Plasma Phys. **21**, 509 (1979).
13. C. Max, J. Arons and A.B. Langdon, Phys. Rev. Lett. **33**, 209 (1974).
14. E.L. Kane and H. Hora, in *Laser Interaction and Related Plasma Phenomena*, ed. by H.J. Schwarz and H. Hora, (Plenum Press, New York, 1977), Vol. 4B, p. 913.
15. K.H. Spatchek, J. Plasma Phys. **18**, 293 (1977).
16. G. Schmidt and W. Horton, Comments Plasma Phys. Controlled Fusion **9**, 85 (1985).
17. P. Sprangle and C.M. Tang, in *Laser Acceleration of Particles*, ed. by C. Joshi and T. Katsouleas, AIP Conf. Proc. No. 130 (Amer. Inst. Phys., New York, 1985), p. 156.
18. G.Z. Sun, E. Ott, Y.C. Lee and P. Guzdar, Phys. Fluids **30**, 526 (1987).
19. P. Sprangle, C.M. Tang and E. Esarey, IEEE Trans. Plasma Sci. **PS-15**, 145 (1987).
20. E. Esarey, A. Ting and P. Sprangle, Appl. Phys. Lett. **53**, 1266 (1988).
21. W.B. Mori, C. Joshi, J.M. Dawson, D.W. Forslund and J.M. Kindel, Phys. Rev. Lett. **60**, 1298 (1988); P. Gibbon and A.R. Bell, Phys. Rev. Lett. **61**, 1599 (1988).
22. C.J. McKinstrie and D.A. Russell, Phys. Rev. Lett. **61**, 2929 (1988).
23. T. Kurki-Suonio, P.J. Morrison and T. Tajima, Phys. Rev. A **40**, 3230 (1989).

24. S.C. Wilks, J.M. Dawson, W.B. Mori, T. Katsouleas and M.E. Jones, Phys. Rev. Lett. **62**, 2600 (1989).
25. E. Esarey, A. Ting and P. Sprangle, NRL memo. report 6541; submitted to Phys. Rev. A (1989).
26. J.B. Rosenzweig, Phys. Rev. A **38**, 3634 (1988).
27. T. Katsouleas and W.B. Mori, Phys. Rev. Lett. **61**, 90 (1988).
28. W.L. Kruer, *The Physics of Laser Plasma Interactions*, (Addison Wesley, Reading, MA, 1988).
29. E. Esarey and A. Ting, NRL memo. report 6542; submitted to Phys. Rev. A (1989).

Rapid Heating of Solids by Ultra-Short Pulse Lasers

R.W. Falcone, M.M. Murnane and H.C. Kapteyn

Department of Physics,

University of California at Berkeley

Berkeley, CA 94720

An ultrashort laser pulse can be focused onto the surface of a solid to produce a short-lived, high-temperature, high-density plasma just inside the surface. The plasma cools extremely rapidly after the laser pulse terminates due to rapid thermal conduction, electron energy loss to ions and expansion of the plasma into the surrounding vacuum.

Experimental work[1,2] has demonstrated the importance of using very short laser pulses (near 100 femtoseconds) with low pre-pulse energy for the production of such plasmas. Energy deposited before the arrival of the short pulse can cause ionized material to be ablated, leading to the formation of a lower density plasma in front of the solid surface which cools more slowly. No significant expansion of the solid occurs during the laser heating pulse if the speed of the plasma expansion times the laser pulse length is on order of the optical skin depth.

Experiments were performed using a high-powered femtosecond dye laser system.[3] The system includes a colliding-pulse mode-locked laser amplified in a series of dye laser amplifiers and yields output pulses with an energy of 3.5 mJ and a pulse length of 160 fsec. These pulses are focused onto a target using an off-axis parabolic reflecting mirror at power-densities on target in excess of 10^{16} W cm^{-2}. Peak plasma temperatures in a silicon target of approximately 350 eV can be produced and x-ray emission from the plasma extends up to photon energies of at least 1 keV.

Reflectivity measurements have confirmed that short-pulse lasers can be coupled into solid or near-solid density plasmas with an associated high, metal-like reflectivity. The experimental reflectivity values are in agreement with the predictions of a Drude model, assuming equilibrium ionization in the plasma.

A modified commercial x-ray streak camera was used to measure the x-ray emission pulse width from the plasma.[4] The temporal response of the streak camera is 2 picoseconds. Assuming an instrumental resolution the true x-ray pulse duration is calculated to be on order of 1 picosecond for low atomic number targets such as silicon. The streak camera trace in Fig. 1 indicates two timing fiducials from the laser bracketing the x-ray pulse.

The major component of the plasma x-ray emission under these conditions is expected to be broadband radiation. At high electron densities, ionization depression reduces the number of bound levels to the first few excited levels and Stark broadening is severe. Line emission will be reduced at near-solid densities and continuum emission (due to recombination, bremsstrahlung and broadened low-lying line radiation) is expected to dominate until the plasma cools and expands. An emission spectrum from a silicon target is shown in Fig. 2.

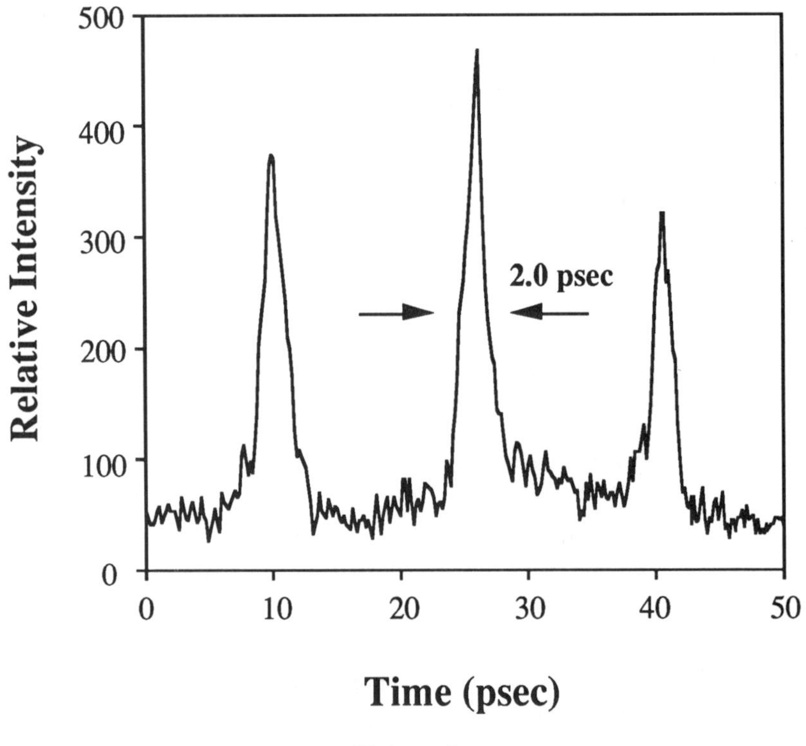

Figure 1

Future directions for this work involve the extension of x-ray measurement techniques to sub-picosecond time scales, the development of more efficient and shorter wavelength laser plasma x-ray sources, and the use of the short pulses for time-resolved x-ray scattering studies from rapidly evolving materials. The use of thin targets having a thickness on the order of the visible laser absorption depth (on order of 100 Å) may also lead to hotter plasmas emitting shorter wavelengths, although cooling may be less efficient, leading to longer x-ray emission pulse lengths. Structured target surfaces may lead to increased laser absorption and subsequent increased plasma temperatures.

This work was supported by the U.S. Air Force Office of Scientific Research, the National Science Foundation, and through a collaboration with Lawrence Livermore National Laboratory under the auspices of the U.S. Department of Energy under contract #W-7405-ENG-48. M.M. Murnane acknowledges support through a University of California President's Postdoctoral Fellowship.

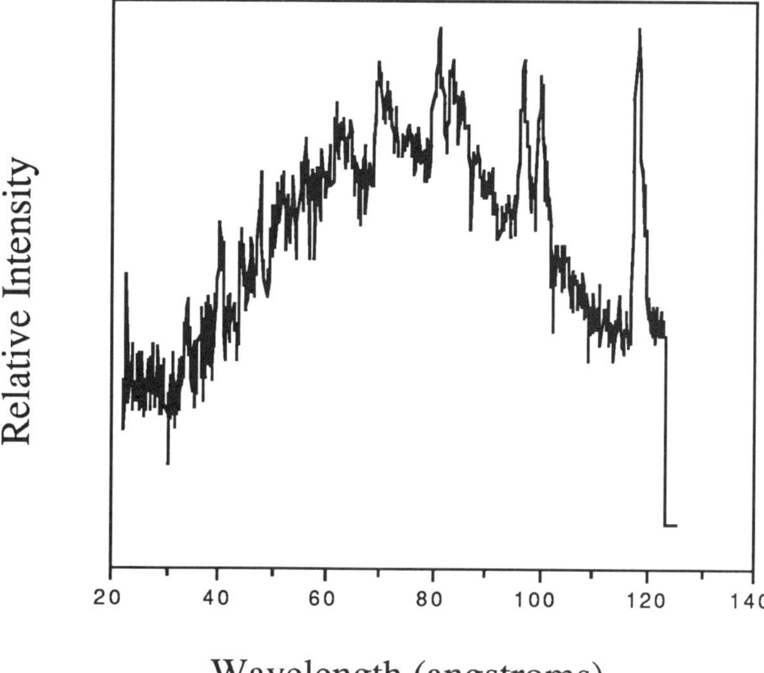

Figure 2

References

1. M.M. Murnane, H.C. Kapteyn, R.W. Falcone, "High-Density Plasmas Produced by Ultrafast Laser Pulses," Phys. Rev. Lett. **62**, 155 (1989).

2. M.M. Murnane, H.C. Kapteyn, R.W. Falcone, "Generation and Application of Ultrafast X-Ray Sources," IEEE Jour. Quant. Elect. **25**, 2417 (1989).

3. M.M. Murnane and R.W. Falcone, "High-power Femtosecond Dye Laser System," J. Opt. Soc. Am. B **5**, 1573 (1988).

4. M.M. Murnane, H.C. Kapteyn, R.W. Falcone, "X-Ray Streak Camera with 2 Picosecond Response," (submitted for publication).

Two-laser Approach to Shorter Wavelength X-ray Lasing

N. Fisch, W. Tighe, E. Valeo and S. Suckewer
Princeton University, Princeton, NJ 08543

Abstract

X-ray lasing in recombining plasmas is becoming attainable at shorter wavelengths. To achieve lasing, new technology, in the form of short-pulse high-power pump lasers, is required. Also required are new theoretical means of describing the interaction of the high power lasers with a dense plasma. This paper reviews selected recent work of the x-ray laser program at Princeton University, with a view towards identifying trends in nonlinear and relativistic effects in recombining plasmas.

I. Introduction

A recombination x-ray laser relies upon the population inversion possible in a recombining plasma. Imagine a plasma that is "supercooled," *i.e.*, it is rapidly cooled to a temperature much less than its ionization energy in a time much shorter than the time it takes for the electrons and ions to recombine. As the plasma recombines, electrons cascade from higher energy states to lower energy states that are unpopulated. If the recombining plasma is dense enough, there is the possibility of amplification of x-rays.

Such amplification has been achieved in highly ionized carbon plasmas at 182 Å, the 3–2 transition in hydrogen-like carbon. At the x-ray laser program at Princeton University (PU), for example, a 300 J commercially available CO_2 laser pumps a carbon plasma, and delivers about 2 mJ of energy in 20 nsec at 182 Å.[1-3] Research at this wavelength is concentrated upon improving further the efficiency of generating the x-ray power, for example, through the use of multilayer mirrors.[4-5]

Quite a different matter, but of great interest, is the extension of these results to shorter wavelengths. For example, the same 3-2 transition in hydrogen-like aluminum, occurs at 39 Å, which is already in the so-called "water window" (24–44 Å), where good contrast can be achieved for biological samples in aqueous solutions. However, it turns out that to achieve lasing in aluminum, or similar materials capable of delivering the shorter wavelength output, severe constraints are placed upon the heating and cooling of the plasma and considerable technological and conceptual advances are required. This paper concentrates on an elucidation of the issues involved, and recent relevant experimental[6-9] and theoretical work[10] carried out at PU.

This paper is organized as follows: Lasing at shorter wavelengths makes such severe demands on the heating and cooling of the plasma that a new approach, ionization without heating[11], is indicated. This second approach is the focus of this paper. In Sec. II, we contrast the two approaches to producing the recombining plasma. In Sec. III, we establish the necessity, in employing the second technique, for a powerful short-pulse (subpicosecond) pump laser. At PU, a two-laser approach[12] is taken in which a conventional long-pulse, high-energy, but relatively low-power laser establishes a target plasma for the second high-power pump laser, and, in Sec. IV, we review the progress at PU in developing the pump laser. These achievements include a pulse compression to 0.3 ps, a small focal spot, control of ASE and control of prepulse energy. In Sec. V, a new theory[10] of nonresonant ionization in powerful fields of the subpicosecond laser is described. This model uses a simple Thomas-Fermi description of the atom to explain a variety of nonresonant ionization data. In Sec. VI, we conclude with an enumeration of other outstanding problems that we expect to dominate the field of recombination lasers in the next several years.

II. Methods of Recombination Lasing

One method of achieving population inversion is ionization by heating. Using a pump laser, say a CO_2 or Nd/Glass laser, electrons are first driven to occupy both higher-state bound levels as well as unbound continuum levels. If the plasma is then cooled rapidly, the higher-state bound electrons decay rapidly, whereas the unbound electrons, relying on 3-body recombination, become bound to the most highly excited levels. During this subsequent recombination, the electrons cascade to lower energy levels, achieving the population inversion. In the case at hand, of rapidly-cooled fully-stripped carbon, the particular transition that is exploited is the 3–2 transition at 182 Å. The inversion occurs primarily here because of the rapid radiative depopulation of the second level, while the third level is populated through strong cascading from the higher levels. The corresponding 3–2 transition in aluminum occus at 39 Å.

An alternative method is to achieve the population inversion without heating the plasma. Rather than to populate all energy levels, bound and unbound, in a thermal distribution, a short-pulse high power laser is directed upon a prepared cold plasma in order to produce a cold electron-ion plasma on a time scale short compared to the recombination time. The electromagnetic field is of sufficient amplitude to rip electrons from atoms by overcoming the coulomb attraction. The electrons oscillate in the field, picking up large oscillatory energy, but in the absence of decorrelation effects, such as a collision, then, as the electromagnetic pulse passes the atom, the freed electrons cool adiabatically in the receding pulse. The result would be ionization without heating, which is a very efficient use of the pump power, and after which the recombination lasing can take place.

The two methods are compared in the schematic figure on the right: In each method a CO_2 laser might provide the initial ionization. In the first method, ionization by heating, the temperature T and the ionization state Z evolve in time as shown in Fig. 1a. As the plasma is heated, the ionization state Z climbs with the temperature T. When the desired ionization state is attained, the challenge then becomes to cool the plasma rapidly, $i.e.$, to cool the plasma in a time that is short compared to the recombination time τ, which, in carbon, might be 5 ns, but, in aluminum, it might be as small as 2 ps. In contrast, in the second method, a low power CO_2 laser might be employed only to provide a low temperature target plasma. When a powerful subpicosecond laser is turned on,

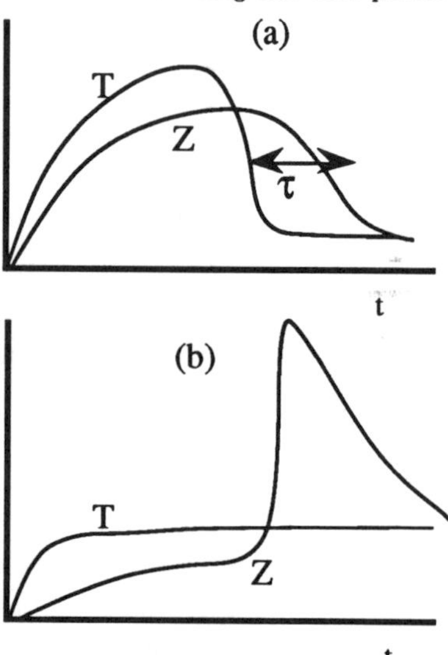

Fig. 1. Recombination lasing by rapid cooling (a) or by ionization without heating (b).

the ion charge state Z is shown (Fig. 1b) to spike higher while the temperature remains flat. Here the problem of rapid cooling is replaced by the challenge of rapid ionization. At PU, a 1/4 μ KrF laser is employed as the second stage laser.

III. Necessity for Powerful Subpicosecond Laser

The necessity in the approach of ionization without heating, for a Powerful SubPicosecond laser (PSP) — $i.e.$ the requirement that the second-stage laser be both powerful and short-pulse — arises from the optimal recombination plasma parameters.

The density must be high, because the upper level number density is proportional to ν_{rec}/ν_{at}, the ratio of the recombination rate to the radiative decay rate of the inverted population. The recombination rate increases as the square of the density, but decreases abruptly with increasing temperature. On the other hand, the collisional deexcitation rate ν_{col} of the inverted population must be small compared to the radiative decay rate, i.e., $\nu_{at} \gg \nu_{col}$, requiring that the density not be too high. This leads to optimal density and maximum allowable temperature. For the 182Å radiation in carbon plasmas this leads to density of 10^{19}cm^{-3}, plasma temperature (at the time of recombination) of 20 eV, and recombination times of 5 ns. For the 39Å radiation in aluminum plasmas this leads to density of $8 \times 10^{21} \text{cm}^{-3}$, plasma temperature of 100 eV, and recombination times of 2 ps. It turns out that for maximal laser gain, the optimum density scales roughly inversely with wavelength. (If the method of ionization by heating were used, for which second-stage PSP laser is not crucial, then the initially heated plasma, prior to the rapid cooling, would have a temperature of about 300 eV in carbon and 1.5 keV in aluminum.)

Consider that the pumping power per ion is given by the x-ray energy E_i times the radiative decay rate ν_{at}, with E_i scaling, for hydrogen-like ions, as Z^2. The x-ray wavelength λ scales as $1/E_i$, and the radiative decay rate ν_{at} scales as $1/\lambda^2$, so the total pumping power per ion, $E_i\nu_{at}$ scales as $1/\lambda^3$. The ion density scales roughly as $1/\lambda$, so the total pumping power scales as $1/\lambda^4$. This indicates the necessity for very high power at short wavelengths.

On the other hand, since ν_{at} scales as $1/\lambda^2$, it is also evident that a short pulse is preferable, since all the useful energy must be delivered within the atomic lifetime. An advantage of the short pulse, therefore, is that the energy only scales as $1/\lambda$. In the PU approach, a KrF UV laser is employed, which, relative to CO_2 or Nd/Glass lasers, can deliver a high intensity to a small focal spot, because of its relatively low wavelength.

IV. POWERFUL SUBPICOSECOND LASER SYSTEM

Having argued for its necessity for producing the x-ray lasing conditions, let us examine the technological achievement of the PSP laser at PU, which takes advantage of recent developments in subpicosecond laser physics.[6-7,13-14] The energy source for this laser is a mode-locked Nd:YAG 1μ laser that pumps a 1 ps dye oscillator at $2/3\ \mu$. The output of the dye oscillator is compresssed to 300 fs, amplified by a factor of 10^5 in a 3-stage dye amplifier, and then frequency doubled to produce $1/3\ \mu$ radiation. This $1/3\ \mu$ radiation is then mixed with power from the YAG pump, at 1μ, to produce $1/4\ \mu$ radiation. After being passed through three successive Kr-F amplifiers, about 130—150 mJ of $1/4\ \mu$ radiation is extracted in a pulse of about 300 fs duration. This pulse, of course, is to be used to ionize (without heating) a plasma, producing the necessary conditions for recombination x-ray lasing.

By splitting the laser output, and then shining the two counterstreaming beams through a XeF cell, the 3-photon fluorescence can be viewed directly in a single shot, showing that a pulse width of 300 fs is indeed obtained. Shining the output onto a 2-D CCD camera, it has been verified that a focal spot of $4\mu \times 10\mu$ has been produced. Thus, even only 150 mJ of energy, concentrated in space and time, produces almost 2×10^{18} W/cm^2.

One of the more difficult technological problems arises from the large amount of amplified spontaneous emission (ASE) in the final stage KrF amplifier. In the first two amplifiers, ASE can be controlled to contain about 0.1% of the total energy, and, what is of greater significance, this emission power is only 10^{-8} of the peak power. Yet, after the final excimer amplification, the ASE may account for as much as 10% of the energy. A number of saturable absorbers have been tested as spatial filters to control the prepulse energy, including anthracene in hexane solution and acridine in a methanol solution. These dyes are effective in reducing ASE, but are difficult to maintain.

Although the ASE can be of substantial energy, the more relevant question is how much of this energy arrives in the prepulse. By advancing the signal relative to the time of maximum gain in the amplifier, it is possible to control the prepulse energy. A delay of a ps signal by 3 ns means that 15% of the energy is in the prepulse, no delay gives 7%, and advancing the ps signal by 3 ns reduces the prepulse energy to only 0.3%.

The control of the prepulse energy may be desirable because a significant amount of prepulse may create a hot plasma, whereas the preferred target may be a cold plasma or even a solid. This heated plasma may then expand, forming a less dense target for the main pulse. As a result, the energy deposition of the main pulse might be less concentrated, resulting in a cooler final plasma temperature. Additionally, there may be additional dissipation mechanisms in preformed plasma that could detract from the hoped for adiabatic ionization in the strong pulse, or the pulse might even be reflected. A greater theoretical understanding is necessary to decide the issues of prepulse control.

In Fig. 2, we show spectra obtained under varying amounts of prepulse in a PSP laser incident upon a teflon target. The top figure shows the cases of no signal delay and a delay of 3 ns; the bottom figure shows the case of signal advance by 3 ns. Noteworthy here are the intensity ratios and the Stark-broadened widths of the F VII lines under the different prepulse conditions, as well as the relative strengths of the C V and C VI resonance lines. The relative broadening is indicative of the density at the time that the ionization state in question is most prevalent, and the data is consistent

with the interpretation offered that cooler plasma conditions may prevail
when there is significant prepulse energy.

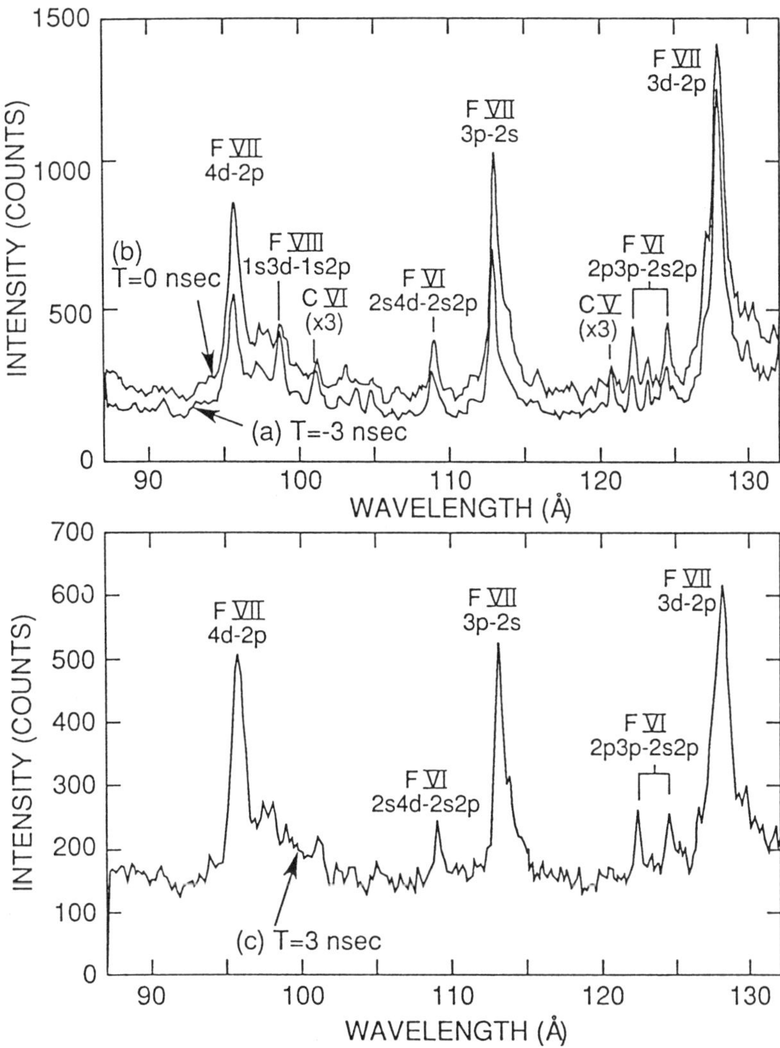

Fig. 2. Spectroscopic data upon irradiation of teflon target with
large prepulses (upper figure) and small prepulse (lower figure).

V. Thomas-Fermi Model

A question of great importance is, given the requirements for the PSP laser, how do huge electric fields act upon high atomic number matter. There is a certain amount of ionization data that has recently become available, and it turns out that this data can be explained rather well with a relatively simple semiclassical theory. This model, spare but successful, retains the screening effect of inner shell electrons in a many-electron atom by treating all the bound electrons as a fermi gas at zero temperature.

The experimental data to which we compare is that of Augst et al.,[15] shown in Fig. 3, in which the threshold laser intensity is plotted as a function of ionization potential. What is meant by threshold intensity is the field intensity at which the appearance of a given charge state has a probability of about 5×10^{-3}. The striking thing about this data is that it is insensitive to details of the atomic structure, which suggests that the statistical treatment of the electrons offered here may be appropriate.

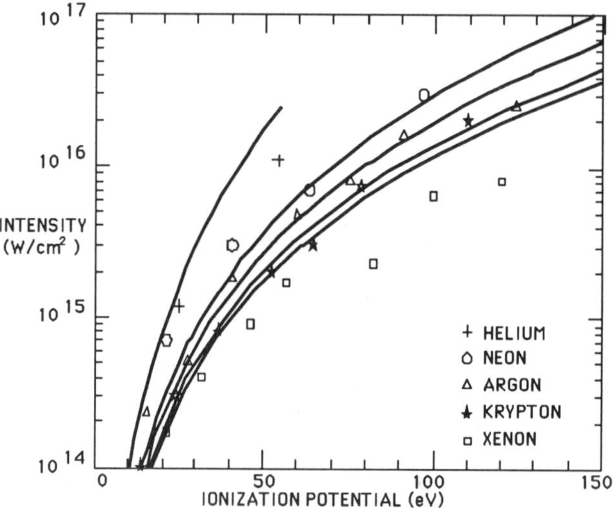

Fig. 3 Intensity vs. ionization potential. Solid lines are the theoretical upper bound.

Following Ref. 10, consider a many-electron atom in an applied laser electric field, $E_l = F \cos \omega t$, directed in the z-direction and where the laser frequency ω is small compared to atomic transition rate, $\omega \ll \nu_{at}$. The applied field can be treated as constant and its maximum value will determine the ionization state of the atom. The potential due to this field is then $\phi = -Fz$. Using a Thomas-Fermi model for high Z atom, we fill up the atom with electrons with momenta

$$p \le p_{max} \equiv [2(\phi - W + Fz)]^{1/2}, \tag{1}$$

where W is the kinetic energy and where the maximum momentum p is spatially dependent on both z, the direction of the applied field, and r, the distance from the nucleus. The density can then be written as

$$n_e = 4\pi \int_0^{p_{max}} dp\, p^2 \frac{1}{4\pi^3} = \frac{p_{max}^3}{3\pi^2} = \frac{2^{3/2}}{2\pi^2} [2(\phi - \epsilon + Fz)]^{1/2} \tag{2}$$

where ϵ is the maximum filled energy. Eq.(2) holds only where n_e is nonzero — *i.e.*, where the electrons are bound. The potential is related to the density n_e by Poisson's equation

$$\nabla^2 \phi = n_e. \tag{3}$$

Substituting in Eq.(3) for n_e from Eq.(2) now yields a nonlinear ordinary integro-differential equation in two-dimensions (there is azimuthal symmetry) for the self-consistent potential ϕ, to be solved with the boundary condition for $r \to 0$ of the unshielded ion potential

$$r\phi(r, u) \to Z \quad \text{as } r \to 0, \tag{4a}$$

and with the boundary condition for $r \to \infty$ of the far field potential of an ionized atom

$$r\phi(r, u) \to Z - Q_{el} + Fr^2\mu \quad \text{as } r \to \infty, \tag{4b}$$

where $\mu = z/r$ and we defined the self-consistent number of bound electrons

$$Q_{el} = \int d^3r\, n_e. \tag{4c}$$

Now note that the following change of variables

$$r' = rZ^{1/3} \tag{5a}$$

$$\phi' = \phi Z^{-4/3} \tag{5b}$$

$$\epsilon' = \epsilon Z^{-4/3} \tag{5c}$$

$$F' = FZ^{-5/3} \tag{5d}$$

leave Eqs.(1-3) and the associated boundary conditions Z-independent. The implication is that it is possible to obtain a universal solution for the ionization state of a multi-electron atom in a strong dc field, in other words, the integro-differential equation here need be solved but once. Superposed on the data of Augst et al. in Fig. 3, we show the analytic result for the intensity that corresponds to a given ionization state (which would be an upper bound to the empirical intensity level, which is deduced from the mere appearance in small quantities of the ionization state), and we note an exceptional fit to the data.

Two points should be noted. First, there are no adjustable parameters in fitting this data, so the essential physics can be assumed to be captured in this very simple model. Second, what has been neglected in this simple presentation is the quantum tunneling, which can be shown, in a more complete calculation,[10] indeed to be negligible.

VI. Conclusions and Trends

There are a number of research trends emerging naturally from the stringent requirements for recombination lasing at short wavelengths that can be identified already as likely to challenge us for several years. Particularly apt for these proceedings, these research areas clearly rely heavily upon nonlinear and, possibly, even relativistic effects. The electric field amplitude in the short pulse VUV fields reported here is already on the order of several 10^{10} V/cm. The quiver energy of electrons in these fields is about 20 keV. For fields large enough to ionize aluminum directly via a multiphoton process, i.e., much greater than 2keV, the quiver energy might rise to more than a hundred keV, at which point relativistic effects become important.

What distinguishes problems here from other areas of laser plasma interaction, for example from laser fusion problems, are the extemely short space and time scales. Thus, even before relativistic effects become important, there are certain new nonlinear effects at play. For example, because of the very short time scale in producing the plasma in femtosecond pulses interacting with a solid, the emerging plasma has little time to move, and gradients are very sharp. At the same time, the large field oscillates electrons across the sharp vacuum-solid interface, resulting in highly nonlinear electron motion even in simple oscillatory fields. Overtaking then takes place for any electrons near the interface. The propagation and reflection of short high power pulses at this interface is an outstanding question. Nonlinear effects also play a large role in propagation and focusing of the short pulses in a preformed plasma. Here, the focusing is particularly important, because the propagation of the high intensity pulse determines the geometry of the recombining plasma.

Apart from the very different physics in laser-plasma interactions that must be understood, there is a challenge to understand and to diagnose these plasmas, again, uniquely because of the short space and time scales. The expanding plasma may have megagauss $\nabla n \times \nabla T$ magnetic fields, but existing only for nanoseconds over distances of only a μm, so that to observe them is a challenge in itself. How these fields might affect the plasma expansion is of concern, and a number of familiar assumptions must be questioned, such as the charge neutrality of the expanding plasma or the utility of the plasma parameter (number of particles in a Debye sphere).

ACKNOWLEDGEMENTS

This contribution is primarily a review of the work done by the many members of the X-ray Laser Project at Princeton University. The work was supported by United States Department of Energy under contract numbers DE–AC02–76–CHO3073 and KC-05-01, and by ONR under Grant N00014-87-K-2006.

REFERENCES

1 S. Suckewer, C. H. Skinner, H. Milchberg, C. Keane and D. Voorhees, Phys. Rev. Lett. **55**, 1753 (1985).

2 S. Suckewer, C. H. Skinner, D. Kim, E. Valeo, D. Voorhees and A. Wouters, Phys. Rev. Lett. **57**, 1004 (1986).

3 S. Suckewer and C. H. Skinner, Science **247**, 1553 (1990).

4 N. M. Ceglio, D. G. Stearns, D. P. Gaines, A. P. Hawryluk, and J. E. Trebes, Opt. Lett. **13**, 108 (1988).

5 D. Kim, C. H. Skinner, G Umesh and S. Suckewer, Opt. Lett. **14**, 665 (1989).

6 W. Tighe, C. H. Nam, J. Robinson, and S. Suckewer, Rev. Sci. Instr. **59**, 2235 (1988).

7 W. Tighe, L. Meixler, C. H. Nam, and S. Suckewer, in *Advances in Laser Science IV*, eds. J.L. Gole, D.S. Heller, M. Lapp, W. C. Stalley (AIP, NY, 1988) p.57.

8 L. Meixler, C. H. Nam, J. Robinson, W. Tighe, K. Krushelnick, S. Suckewer, and J. Goldhar, in *Short Wavelength Coherent Radiation: Generation and Application*, eds. R. Falcone and J. Kirz (Opt. Soc. Am., Wash. Dc, 1989) Vol. 2, p.106.

9 C. H. Nam, W. Tighe, E. J. Valeo, and S. Suckewer, App. Phys. B (1990).

10 S. Susskind, E. J. Valeo, C. R. Oberman and I. B. Bernstein, submitted to Phys. Rev. Lett. (1989).

11 N. H. Burnett and P. B. Corkum, J. Opt. Soc. Am. B **6**, 1195 (1989).

12 C. W. Clark, M. G. Littman, R. Miles, T. J. McIlrath, C. H. Skinner, S. Suckewer, and E. Valeo, J. Opt. Soc. Am. B **3**, 371 (1986).

13 A. P. Schwarzenbach, T. S. Luk, I. A. McIntyre, U. Johann, A. McPherson, K. Boyer, and C. K. Rhodes, Opt. Lett. **2**, 499 (1986).

14 S. Szatmari, F. P. Schafer, E. Muller-Horsche, and W. Muckenheim, Opt. Commun. **63**, 305 (1987).

15 S. Augst, D. Strickland, D. D. Meyerhofer, S. L. Chim, J. H. Eberly, Phys. Rev. Lett. **63**, 2212 (1989).

THEORY OF GENERATION OF RADIATION BY A BEAM-PLASMA SYSTEM*

Nicholas A. Krall
and
Marlene Rosenberg**

Krall Associates
Del Mar, CA 92014

We describe a phenomenological turbulence model which predicts that the efficiency of generation of radiation from a device which injects counterstreaming electron beams into an unmagnetized, plasma-loaded waveguide should increase strongly with the beam energy density.

This model is consistent with experimental results reported elsewhere.[1,2] The observed power levels, scaling of the radiated power with beam current, optimum beam energy for maximum power output, and modulation of the radiation at ω_{pi} are all consistent with the generation of counterstreaming plasma waves by the beams, which mix by resonant three-wave coupling to produce radiation at $2\omega_{pe}$. The optimum energy of the beam is determined by saturation conditions, the scaling with current by the mechanics of three-wave coupling, and the observed modulation of the emitted radiation by the nonlinear generation of ion density fluctuations by the saturated plasma wave spectrum.

More specifically, the beam convergence observed in the experiment is consistent with the Bennett pinch conditions, with self focussing for currents $I_b^2 > 3.2 \times 10^{-10} n_b A_b T_b \text{Amps}^2$, where $n_b(\text{cm}^{-3})$ is the beam density, $A_b(\text{cm}^2)$ is the beam cross-sectional area, and $T_b(\text{eV})$ is the beam transverse temperature. Further, each beam is predicted to be unstable as it propagates through the background plasma, and to produce a spectrum of electron plasma waves at a frequency $\omega \simeq \omega_{pe}$ and axial wavenumber $k \simeq \omega_{pe}/v_b$, where v_b is the beam speed. This spectrum saturates by electron trapping at a level $W_{sat} = n_b E_b (n_b/n_e)^{1/3}$, where E_b is the beam directed energy and n_e is the density of the background plasma. The distance the beam travels before the wave saturates is a few times v_b/δ, where δ is the instability growth rate, $\delta \sim \omega_{pe}(n_b/n_e)^{1/3}$. If this distance is comparable to the length of the waveguide, the waves produced by both the beams overlap spatially and saturate before leaving the system. This gives an optimum coupling energy E_b of order $(m_e/2)(L\omega_{pe}/2N)^2 (n_b/n_e)^{2/3}$, where N is the number of e-foldings of the unstable wave before saturation. Resonant three-wave interactions $\omega_3 = \omega_1 + \omega_2$, $k_3 = k_1 + k_2$ couple the two counterpropagating electrostatic beam waves to a TM mode of the circular waveguide at $\omega = 2\omega_{pe}$ which radiates out of the plasma.

This model explains why the experiment sees radiation only when the background plasma and both beams are present. The requirement that the beams be of comparable energy is consistent with the resonant three-wave conditions, since the radiating wave is longer wavelength than the beam waves. The maximum in power output as a function of beam energy is consistent with the energy for optimum coupling determined from wave-saturation length arguments. Finally, the spiky structure of the radiation, modulated at the ion plasma frequency, can be understood by noting that the saturated beam waves can drive a modulational instability on the time scale of the ion plasma frequency.

We conclude that high efficiency generation of mm-waves from a beam plasma system requires the use of higher current beams than previously used, and the use of optimum beam energy for the actual parameters of the waveguide and beam.

*Work supported by the AFOSR.

**Consultant; presently at the Center for Astrophysics and Space Sciences, University of California at San Diego, La Jolla, CA 92037.

[1] R. W. Schumacher and J. Santoru, Bull. Am. Phys. Soc. **32**, 1885 (1987).

[2] R. W. Schumacher, J. Santoru, M. Rosenberg, and N. A. Krall (paper in preparation).

On the Eulerian Analysis of Large Amplitude Relativistic Plasma Waves

W. P. Leemans and C. Joshi
Electrical Engineering Department
University of California Los Angeles, Los Angeles CA 90024

Introduction

In the last decade the physics of non-linear systems has received a considerable amount of interest. Recently we have presented[1] a plasma based system whose model equation can be reduced, under the appropriate conditions, to a driven anharmonic oscillator. The physical system consists of a large amplitude relativistic plasma wave excited through collinear optical mixing, in the presence of a spatial density modulation.

Previous authors[2,3] have investigated the possibility of this system exhibiting bistability and evolving into a chaotic state under appropriate conditions. However their analysis was based upon the equation of motion for the longitudinal electric field in an Eulerian frame, using the weakly relativistic approximation. This approximation is valid when α_i, the non-relativistic quiver velocity of an electron in a laser field, is much less than 1 and the amplitude of the plasma wave satisfies the condition $\frac{n_1}{n_0} \ll 1$ where n_1 is amplitude of the density modulation and n_0 is the background plasma density.

In the present paper we re-examine the consequences of making the weakly relativistic approximation when investigating chaotic solutions. We rederive an equation of motion for the longitudinal electric field using this weakly relativistic approximation and then make the connection with the well known Duffing equation[4] for a non-relativistic particle moving inside a potential well $V(\Psi) = \frac{\Psi^2}{2} - \frac{\Psi^4}{4}$, where Ψ is the amplitude of the displacement. Using a rotating Van der Pol plane[5] we calculate the steady state amplitude of the displacement as a function of driving strength and detuning ratio. The Duffing like equation is then solved numerically and the results compared to the analytically obtained frequency response function. The validity of the Duffing model is then assessed and it is found that it can not be used to study the interesting regime of large driving strengths where the non-linear dynamics dominate the system behavior.

Eulerian Analysis of Electron Plasma Waves

Consider the fluid equations for a plasma in which the density is changing in time due to collisional ionization at a rate λ. The equation of continuity and the equation of motion for a fluid element are respectively given by

$$\frac{\partial n(\mathbf{r},t)}{\partial t} + \nabla \cdot [\, n(\mathbf{r},t) \frac{P}{m_0 \gamma} \,] = \lambda\ n(\mathbf{r},t) \tag{1},$$

and

$$\frac{\partial}{\partial t} [\, \mathbf{P}(\mathbf{r},t) \,] + v \frac{\partial \mathbf{P}(\mathbf{r},t)}{\partial r} - q [\, \mathbf{E} + \frac{\mathbf{p} \times \mathbf{B}}{\gamma\ m_0 c} \,] n(\mathbf{r},t) + \lambda\ \mathbf{P}(\mathbf{r},t) = 0 \tag{2}.$$

The two fluid equations (1) and (2), complemented by the Maxwell's equations

$$\nabla \cdot \mathbf{E} = -4\pi e\ (n_e - n_i) \tag{3}$$

$$\nabla \times \mathbf{B} = \frac{1}{c}\frac{\partial \mathbf{E}}{\partial t} - \frac{4\pi e}{c}\ (n_e v_e - n_i v_i) \tag{4}$$

$$\nabla \times \mathbf{E} = -\frac{1}{c}\frac{\partial \mathbf{B}}{\partial t} \tag{5}$$

are used to derive the equation which describes the evolution of the longitudinal electric field generated by beating two transverse linearly polarized electromagnetic waves in a plasma.   In these equations n_e (n_i) and v_e (v_i) are the electron (ion) fluid density and velocity, p_e is the electron fluid momentum and all other quantities have their usual meaning.

Let us assume that the ions are immobile and analyze the set of equations in an Eulerian coordinate system.   Combining equations (4) and (5) we obtain:

$$\left(\nabla^2 - \frac{1}{c^2}\frac{\partial^2}{\partial t^2} - \nabla\nabla.\right) \mathbf{E} = -\frac{4\pi e}{c^2}\frac{\partial}{\partial t}(N_e \mathbf{v}_e) \tag{6}$$

We now define $N_e = N_o + n_e$, where N_o is a background steady state value and n_e is oscillatory. Equation (6) becomes

$$\left(\nabla^2 - \frac{1}{c^2}\frac{\partial^2}{\partial t^2} - \nabla\nabla.\right) \mathbf{E} = -\frac{4\pi e}{c^2}\left(N_o \frac{\partial}{\partial t}\mathbf{v}_e + \frac{\partial}{\partial t}(n_e \mathbf{v}_e)\right) \tag{7}$$

From the equation of motion we obtain

$$\frac{\partial \mathbf{v}_e}{\partial t} + \mathbf{v}_e.\nabla \mathbf{v}_e = -\frac{e}{\gamma m_o}\left(\underline{\underline{\mathbf{I}}} - \frac{\mathbf{v}_e \mathbf{v}_e}{c^2}\right)\left(\mathbf{E} + \frac{\mathbf{v}_e \times \mathbf{B}}{c} - \lambda \gamma m_o \mathbf{v}_e\right) \tag{8}$$

where $\underline{\underline{\mathbf{I}}}$ is the unit tensor.

Substituting equation (8) into equation (7) we find

$$\left(\nabla^2 - \frac{1}{c^2}\frac{\partial^2}{\partial t^2} - \nabla\nabla.\right) \mathbf{E} = -\frac{4\pi e}{c^2}\left\{\frac{\partial}{\partial t}(n_e \mathbf{v}_e) - N_o \mathbf{v}_e.\nabla \mathbf{v}_e - \right.$$

$$\left. \frac{N_{oj}\, e}{\gamma m_o}\left(\underline{\underline{\mathbf{I}}} - \frac{\mathbf{v}_e \mathbf{v}_e}{c^2}\right)\left(\mathbf{E} + \frac{\mathbf{v}_e \times \mathbf{B}}{c} - \nu \gamma m_o \mathbf{v}_e\right)\right\} \tag{9}$$

For the weakly relativistic case we expand γ^{-1}:

$$\gamma^{-1} = \sqrt{1 - \left(\frac{v}{c}\right)^2} \approx 1 - \frac{1}{2}\left(\frac{v}{c}\right)^2 \tag{10}$$

The wave equation becomes

$$\left(\nabla^2 - \frac{1}{c^2}\frac{\partial^2}{\partial t^2} - \nabla\nabla.\right) \mathbf{E} = -\frac{4\pi e}{c^2}\left\{\frac{\partial}{\partial t}(n_e \mathbf{v}_e) - N_o \mathbf{v}_e.\nabla \mathbf{v}_e - \right.$$

$$\frac{N_o\, e}{\gamma m_o}\left\{\left[\left(\underline{\underline{\mathbf{I}}} - \frac{\mathbf{v}_e \mathbf{v}_e}{c^2}\right)\frac{1}{2}\left(\frac{v}{c}\right)^2 + \frac{\mathbf{v}_e \mathbf{v}_e}{c^2}\right].\mathbf{E} - \left(1 - \left(\frac{v}{c}\right)^2\right)\frac{\mathbf{v}_e \times \mathbf{B}}{c}\right\} +$$

$$\left. \lambda\, N_o\, e\left(\underline{\underline{\mathbf{I}}} - \frac{\mathbf{v}_e \mathbf{v}_e}{c^2}\mathbf{v}_e\right)\right\} \tag{11}$$

The evolution of the longitudinal component of the electric field is then determined by

$$\left\{ \frac{\partial^2}{\partial t^2} + \omega_p^2 [1 - \frac{3}{2} (\frac{v_x}{c})^2 - \frac{1}{2} (\frac{v_z}{c})^2] \right\} E_x = 4\pi e \left\{ \frac{\partial}{\partial t} (n_e v_{ex}) - \right.$$

$$N_0 \, \mathbf{v_e} \cdot \nabla \, v_{ex} + \frac{N_0 \, e}{\gamma \, m_0} \frac{v_x v_z}{c^2} E_z - \frac{N_0 \, e}{\gamma \, m_0} (\frac{\mathbf{v_e} \times \mathbf{B}}{c})_x +$$

$$\left. \lambda \, N_0 \, e \, v_{ex} (1 - \frac{v_x^2 + v_z^2}{c^2}) \right\} \tag{12}$$

Following Mori's[6] work we use the Mitropolsky-Bogoliubov perturbation technique to obtain an equation of motion for the longitudinal electric field value.   The wave equation becomes:

$$\left\{ \frac{\partial^2}{\partial t^2} + [\, \sigma + \lambda \, [1 - \frac{1}{4} (\frac{v_\phi}{c})^2 \, |e_0|^2 - \frac{1}{2} (\alpha_1^2 + \alpha_2^2)] \,] \, \frac{\partial}{\partial t} + \right.$$

$$\left. \omega_p^2 \, [1 - \frac{3}{8} (\frac{v_\phi}{c})^2 \, |e_0|^2 - \frac{1}{4} (\alpha_1^2 + \alpha_2^2)] \, \right\} \, \boldsymbol{\varepsilon}_0 =$$

$$\frac{c}{v_\phi} \frac{\alpha_1 + \alpha_2}{2} \, \omega_p^2 \, \sin (\Delta k \, x - \Delta \omega \tau) \tag{13}$$

where v_ϕ is the phase velocity of the wave, $\alpha_i = \dfrac{v_{osc}}{c} = \dfrac{eE_i}{m\omega_i c}$ is the normalized non-relativistic oscillatory velocity of the electron in the laser field and σ is a phenomenological damping term to accommodate for such effects as mode coupling to slow waves. Equation (13) can be rewritten as :

$$\frac{\partial^2 \Psi}{\partial t^2} + (\, c_1 + c_2 |\Psi|^2) \frac{\partial \Psi}{\partial t} + (c_3 + c_4 |\Psi|^2) \, \Psi = F \sin(\Delta k \, x - \Delta \omega \, \tau) \tag{14}$$

with $c_1 = \sigma + \lambda \, (1 - \frac{1}{2} (\alpha_1^2 + \alpha_2^2))$

$$c_2 = -\lambda \frac{1}{4} \left(\frac{v_\phi}{c}\right)^2$$

$$c_3 = \omega_p^2 \left[1 - \frac{1}{4}(\alpha_1^2 + \alpha_2^2) \right]$$

$$c_4 = \omega_p^2 \left[\frac{3}{8} \left(\frac{v_\phi}{c}\right)^2 |e_0|^2 \right]$$

Under this form one recognizes the equation of a driven oscillator with both non-linear damping (a Van der Pol type oscillator) and non-linear spring constant (a Duffing type oscillator). In the next section we focus on the case in which $\sigma \gg \lambda$, i.e. the Duffing oscillator.

Duffing Model of Electron Plasma Waves

It is well known that the Duffing equation exhibits the jump phenomenon and that a period doubling route to chaos can exist[7]. In order to verify that these same phenomena can be observed for large amplitude relativistic plasma waves in an actual collinear optical mixing experiment, we need to establish a parameter regime which is accessible in the laboratory.   Therefore it is necessary to develop an understanding of the parameter space and fundamental characteristics of the canonical Duffing equation. Furthermore we have to ascertain that the aforementioned effects occur for weakly relativistic waves, as assumed in our derivation of the equation of motion, and for wave amplitudes less than the wave breaking limit.

In what follows we assume that the fluid element excursion is small compared to the beat-wave wavelength and apply a capacitor model for the driver term.   The beat-wave equation can then be written under the standard Duffing form with a cosinusoidal driver term :

$$\frac{d^2\Psi}{dt^2} + \Gamma \frac{d\Psi}{dt} + \alpha\,\omega_p^2\,\Psi + \beta\,\omega_p^2\,\Psi^3 = F \cos \Omega t \qquad (15)$$

where all parameters are real and $\Gamma \equiv \sigma$. This equation models the motion of a particle with charge ω_p^2 moving in a potential of the form

$$V(\Psi) = \alpha \frac{\Psi^2}{2} + \beta \frac{\Psi^4}{4} \tag{16}$$

The turning point of this potential is found from

$$\frac{dV}{d\Psi} = 0 \text{ or } \Psi_0 = \sqrt{\frac{-\alpha}{\beta}}$$

Renormalizing time with respect to $\sqrt{\alpha}\ \omega_p^2$ and space with respect to the turning point distance Ψ_0 we obtain

$$\frac{d^2\eta}{d\tau^2} + \Gamma \frac{d\eta}{d\tau} + \eta - 4\eta^3 = F \cos \omega\tau \tag{17}$$

where $\eta = \dfrac{\Psi}{2\Psi_0}$ and $\tau = \omega_p t$.

We rewrite equation (17) as a set of first order non-linear differential equations where the position x_1, velocity x_2 and $\omega\tau$ are taken as the independent variables:

$$\begin{aligned}
\overset{\circ}{x}_1 &= x_2 \\
\overset{\circ}{x}_2 &= -x_1 + 4x_1^3 - \Gamma x_2 + F \cos \iota \\
\overset{\circ}{\iota} &= \Omega
\end{aligned} \tag{18}$$

where $|\lambda|$ is assumed small, $\lambda = (\Gamma, F, \rho)$; $\rho = 1 - \omega^2$ $\tag{19}$
Let us first set F equal to zero so that equation (18) reduces to an autonomous equation. The fixed points are given by

$$x_{1s} = 0 \ , \ x_{2s} = 0 \text{ and } x_{1s} = \pm\frac{1}{2} \ , \ x_{2s} = 0 \tag{20}.$$

Performing a local stability analysis it is straightforward to show that , for $\Gamma \geq 0$ the first fixed point is a sink while the two others are saddle points.

We now return to the non-autonomous equations: in case $|\lambda|$ small one can use the method of averaging to analyze the system.

Consider therefore a Van der Pol plane[5] (z_1, z_2) rotating at the frequency ω :

$$z_1 = x_1 \cos \omega\tau - \frac{x_2}{\omega} \sin \omega\tau$$

$$z_2 = - x_1 \sin \omega\tau - \frac{x_2}{\omega} \cos \omega\tau \tag{21}$$

or equivalently

$$x_1 = z_1 \cos \omega\tau - z_2 \sin \omega\tau$$

$$x_2 = -\Omega \, [\, z_1 \sin \omega\tau + z_2 \cos \omega\tau] \tag{22}$$

Differentiating equation (21) with respect to time, substituting equations (18) and (22) and time averaging over one period, i.e $\tau = 0 \rightarrow \dfrac{2\,\pi}{\omega}$ results in

$$\overset{\circ}{z_1} = \frac{1}{2\omega} [\, - \rho \, z_2 - \Gamma \, \omega \, z_1 + 3 \, z_2 \, (\, z_1{}^2 + z_2{}^2) \,]$$

$$\overset{\circ}{z_2} = \frac{1}{2\omega} [\, \rho \, z_1 - \Gamma \, \omega \, z_2 - 3 \, z_1 \, (\, z_1{}^2 + z_2{}^2) - F] \tag{23}$$

Using a polar coordinate system

$$z_1 = r \cos \theta$$

$$z_2 = r \sin \theta \tag{24}$$

equation (23) becomes

$$\overset{\circ}{r} = \overset{\circ}{z_1} \cos \theta + \overset{\circ}{z_2} \sin \theta$$

$$= \frac{1}{2\Omega} (\, - \Gamma \, \omega \, r - F \sin \theta)$$

$$\tag{25}$$

$$r\overset{\circ}{\theta} = - \overset{\circ}{z_1} \sin \theta + \overset{\circ}{z_2} \cos \theta$$

$$= \frac{1}{2\Omega} (\, \rho \, r + 3 \, r^3 - F \cos \theta)$$

In order to find the steady state amplitude A and phase Φ of the oscillation we have to evaluate the fixed points of equations (25) :

$$\overset{\circ}{r} = 0 \quad \Rightarrow \quad F \sin \Phi = - \Gamma \, \omega \, A$$

$$\overset{\circ}{r\theta} = 0 \quad \Rightarrow \quad F \cos \Phi = \rho \, A - 3 \, A^3 \tag{26}$$

Therefore, the amplitude of the steady state oscillation is found from

$$A^6 + \frac{2}{3} \rho \, A^4 + \frac{1}{9} (\rho^2 + \Gamma^2 \, \omega^2) \, A^2 - \frac{F^2}{9} = 0 \tag{27}.$$

and the phase with respect to the driver is given by

$$\tan \Phi = - \frac{\Gamma \omega}{\rho - 3 \, A^2} \tag{28}$$

The roots of this cubic polynomial in A^2 as a function of ω are being plotted for $\Gamma = 0.4$ and F respectively equal to 0.10, 0.11, 0.12 in figure 1.

The upper branch is associated with a steady state motion around the points $x_1 = \pm \frac{1}{2}$ while the lower branch is associated with motion around the point $x_1 = 0$. As F increases for a given damping rate Γ we notice that the two branches connect allowing the steady state amplitude of a particle with equilibrium position around $x_1 = 0$ to build up to amplitudes close to the unstable points.

To study the detailed dynamics of the system , i.e. not just follow the motion of the particle with frequency equal to the driver frequency, we solved equation (22) using a fourth order Runge-Kutta method. For a given set of parameters the obtained detuning curve is shown in figure 2. It shows the existence of a hysterectic loop. This hysteresis effect occurs whenever the lower branch of the steady state solutions is multivalued for a given detuning ratio.

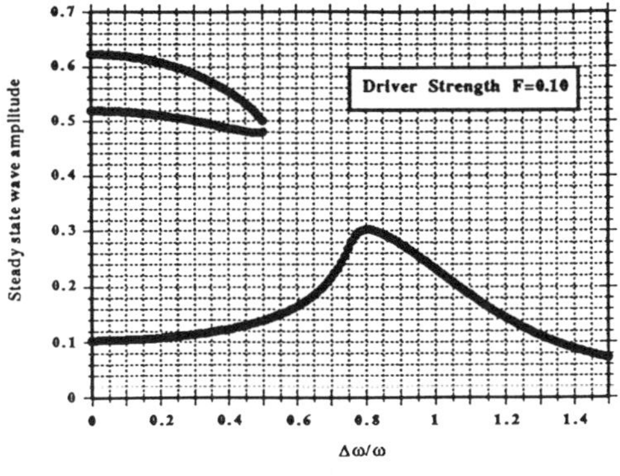

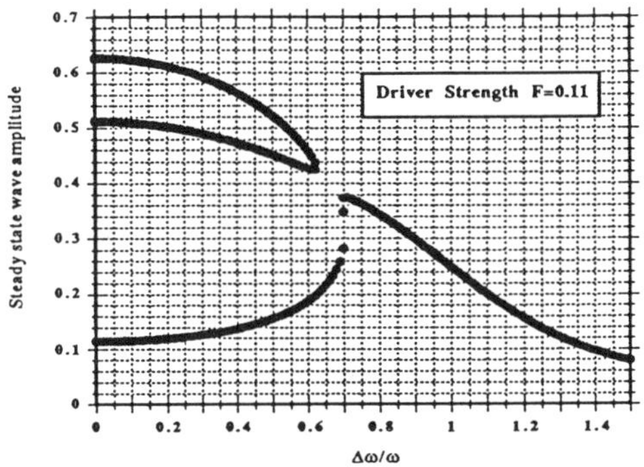

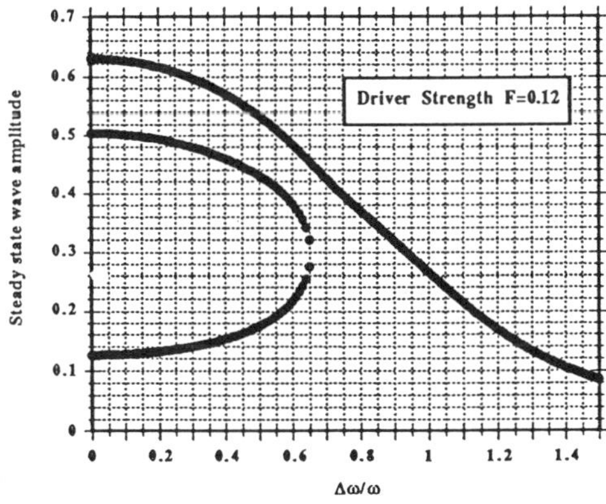

Fig. 1 Steady state amplitude vs. detuning ratio for the Duffing oscillator. The damping rate was fixed at $\Gamma = 0.4$ while the driver strength is F = 0.1, 0.11, 0.12 respectively.

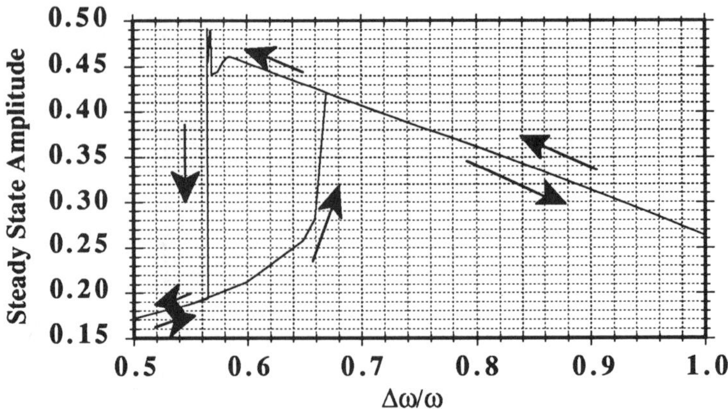

Fig. 2   Detuning curve for the Duffing oscillator. The driver strength was fixed at F = 0.12 and the damping rate was Γ = 0.4. The arrows denote the direction in which the detuning ratio is swept.

A completely different phenomenon has been observed when, for a given damping rate, the driver strength is large enough for the two branches to be connected. It is then possible to build up the amplitude of the oscillation to values close to the unstable limit. Using a value for the driver strength for which the branches are connected, the Duffing oscillator was started with both initial displacement and velocity zero. We then reduced the ratio of driver frequency to natural frequency of the system in a step wise manner, each time using the final displacement and velocity before the change as initial conditions for after the parameter change. As most clearly witnessed by the FFT-spectrum, period doubling occurred when the amplitude of the displacement approached the turning point amplitude, for driver frequencies between ω= 0.5 and ω=0.6. This period doubling of the Duffing oscillator eventually results in chaotic behavior[7]. To summarize, it is found that a hysterectic loop develops whenever the lower branch is multivalued at a given detuning ratio, and that a period doubling route to chaos only develops when the branches are connected. The amplitude associated with the onset of both phenomena is close to the turning point amplitude for this potential.

At this stage we have to asses the validity of our model. The wave equation was derived in the weakly relativistic limit i.e. the Taylor expansion of the Lorentz factor was terminated after the second term. The resulting restoring force in the wave equation is then indeed equivalent to the restoring force for a softening spring. In a Lagrangean frame it is straightforward to show that the equation of motion in the absence of damping for the momentum of relativistic plasma waves is given in it simplest form by

$$\frac{d^2p}{dt^2} + \omega_p{}^2 \frac{p}{\sqrt{1 + p^2}} = \frac{d}{dt} F_{NL} \tag{29}$$

where F_{NL} is the ponderomotive force. The restoring force can now be derived from a potential

$$V(p) = \sqrt{1 + p^2} - 1 \tag{30}$$

In figure 3 we show the exact potential as given by equation (29) and its Taylor expansion up to 2nd order. As can be seen from figure 3, the two potentials start differing significantly beyond $|p|$ = 1 and the dynamics associated with the two potentials is completely different : while the exact potential has only one stable equilibrium point, its Taylor expansion has one stable and two unstable equilibrium points. And, as seen from our previous analysis the period doubling occurs for amplitudes close to the unstable points.

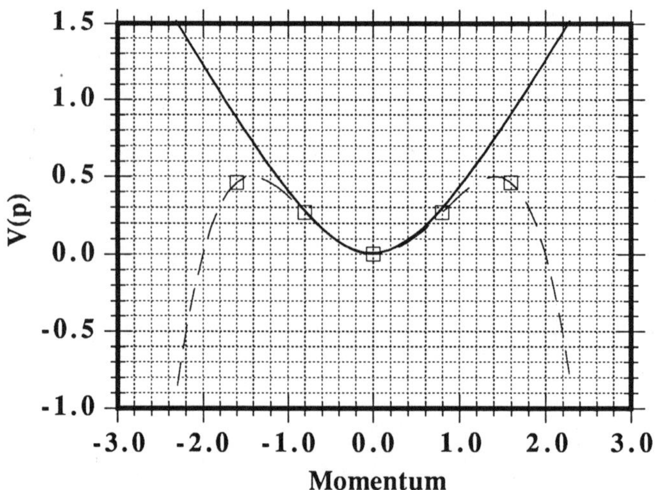

Fig. 3   Potential $V(p) = \sqrt{1 + p^2} - 1$ (solid curve) and its Taylor expansion $\dfrac{p^2}{2} - \dfrac{p^4}{8}$ (dashed curve) as a function of fluid momentum p.

To summarize, the rich non-linear behavior (i.e. period doubling route to chaos and bistability) exhibited by the Eulerian model-equation of the longitudinal electric field, is an artefact of the weakly relativistic approximation used in the derivation of equation (14).   Details of a Lagrangean analysis, in which the relativistic effects on the fluid element have been treated exactly, have been published[1] elsewhere.   The main conclusion is that the aforementioned non-linear effects do indeed occur.   Parameter regimes have been established in which the electron plasma wave shows bistable behavior and follows a period doubling route to chaos[8].   The bistable behavior has been obtained even in the absence of a density ripple but the necessity of these ripples to observe bifurcations has been determined both numerically and by the use of self-consistent computer simulations using the particle in cell code WAVE[9].

Acknowledgements

This work was supported by DOE contract DE-AS03-83-ER 40120 and DOE grant DE-FG03-88-ER 40474. We would like to thank Drs. C. E. Clayton and W. B. Mori for many helpful discussions and suggestions.

References

1. W. P. Leemans, C. Joshi, C. E. Clayton, and W. Mori, submitted to Phys. Rev. A.
2. J. X. Ma and Z. Z. Xu, J. Appl. Phys. **65**, 9 (1989)
3. J. T. Mendonca, J. Plasma Physics **34**, 115 (1985)
4. G. Duffing, Erzwungene Schwingungen bei Veranderlicher Eigenfrequenz, F. Vieweg u. Sohn: Braunschweig
5. J. M. T. Thompson and H. B. Stewart, Nonlinear Dynamics and Chaos, J. Wiley&Sons Ltd. (1986)
6. W. B. Mori, Ph. D. dissertation, UCLA (1987)
7. P. Holmes and D. J. Rand, J. of Sound and Vibration, **44**, 237 (1976);
   B. Huberman and J. P. Crutchfield, Phys. Rev. Lett. **43**, 1743 (1979)
8. M. J. Feigenbaum, J. Stat. Phys. **19**, 25 (1978)
9. R. L. Morse and C. W. Neilson, Phys. Fluids **14**, 830 (1971)

MODE COMPETITION EFFECTS IN
FREE ELECTRON LASERS AND GYROTRONS*

B. Levush and T. M. Antonsen

Laboratory for Plasma Research, University of Maryland,

College Park, MD 20742-3511

ABSTRACT

In many cases in high frequency, high power coherent radiation generators (such as free electron laser and gyrotrons) the linear gain is positive for many modes and therefore these modes will grow and compete for the beam energy. The questions related to mode competition, coherency of the radiation and maximization of the interaction efficiency are of great importance. To address these issues simple multi-mode models have been formulated. This paper is a short review of the recent results from both simulation and analyses of these models.

I. INTRODUCTION

One of the most important problems in the design of high power, high frequency coherent radiation generators such as free electron lasers and gyrotrons is insuring that the device operates in the desired mode. In many cases the linear gain is positive for many modes and therefore these modes will grow and compete for the beam energy. One is then led to ask the following questions. Is operation in a single mode possible? If so, how long will it take to reach a desired coherency? What steps must be taken to maximize the electronic efficiency of the device while ensuring single mode operation?

In our recent theoretical studies an attempt was made to address these questions regarding the stability of the single mode operation, the time scale for establishing a desired coherency and the control of the operating mode. Here we will report briefly the results of these studies and refer readers to the existing publications for the details.

II. MODEL DESCRIPTION AND STABILITY ANALYSES

The present theoretical models[1-4] are limited to the low gain regime. By this it is meant that the radiation field in the resonator can be expressed as a superposition of empty cavity modes whose amplitudes and phases change slowly in time compared with the transit time of the moving electrons with velocity v_z, through the interaction region, length L, or the time of flight of radiation through the cavity size, L_c. Thus, two distinct time scales are introduced in the model; $\tau_0 = t/(L_c/v_g)$ and $\tau_s = \left(\frac{t\omega_0}{Q}\right)$, where τ_0 is the fast time (v_g is group velocity of the radiation) and τ_s is the slow time. (Here ω_0 is the central frequency of the radiation and Q is the quality factor of the cavity).

The radiation field is periodic in fast time variable, with period 2. Equivalently we have assumed that the frequencies of the competing cavity modes are equally spaced.

Thus the normalized field in the cavity has the multiple time scale representation

$$a(\tau_s, \tau_0) = \sum_n a_n(\tau_s) \exp[-\pi n(\epsilon\xi + \tau_0)], \tag{1}$$

where ξ z/L and the slippage parameter is given by

$$\epsilon = \frac{L}{L_c}\left(\frac{v_g}{v_z} - 1\right). \tag{2}$$

This approach leads to the following simple set of equations for FEL oscillator. Each particle is described by a pendulum equation

$$\frac{dp}{d\xi} = \frac{d^2\psi}{d\xi^2} = Im\left(\sum_n a_n(\tau_s)e^{i[\psi - n\pi(\epsilon\xi + \tau_0)]}\right), \tag{3}$$

where ψ is the particle phase, p is proportional to the energy deviation of the particle. The quantity a_n is the complex amplitude of the nth mode and it depends on a slow time variable τ_s. The evolution of a_n is described by the equation

$$\left(\frac{d}{d\tau_s} + \frac{1}{2}\right) a_n(\tau_s) = -i\hat{I}\int_0^2 \frac{d\tau_0}{2}\int_0^1 d\xi < e^{-i[\psi - n\pi(\epsilon\xi + \tau_0)]} >, \tag{4}$$

where $\hat{I}$ is the normalized current and the angular average represents the projection of the beam current on the nth mode is over initial entrance phases of the particles. In the model it is assumed that all cavity modes have the same damping rate, which in normalized units is $1/2$. Similar equations have been derived for a gyrotron oscillator, see Refs. 5 and 6.

It has been recognized that by using simple models such as the ones presented here the number of parameters which are required to describe mutli-frequency FEL and gyrotron low gain oscillators can be reduced to three and four, respectively. The first parameter is the normalized detuning. In the case of gyrotrons the detuning parameter is defined to be $\delta_n = \frac{1}{2}(\omega_n - \Omega_0/\gamma)T$ where ω_n is the mode frequency of the nth mode, Ω_0/γ is the initial relativistic cyclotron frequency and T is the electron's time of flight through the interaction region, $T = L/v_{z0}$. For free electron lasers the corresponding parameter is $p_{inj,n} = ((k_n + k_w)v_{z0} - \omega_n)T$, where k_w is the wiggler wave number, $k_n = \omega_n/c$ and v_{z0} is the injected axial velocity. In a low gain oscillator the frequency and wave number must correspond approximately to the cavity mode and we have assumed uniform spacing of modes.

In the FEL case

$$\omega_n = \omega_0 + \frac{n\pi}{L_c}v_g, \tag{5}$$

where $n = 0, \pm 1, \pm 2$, labels the mode, ω_0 is the frequency of some arbitrarily chosen reference mode. In the gyrotron case

$$\omega_n = \omega_0 + \frac{2\pi}{T_R}n, \tag{6}$$

where T_R is the repetition time for the high frequency field. When many modes are present in the system each mode has its own detuning. Thus, for a given injection velocity, in the FEL case,

$$p_{inj,n} = p_{inj,0} - \epsilon\pi n, \tag{7}$$

where $p_{inj,0}$ is the detuning for the $n = 0$ mode. In the gyrotron case, the detuning

$$\delta_n = \delta_0 + \frac{2\pi}{T_0}n, \tag{8}$$

where $T_0 = T_R/T$. Thus, the parameters ϵ and T_0^{-1} measure the spectral density in FEL's and gyrotrons, respectively. To describe the gyrotron operation, an additional parameter $\mu = T\Omega_0\beta_\perp^2/2\gamma_0$ is introduced.

Finally, the last parameter is a dimensionless current, $\hat{I}$. The actual expressions for the dimensionless current for the FEL and gyrotron oscillators are quite complicated and are not presented here. One can find the appropriate formulas in Ref. 2 and Ref. 6, respectively.

In the single mode theory the number of parameters is reduced: $p_{inj,0}$ and $\hat{I}_{\text{FEL}}$, in the FEL case, and δ_0, μ and $\hat{I}_{gyr}$, in the gyrotron case, and one obtains for a given normalized current the normalized electric field strength, a_0, needed to maintain the mode in steady state. Therefore, often the normalized current is replaced by normalized electric field as a parameter.

A general feature of these devices is that for a given current modes with a range of detunings are potential stable single mode equilibria. Figure 1 shows the region of stable values of detuning p_{inj} and normalized electric field, a_0, for FEL. Also shown are the constant normalized efficiency contours and the constant dimensionless current contours. Here current, χ, is normalized to the minimum start current. As can be seen, for a given current, say $\chi = 3$ detunings p_{inj} between 2 and 5.7 are stable. The number of modes which could be stable is therefore $N \simeq 3.7/(\epsilon\pi)$.

Stability regions have been generated also for the gyrotron oscillator.[5,6] A similar conclusion, that for a given current a number of detunings are stable, is derived from that analyses. However, in the gyrotron case the picture is more

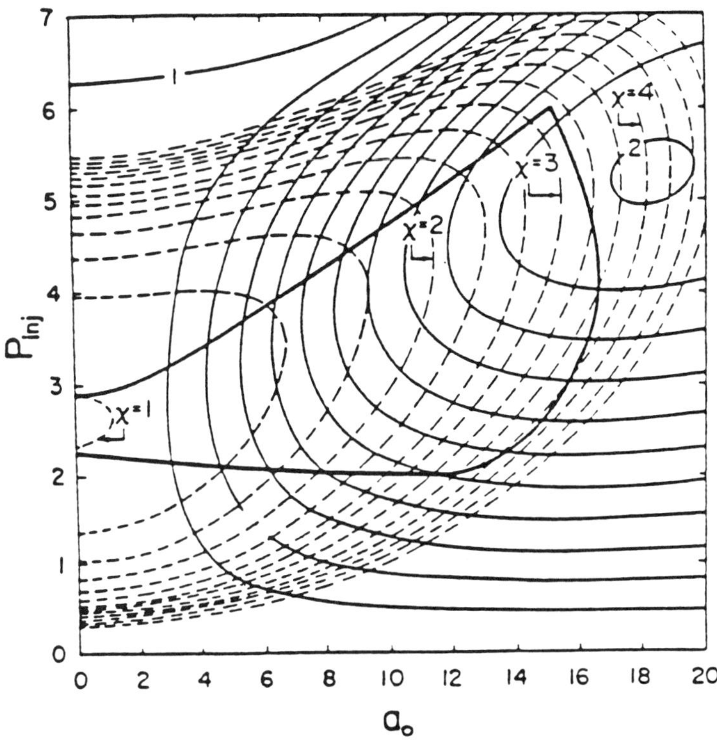

Fig. 1. Solid lines are the equal energy extraction (Δp) curves in the (p_{inj}, a_0) plane. The curve labeled 1 corresponds to $\Delta p = 0$, the curve 2 corresponds to $\Delta p = 5.5$. The difference between the neighboring level curves is 0.5. The dashed lines are the level curves of $\hat{I}/v$, which is obtained from energy balance at a particular value of a_0 and p_{inj}. The numbers on the curve indicate the value of $\chi = \hat{I}/\hat{I}_{start}$. Only inside of the triangular-shaped region is stable single-mode operation possible.

complicated, since the shape of the stability region depends strongly on the parameter μ. Note, that in gyrotron theory the mode separation is $(2\pi/T_0)$.

In both FEL and gyrotron oscillators, if the normalized current χ is less than some critical value χ_{cr}, single mode operation is possible. In the FEL case $\chi_{cr} = 4$ and in the gyrotron case $\chi_{cr} = \chi_{cr}(\mu)$; for example, $\chi_{cr}(\mu = 13) \simeq 11$ and $\chi_{cr}(\mu = 22) \simeq 5$. For a given $\chi < \chi_{cr}$, which of these equilibria is reached depends mainly on the time history of the detunings but also to a degree on the noise level.

III. COHERENCY TIME SCALES

To answer the second question, how long it will take to reach a desired coherency, we calculate the damping rate, γ_n, of the sideband modes, a_n with $n \neq 0$ in the presence of a large, nonlinearly saturated, equilibrium mode a_0. We found that, for FEL's, $\gamma_n \sim \frac{\omega_0}{Q}(\epsilon n)^2$, and for gyrotrons, in the case of large μ, $\gamma_n \sim \frac{\omega_0}{Q}\left(\frac{2n}{T_0}\right)^2$. Thus in a low gain oscillator the time t required for the Nth satellite to decay is

$$t \sim \frac{t_d}{(\epsilon N)^2}. \tag{9}$$

Solving for N gives half the number of modes present after a time t. Using the separation in frequency between the modes, the definition of ϵ and the resonance condition to express the spectral half-width, $(\Delta f/f_0)$ we find

$$\frac{\Delta f}{f_0} \sim \frac{1}{2N_W}\sqrt{\frac{t_d}{t}}, \tag{10}$$

where N_W is the number of wiggler periods. Similarly, for gyrotrons we obtain

$$\frac{\Delta f}{f_0} \sim \frac{1}{2N_c}\sqrt{\frac{t_d}{t}}, \tag{11}$$

where N_c is the number of cyclotron periods in the interaction region. Thus, Eqs. (10) and (11) indicate that at saturation the spectral width is roughly equal to the gain bandwidth and then decreases slowly in time.

To illustrate this effect we show the results of a FEL oscillator, simulation with $\epsilon = 0.05$. The spectrum of mode amplitudes at $\tau_s = 60$ and fast time dependence, τ_0, of the amplitude $|a(\tau_s, \tau_0)|$ and the phase $\phi(\tau_s, \tau_0)$ of the radiation field at the entrance to the cavity are shown in Fig. 2. At $\tau_s = 60$ the amplitude $|a(\tau_s, \tau_0)|$ is almost a constant on the fast time scale, while the phase $\phi(\tau_s, \tau_0)$ is strongly modulated. The spectrum is still brood, the number of modes at half maximum is about 11 modes.

Figure 3 exhibits the situation of the same simulation at $\tau_s = 946.1$. The number of modes at half maximum reduced to three modes, the amplitude of

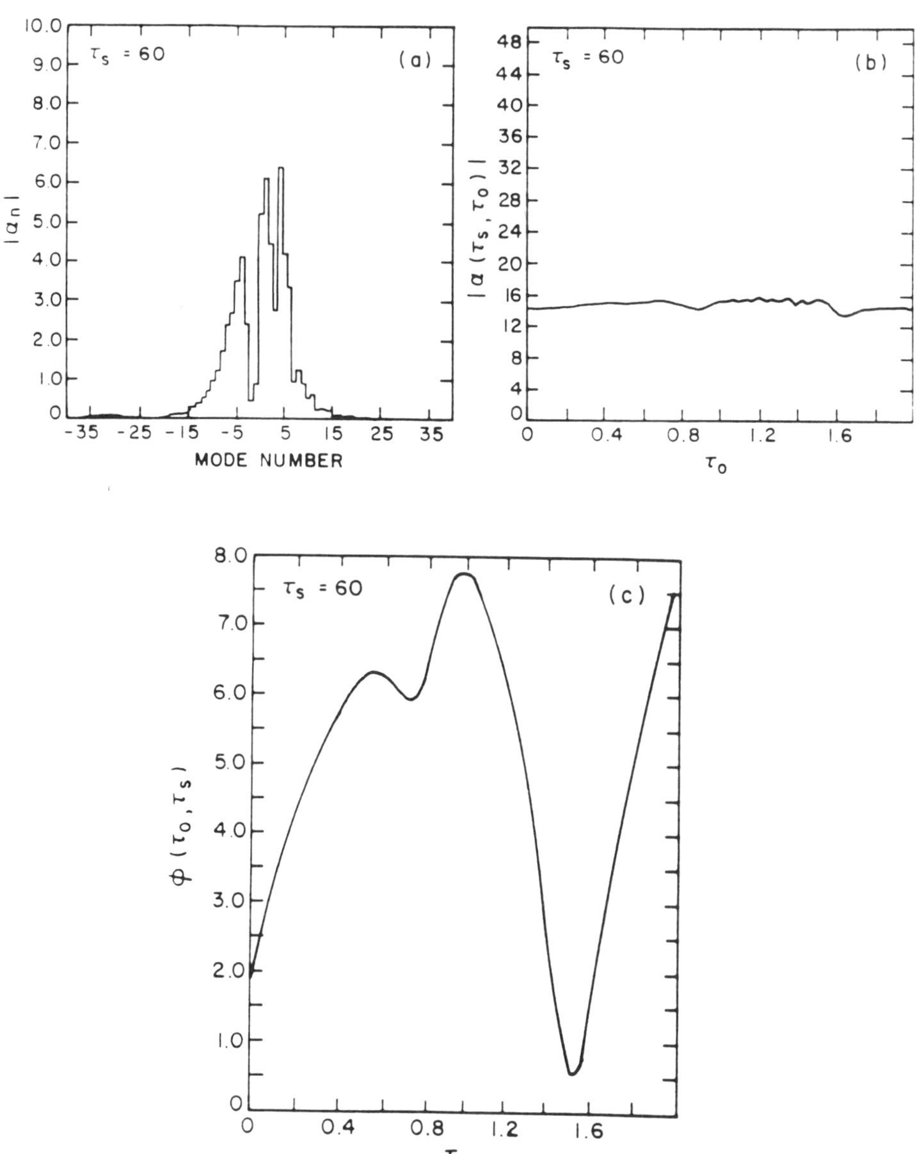

Fig. 2. Sample of a multimode simulation with 81 modes, $\epsilon = 0.05$, $\chi = 3$. Frames: (a) spectrum at $\tau_s = 60$ and fast time, τ_0, dependence of (b) field amplitude $|a(\tau_s, \tau_0)|$ and (c) field phase $\phi(\tau_s, \tau_0)$.

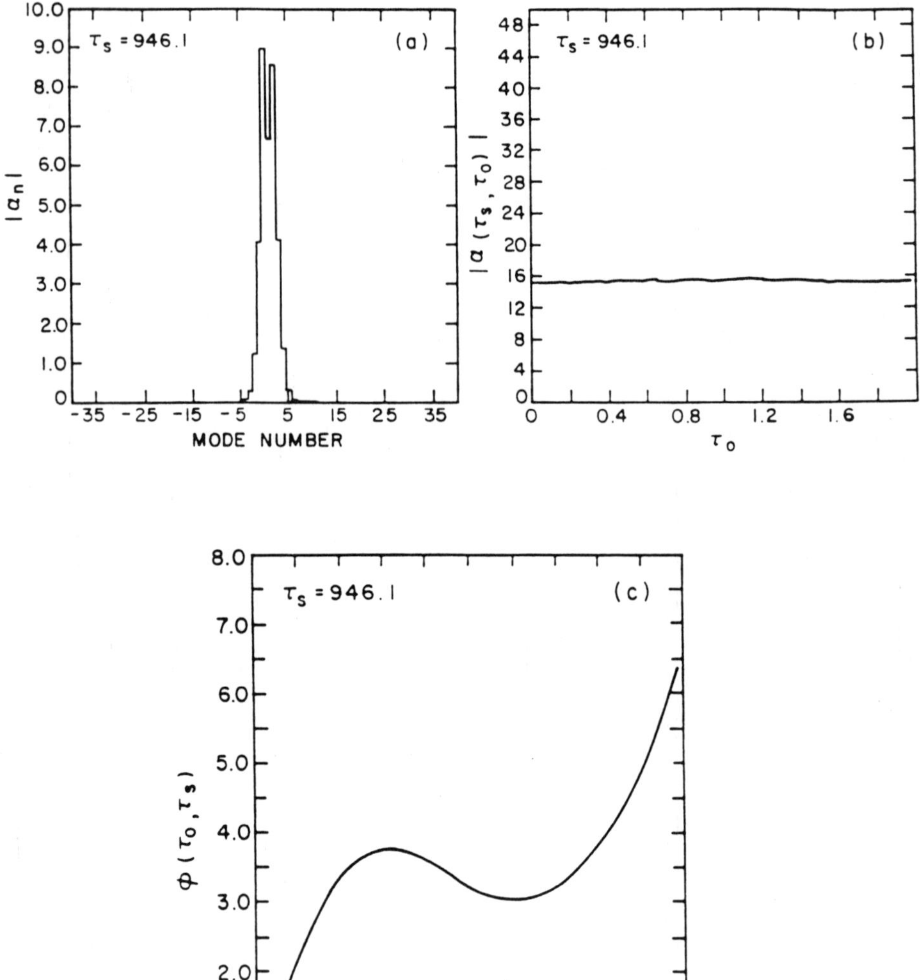

Fig. 3. The same as Fig. 2 at $\tau_s = 946.1$.

the field as a function of τ_0 is practically constant and the phase is much more relaxed. Thus, the neighboring satellites do not produce strong modulation in the amplitude of the radiation, only in its phase.

IV. MODE CONTROL

Finally, recently an attempt has been made to address the third question regarding control of the operating mode.[4,6] A number of methods of mode control such as priming and mode locking suggest themselves immediately. However, we focused on the effective mode control that occurs due to the time dependence of various system parameters during the start up phase of the oscillator. If all parameters instantly achieved their final values and the noise level was low, then the mode with the largest linear gain would grow the fastest and eventually suppress its neighbors to become the single final mode.

The opposite extreme is the case in which the system parameters attain their final values on a time scale long compared with the cavity decay time. In this case the device passes through a sequence of mode "hoppings" until a final mode is reached. Mode hopping occurs when the time dependence of the system parameters carries the detuning of the dominant mode outside the range of stable operation across the stability boundary. For slow variation of parameters the hopping occurs between adjacent modes in frequency. However, in gyrotron cases we have found that the condition on the "slowness" can be severe such that hopping by more than one mode is not uncommon. Effective mode control occurs in this adiabatic case in that frequency the desired mode is one with the largest positive detuning. Thus, by programming the voltage to rise during the start up phase (which generally causes detunings to rise with time) one can insure that the final mode is within one or two modes of having the maximum stable detuning. In the case of gyrotrons for example this provides a means of accessing "hard excitation" equilibria starting from noise.

Figure 4 shows the time history of the normalized (e_n) amplitudes of a number of modes obtained from a nonlinear, numerical simulation of a gyrotron with a rising voltage pulse. The mode numbers correspond to modes whose final detunings are given by $\delta_n = 3.1 + 2\pi n/6.5$. Thus, the separation in detunings between adjacent modes is 0.97. Time is normalized to the cavity decay time, the rise time of the voltage was 190 cavity decay times, and the current was held constant at $\chi = 9.1$. As can be seen, the system evolves through a sequence of single mode equilibria, jumping by two modes at a time. If the voltage rise time was made still larger we anticipate that the jumping could be reduced to single mode increments allowing for accessing the hard excitation equilibrium at $\delta = 3.1$. For a voltage rise which is faster (25 cavity decay times) the system settles directly into a mode with this detuning. Whereas for still shorter rise times the final detuning is lower.

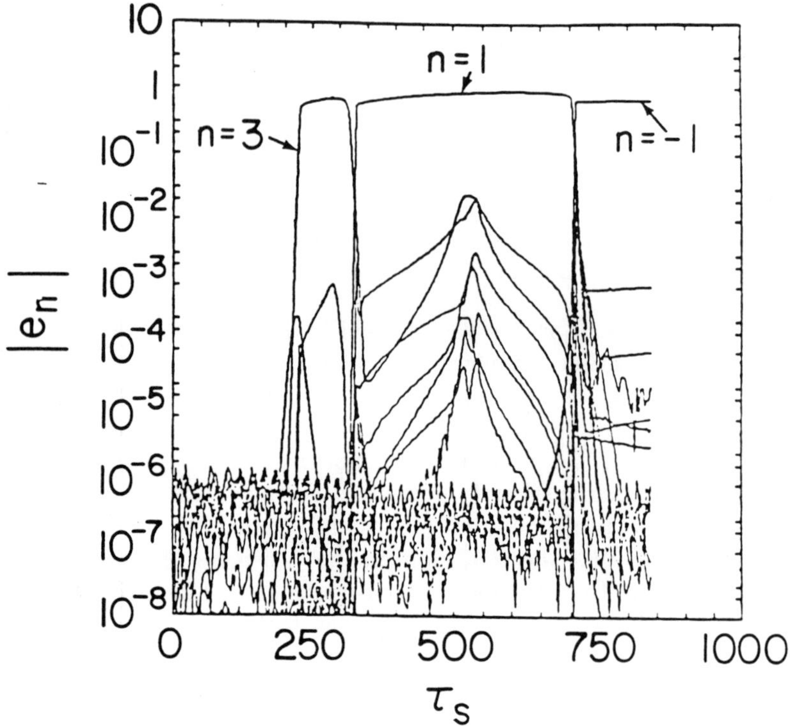

Fig. 4. Time history of the normalized amplitudes $|e_n|$ of a number of modes. The time is in units of the decay time of radiation in the empty cavity.

SUMMARY

For high power and short wavelength FEL and gyrotron oscillators the effective mode density is high in that the current is above threshold for many modes. One then is led to ask the following questions. Is operation in a single mode possible? How long will it take to reach a desired coherency? What steps must be taken to select the operating mode? The answer to the first question has been determined to be yes, provided that certain conditions are met. Regarding the second question, theory indicates that in a low gain continuous beam FEL and gyrotron oscillators with a dense spectrum the process of establishing a coherent radiation has two stages. First, the amplitude of the radiation reaches a stationary state, while the phase of the radiation is still strongly modulated. Next, the phase modulation is slowly relaxing, leading to the narrowing of the spectrum. The corresponding time scale is inversely proportional to the mode separation squire. Therefore, the coherency time can be very long if, in the FEL case, the slippage parameter is small or if, in the gyrotron case, the repetition rate, T_R, of the radiation field is much larger than the electron transit time, T.

Finally, it is suggested that by programming the voltage to rise during the start up phase (which generally causes detunings to rise with time) one can insure that the final mode is within one or two modes of having the stable detuning corresponding to the state with maximum electronic efficiency.

REFERENCES

1. N. S. Ginzburg and M. I. Petelin, Int. J. Electron. **59**, 291 (1985).

2. T. M. Antonsen, Jr. and B. Levush, Phys. Fluids B **1**, 1097 (1989).

3. T. M. Antonsen, Jr and B. Levush, Phys. Rev. Lett. **62**, 1488 (1989).

4. B. Levush and T. M. Antonsen, Jr., Nucl. Instrum. Meth. Phys. Res. **A285**, 136 (1989).

5. T. M. Antonsen, Jr., B. Levush and W. Manheimer, Phys. Fluids B **2**, 419 (1990).

6. B. Levush and T. M. Antonsen, Jr., IEEE Transactions on Plasma Science, **18**, 260 (1990).

*Work supported by US DOE and Plasma Physics Division NRL.

Optical Guiding in the High Brightness Limit

John M.J. Madey
Duke University

ABSTRACT

Based on the analytic small signal theory of Xie, the gain and optical mode radius in a high gain FEL approach well-defined asymptotic limits as the electron beam emittance is reduced to zero. It should be possible to approach these limits by improving the brightness of existing electron injectors and in the infrared and visible accelerators.

NONLINEAR PLASMA AND BEAM
PHYSICS IN PLASMA WAKE-FIELDS

J. B. ROSENZWEIG[†]

Fermi National Accelerator Laboratory
P.O. Box 500, M.S. 306,
Batavia, Illinois 60510

ABSTRACT

In experimental studies of the Plasma Wake-field Accelerator performed to date at the Argonne Advanced Accelerator Test Facility, significant nonlinearities in both plasma and beam behavior have been observed. The plasma waves driven in the wake of the intense driving beam in these experiments exhibit three-dimensional nonlinear behavior which has as yet no quantitative theoretical explanation. This nonlinearity is due in part to the self-pinching of the driving beam in the plasma, as the the denser self-focused beam can excite larger amplitude plasma waves. The self-pinching is a process with interesting nonlinear aspects: the initial evolution of the beam envelope and the subsequent approach to Bennett equilibrium through phase mixing.

Submitted to the Proceedings of the LaJolla Topical Conference on Research Trends in Nonlinear and Relativistic Effects in Plasmas, LaJolla, California, Feb. 5–8, 1990.

[†]Also at Argonne National Laboratory, Argonne, IL 60439.

1. Introduction

The Plasma Wake-field Accelerator,[1-4] (PWFA), a scheme in which charged particles are accelerated in the potentially ultra-high gradient fields supported by plasma waves driven in the wake of an intense particle beam, has been the subject of experimental investigation at the Argonne Advanced Accelerator Test Facility[5,6] (AATF). In course of these experiments, many aspects of nonlinear plasma and beam physics have been observed. These include the driving of non-linear plasma waves in the wake of the beam, and the behavior of the beam itself under the influence of its strong, nonlinear self-focusing wake-fields. This self-focusing effect yields considerably larger lens strengths than conventional methods, a fact which has led to the proposal of employing plasma wake-fields to create a powerful final focusing lens for use in a future linear e^+e^- collider.[7-11] The first compelling evidence for electron beam self-focusing in the PWFA tests at the AATF was in fact the observation of nonsinusoidal plasma wave supported wake-fields left behind the driving beam. At high driving beam currents the accelerating wake-fields were enhanced in both amplitude and in harmonic content beyond what was expected assuming the driving beam did not pinch. The degree of nonlinearity in the wake-fields provided an estimate on the self-pinched driving beam radius; this estimate agreed well with what was calculated from the theory of plasma focusing.[6]

The experimentally observed nonlinear plasma waves represent one aspect of nonlinear physics, the deviation from linear dynamics with increasing amplitude. The theoretical problem of describing nonlinear electron plasma waves excited by relativistic electron beams has been examined thoroughly the in one-dimensional limit.[12-17] On the other hand, due to the mathematical difficulty of the analysis, not much progress has been made in treating three-dimensional nonlinear plasma waves. Theoretical estimates of the expected degree of wake-field wave nonlinearity in the experiments at the AATF were based on a rough synthesis of three-dimensional linear[7] and one-dimensional nonlinear PWFA theories. This estimation method and its limitations are outlined below, and the results of this analysis are compared with the experimental data. The present status of and future prospects for theoretical work on the three-dimensional theory of nonlinear plasma waves are examined.

After the discussion of nonlinear plasma waves in PWFA research, the investigation of nonlinear dynamics in the self-pinching electron beam will be undertaken in detail. The nonlinear physics contained in the driving beam dynamics has some similarity to the nonlinear plasma waves in that the nonsinusoidal envelope motion results in a steep density spike, although this motion rapidly dissipates. Even ideal envelope wave motion requires an intrinsically nonzero transverse beam temperature, in contrast to the essentially cold fluid behavior of the nonlinear wake plasma waves. The combined effects of the beam temperature and the nonlinearity of the self-focusing forces cause the coherent envelope motion to be collisionlessly damped to approach an equilibrium through the mechanism of fast phase mixing. This equilibrium corresponds to a Bennett density profile whose minimum width is determined by the beam current and initial temperature.

The beam dynamics is treated both analytically and computationally. Because in a certain limit the focusing wake-fields become nearly independent of the plasma density and depend only on the beam profile, the self-consistent evolution of the beam distribution can be treated as an issue approximately separate from the plasma dynamics. Analytical models of the self-pinch process are employed below: laminar flow is assumed in calculating the initial focusing dynamics, and the Maxwell-Vlasov equation is utilized to discuss the asymptotic approach to self-pinched equilibrium. Particle-in-cell computer simulations are shown to complement and clarify the conclusions of the analysis, and to examine the collisionless approach to equilibrium more completely. The physics of the self-pinched beam is then compared to the related process of emittance growth of a space charge dominated beam in a linear focusing channel. After the theoretical context is established, results from the AATF experimental measurements are presented and compared with the theory.

2. Linear Plasma Wake-field Theory

In the linear theory of plasma wake-fields the function of the bunched beam is to provide an impulse to the plasma electron fluid, which causes electron density oscillations to be excited at the plasma frequency $\omega_p = \sqrt{4\pi e^2 n_0/m_e}$, where n_0 is the equilibrium ambient electron density. Linearization of the fluid equations requires that the perturbed electron density $n_1 \equiv n - n_0$ be a small quantity compared to n_0. The linearized fluid equation for plasma oscillations excited by an electron beam is given by

$$\frac{\partial^2 n_1}{\partial t^2} + \omega_p^2(n_1 + n_b) = 0, \qquad (2.1)$$

where n_b is the beam density. If one assumes a steady state condition where the plasma response in time t and longitudinal coordinate z is given only in the combination $\xi = z - v_b t$, with the beam velocity $v_b = \beta_b c$ taken to be constant (in practice we are concerned almost entirely with ultra-relativistic beams with $v_b \simeq c$), then Eq. (2.1) can be written as

$$\frac{\partial^2 n_1}{\partial \xi^2} + k_p^2(n_1 + n_b) = 0, \qquad (2.2)$$

where $k_p = \omega_p/v_b$. This linear model of plasma oscillations yields a picture of independent local oscillators with natural frequency ω_p excited by a source moving with velocity v_b, giving rise to a simple dispersionless wave with phase velocity $v_\phi = v_b$.

If we can write the driving beam density for a cylindrically symmetric beam as a product of longitudinal and transverse distributions, $n_b = g(\xi)f(r)$, with g

normalized to unity, then the solution to Eq. (2.2) can is

$$n_1(\xi, r) = f(r) \int\limits_{\xi}^{\infty} d(k_p \xi') g(\xi') \sin\left[k_p(\xi - \xi')\right]. \tag{2.3}$$

The radial profile of the plasma wave amplitude is the same as that of the beam; a convolution integral over beam's current profile gives the the longitudinal response. This convolution over the beam pulse gives a maximum excited wave if the beam scale length is shorter than a plasma skin-depth k_p^{-1}, approaching $\|n_1(r)\| = k_p f(r)$. If the beam has a Gaussian current profile, then evaluation of the convolution integral gives the degradation of the wave amplitude with rms beam length σ_z explicitly; $\|n_1(r)\| = k_p f(r) \exp\left[-(k_p \sigma_z)^2/2\right]$.

Notice that in the limit that the beam bunch is much wider than it is long the quantity $f(r)$ can be interpreted as a surface charge density. If the beam is also much wider than the plasma skin-depth, then the electric field can be easily approximated as arising from a one-dimensional charge distribution –

$$E_z \simeq 4\pi e \int\limits_{\xi}^{\infty} [n_b(\xi') + n_1(\xi')] d\xi' \simeq 4\pi e f(r) \cos(k_p \xi). \tag{2.4}$$

Inclusion of the effects of the wave's finite transverse geometry requires substitution a function $F(k_p r)$ (which is a convolution over the transverse profile using the radial Green's function[7]) for $f(r)$ in Eq. (2.4). This function which is nearly proportional to $f(r)$ if the beam is wide compared to k_p^{-1}. In the opposite limit, the longitudinal field becomes logarithmically small and nearly constant inside the beam profile if the beam is narrow with respect to k_p^{-1}. This is due to the fact that the plasma motion, and the related electric field, becomes predominantly radial in this limit. To quantify this effect, we define the radial field efficiency $\eta_r(k_p \sigma_r)$ of the wave, which is the ratio of the convolution integral on axis $F(0)$ to its value in the limit that $k_p \to \infty$. The dependence of the on-axis longitudinal field on the width of the beam distribution σ_r is explicit in this definition.

The efficiency $\eta_r(k_p\sigma_r)$ is a monotonically increasing function of its argument, approaching unity at large values of $k_p\sigma_r$.

The most relevant remaining facet of the linear theory concerns the strong transverse wake-fields which act upon the beam itself. In general, if the longitudinal/temporal dependence of the plasma response can be expressed as a function of ξ alone, as we have assumed, then the wake-field in the limit of an ultra-relativistic beam

$$\mathbf{W} \equiv \mathbf{E} + \hat{\mathbf{z}} \times \mathbf{B} \tag{2.5}$$

can be derived from a potential,

$$\mathbf{W} = \nabla(A_z - \phi) \tag{2.6}$$

This is a differential form of the Panofsky-Wenzel theorem,[18] which will explicitly manifest itself in the nonlinear wake-field data we will discuss below. In the wake plasma wave, the longitudinal component of the electromagnetic vector potential A_z vanishes and the wave is electrostatic. Inside the beam itself, however, A_z does not vanish in general, and can give rise to magnetic self-pinching forces.

The most interesting regimes occur in the limit that the beam is long compared to the plasma skin-depth, in which case the beam can be nearly charge neutralized, *i.e.* $n_1 = -g(\xi)f(r)$. If in addition the beam is wide compared to the skin-depth, then the plasma electron return current will flow inside of the beam, and the net force on the beam is approximately nullified, as the beam charge and current densities are neutralized. In such a case, charged particles moving in the direction opposite to the beam also feel no net force. This scheme, called plasma compensation, has been proposed as a method of alleviating problems associated with the strong beam-beam interaction in high energy linear colliders.[19]

In practice, high energy electron beams tend to be long and narrow, thus allowing choice of plasma density which gives a skin-depth small compared to the beam length, but large compared to its width. In this case, the plasma return current flows in a disk of radius $\sim k_p^{-1}$ and the beam current density is not neutralized. The net force felt by the beam particles is thus the self-focusing effect of the magnetic fields arising from the beam current distribution. Since the beam is assumed to be ultra-relativistic, this force can be estimated simply from Gauss' law,

$$F_r = -\frac{4\pi e^2 g(\xi)}{r} \int\limits_0^r r' f(r') dr'. \tag{2.7}$$

In this approximation the force depends only on the enclosed current density at a given point (r, ξ). Since the current density is not uniform inside the beam in general, the self-focusing wake-fields are not linear in r and independent of ξ as one would desire for aberration-free optics. The effects of the focusing nonlinearities on the performance of a plasma lens used near the interaction point of a linear collider has been examined extensively in Refs. 8-11.

From the discussion above, it would seem that there are two distinct regimes where the plasma wake-fields can provide large fields which may be of use in accelerator physics. If one utilizes an appropriate density plasma the short, wide beam can be used to drive large amplitude waves with high-gradient electric fields useful for accelerating other particle bunches, as in the PWFA. On the other hand, the long, narrow beam can be strongly focused by its self-magnetic fields which are left unbalanced when the plasma response neutralizes the beam's space charge density. It is in fact at the border of these two regimes that the most interesting experimental situation has been encountered, as will be seen below.

3. Nonlinear Plasma Wake-field Theory

The nonlinear theory of plasma wake-fields has not been developed to the point that the linear theory has, with the facility for calculating multi-dimensional effects. For this reason the theory is of limited practical use in quantitatively predicting or understanding experimental results. The one-dimensional treatments have, however, examined several relevant aspects of the nonlinear regime of the PWFA: modulational, convective and relativistic effects on the plasma waves,[13] improved transformer ratio (the ratio of the maximum accelerating field in the wake to the maximum decelerating field inside the driving beam) in the PWFA,[12] thermal limits on wave amplitude (wave breaking),[14],[15] and ion motion.[17] Despite the limitations of the theory, it is of instructive value to review the major results of the nonlinear treatments at this point.

The first nonlinear corrections to the linear theory arise through the effects of the convective derivative and the modulation of the bulk plasma electron density in the fluid equations. The nonrelativistic theory developed for this case entails keeping the full nonlinear terms in the fluid equation of motion, excluding the relativistically correct expression for the momentum. In this one-dimensional model the wave characteristics are dependent only on ξ. A useful form of the solution to nonrelativistic nonlinear fluid equation is in terms of harmonics of the plasma frequency, suppressing phase factors,[20]

$$n - n_0 = \sum_{m=1}^{\infty} \frac{m^m e^{ik\xi}}{2^{m-1} m!} \left(\frac{n_1}{n_0}\right)^m. \tag{3.1}$$

The quantity n_1 in this expression now represents the amplitude of the fundamental component of the plasma wave. The longitudinal electric field associated with this charge distribution is thus

$$E = 4\pi e k_p n_0 \sum_{m=1}^{\infty} \frac{m^{m-1} e^{ik\xi}}{2^{m-1} m!} \left(\frac{n_1}{n_0}\right)^m, \tag{3.2}$$

again ignoring phase factors. Note that if $n_1 = n_0$ in this expression that the elec-

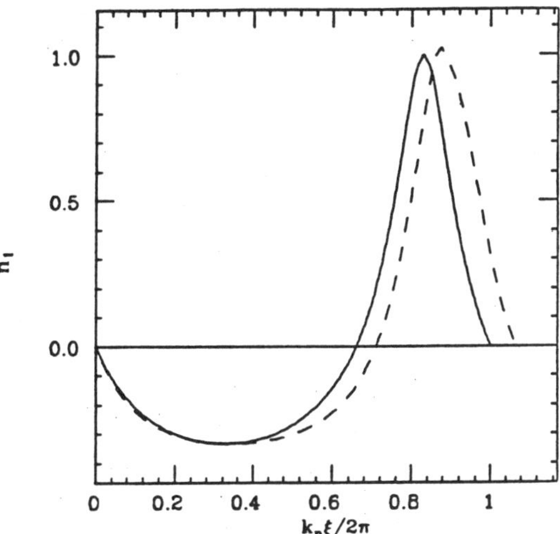

Figure 1. Perturbed plasma electron densities n_1/n_0 for one cycle of a nonlinear plasma wave, with maximum plasma fluid velocity $\beta_m = 0.5c$. The solid line represents the nonrelativistic solution; the dotted line shows the correct, period lengthened relativistic solution.

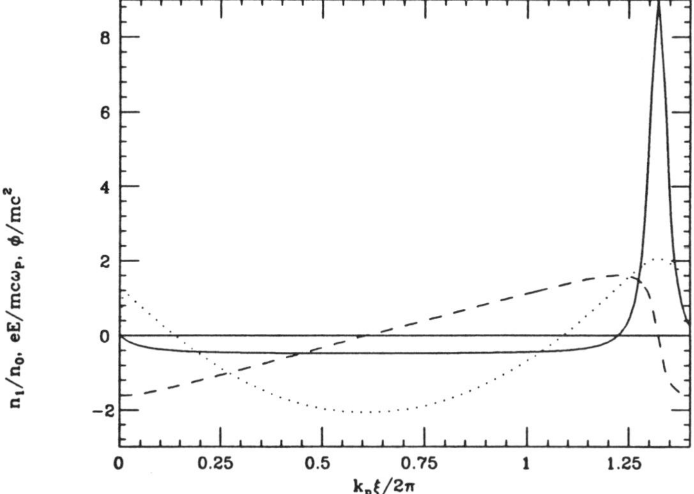

Figure 2. Nonlinear plasma wave of maximum fluid velocity $\beta_m = 0.9c$. Solid line shows n_1/n_0, dashed line indicates the electric force $-eE/m_ec\omega_p$, and dots represent the electrostatic potential $-e\phi/m_ec^2$.

tric field associated with the fundamental is $E = m_e c \omega_p / e$, which is referred to, somewhat inaccurately, as the "wave-breaking" field.[21] This harmonic decomposition of the electric field can be used to estimate the amplitude of the nonlinear waves in experiments if the harmonics in the longitudinal field are known. To incorporate the beam's role in the plasma wave excitation and the finite transverse dimension of the excited wave we add a factor to account for the radial efficiency

$$E = 4\pi e k_p n_0 \sum_{m=1}^{\infty} \frac{m^{m-1} e^{ik\xi} \eta_r(mk_p \sigma_r)}{2^{m-1} m!} \left(\frac{n_1}{n_0}\right)^m, \tag{3.3}$$

and we equate our present definition n_1 with our previous usage, taking the amplitude of the fundamental to be that given by the linear theory, which for a cylindrically symmetric bi-Gaussian beam profile is

$$n_1 = \frac{k_p N \exp\left[-(k_p \sigma_z)^2/2\right]}{2\pi \sigma_r^2}, \tag{3.4}$$

where N is the number of particles per bunch, and $\sigma_{r(z)}$ is the rms beam radius (length).

The plasma waves that develop in the nonlinear nonrelativistic theory show steepening and an increasingly narrow spike in positive electron density as the amplitude is increased. This steepening, which is illustrated in Fig. 1, generates the harmonics which are contained in Eq. (3.1).

The addition of relativistic effects into the fluid treatment introduces an additional feature to the wave dynamics, that of period lengthening with increased amplitude. This phenomenon is shown for a moderately relativistic case in comparison to the nonrelativistic solution for the same maximum fluid electron velocity $v = \beta c$ in Fig. 1. The equation for one-dimensional plasma oscillations in the limit $v_b \to c$ with functional dependence only on ξ can be cast in the form[12]

$$\frac{d^2 x}{d(k_p \xi)^2} = \frac{1}{2}\left[\frac{1}{x^2} - 1 + \frac{2n_b}{n_0}\right] \tag{3.5}$$

where $x = \sqrt{(1 - \beta)/(1 + \beta)}$. It should be noted that Eq. (3.5) is equivalent to the Poisson equation, as the electrostatic potential is given by $-\phi = m_e c^2 x$ and the perturbed electron density $n_1/n_0 = (x^{-2} - 1)/2$.

Some qualitative remarks on the solution to Eq. (3.5) are in order. The first is that the driven equation ($n_b > 0$) has a bifurcation between oscillatory and monotonically increasing behavior at $n_b = n_0/2$. This is in fact the required beam density for an enhanced transformer ratio in the proposed nonlinear PWFA scheme.[12] The undriven waves show oscillatory behavior, which in the large amplitude limit tends to exhibit long distances over which the perturbed plasma electron density approaches $-n_0/2$, interrupted by very narrow spikes of high positive density. The corresponding electric field becomes saw-toothed, with long linear rises followed by sharp drops, as illustrated in Fig. 2. The period expansion with amplitude physically comes from the relativistic mass increase of the plasma electrons; mathematically one can see that while for nearly linear waves ($|x - 1| \ll 1$) there is a linear restoring force in Eq. (3.5), for large amplitude waves where $x \gg 1$ the restoring force in the amplitude dependent oscillator saturates, forcing the local oscillation frequency down over much of the cycle and thus expanding the period. The expansion of the period allows for the possibility of electric fields larger than "wave-breaking".[14]

4. Nonlinear Plasma Wake-fields: Experiment

Much of the phenomena we have outlined above have been observed to some degree in the nonlinear PWFA experiments[6] at the AATF. We now examine a few of the most relevant observations from these experiments. The 21 MeV driving electron beam pulse used in the nonlinear experiments has the following characteristics: number of electrons per pulse $N = 2.5 \times 10^{10}$ ($Q = 4$ nC), rms pulse length $\sigma_z = 2.1$ mm, and initial rms radius $\sigma_r = 1.4$ mm. The plasma source length was set at $L_p = 33$ cm and plasma densities were variable between $n_0 = 0.5 - 8.0 \times 10^{13}$ cm^{-3}. The wake-fields in these experiments were directly measured by use of a low-intensity 15 MeV test electron bunch of similar dimensions to the driving beam, which could be delayed to travel a variable distance behind the driver. The test beam could also be misaligned to measure transverse wake-fields. Both the energy spectrum and the transverse deflections of the test bunch were measured in a broad-range high resolution magnetic spectrometer. The data from these experiments is presented by taking the energy and transverse centroids of the test beam distribution a given time delay, and then stepping the delay sequentially with small time steps, repeating the centroid measurements at each time step. A plot of the test bunch energy or deflection centroid versus delay time is referred to as a wake-field scan.

To illustrate the most striking nonlinear PWFA phenomena observed we show two high resolution wake-field scans. The scan shown in Figs. 3-4, is taken with a relatively low plasma density $n_0 = 7.3 \times 10^{12}$ cm^{-3}. The witness and driver beams are horizontally misaligned in this scan to allow observation of the longitudinal dependence of the transverse wake-fields in this case. Several qualitative remarks can be made upon inspection of Figs. 3(a) and 4. The first is that both the longitudinal and transverse wake-fields are stable, oscillatory functions of the distance behind the driving beam ξ. In fact, the wake-fields suffer little degradation in form or amplitude out to 18 wavelengths behind the beam in this scan. This is surprising, as narrow nonlinear plasma wave may be subject to differences in frequency as a function of transverse position. Secondly, the longitudinal wake-fields W_z have taken on a more saw-tooth appearance, as we would be naively expect from the one-dimensional nonlinear theory. The transverse wake-fields show a form consistent with the Panofsky-Wenzel theorem, which implies that $W_z = \partial_z \int_0^r dr W_r$ for a cylindrically symmetric driver. It is apparent from inspection that the measured longitudinal wake-fields are to a good approximation proportional to the longitudinal derivative of the measured transverse wake-fields.

In addition to these qualitative remarks on the wave-forms, we have can examine the Fourier spectrum of the longitudinal wake-fields to attempt to quantify the physical basis for the nonlinearity of these waves using our perturbation treat-

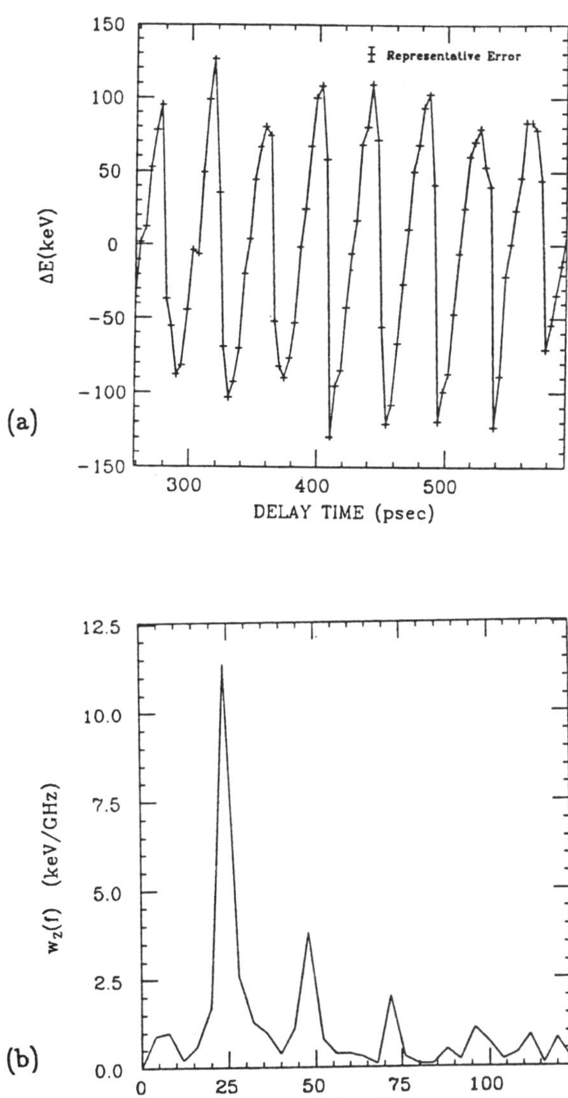

Figure 3. (a) Longitudinal wake-field scan – test beam energy centroid ΔE motion vs time delay, with plasma density of $n_0 = 7.3 \times 10^{12}$ cm^{-3}. (b) FFT amplitude function $w_z(f)$ for longitudinal wake-fields.

ment. The FFT of the longitudinal wake-field shown in Fig. 3(a) is displayed in Fig. 3(b); note that the ratio of first harmonic to the fundamental amplitude in the wake-fields is about 0.3. At a higher plasma density, which for the scan displayed in Fig. 5 is $n_0 = 2.8 \times 10^{13}$ cm^{-3} the longitudinal wake-fields display even more steepening. In this scan the the ratio of first harmonic E_2 to fundamental E_1 amplitude is 0.48. This wave amplitude is not consistent with the prediction excitation amplitude from linear theory if one ignores possible pinching of the driver beam. This is easily seen by using Eq. (3.4) to evaluate the the present case, with σ_r taken as its initial value to obtain $n_1/n_0 = 0.04$, which is much smaller than the estimate from harmonic content. This fact, along with other evidence available from the PWFA measurements,[6] led to the suggestion that significant self-pinching of the driver must have occurred, an assertion which was directly verified in later experiments.

The plasma skin-depth in the scan shown in Fig. 5 is 0.1 mm, or less than one-half of the rms beam length. This case is geometrically in the regime of the plasma lens. The dynamics and equilibration of the beam self-pinch will be explored in much greater detail below, so a simplified explanation of the the self-focusing effect on the nonlinear PWFA experiment is offered presently. Calculation using thick plasma lens theory[8] predict that the beam in this case will focus inside the plasma, and give an approximate equilibrium self-pinched beam radius is $\sigma_{eq} \simeq \epsilon\sqrt{\gamma/\nu}$, where ϵ is the beam transverse emittance, γ is the beam relativistic Lorentz factor and ν is Budker's parameter, the number of particles per unit length in units of a classical particle radius. The maximum current as a function of ξ in our beam corresponds to $\nu(0) = \nu_0 = 1.3 \times 10^{-2}$. Thus, with $\epsilon = 7 \times 10^{-6}$ m-rad we have $\sigma_{eq} = 0.44$ mm, or approximately one-third the original beam radius. This pinching is not uniform along the length of the beam, as in our model it depends on the enclosed current at a given point in ξ, a fact we can reflect by writing $\nu(\xi) = \nu_0 \exp\left(-\xi^2/2\sigma_z^2\right)$. We then have $\sigma_{eq}(\xi)^{-2} = (\nu_0\epsilon)\exp\left(-\xi/2\sigma_z^2\right).$, and the equilibrium density profile on axis is proportional to $\nu(\xi)/\sigma_{eq}^2 \sim \exp\left(-\xi/\sigma_z^2\right)$, $i.e.$ the bunch is effectively shortened

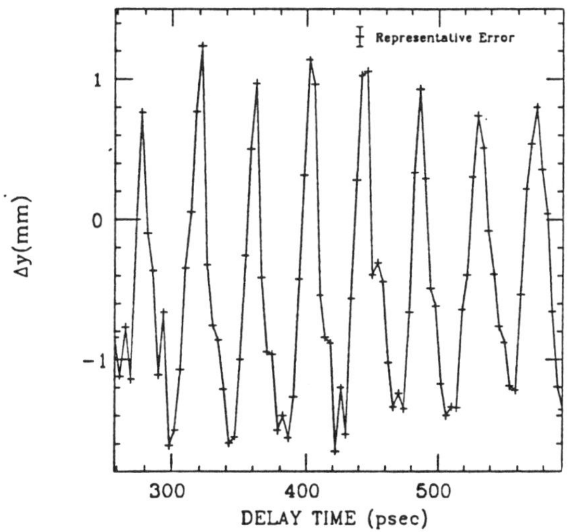

Figure 4. Transverse wake-field scan – test beam deflection plane centroid Δy vs time delay for the same scan as Fig. **3**.

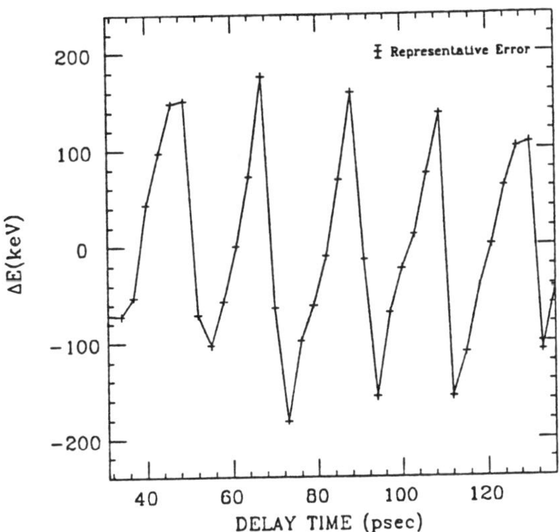

Figure 5. Longitudinal wake-field scan, with plasma of density $n_0 = 2.8 \times 10^{13}$ cm^{-3}.

on axis by a factor of $\sqrt{2}$. Both the beam pinching and the effective on axis pulse shortening serve to drive larger amplitude wake plasma waves, as can be seen from Eq. (3.4).

Taking into account the beam self-pinching the estimated ratio of the first harmonic to the fundamental in the second scan, using $\sigma_r = \sigma_{eq} = 0.44$ mm and $\sigma_z = 2.1/\sqrt{2}$ mm in the linear response formula, is $E_2/E_1 = 0.38$, in much better agreement with observation. If we apply the deconvolution analysis of Ref. 12 to include the effects of the finite resolution of the test beam, we obtain for this case a peak longitudinal field of $W_m = 5.3$ MeV/m. This is in reasonable accord with the value from Eq. (3.3) of 6 MeV/m, and with simulations performed at UCLA,[22] which indicate a maximum average accelerating gradient for this case of 7 MeV/m. The observed steepened form of the wake-fields is also reproduced well in the simulations. In regard to the maximum longitudinal field, it should be noted that no resolvable period lengthening has been observed in the PWFA experiments at the AATF, despite the large density nonlinearities. This is reflected in the fact that the wave-breaking field E_{wb} for this plasma density is over 500 MeV/m, and we have measured the field to only one percent of this value. In the one-dimensional limit E/E_{wb} is approximately the same as n_1/n_0. This does not occur in our case because the plasma electron motion becomes more radial due to the narrowness of the driving beam. This effect lowers the longitudinal field, and the plasma electrons do not attain relativistic velocities to support large density perturbations. Thus one would not expect period expansion in a narrow wave.

In these nonlinear PWFA measurements, the beam is narrower than a plasma skin-depth, and thus the plasma electron response and associated wake-fields cannot be well explained by a one-dimensional theory. Thus the theoretical arguments presented above should not be trusted to give quantitative predictions; we have been fortunate that the harmonic decomposition analysis agreed as well as it did with the measurements, and cannot expect this to be so in more nonlinear or narrower waves. A more rigorous approach to this problem would be desirable. Exact analysis remains an intractable problem; currently, perturbation treatments of nonlinearities in relativistic phase velocity plasma waves are being developed by several investigators.[23] Particle-in-cell simulations have proven to be valuable so far in modelling nonlinear effects in the PWFA.[24] It is reasonable to suspect that simulations will continue to play an important role in understanding nonlinear plasma wake-fields. There are many questions that must still be addressed in this area, including the effect of multi-dimensional nonlinear effects on the transformer ratio in the PWFA, the stability of the nonlinear plasma oscillations – both in the one-dimensional limit and including transverse variations – and the maximum wake-field that can be expected in the limit that the beam is very narrow.

5. Nonlinear Beam Dynamics: Initial Self-Pinch

Much of the previous theoretical work on plasma wake-field focusing concerns the effect of a thin plasma lens on the transverse profile of a particle beam. In the present experiments, however, the plasma column is long compared to the focal length of the lens, and the beam dynamics are much more complex. In Ref. 8, there is a treatment of the problem of the thick lens correction to the thin lens which includes the effects of finite beam emittance and the raising of the focusing strength as the beam becomes more dense inside the lens. This analysis is not strictly applicable to the if the beam is not uniform in density, but shows interesting nonlinear characteristics. A phase-space distribution which allows use of this approach is the microcanonical distribution of Kapchinskii and Vladimirskii[25] (K-V distribution). In this distribution, in which the beam density is uniform out to the beam envelope, the emittance ϵ is defined to be four times the rms emittance, and equation for the beam envelope A can be written

$$A'' + \frac{2\nu}{\gamma A} = \frac{\epsilon^2}{A^3},\tag{5.1}$$

where $'$ indicates a derivative with respect to z, The oscillations described by this equation are approximately simple harmonic if the envelope is not greatly perturbed from its equilibrium value $A = \sqrt{\gamma/2\nu\epsilon}$. If the beam envelope is not closely matched to this value, then the solution to Eq. (5.1) predicts a large density spike ($n_b = \nu/\pi r_e A^2$) and period expansion with amplitude, as in the case of nonlinear plasma waves.

This coherent envelope motion does not exist for a non-K-V distribution, but is approximated only during the initial self-pinching for a beam which is initially larger than its equilibrium radius. In this case, the envelope dynamics can be simplified further by assuming laminar flow of the beam particles, an approach which is equivalent to setting $\epsilon = 0$ in Eq. (5.1) for a K-V beam. This approximation breaks down at the same point as the K-V treatment, as it also ignores the details of the self-consistent evolution of the beam phase space, but it allows us to simply estimate the distance to the first envelope minimum. Since the transverse wake-fields felt by a particle depend only on the enclosed current, which is conserved in laminar flow, and on the radius of the particle, each particle obeys an equation of motion for paraxial trajectories derivable from a logarithmic potential,

$$r'' + \frac{C(r_0)}{r} = 0, \tag{5.2}$$

where $C(r_0)$ is a constant dependent on initial radius r_0. Near the center of a cylindrical Gaussian this constant is given by $C(r_0) \simeq (\nu/\gamma)(r_0/\sigma_r)^2$, where γ is the Lorentz factor and we have introduced the Budker parameter ν, which is the number of particles per unit length measured in classical electron radii r_e. Integrating Eq. (5.2), the distance from the plasma boundary to the first focus is calculated,

$$s = \frac{1}{\sqrt{2C(r_0)}} \int_0^{r_0} \frac{dr}{\sqrt{\log(r_0/r)}} = r_0 \sqrt{\frac{\pi}{2C(r_0)}}. \tag{5.3}$$

This expression shows the linear dependence of this approximate half-oscillation period on initial amplitude. For a Gaussian initial profile one has $s \simeq \sigma_r \sqrt{\pi\gamma/\nu}$. In the AATF experiments where the transverse beam profiles are measured, described further below, the number of particles per bunch was $N = 3.2 \times 10^{10}$, the rms bunch length $\sigma_z = 2.1$ mm and $\gamma = 42$, so the Budker parameter is $\nu = 1.6 \times 10^{-2}$. The beam is thus expected to come to its initial focus $s \simeq 8$ cm, which is well before the end of the plasma column of length $L = 35$ cm.

It is difficult to analytically estimate the minimum pinched beam radius in the presence of aberrations and finite initial emittance, because the transverse profile of the beam changes so dramatically as it focuses under the influence of nonlinear fields. The evolution of the rms beam envelope can be calculated by use of an equation which is formally identical to Eq. (5.1),[26] but in order to solve it one must known the evolution of the emittance, which is not a constant for non-K-V distributions. In order to solve for the emittance one must know the details of the phase space distribution. This subject can be addressed most straightforwardly by use of computer simulations. Computational treatments using a particle-in-cell code of this and related aspects of the self-pinching process will be presented below. Before proceeding to the computational work, an analytical model of the beam's approach to equilibrium is explored.

6. Nonlinear Beam Dynamics: Approach to Equilibrium

The transverse phase space dynamics of the beam under the influence of its self-focusing magnetic fields can be calculated in principle from the Maxwell-Vlasov equations, the self-consistent combination of the Maxwell electromagnetic field equations and the Vlasov equation, which is written

$$\frac{\partial f}{\partial t} + (\gamma m_e)^{-1}\mathbf{p}_\perp \cdot \nabla_{\mathbf{r}_\perp} f + \mathbf{F} \cdot \nabla_{\mathbf{p}_\perp} f = 0. \tag{6.1}$$

Here $f(\mathbf{r}_\perp, \mathbf{p}_\perp)$ is the beam's transverse distribution function and $\mathbf{F}$ is the Lorentz force arising from the charge and current distribution under consideration. In this case the force is due to the transverse wake-fields given by Eq. (2.7), which depend on the distribution of the beam particles in configuration space.

At this point it can be remarked that, because of the large nonlinearities in the radial focusing force and the effects of the individual particles' angular momentum, all particles have different effective betatron wave-numbers describing the periodicity of their orbits. Thus the initial coherent self-focusing motion of a set of particles decoheres as the particle motion becomes out of phase with each other, and the situation is approached where the distribution in configuration space becomes uncorrelated with the distribution in momentum space, *i.e.* the distribution seeks an equilibrium through collisionless or Landau damping. This approach to equilibrium through phase decoherence in the particle oscillations, whereby the coherent radial beam motion dissipates, was in fact initially sketched out by by Bennett in 1955, who termed the effect 'mixing'.[27] This effect is similar to a phenomenon, also referred to as decoherence, observed in the experimental study of nonlinear transverse dynamics in synchrotrons.[28] In our case, however, the nonlinear fields are provided by the beam distribution itself, and thus feedback is provided which allows us to view the dissipation of the coherent radial motion as a form of Landau damping.

As a specific type of stationary distribution is of present interest, the Maxwell-Vlasov equation is now written in cylindrincally symmetric equilibrium as

$$\frac{p_r}{\gamma m}\frac{\partial f}{\partial r} + W_r\frac{\partial f}{\partial p_r} = 0, \tag{6.2}$$

and the distribution function is assumed separable in coordinate and momentum dependence, $f(r, p_r) = R(r)P(p_r)$, because of the decorrelating effects of the nonlinearities. This assumption will be validated later by our computer simulations.

Upon substitution of the radial dependence of the magnetic self-force W_r from Eq. (2.7), and separation of variables, one obtains the momentum equation

$$\frac{\partial P}{\partial p_r} = -\frac{\alpha p_r}{\gamma m}P, \tag{6.3}$$

where $-\alpha$ is the separation constant, and the radial equation

$$\frac{\partial R}{\partial r} = -R\frac{2\alpha e^2 \beta^2}{r} \int\limits_0^r R(r')r'\,dr'.$$ (6.4)

The solution to the Eq. (6.3) is, of course, a Gaussian (thermal) distribution,

$$P = \sqrt{\frac{\alpha}{2\pi\gamma m}}\, \exp\left[-\alpha p_r^2/2\gamma m\right].$$ (6.5)

The solution to the radial equation corresponding to this thermal equilibrium in momentum space is a Bennett profile[27] which has the form

$$R(r) \sim n_b(r) = \frac{\rho_b}{[1 + (r/a)^2]^2}\quad,$$ (6.6)

where a is the Bennett radius. The usual expression for the Bennett radius relates it to the beam Debye length λ_D, with $\beta \simeq 1$,

$$a^2 = 8\lambda_D^2 = \frac{2kT_\perp}{\pi e^2 \rho_b} = \frac{2kT_\perp}{\nu mc^2}a^2\quad,$$ (6.7)

which only specifies the relationship between the beam transverse temperature and current, not the radius, which cancels out of the equation. This uncertainty can be be removed by invoking an approximate constraint on the asymptotic form of the distribution function.

To derive this constraint, note that from the form of the self-fields of a cylindrically symmetric beam and the associated Maxwell-Vlasov equation that the phase space density must be constant at $(\mathbf{r}_\perp, \mathbf{p}_\perp) = (\mathbf{0}, \mathbf{0})$, as from Eq. (6.1) we see that $\partial f/\partial t = 0$ there. If one takes an original four-dimensional phase space density corresponding to a cylindrically symmetric bi-Gaussian profile, and equates its initial value of $f(\mathbf{0}, \mathbf{0})$ to the final value of $f(\mathbf{0}, \mathbf{0})$ associated with the Bennett equilibrium, the Bennett radius is given by

$$a^2 = \frac{4\epsilon_n^2}{\gamma\nu},$$ (6.8)

where $\epsilon_n = \beta\gamma\epsilon$ is the initial normalized emittance.

This argument is not strictly rigorous, however. Even though $f(\mathbf{0},\mathbf{0})$ is a constant of the motion, the form of the asymptotic state that we have assumed is not completely correct. It has been known for some time that the Bennett profile, being a state in thermal equilibrium, can evolve from a different initial state due to the thermalizing influence of multiple scattering of beam particles off the background plasma ions.[29] The distribution function can become smooth during this collisional process and thus approach the equilibrium solution to the Maxwell-Vlasov equation, which is the product of two smooth functions, a Bennett radial profile and Gaussian momentum profile.

In the case of the collisionless damping, however, the near-equilibrium state evolves from the initial pinch by filamentation of the beam phase space as it spirals under the influence of its own nonlinear self-fields. This filamentation process does not greatly affect the center of the distribution in phase space if the self-focusing forces there are nearly linear, which is the case for conditions not too far from equilibrium. This is because near equilibrium the small amplitude orbits in phase space are well behaved rotations about the fixed point of simple harmonic motion, $(\mathbf{0},\mathbf{0})$. On the other hand, large amplitude orbits experience very nonlinear fields, and the filamentation is quite pronounced in these regions of phase space. Since the derivation given above of the asymptotic Bennett radius is concerned with the final values of $f(\mathbf{r}_\perp,\mathbf{p}_\perp)$ at or near $(\mathbf{0},\mathbf{0})$, the fact that the motion near this point is well behaved makes the argument, which depends critically on approximation of the distribution as a smooth function at small amplitudes, quite accurate for cases relatively close too equilibrium.

If the motion is not well behaved even for small amplitude particles, as happens if the initial conditions are too far from equilibrium (e.g. $a \ll \sigma_r$), then the distribution function filaments near the origin in phase space, and the approximation of the actual distribution function as a product of smooth continuous functions is not as good. Filamentation at small amplitudes thus causes the asymptotic Bennett radius to be larger, and we rewrite Eq. (6.8) as an inequality, $a \geq 2\epsilon_n/\sqrt{\gamma\nu}$. In fact, other effects, such as a deviation from the assumed initial cylindrical symmetry, spurious plasma oscillations or nontrivial return current density inside the beam radius will serve to strengthen this inequality. All of these effects are present in experiments to some extent.

As an aside, we note that the filamentation of phase space and associated emittance growth is closely related to the degree to which the beam is initially near its Debye shielded equilibrium. In the case of the magnetically self-constricted beam equilibrium, the Debye shielding length is essentially just the Bennett radius, as is seen from Eq. (6.7). The beam distribution is inherently thermal, and no fluid description is needed to describe the beam phase space. If the initial beam radius is much larger than the Bennett radius (Debye length), then the fluid motion of the beam is important, however, and filaments will form in phase space, effectively converting fluid motion into thermal motion. This behavior is also displayed in the related system of a space-charge dominated beam propagating in a constant focusing channel.[30-32] If a space-charge dominated beam is matched to a focusing in the rms sense, then the final emittance is given by[32]

$$\epsilon^2 = \epsilon_0^2 + \frac{K}{16} R^2 U_n \tag{6.9}$$

where R is the rms beam radius, K is the generalized beam perveance, and U_n is the normalized nonlinear field energy of the initial beam distribution, a dimensionless factor that is a measure of the degree of fluid motion in phase space necessary to shield the external focusing forces of the channel. As a space charge-dominated beam reconfigures to minimize its nonlinear field energy, it of course becomes more uniform, except for a region of a few Debye lengths near the beam edge where the density rapidly vanishes. It is seen from Eq. (6.9) that this process results in more phase space filamentation and emittance growth in the final state as the rms beam radius R becomes large compared to the Debye length $\lambda_D = (2K)^{-1/4}\sqrt{R\epsilon}$, or equivalently, as R is increased in comparison to the emittance. This corresponds to weaker focusing, or more laminar (fluid-like) initial flow of the beam. In this limit the spiralling beam fluid filaments in phase space can occupy a region whose area is proportional to the rms beam radius, giving rise to the dependence seen in Eq. (6.9).

In the limit of applicable beam-plasma parameters, $k_p\sigma_z \gg 1 \gg k_p\sigma_r$, some predictions can now be made concerning the outcome of experiments. In the present experiments the normalized emittance $\epsilon_n = 3 \times 10^{-4}$ rad-m. Since the maximum in the beam current at $\xi = 0$ corresponds to a maximum Budker parameter of $\nu_0 = 1.6 \times 10^{-2}$, the minimum Bennett radius predicted from the Vlasov analysis is $a_0 \geq 0.72$ mm. This pinching is not uniform along the length of the beam, as in the model it depends on the current at a given point in ξ, a fact that can be reflected by writing $\nu(\xi) = \nu_0 \exp\left(-\xi^2/2\sigma_z^2\right)$. One then has $a(\xi)^{-2} = a_0^{-2} \exp\left(-\xi/2\sigma_z^2\right)$, and the equilibrium density profile on axis is proportional to $\exp\left(-\xi/\sigma_z^2\right)$, $i.e.$ the bunch is effectively shortened on axis by a factor of $\sqrt{2}$.

7. Nonlinear Beam Dynamics: Simulation

In order to examine the deviations from the approximate theoretical treatments presented to this point, particle-in-cell simulations of the beam motion have been performed. In these calculations, which employed a modified version of the code EMMA,[33] written by Noble to calculate emittance growth in a space-charge dominated transport channel, cylindrical symmetry is assumed, longitudinal effects are ignored, and the wake-fields are calculated simply by using Eq. (2.7). The evolution of the beam distribution at one point in ξ can be quickly followed under these assumptions. Some relevant aspects of these computational results are presented here, which clarify the analytical work and provide more concrete predictions for the experimental data. These results are specialized in that the spurious effects of plasma oscillations are ignored in order to concentrate on the beam dynamics; fully electromagnetic $2\frac{1}{2}$-dimensional particle-in-cell simulations of beam self-focusing by plasma wake-fields have been performed previously.[11]

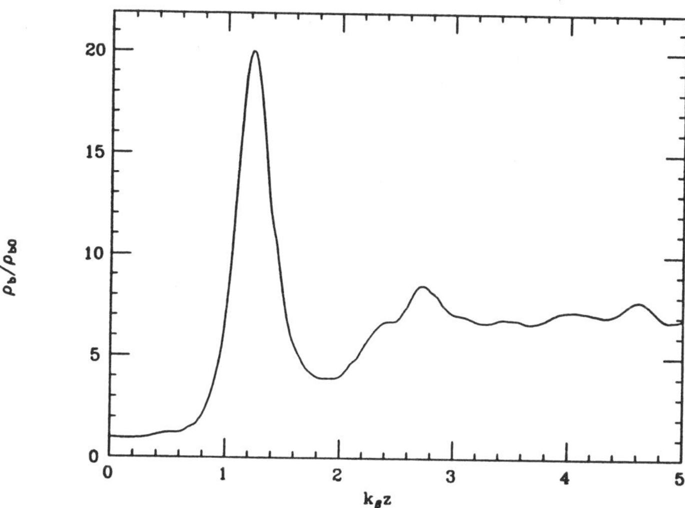

Figure 6. Peak calculated beam density ρ_b/ρ_{b0} as a function of $k_\beta z$, for initial condition $a/\sigma_r = 0.51$. For minimum equilibrium Bennett radius $\rho_b/\rho_{b0} = 7.85$.

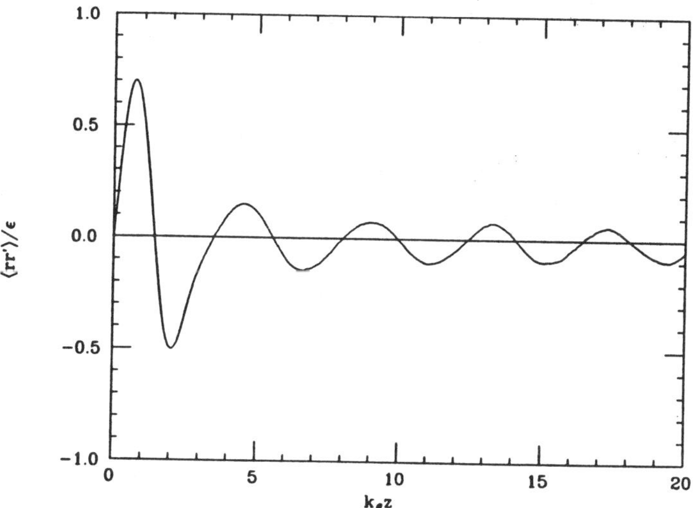

Figure 7. Calculated correlation parameter $\langle rr' \rangle / \epsilon_r$ as a function of $k_\beta z$, for initial condition $a/\sigma_r = 0.51$.

To begin, we examine the evolution of the peak beam density ρ_b/ρ_{b0} as it self-pinches, which is shown in Fig. 6. The longitudinal distances are given in units of k_β^{-1}, where $k_\beta = \sqrt{\nu/\gamma\sigma_r^2}$ is the initial small amplitude betatron wavenumber in the beam's self-focusing wake-field. All parameters in the computations shown correspond to the strongest focusing experimental case, in which $a/\sigma_r = 0.51$, where the initial beam distribution function is Gaussian in momentum and coordinate and the beam is at a waist. The initial pinching occurs in $k_\beta s = 1.27$, which is in excellent agreement with the value of $k_\beta s = \sqrt{\pi/2} \simeq 1.25$ derived in Eq. (5.3). The peak beam density is approximately 20 times the initial peak density at this point; after slight defocusing, it subsequently does not reach such large densities, but tends rapidly towards equilibrium. The maximum predicted equilibrium density associated with $a/\sigma_r = 0.51$ is $\rho_b/\rho_{b0} = 2(\sigma_r/a)^2 = 7.85$. It appears that the equilibrium which develops is slightly less dense than this.

The fluctuations in the peak density have nearly damped after $k_\beta z = 3$, or one beam envelope oscillation, which is extremely quick. The degree to which the distribution function comes into equilibrium, *i.e.* approaches a separable Bennett-coordinate/Gaussian-momentum form, can be quantified by examining the correlation parameter $\langle rr' \rangle /\epsilon_r$. Here $\epsilon_r = \sqrt{\langle r^2 \rangle \langle (r')^2 \rangle - \langle rr' \rangle^2}$ is the radial rms emittance. The evolution of this parameter is shown in Fig. 7. After initial large excursions associated with the first focus and defocus, the correlations then damp more slowly while oscillating with period $k_\beta \lambda \simeq 4$. These oscillations do not in large part reflect oscillation of the core of the beam, as these would have a period $k_\beta \lambda = 2\pi(a/\sigma_r) \simeq 3.2$, and would show up also in Fig. 6. The correlations are instead due to the spiralling 'arms' of the distribution, which occur at large amplitude and thus have a lower average wave-number. These spirals, which come into equilibrium more slowly than the beam core due to the decrease in focusing strength with amplitude and the larger initially empty regions of phase space with which they must mix, have been observed in the simulations; they are not shown in the interests of brevity. These results support the conclusion that the phase space correlations in the beam core indeed dissipate quickly, after one radial oscillation, as is necessary to apply the results of the Vlasov analysis.

It is apparent from Fig. 6 that the maximum beam density has stabilized by $k_\beta z = 5$, which corresponds to the length of the plasma column in these experiments. It is reassuring to plot the transverse profile of the beam density at this point, as is shown in Fig. 8, along with the Bennett profile from the theoretical prediction of the Vlasov treatment. The calculated profile is slightly less dense and has a marginally wider core than the minimum-radius Bennett profile, indicating small effects due to phase space filamentation.

Thus the simulations confirm the major conclusions of the analytical approaches that have been developed, and give additional insight into the expected experimental results. The observation of core equilibration within $k_\beta z = 3$ allows use of the results of the Vlasov treatment to predict the self-pinched beam radius, as this equilibration length is shorter than the plasma column in the measurements, which are presently described.

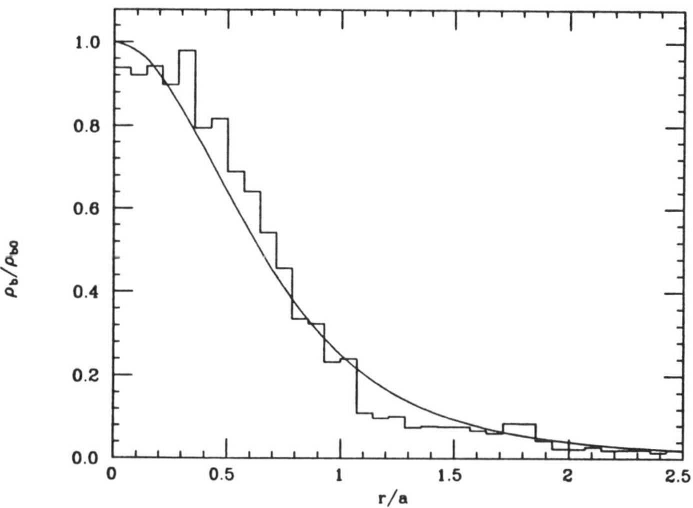

Figure 8. Histogram of the calculated beam density at $k_\beta z = 5$ (the end of the plasma column in the experiments), for initial condition $a/\sigma_r = 0.51$. Solid line is the Bennett profile predict by the Vlasov treatment.

8. Measurement of Self-focused Beam Profiles in Plasma

In order to further investigate the mechanisms behind the wake-field excitation of nonlinear plasma waves and to test aspects of plasma focusing theory, experimental measurements of the self-focused beam profiles at the end of a plasma column have been recently performed at the AATF.[34] The beam pulse delivered to the plasma source has, to restate, the following parameters; $N \simeq 3.2 \times 10^{10}$ electrons, rms length $\sigma_z \simeq 2.1$ mm transverse emittance $\epsilon \simeq 7 \times 10^{-6}$ m-rad, and $\sigma_r \simeq 1.4$ mm. This experiment used a one-picosecond resolution streak camera as the basis of a diagnostic that allowed measurement of the transverse profile of the beam as a function of longitudinal position in the beam. We summarize the results of these measurements here; for a detailed discussion see Ref. 34.

The time resolved pictures of the beam intensity profile obtained in this manner confirmed that at high plasma density ($k_p \sigma_z \geq 2$), the beam profile was self-pinched with equilibrium radius dependent on ξ, with the narrowest profile occurring at the current maximum, as expected. This self-pinched distribution led to effective pulse shortening on axis, also validating a theoretical prediction. As an example of a self-pinched profile we will show a case where the plasma density $n_0 = 6.0 \times 10^{13}$ cm^{-3}. In terms of the initial beam profile, this corresponds to a beam length in plasma radians $k_p \sigma_z \simeq 3$, and is thus inside the regime where the plasma space charge neutralizes the beam and magnetic self-focusing occurs.

The plots in Fig. 9 shows the transverse cross-section (a projection of the most intense 0.5 mm long transverse strip) of the beam intensity streak image for the dense plasma present and absent. In the case of plasma absent shown in Fig. 9(b), we also include a best fit of the data, to a Gaussian of width $\sigma_r = 1.4$ mm, with a dashed line. When the dense plasma is present the self-pinched profile shown in Fig. 9(a) is obtained, with a best fit of the data, to a Bennett profile of radius $a = 0.91$ mm, indicated by the dashed line. The agreement in form is about as good as would be expected on the basis of the profile shown in Fig. 8. Recall that the minimum predicted Bennett radius for this case is $a_0 = 0.72$

mm. The experimentally measured value is larger by about 25% due to the effects previously mentioned; the beam does not have perfect cylindrical symmetry initially, and deviations from this ideal undoubtedly allow more filamentation and effective dilution of the phase space density of the beam. In addition, there are effects due to the plasma response; even in the high plasma density case considered here the product $k_p \sigma_z$ which results on axis after pinching (due to the effective pulse shortening described above) is about 2.2, which is just on the border of the adiabatic regime of the plasma electron fluid motion where one can assume approximate charge neutralization of the beam. The focusing in the core of the beam can be lowered by the lag of the plasma electron response to the larger gradients in the beam charge density.

The effects of this response lag are displayed in Fig. 10. The on-axis beam profiles – projection of the most intense 0.2 mm wide longitudinal strip – of the beam image are plotted for four different plasma densities, $n_0 = 0.9, 1.5, 2.9,$ and 6.0×10^{13} cm^{-3} (dashed, dotted, dot-dash and solid lines, respectively). At low density the plasma skin-depth is longer than the beam scale length and the maximum focusing, reflected by the maximum in beam intensity, lags the current maximum. At the highest plasma density, the plasma response is nearly adiabatic and the on-axis profile regains its symmetric, Gaussian-like shape. The rms beam length for the highest density case is $\sigma_z = 1.5 \simeq 2.1/\sqrt{2}$ mm, in agreement with our naive prediction concerning the on-axis pulse shortening. This observation is important, as it is necessary to invoke pulse shortening to attempt explanation of the wake-field nonlinearity observed in Fig. 5. More detailed presentation and discussion of the results of the plasma wake-field focusing experiments at the AATF can be found in Ref. 34.

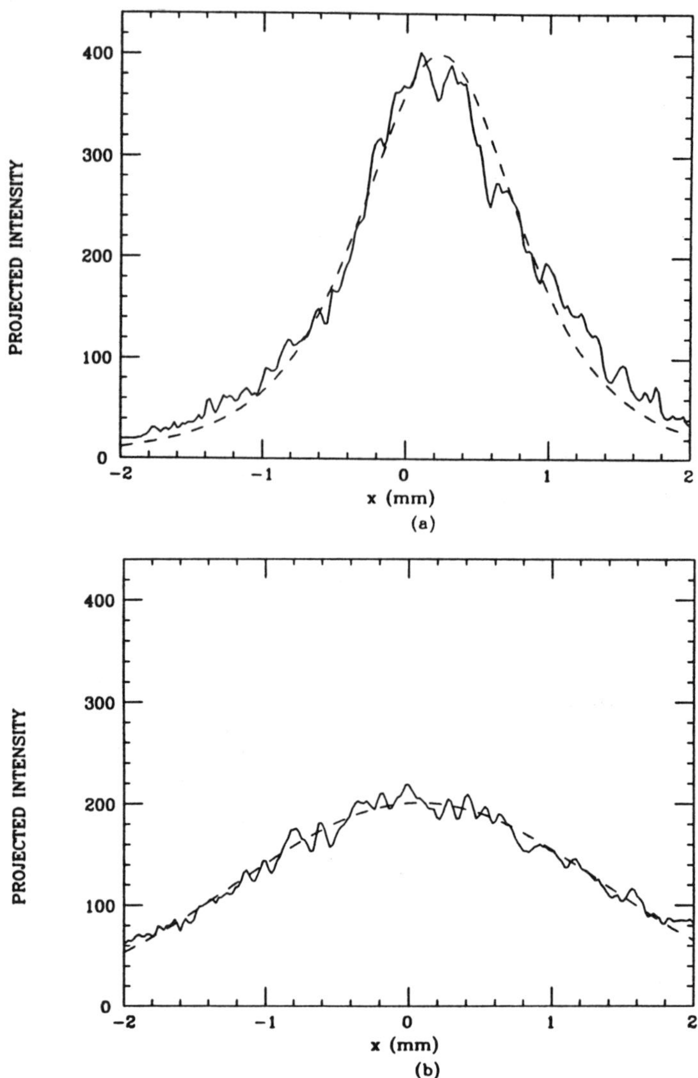

Figure 9. Plots of beam transverse intensity profiles (a projection of the most intense 0.5 mm long transverse strip in the streak image) (a) With dense plasma present – the self-pinched profile (solid line) with a best fit of the data to a Bennett profile of radius $a = 0.91$ mm (dashed line). (b) With no plasma – the unpinched profile (solid line) with a best fit of the data to a Gaussian profile of width $\sigma_r = 1.4$ mm (dashed line).

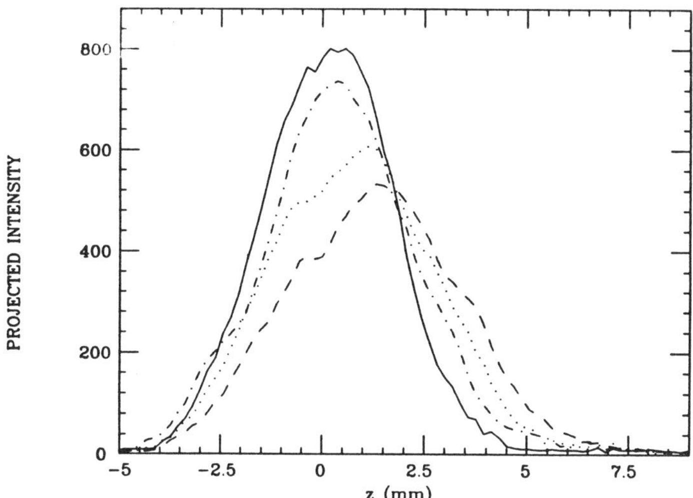

Figure 10. The on-axis beam profiles – projection of the most intense 0.2 mm wide longitudinal strip – of the beam image, plotted for four different plasma densities, $n_0 = 0.9, 1.5, 2.9,$ and 6.0×10^{13} cm^{-3} (dashed, dotted, dot-dash and solid lines, respectively).

9. Conclusions

In this paper we have examined experimental and theoretical aspects of nonlinear beam and plasma physics in plasma wake-fields. The beam and plasma behavior are, of course, intimately related experimentally, and this presentation has partially been an attempt to reflect that relationship. As a whole, the theoretical picture we have drawn explains the experimental situations encountered quite well. The experimental results to form a fairly self-consistent and nearly complete set, in that they investigated the obvious physical questions in beam and plasma parameters accessible with present experimental facilities.

Having understood past and present experiments with their challenging nonlinear attributes to a fair degree of satisfaction, we can look with some trepidation and excitement towards the future. The planned upgrade of the AATF[35] at Argonne will include a 150 MeV low emittance electron beam, with 100 nC in 5 psec pulse. The peak current in this beam is 75 times that used in the present nonlinear wake-field experiments. With this beam one can access a regime where the beam will actually be denser than the plasma ($n_0 \sim 2 \times 10^{14}$ cm^{-1}). None of the approximations we have used to explain the plasma response will be of reasonable validity in this limit. While the plasma physics in this regime becomes much more nonlinear, the transverse beam behavior becomes more linear – inside the beam the plasma electrons will be completely ejected, leaving a uniform density ion column which gives rise to linear focusing forces. This is the regime of the underdense plasma lens,[10,11] which is superior to the overdense case we have considered as a final focusing linear-collider lens. This is due to the background problem that can arise from the hard beam lepton-plasma ion collisions. Demonstration of underdense plasma focusing is one interesting result that may arise from future AATF experiments. Depending on the details of the plasma response, one can also conceive of observing accelerating gradients in these nonlinear wake-fields which are approximately "wave breaking" amplitude, which is greater than 1 GeV/m for projected plasma densities. More computational investigation of these experimental scenarios is necessary to obtain more rigorous predictions concerning the nonlinear wake-fields driven by very intense electron beams. At the same time, it would also be most beneficial to have useful analytical models of the three-dimensional nonlinear plasma response, in order to better understand the fundamental physics of the plasma wake-field interaction.

REFERENCES

1. P. Chen, J. M. Dawson, R. W. Huff, and T. Katsouleas, *Phys. Rev. Lett.*, **54,** 693 (1985).

2. T. Katsouleas, *Phys. Rev. A* **33,** 2056 (1986).

3. R. D. Ruth, A. Chao, P. L. Morton, and P. B. Wilson, *Particle Accelerators* **17,** 171 (1985) .

4. R. Keinigs and M. Jones, *Physics of Fluids* **30,** 252 (1987).

5. J. B. Rosenzweig, D. B. Cline, B. Cole, H. Figueroa, W. Gai, R. Konecny, J. Norem, P. Schoessow, and J. Simpson, *Phys. Rev. Lett.* **61,** 98 (1988).

6. J. B. Rosenzweig, P. Schoessow, B. Cole, W. Gai, R. Konecny, J. Norem and J. Simpson, *Phys. Rev. A – Rapid Comm.,* **39,** 1586 (1989).

7. P. Chen, *Particle Accelerators* **20** (1985) 171.

8. J. B. Rosenzweig and P. Chen, *Phys. Rev. D* **39,** 2039 (1989).

9. J. B. Rosenzweig, B. Cole, D. B. Cline, and D. J. Larson, *Particle Accelerators,* **24,** 11 (1988).

10. P. Chen, S. Rajagoplan, and J. B. Rosenzweig, *Phys. Rev. D* **40,** 923 (1989).

11. J.J. Su, T. Katsouleas, J. Dawson and R. Fedele, UCLA Report PPG-1177, 1988, submitted to *Phys. Rev. A.*

12. J. B. Rosenzweig, *Phys. Rev. Letters,* **58** (1987) 555.

13. R.J. Noble, *Proceedings of the Twelfth International Conference on High Energy Accelerators,* (Fermilab, Batavia, IL, 1984)

14. J. B. Rosenzweig, *Phys. Rev. A,* **38,** 3634 (1988).

15. T. Katsouleas and W. B. Mori, *Phys. Rev. Lett.,* **61,** 90 (1988).

16. A.Ts. Amatuni, S.S. Sekhpossian and E.V. Elbakian, Yerevan Preprint YerPhi-935(86)-86 Yerevan, USSR (1986).

17. J. B. Rosenzweig, *Phys. Rev. A*, **40,** 5249 (1989).

18. W.K.H. Panofsky and W.A. Wenzel, *Ref. Sci. Instrum.* **27,** 967 (1956).

19. B. Autin, A.M. Sessler and D.H. Whittum, *Proceedings of the 1989 IEEE Particle Accelerator Conference*, 1812 (IEEE, New York, 1989).

20. D. Umstadter, R. Williams, C. Clayton, and C. Joshi, *Phys. Rev. Lett.* **59,** 292 (1988).

21. J.M. Dawson, *Phys. Rev.* **113,** 383 (1959).

22. J.J. Su, presented at the Lake Arrowhead Workshop on Advanced Accelerator Concepts (Lake Arrowhead, CA 1989).

23. W.B. Mori, J.J. Su, and G. Miano, private communication.

24. J.J. Su, T. Katsouleas, J. Dawson, P. Chen, M. Jones and R. Keinigs, *IEEE Trans. Plasma Sci.* **PS-15,**192 (1987).

25. I.M. Kapchinskii and V.V. Vladimirskii, *Proc. of International Conference on High Current Accelerators*, 274 (CERN, Geneva, 1959)

26. F. Sacherer, *IEEE Trans. Nucl. Sci.* **18,** 1105 (1971).

27. W. H. Bennett, *Phys. Rev.* **45,** 890 (1934), and *Phys. Rev.* **98,** 1584 (1955).

28. A. Chao, D. Johnson, S. Peggs, J. Peterson, C. Saltmarsh, L. Sachinger, R. Meller, R. Siemann, R. Talman, D. Edwards, D. Finley, R. Gerig, N. Gelfand, M. Harrison, R. Johnson, N. Merminga, and M. Syphers, *Phys. Rev. Lett.* **61,** 2752 (1988).

29. E.P. Lee, *Phys. Fluids* **19,** 60 (1976).

30. T.P. Wangler, K.R. Crandall, R.S. Mills and M. Reiser, *IEEE Trans. Nucl. Sci.* **32,** 2196 (1985).

31. J. Struckmeier, J. Klabunde and M. Reiser, *Particle Accelerators* **15,** 47 (1984).

32. O.A. Anderson, *Particle Accelerators* **21,** 197 (1987).

33. R. J. Noble, *Proceedings of the 1989 Particle Accelerator Conference,* 1067 (IEEE, New York, 1989).

34. J. B. Rosenzweig, W. Gai, C.H. Ho, R. Konecny, S. Mtingwa, J. Norem, M. Rosing, P. Schoessow and J. Simpson, accepted for publication in *Phys. Flu. B.*

35. W. Gai, C. Ho, R. Konecny, S. Mtingwa, J. Norem, J. Rosenzweig, J. Simpson, B. Cole and M. Rosing, *Proceedings of the 1989 Particle Accelerator Conference,* 612 (IEEE, New York, 1989).

A CONTINUOUS PLASMA FINAL FOCUS

D. H. Whittum

Lawrence Berkeley Laboratory, Berkeley, California 94720

ABSTRACT

Scaling laws are set down for a plasma cell used for transport, focussing and current neutralization of fine, intense, relativistic electron beams. It is found that there exists a minimum beam spot size, $\sigma_{min} \sim \varepsilon_n (I_A / \gamma I)^{1/2}$, in such a focussing system. Propagation issues, including channel formation, synchrotron radiation, beam ionization and instabilities, are discussed. Three numerical examples are considered.

PACS numbers: 41.80Ee, 52.40Mj, 07.77.+p

INTRODUCTION

A relativistic electron beam (REB) injected into a plasma less dense than the beam expels plasma electrons from the beam volume, producing an "ion-channel." The radial electric field due to the ions then focusses the beam. A plasma more dense than the beam will neutralize the beam charge, so that the REB is focussed by its own magnetic field. A still denser plasma may partially neutralize the current. Over the last twenty years, these and other features of REB propagation in plasmas have been studied extensively, theoretically and experimentally.[1] In recent years, ion-channel focussing has been successfully employed in the transport of high current beams for accelerator research.[2,3]

The "adiabatic focusser", proposed by Chen *et al.*[4] extends the underdense-plasma ion-focussing mechanism, for use in a TeV linear electron-positron collider. They propose to increase the plasma density along the direction of beam propagation, so as to focus the beam continuously to a spot size smaller than can be achieved through conventional magnetic optics. They observe that continuous focussing is a means of circumventing the Oide limit on the spot size in a discrete focussing system.[5]

At the same time, a subject of ongoing interest, in TeV linear electron-positron collider design, is the reduction of coherent beam-beam effects: beamstrahlung and disruption.[6,7] One method which has been proposed is current neutralization in an overdense plasma at the interaction point (IP).[8,9] Beamstrahlung and disruption are suppressed due to plasma return currents which reduce the magnetic pinch forces seen by the two colliding beams.

In this note, a plasma final focussing system, consisting of an underdense adiabatic focussing cell, as proposed by Chen *et al.*, followed by an overdense current neutralization cell at the IP, is considered (Fig.1). The parameter range of interest consists of electron bunches which are short (1-10 ps), fine (mm-μm radius), high current (100 A - 1000 A) and highly relativistic (100 MeV - 1 TeV), propagating through a plasma of density 10^{12} cm^{-3} - 10^{22} cm^{-3}. Parameters in this range have attracted growing interest in recent years, in connection with the plasma wakefield accelerator,[10,11] the plasma lens,[12] and the beat-wave accelerator.[13]

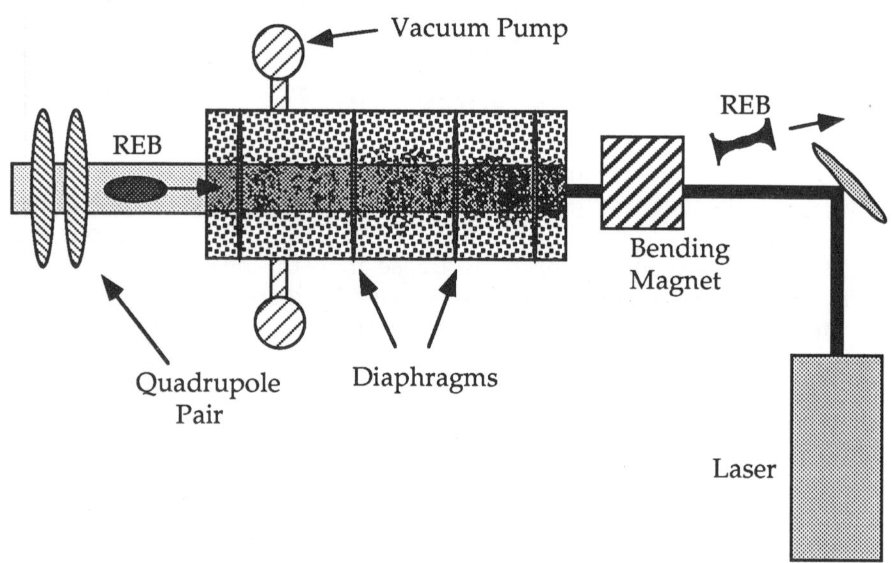

Figure 1. Set-up for a "proof-of-principle" continuous plasma focus experiment.

In the following sections, the basic scaling laws for such a "continuous plasma final focus" are set down. Issues discussed are scattering, ion channel formation, radiative losses, beam ionization and instabilities. Numerical examples are given and conclusions are offered.

SCALING LAWS

In the adiabatic focusser, an axial density gradient in a neutral gas is maintained prior to ionization, through differential pumping. An ionizing laser pulse then produces an axially increasing plasma density. Within less than a recombination time, an REB is injected. The gradient in plasma density results in an axially increasing electrostatic force, due to ion space-charge, on beam electrons as they traverse the cell. Consequently, the beam spot size is continuously reduced; the beam is "adiabatically focussed."

The continuous plasma final focus consists of such an adiabatic focussing cell, terminated with an abruptly increased plasma density extending through the IP. As the beam enters this overdense plasma, return currents are induced within the beam volume, reducing the azimuthal magnetic field that would otherwise disrupt the two beams in collision.

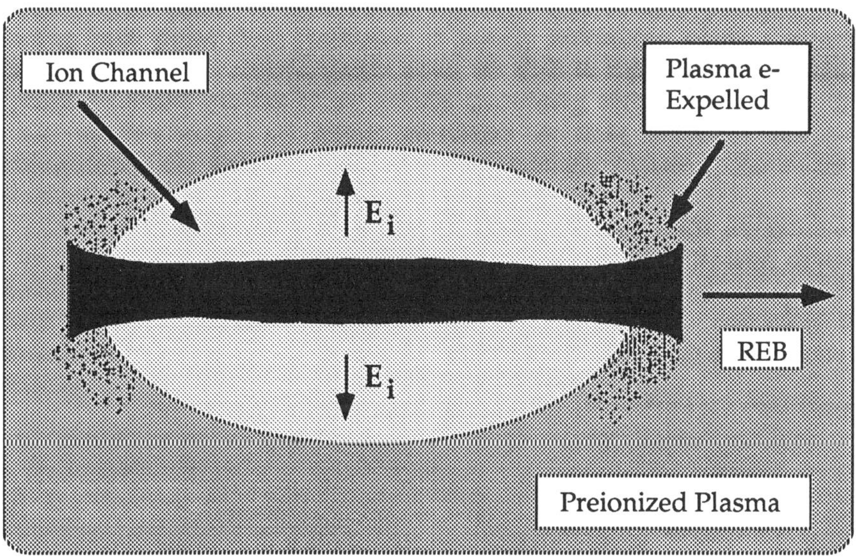

Figure 2. The radial electric field of the beam expels plasma electrons from a large volume, or "channel". Beam electrons are then focussed by the radial electric field of the relatively immobile ions.

In this section, focussing near the axial center of a long cigar-shaped beam, in a perfectly rigid channel, is considered (Fig. 2), and the relevant scaling laws are noted. Discussion of channel formation is taken up in the next section.

As the beam head propagates through the underdense plasma, it continuously expels plasma electrons from the beam volume, forming an "ion-channel", or volume from which plasma electrons have been completely ejected by the beam charge. For a very underdense plasma, the radius of this channel is given by $R \sim \sigma(2n_b/n_p)^{1/2}$, where σ is the rms beam radius, n_p is the plasma density prior to channel formation, and n_b is the beam density on axis.[14]

For effective focussing, the restoring force due to the ion charge should be much larger than the transverse Lorentz force on the beam due its self-fields. This requires $n_b > n_p >> n_b(1/\gamma^2 + \beta_\perp^2)$, where $\beta_\perp = v_\perp/c$, with $v_\perp << c$ the transverse velocity. The speed of light is c and γ is the Lorentz factor. The transverse Lorentz force seen by a beam electron in the channel is then, just the radial electric field due to the ion charge, $E_i \sim 2\pi e n_p r$, with r the radial coordinate. The electron charge is $-e$ and its mass is m. In the potential well of the ion-charge, beam electrons oscillate transversely with wavenumber $k_\beta = k_p/(2\gamma)^{1/2}$, where $k_p = \omega_p/c$ and ω_p is the plasma frequency, $\omega_p^2 = 4\pi n_p e^2/m$.

To transport the beam into the plasma without excessive emittance growth, k_β should always vary adiabatically. This determines the initial plasma electron density, n_{pi}, in terms of the initial beam spot size, σ_i: $n_{pi} = \varepsilon_n^2/(2\pi\gamma r_e \sigma_i^4)$. Here, $\varepsilon_n = \gamma k_\beta \sigma^2$ is the normalized emittance, r_e is the classical electron radius and σ is the rms spot size. Neglecting radiation, scattering, and self-fields, ε_n is an adiabatic invariant. Thus an adiabatic increase in k_p, increases k_β, and decreases σ. This is the principle of the adiabatic focusser.

This adiabaticity requires that the plasma density be tapered over a length of order the initial betatron wavelength, and provides an estimate of the overall length of the plasma cell, $L_p \sim 2\pi\gamma\sigma^2/\varepsilon_n$.

As the beam is focussed to an ever smaller spot, the plasma density approaches the beam density and the character of the focussing changes. In the overdense regime, the ion space-charge is sufficient to neutralize the beam charge, so that the beam is focussed by its own magnetic field. This transition, from the underdense regime and ion space-charge focussing, to the overdense regime and beam self-pinching, occurs for a minimum beam radius, σ_{min}, determined by setting $n_p = n_b$: $\sigma_{min} = \varepsilon_n(I_A/\gamma I)^{1/2}$. The quantity $I_A = mc^3/e = 17.05$ kA, is the Alfven current, and $I = Nec/(2\pi)^{1/2}\sigma_z$ is the peak beam current. The number of electrons per bunch is N. The density at this transition is $n_{pt} = \gamma(I/I_A)^2/(2\pi r_e \varepsilon_n^2)$, and the betatron wavenumber at this density is $k_{\beta max} \sim (I/I_A)\varepsilon_n^{-1}$.

In the overdense regime, the effective betatron wavenumber provided by the beam magnetic field is $k_\beta \sim (I_{net}/2\gamma I_A)^{1/2}/\sigma$, where the net current, I_{net}, is the sum of the beam current and the plasma return current within the beam volume.[15] Since $I_{net} \leq I$, the maximum focussing strength is bounded: $k_\beta \leq k_{\beta max}$.

Therefore, once the beam spot size is focussed to σ_{min}, the adiabatic focussing is complete, and the overdense ion-focussed regime relied on in conventional ion-focussing experments obtains. This establishes a limit on spot size in the adiabatic focusser, neglecting radiation damping. However, for the low emittance, high energy beams of a TeV collider, this limit is far smaller than the beam spot size required. Therefore, the design final spot size, σ_f, will usually be larger than the minimum possible spot size, σ_{min}. In this case, the final density in the focussing section, $n_{pf} = \varepsilon_n^2/(2\pi\gamma r_e \sigma_f^4)$, will usually be much less than n_{pt}.

Current neutralization requires a plasma skin depth short compared to the beam radial size, and a magnetic diffusion time long compared to the beam length. It is shown in Ref. 8 that the magnetic field reduction associated with an REB in a collisionless plasma scales as a function only of $k_p\sigma$. Taking a reduction of 70% as a figure of merit, $k_p\sigma_f \sim 1.4$ is required. To obtain partial current neutralization, without an increase in beam spot size, the adiabatic focussing cell should then be terminated within a distance $\lambda_{\beta f}$ of the IP with a nonadiabatic increase in plasma density to a value, n_{pc}, such that, $n_{pc} \sim 1/(2\pi r_e \sigma_f^2)$.

The length of this cell should be of order a few bunch lengths, and, to avoid defocussing due to plasma return currents, it should be less than the final betatron wavelength at the focusser exit. This implies, $\sigma_z < \lambda_{\beta f}$. If the adiabatic focusser is terminated with $n_{pf} < n_{pt}$, the beam may pinch as it enters the current neutralization cell. Pinching is ngeligible if the cell length is much less than $\lambda_{\beta min}$; this requires $\sigma_z < \lambda_{\beta min}$.

BEAM PROPAGATION

The simplified analysis of the last section considered focussing of a long cigar shaped bunch, neglecting the details of channel evolution at the bunch head and tail. However, these details are known to limit propagation in many beam-plasma applications and, in this section, their effect in the continuous plasma focus is considered.

Emittance growth due to scattering has been calculated by Montague and Schnell.[16] Applying their result, $\Theta_{rms}(z)$, the rms scattering angle after traversing a length, z, of gas, varies according to,

$$\frac{d}{dz}\Theta^2_{rms} = 8\,\pi\,n_0\,\frac{Z^2 r_e^2}{\gamma^2}\,\ln\!\left(\frac{\theta_{max}}{\theta_{min}}\right),$$

(1)

where n_0 is the density of neutral atoms. For a partially ionized gas from which plasma electrons have been ejected, $\theta_{min}\sim\hbar/(Rmc\gamma)$, for scattering from ions, and $\theta_{min}\sim\hbar/(amc\gamma)$ for scattering from neutral atoms. However, it will be assumed that the ionization fraction, f, is sufficiently low that scattering with neutral atoms dominates. The atomic number is Z and $a\sim 1.4\,a_B\,Z^{-1/3}$, is the screening radius in the Thomas-Fermi model. Planck's constant is $\hbar$ and a_B is the Bohr radius. The maximum scattering angle is $\theta_{max}\sim\hbar/(r_n mc\gamma)$ where $r_n\sim 0.5\,r_e\,A^{1/3}$ is the nuclear radius and A is the atomic weight. This gives, $\theta_{max}/\theta_{min}\sim 5.26\ 10^4/(AZ)^{1/3}$.[17] Emittance growth is given by,

$$\frac{d\varepsilon_n}{dz} = \frac{\gamma}{2\,k_\beta}\,\frac{d}{dz}\,\Theta^2_{rms},$$

(2)

The change in normalized emittance in passing through the cell is then

$$\Delta\varepsilon_n \approx \frac{r_e Z^2}{\alpha_0 f}\,\ln\!\left(\frac{\theta_{max}}{\theta_{min}}\right)\ln\!\left(\frac{\lambda_{\beta i}}{\lambda_{\beta f}}\right).$$

(3)

In the overdense regime, envelope expansion is qualitatively different because the quasistatic beam equilibrium is maintained by the beam magnetic field, rather than the (external) field of the ion charge. As the beam expands, the focussing is reduced, with the result that the beam envelope exponentiates, on the scale of the Nordsieck length,[18]

$$L_N = \frac{1}{4\,\pi\,n_0\,r_e^2}\,\frac{\gamma\,I}{I_A}\,\frac{1}{Z^2\ln\!\left(\dfrac{\theta_{max}}{\theta_{min}}\right)},$$

(4)

where channel radiation has been neglected. The Nordsieck length is always much longer than the current neutralization cell.

As the beam moves through the plasma, the beam head must eject electrons from the channel. The ion charge thus exposed provides focussing for electrons to the rear. In the meantime, electrons at the front are not strongly focussed, and expand due to emittance. These and other issues have been discussed in connection with "beam head erosion" of long pulses injected into an unionized gas,[19] and for long pulses in a preionized plasma of radial extent comparable to the beam.[20] The regime of interest here has not been extensively studied, but it is expected that erosion should be negligible, since the plasma is preionized, the electron bunch is short, the emittance is low, the energy is high, and the propagation length is short.[21]

Specifically, regions at the beam head and tail will be less dense than the plasma and will be magnetically focussed by much less than the peak azimuthal magnetic field. Thus a realistic beam profile would appear flared at each end. For best focussing the beam current rise should be adiabatic on the ω_p^{-1} time scale, i.e.,[22] $\sigma_z > 1/(4\pi n_{pi} r_e)^{1/2}$.

RADIATION IN THE ION-CHANNEL

Radiation in the ion-channel is of interest as a diagnostic, and of possible concern for its effect on beam optics. Two types of radiation are considered: bremsstrahlung and synchrotron radiation due to the betatron motion.

Bremsstrahlung may be characterized by the radiation length λ_R,[23]

$$\lambda_R^{-1} = \frac{16}{3}\alpha\, n_0\, r_e^2\, Z^2 \ln\left(\frac{233}{Z^{1/3}}\right), \tag{5}$$

where α is the fine structure constant. The fractional energy loss is then

$$\left(\frac{\Delta\gamma}{\gamma}\right)_B \approx \int_0^{L_p} \frac{dz}{\lambda_R} \approx \frac{4\alpha}{3\pi\alpha_0} r_e^2 \lambda_{\beta i}\left(\sqrt{n_{0f} n_{0i}} - n_{0i}\right) Z^2 \ln\left(\frac{233}{Z^{1/3}}\right), \tag{6}$$

where n_{0i} and n_{0f} are the initial and final neutral densities, respectively. This loss is typically very small.

Radiation due to the betatron motion takes on the character of wiggler radiation, for strong focussing $(\gamma\beta_\perp \geq 1)$.[24] In principle, the exact single particle trajectory should be used to compute the radiation fields; but for estimates here, it is enough to note the main features.

The spectrum on axis will extend through the frequency range $\omega \sim 2\gamma^2 c k_\beta/(1+\gamma^2\beta_\perp^2)$ to $\omega \sim 2\gamma^2 c k_\beta$, due to the spread in $\beta_\perp$ within the beam. Integrated over all angles, the spectrum is characterized by the critical frequency, $\omega_c = 3\gamma^3 c/\rho$, where $\rho = 1/(k_\beta^2 \sigma)$, is the effective bending radius. The angular distribution extends to angles of order $\beta_\perp$. Quantum effects are small provided $\Upsilon < 0.2$, where, $\Upsilon = \gamma^2 \lambda_c / \rho$, and λ_c is the Compton wavelength.[25]

As in a damping ring, synchrotron radiation can decrease the normalized emittance of the beam.[26] However, for the continuous plasma focus it is desirable to limit radiation losses to a small fraction of the beam energy. Fractional energy loss is computed in Ref. 4 and the result, for Υ small, is

$$\left(\frac{\Delta\gamma}{\gamma}\right)_s = -\frac{2\pi^2}{3}\, \gamma^2 \varepsilon_n r_e \frac{\left(1+\alpha_0^2\right)^2}{\alpha_0}\left(\frac{1}{\lambda_{\beta f}^2} - \frac{1}{\lambda_{\beta i}^2}\right). \tag{7}$$

Here, $\Delta\gamma$ is the change in γ and a linear variation in λ_β is assumed: $\lambda_\beta = \lambda_{\beta i} - 4\pi\alpha_0 z$. For the examples, $\alpha_0 \sim 1/4\pi$, corresponding to a length, $L_p \sim \lambda_{\beta i}$.

BEAM IONIZATION

Ionization by the beam is of concern in determining the actual axial plasma density profile. Ionization is produced by the beam through electron impact, gas breakdown, and stripping of atoms and ions in the strong radial electric field at the beam "edge". To accurately compute the net volume rate of ionization requires numerical solution of detailed rate equations, and modelling of the chemistry of the particular gas used. To estimate the effect of impact ionization, a phenomenological estimate must be made for the effective area into which secondary electrons are ejected.[27] In this section, only a few simple estimates are made.

The time scale for ionization in the overdense regime via impact ionization of neutrals by beam electrons and secondaries is $\tau_b \sim 1/(n_0 \sigma_{bi} c)$, where σ_{bi} is an effective ionization cross-section of order 10^{-18} cm^2.[28] This ionization time is ~ 1 ps at a density of 3×10^{19} cm^{-3}.

The character of breakdown produced by long pulses is determined by the value of E/p, the ratio of radial electric field to pressure.[29] For very fine beams, E/p will be sufficiently large that secondary electrons are ejected far beyond the beam volume before they create additional ionization.

In addition, for short pulses, a key limitation is the formative time required for breakdown. This is roughly the time for one secondary electron accelerated in the beam field, to ionize one neutral, $\tau_e \sim 1/(n_0 \sigma_{ei} v_e)$, where σ_{ei} is the cross-section for ionization by secondaries and v_e is the secondary velocity. The quantity $\sigma_{ei} v_e$ peaks at secondary electron energies of order ~ 100 eV, with $\sigma_{ei} v_e \sim 10^{-7} - 10^{-8}$ cm^3/sec, depending on the gas.[30] The time scale τ_e is then of order ~1 ps at a density of $3 \cdot 10^{19}$ cm^{-3}. Based on this estimate, energetic secondaries, and significant ionization beyond the beam volume may be expected depending on the particular parameters.

The radial electric field at the beam edge will be adequate to strip an atomic electron with ionization potential, $\Delta\varepsilon$, for currents of order

$$I \approx \alpha^4 \frac{\sigma}{r_e} \left(\frac{\Delta\varepsilon}{e^2/a_B} \right) I_A .$$
(8)

For very fine beams, this mechanism may fully ionize a channel larger than the beam, with some multiple ionization.

When field stripping may be neglected, plasma electrons are also lost through recombination on a time scale $\tau_r \sim 1/(\alpha_r n_p)$, and through attachment on a time scale $\tau_r \sim 1/(\alpha_a n_0)$. Here, α_r and α_a are the recombination and attachment coefficients, respectively.[31] Taking recombination in N_2 as an example, $\alpha_r \sim 2 \cdot 10^{-7}$ cm^3/sec, at electron energies ~1 eV.[31] At a density of $3 \cdot 10^{19}$ cm^{-3}, $\tau_r \sim 0.2$ ps and this is quite short. However, α_r will be lower for more energetic electrons. In addition, despite recombination and attachment, the beam volume will become depleted of plasma electrons, provided the impact ionization time scale is short enough. This occurs because, as electrons go through successive ionizations and recombinations, they diffuse away from the beam center.

Any realistic model of beam ionization, for TeV collider parameters, will have to incorporate all of these effects.

INSTABILITIES

A number of instabilities complicate the equilibrium outlined above, and in this section, the growth rates are noted.

In the focussing cell, the equilibrium discussed so far, consisting of a beam travelling down a static channel, is maintained only to the extent that ions are immobile. In fact, ions at radius R collapse inward, neutralizing the beam charge, on a time scale, $\tau_{ion} \sim (m_i/m)^{1/2} (I_A/I)^{1/2} (R/4c)$, where m_i is the ion mass. For pulses longer than τ_{ion}, focussing is stronger for the beam tail than the head. This should be avoided since it may result in disruption and emittance growth.[32]

In addition, in the underdense regime, it has been suggested that a "transverse two-stream instability" may develop,[33] whereby a displacement of the beam centroid perturbs the channel wall, which then acts back on the beam. However, only preliminary work has been performed on this problem and a growth rate has not yet been derived.

In the current neutralization cell, significant current cancellation requires a low collision rate. However, in the collisionless limit, instabilities may replace collisions in dissipating the energy of the secondaries.[34] In particular, the two-stream (Buneman) instability will couple the electron motion to the ions on a time scale v_{eiTS}^{-1}, where $v_{eiTS}/\omega_p \sim (3^{1/2}/2)(m/2m_i)^{1/3}$. This time scale can be quite short in the current neutralization cell. On the other hand, this instability convects away from the beam, and the carriers of the return current are constantly being replaced with an unperturbed flow of plasma electrons. A thorough analysis of the effect of this instability on current neutralization, including the effects of energy spread, has not been performed. However, numerical simulations performed in Ref. 8 show no evidence of return current disruption due to this effect, indicating that the plasma electron energy spread is probably sufficient to damp growth.

In addition, in the overdense regime, significant return currents flow within the beam volume and two adjacent plasma electron return current filaments attract. Filaments form and disrupt the intended current neutralization.[35] The growth rate for the Weibel or filamentation instability is $v_w \sim \omega_p (n_b/\gamma n_p)^{1/2}$, and typically a few e-folds may develop.[36,37]

Finally, the ion-hose instability[20] requires numerical study; however, a simple estimate using the rigid beam model indicates that for the examples considered here, at most a few e-folds can be expected.

EXAMPLES

In this section, three applications of a continuous plasma final focus are considered: a proof-of-principle experiment at the TRISTAN injector at the National Laboratory for High Energy Physics (KEK), an application for luminosity enhancement at the Stanford Linear Collider (SLC), and a hypothetical TeV electron-positron collider. A summary of parameters is given in Table I.

The TRISTAN injector offers the possibility of doing single beam focussing and current neutralization experiments as a "proof-of-principle." A cell of length $L_p \sim 3$ m, with initial density, $n_i \sim 1\ 10^{11}$ cm^{-3}, and final density, $n_f \sim 2\ 10^{13}$ cm^{-3} would focus the spot size from, $\sigma \sim 1$ mm, to $\sigma \sim 0.3$ mm. An increase in plasma density up to $n_c \sim 6\ 10^{14}$ cm^{-3}

over a length of a few millimeters would produce partial current neutralization.

One complication with these parameters is that the adiabatic current rise condition is not satisfied at injection. Thus nonlinear plasma oscillations would be excited and may cause emittance growth.

Table I. Parameters for applications of a continuous plasma focus.

	TRISTAN	SLC	TeV-LC	(Units)
Beam Parameters				
$mc^2\gamma$	0.25	46	1000	(GeV)
I	0.4	0.25	1.0	(kA)
ε_n	10^{-3}	10^{-5}	10^{-5}	(m-rad)
N	$6\ 10^{10}$	10^{10}	$5\ 10^{10}$	(---)
σ_z/c	10	2.5	1.0	(ps)
Plasma Parameters				
n_{pi}	$1\ 10^{11}$	$1\ 10^{14}$	$3\ 10^{15}$	(cm^{-3})
n_{pf}	$2\ 10^{13}$	$1\ 10^{18}$	$3\ 10^{19}$	(cm^{-3})
n_{pc}	$6\ 10^{14}$	$2\ 10^{20}$	$6\ 10^{21}$	(cm^{-3})
Focussing Cell Parameters				
σ_i	1000	5.0	1.0	(μm)
σ_f	300	0.5	0.1	(μm)
σ_{min}	300	0.3	0.03	(μm)
$\lambda_{\beta f}$	27	1.4	1.2	(cm)
L_p	3.0	1.4	1.3	(m)
Radiation & Scattering Parameters				
$(\Delta\gamma/\gamma)_S$	$8\ 10^{-10}$	$1\ 10^{-4}$	0.1	(---)
Υ	$2\ 10^{-8}$	$3\ 10^{-4}$	$6\ 10^{-2}$	(---)
$\Delta\varepsilon_n$	$2\ 10^{-11}\,Z^2/f$	$8\ 10^{-13}\,Z^2/f$	$2\ 10^{-12}\,Z^2/f$	(m-rad)

At the SLC, a continuous plasma focus could be employed for a proof-of-principle experiment, and to significantly enhance the luminosity in a working collider.[38] The initial plasma electron density would be $n_i \sim 1\ 10^{14}$ cm^{-3} and the length of the focusser would be $L_p \sim 1.4$ m. For a final density $n_{pf} \sim 1\ 10^{18}$ cm^{-3}, the spot size would be $\sigma \sim 0.5$ μm. The density required for partial current neutralization would be $\sim 2\ 10^{20}$ cm^{-3}. As a proof of principle, current neutralization experiments would be interesting; however, for the SLC, beamstrahlung and disruption are small and current neutralization is not required.

With A~100, the time scale for ion motion at the focusser exit is τ_{ion}~3 ps and this is probably acceptable. In the current neutralization section, the time scale for filamentation is τ_w ~2 ps. The time-scale for the electron-ion two-stream instability is τ_{eiTS} ~0.1 ps and this is short. However, plasma electron energy spread will likely damp growth.

At a density of 10^{18} cm^{-3} the formative time for breakdown will be τ_e ~10 ps, so that breakdown will be marginal. In the neutralization cell, τ_e ~ 0.03 - 0.3 ps depending on the gas, while the impact ionization time scale is τ_b ~0.1 ps. Field stripping of atoms will be significant for ionization potentials less than ~ 5 eV. Otherwise, recombination may be significant since a simple estimate gives τ_r ~10^{-2} ps.

The last example is a hypothetical TeV linear collider. The normalized emittance used for this example was an order of magnitude larger than in conventional TeV collider designs, and the charge per bunch was taken to be 5 10^{10}, which is a bit higher than is typical. The initial plasma electron density would be n_i ~ 3 10^{15} cm^{-3}. The length of the focusser would be L_p ~ 1.3 m. For a final density of 3 10^{19} cm^{-3}, the spot size would be σ ~ 0.1 μm. The betatron wavelength would be ~ 1.2 cm and the density required for partial current neutralization would be ~ 6 10^{21} cm^{-3}. Densities of 10^{22} cm^{-3} would be desirable.

Taking A ~ 100, the time scale for ion motion near the focusser exit is τ_{ion} ~0.5 ps and this is shorter than a bunch length. In the current neutralization section, the time scale for filamentation is τ_w ~1 ps. The time-scale for the electron-ion two-stream instability is τ_{eiTS} ~4 10^{-3} ps. Further work remains to assess the effect of these instabilities on current neutralization.

At a density of 3 10^{19} cm^{-3} the formative time for breakdown will be τ_e ~0.3 ps,   while the impact ionization time is τ_b ~1 ps. The recombination time will be of order τ_r ~0.2 ps. Therefore it is likely that beam ionization will be significant at the focusser exit. In the neutralization cell, τ_e ~ 2 - 20 fs depending on the gas, while τ_b ~5 fs. Field stripping of atoms will be significant for ionization potentials less than ~ 30 eV, so that an annulus will be cleared around the beam edge, in which all atoms are at least singly ionized. It is evident from these simple estimates that copious ionization will be produced by the beam in the current neutralization section.

Fractional energy loss is ~10% and this is roughly the beamstrahlung energy loss in conventional TLC designs. However, energy loss of this size is not an intrinsic feature of the continuous plasma focus and it can be reduced by reducing the emittance.

CONCLUSIONS

The concept of a continuous plasma final focus, consisting of an adiabatic plasma focussing cell, followed by a short, dense current neutralization cell has been outlined. The scaling laws for such a device have been set down, together with three numerical examples.

Further work remains to assess the effect of ion-motion at the focusser exit in a TeV collider design, as well as the Weibel and Buneman instabilities in the current neutralization cell.

Much analytical and numerical work remains to be done for a practical experiment. Interesting problems include: (1) studies of the high energy products of beam-plasma collisions, (2) studies of continuous plasma focussing of positron beams, (3) design of the vacuum system, (4) studies of realistic beam ionization profiles and their effect on focussing, (5) matching of weakly focussed beams (nonlinear wakefield theory in a very underdense plasma), (6) the effects of ion motion at the focusser exit, and (7) numerical simulation of channel formation, including ion-motion, collisions, and dipole perturbations to the beam centroid.

ACKNOWLEDGMENTS

Discussions with Andrew M. Sessler were of great help. Conversations with William M. Sharp and Simon S. Yu are greatly appreciated. Comments by John J. Stewart and Yong Ho Chin were quite useful. Thanks go to Atsushi Ogata for encouraging this note.

Work supported by the Office of Energy Research, U.S. Dept. of Energy, under Contract No. DE-AC03-76SF00098. Work at KEK was supported in part by the XIV International Conference on High Energy Accelerators and The National Laboratory for High Energy Physics (KEK), Tsukuba, Japan.

---

[1] G. Wallis, K Sauer, D. Sunder, S. E. Rosinskii, A. A. Rukhadze and V. G. Rukhlin, Sov. Phys.-Usp. **17**, 492 (1975); R. Okamura, Y. Nakamura, and N. Kawashima, Plasma Phys. **19**, 997 (1977); P. C. de Jagher, F. W. Sluijter, and H. J. Hopman, Physics Reports **167**, 177 (1988).

[2] W. E. Martin, G. J. Caporaso, W. M. Fawley, D. Prosnitz, and A. G. Cole, Phys. Rev. Lett. **54**, 685 (1985).

[3] R. B. Miller, <u>Physics of Particle Accelerators,</u> edited by Melvin Month and Margaret Dienes, AIP Conf. Proc. **184**, (New York, 1989), Vol. 2, p. 1730.

---

[4]P. Chen, K. Oide, A.M. Sessler, S. S. Yu, in <u>Proceedings of the XIV International Conference on High Energy Accelerators</u>, (Tsukuba, 1989).

[5]K. Oide, Phys. Rev. Lett. **61**, 1713 (1988). The advantage of the ion-channel in this regard is that focussing is continuous, and, for multi-GeV beams, much stronger than is attainable through continuous magnetic focussing. The features of the continuous plasma focus are then rather different from the discrete plasma focus, although the physics is closely related.

[6]R. Hollebeek, Nucl. Instrum. Methods **184**, 333 (1981); G. Bonvicini, E. Gero, R. Frey, W. Koska, C. Field, N. Phinney, A. Minten, Phys. Rev. Lett. **62**, 2381 (1989).

[7]R. B. Palmer, "The Interdependence of Parameters for a TeV Collider," SLAC-PUB-4295; S. van der Meer, "The CLIC Project and the Design for an e^+e^- Collider, CLIC Note 68, (CERN, 1988).

[8]D. H. Whittum, A. M. Sessler, S. S. Yu, and J. J. Stewart, "Plasma Suppression of Beamstrahlung," LBL Report No. 25759, (to be published in Part. Acc.).

[9]Use of two-additional beams has also been studied. J. B. Rosenzweig, B. Autin, and P. Chen, in <u>Proceedings of the Lake Arrowhead Workshop on Advanced Accelerator Concepts</u>, (UCLA, 1989), LBL No. 27058.

[10]P. Chen, J. M. Dawson, R. W. Huff, and T. Katsouleas, Phys. Rev. Lett. **54**, 693 (1985); T. Katsouleas, Phys. Rev. A **33**, 2056 (1986).

[11]J. B. Rosenzweig, D. B Cline, B. Cole, H. Figueroa, W. Gai, R. Konecny, J. Norem, P. Schoessow, and J. Simpson, Phys. Rev. Lett. **61**, 98 (1988).

[12]P. Chen, Part. Acc. **20**, 171 (1987); P. Chen, J. J. Su, T. Katsouleas, S. Wilks, and J. M. Dawson, IEEE Trans. Nucl. Sci. **PS-15**, 218 (1987).

[13]T. Tajima and J. M Dawson, Phys. Rev. Lett. **43**, 267 (1979).

[14]In terms of $s = z - v_z t$, the beam density is assumed to take the form:

$$n_{\text{beam}}(r,s) = n_b \exp\left(-\frac{r^2}{2\sigma^2} - \frac{s^2}{2\sigma_z^2}\right),$$

[15]Focussing is non-linear; electrons at the beam edge see weaker focussing.

[16]B. W. Montague and W. Schnell, in <u>Laser Acceleration of Particles</u>, edited by Chan Joshi and Thomas Katsouleas, AIP Conf. Proc. **130**, (AIP, New York, 1985), p. 146.

[17]Scattering with neutral atoms dominates for $f < \ln(a/r_n)/\ln(R/r_n) \sim 10\%$.

[18]T. P. Hughes and B. B Godfrey, Phys. Fluids **27**, 1531 (1984).

[19]W. M. Sharp and M. Lampe, Phys. Fluids **23**, 2383 (1980).

[20]H. L. Buchanan, Phys. Fluids **30**, 221 (1987).

[21]W. M. Sharp and W. M. Fawley, (private communication).

[22]For shorter pulses, nonlinear, plasma oscillations are driven by the rapidly rising beam current, as in a nonlinear plasma wake-field accelerator;J. B. Rosenzweig, Phys. Rev. Lett. **58**, 555 (1987). If turbulence results in a plasma electron temperature, T_e, the channel edge will have a finite thickness of order the corresponding Debye wavelength. The discussion assumes $\lambda_D < r_c$, and this requires $k_B T_e < 4mc^2 I/I_A$. For I ~ 1 kA, this is $k_B T_e < 100$ keV.

[23]J. D. Jackson, Classical Electrodynamics, 2nd ed. (Wiley, New York, 1975).

[24]A. Hofman, Physics Reports **68**, 253 (1980).

[25]A. A. Sokolov and I. M. Ternov, Radiation from Relativistic Electrons, (AIP, New York, 1986).

[26]W. A. Barletta, in Proceedings of the Workshop on New Developments in Particle Acceleration Techniques, (Orsay, 1987) and LLNL No. 96947; E. P. Lee, "Radiation Damping of Betatron Oscillations," UCID-19381 (1982).

[27]D.P. Murphy, M. Raleigh, R.E. Pechacek, and J. R. Grieg, Phys. Fluids **30**, 232 (1987).

[28]A. E. S. Green, Radiation Research **64**, 119 (1975).

[29]P. Felsenthal, J. M. Proud, Phys. Rev. **139**, 1796 (1965).

[30]M. Mitchner and C. H. Kruger, Jr., Partially Ionized Gases, (Wiley, New York).

[31]F. J. Mehr and M. A. Biondi, Phys. Rev. **181**, 264 (1969).

[32]The beam-ion longitudinal two-stream instability growth rate, $\nu_{biTS}/\omega_i=3^{1/2}/2(\omega_b^2/2\gamma^3\omega_i^2)^{1/3}$, is typically small (the limit $\omega_i^2 > \omega_b^2/\gamma^3$ is assumed). The ion plasma frequency is ω_i and $\omega_b^2=4\pi n_b e^2/m$.

[33]W. M. Sharp and S. S. Yu (private communication).

[34]D. Prono, B. Ecker, N. Bergstrom, and J. Benford, Phys. Rev. Lett. **35**, 438 (1975); D. A. McArthur and J. W. Poukey, Phys. Rev. Lett. **27**, 1765 (1971).

[35]R. B. Miller, Intense Charged Particle Beams, (Plenum, New York, 1982).

[36]Resistive instabilities are neglected in the collisionless limit ($\nu \tau < 1$).

[37]The growth rate for the beam-electron-plasma-electron two-stream instability is typically small: $\nu_{beTS}/\omega_p \sim 3^{1/2}/2(n_b/2\gamma^3 n_p)^{1/3}$.

[38]Plasma focussing of positron beams is not addressed here. To estimate luminosity enhancement, positron beams focussed conventionally or by a discrete plasma lens must be considered.

NOVEL PLASMA-BASED FREQUENCY UPSHIFT METHODS FOR SHORT PULSE LASERS

S.C. Wilks

Lawrence Livermore National Laboratory, Livermore, CA 94550

J.M. Dawson and W.B. Mori

Dept. of Physics, University of California, Los Angeles, California 90024

ABSTRACT

We discuss various novel methods of frequency upshifting short (≤ 1 picosecond) pulses of laser light. All of these methods make use of either the sudden creation of a plasma or relativistic plasma waves. The first method discussed is known as photon acceleration. This method makes use of the fact that a laser pulse moving in a plasma can be thought of as a packet of photons, each possessing an effective mass of $m_\gamma = \hbar\omega_{pe}/c^2$ and moving with the group velocity of the laser pulse. These photons experience a force acting on them when in the presence of a gradient in the plasma density. By using a relativistic plasma wave (i.e., a moving density gradient) traveling with the photons, the energy of the photons (thus the frequency) can be continuously increased. We then discuss the sudden creation of a plasma in a region where there exists an electromagnetic wave. This results in a frequency shift of the wave. A similar method is the creation of an ionization front moving near the speed of light, whereby the interaction of this plasma front with an EM wave also results in a frequency upshift of the original wave.

PHOTON ACCELERATOR METHOD OF FREQUENCY UPSHIFTING

The two plasma-based particle acceleration schemes, the plasma beat wave accelerator (PBWA)[1] and the plasma wake field accelerator (PWFA)[2], employ relativistic plasma waves [3] to generate large electric field gradients (~ 1 GeV/cm)[4] for the acceleration of charged particles. A third method, (which is actually the proposed method in the original beat wave article by Tajima and Dawson [1], and will be implicitly included in our references to the PBWA), relies on a short ($\leq \lambda_p/2$) powerful laser pulse to create the plasma wave. Each scheme relies on using a finite-length, electrical disturbance (in the PBWA, an intense laser pulse; in the PWFA, a relativistic electron beam) propagating through a plasma to set up plasma waves that have a phase velocity of nearly the speed of light in vacuum. Once generated, one then injects a trailing bunch of electrons [2],[5] into the accelerating phase of the plasma wave, so that energy can be transferred from the wave to the trailing electrons (this is referred to as "beam loading"). In this article, we report on the results of replacing this trailing bunch of electrons with a short (less than 1/2 plasma wavelength) pulse of electromagnetic (EM) radiation. It will be shown that the frequency of the radiation is continuously upshifted, in analogy with the energy gain of the trailing bunch of electrons in the original concept of plasma accelerators.

The idea of "loading" the plasma wave with a laser pulse can be explained conceptually as follows. A light pulse traveling through a plasma leaves behind it a "wake", or density perturbation, of amplitude $\frac{\delta n}{n} \approx \frac{eE}{m_e \omega_p c}$. Now consider a second, identical, pulse placed $1\frac{1}{2}$ plasma wavelengths behind the first pulse. This second pulse will create an identical wake that is 180 degrees out of phase with the first wake. The superposition of the two wakefields behind the second pulse results in a lowering of the amplitude of the plasma wave. It is clear that the second laser pulse has absorbed some fraction of the energy stored in the wake created by the first laser pulse. Fig. 1 shows the results of a 1-D, particle-in-cell (PIC) computer simulation of this loading. Assuming photon conservation, this increase in energy implies that the frequency of the second pulse can be upshifted. This follows from the fact that the energy of the pulse is $U = N\hbar\omega$, N being the total number of photons in the packet.

We now obtain an estimate on the rate of frequency upshift possible in the presence of a relativistic Langmuir wave, of amplitude δn in an otherwise homogeneous plasma with density n_0, that moves with the velocity of the driver (v_p), in the positive x direction. The plasma frequency associated with this density perturbation can therefore be written as

$$\omega_p^2(x,t) = \omega_{p0}^2(1 - \epsilon \sin(k_p(x - v_p t))) \tag{1}$$

where $\epsilon = \delta n/n_0, \omega_{p0}^2 = 4\pi n_0 e^2/m_e$, and $k_p = 2\pi/\lambda_p$. We now assume that a laser pulse, of width $\Delta x \leq \lambda_p/2$, and initial frequency $\omega > \omega_p$, has been injected into the accelerating phase of the wave. (The group velocity of the pulse is $v_g = c\sqrt{1 - \omega_p^2/\omega^2}$.) The plasma dispersion relation for this EM wave in the plasma is

$$\omega^2 = \omega_p^2(x,t) + c^2 k^2. \tag{2}$$

The variation of this is simply

$$2\omega\delta\omega - 2c^2 k\delta k = \delta\omega_p^2. \tag{3}$$

If we transform into the plasma waves frame (i.e., the primed frame) we obtain $\omega' = \gamma_p(\omega - v_p k)$. Using this, and the fact that in the wave frame $\omega\prime$ is constant, we can rewrite Eq (3) as

$$\delta\omega_p^2 = 2\omega\delta\omega\left(1 - \frac{c^2 k}{\omega v_p}\right). \tag{4}$$

In the lab frame, the variation of $\delta\omega_p^2$ is

$$\delta\omega_p^2 = \frac{\partial\omega_p^2}{\partial x}(\delta x - v_p\delta t) \tag{5}$$

where ω_p^2 is assumed to be a function of $(x - v_p t)$. Using Eq. (1) and the fact that $\delta x/\delta t$ is the velocity of the pulse (v_g), Eqs. (4) and (5) can be combined to give

$$\frac{\delta\omega}{\delta x} = \frac{\omega_{pe}^2 \epsilon k_p}{2\omega}. \tag{6}$$

It is found that this increase in frequency of the frequency of the accelerated laser packet exactly accounts for the loss in the energy if the accelerating plasma wave.

For the laser packet to gain energy as given by Eq. (6) requires the photons to feel the same gradient for the entire time they are accelerated. However, since the accelerating wave has a velocity $v_p < c$ (similar to the way the acclerating wave in a linac must be less than c) which we take to be constant, we see that as the photons accelerate, they phase slip with respect to the wave, and cannot continue to gain energy indefinitely. To estimate the limit on the frequency upshift possible using PWFA or PBWA - generated Langmuir waves in a homogeneous plasma due to this phase slippage effect, we now make a Lorentz transformation to a frame moving with the phase velocity of the plasma wave. Since the dispersion relation for an electromagnetic wave in a plasma is invariant under a Lorentz transformation, it is obvious that by going to this frame,

$$c^2 k\prime^2 = \omega\prime^2 - \omega_{p0}^2(1 - \epsilon\sin k_p\prime x\prime), \tag{7}$$

where primes denote moving frame quantities. In order to find the maximum frequency upshift of the laser pulse, we note that if the pulse is injected into the trough of the plasma wave at t=0, then the maximum increase in $k\prime$ will be when the pulse is just turned around at the peak density of the wave, then travels back until it reaches the trough once more. The point at which it will be turned around (the peak of the density) is also the point where $k\prime = 0$, or equivalently $\omega\prime = \omega_{p0}\sqrt{1+\epsilon}$. This will necessarily be the frequency of the pulse in the moving frame for all times because nothing in the wave frame is changing in time, except position in the wave. From these considerations and Eq. (7), we find that the initial wavenumber in the moving frame is given by $k_i\prime = -\frac{\omega_{p0}}{c}\sqrt{2\epsilon}$. In the photon picture, this is equivalent to the photons initially traveling backwards in the wave frame. They are gradually accelerated as they slip back in the wave, until they are reflected at the peak, and when they reach the bottom of the trough, the photon momentum has simply changed sign, to $k_f\prime = +\frac{\omega_{p0}}{c}\sqrt{2\epsilon}$. By Lorentz transforming the frequency back to the lab frame, we find that this change in $k\prime$ gives a new frequency of

$$\omega_f \approx \omega_i(1 + 2\sqrt{2\epsilon} + 4\epsilon + O(\epsilon^{3/2})), \tag{8}$$

where it has been assumed that $\sqrt{1+\epsilon} \approx 1+\epsilon/2$ and $\beta_\phi \approx 1$ (i.e., the plasma wave phase velocity is roughly c.) Note that this maximum final frequency can be achieved only if the wave phase velocity is chosen such that $\gamma_{wave} \leq (\omega_i/\omega_p)(\sqrt{1+\epsilon}-\sqrt{2\epsilon})^{-1}$. This can be found from the above condition on $\omega\prime$ at the turning point and the Lorentz transformation $\omega_i = \gamma_{wave}(\omega_i\prime + \beta k_i\prime)$.

For realistic density perturbations ($\delta n/n_0 \sim 0.1 - 0.5$), Eq. (8) predicts no more than about a factor of five increase in frequency. However, this restriction simply arises from the dephasing of the accelerated photons. One possible solution to this is to continuously change the phase velocity of the plasma wave as the photons are accelerated so that they remain at the same phase, in the same manner that wigglers are tapered to keep the electrons in the deaccelerating phase for a longer distance. A ramped plasma density can accomplish this, provided that the density-gradient scale length is roughly [6]

$$L_n = (\frac{1}{n}\frac{dn}{dx})^{-1} \approx m\lambda_p\gamma_{wave}^2\frac{1+\sqrt{\epsilon}}{\epsilon}, \tag{9}$$

where $m=$ the number of wavelengths behind the back of the driver (in this case, a relativistic electron beam with peak density ϵ and energy γ_{wave}) the packet sits. (The fact that driver is slowing down, and ϵ is decreasing is neglected in this estimate.) If density ramps such as this could be produced, continuous upshifts at the rate given by Eq. (6) would result, allowing for upshifts of factors of ten or more. However, various limiting factors other than phase slippage become important before these wavelengths can be achieved. One important limitation is the diffraction length, which is a measure of how far the pulse can propagate in vacuum, before the diameter of the beam becomes unacceptably wide. For the simple case of a laser pulse focused into a plasma in the absence of a plasma wave, this length is roughly given by $L_{diff} \sim \pi r_0^2/\lambda$, where r_0 is the laser spot size at the focus. We actually expect this limitation to be less restrictive when the plasma wave is present, due to a region of both focusing and accelerating in the plasma wave [2],[3].

As an example of this upshift, Eq. (6) is plotted in Fig. 2 for two different lasers. The initial laser pulses come from KrF ($\lambda = 0.26\mu$m) and a Nd^{3+}:glass, or YAG ($\lambda = 1.06\mu$m) lasers that are injected into a plasma with density $n_0 = 10^{18}$cm^3 containing a 30 % plasma wave ($\epsilon = 0.3$). Fig. 2 shows the amount of wavelength decrease (corresponding to frequency upshift) that can be expected if the plasma density is tailored according to Eq. (9). (This corresponded to a 2% density increase in the case of KrF, and a 28 % increase for Nd:glass.) The pulse lengths of both were chosen to be $\lambda_p/4 = 1.4\mu$m. The characteristic length for dispersion to become a problem is $L_{disp} \approx 5$m for KrF, and $L_{disp} \approx 0.9$m for Nd:glass. Diffraction of the pulses become a problem when $L_{diff} \approx 3$m for KrF, and $L_{diff} \approx 0.74$m for Nd:glass (where a spot size radius of 5mm was used.) As mentioned aboved, 2-D effects should relax this constraint. Notice that as the distance the pulse travels through the plasma approaches ~ 1 m, the final wavelength becomes essentially independent of the initial wavelength.

Computer simulations have been performed with the plasma simulation codes WAVE [7] and ISIS, to test these predictions. Fig. 3 shows the results of a 1 and $\frac{2}{2}$ dimensional, fully relativistic, electromagnetic PIC simulation designed to show frequency upshift . The plasma wave was set up by a wakefield driver [2] with initial $\gamma = 22$. The density perturbation, $\epsilon \sim 0.25$, set up by this driver produced a wake into which was injected a laser pulse of width $L_{pulse} \approx 2.5c/\omega_p$, initial frequency $\omega_i = 18\omega_p$. After a distance of $237c/\omega_p$, we found that there was an upshift of 10% in the laser pulse, or a final frequency of $\approx 19.8\omega_p$. Eq. (6) predicts an upshift to $\approx 19.5\omega_p$. This slight difference is attributed to the fact that even at density perturbations of 25%, the wake shows nonlinear wave steepening. This implies a larger electric field (or equivalently, a larger density gradient) than expected for a density perturbation of 25%, which the simple model above assumed to be linear. Thus, this nonlinearity would cause the rate at which the pulse is upshifted to be slightly higher than what the linear theory predicts.

In this section we have presented a novel method of continously upshifting short ($L_{pulse} \leq \lambda_p/2$) pulses of electromagnetic radiation by use of relativistic electron plasma waves. By choosing the length of time the packet interacts with the plasma wave, any frequency between the initial frequency and the final frequency (Eq. (8), for a homogeneous plasma,) can be obtained. It is found that if the plasma density profile satisfies Eq. (9), the frequency can be increased by almost a factor of ten.

SUDDEN IONIZATION METHODS

We will now discuss a different type of laser plasma interaction that results in an upshifting of the laser frequency. Some early efforts to understand frequency upshifts of high power lasers ($I > 10^{11}$ W/cm^2) due to an interaction with a time dependent plasma density were made by Yablonovitch [8] and Bloembergen [9] in the early 1970's. Yablonovitch began by carrying out some experimental studies with a CO_2 laser ($I= 10^{12}$ w/cm^2, r=9μm, and a pulse length of 100 nanoseconds) which he focused into 1 atm. of nitrogen gas. He observed a significant transfer of energy from the Stokes to the anti-Stokes side of the frequency spectrum for the pulse. He proposed a model whereby the frequency upshift was attributed to the generation of plasma in the focal region of the laser pulse. Using a simple Drude electron model for the plasma, he found that plasma generation caused a phase shift in the laser pulse given by

$$\phi = -\Delta n(t)z\omega/c \qquad (10)$$

where $\Delta n(t)$ is the change in complex index of refraction, which is proportional to the plasma density as a function of time, z is the distance the beam propagates and ω is the laser frequency. Alfano and Shapiro [10] observed filamentation and self-phase modulation of intense short pulse lasers focused into glass and liquids. (Recently Corkum [11] has also done work on this "supercontinuum generation" in gases.) Bloembergen [9] put forth a theoretical explaination for the anti- Stokes broadening of these pulses by including plasma formation as a possible upshift mechanism. However, it was not until the advent of subpicosecond lasers with the necessary pulse length and power that definite frequency upshifts were observed. Recently, Downer *et al.* [12] has re-examined this upshifting mechanism with the following set - up. A 1 millijoule, 90 femtosecond, $\lambda = 626$ nm laser pulse is sharply focussed into a gas cell with a pressure of approximately 1 atmosphere (this can be varied). At some point, the intensity increases to a point ($\approx 10^{16}$ W/cm^2) where breakdown occurs, thus creating a plasma in the cell. He has found that not only is the anti-Stokes side of the pulse increased in energy, relative to the Stokes side, but that the actual peak of the laser pulse is blue shifted. In fact, they find that the entire central wavelength of the resulting pulse has been blue shifted, sometimes by as much as 11 nm (615 nm). In order to explain this upshift, he uses the Drude model discussed above, and obtains

$$\Delta\lambda = \frac{\lambda^3 e^2}{2\pi mc^3}\frac{d}{dt}\int^L N(x)dx \tag{11}$$

where $N(x)$ is the plasma density along the propagation direction x, and the interaction length l is approximately the confocal beam parameter. Reasonable agreement is found if certain assumptions about the plasma are assumed. These experiments clearly demonstrate the basic idea that an electromagnetic wave traveling through a time-dependent dielectric medium will suffer a frequency change. However, as the same pulse that does the ionizing is also being upshifted, things become complicated when one begins to compare to theory.

IONIZATION FRONT METHODS

This method, studied by Semenova [13], Borisov [14], Lampe *et al.* [15] and more recently by Mori [16], is being experimentally investigated by Joshi and co-workers at UCLA. This experiment provides a concrete example of how the ionization front upshifting method works. An ionization front is produced, in a region of waveguide, where a propagating wave exists in the cavity. The cavity is initially filled with a gas, which has an effective index of refraction of 1, and a standing electromagnetic wave of frequency ω_i. The standing wave can also be represented as the superpostion of two wave, each propagating in opposite directions. An ionizing laser pulse (possibly a high power, 1 μm laser) is then sent into one end of the cavity, creating an ionization front moving at slightly less than the speed of light.

There are two cases of interest. The first is the overdense case, where the plasma is overdense to the radiation in the frame moving with the front. The case where the plasma is opaque to the radiation in the frame of the front is called the underdense case. For the overdense case, Lampe *et al.* [15] have found that the plasma acts like a mirror and one simply obtains the doubly Doppler shifted frequency

$$\omega_f = \omega_i \frac{1 + v_f/c}{1 - v_f/c} = 4\gamma_0{}^2 \omega_i \tag{12}$$

where v_f is the velocity of the front. They also find that the reflected pulse is compressed. For the underdense case, by matching the fields in the plasma and in vacuum at the ionization front (as well as including the constraint that the current vanish on the front boundary,) Mori [16] has found that the upshifted wave has a frequency given by

$$\omega_f = \omega_i \left(1 + \frac{1}{4} \frac{\omega_p^2}{\omega_i^2} \right) \tag{13}$$

This upshift is also accompanied by a pulse compression. Mori [17] and Sprangle [18] have proposed that this mechanism be studied as a possible source of short (sub-femtosecond) soft x-ray pulses. It is also found that the ionization front produces a static magnetic field at $2k_0$ that is trapped in the plasma that was created behind the front. This will be discussed in more detail in the following section. In this case, the stationary magnetic field is given by

$$B(x) = E_i 2\beta_0 \frac{1 - \sqrt{\epsilon_f}}{1 - \beta_0 \sqrt{\epsilon_f}} \tag{14}$$

where E_i is the initial amplitude of the EM wave, $\beta_0 = v_0/c$, and $\epsilon = (1 - \omega_p^2/\omega_f^2)$. This equation is valid for both the overdense and underdense cases.

We will now consider a related frequency upshifting mechanism known as the flash ionization method, which can actually be thought of as an ionization front with a front velocity of infinity.

FLASH IONIZATION METHOD

We will now discuss the effects of quickly creating a plasma around a monochromatic electromagnetic source wave, on time-scales on the order of a cycle of the wave. It is found that this results in an upshifting of the wave frequency, which can be varied by changing the plasma density. In fact, if the density of the gas is chosen so that the plasma density is larger than the critical density of the source wave, a large fraction of the wave energy is converted to a wave with frequency above the plasma frequency. It is also found that a substantial fraction of the B-field associated with the initial wave can be frozen in the plasma as a time independent B-field. For fields generated at the focus of a powerful CO_2 laser, these fields can be in the megagauss range, making them attractive as wigglers for FELs or other related applications. Computer simulations have been used to study this process in detail, including the effects of finite ionization time.

Recent advances in high power lasers capable of producing 10-100 femtoseconds long pulses [19] of photons with energies of between 2 and 4 eV have made possible the ionization of small (1 mm^3 $\sim$ 10 cm^3) regions of gas in a time on the order of the pulse duration. We will consider what happens to an EM wave, which we will refer to as the source wave (which might be a 1 μm laser beam, for example) propagating through a gas which is ionized during a single cycle of this wave. It is predicted from linear theory that the frequency of the resultant radiation is upshifted past the plasma frequency and that a fraction of the original wave energy remains in the plasma as a steady state magnetic field. These predictions, which are based on the ideal case of instantaneous plasma creation, are found to agree with results given by 1 and 2D particle-in-cell (PIC) computer simulations. We also use the simulations to investigate how the quality of the resulting radiation is altered when the gas is slowly ionized over a number of cycles (1 to 10) of the source wave.

The linear analysis begins with a purely right-going laser pulse, propagating in the as yet unionized gas. Experimentally, it will be required that the source laser power be sufficiently low such that this laser does not ionize the gas; the gas at this point has an index of refraction of $\sim$ 1, for a laser of this frequency. As a first approximation, the source laser is represented by plane polarized wave solutions to Maxwell's equations in vacuum:

$$\vec{E} = E_0 \cos(k_0 x - \omega_0 t)\hat{e}_y \tag{15}$$

$$\vec{B} = B_0 \cos(k_0 x - \omega_0 t)\hat{e}_z \tag{16}$$

where $E_0 = B_0$ and $\omega_0 = k_0 c = 2\pi\nu_0$ is the frequency of the source laser. Suddenly, in a time interval short compared to ν_0^{-1}, we create a plasma around a portion of this laser pulse at t=0. We now consider what effect the introduction of this plasma has on the frequency spectrum of the pulse. At time t=0, inside the plasma the fields have the form of Eqs. (15) and (16) evaluated at t=0. At subsequent times, the fields still have the same spatial periodicity. However, the field now evolves in time with a final frequency ω_f, given by the dispersion relation [20],[21]

$$\omega_f{}^2 = k_0^2 c^2 + \omega_p{}^2. \tag{17}$$

Therefore the upshifted frequency can be given as

$$\omega_f = \omega_i \sqrt{1 + \frac{\omega_p^2}{\omega_i^2}}. \tag{18}$$

Note that although this is true for any plasma density, a large up-shift in frequency is possible if the density is chosen to be greater than the critical density, $n_{crit} = \omega_0^2 m/(4\pi e^2)$. What is unusual about this situation is that electromagnetic waves of the original frequency do not ordinarily exist inside an overdense plasma. The more typical case considered is that of a vacuum-plasma boundary; and if a laser is fired from the vacuum at the overdense plasma, it penetrates only a skin depth ($\sim c/\omega_p$) and is reflected back into vacuum. This arises from the fact that the discontinuity in space (the vacuum-plasma interface) keeps ω fixed (time independence of the medium) but allows for a change in k and k is imaginary in an overdense plasma. However, for the case of flash ionization, we are now creating a discontinuity in time, which means that ω can change, but the wavelength must remain the same before and after the ionization takes place. Therefore, the frequency of the wave shifts to satisfy Eq. (17), keeping k at it's original value, k_0. Mathematically, the initial value problem allows these solutions inside the plasma, whereas the boundary value problem does not.

A further difference from the standard vacuum-plasma interface solution can be found from the wave equation for the B-field. In addition to the two solutions (the right- and left-going waves) implied in Eq. (17), there exists a solution for which $\omega = 0$ and $k = k_0$. This additional solution is required because introducing the plasma adds an additional degree of freedom (the motion of the electrons) and the additional wave amplitude is required to solve the initial value problem with given $\vec{E}, \vec{B}$, and $\vec{v}$. This stationary, sinusoidally-varying magnetic field remains in the plasma, even after the upshifted light has been radiated out of the plasma. All three solutions are shown in the $\omega - k$ diagram in Fig. 4. At $t < 0$, there is only the source wave at $\omega = \omega_0$ and $k = k_0$. After the plasma has been created, this solution (lying on the light cone) is no longer allowed. However, solutions on the parabolic curves (i.e., the electromagnetic dispersion curves for a light wave in a plasma) are allowed, as are zero frequency modes. In the figure, the "jump" to these solutions is vertical because the change in the dielectric constant, ϵ, due to the sudden creation of the plasma is a change in time, and not space. Therefore, the wave jumps to the dots shown on the curves. Finally, when the light exits the plasma, it now jumps horizontally, because of the discontinuity in space; i.e., the vacuum – plasma interface. These solutions can be understood from the electrons response as follows. When the electrons are suddenly set free and are accelerated, they radiate in all directions such that the initially right-going wave is now broken into both right and left-going components. In addition, by following the electric field of the initial wave, the electrons immediately create a transverse current in the plasma, allowing for the static B-field, or zero-frequency, solution. In fact, it is easy to see why the wavelength of the static B-field is the original laser wavelength from this argument. It is found [21] that this stationary B-field is given as

$$B_s \cos k_0 x = \frac{\omega_p^2 E_0}{\omega_p{}^2 + \omega_0{}^2} \cos k_0 x. \tag{19}$$

Note that as the plasma density is increased, such that $\omega_p \gg \omega_0$, the amplitude of the field approaches the value of the original field, which can be on the order of a megagauss at the focal spot of a CO_2 laser with an intensity of 2.5×10^{14} W/cm^2. For this laser, the magnetic field will vary sinusiodally in space with a wavelength of $10 \mu m$. This field may have applications as an undulator; in fact, tapering could easily be achieved by varying the neutral pressure.

In order to test this model, we have performed a series of computer simulations using the plasma code WAVE [7]. For the case of instantaneous turn-on (top of Fig. 5, $\Delta t = 0$), the frequency and amplitude of the output radiation agree with what the above model predicts. An important question that can now be asked is this: How is this phenomenon affected when finite ionization time is considered? Although this new problem is analytically solvable in certain limits, the simulations have allowed us to determine what the effects of slowly creating the plasma, over periods of 1 to 10 cycles of the incident wave, are on the quality of the resulting radiation. Fig. 5 compares the power spectrum of the ideal, instantaneously-created plasma case with two other cases where the density was ramped linearly over times of 1 and 10 ν_0^{-1}. As might be expected, increasing the ionization time still results in upshifted light albeit at progressively lower powers. However, with slow ionization some radiation leaves the plasma as the frequency (and plasma density) is increasing. This accounts for much of the broad, flat spectrum. A non-linear treatment must still be done to completely explain the rich spectrum seen for finite ionization times.

One possible application of this method is to imagine upshifting a laser pulse by at least a factor of 2. To show that creation of plasmas with densities near $2n_{crit}$ for CO_2 in tens of femtoseconds may be possible with existing technology, consider that to create a xenon plasma of density 2×10^{19} cm^{-3} in a volume of dimensions $100\mu m \times 50\mu m \times 50\mu m$ with a 0.3μm laser requires only about 10^{-2} milliJoules of energy in the ionizing pulse. Here we have taken multiphoton ionization to be the mechanism responsible for plasma creation, and have assumed that complete ionization takes place only over the length of the pulse (i.e., no cascading). Currently, XeCl excimer lasers are capable of 3.5 mJ in 100 femtoseconds (about 3 oscillations of 10μ radiation) which, when focused into a $50\mu m^2$ area gives an intensity of 10^{15} W/cm^2. Such a pulse both exceeds the multiphoton ionization threshold and contains enough energy to create a sizable region of plasma adaquate for producing a signifigant amount of upshifted radiation. One possible senario would be to send a single ionizing laser perpendicular to the source wave. In this case, an ionization front propagates across the source wave. Preliminary 2-D simulations show that the only modification to our analytical results is that the wave fronts become curved. Our studies also show that with two or more time-tailored laser pulses, it is possible to (approximately) instantaneously ionize a finite region of gas without violating causality.

To conclude this section, we have studied a possible method of frequency upshifting EM waves, provided one can quickly create a plasma with a higher power laser provided that the ionization time is short compared to a cycle of the original EM wave to be uposhifted. This upshifted light is also tunable in the sense that by varying the plasma density, one is actually varying the output frequency given in Eq. (17). In addition, the model discussed above also predicts a residual, static magnetic field, given by Eq. (19). 1 and 2-D computer simulations of this phenomena show all of these effects to occur as predicted from linear theory. In addition, we have investigated the effects of finite ionization time, and nonlinear longitudinal and transverse motion, with the help of computer modeling. Although we have commented on only two applications of this phenomenon (conversion of a EM wave of a given frequency to a new tunable, upshifted frequency and the use of the trapped, static magnetic field as the wiggler field for a free electron laser production of kilovolt plasmas) other possible applications are currently being investigated. For instance, by embedding the wave in a gas of nonuniform pressure before ionization, any desired frequency time history for the radiation can be generated.

ACKNOWLEDGEMENTS

Work performed under the auspices of the United States Department of Energy by the Lawrence Livermore National Laboratory under contract number W-7405-ENG-48.

REFERENCES

[1] T. Tajima and J.M. Dawson, Phys. Rev. Lett. **43**, 267 (1979).

[2] R. Ruth, A. Chao, P. Morton, and P. Wilson, Particle Accelerators, **17**, 171 (1985); R. Keinigs and M. E. Jones, Phys. Fluids **30**, 252 (1987).

[3] "Laser Acceleration of Particles", Eds. C. Joshi and T.Katsouleas, AIP Conf. Proc. No. 130, Am. Inst. of Phys., New York, 1985.

[4] C. Joshi *et al.*, *Nature*(London), **311**, 525 (1984).

[5] S. Van der Meer, CLIC Note No. 3, CERN/PS/85-65 (AA) (Geneva, Switzerland, November, 1985); T. Katsouleas, *et al.*, Particle Accelerators, **22**, 81 (1987); S. Wilks, *et al.*, IEEE Transactions on Plasma Science, vol. PS-15, no. 2, April, 1987.

[6] T. Katsouleas, Phys. Rev. A, **33**, 4412, (1986).

[7] R.L. Morse and C.W. Nielson, it Physics of Fluids, **14**, 830, (1971).

[8] E. Yablonovitch, *Phys. Rev. A* **10**, 1888 (1974).

[9] N. Bloembergen, Opt. Comm. **8**, 285 (1973).

[10] R.R. Alfano and S.L. Shapiro, Phys. Rev. Lett. **24**, 592 (1970).

[11] P.B. Corkum, C. Rolland, and T. Srinivasan-Rao, Phys. Rev. Lett. **57**, 2268 (1986).

[12] M.C. Downer, G. Focht, D.H. Reitze, W.M. Wood, and T.R. Zhang, in Springer Series in Chemical Physics, **48**, *Ultrafast Phenomena VI*, Eds. T. Yajima, *et.al.*, Springer - Verlag, Berlin, p. 128 (1988); and also W.M. Wood, G. Focht, and M.C. Downer, Optics Letters, **13**, 984 (1988).

[13] V.I. Semenova, Izv. V.U.V. Radiofiz. **10**, 1077 (1967); **15**, 665, 1793 (1972) [Sov. Radiophys. **10**, 599 (1967); **15**, 505, (1972)].

[14] V.V. Borisov, Izv. V.U.V. Radiofiz. **13**, 1376 (1970) [Sov. Radiophys. **13**, 1059, (1970)].

[15] M. Lampe, E. Ott, and J.H. Walker, Phys. Fluids **21**, 42 (1978).

[16] W.B. Mori, submitted to Phys. Rev. Lett.

[17] W.B. Mori, presented at 19th Annual Anomalous Conference, Durango, Colorado, June 1989; Bulletin of the APS **34**, (9), 2069 (1989).

[18] P. Sprangle, E. Esarey and A. Ting presented at 13th Conference on the Numerical Simulation of Plasmas, Santa Fe, New Mexico, Sept. 17-20 (1989).

[19] P.B. Corkum, *IEEE Journal of Quantum Electronics*, **QE-21**, No. 3, 216 (1985); A. Migus, A. Antonetti, J. Etchepare, D. Hulin, and A. Orszag, J. Opt. Soc. Am. B, Vol. 2 No. 4, (1984), and references therein.

[20] D.K. Kalluri, *IEEE Trans. Plasma Sci.* **16**, 11 (1988).

[21] S.C. Wilks, J.M. Dawson, and W.B. Mori, Phys. Rev. Lett. **61**, 337 (1988).

FIGURE CAPTIONS

FIG. 1. Superposition of wakes from two identical laser pulses spaced $1\frac{1}{2}$ plasma wavelengths apart, shown at t=15.0 ω_p^{-1}. (a) Electric field components of the laser pulses. (b) Longitudinal electric field showing partial absorbtion of plasma wave energy by the second pulse. Both laser pulses have frequencies of $\omega = 6\omega_p$ and were injected (with a delay between pulses) into a homogeneous plasma from the left hand side of the simulation box.

FIG.2. Amount of upshift possible for two different types of lasers,(KrF and Nd:glass,) assuming appropriately ramped densities.

FIG. 3. Frequency spectrum of laser packet showing frequency upshift. The initial pulse had a frequency of $\omega_i = 18\omega_p$ (solid line), and the final pulse (dotted line) has a frequency of $\omega_f \approx 19.8\omega_p$.

FIG. 4. Dispersion relation for electromagnetic waves in an instantaneously created overdense plasma. Note that the discontinuity in time (caused by ionizing the plasma) not only converts the original wave (ω_0, k_0) into a left- and right-going wave $(\pm\omega_f, k_0)$, but also allows for a zero frequency solution which corresponds to the trapped B-field $(0, -k_0)$.

FIG. 5. Power spectrum for radiation leaving the right and left boundaries of the plasma for increasing ionization times, Δt. The plasma (20 source wavelengths long) was created instantaneously in (a); over a time interval ν_0^{-1} for (b); and over a time interval $10\nu_0^{-1}$ in (c). Plasma density is $2n_{crit}$ for each run; thus, the peak frequency has been shifted up according to Eq. (17). The insets show the actual field that was transformed to give the spectra.

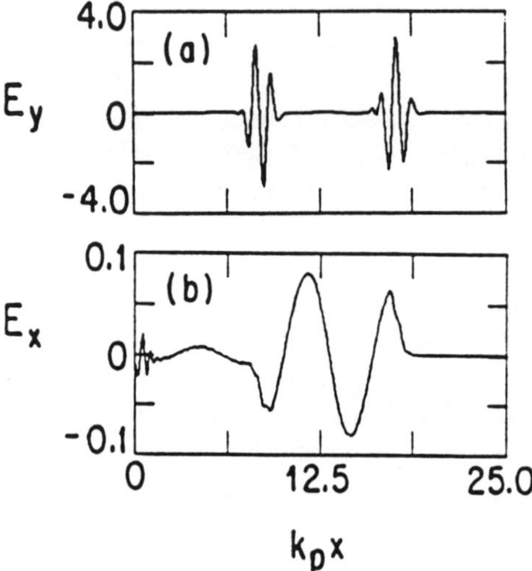

FIG. 1.

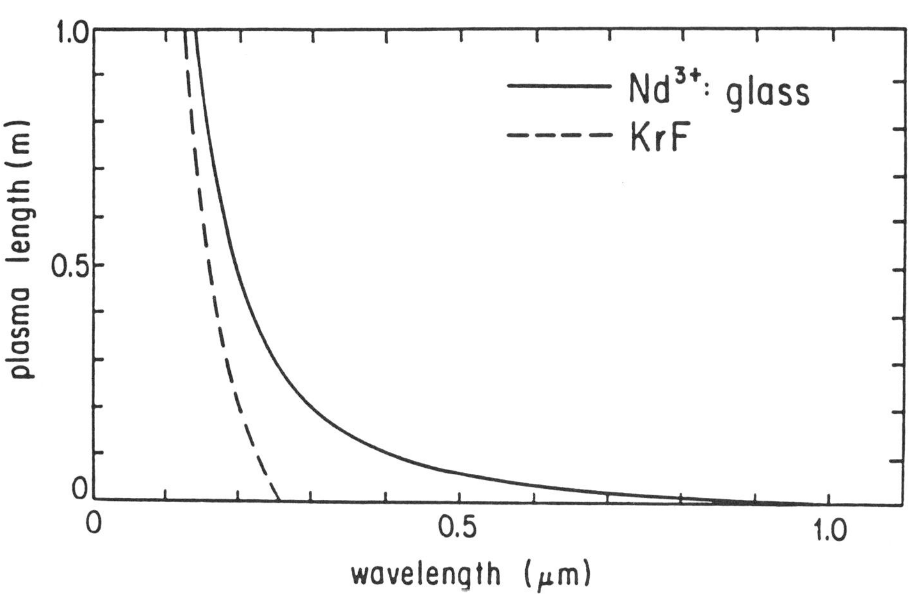

FIG.2.

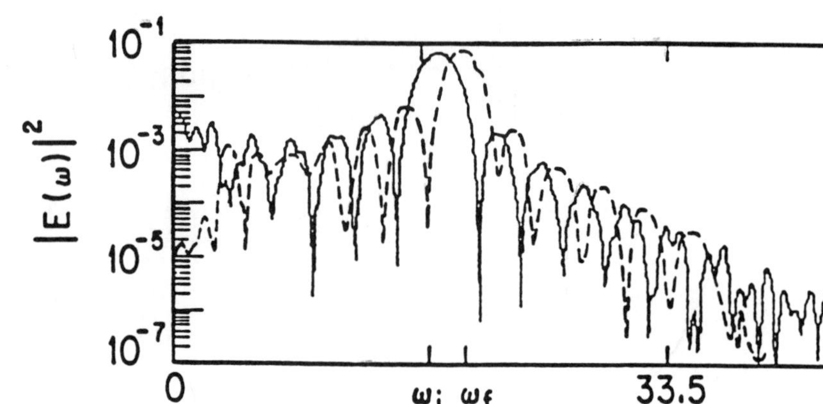

FIG. 3.

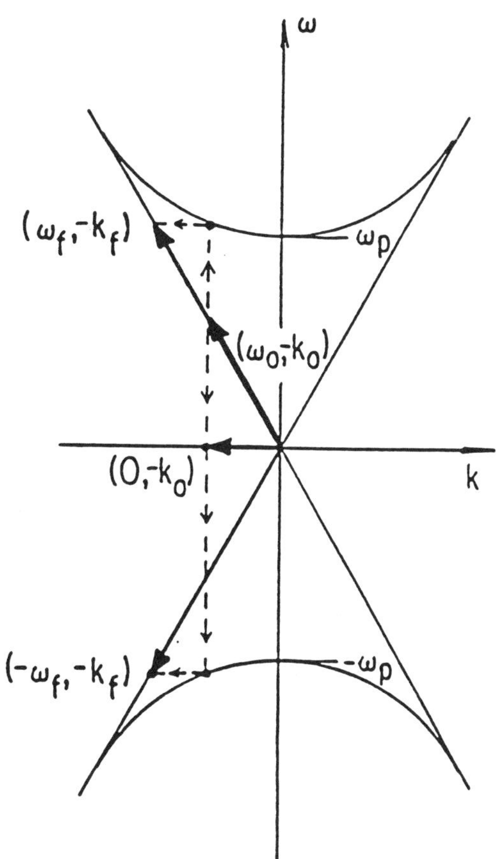

FIG. 4.

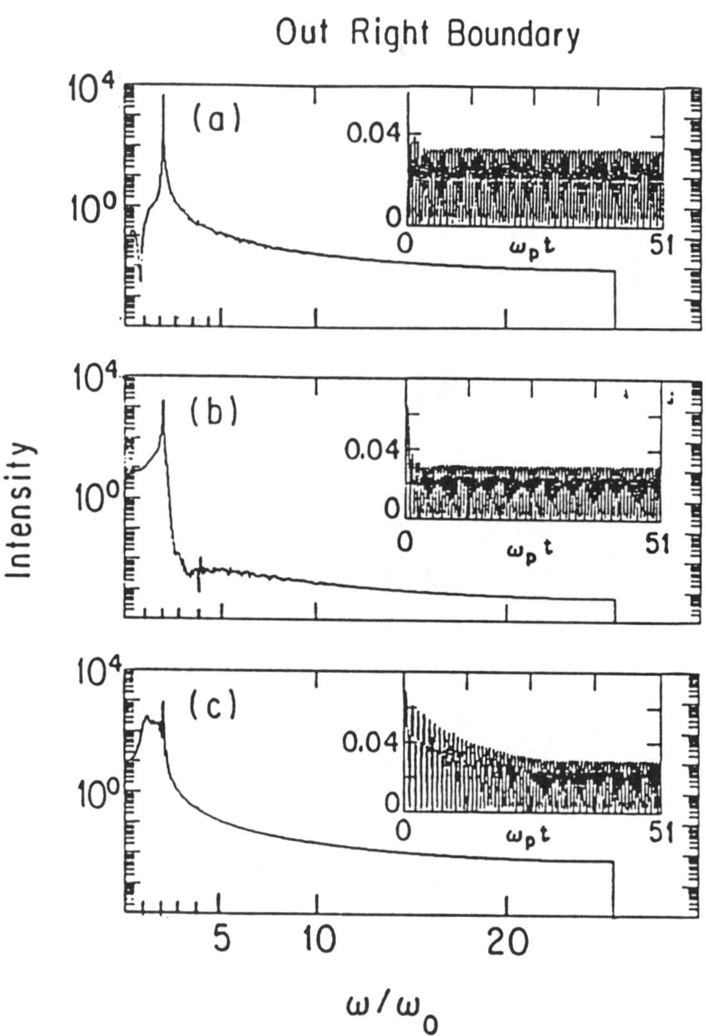

FIG. 5.

Part V

NONNEUTRAL PLASMAS

LARGE-AMPLITUDE COHERENT STRUCTURES
IN NONNEUTRAL PLASMAS WITH CIRCULATING ELECTRON FLOW[†]

Ronald C. Davidson, Hei-Wai Chan, Chiping Chen and Steven Lund
Plasma Fusion Center
Massachusetts Institute of Technology
Cambridge, Massachusetts 02139

ABSTRACT

The nonlinear dynamics of nonneutral plasmas with circulating electron flow is often characterized by large-amplitude coherent structures. Use is made of a cold-fluid guiding-center model to investigate the properties of stationary, two-dimensional, large-amplitude vortex structures in a low-density ($s_e = \omega_{pe}^2/\omega_{ce}^2 \ll 1$) nonneutral plasma column confined by an axial magnetic field $B_0\hat{e}_z$. In addition, particle-in-cell computer simulation studies are presented which describe the nonlinear evolution of a high-density ($s_e = \omega_{pe}^2/\omega_{ce}^2 \sim 1$) nonneutral electron layer in a relativistic cylindrical magnetron, including the formation of a large-amplitude "spoke" structure in the circulating electron density.

I.   INTRODUCTION

The formation and evolution of large-amplitude coherent structures play an important role in describing the nonlinear dynamics of nonneutral plasmas[1] with circulating electron flow. This is true in systems ranging from low-density ($s_e = \omega_{pe}^2/\omega_{ce}^2 \ll 1$) rotating nonneutral plasmas initially subject to the diocotron instability, to high-density ($s_e = \omega_{pe}^2/\omega_{ce}^2 \sim 1$) circulating nonneutral electron layers in conventional and relativistic magnetrons. Here, ω_{pe} is the electron plasma frequency, ω_{ce} is the electron cyclotron frequency associated with the confining magnetic field, and $s_e = \omega_{pe}^2/\omega_{ce}^2$ is a measure of the self-field intensity.

This paper makes use of a cold-fluid guiding-center model to investigate the properties of stationary, two-dimensional,

---

[†]Research supported in part by the Office of Naval Research, the Department of Energy High Energy Physics Division, and the Naval Research Laboratory Plasma Physics Division.

large-amplitude vortex structures in a low-density nonneutral plasma column (Sec. II). In addition, particle-in-cell computer simulations are presented which describe the nonlinear evolution of a high-density nonneutral electron layer in a relativistic cylindrical magnetron, including the formation of a large-amplitude "spoke" structure in the circulating electron density (Sec. III).

By way of background, one of the most ubiquitous properties of low-density nonneutral plasma initially subject to the diocotron instability[2-12] is the development of long-lived, rotating vortex structures[13] during the nonlinear evolution of the system. This has been observed experimentally in annular electron layers,[14,15] intense propagating annular electron beams,[16] nonneutral plasma columns with[17,18] and without[19-21] central conductors, and in computer simulation studies.[18] The fact that long-lived coherent structures exist in these systems for many rotation periods and thousands of cyclotron periods suggests the existence of large-amplitude solutions which are stationary in the rotating frame. In Sec. II, use is made of a cold-fluid guiding-center model to investigate the properties of large-amplitude stationary vortex structures in a low-density nonneutral plasma column with $s_e = \omega_{pe}^2/\omega_{ce}^2 \ll 1$.

In contrast, magnetrons operate with dense nonneutral electron layers characterized by $s_e = \omega_{pe}^2/\omega_{ce}^2 \sim 1$. In relativistic magnetrons,[22-30] pulsed high-voltage diodes are used to generate microwaves at gigawatt power levels. Although magnetrons are widely used as microwave sources, a fundamental understanding of the underlying interaction physics is still being developed,[31] particularly in the nonlinear regime. Part of the theoretical challenge is associated with the fact that the electrons emitted from the cathode interact with the electromagnetic waves excited in the anode-cathode gap in a highly nonlinear way. This is manifest through a strong azimuthal bunching of the electrons and the formation of a large-amplitude spoke structure in the circulating electron density. In this regard, computer simulation studies[32-35] provide a particularly valuable approach to analyze the interaction physics and nonlinear electrodynamics in magnetrons. In Sec. II, we summarize recent

computer simulations[35] of the multiresonator, cylindrical A6 magnetron configuration[26] using the two-dimensional particle-in-cell
code MAGIC,[36] including the formation of a coherent, large-amplitude
spoke structure in the circulating electron density.

II.   LARGE-AMPLITUDE VORTEX STRUCTURES
IN LOW-DENSITY NONNEUTRAL PLASMA

In this section, use is made of a cold-fluid guiding-center
model (Sec. II.A) to investigate the properties of stationary
coherent structures in a low-density ($s_e = \omega_{pe}^2/\omega_{ce}^2 \ll 1$) nonneutral
plasma column (Sec. II.B). Particular examples of large-amplitude
$\ell = 1$ and $\ell = 2$ vortex structures are then presented (Sec. II.C).

A.   Nonrelativistic Guiding-Center Model

We consider a low-density nonneutral electron plasma in cylindrical geometry with

$$\frac{\omega_{pe}^2}{\omega_{ce}^2} = \frac{4\pi\, n_e(\underset{\sim}{x},t)m_e c^2}{B_0^2} \ll 1 \; . \tag{1}$$

Here, m_e is the electron mass, c is the speed of light <u>in vacuo</u>, and
$n_e(\underset{\sim}{x},t)$ is the electron density. The electrons are confined radially by a uniform axial magnetic field $B_0\hat{\underset{\sim}{e}}_z$, and cylindrical conducting walls are located at $r = a$ and $r = b$ (Fig. 1). The case in
which the central conductor is absent is treated by setting $a = 0$.
In the present analysis, a cold-fluid guiding-center model[2] is
adopted in which electron inertial effects are neglected ($m_e \to 0$),
and the motion of a strongly magnetized electron fluid element is
determined from

$$0 = -en_e(\underset{\sim}{x},t)\left[\underset{\sim}{E}(\underset{\sim}{x},t) + \frac{1}{c}\,\underset{\sim}{V}_e(\underset{\sim}{x},t) \times B_0\hat{\underset{\sim}{e}}_z\right] \, , \tag{2}$$

where $-e$ is the electron charge, and $\underset{\sim}{V}_e(\underset{\sim}{x},t)$ is the average flow
velocity. In the electrostatic approximation, the electric field
can be expressed as $\underset{\sim}{E}(\underset{\sim}{x},t) = -\nabla\phi(\underset{\sim}{x},t)$. Therefore, Eq.(2) gives

$$\underset{\sim}{V}_e(x,t) = -\frac{c}{B_0}\,\nabla\phi(\underset{\sim}{x},t) \times \hat{\underset{\sim}{e}}_z \tag{3}$$

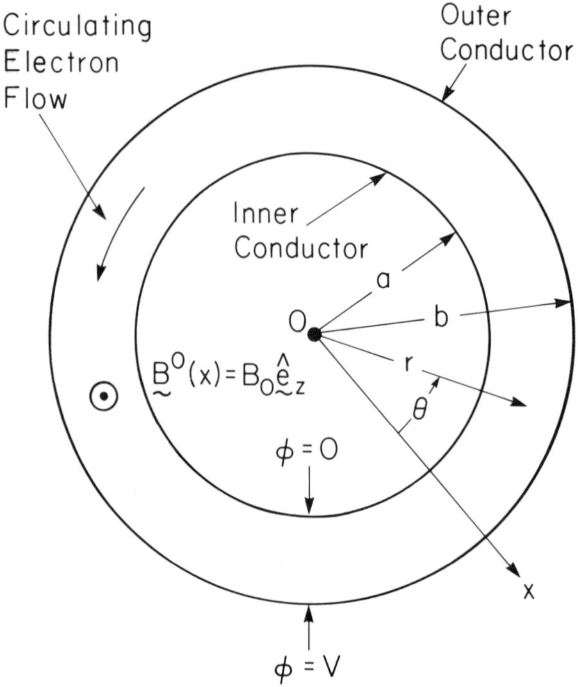

Fig. 1. A low-density nonneutral electron plasma is immersed in a strong axial magnetic field $B_0\hat{e}_z$ between two cylindrical conductors at $r = a$ and $r = b$. In the electrostatic approximation, the perpendicular flow velocity is $\underset{\sim}{V}_e = -(c/B_0)\nabla\phi(\underset{\sim}{x},t) \times \hat{e}_z$.

for the perpendicular fluid motion. In cylindrical geometry, Eq.(3) reduces to

$$V_{re}(r,\theta,t) = -\frac{c}{B_0 r}\frac{\partial}{\partial\theta}\phi(r,\theta,t) , \qquad (4)$$

$$V_{\theta e}(r,\theta,t) = \frac{c}{B_0}\frac{\partial}{\partial r}\phi(r,\theta,t) , \qquad (5)$$

where $\partial/\partial z = 0$ is assumed. Because $\nabla \cdot \underset{\sim}{V}_e = 0$ follows from Eq.(3), the continuity equation can be expressed as

$$\left(\frac{\partial}{\partial t} - \frac{c}{B_0 r}\frac{\partial\phi}{\partial\theta}\frac{\partial}{\partial r} + \frac{c}{B_0 r}\frac{\partial\phi}{\partial r}\frac{\partial}{\partial\theta}\right)n_e(r,\theta,t) = 0 .$$

(6)

Of course, Eq.(6) must be supplemented by Poisson's equation

$$\left(\frac{1}{r}\frac{\partial}{\partial r}r\frac{\partial}{\partial r} + \frac{1}{r^2}\frac{\partial^2}{\partial\theta^2}\right)\phi(r,\theta,t) = 4\pi en_e(r,\theta,t) ,$$

(7)

which relates self-consistently the electrostatic potential $\phi(r,\theta,t)$ to the electron density $n_e(r,\theta,t)$.

Equations (6) and (7) constitute a fully nonlinear description of the two-dimensional evolution of the system in the cold-fluid guiding-center approximation with $m_e \to 0$. Although the ratio $\omega_{pe}^2/\omega_{ce}^2 = 4\pi n_e m_e c^2/B_0^2$ approaches zero in the limit of zero electron mass, note that the effective diocotron frequency defined by $\omega_D = \omega_{pe}^2/\omega_{ce} = 4\pi n_e ec/B_0$ remains finite as $m_e \to 0$. Assuming that the cylinders at $r = a$ and $r = b$ in Fig. 1 are perfect conductors, it is required that

$$E_\theta(r,\theta,t) = -\frac{1}{r}\frac{\partial}{\partial\theta}\phi(r,\theta,t) = 0, \text{ at } r = a \text{ and } r = b ,$$

(8)

which corresponds to zero tangential electric field at the conducting walls. Making use of $V_{re}(r,\theta,t) = -(c/B_0 r)(\partial/\partial\theta)\phi(r,\theta,t)$, it follows from Eqs.(4) and (8) that

$$V_{re}(r,\theta,t) = 0, \text{ at } r = a \text{ and } r = b ,$$

(9)

which corresponds to zero radial flow of the electron fluid at $r = a$ and $r = b$. The nonlinear equations (6) and (7) possess certain global conservation constraints[2,5] which provide important insights regarding the nonlinear evolution of the system. In particular, it is convenient to introduce the density-weighted mean-square radius of guiding-center locations defined by

$$U_r = \int_a^b drr\int_0^{2\pi} d\theta\, r^2 n_e(r,\theta,t) ,$$

(10)

and the generalized entropy defined by

$$U_G = \int_a^b dr\, r \int_0^{2\pi} d\theta\, G(n_e) \ . \tag{11}$$

Here, $G(n_e)$ is a smooth, differentiable function with $G(n_e \to 0) = 0$. Making use of Eqs.(6) and (7) and the boundary conditions in Eq.(8), it is readily shown that

$$\frac{d}{dt} U_r = 0 \ , \tag{12}$$

and

$$\frac{d}{dt} U_G = 0 \ . \tag{13}$$

That is, U_r = const. and U_G = const. are globally conserved quantities no matter how complicated the nonlinear evolution of the system described by Eqs.(6) and (7). In this regard, note that U_r = const. is a statement of the conservation of canonical angular momentum, $\int d^2x(m_e rV_{\theta e} - eB_0 r^2/2c)n_e$ = const., in the limit of zero electron mass ($m_e \to 0$).

Not only are Eqs.(12) and (13) useful conservation relations for describing the nonlinear evolution of the system, these constraint conditions can also be used to derive a sufficient condition for azimuthally symmetric equilibria $n_e^0(r)$ to be stable to small-amplitude perturbations $\delta n_e(r,\theta,t)$. In particular, for mono-tonically decreasing profiles with

$$\frac{1}{r}\frac{\partial}{\partial r} n_e^0(r) \leq 0 \ , \quad \text{for } a \leq r \leq b \ , \tag{14}$$

it can be shown that the density perturbation $\delta n_e(r,\theta,t)$ cannot grow without bound, and the system is linearly stable.[2,5,6] That is, Eq.(14) is a <u>sufficient condition for stability</u> to small-amplitude perturbations in the context of the cold-fluid guiding-center model based on Eqs.(6) and (7). Therefore, a <u>necessary condition for instability</u> is that the density profile $n_e^0(r)$ have a maximum at some radius $r = r_M$ intermediate between $r = a$ and $r = b$. An example of a profile subject to the diocotron instability is a sufficiently thin

annular electron layer[2,3,10] in which the inner and outer radii of
the layer (at $r = r_b^-$ and $r = r_b^+$, say) are not in contact with the
conductors at $r = a$ and $r = b$. A smooth density profile $n_e^0(r)$ with
a sufficiently large density depression $n_e^0(r = a)/n_e^0(r = r_M) < 1$ is
also subject to the diocotron instability.[4]

B. Nonlinear Stationary Structures in the Rotating Frame

A thorough review of the linear properties of the diocotron
instability is presented in Chapter 6 of Ref.1 and will not be
repeated here. Rather, we focus on the application of Eqs.(6) and
(7) to describe large-amplitude rotating structures in nonneutral
plasma. As indicated earlier, one of the most ubiquitous properties
of nonneutral plasma initially subject to the diocotron instability
is the development of long-lived vortex structures[13] during the
nonlinear evolution of the system. This has been observed experimentally in annular electron layers,[14,15] intense propagating
annular electron beams,[16] nonneutral plasma columns with[17,18] and
without[19-21] central conductors, and in computer simulation
studies.[18] The fact that long-lived coherent structures exist in
these systems for many rotation periods and thousands of cyclotron
periods suggests that the nonlinear equations (6) and (7) support
large-amplitude solutions which are stationary in the rotating
frame.

To investigate this possibility, we look for solutions to
Eqs.(6) and (7) which depend on θ and t solely through the linear
combination, $\theta - \omega_r t$, where $\omega_r = $ const. is the angular rotation
velocity of the disturbance. In particular, we introduce the
coordinate transformation

$$\theta' = \theta - \omega_r t \ ,$$

$$r' = r \ , \tag{15}$$

$$t' = t \ .$$

For stationary solutions $n_e(r', \theta')$ and $\phi(r', \theta')$ with $\partial/\partial t' = 0$, the
continuity and Poisson equations (6) and (7) can be expressed in the
rotating frame as

$$\left(-\omega_r + \frac{c}{B_0 r'}\frac{\partial\phi}{\partial r'}\right)\frac{\partial n_e}{\partial\theta'} - \frac{c}{B_0 r'}\frac{\partial\phi}{\partial\theta'}\frac{\partial n_e}{\partial r'} = 0 \ , \qquad (16)$$

and

$$\left(\frac{1}{r'}\frac{\partial}{\partial r'} r'\frac{\partial}{\partial r'} + \frac{1}{r'^2}\frac{\partial^2}{\partial\theta'^2}\right)\phi = 4\pi e n_e \ . \qquad (17)$$

It is convenient to introduce the stream function $\psi(r',\theta')$ defined by

$$\psi(r',\theta') = \frac{c}{B_0}\phi(r',\theta') - \frac{1}{2}\omega_r r'^2 \ . \qquad (18)$$

Equations (16) and (17) then become

$$\frac{\partial\psi}{\partial r'}\frac{\partial n_e}{\partial\theta'} - \frac{\partial\psi}{\partial\theta'}\frac{\partial n_e}{\partial r'} = 0 \ , \qquad (19)$$

$$\left(\frac{1}{r'}\frac{\partial}{\partial r'} r'\frac{\partial}{\partial r'} + \frac{1}{r'^2}\frac{\partial^2}{\partial\theta'^2}\right)\psi = \frac{4\pi e c}{B_0} n_e - 2\omega_r \ . \qquad (20)$$

Note that introducing the term $-\omega_r r'^2/2$ in the definition of the stream function $\psi(r',\theta')$ in Eq.(18) is equivalent to reducing the density by a constant amount $(B_0/2\pi e c)\omega_r$ in Poisson's equation (20).

The general stationary solution to the continuity equation (19) in the rotating frame is

$$n_e(r',\theta') = n_e[\psi(r',\theta')] \ , \qquad (21)$$

where $n_e(\psi)$ is a (yet unspecified) function of ψ. Substituting Eq.(21) into Eq.(20) then gives

$$\left(\frac{1}{r'}\frac{\partial}{\partial r'} r'\frac{\partial}{\partial r'} + \frac{1}{r'^2}\frac{\partial^2}{\partial\theta'^2}\right)\psi = \frac{4\pi e c}{B_0} n_e(\psi) - 2\omega_r \ . \qquad (22)$$

It is evident from Eq.(22) that there is considerable latitude in determining stationary solutions that depend on both r' and θ' in the rotating frame. Once the functional form of $n_e(\psi)$ is specified, then Eq.(22) is solved numerically or analytically, as appropriate, for the stream function $\psi(r',\theta) = c\phi(r',\theta)/B_0 - \omega_r r'^2/2$. The boundary conditions consistent with zero tangential electric field at the perfectly conducting walls in Fig. 1 are given by [see Eq.(8)]

$$\left[\frac{\partial}{\partial\theta'}\,\psi(r',\theta)\right]_{r'\,=\,a} = 0 = \left[\frac{\partial}{\partial\theta'}\,\psi(r',\theta')\right]_{r'\,=\,b} . \tag{23}$$

In addition, it is assumed that the inner and outer conductors are maintained at a constant potential difference corresponding to $\phi(r' = a,\theta) = 0$ and $\phi(r' = b,\theta) = V$, or equivalently,

$$\psi(r' = a,\theta') = -\frac{1}{2}\omega_r a^2 ,$$

$$\psi(r' = b,\theta') = \frac{cV}{B_0} - \frac{1}{2}\omega_r b^2 , \tag{24}$$

where V is the potential difference.

C.   Example of Large-Amplitude Vortex Solutions

Depending on the choice of $n_e(\psi)$, there is considerable latitude in determining stationary solutions to Eq.(22) that depend on both r' and θ' in the rotating frame. In this section, we consider a simple example which is analytically tractable and corresponds to a large-amplitude vortex solution. In particular, it is assumed that $n_e(\psi)$ is specified by the linear function

$$n_e(\psi) = \hat{n}_e(C_0 + C_1\psi) , \tag{25}$$

where C_0 and C_1 are constants, and $\hat{n}_e$ = const. is a measure of the characteristic electron density in the interval $a \leq r \leq b$. Introducing the effective diocotron frequency ω_D = const. defined by

$$\omega_D = \frac{4\pi\,\hat{n}_e ec}{B_0} , \tag{26}$$

Poisson's equation (22) becomes

$$\left(\frac{1}{r'} \frac{\partial}{\partial r'} r' \frac{\partial}{\partial r'} + \frac{1}{r'^2} \frac{\partial^2}{\partial \theta'^2} \right) \psi = \omega_D C_0 - 2\omega_r + \omega_D C_1 \psi . \tag{27}$$

The (linear) differential equation (27) can be solved exactly for $\psi(r',\theta')$ subject to the boundary conditions in Eqs.(23) and (24). We examine solutions to Eq.(27) of the form

$$\psi(r',\theta') = \psi_0(r') + a_\ell \psi_\ell(r')\cos(\ell\theta') , \tag{28}$$

where $\ell \neq 0$ is an integer and a_ℓ is a constant amplitude factor. Substituting Eq.(28) into Eq.(27) then gives

$$\left(\frac{1}{r'} \frac{\partial}{\partial r'} r' \frac{\partial}{\partial r'} + k^2 \right) \psi_0(r') = \omega_D C_0 - 2\omega_r , \tag{29}$$

and

$$\left(\frac{1}{r'} \frac{\partial}{\partial r'} r' \frac{\partial}{\partial r'} + k^2 - \frac{\ell^2}{r'^2} \right) \psi_\ell(r') = 0 , \tag{30}$$

where $k^2 \equiv -\omega_D C_1 > 0$ is assumed. The solutions to Eqs.(29) and (30) in the interval $a \leq r \leq b$ are linear combinations of $J_0(kr')$ and $N_0(kr')$, and $J_\ell(kr')$ and $N_\ell(kr')$, respectively. Here, $J_\ell(x)$ is the Bessel function of the first kind of order ℓ, and $N_\ell(x)$ is the Neumann function of order ℓ. For present purposes, we also consider the class of solutions in which the electron density is equal to zero at the outer conductor, i.e., $n_e(r' = b, \theta') = \hat{n}_e[C_0 + C_1\psi(r' = b,\theta')] = 0$. Making use of $\psi(r' = b,\theta') = cV/B_0 - \omega_r b^2/2$ [Eq.(24)], this condition relates the constants C_0 and $C_1 = -k^2/\omega_D$ by

$$C_0 = -C_1 \left(\frac{cV}{B_0} - \frac{1}{2} \omega_r b^2 \right) . \tag{31}$$

We solve Eqs.(29) and (30) for $\psi_0(r')$ and $\psi_\ell(r')$ and substitute into Eq.(28). Enforcing the boundary conditions in Eqs.(23) and (24) then gives the desired solutions for the electrostatic potential $c\phi(r',\theta')/B_0 = \psi(r',\theta') + \omega_r r'^2/2$ and the electron density

$n_e(r',\theta') = n_e[C_0 + C_1\psi(r',\theta')]$. In laboratory-frame variables (r,θ,t), some straightforward algebra gives

$$\frac{c}{B_0}\phi(r,\theta,t) = \left[\frac{cV}{B_0} - \frac{\omega_r}{2}\left(b^2 - r^2 + \frac{4}{k^2}\right)\right]$$

$$- \left[\frac{cV}{B_0} - \frac{\omega_r}{2}\left(b^2 - a^2 + \frac{4}{k^2}\right)\right]\left[\frac{J_0(kb)N_0(kr) - N_0(kb)J_0(kr)}{J_0(kb)N_0(ka) - N_0(kb)J_0(ka)}\right]$$

$$+ \frac{2\omega_r}{k^2}\left[\frac{N_0(ka)J_0(kr) - J_0(ka)N_0(kr)}{N_0(ka)J_0(kb) - J_0(ka)N_0(kb)}\right]$$

$$+ a_\ell\left[J_\ell(kr) - \frac{J_\ell(ka)}{N_\ell(ka)}N_\ell(kr)\right]\cos[\ell(\theta - \omega_r t)] \; ,$$

(32)

and

$$n_e(r,\theta,t) = \hat{n}_e\frac{2\omega_r}{\omega_D}\left[1 - \frac{N_0(ka)J_0(kr) - J_0(ka)N_0(kr)}{N_0(ka)J_0(kb) - J_0(ka)N_0(kb)}\right]$$

$$+ \hat{n}_e\frac{k^2}{\omega_D}\left[\frac{cV}{B_0} - \frac{\omega_r}{2}\left(b^2 - a^2 + \frac{4}{k^2}\right)\right]\left[\frac{J_0(kb)N_0(kr) - N_0(kb)J_0(kr)}{J_0(kb)N_0(ka) - N_0(kb)J_0(ka)}\right]$$

(33)

$$- \hat{n}_c\frac{k^2 a_\ell}{\omega_D}\left[J_\ell(kr) - \frac{J_\ell(ka)}{N_\ell(ka)}N_\ell(kr)\right]\cos[\ell(\theta - \omega_r t)] \; .$$

Here, a_ℓ is the constant amplitude factor, $\ell = 1,2,3,\cdots$ is an integer, and the parameter k is chosen to satisfy

$$J_\ell(kb)N_\ell(ka) - J_\ell(ka)N_\ell(kb) = 0 \; .$$

(34)

From Eqs.(32) and (34), it follows trivially that the electrostatic potential $\phi(r,\theta,t)$ satisfies the boundary conditions $[\phi]_{r=a} = 0$, $[\phi]_{r=b} = V$, $[\partial\phi/\partial\theta]_{r=a} = 0$, and $[\partial\phi/\partial\theta]_{r=b} = 0$, consistent with Eqs.(23) and (24). In addition, from Eqs.(33) and (34), the electron density is equal to zero at the outer conductor in Fig. 1,

i.e., $n_e(r = b, \theta, t) = 0$. On the other hand, evaluating Eq.(33) at
$r = a$ gives the steady value $n_e(r = a, \theta, t) = n_e(a)$, where

$$n_e(a) \equiv \hat{n}_e \, \frac{k^2}{\omega_D} \left[\frac{cV}{B_0} - \frac{\omega_r}{2} (b^2 - a^2) \right] . \tag{35}$$

Evidently, Eqs.(32) and (33) describe structured potential and
density profiles which rotate azimuthally about the z-axis in Fig. 1
with angular velocity ω_r = const. Note that the profiles are
stationary (independent of time) in the rotating frame. Furthermore, for specified integer ℓ, the profiles described by Eqs.(32)
and (33) have azimuthal periodicity $2\pi/\ell$ in the rotating frame. In
particular, insofar as Eqs.(32) and (33) represent a coherent vortex
structure, $\ell = 1$ corresponds to one vortex, $\ell = 2$ corresponds to two
vortices, etc. As a further important point, for the solution in
Eq.(33) to be physically acceptable, it is required that the
electron density profile satisfy

$$n_e(r, \theta, t) \geq 0 \tag{36}$$

in the entire region $a \leq r \leq b$ and $0 \leq \theta \leq 2\pi$ between the conducting
cylinders in Fig. 1. This places restrictions on the allowed values
of the dimensionless parameters kb, ω_r/ω_D and $k^2 cV/\omega_D B_0$, and the
dimensionless amplitude $k^2 a_\ell/\omega_D$ of the oscillatory (θ-dependent)
term in Eq.(33).

Figures 2–5 illustrate properties of the solutions for ϕ and n_e
in Eqs.(32) and (33) for disturbances with azimuthal mode number
$\ell = 1$ (Figs. 2 and 3) and $\ell = 2$ (Figs. 4 and 5). Because the
structures in Eqs.(32) and (33) rotate azimuthally about the z-axis
with angular velocity ω_r = const., the information in Figs. 2–5 is
displayed at time $t = 0$ without loss of generality.

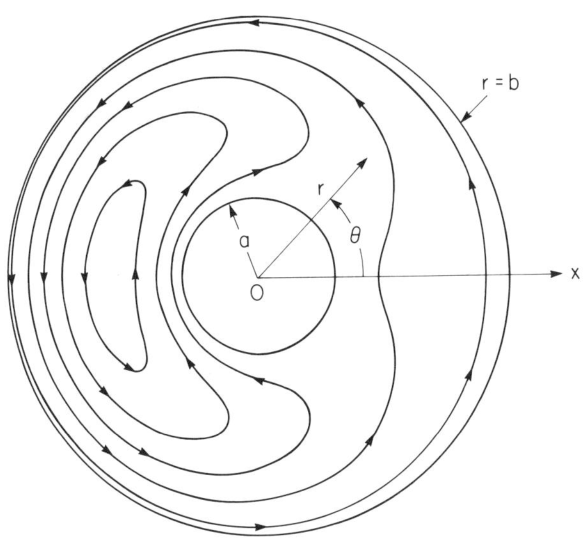

Fig. 2. Plots of the equipotential contours $\phi(r,\theta,t=0)$ = const. calculated from Eq.(32) for $\ell = 1$ and the system parameters in Eq.(37). The arrows indicate the direction of the circulating electron flow $\underset{\sim}{V}_e = -(c/B_0)\nabla\phi \times \hat{\underset{\sim}{e}}_z$.

The choice of system parameters in Figs. 2 and 3 corresponds to $\ell = 1$ and

$$\frac{a}{b} = 0.3 \;,\quad kb = 4.7058 \;,\quad \frac{\omega_r}{\omega_D} = 0.2 \;,$$

$$\frac{k^2 cV}{\omega_D B_0} = 2.2144 \;,\quad \frac{k^2 a_\ell}{\omega_D} = 4.4289 \;. \tag{37}$$

Here, $kb = 4.7058$ is the first zero of Eq.(34) for $\ell = 1$ and $a/b = 0.3$. Figure 2 shows plots of the equipotential contours, $\phi(r,\theta,t=0)$ = const., calculated from Eqs.(32) and (37). From Eqs.(3)–(5), the flow velocity $\underset{\sim}{V}_e = -(c/B_0)\nabla\phi \times \hat{\underset{\sim}{e}}_z$ is tangential to the contours ϕ = const. Therefore, the local flow velocity

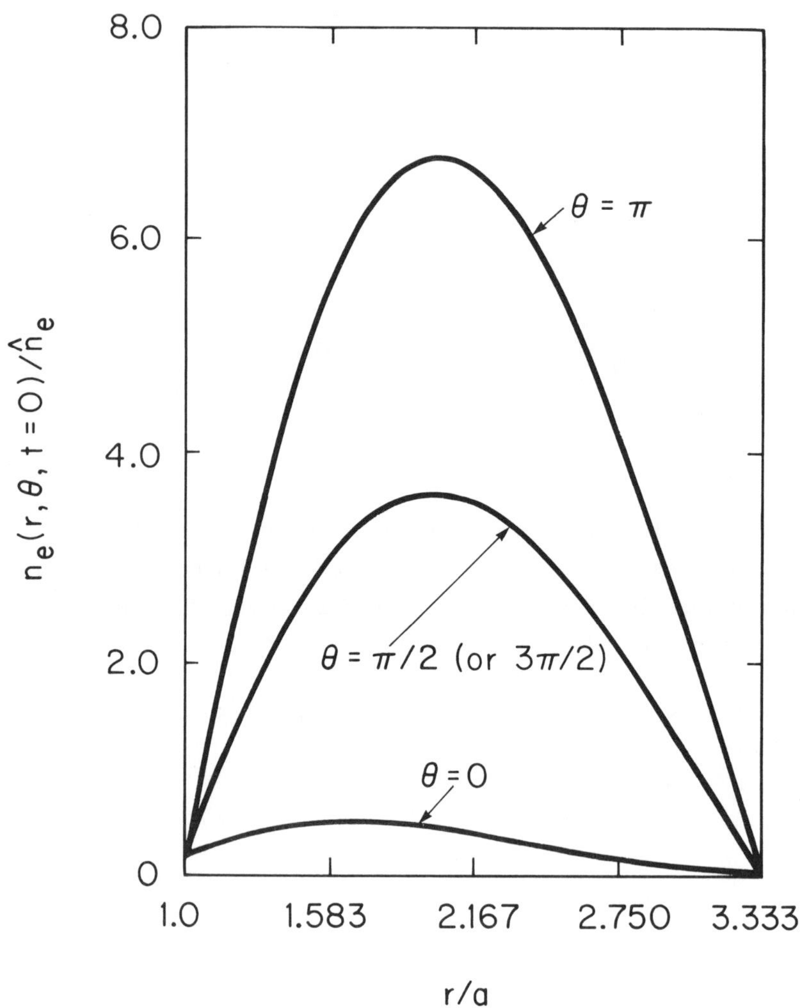

Fig. 3.   Plots versus r/a of the density profile $n_e(r,\theta,t=0)$ calculated from Eq.(33) for $\ell = 1$ and values of θ corresponding to $\theta = 0$, $\theta = \pi/2$ (or $\theta = 3\pi/2$), and $\theta = \pi$.   The choice of system parameters is the same as in Eq.(37) and Fig. 2.

circulates in the direction indicated by the arrows in Fig. 2. Evidently, the structure centered around $\theta = \pi$ in Fig. 2 corresponds to a large-amplitude vortex localized between $r = a$ and $r = b$. The corresponding radial dependence of the density profile $n_e(r, \theta, t = 0)$ calculated from Eqs.(33) and (37) is illustrated in Fig. 3 for $\theta = 0$, $\theta = \pi/2$ (or $\theta = 3\pi/2$), and $\theta = \pi$. It is evident from Fig. 3 that the density compression is large at the center of the vortex. Note also from Eq.(33) that the profile for $n_e(r, \theta = \pi/2, t = 0)$ plotted in Fig. 3 is the same as the azimuthally averaged density profile

$$\langle n_e \rangle (r, t = 0) = \frac{1}{2\pi} \int_0^{2\pi} d\theta \, n_e(r, \theta, t = 0) \ .$$

The choice of system parameters in Figs. 4 and 5 corresponds to $\ell = 2$ and

$$\frac{a}{b} = 0.3 \ , \quad kb = 5.4702 \ , \quad \frac{\omega_r}{\omega_D} = 0.4 \ ,$$

$$\frac{k^2 cV}{\omega_D B_0} = 5.9847 \ , \quad \frac{k^2 a_\ell}{\omega_D} = 2.9923 \ . \tag{38}$$

Here, $kb = 5.4702$ is the first zero of Eq.(34) for $\ell = 2$ and $a/b = 0.3$. Figure 4 shows plots of the equipotential contours, $\phi(r, \theta, t = 0) = \text{const.}$, calculated from Eqs.(32) and (38). The direction of the local flow velocity is indicated by the arrows in Fig. 4. Evidently, for $\ell = 2$, there are two large-amplitude vortices centered around $\theta = \pi/2$ and $\theta = 3\pi/2$. Moreover, the density compression is large at the center of the vortices. This is evident from Fig. 5 which shows the radial dependence of the density profile $n_e(r, \theta, t = 0)$ calculated from Eqs.(33) and (38) for $\theta = 0$ (or $\theta = \pi$) and $\theta = \pi/2$ (or $\theta = 3\pi/2$).

To summarize, Figs. 2–5 and the analysis in Secs. II.B and II.C demonstrate that a simple cold-fluid guiding-center model of a low-density nonneutral plasma supports large-amplitude vortex solutions which are stationary in the rotating frame. What is most striking

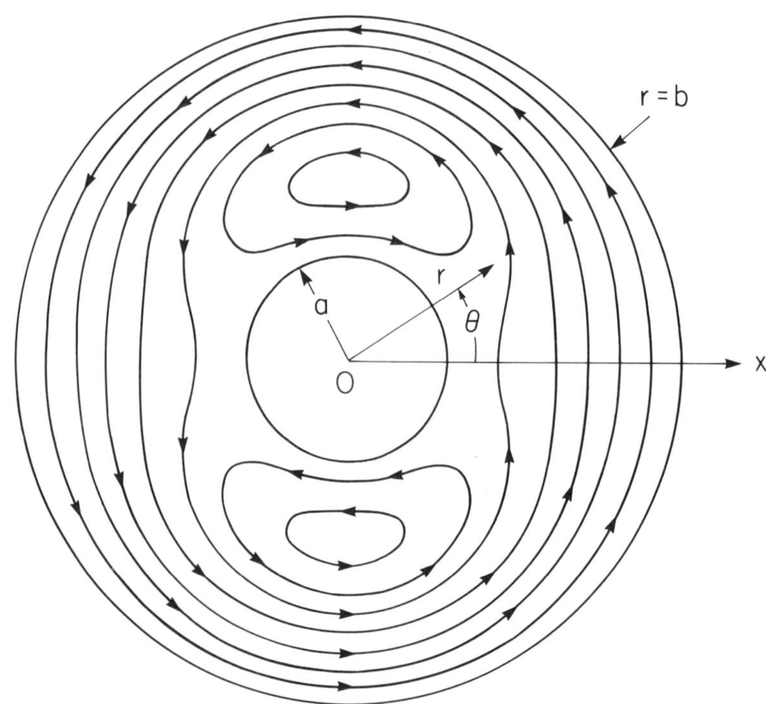

Fig. 4.   Plots of the equipotential contours $\phi(r,\theta,t=0)$ = const. calculated from Eq.(32) for $\ell = 2$ and the choice of system parameters in Eq.(38).   The arrows indicate the direction of the circulating electron flow $\underset{\sim}{V}_e = -(c/B_0)\nabla\phi \times \hat{\underset{\sim}{e}}_z$.

is that the coherent structures described by Eqs.(32) and (33) are very rich in detail for the case in which $n_e(\psi)$ is assumed to have a simple linear dependence on ψ with $n_e(\psi) = \hat{n}_e(C_0 + C_1\psi)$.   Even more structure would be present in the nonlinear case where a quadratic term is included with $n_e(\psi) = \hat{n}_e(C_0 + C_1\psi + C_2\psi^2)$.   Finally, it should be emphasized that the present analysis addresses only the existence of coherent structures which are stationary in the rotating frame.   The question of accessibility of such solutions from prescribed initial conditions $n_e(r,\theta,t=0)$ is not addressed by the present analysis, although the constants ω_r/ω_D, $k^2 a_\ell/\omega_D$, etc., occurring in Eqs.(32) and (33) can be related to the initial

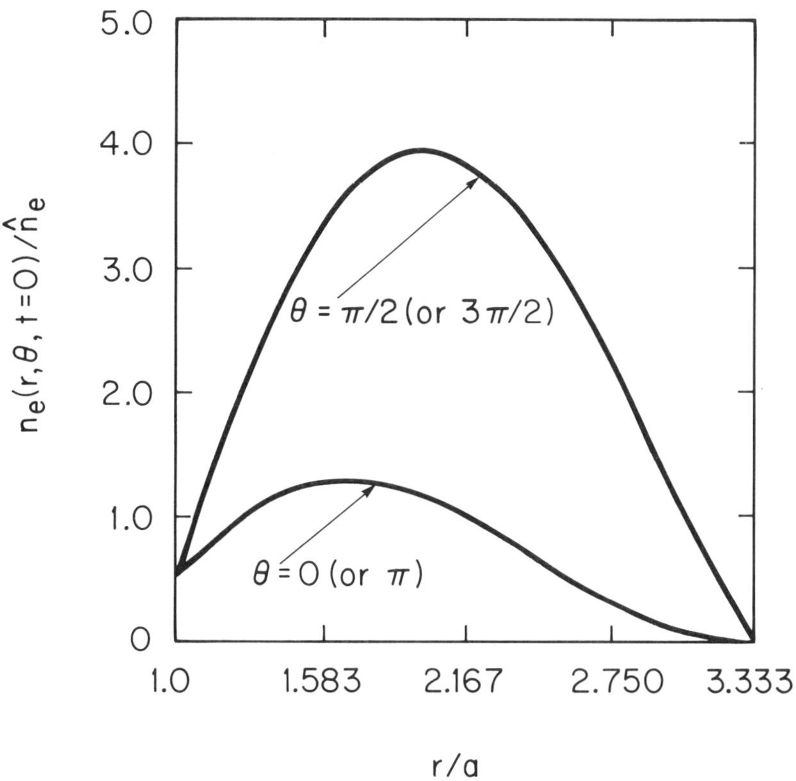

Fig. 5. Plots versus r/a of the density profile
$n_e(r,\theta,t=0)$ calculated from Eq.(33) for $\ell = 2$ and values
of θ corresponding to $\theta = 0$ (or $\theta = \pi$), and $\theta = \pi/2$ (or
$\theta = 3\pi/2$). The choice of system parameters is the same as
in Eq.(38) and Fig. 4.

conditions by the global conservation constraints in Eqs.(10) and
(11). In addition, the stability of such large-amplitude structures
requires an analysis of the evolution of small-amplitude perturbations, δn_e and $\delta\phi$, about the solutions in Eqs.(32) and (33).

III. <u>COMPUTER SIMULATION OF THE NONLINEAR ELECTRODYNAMICS</u>
<u>OF RELATIVISTIC MAGNETRONS</u>

In relativistic magnetrons,[22-30] pulsed high-voltage diodes (operating in the several hundred kV to MV range, say) are used to generate microwaves at gigawatt power levels. Although magnetrons are widely used as microwave sources, a fundamental understanding of the underlying interaction physics is still being developed,[31] particularly in the nonlinear regime. Much of the theoretical challenge in describing multiresonator magnetron operation arises from the complexity introduced by the corrugated anode boundary[30] and the fact that the electrons emitted from the cathode interact with the electromagnetic waves in the anode-cathode gap in a highly nonlinear way. This is manifest through strong azimuthal bunching of the electrons and the formation of a large-amplitude "spoke" structure in the circulating electron density. In this regard, computer simulation studies[32-35] provide a particularly valuable approach to analyze the interaction physics and nonlinear electro-dynamics in magnetrons. In this section, we summarize recent computer simulations[35] of the multiresonator A6 magnetron configura-tion of Palevsky and Bekefi[26] using the two-dimensional ($\partial/\partial z = 0$) particle-in-cell code MAGIC.[36]

A. <u>Simulation Model and A6 Magnetron Configuration</u>

The MAGIC simulation code includes cylindrical effects, and relativistic and electromagnetic effects in a fully self-consistent manner. Unlike previous computer simulations,[32-34] the magnetron oscillations in the present studies[35] are excited from noise, i.e., without preinjection of a finite-amplitude rf signal or preferential excitation of 2π-mode or π-mode oscillations. In addition, the present simulations[35] are carried out in cylindrical rather than planar[32,33] magnetron geometry. Figure 6 shows the cross section of the A6 magnetron diode.[26] The cold, field-emission, graphite cathode is located at radius a = 1.58 cm. The inside radius of the anode block is b = 2.11 cm, and six vane-type resonators with outer radius d = 4.11 cm are used, with angle $\psi = 20°$ subtended by the resonators on axis. The axial magnetic field $B_f\hat{e}_z$ prior to

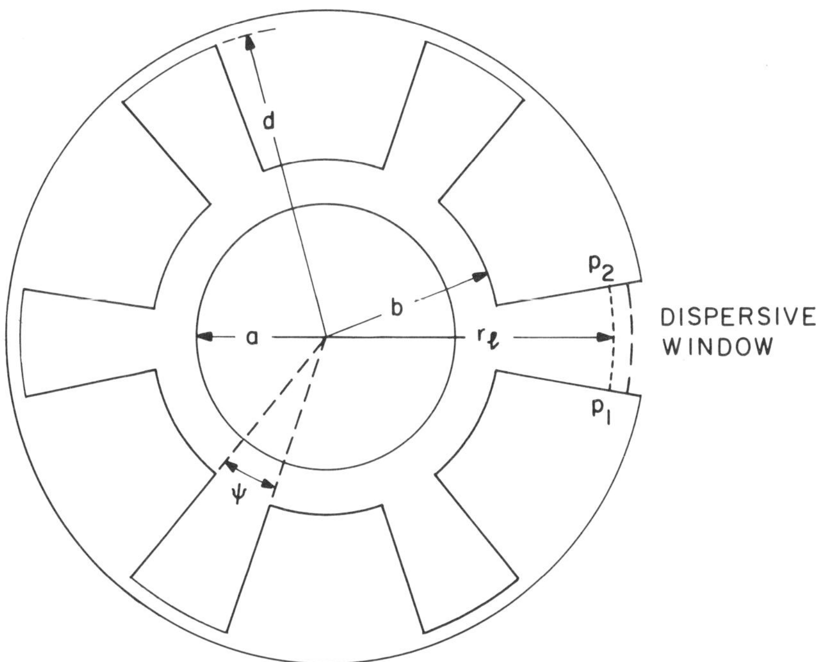

Fig. 6.   Schematic of the A6 magnetron used in the computer simulations. The rf power is partially absorbed by
the dispersive window (dashed line) located at $r = d =$
4.11 cm in the open resonator. Here, $a = 1.58$ cm,
$b = 2.11$ cm, $d = 4.11$ cm, and $\psi = 20°$.

formation of the circulating electron layer (the so-called "fill"
magnetic field) ranges from 4–10 kG in typical operation.[26] The
axial length of the anode block is $L = 7.2$ cm, and the operating
voltage is 300–400 kV. The annular interaction space between $r = a$
and $r = b$ together with the periodically-spaced vanes can be viewed
as a coaxial microwave resonator. For transverse electric (TE)
modes with $\delta \underset{\sim}{E}$ perpendicular to $B_f \hat{\underset{\sim}{e}}_z$ and $\delta \underset{\sim}{B}$ parallel to $B_f \hat{\underset{\sim}{e}}_z$, the
vacuum electric field pattern is such that δE_θ is in phase in
adjacent resonators for the 2π mode, whereas δE_θ is out of phase in
adjacent resonators for the π mode.

In circumstances where the magnetically insulated electron flow
is described by an ideal Brillouin flow model,[22] the cylindrical

expressions for the Hull cut-off voltage V_H and the Buneman-Hartree threshold voltage V_{BH} can be expressed relativistically as[31]

$$\frac{eV_H}{m_e c^2} = \left[1 + \frac{e^2 B_f^2}{m_e^2 c^4} \left(\frac{b^2 - a^2}{2b} \right)^2 \right]^{1/2} - 1 \; , \tag{39}$$

and

$$\frac{eV_{BH}}{m_e c^2} = \frac{eB_f}{m_e c^2} \left(\frac{b^2 - a^2}{2b} \right) \beta_p - \left[1 - \left(1 - \beta_p^2 \right)^{1/2} \right] \; . \tag{40}$$

Here, $\beta_p c$ is the phase velocity of the electromagnetic wave excited in the interaction region. For steady Brillouin flow in a cylin-drical diode with specified fill field B_f, the inequality $V < V_H$ is required to assure magnetic insulation of the azimuthal flow from contact with the anode at $r = b$, whereas $V > V_{BH}$ is required for interaction of the outermost electrons with the electromagnetic wave field.

In the simulations, Maxwell's equations and the particle orbit equations are solved relativistically and electromagnetically, using (typically) more than 3000 macroparticles and a nonuniform, two-dimensional grid consisting of approximately 3000 cells. When a voltage $V_D(t)$ is applied across the diode shown in Fig. 6, the elec-trons are emitted from the cathode through a space-charge-limited emission process in which the instantaneous electric field normal to the cathode surface vanishes. The radial momenta of the emitted electrons are randomly distributed from 0 to 0.02 $m_e c$. On the other hand, electrons are absorbed by the anode or cathode whenever they strike the surface. The simulations are carried out with one open resonator, which is modeled by a dispersive window placed along the dashed line in Fig. 6 at $r = d = 4.11$ cm. At the window, the boundary condition is such that most of the electromagnetic wave energy is absorbed, while a small fraction of the wave energy is reflected back into the cavity. Such a window yields a finite Q factor for the system. The rf power output is determined from the net flow of electromagnetic energy expressed as an area-integral of the Poynting flux over the window surface shown in Fig. 6.

A quasi-static model is used in the simulations to describe the high-voltage pulse applied to the diode. In such a model, the diode voltage is given by $V_D(t) = Z(t)V_0(t)/[Z_0 + Z(t)]$, where $V_0(t)$ is the voltage pulse provided by the power supply, $Z_0 = $ const. is the impedance of the power supply, and $Z(t)$ is the magnetron impedance. In the simulations, the voltage pulse $V_0(t)$ is assumed to have the form

$$V_0(t) = \begin{cases} 0 \, , & t < 0 \, , \\[2ex] (t/t_0)V_m \, , & 0 \leq t < t_0 \, , \\[2ex] V_m \, , & t \geq t_0 \, , \end{cases} \tag{41}$$

where t_0 and V_m are the rise time and maximum value of the voltage pulse, respectively. The rise time assumed in the simulations is $t_0 = 4.0$ ns, corresponding to the experimental value.[26] Although the applied high-voltage pulse in the simulations is described by Eq.(41), it should be emphasized that all extraordinary-mode rf excitations ($\underset{\sim}{\delta B} = \delta B_z \hat{e}_z$ and $\underset{\sim}{\delta E} = \delta E_r \hat{e}_r + \delta E_\theta \hat{e}_\theta$) are treated fully electromagnetically.

B.   Simulation Results

Typical numerical results are presented in Figs. 7-10 for the case of an ideal power supply where $Z_0 = 0$ and $V_D(t) = V_0(t)$.

Figure 7(a) shows the time history of the integrated rf field profile

$$V_\theta(t) = \int_{P_1}^{P_2} d\theta \; r_\ell \, \delta E_\theta(r_\ell, \theta, t) \tag{42}$$

for the choice of system parameters $B_f = 7.2$ kG, $V_m = 350$ kV and $t_0 = 4.0$ ns. In Fig. 7(b), the Fourier transform of the signal in Fig. 7(a) is plotted versus frequency f. Here, the integration path corresponds to the dotted line in Fig. 6 from P_1 to P_2 at $r = r_\ell = 3.7$ cm. In Fig. 7(a), the nonlinear saturation of the magnetron oscillations occurs at $t \simeq 10$ ns, where the ratio of the saturated amplitude $V_{\theta s}$ and the applied diode voltage $V_D = V_m$ is $V_{\theta s}/V_D \simeq 0.85$.

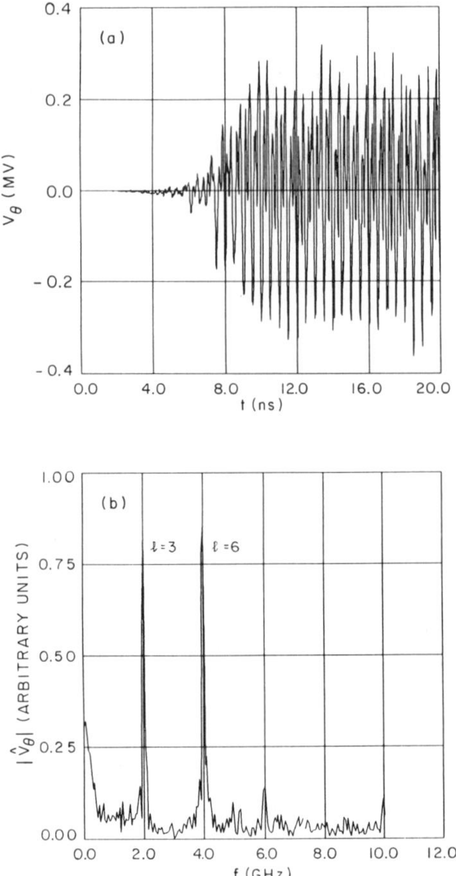

Fig. 7.   Shown in Fig. 7(a) is the time history of the integrated rf field profile $V_\theta(t) = \int_{P_1}^{P_2} d\theta\, r_\ell\, \delta E_\theta(r_\ell, \theta, t)$ obtained in the simulations at radius $r = r_\ell = 3.7$ cm in the open resonator for the choice of system parameters $B_f = 7.2$ kG, $V_m = 350$ kV, $t_0 = 4.0$ ns, and $Z_0 = 0$. Figure 7(b) shows the magnitude of the Fourier transform, $|\hat{V}_\theta(f)|$, of the signal in Fig. 7(a). The two distinct peaks at $f = 2.0$ GHz and $f = 4.0$ GHz correspond to π–mode ($\ell = 3$) and 2π–mode ($\ell = 6$) oscillations, respectively.

In Fig. 7(b), the two distinct peaks at the frequencies $f = 2.0$ GHz and $f = 4.0$ GHz correspond to π-mode and 2π-mode oscillations, respectively. The 2π-mode oscillation frequency $f = 4.0$ GHz is 14% lower than the frequency $f = 4.55$ GHz observed in the experiment,[26] which may be due to the absence of finite-axial-length effects in the simulations (where $\partial/\partial z = 0$ is assumed). Although both the π-mode and 2π-mode excitations have nearly the same wave amplitudes in Fig. 7(b), the (higher frequency) 2π mode delivers more rf power than the π mode.

By evaluating the area-integral of the outward Poynting flux $(c/4\pi)\delta E_\theta \delta B_z$ over the surface of the dispersive window at $r = d = 4.11$ cm in Fig. 6, the peak rf power output in the simulations is calculated for various values of the applied magnetic field B_f. The dependence of the normalized rf power on magnetic field is shown in Fig. 8. Here, the dots correspond to the experimental values,[26] and the triangles are obtained from the simulations with $Z_0 = 0$. In Fig. 8, the normalization is chosen such that the maximum values of the rf power in both the simulations and the experiment are equal to unity. The actual value of the maximum rf output per unit axial length in the simulations is 4.3 GW/m. For the A6 magnetron, with axial length $L = 7.2$ cm, this corresponds to $P = 0.3$ GW, which is somewhat less than the maximum rf power $P = 0.45$ GW measured in the experiment.[26] (The values of power quoted here are rms values.) Apart from a constant scale factor, it is evident from Fig. 8 that the simulations are in excellent agreement with the experimental results. The difference in scale factor may be due to the fact that a larger fraction of the rf power in the simulations is reflected back into the cavity, and the effective Q-value in the simulations ($Q \sim 100$) is greater than in the experiment ($Q \sim 20\text{-}40$). As B_f is decreased below 6 kG, crossing the Hull cut-off curve[22] at a diode voltage corresponding to $V_D = 350$ kV, it is found that the $5\pi/3$-mode ($\ell = 5$) becomes the dominant rf excitation in the simulations.

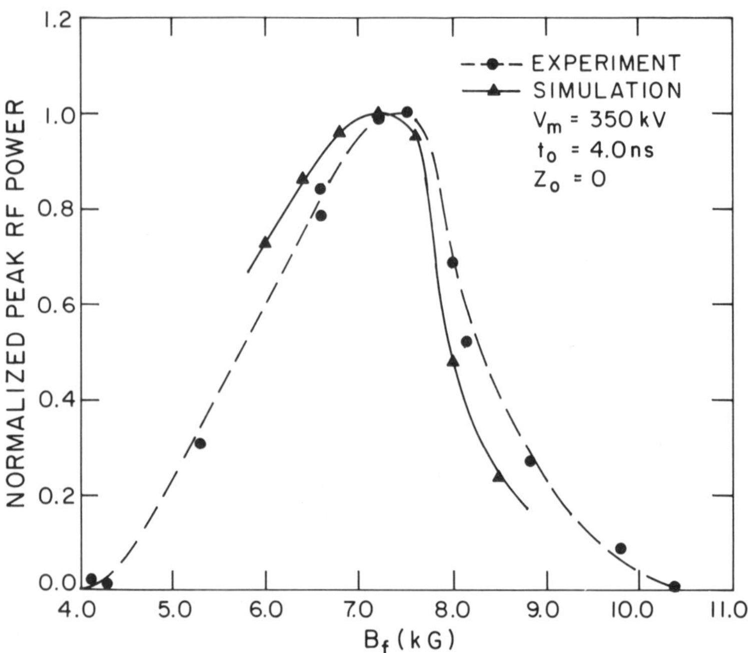

Fig. 8.   Plots of the normalized peak rf power versus the applied magnetic field B_f.   The dots correspond to the experimental results [A. Palevsky and G. Bekefi, Phys. Fluids 22, 986 (1979)], and the triangles correspond to the simulation results for V_m = 350 kV, t_0 = 4.0 ns, and Z_0 = 0.   The maximum rf power is 0.45 GW per open port in the experiment, and 0.3 GW in the simulations.

Figure 9 shows radial plots of the charge density,

$$e\langle n_e \rangle (r,t) = \frac{1}{2\pi} \int_0^{2\pi} e n_e (r,\theta,t) d\theta \ , \tag{43}$$

averaged over the azimuthal angle θ, at several instants of time for the same values of system parameters as in Fig. 7.   In Fig. 9, the outer radius of the electron layer [$r = r_b(t)$], designated by the

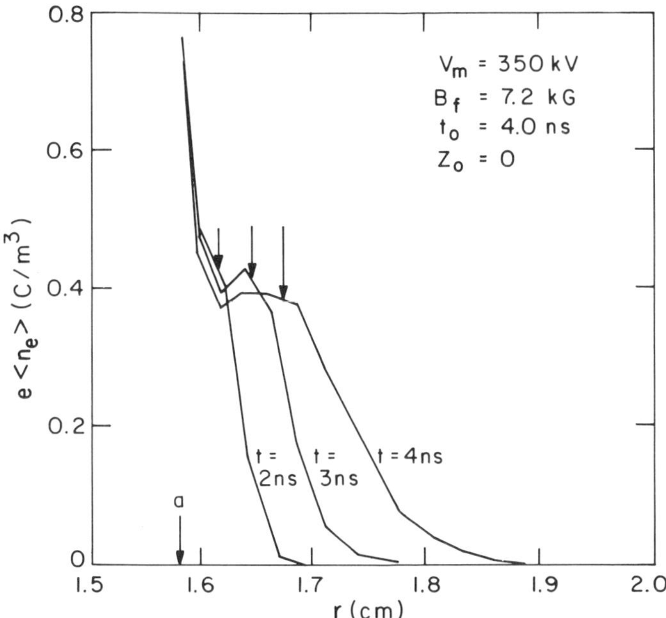

Fig. 9.   Plots of the azimuthally averaged charge density
$e\langle n_e\rangle(r,t)$ versus radial distance r obtained in the simulations at times t = 2.0, 3.0, 4.0 ns for system parameters
the same as in Fig. 7.   Here, the arrows designate the
location of the outer envelope (r = r_b) of the electron
layer calculated from a simple Brillouin flow model.

arrows, is calculated from a simple Brillouin flow model[22] for B_f =
7.2 kG and diode voltages $V_D(t)$ = 0.5 V_m, 0.75 V_m, 1.0 V_m, corresponding to t = 2.0, 3.0, 4.0 ns.   It is clear from Fig. 9 that a
substantial fraction of the electrons occupy the region between
r = r_b and the anode (r = b).   The existence of a long tail in the
electron density profile indicates that the electron flow differs
significantly from the ideal Brillouin flow model.   For the A6
magnetron operating at V_m ≃ 350 kV, cylindrical and relativistic
effects are relatively mild.   For example, at t = t_0 = 4.0 ns, the

layer aspect ratio is $A = a/(r_b - a) \simeq 1.15$. We define the local self-field parameter $s_e(r)$ by

$$s_e(r) = \frac{\omega_{pe}^2(r)/\gamma_e(r)}{\omega_{ce}^2(r)/\gamma_e^2(r)}, \qquad (44)$$

where $\omega_{ce}(r) = e\langle B_z\rangle/m_e c$ is the nonrelativistic cyclotron frequency, $\omega_{pe}^2(r) = 4\pi\langle n_e\rangle e^2/m_e$ is the nonrelativistic plasma frequency-squared, and $\gamma_e(r) = (1 - V_{\theta e}^2/c^2)^{-1/2}$ is the relativistic mass factor of an electron fluid element. Under ideal Brillouin flow conditions,[22] the self-field parameter satisfies $s_e = 1$ (in the planar approximation). In the simulations, however, it is found that $s_e(r)$ decreases considerably as r increases from $r = a$ to $r = r_b$ and beyond. For example, at $t = t_0 = 4.0$ ns in Fig. 9, the self-field parameter decreases from $s_e(r = a) \simeq 1$ at the cathode, to $s_e(r = r_b) \simeq 0.5$ at $r = r_b$.

Although the azimuthal bunching of the electrons is relatively small for times up to 4 ns, by $t \simeq 6$ ns the system begins to enter a nonlinear regime characterized by large-amplitude spoke formation. Highly developed spokes are evident in Fig. 10(b) which shows density contour plots at $t = 8$ ns for the choice of system parameters $B_f = 7.2$ kG, $V_m = 350$ kV and $t_0 = 4$ ns (similar to the conditions in Fig. 7, and the maximum power simulation point in Fig. 8). As the system evolves, the spokes rotate as coherent nonlinear structures in the azimuthal direction for hundreds of electron cyclotron periods. In addition, by $t = 7$ ns, there is current flow from the cathode to the anode. At saturation, which occurs at $t \simeq$ 10 ns, the time-averaged diode current per unit axial length is $I_D \simeq$ 100 kA/m, and the amplitude of the integrated rf field profile $\int_a^b dr\, \delta E_r(r, \theta, t)$ is comparable with the diode voltage $V_D \simeq V_m = 350$ kV.

To summarize, with regard to the dependence of rf power on magnetic field, the simulation results are in excellent agreement with experiment (within a constant scale factor). Also, in terms of rf power output, the simulations confirm that the A6 magnetron oscillates preferentially in the 2π mode. In the preoscillation regime, it is found that the electron flow differs substantially from

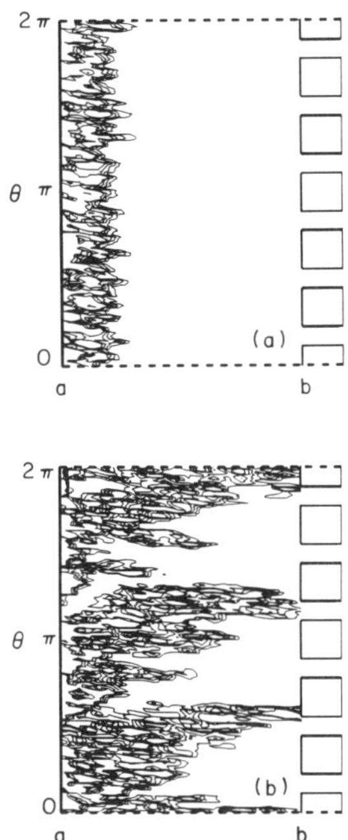

Fig. 10.   Density contour plots for $n_e(r,\theta,t)$ obtained in
the simulations at (a) t = 3.0 ns and (b) t = 8.0 ns for
the same system parameters as in Fig. 7.

Brillouin flow conditions. In the nonlinear regime, the saturation
is dominated by the formation of a large-amplitude spoke structure
in the circulating electron density. The simulations also show that
the magnetron performance and rf power generation are degraded when
the impedance Z_0 of the external power supply is increased (in
agreement with experiment) from the ideal value $Z_0 = 0$. As a
general conclusion, based on the results presented here, it is
expected that the MAGIC simulation code can be used as an effective
tool for developing a fundamental understanding of the large-
amplitude spoke dynamics and saturation in magnetrons, as well as
for experimental magnetron design.

IV.   <u>REFERENCES</u>

1. R.C. Davidson, <u>Physics of Nonneutral Plasmas</u> (Addison-Wesley, Reading, Massachusetts, 1990).

2. Ibid., Chapter 6.

3. R.C. Davidson, K.T. Tsang and H.S. Uhm, Phys. Fluids <u>31</u>, 1727 (1988).

4. R.C. Davidson, Phys. Fluids <u>28</u>, 1937 (1985).

5. R.C. Davidson, Phys. Fluids <u>27</u>, 1804 (1984).

6. R.J. Briggs, J.D. Daugherty and R.H. Levy, Phys. Fluids <u>13</u>, 421 (1970).

7. J.D. Daugherty, J.E. Eninger and G.S. Janes, Phys. Fluids <u>12</u>, 2677 (1969).

8. R.H. Levy, Phys. Fluids <u>11</u>, 920 (1968).

9. O. Buneman, R.H. Levy and L.M. Linson, J. Appl. Phys. <u>37</u>, 3203 (1966).

10. R.H. Levy, Phys. Fluids <u>8</u>, 1288 (1965).

11. O. Buneman, J. Electron. Control <u>3</u>, 507 (1957).

12. C.C. MacFarlane and H.G. Hay, Proc. Phys. Soc. (London) <u>63B</u>, 409 (1950).

13. S.A. Prasad and J.H. Malmberg, Phys. Fluids <u>29</u>, 2196 (1986).

14. R.L. Kyhl and H.F. Webster, IRE Trans. Electron Devices <u>ED-3</u>, 172 (1956).

15. J.R. Pierce, IRE Trans. Electron Devices <u>ED-3</u>, 183 (1956).

16. C.A. Kapetanakos, D.A. Hammer, C. Striffler and R.C. Davidson, Phys. Rev. Letters <u>30</u>, 1303 (1973).

17. G. Rosenthal, G. Dimonte and A.Y. Wong, Phys. Fluids <u>30</u>, 3257 (1987).

18. G. Rosenthal and A.Y. Wong, "Localized Density Clumps and Potentials Generated in a Magnetized Nonneutral Plasma," UCLA Report No. PPG1282 (1989).

19. K.S. Fine, C.F. Driscoll and J.H. Malmberg, Phys. Rev. Lett. <u>63</u>, 2232 (1989).

20. J.H. Malmberg, C.F. Driscoll, B. Beck, D.L. Eggleston, J. Fajans, K. Fine, X.-P. Huang and A.W. Hyatt, in <u>Nonneutral Plasma Physics</u>, eds., C.W. Roberson and C.F. Driscoll, AIP Conference Proceedings <u>175</u>, 28 (1988).

21. C.F. Driscoll, J.H. Malmberg, K.S. Fine, R.A. Smith and X.-P. Huang, in <u>Plasma Physics and Controlled Nuclear Fusion Research</u>, Nice (IAEA, Vienna, 1989), Vol. 3, p. 507.

22. R.C. Davidson, Chapter 8 of Ref. 1.

23. J. Benford, in <u>High-Power Microwave Sources</u>, eds., V. Granatstein and I. Alexeff (Artech House, Boston, Massachusetts, 1987) p. 351.

24. J. Benford, H.M. Sze, W. Woo, R.R. Smith and B. Harteneck, Phys. Rev. Lett. <u>62</u>, 969 (1989).

25. G. Bekefi and T.J. Orzechowski, Phys. Rev. Lett. <u>37</u>, 379 (1976).

26. A. Palevsky and G. Bekefi, Phys. Fluids <u>22</u>, 986 (1979).

27. T.J. Orzechowski and G. Bekefi, Phys. Fluids <u>22</u>, 978 (1979).

28. A.G. Nokonov, I.M. Roife, Yu.M. Savel'ev and V.I. Engel'ko, Sov. Tech. Phys. <u>32</u>, 50 (1987).

29. I.Z. Gleizer, A.N. Didenko, A.S. Sulakshin, G.P. Fomenko and V.I. Tsvetkov, Sov. Tech. Phys. Lett. <u>6</u>, 19 (1980).

30. H.S. Uhm, H.C. Chen and R.A. Stark, Proc. SPIE <u>1061</u>, 170 (1989).

31. Y.Y. Lau, in <u>High-Power Microwave Sources</u>, eds., V. Granatstein and I. Alexeff (Artech House, Boston, Massachusetts, 1987) p. 309.

32. S.P. Yu, G.P. Kooyers and O. Buneman, J. Appl. Phys. <u>36</u>, 2550 (1965).

33. A. Palevsky, G. Bekefi and A.T. Drobot, J. Appl. Phys. <u>52</u>, 4938 (1981).

34. A. Palevsky, et al., in <u>High-Power Beams</u>, eds., H.J. Doucet and J.M. Buzzi (Ecole Polytechnique, Palaiseau, France, 1981) p. 861.

35. H.-W. Chan, C. Chen and R.C. Davidson, "Computer Simulation of Multiresonator Cylindrical Magnetrons," submitted for publication (1990).

36. B. Goplen and J. McDonald, private communication (1989). The MAGIC simulation code was developed by researchers at Mission Research Corporation. The simulation results presented in this paper use the code version dated 1988.

WAVE AND VORTEX DYNAMICS IN PURE ELECTRON PLASMAS

C.F. Driscoll

Physics Department, Univ. of Calif. at San Diego, La Jolla, CA 92093

ABSTRACT

Magnetically confined columns of electrons are excellent experimental manifestations of 2 dimensional vortices in an inviscid fluid. Surface charge perturbations on the electron column (diocotron modes) are equivalent to surface ripples on extended vortices; and unstable diocotron modes on hollow electron columns are examples of the Kelvin-Helmholtz instability. Additionally, vortex interaction processes can be studied using two or more isolated electron columns. The experiments can include nonlinear, finite size, and finite Larmor radius effects, and point out deficiencies in simple fluid theories.

ELECTRON SYSTEM

Pure electron plasmas are one of the simplest systems on which one can study plasma wave and transport effects. This talk will point out a few of the areas where interesting comparisons between theory and experiment are occurring. More complete descriptions can be found in the references.

A schematic of the experimental apparatus[1,2,3] is shown in Figure 1. The electrons are contained in a grounded conducting cylinder, with a uniform axial magnetic field B_z providing radial confinement, and negative voltages applied to end cylinders providing axial confinement. The electron column rotates due to the radial electric field E from the unneutralized charge.

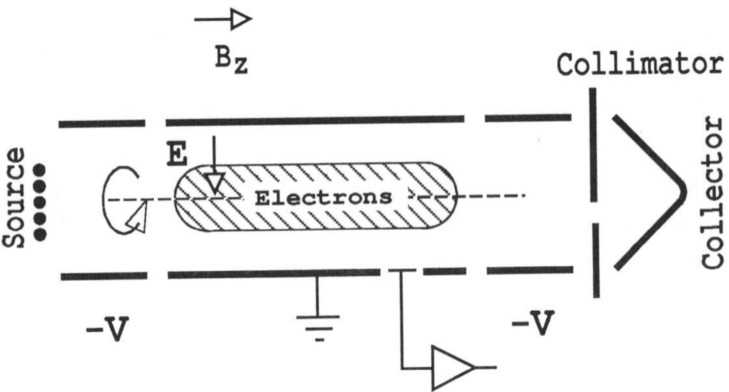

Figure 1. The cylindrical containment apparatus.

The trapped electrons can be manipulated in a variety of ways to create the desired "initial conditions" of a rotating column with specified density profile. The resulting evolution can then be measured by wall antennas, or by axially dumping the electrons to obtain the (z-integrated) density $n(r,\theta,t) \equiv \int dz\, \tilde{n}(r,\theta,z,t)/L$. For most processes of interest here, the axial bounce motions of the electrons are fast compared to the drift rotational motions.

2D FLUID EVOLUTION

The basic 2 dimensional drift-Poisson equations governing a magnetized electron column are isomorphic to the 2D Euler equations governing a constant density inviscid fluid.[2,4,5] Here, the electron dynamics is approximated by 2D guiding center theory: the axial bouncing of the electrons averages over any z-variations, and we consider only the $(r-\theta)$ drift velocity $\mathbf{v}(r,\theta,t)$. Both systems evolve such that the vorticity $\Omega(r,\theta,t)$ of the flow has zero convective derivative, that is, $(\partial/\partial t + \mathbf{v}\cdot\nabla)\Omega = 0$. In the electron system only, the vorticity is proportional to the measured electron density, i.e. $n(r,\theta,t) \propto \Omega(r,\theta,t)$.

This isomorphism implies that surface charge perturbations on electron columns, called diocotron modes,[1,2,6] are equivalent to the surface ripples on extended vortices first studied by Kelvin.[7] When the radial density profile (or vorticity profile) is monotonically decreasing, the Rayleigh stability criterion[6,8] demonstrates that all modes are stable. When the density profile is non-monotonic (i.e. "hollow"), the unfavorable shear in the rotation velocity gives rise to unstable diocotron modes, which are examples of the Kelvin-Helmholtz instability. More generally, the nonlinear interaction of two or more electron columns is the same as the interaction of two or more fluid vortices.[2]

The electron system can thus be considered a test of 2D fluid theory in a realistic system. The "realism" consists of effects such as a continuous density profile, finite system length, finite Larmor radius, large wave amplitude, and trapped particle distributions; some of these effects are within the 2D fluid model, others outside it. We find that considerable re-evaluation and extension of theory is required to understand the experiments.

DIOCOTRON MODES

One basic experimental "realism" is that of a continuous density profile. Theory approximations based on "step profiles" (as in Fig. 2) are useful for obtaining simple answers, but suggest that there are as many modes varying as $e^{il\theta}$ as there are (unphysical) steps, and that the modes occur in complex conjugate pairs. This has lead to the historical misconception that the stable mode becomes unstable as the profile becomes hollow. Experimentally, we observe that the stable and unstable modes are distinct and may propagate simultaneously.[1,2,9,10] The measurements completely characterize the modes, showing they have different radial eigenfunctions as well as different frequencies.

The step profile approximation also gives poor quantitative predictions of instability growth rates. For example, the step profile of Fig. 2c is predicted to be stable, whereas the continuous profile is predicted and observed to be unstable. For azimuthal mode number $l=2$, numerical eigenvalue solutions with continuous hollow profiles predict growth rates within 50% of the measured growth rates.[9] While this can be considered to be fair agreement, the source of the discrepancy is unknown at present.

For $l=1$, theory and experiment differ substantially. Linear mode theory predicts that there are no exponentially unstable diocotron modes in our geometry (which has no conductor at $r=0$).[4,6] However, a recent linear initial value analysis based on the Laplace transform predicts that there is a subtle instability which grows asymptotically as $\sqrt{t}$.[11] Experimentally, we observe a robust exponential instability which is basically similar to the unstable $l=2$ mode.[1,2] Figure 3 shows the growth of the unstable $l=1$ mode for two different initial amplitudes, and the prediction of the initial value analysis (dashed). This

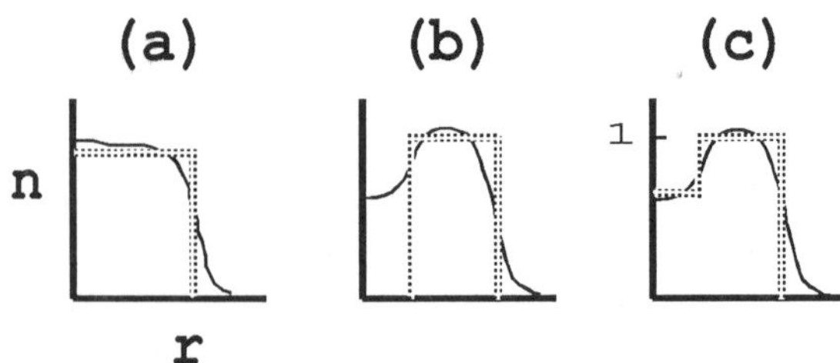

Figure 2. Typical monotonic and hollow density profiles, with step approximations.

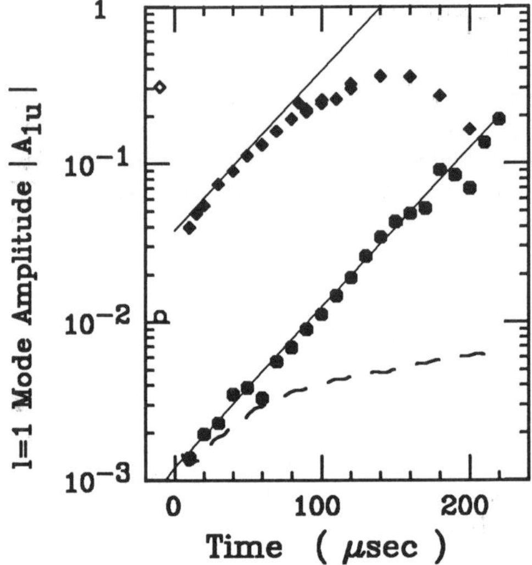

Figure 3.　Unstable $l = 1$ mode amplitude vs time for two different initial conditions, compared to initial value theory (dashed).

discrepancy may indicate that the theory is inadequate, or that experimental subtleties such as finite system length or finite electron Larmor radius must be considered. In either case, this new instability is an important effect for typical experimental systems.

FINITE LENGTH EFFECTS

When the electron column is made axially short, finite length effects can completely eliminate these instabilities. In particular, profiles with non-monotonic *local* densities $\tilde{n}(r, z = 0)$ and a length to diameter ratio of about 2:1 show no instabilities even on a time scale of 10^5 rotations.[3] Finite length effects can also have dramatic effects on the stable modes on monotonic density profiles. For example, the frequency of the $l = 1$, $k_z = 0$ diocotron mode can more than double as the electron column is made axially short,[12] becoming the containment-potential-dependent "magnetron frequency" of hyperbolic Penning traps.[13]

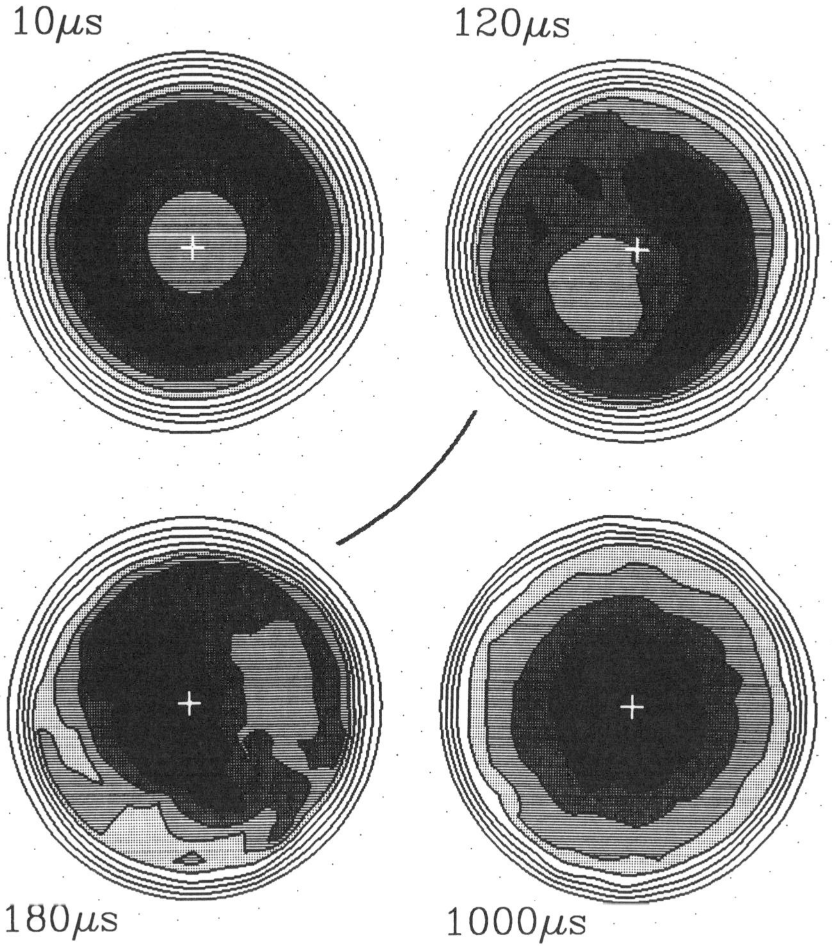

Figure 4.   Experimentally measured contour plots of $n(r,\theta)$ at 4 times during the $l = 1$ instability. The 8 contours are on a linear scale spanning 1 to 1/8.

These finite length corrections are not well-approximated by calculating finite k_z wave numbers on an a periodic system; rather, the boundaries of the plasma must be considered explicitly. Fortunately, the axial density dependence $\tilde{n}(r,z)$ can be calculated from the measured z-averaged density $n(r)$, the measured temperature $T(r)$, and the known boundary conditions at the walls, presuming only local thermal equilibrium along any given field line.[3] Thus, one has all the data required for a realistic description of finite length modes.

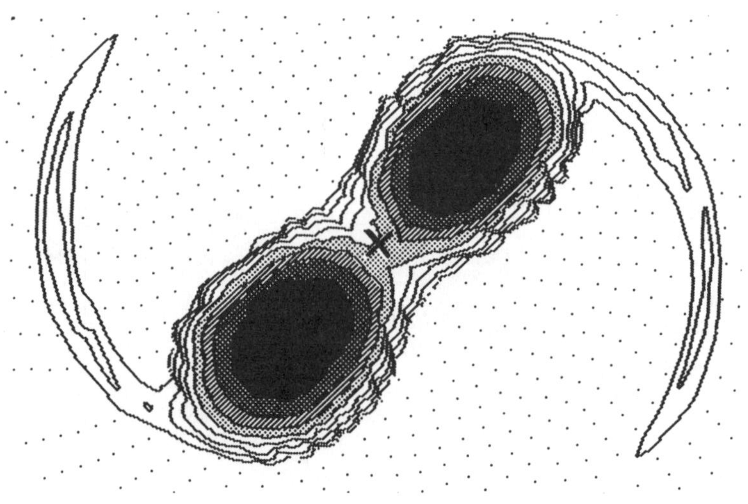

Figure 5.   Experimentally measured contour plots of $n(r,\theta)$ during 2-vortex merger. The 8 contours are on a log scale spanning 1 to .01.

VORTICES

The diocotron instabilities on hollow columns saturate with the formation of smaller vortices and filamentary structures.[1,2,9] When one unstable l mode dominates, there will be l vortices; when many modes grow, the nonlinear saturation may be more complex. Radial transport to a stable, monotonically decreasing density profile is in general the result of nonlinear interactions among these various vortices, with a prominent interaction being the vortex pairing instability.[14,15] There are also interesting turbulence and noise questions associated with this process.[9]

Figure 4 shows contours of $n(r,\theta)$ at 4 times when this evolution is dominated by an unstable $l=1$ mode,[1] corresponding to the upper data of Fig. 3. When the $l=2$ mode dominates, two vortices form, and particle transport to a stable density profile is the result of a nonlinear vortex pairing instability.[2,9]

We are able to study the vortex pairing instability in detail by starting with two well formed, isolated vortices.[2,16] We find that the time required to merge varies dramatically from less than one orbit period to over 10^4 orbit periods as the separation between the vortices varies from 1.8 to 2.0 vortex diameters (FWHM). Figure 5 shows $n(r,\theta)$ for two vortices beginning to merge, with concomitant filamentary tail formation.

One important advantage of the electron system is that the internal electron viscosity is very low, and there are no radial or axial boundary layers to dissipate the vortices. Unlike conventional fluids which "spin down" in 10 to 30 orbit periods,[15] the electron system shows little dissipation even on time scales of 10^4 orbits. The electron system may thus offer the best experimental data on this fundamental vortex interaction process.

ACKNOWLEDGMENTS

The experiments described here were performed in collaboration with Kevin Fine, John Malmberg, Tim Mitchell, and X.-P. Huang; and theory discussions with Ralph Smith and Marshall Rosenbluth are gratefully acknowledged. This research was supported by ONR N-00014-82-K-0621, NSF PHY87-06358, and DOE DE-FG03-85ER53199.

REFERENCES

1. C.F. Driscoll, Phys. Rev. Lett. **64**, 645 (1990).

2. C.F. Driscoll and K.S. Fine, "Experiments on Vortex Dynamics in Pure Electron Plasmas," submitted to Phys. Fluids (1989).

3. C.F. Driscoll, J.H. Malmberg and K.S. Fine, Phys. Rev. Lett. **60**, 1290 (1988).

4. R.H. Levy, Phys. Fluids **8**, 1288 (1965); Phys. Fluids **11**, 920 (1968).

5. R.J. Briggs, J.D. Daugherty and R.H. Levy, Phys. Fluids **13**, 421 (1970).

6. R.C. Davidson, *Theory of Nonneutral Plasmas* (Benjamin, Reading, Mass., 1974) Sec. 2.10.

7. W. Kelvin, Phil. Mag. (5), x. 155 (1880) [*Papers*, iv. 152].

8. J.W.S. Rayleigh, Proc. London Math. Soc. **11**, 57 (1880).

9. C.F. Driscoll, J.H. Malmberg, K.S. Fine, R.A. Smith, X-P. Huang and R.W. Gould, *Plasma Physics and Controlled Nuclear Fusion Research 1988* (IAEA, Vienna, 1989), Vol. 3, pp. 507-514.

10. K.S. Fine, C.F. Driscoll and J.H. Malmberg, Phys. Rev. Lett. **63**, 2232 (1989).

11. R.A. Smith and M.N. Rosenbluth, Phys. Rev Lett. **64**, 649 (1990).

12. K.S. Fine, Ph.D. Thesis, Univ. of Calif. at San Diego (1988), Ch. 5. UMI AAD89–25064.

13. J.J. Bollinger and D.J. Wineland, Phys. Rev. Lett. **53**, 348 (1984).

14. M.V. Melander, N.J. Zabusky and J.C. McWilliams, J. Fluid Mech. **195**, 303 (1988).

15. R.W. Griffiths and E.J. Hopfinger, J. Fluid Mech. **178**, 73 (1987).

16. K.S. Fine, C.F. Driscoll, T.B. Mitchell and J.H. Malmberg, Bull. Am. Phys. Soc. **34**, 1932 (1989).

PURE ION PLASMAS, LIQUIDS AND CRYSTALS

D.H.E. Dubin and T.M. O'Neil
Physics Department, Univ. of Calif. at San Diego, La Jolla, CA 92093

ABSTRACT

A brief review of aspects of the behavior of strongly correlated trapped non-neutral plasmas is presented. Unlike neutral plasmas, non-neutral plasmas can come to a state of confined thermal equilibrium and can be cooled to low temperature at sufficiently high density so that strong correlation effects become important. Under these conditions analytic theory and computer simulations for small clouds of trapped charges predict the formation of novel non-neutral liquid and crystalline states whose structure is strongly influenced by finite size effects. Some of these effects have been observed in experiments.

CONFINEMENT AND THERMAL EQUILIBRIUM

In experiments at the National Institute of Standards and Technology,[1] an un-neutralized cloud of N ions ($N \sim 10^2 - 10^4$) is trapped for long periods of time. The density n_0 is sufficiently high and the temperature T is sufficiently low so that the correlation parameter Γ, defined by $\Gamma = e^2/akT$, is larger than unity (here e is the ion charge and a is the inter-ion spacing, given by $4\pi a^3 n_0/3 = 1$). In this regime the system becomes strongly correlated and transitions to non-neutral liquid or even non-neutral crystalline states are observed. This note briefly reviews some aspects of the theory and experiments involving these novel strongly correlated plasmas.

Plasmas which consist of a single charge species—i.e. non-neutral plasmas—have many properties in common with neutral plasmas. For instance, non-neutral plasmas exhibit the phenomenon of Debye shielding and also exhibit collective effects such as plasma waves. However, there are also important differences between neutral and non-neutral plasmas.

For instance, unlike neutral plasmas, non-neutral plasmas can be confined for long periods of time using only static electric and magnetic fields. A typical confinement geometry used in experiments is shown in Fig. 1—the "cylindrical Penning trap." Axial confinement is provided by an electrostatic axial potential well induced by a voltage difference between the central cylinders and end cylinders. This potential must satisfy Laplace's equation; and since solutions of this equation cannot have absolute minima the axial well is actually a "saddle," so it is deconfining in the radial direction. Radial confinement is provided by a uniform axial magnetic field $\mathbf{B}$. The plasma rotates on axis through this field, providing an inward $\mathbf{v} \times \mathbf{B}$ force which balances the radial electric field. Confinement properties of such traps have been studied in detail using electron plasmas.[2,3]

Confinement times of several days have been achieved in this type of trap,[4] so states of confined thermal equilibrium are possible. The shape of the cloud of charges can be uniquely determined under the assumption of thermal equilibrium[5]; for the case of small clouds used in the experiments this shape is an ellipsoid of revolution[6,7] with approximately constant overall density. Correlations between charges are then set up within this shape. A consequence of thermal equilibrium is that these correlations can be shown to be the same as those in a one-component plasma (OCP)[8,9]—a system of charges confined by a uniform neutralizing background charge. (In the case of the trapped single species plasma, rotation through the magnetic field acts like a uniform neutralizing charge.)

The OCP is a venerable paradigm of plasmas, dating as far back as J.J. Thompson's "plum pudding" model of the atom. The properties of the infinite homogeneous OCP have been studied extensively in the strongly correlated regime.[10] Computer simulations and analytic theory for an infinite homogeneous OCP predict that for $\Gamma \gtrsim 2$ the system of charges begins to exhibit local order characteristic of a liquid, and for $\Gamma \gtrsim 172$ there is a first-order phase transition to a bcc crystal.[11] In cryogenic electron plasma experiments, $\Gamma \sim 1-2$ have been achieved[4]; in ion plasma experiments Γ values in the range of several hundred have been measured, putting the system well into the regime of strong correlation. However, these latter experiments have up to the present involved relatively small numbers of ions ($N \lesssim 10^4$) so theoretical studies of infinite systems cannot be trusted. On the other hand, these small systems are ideal for numerical simulations with realistic boundary conditions. Such simulations have been carried out by various authors,[12–14] and correlation properties have been observed which are quite different than those of the infinite homogeneous OCP.

SIMULATION RESULTS

For moderate values of Γ ($2 < \Gamma \lesssim 100$) one finds that the density of the cloud exhibits spatial oscillations. These oscillations have maximum amplitude at the cloud surface and decay back to the background density n_0 with increasing distance from the surface. The oscillations are evidence of local order, and the damping length is a measure of the correlation length. As Γ increases, the correlation length increases and the oscillations increase in amplitude until finally the density approaches zero between peaks.[12] For a spherical cloud of 100 charges this occurs at $\Gamma \sim 140$ (see Fig. 2). Thus, the cloud separates into concentric spheroidal shells. In this regime charges rarely move from shell to shell, but they diffuse freely within the shell surfaces. Also, inter-ion correlations within the shells also exhibit decaying oscillations characteristic of a liquid. Thus, the system might be characterized as a liquid within the shells, and as a solid in the direction perpendicular to the shells, as in a smectic liquid crystal.[12]

For still larger values of Γ particle diffusion within the shells also approaches zero and an imperfect 2-D hexagonal crystal is formed within each shell (see Fig. 3).[12–13] Correlation functions and diffusion coefficients have been measured for a range of cloud shapes and Γ values in both the smectic and crystalline regimes.[15]

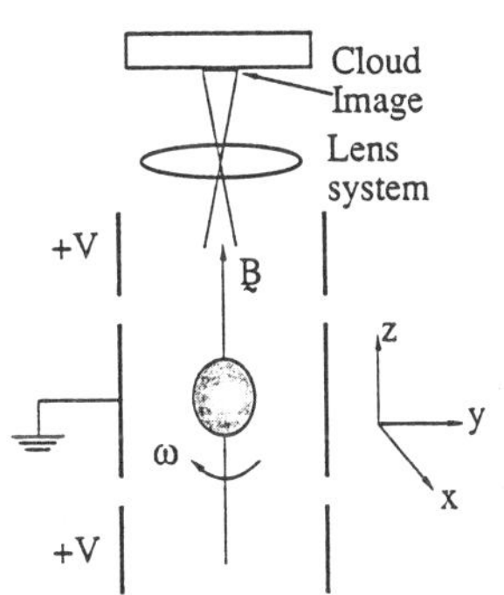

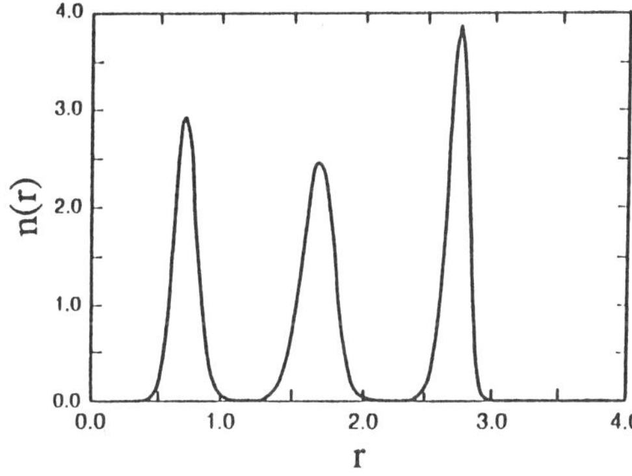

Fig 1. Schematic of cylindrical Penning trap and imaging system.

Fig 2. Density in a spherical cloud with $N = 100$ and $\Gamma = 140$.

The lattice structure in the crystalline regime is quite different from the bcc lattice predicted for an infinite homogeneous OCP. Simulations have been performed on up to 2000 ions, and no evidence of bcc structure has been observed. However as $N \to \infty$ the system becomes infinite and homogeneous and one would expect to observe bcc structure within the cloud. Recent theoretical work based on the extremely oblate "slab" limit of the bounded crystal predicts that the system may have to be quite large before the minimum energy state displays bcc symmetry, perhaps requiring as many as 60 shells before the infinite homogeneous structure is recovered.[16] The calculation is based on the fact that the bulk free energy of the infinite bcc lattice is only a few parts in 10^4 less than that of other lattice configurations such as hcp or fcc. Thus, the addition of a surface term to the free energy can change the thermal equilibrium lattice structure, even if this surface term is relatively small compared to the bulk term.

EXPERIMENTS

We now briefly describe some of the experimental results. Clouds of up to 15,000 Be^+ ions are trapped at densities of 10^7–10^8 cm^{-3} and cooled to temperatures as low as 10 mK using the technique of "laser cooling."[17] These temperatures and densities correspond to values of Γ in the range of several hundred, so strong-correlation behavior is expected. In particular, the prediction that the cloud forms concentric spheroidal shells has been confirmed by using an optical imaging system to probe the cloud density (see Fig. 1). A "probe" laser beam is passed through the cloud in the x-y plane. The laser frequency is adjusted to an internal resonance in the ions, which therefore fluoresce. The fluorescence is focused by a series of lenses onto the photocathode of a photon counting imaging tube to provide a cross-sectional image of the cloud density. If the cloud consists of concentric shells, the fluorescence shows intensity peaks at the density maxima. Such an image is shown in Fig. 4 (taken from Ref. 1), which shows the cloud fluorescing in the light provided by a probe beam and two laser cooling beams. The spacing between shells, D, is given by $n_0 D^{1/3} \cong 1$ which is within experimental error of the values $n_0 D^{1/3} \sim .92$–.96 predicted by theory and simulation.[14,15] In some cases, rather than forming closed shells,

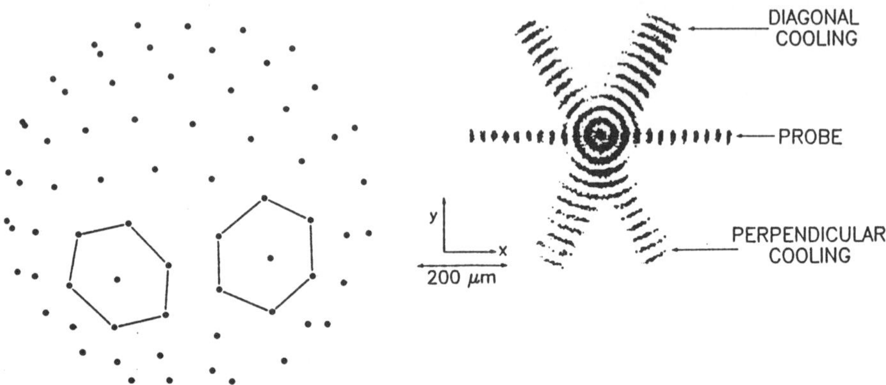

Fig 3. Projection onto plane of T=0 equilibrium state for ½ of outer shell of $N = 256$ spherical cloud.

Fig 4. Image of Be^+ cloud of 15,000 ions (from Ref. 1).

the clouds condense into nested concentric cylinders with open ends. This behavior is not yet understood and may be due to shear in the cloud rotation induced by external torques, the presence of impurity ions, etc.

Experiments on small numbers of ions have also been carried out.[18,19] When only a few charges ($N \sim 1$–10) are trapped and cooled they condense into various simple geometrical configurations. "Phase transitions" between different minimum energy configurations are predicted to occur as the trap voltages are varied.[20] Nonlinear behavior of these few degree of freedom systems has also been studied.[21] The experimental work on Coulomb clusters has so far been carried out in radio frequency "Paul" traps rather than in Penning traps. However, one can show that the equilibrium configurations are almost identical in the two types of trap.

Some experiments have also been performed in which two or more types of ion are confined. The ions' different charge to mass ratios should then lead to centrifugal separation of the ions in the spinning cloud[22]; this has been observed in clouds of Be^+ and Hg^+ ions.[23]

ACKNOWLEDGMENT

The authors gratefully acknowledge useful discussions with D. Wineland, J.J. Bollinger, and W. Itano. This work is supported by NSF PHY87-06358, ONR N00014-82-K-0621 and a grant of computer time from the San Diego Supercomputer Center.

REFERENCES

1. S. Gilbert, J. Bollinger and D. Wineland, Phys. Rev. Lett. **60**, 2022 (1988).

2. J.H. Malmberg and C.F. Driscoll, Phys. Rev. Lett. B44, 654 (1980).

3. C.F. Driscoll and J.H. Malmberg, Phys. Rev. Lett. **50**, 167 (1983).

4. J.H. Malmberg, T.M. O'Neil, A.W. Hyatt and C.F. Driscoll, Proceedings of the 1984 Sendai Symposium on Plasma Nonlinear Phenomena, Tohoku Univ., Sendai, Japan, p. 31 (1984).

5. S.A. Prasad and T.M. O'Neil, Phys. Fluids **22**, 278 (1979).

6. J. Bollinger and D. Wineland, Phys. Rev. Lett. **53**, 348 (1984).

7. L. Turner, Phys. Fluids **30**, 3196 (1987).

8. J.H. Malmberg and T.M. O'Neil, Phys. Rev. Lett. **39**, 1333 (1977).

9. D.H.E. Dubin and T.M. O'Neil, Phys. Fluids **29**, 11 (1986).

10. S. Ichimaru, H. Iyetomi and S. Tanaka, Phys. Rep. **149**, 91 (1987).

11. S. Brush, E. Sahlin and E. Teller, J. Chem. Phys. **45**, 2102 (1966); E.L. Pollock and J.P. Hansen, Phys. Rev. A **8**, 3110 (1973); W.L. Slattery, G.D. Doolen and H.E. DeWitt, Phys. Rev. A **26**, 2255 (1982); S. Ogata and S. Ichimaru, Phys. Rev. A **36**, 5451 (1987); D.H.E. Dubin, Phys. Rev. A (submitted, 1989).

12. D.H.E. Dubin and T.M. O'Neil, Phys. Rev. Lett. **60**, 511 (1988).

13. J. Schiffer, Phys. Rev. Lett. **61**, 1843 (1988).

14. H. Totsuji, in Strongly Coupled Plasma Physics, F.J. Rogers and H.E. DeWitt, editors (Plenum, NY, 1986), p. 19.

15. D.H.E. Dubin and T.M. O'Neil, To be published in *Proceedings of the XXIV Yamada Conference*, Lake Yamanaka, Japan (August 1989).

16. D.H.E. Dubin, Phys. Rev. A **40**, 1140 (1989).

17. D.J. Wineland and W. Itano, Physics Today, p. 2 (June 1987).

18. F. Deidrich, E. Peik, J. Chen, W. Quint and H. Walther, Phys. Rev. Lett. **59**, 2931 (1987).

19. D. Wineland, J. Berquist, W. Itano, J. Bollinger and C. Manney, Phys. Rev. Lett. **59**, 2935 (1987).

20. D.H.E. Dubin (in preparation).

21. J. Hoffnagle, R.G. DeVoe, L. Reyna and R.G. Brewer, Phys. Rev. Lett. **61**, 255 (1988).

22. T.M. O'Neil, Phys. Fluids **24**, 1447 (1981).

23. D.J. Larson, J.C. Berquist, J.J. Bollinger, W.M. Itano and D.J. Wineland, Phys. Rev. Lett. **57**, 70 (1986).

Use of the Positron as a Plasma Particle[*]

C.M. Surko and T.J. Murphy
Physics Department
University of California, San Diego
La Jolla, CA 92093

ABSTRACT

The use of positrons in laboratory plasma physics experiments is considered. Recent progress in this area is discussed, including the creation of a single-component positron plasma in the laboratory. Specific applications of such antimatter plasmas are also discussed, with emphasis on areas where existing plasma physics technology and that currently under development are likely to produce results in the next few years.

PACS numbers: 52.25.Wz, 52.40.-w, 71.60 + z

---

[*]Published in <u>Phys. Fluids</u> B2, 1372 (1990)

I.   INTRODUCTION

The positron is an attractive choice as a plasma particle. It is the antiparticle of the electron and has the same mass and magnitude of charge as the electron but is positively charged. Thus a plasma composed of positrons and electrons is unique. It is also relatively ideal, in the sense that neither of the plasma particles has complicating internal structure such as excited or metastable states. Finally, over a wide range of parameters, annihilation of electrons and positrons, which is the analog of recombination in plasmas composed of ions and electrons, is relatively unimportant: even at an electron density of 1×10^{12} cm^{-3} and a temperature as low as 1 eV, the positron annihilation time is greater than 1 sec.

The positron has a considerable history in plasma physics. Alfvén has discussed the role which electron-positron plasmas might play in astrophysical situations.[1] The annihilation of positrons in a plasma has also been considered in detail.[2] Waves and instabilities in electron-positron plasmas were discussed theoretically by Tsytovich and Wharton.[3] Finally, in a novel laboratory experiment in the 1960s, particle confinement in a magnetic mirror was studied using fast positrons with energies in the range of several hundred keV. In this case, small amounts of a radioactive isotope of neon were introduced into the device, and the positrons were then produced by radioactive decay of the neon atoms.[4] That positrons have not yet become a common tool in laboratory plasma physics is due to the fact that slow positrons are relatively difficult to produce and handle.

The main point of the present paper is that our ability to accumulate and store positrons in the laboratory has improved significantly in the last decade, and we expect considerable progress in this area in the near future. This progress has come in two areas. There is now a relatively large scientific community interested in laboratory positron physics.[5] In addition, there are now relatively efficient methods available to produce slow positrons. The key to this technology was the development of materials which are capable of converting fast positrons from either radioactive decay or from pair production (which have energies of several hundred keV) to slow positrons with energies of a few eV. For example, single-crystal tungsten[6,7] moderators have now been developed with efficiencies of 0.1%. The other area in which progress has been made is the confinement and manipulation of single-component plasmas. In particular, techniques are now available to produce dense, quiet, and cold single-component electron plasmas, and many aspects of the confinement and cooling of these plasmas are now understood.[8,9] These techniques can be utilized more or less directly in the production of positron plasmas.

We have recently developed a method of accumulating slow positrons.[10,11] Using this trapping scheme, we have been able to accumulate and store 3×10^5 positrons for 60 sec.[11] The positrons cool to room temperature in about 3 sec. The resulting Debye length is slightly smaller than the size of the charge cloud, and thus the charge cloud is a single-component, positron plasma.[11] In this article, we review the essential features of this trapping scheme. We discuss areas in which progress might be made in the near future and the characteristics of the positron plasmas one can expect to be able to create. Finally, we discuss potential applications of such accumulations of large numbers of positrons and trapped positron plasmas.

II.   POSITRON PLASMAS IN THE LABORATORY

A.   The Positron Trapping Experiment

We have created a single-component positron plasma in a modified version of a Penning trap.[11]   The trap consists of a set of cylindrical electrodes coaxial with an 800 G magnetic field.   The electrodes create a confining potential well in the axial direction, and the magnetic field provides radial confinement.   Positrons from a radioactive ^{22}Na source were moderated by a single-crystal tungsten moderator and then guided into the region of the potential well.   To become trapped in the well, the positrons must lose energy.   As shown in Fig. 1, this is accomplished by inelastic scattering in low-pressure nitrogen gas.   This energy loss must be sufficiently rapid to trap the positrons before they transit the region of the well.   However, scattering of the positrons by the nitrogen gas also tends to produce radial diffusion to the walls of the trap.   In order to minimize this diffusion loss, three stages with progressively lower nitrogen pressures were used.   The trap is operated with these three stages at pressures of approximately 10^{-3} Torr, 10^{-4} Torr, and 10^{-6} Torr, with each stage at a progressively lower electrostatic potential.

Using an 80 mCi ^{22}Na source, as many as 3×10^{5} positrons were accumulated.   The measured confinement time of 60 sec was limited by the annihilation of the positrons with the background nitrogen gas in the final stage of the trap.   The radial distribution of the positron density was measured by dumping the positrons on an annular set of metal collector plates and measuring the resulting annihilation radiation.   The temperature of the positron gas was measured using a "magnetic beach" energy analyzer.[11] This involves measuring the component of the energy of the positrons perpendicular to the magnetic field.   It is accomplished by measuring the shift in the parallel energy distribution of the positrons at the collector as a function of an additional magnetic (mirror) field in the region near the collector.   As shown in Fig. 2,   the positrons were found to quickly lose energy by collisions with the nitrogen molecules, cooling to within 10% of room temperature in 3 sec.

The extent of the charge cloud along the magnetic field was calculated,[9] using the measured radial distribution of the positrons and the applied electrostatic potential, and assuming that the positrons are in thermal equilibrium at room temperature at each radius.   Shown in Fig. 3 is a two-dimensional plot of the density of the resulting positron gas.   The 25% of peak density contour extends approximately 2.3 cm in the direction along the magnetic field and is 3.0 cm in diameter in the direction perpendicular to the magnetic field.

The shape and extent of the density profile in the direction perpendicular to the field is determined both by the initial positron deposition profile and by radial diffusion.   The dominant radial transport process is diffusion in the second stage of the positron trap;   in the final stage, the gas pressure is sufficiently low that the effect of diffusion is small on the scale of the measured positron confinement time.   For the conditions of the experiment, this time is limited by direct annihilation of the positrons with the nitrogen gas.

B.   Work in Progress and a Look to the Future

Several relatively straightforward improvements can be made in the current operation of the trap.   After a positron density of $\sim 1 \times 10^4$ cm^{-3} is achieved operating the trap at the values of background nitrogen pressure described above,   the N_2 pressure can be further reduced by more than one order of magnitude in the third stage of the trap.   Then the positrons coming into the third stage will be trapped by collisions with previously-trapped positrons.   The resulting trapped positron gas would continue to be cooled by collisions with the nitrogen molecules, but now at a slower rate.   In this operating regime, the annihilation time will be increased, so that the resulting confinement time is expected to be $\gtrsim 10^3$ sec.   In the present trap, there are also losses (of a factor of three or so) of the incoming positrons from the source.   This loss is apparently due to magnetic mirroring, caused by the excess perpendicular energy given to the positrons near the moderator by radial electric fields present in the current version of the apparatus.   An additional loss of a factor of two is also present due to less than optimum source/moderator arrangement,   as well as a factor of four loss of positrons going from the first to the final stage of the trap.

It appears that most of these losses can be either greatly reduced or eliminated.   If this is done, the maximum capture rate of the trapping scheme, starting with 2 eV positrons from the moderator, is expected to be limited to 0.2 - 0.25 by positronium atom formation in the first stage of the trap.   We would then expect to be able to accumulate and store $\sim 1 \times 10^9$ e$^+$ for more than 10^3 sec. The plasma temperature is expected to be 300 K and it is estimated to have a characteristic dimension of 10 cm and a volume of 10^3 cm^3, resulting in a Debye wavelength of 2 mm. Such a plasma would then be sufficiently large that a wide variety of plasma physics experiments could be considered.

We have begun to develop techniques to measure the dielectric properties of the positron plasma by exciting and measuring the resonant modes near the plasma frequency. In principle, this could provide a nonperturbative measure of quantities of interest, such as plasma density. In the next step, we envision injecting a passing electron beam to excite waves and instabilities in this unique, equal-mass plasma system. However, such a plasma experiment amounts to a large extrapolation from previous experience. We need to try this experiment in order to determine how easy it will be to study either the linear or nonlinear phenomena in this system.

Alternative trapping schemes have been discussed elsewhere. For example, in the paper by Tsytovitch and Wharton[3], they discuss the production of positrons by a high-current, 4-MeV electron beam, and they propose trapping the resulting electron-positron plasma in a magnetic mirror. A positron trapping scheme has also been proposed which uses RF at the magnetron frequency to pump positrons radially inward.[12] In this scheme, the positron source is located off axis, and the inward motion must be sufficiently rapid that the positrons do not return to the source. Another scenario using a multiple-stage trap has also been discussed.[13] In this scheme, the moderator is ramped in potential and the particles are compressed into another stage by a peristaltic action.[13] They are then cooled by other means, such as a remoderator.

There is much to be gained if one could efficiently capture fast positrons directly from a radioactive source (i.e., thereby avoiding use of the 0.1% efficient moderator).   However, the fast positrons are distributed in phase space and have a broad spectrum of energies up to 540 keV.   At present, we are not aware of a scheme which will do this, but we know of no reason in principle that this cannot be done.

III.   OTHER APPLICATIONS OF TRAPPED POSITRONS

There are many other potential uses of large numbers of trapped positrons.   For example, we are developing techniques to use positrons to study transport in hot plasmas such as those in tokamaks.[14,15]   As shown schematically in Fig. 4, we envision that this can be done by converting the trapped positrons to fast positronium atoms.   The atoms are then injected into the plasma and become ionized by collisions with the plasma particles, thus depositing the positrons in the core of the plasma.   The positrons will quickly thermalize with the plasma electrons.   They then act as a thermalized, identifiable, electron-mass test particle.   Such positron transport experiments are likely to shed light on important aspects of the observed anomalous transport in tokamak plasmas which are not conveniently accessible by other means.   For example, a key question that such an experiment might address is the importance of magnetic braiding in anomalous energy transport.   These experiments will require the trap improvements described above.   For a device the size of the TEXT tokamak[16], the optimum energy of the positronium atom beam is of the order of 100 eV. We are currently in the process of building the apparatus to efficiently create such a positronium beam.   This will be done by passing the trapped positron plasma through either a thin (30-50A) carbon foil or a H_2 gas cell.

Another application of trapped positron plasmas is modelling astrophysical processes in the laboratory. Of particular interest is the effect of free electrons on the slowing of positrons in the galactic medium and the resulting annihilation linewidth. If the rate of slowing of the positrons is increased, the positrons spend less time above the threshold for positronium atom formation, and the Ps fraction is reduced.[17] Experiments in which positrons are injected into partially-ionized hydrogen plasmas in order to determine the relevant annihilation schemes would help to establish the nature of the annihilation media near the center of the galaxy.[18]

Also of interest is the annihilation of positrons in the envelopes of novae.[19,20] It has been suggested that the annihilation of positrons produced in the radioactive decay of ^{13}N and ^{18}F might be detectable.[19] Experimental studies of the annihilation of positrons in partially ionized plasmas could answer questions about the fraction of positrons going into specific modes of annihilation as a function of plasma parameters such as electron density, electron temperature, and the degree of ionization of the gas.

From experience with single-component electron plasmas, it is known that one can cool such plasmas to low temperatures (e.g., $T \lesssim 50K$). In this case, one could envision creating high emittance positron sources which might be useful in a number of applications, such as a e^+ injector for particle accelerators. A related use of bright e^+ sources would be the production of hard x rays by channeling positrons in crystals.

Trapped positron plasmas are also likely to be useful in the production of antihydrogen.[21]   Recently, antiprotons have successfully been trapped,[22] and antihydrogen can be formed in three-body collisions of the antiprotons with positrons.   It is expected that this will occur with reasonable efficiency if a cold (i.e., 4 K) positron plasma is combined with the trapped antiprotons.[23]

IV.   CONCLUDING REMARKS

We are now at the stage where we can begin to produce very modest e^+ plasmas in the laboratory reliably and relatively efficiently.   We envision that improvements in this capability will allow us to address a range of new and interesting scientific problems.

V.   ACKNOWLEDGEMENTS

We wish to acknowledge the collaboration of M. Leventhal, A. Passner, and F.J. Wysocki in the work described here.   We also wish to acknowledge useful conversations with C.F. Driscoll, J.H. Malmberg, A.P. Mills Jr., T.M. O'Neil, and W.L. Rowan.   This work is supported by the Office of Naval Research,   and the fusion diagnostic experiment is supported by the U.S. Department of Energy.

1.   H. Alfvén, <u>Cosmic Plasma</u> (D. Reidel Co., Boston, 1981), and references therein.

2.   W. Wolfer, Ph.D. Thesis, U. of Florida, 1969 (available from University Microfilms, Ann Arbor, MI); R.J. Gould, Physica **60**, 145 (1972) and Ap. J. **344**, 232 (1989); and references therein.

3.   V. Tsytovitch and C.B. Wharton, Comments on Plasma Physics and Controlled Fusion **4**, 91 (1978).

4.   G. Gibson, W.C. Jordan, and E.J. Lauer, Phys. Rev. Lett. **5**, 141 (1960); Phys. Fl. **6**, 116 (1963), and ibid. **6**, 133 (1963).

5.   See, for example, <u>Positron Annihilation</u>, by L. Dorikens, M. Dorikens, and D. Segers (World Scientific, Singapore, 1989).

6.   E. Gramsch, J. Throwe, and K.G. Lynn, Appl. Phys. Lett. **51**, 1862 (1987).

7.   N. Zafar, J. Chevallier, F.M. Jacobsen, M. Charlton, and G. Larrechia, Appl. Phys. **A47**, 409 (1988).

8.   J.H. Malmberg and C.F. Driscoll, Phys. Rev. Lett. **44**, 654 (1980).

9.   C.F. Driscoll, J.H. Malmberg and K.S. Fine, Phys. Rev. Lett. **60**, 1290 (1988).

10. C.M. Surko, A. Passner, M. Leventhal, and F.J. Wysocki, Phys. Rev. Lett. **61**, 1832 (1988).

11. C.M. Surko, M. Leventhal and A. Passner, Phys. Rev. Lett. **62**, 901 (1989).

12. G. Gabrielse and B.L. Brown, in <u>The Hydrogen Atom</u>, G. Bassini, M. Inguscio, and T.W. Hansch, eds. (Springer Verlag, New York, 1989), p. 196.

13. R. Conte, A. Rich, D. Gidley, M. Skalsey, J. Van House, H. Poth, W. Schwab, B. Seligmann, M. Wortge, and A. Wolf, Hyperfine Interactions **44**, 201 (1988).

14. C.M. Surko, M. Leventhal, W.S. Crane, A. Passner, F.J. Wysocki, T.J. Murphy, and J. Strachan, and W.L. Rowan, Rev. Sci. Instrum. **57**, 1862 (1986).

15. T.J. Murphy, Plasma Physics and Controlled Fusion **29**, 549 (1987).

16. K.W. Gentle, Nuclear Technology/Fusion **1**, 479 (1981).

17. R.W. Bussard, R. Ramaty and R.J. Drachman, Ap. J. **228**, 928 (1979).

18. B.L. Brown and M. Leventhal, Phys. Rev. Lett. **57**, 1651 (1986).

19. D.D. Clayton and F. Hoyle, Ap. J. (Letters) **187**, L101(1974).

20. M.D. Leising and D.D. Clayton, Ap. J. **323**, 159 (1987).

21. G. Gabrielse, S.L. Rolston, L. Haarsma, and W. Kells, Phys. Lett. **A129**, 38 (1988).

22. G. Gabrielse, X. Fei, L.A. Orozco, R.L. Tjoelker, J. Haas, H. Kalinowsky, T.A. Trainor, and W. Kells, Phys. Rev. Lett. **63**, 1360 (1989).

23. M.E. Glinsky and T.M. O'Neil, Bull. Am. Phys. Soc. **34**, 1934 (1989), and private communication.

Fig. 1.   The three-stage positron trap.   Shown schematically is the electrostatic potential V(z) along the direction of the applied magnetic field. Positrons incident from the right make an inelastic collision, A, with N_2 gas in the high pressure (10^{-3} Torr) region in their first transit through the potential well, and they are then trapped.   They then make similar collisions, B and C, to finally become trapped near the bottom of the potential well at an N_2 gas pressure of 10^{-6} Torr, for which cross-field diffusion is relatively unimportant.   The time spent in the trap before transition B is typically less than 1 msec, and the time spent before transition C is of the order of 30 msec.

Fig. 2.   The average positron energy <E> minus that corresponding to room temperature (0.038 eV) is shown as a function of the positron storage time. The positrons cool to within 10% of room temperature in 3 sec.   The N_2 gas pressure was 1.5 x 10^{-6} Torr.

Fig. 3.   The density contours of the trapped positron gas in units of the central density, $n_0 = 2$ x 10^4 cm^{-3}.   The Debye lengths corresponding to the contours shown are 1.0 cm (0.75 n_0), 1.2 cm (0.5 n_0) and 1.7 cm (0.25 n_0).

Fig. 4.   A schematic diagram of the positronium atom beam diagnostic for hot plasmas.   The positrons are deposited in the plasma in a pulse in time, and the subsequent positron transport is measured by the arrival times of the positrons at a limiter or divertor plate.

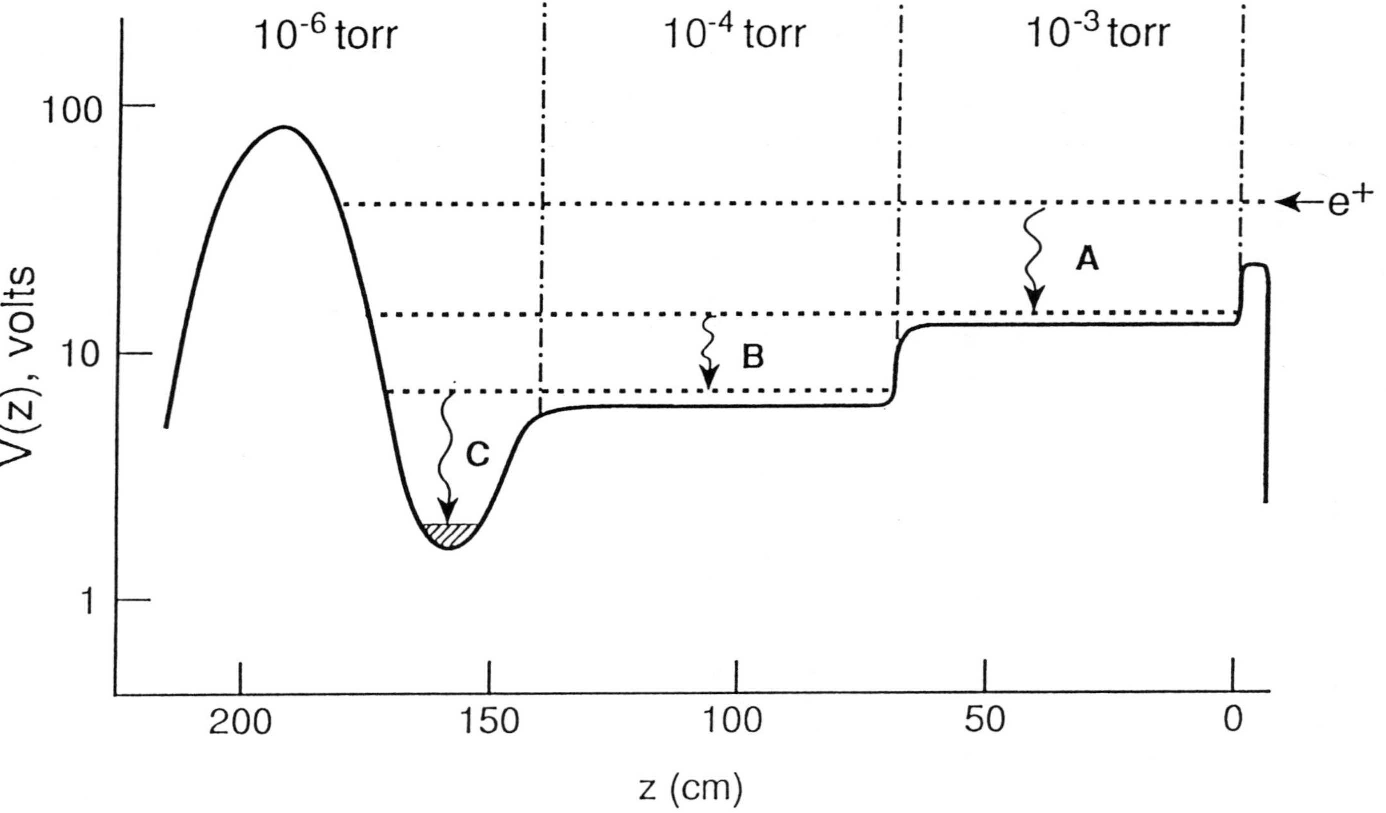

10^{-6} torr
10^{-4} torr
10^{-3} torr
100
10
1
V(z), volts
e+
A
B
C
200
150
100
50
0
z (cm)

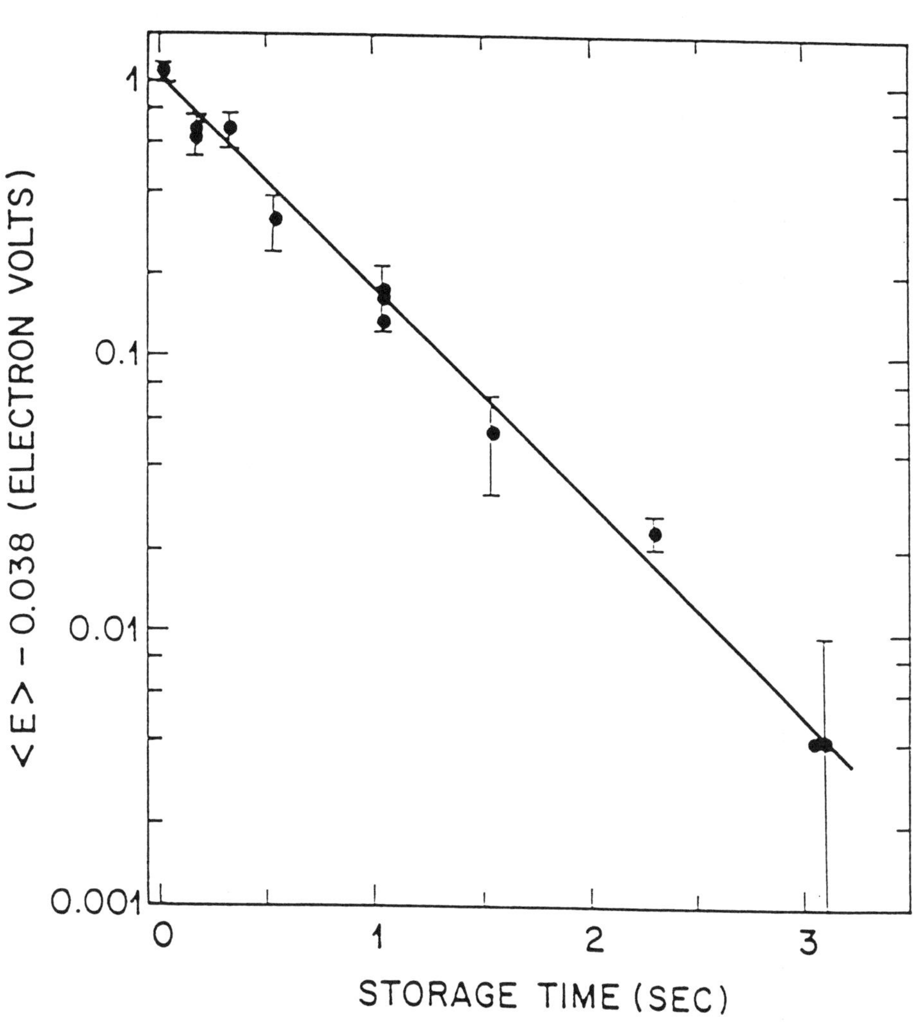

FIG. 2

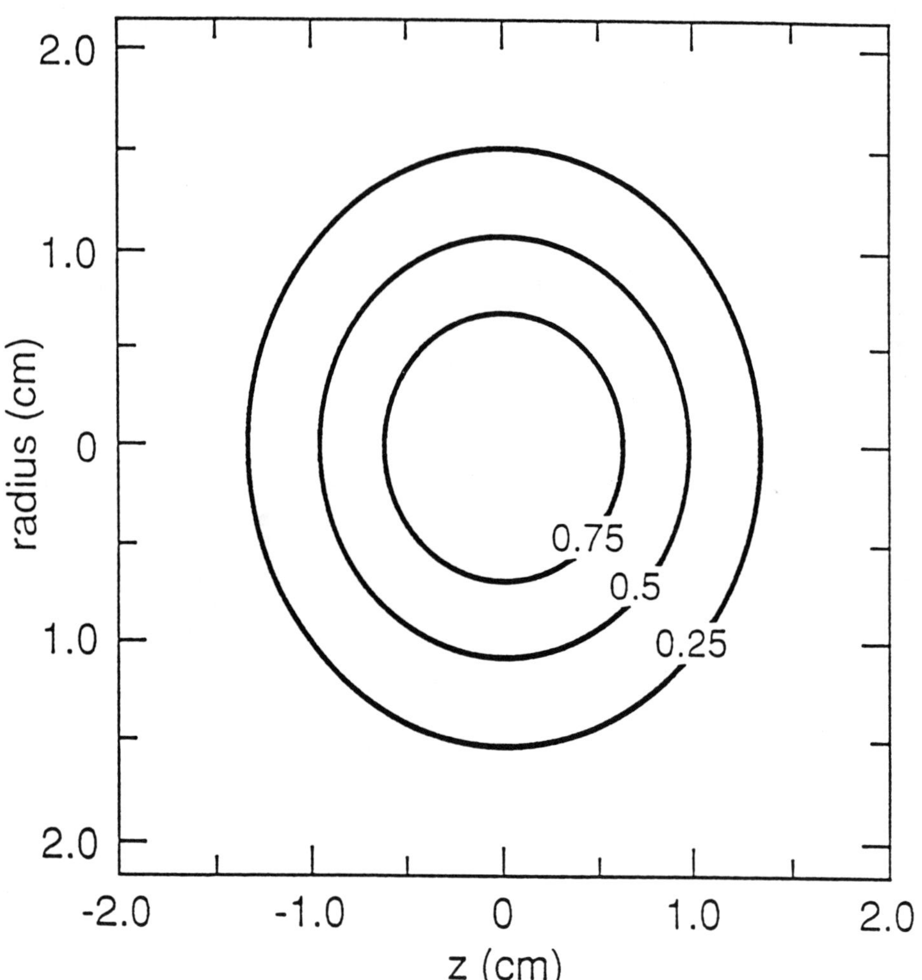

FIG. 3

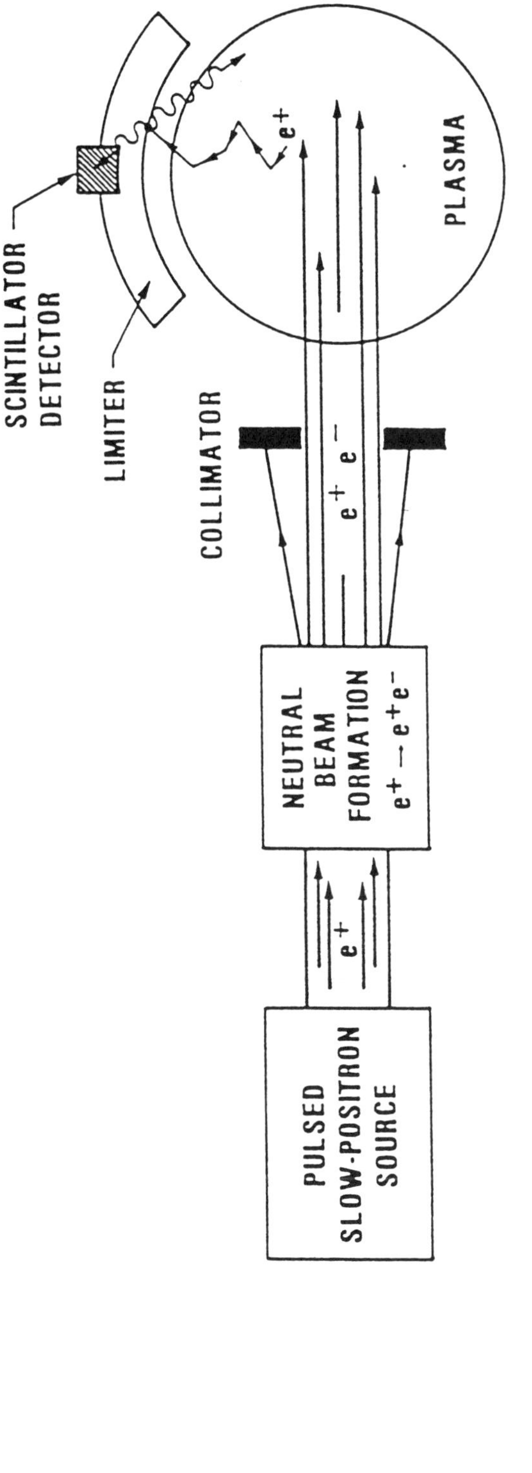
SCINTILLATOR DETECTOR
LIMITER
COLLIMATOR
PLASMA
e+
e+ e−
NEUTRAL BEAM FORMATION e+ → e+e−
e+
PULSED SLOW-POSITRON SOURCE

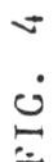

FIG. 4

Part VI

UPPER ATMOSPHERIC, SPACE AND ASTROPHYSICAL PLASMAS

Theory and Simulation of Discrete Packets $F_s(t)$ of Waves of Pc-1 Micropulsations Generated in the Magnetosphere

Ya. L. Alpert

Harvard–Smithsonian Center for Astrophysics, Cambridge, Massachusetts

Results and a method of simulation of Pc-1 micropulsations are given, based on the joint solution of a linear oscillating equation and a system of nonlinear Korteweg, de Vries, and Burgers differential equations. The latter describe slow, weak oscillatory and monotonic shock waves. Physically, it is a modulation of the fast oscillations, excited in the magnetosphere, due to the cyclotron instability, by these slow shock waves. They form discrete signals of various shapes. Diagnosis of some local properties of the magnetosphere is achieved by these calculations. The simulation reproduces well the shape of the experimental signals and permits us to estimate the parameters of the region where the signals are excited. They are in good agreement with the known data. This method may help to understand a variety of discrete processes observed in nature.

INTRODUCTION

About three decades ago, very interesting electromagnetic phenomena were discovered in the magnetosphere, namely, the excitation of geomagnetic micropulsations Pc of various kinds. Some of the most unique of them are the so-called Pc-1 pulsations in the frequency range $0.2\,\mathrm{Hz} < f < (2-3)\,\mathrm{Hz}$, $f < F_H$; F_H is the gyrofrequency of the ions of the magnetoplasma [see, for example, *Jacobs, 1970; Gulelmi and Troitskaya, 1973*].

The adequate physically clear term of Pc-1 micropulsations, namely, "hydromagnetic whistlers," was introduced, it seems, first by *Tepley* [1961]. This term realizes the fact that these ion whistler waves are propagating like whistling atmospherics – electron whistler waves in the frequency range $f_L \ll f < f_H$, being guided along the Earth's magnetic force lines that pass through the source location in the magnetosphere [*Storey*, 1957; f_L and f_H are the low-hybrid and gyrofrequencies].

Since the Pc-1 pulsations are multiply reflected in the magnetically conjugate points, they form trains of signals recorded in the observation point. One of the best published dynamical spectrograms of such a train is given in Figure 1 [*Liemohn*, 1969]. Fifteen dynamical spectra of multiple reflected micropulsations Pc-1 are recorded in this figure; the time of the process was about one hour; the central frequency of these signals $f_{os} = 0.52\,\mathrm{Hz}$. Such processes last for two or more hours in many cases, and dozens of Pc-1 signals are recorded. Of course, during such long intervals of time, the conditions in the magnetosphere are changing, and some properties of the micropulsations should also be changed. Namely, the study of the behavior and of the evolution of the dynamical spectra of such trains of Pc-1 micropulsations in different conditions and their connection with the state of the magnetosphere was and is the main aim and the experimental method of this wide and interesting field of investigations until now.

The mechanism of the excitation of Pc-1 signals is in general known. It is the gyroresonance instability of the magnetoplasma under the influence of hot anisotropic beams of protons [see *Kennel and Petschek, 1966; Gendrin et al., 1971; Alpert, 1983*]. But until now, the discreteness of Pc-1 micropulsations and the very diverse shape of these packets of waves $F_s(t)$ remained an enigma. The shape of $F_s(t)$ is even changing sometimes from signal to signal in one train.

Some records $F_s(t)$ of such packets of waves of different kind are shown for illustration in Figures 2 and 3. The most simple of them with one maximum are shown in Figures $2a, b, c$. They were recorded at Sogra (Russia, geomagnetic latitude 57.9°N). Two signals $\langle a \rangle$ and $\langle c \rangle$ were recorded at conditions when the level of the "noise" – the background of oscillations of the magnetoplasma – was very low. The shape of $F_s(t)$ of all of these trains was not changed. But the behavior of $F_s(t)$ of another train, shown in Figure $2d$, was more complicated. The envelopes of these signals were also smooth, but their shape was changing signal by signal. One of them ($s = 1$) has two maxima, another ($s = 10$) has three maxima, and two ($s = 16$ and 19) have one maximum. Packets of waves $F_s(t)$ of very complicated, even irregular, shape, are shown in Figure 3 (France, Husafell, geomagnetic latitude 70.2°N). This train of 20 signals lasted about two hours. No signals of similar shape were in this train.

The consideration of experimental data of Pc-1 packets of waves $F_s(t)$ like those noted above induces the following ideas.

Firstly, we have to learn more deeply the nature of the diversity of these signals and of their properties. The methods of recording and analyzing the experimental data should be correspondingly developed and expanded in comparison with those that were and are still used. The time behavior of the records of the wave packets $F_s(t)$ should be studied in more detail. One of these methods may be the detail analysis of their Fourier spectra. However, in order to find new characteristics of this process, the time resolution of the records and the frequency resolution of the Fourier analysis should be sufficiently high.

Secondly, can we find a general physical mechanism and adequate mathematics that simulate on the whole the process of generating and shaping the variety of discrete packets of waves $F_s(t)$ discussed here?

The Fourier spectra of trains of Pc-1 wave packets $F_s(t)$ of different kinds were studied in the past by *Alpert and Fligel* [1985, 1987], *Alpert* [1987a], and *Alpert et al.* [1989]. It is appropriate to point out here very briefly two of the main new results of these investigations. They are the following.

1. Symmetric *side bands* – satellites $f_{sk}^{\mp}$ appear in the Fourier spectra of the wave packets $F_s(t)$. Their frequencies seem to determine the *resonance* frequencies $\Delta f_{sk}^{\mp}$ of the magnetospheric cavity where these wave packets are generated. Namely,

$$\overline{|f_{sk}^{\mp} - f_{so}|} = \overline{\Delta f_{sk}^{\mp}} = (0.011\text{–}0.012, \quad 0.022\text{–}0.024,$$

$$0.034\text{–}0.035, \quad 0.045\text{–}0.047) \;\; \text{Hz} \;\; , \qquad k = 1,2,3,4$$

where f_{so} are the central frequencies of the spectra. The frequencies $\Delta f_{sk}^{\mp}$ closely coincide with the frequencies of Pc-3 and Pc-4 geomagnetic pulsations.

2. The spectra of Pc-1 signals often have intensive maxima at the double frequency $f = 2f_{so}$ and weaker maxima at $f = 3f_{so}$. This means that the process of generation of these signals has *nonlinear characteristics* – it is accompanied by the generation of *twofold* and *threefold* Pc-1 signals.

Relating to the simulation of the Pc-1 wave packets $F_s(t)$ of different shape, the aim of this paper is to describe an appropriate sufficiently method that was used by the author. In a few words, the main idea of this method is the following.

The *harmonical oscillations* $Y(t)$ generated at the frequency f_{so}, due to the gyroresonance instability of the magnetoplasma, are *modulated* by the other kind of slower oscillatory waves described by the function $Z(t)$. The interaction of both functions $Y(t)$ and $Z(t)$ forms the wave packets $F_s(t)$. The modulator $Z(t)$ should be of a nonlinear nature. This necessary condition is self-evident. Indeed, for example, a smooth signal with one maximum, as shown in Figure 1*a*, may be formed by a tangent hyperbolic function $Z(t) = th(t)$. Then the *"modulation function"* of the oscillatory linear differential equation of the oscillations $Y(t)$ is equal to $[1 - a_1 Z(t/a_2)]$, where $\langle a_1 \rangle$ and $\langle a_2 \rangle$ are numerical coefficients.

The first step of this study to simulate Pc-1 signals of simple shape was done with $Z(t) = th(t/a_2)$ [*Alpert*, 1986]. The results of this study gave hope and stimulated the search for a more general and flexible mathematical method, based on clear physical ideas, which may be used for simulation of $F_s(t)$ functions of more complicated shape. It was found [see *Alpert*, 1987b] that such functions $Z_i(t)$ may be solutions of the nonlinear Korteweg, de Vries, and Burgers equations for stationary running waves. These functions describe nonlinearly weak, slow oscillatory shock waves. In the coordinate system, moving with the velocity V_o, the velocity of these waves is $V_{oi} \ll V_o$. In our case, $V_o = V_A$, where V_A is the Alfven velocity. This method was used to simulate Pc-1 pulsations of more complicated shape, as shown in Figure 2d. By making relatively small changes in the velocity V_{oi}, the appropriate functions $Z_i(t)$ reproduce well the envelopes of all the signals $F_s(t)$ of this train. Furthermore, the results of this study may be used for diagnosis of the properties of the medium where the signals were excited. As a result, the characteristics of the hot anisotropic beam – the source of the instability of the magnetoplasma – were estimated. They are in good agreement with the data obtained in different experiments, particularly on satellites (see §3 below).

The results of the processing of these experimental data and of signals of more complicated shape, similar to those shown in Figure 3, are described in detail in this paper. Some of them may be simulated by a sum of two oscillatory weak shock waves, namely, by functions $Z(t) = \alpha_1 Z_1(t) + \alpha_2 Z_2(t)$. It seems that more than two functions $Z_i(t)$ should often be used. In addition, although the weak slow shock waves play the main role in the shaping of the functions $F_s(t)$, some statistical mechanisms may also occur and complicate this process. However, these problems should be the subject of another study. It should also be pointed out that the method used in this study may provide a clue to the understanding of discrete process of other kinds. Such processes are observed rather frequently in nature.

1. Excitation and Shaping of Discrete-Isolated Packets of Waves

The generation of discrete packets of waves $F_s(t)$ in different media may be described by the linear oscillatory differential equation

$$\frac{d^2 Y}{dt^2} - 2\gamma_{io}\left[1 - \alpha_o \cdot Z(t)\right]\frac{dY}{dt} + \omega_o^2 = 0 \ , \tag{1}$$

where $\omega_o = 2\pi f_o$ is the angular frequency of the harmonical oscillations $Y_o(t)$ when $\gamma_{io} = 0$.

Let us now assume that $\gamma_{io} > 0$. Then, depending on the sign of the multiplier of the first derivative of equation (1), the value

$$\gamma_i(t) = 2\gamma_{io}[1 - \alpha_o \cdot Z(t)] \tag{2}$$

may be > 0 or < 0, i.e., it may represent the *growth rate* or the *attenuator factor* of equation (1). As a result, depending on the behavior of the function $Z(t)$ in time $\langle t \rangle$, the solution $Y(t)$ of equation (1) is changing its shape. If from the very beginning at $t = t_o$ the value $\gamma_i(t) > 0$, the function $Y(t)$ will increase and the process of excitation of a signal $F_s(t)$ will have begun. Then, following the behavior of $Z(t)$, its envelope may be smooth or become an oscillation function, and finally it will attenuate exponentially up to zero if $\alpha_o \cdot Z(t) > 1$, when $t > \tau_s$, $t \to \infty$. A discrete packet of oscillations $F_s(t)$ is formed

$$F_s(t) \neq 0 \ , \qquad \text{when } t > t_o \ , \quad t \underset{\sim}{<} \tau_s \ ,$$
$$F_s(t) = 0 \ , \qquad \text{when } t < t_o \ , \quad t > \tau_s \ . \tag{3}$$

The central frequency of this packet of waves $f_{so} < f_o$ and its duration τ_s both depends on the character of the attenuation of the solution of equation (1). That is self-evident: it is assumed that the oscillatory character of the function $Z(t)$ is much slower than the oscillations of $Y(t)$. Schematically, the described process is shown in Figure 4. It is clear that if $Z(t)$ is similar to the tangent-hyperbolic function, it forms a smooth signal $F_s(t)$ with one maximum (see Fig. 4a). But oscillatory functions $Z(t)$, similar to the one shown in Figure 4b, form oscillatory envelopes of $F_s(t)$ with the appropriate numbers of maxima and minima of $Z(t)$.

Physically, the described theoretical method of formation of discrete-isolated signals may be considered as modulation of fast oscillations by slower oscillations, i.e., $Z(t)$ is the "modulator" and $[1 - \alpha_o \cdot Z(t)]$ is the "modulation function" of this process. The constant α_o regulates the shape of the envelope of $F_s(t)$ and satisfies the obvious conditions

$$\alpha_o \cdot Z_{min} < 1 \ , \qquad\qquad \alpha_o \cdot Z_{max} > 1 \ , \tag{4}$$

where Z_{min} and Z_{max} are the minimum and maximum values of the modulator $Z(t)$.

It is clear that the problem of simulation – of shaping any discrete signals – is reduced to finding appropriate modulators $Z(t)$ and to choosing the most suitable constant α_o. If the modulator is the sum of several functions $Z_i(t)$, namely $Z(t) = \sum_{i=1,2,\dots} \alpha_i \cdot Z_i(t)$, then the influence of each of these functions should be weighted by the choice of the constants α_1, α_2, etc. (see §3 below).

Certainly, the modulators $Z_i(t)$ should describe a congruous physical mechanism that can simulate, on the whole, the main properties of the variety of the adequate type of discrete signals of different kinds excited in different conditions. It was already noted in the introduction that the slow, weak oscillatory shock waves $Z_i(t)$ described by the nonlinear Korteweg, de Vries, and Burgers equations are very suitable and flexible to solve such a problem. Let us now describe briefly the main properties of this equation and of the equation obtained from it, which is used for shaping the envelopes of discrete packets of waves $F_s(t)$ [see *Korteweg and de Vries, 1895; Burgers, 1940; Karpman, 1970; Kadomtsev, 1976*].

In the coordinate system $\xi = x - V_x \cdot t$, moving with the velocity V_x relative

494   Discrete Packets $F_s(t)$ of Waves of Pc-1 Micropulsations

to the medium (x is the space coordinate), this equation is

$$\beta_o \frac{\partial^3 z}{\partial \xi^3} - \mu_o \frac{\partial^2 z}{\partial \xi^2} + z \frac{\partial z}{\partial \xi} + \frac{\partial z}{\partial t} = 0 \ . \tag{5}$$

The stationary solution of equation (5), $z = z(x - v_o \cdot t)$, $v_o \ll V_x$, is the following:

$$\beta_o \frac{d^3 z}{dx^3} - \mu_o \frac{d^2 z}{dx^2} + (z - v_o) \frac{dz}{dx} = 0 \ . \tag{6}$$

For these waves, equation (5) becomes an ordinary differential equation. This equation is invariant relative to the transformations $z = z + \text{const}$, $v_o = v_o + \text{const}$. Therefore, one step of integration reduces equation (6) to

$$\beta_o \frac{d^2 z}{d\xi^2} - \mu_o \frac{dz}{d\xi} - \left(v_o - \frac{z}{2} \right) z = 0 \ . \tag{7}$$

Equation (7) describes running weak, slow oscillatory shock waves. The values β_o and μ_o in equation (7) are the coefficients of the complex dispersion equation of these waves

$$\omega(k) = kV_x - k^3 \cdot \beta_o + i\mu_o k^2 \ , \tag{8}$$

where $\beta_o = \frac{V_x}{2k^2} = \frac{V_x \cdot \Lambda^2}{8\pi^2}$ cm^3 s^{-1}, μ cm^2 s^{-1} is the viscosity of the medium, $k = \frac{2\pi}{\Lambda} \ll 1$ determines the so-called "characteristic length" – the wave length Λ of the nonlinear waves, and the wave number $k \ll 1$. The behavior of the solution of equation (7) may be understood qualitatively by the analogy with the behavior of a nonlinear oscillator, and in general it can be studied only numerically. The main properties of $z(\xi)$, interesting for our study, are the following:

The shock wave $z(\xi)$ is oscillating, if $\mu_o < \sqrt{4\beta_o v_o}$; $\tag{9}$

its maximum amplitude $z_{max} = 3v_o$, when $z(+\infty) \to 0$.

It is a monotonic shock wave when $\mu_o > \sqrt{4\beta_o v_o}$ and $z(-\infty) \to 2v_o$, $\qquad$ (10)

The Mach number of these waves is $M = 1 + \dfrac{v_o}{V_x}$, $\dfrac{v_o}{V_x} \ll 1$; $\qquad$ (11)

their velocity in the laboratory coordinate system is $V = V_x + v_o$.

When $\mu_o \ll \mu_{cr} = \sqrt{4\beta_o \cdot v_o}$, we have

$$z(\xi) = 2v_o + \text{const} \cdot \exp\left(\frac{\mu_o}{2\beta_o}\xi\right) \cdot \cos\left(\sqrt{\frac{2v_o}{\beta_o}}\xi\right) . \qquad (12)$$

For simulating signals $F_s(t)$ of a different shape, it is very important

that the oscillatory character of $z(\xi)$ be mainly determined by the velocity v_o.

This property of $z(t)$ greatly simplifies the shaping method and it promotes

its flexibility. The viscosity coefficient μ_o at most regulates the damping of the

oscillations of the shock waves. As was pointed out above, equations (7) and

(5) are written in the space coordinate system, $\xi = (x - V_x \cdot t)$. However, the

process of generation and modulation of the wave packets should be considered as

a local process. The signal is produced at a local region around the origin of the

laboratory coordinate system $x = 0$. Therefore, for shaping $F_s(t)$, equation (7)

should be transformed into the time coordinate system t, using the relations

$d\xi = -V_x dt$, $d\xi^2 = V_x^2 dt^2$. Furthermore, a new variable $Z(t) = z(t)/\beta$ s^{-2} may

be used. As a result of these two operations, equation (7) is reduced to

$$\frac{d^2 Z}{dt^2} + \mu \frac{dZ}{dt} - \left(V_o - \frac{Z}{2}\right) \cdot Z = 0 , \qquad (13)$$

where $\mu = \frac{\mu_o}{\beta V_A}$ s^{-1}, $V_o = \frac{v_o}{\beta}$ s^{-2}, and $\beta = \frac{\beta_o}{V_A^2}$ cm $\cdot$ s $= 1$. This equation depends only on two parameters. Below it will be shown that the equality $\beta = 1$ may be satisfied in the problem discussed here.

Thus, the problem of simulation of packets of waves $F_s(t)$ is reduced to the joint solution of two ordinary second-order differential equations. One is equation (1), and the other is a system of nonlinear equations (13), if the modulator $Z(t) = \sum\limits_{i=1,2,\ldots} \alpha_i \cdot Z_i(t)$. An alternative system of interconnected nonlinear equations (13) can also be used, namely,

$$\frac{d^2 Z_i}{dt^2} + \mu_i \frac{dZ_i}{dt} - \left(V_{oi} - \frac{Z_i}{2} \right) Z_i = \sum_{i \neq j} \alpha_{ij} \cdot Z_j \,, \tag{14}$$

where α_{ij} are correlation coefficients of this system and $i, j = 1, 2, 3, \ldots$. Both of these methods were used in this study.

At the end of this section, the evolution of the solution $Z(t)$ of equation (13) for $V_o = $ const and different values of μ is shown in Figure 5.* The maximum values of $Z(t) = 1$, and the time t, μ, and V_o are normalized in this figure and have conventional values. It is seen that by changing μ, the oscillatory character of $Z(t)$ is not changed. Certainly, only the amplitudes of the oscillations are diminished and the number of oscillations is diminished because of their damping with the increase of μ. Other details of the behavior of the $Z(t)$ function will be shown later.

---

* Equation (13) was solved numerically by the Sharp Pocket Computer PC-1500 and plotted on its printer.

2. SIMULATION OF Pc-1 MICROPULSATIONS

The excitation of the plasma oscillations that form the Pc-1 packets of waves is due to the gyroresonance instability of the magnetoplasma. The source of this instability is a beam of hot ions (protons), characterized by the ion density N_{ib}, by the average longitudinal component at the thermal velocity of the beam $V_b \cos \vartheta = V_{bo} = \overline{v_{i\parallel}} = \sqrt{\frac{2T_{i\parallel}}{M}}$, and by the anisotropy coefficient of this beam

$$\Theta_{\perp,\parallel} = \frac{T_{i\perp}}{T_{i\parallel}} = \left(\frac{\overline{v_{i\perp}}}{\overline{v_{i\parallel}}} \right)^2 \quad , \tag{15}$$

where $T_{i\perp}$ and $T_{i\parallel}$ are the transverse and longitudinal temperatures of the beam, relative to the direction of the magnetic field $\overline{H_o}$. This mechanism is described by the gyroresonance equation

$$\omega = (\overline{k}_o \cdot \overline{V}_b) - \Omega_H = \frac{\omega}{c} n_o V_{bo} - \Omega_H \quad , \tag{16}$$

for $\omega <$ or $\ll \Omega_H$, and by the values

$$\gamma_{io} = \sqrt{\pi} \Omega_H \frac{N_{ib}}{N_{io}} \frac{\left(1 - \frac{\omega}{\Omega_H}\right)^{7/2}}{\left(\frac{\omega}{\Omega_H}\right)^2 \left(2 - \frac{\omega}{\Omega_H}\right) \frac{V_{bo}}{V_A}} \exp\left\{ -\left(1 - \frac{\omega}{\Omega_H}\right)^2 \left(\frac{\omega}{\Omega_H} \cdot \frac{V_{bo}}{V_A}\right)^{-2} \right\} \quad ,$$

$$\gamma_i = \gamma_{io} \cdot \left(\frac{T_{i\perp}}{T_{i\parallel}} - \frac{\Omega_H}{\Omega_H - \omega} \right) \quad , \tag{17}$$

[see *Kennel and Petchek, 1966*].

Let us now suppose, following the notation of equation (1) above in §1, that the anisotropy coefficient is varied in time. Then it follows from equation (17) that $\gamma_i(t)$ is the growth rate of this instability only when

$$\Theta_{\perp\parallel}(t) > \frac{\Omega_H}{\Omega_H - \omega} \quad , \tag{18}$$

since $\gamma_{io} > 0$ and does not change its sign [see eq. (17)]. The value $\gamma_i < 0$, i.e., it

becomes an attenuation factor when $\Theta_{\perp\|}(t) < \frac{\Omega_H}{\Omega_H - \omega}$. Furthermore, by shaping

the discrete Pc-1 wave packets [see eqs. (1) and (16)], the modulation function is

$$[1 - \alpha \cdot Z(t)] = -\left[\Theta_i(t) - \frac{\Omega_H}{\Omega_H - \omega}\right] , \tag{19}$$

and

$$\Theta_{\perp\|}(t) = \alpha_o Z(t) + \left(\frac{1}{1 - \frac{\omega}{\Omega_H}} - 1\right) . \tag{20}$$

The modulator $Z(t)$ determines the time behavior of the anisotropy coefficient

$T_{i\perp}/T_{i\|}$, and formula (17) determines the density of the beam N_{ib}.

Thus, by the theoretical calculations of the experimental packets of waves

$F_s(t)$ of these signals by the joint solution of equations (1) and (13), the formulas

(16), (17), and (20) should be used. The meaning of the notations of these

formulas and of other equations used in this case is given below.

The complex dispersion equation of the magnetosphere for $\frac{\omega}{\Omega_H} \ll \cos\Theta$,

$kc \ll \Omega_o$ in powers of the small wave number $k = \frac{\omega}{c} n_o = \frac{2\pi}{\Lambda}$ (n_o is the refractive

index) is

$$\omega * (k) = kV_A \cdot \cos\Theta - k^3 \cdot \beta_o \cdot \cos\Theta + i\mu_o k^2 , \qquad k = \frac{\omega}{c} n_o , \tag{21}$$

as the wave length Λ of these waves is large. This equation contains the first

nonlinear term in the expansion of the general dispersion equation [see *Alpert*,

1983]. It is the nonlinear correction to the linear dispersion law, as it is used by

derivation of the Korteweg, de Vries, and Burgers equations. The dissipation in these equations is taken into account only to a linear approximation.

Equation (21) is similar to equation (8). Only in equation (21), the velocity $V_x = V_A$, V_A is the Alfven velocity and the direction (angle Θ) between the wave vector $\overline{k}$ and the direction of the Earth's magnetic field $\overline{H}_o$ is taken into account. The other values used in equation (21) and in equations (16) to (20) are the following.

The refractive index of these waves is

$$n_o = \frac{n_A}{\cos\Theta}\left(1 - \frac{\omega^2}{\Omega_H^2}\right)^{-1/2}, \qquad \frac{\omega^2}{\Omega^2} \ll \frac{1}{2}\,tg^2\theta\,, \qquad (22)$$

where $n_A = \frac{\Omega_o}{\Omega_H}$ is the Alfven refractive index, $\Omega_o^2 = \frac{4\pi N_{oi}^2}{M}$, $\Omega_H = \frac{eH}{Mc}L$ and N_{oi} are the ions Langmuir and gyrofrequencies and the ion density of the magnetoplasma, M is the mass of ions, and c is the velocity of light. Equations (22) and (16) determine the velocity of the hot beam:

$$V_{bo} = V_A\left(1 + \frac{\Omega_H}{\omega}\right)\left(1 - \frac{\omega}{\Omega_H}\right)^{1/2}. \qquad (23)$$

The coefficient β_o in equation (21) is determined from the expansion of the general dispersion equation [see *Alpert*, 1983] in powers of the small values of k_o. Namely,

$$\beta_0 = \frac{c^2 V_A}{2\Omega_0^2}\cdot\cos^2\Theta \quad\text{if}\quad tg^2\Theta \gg 1. \qquad (24)$$

Large angles between the vectors $\overline{k}$ and $\overline{H}_o$ are indeed adequate for the process of   formation of Pc-1 signals. To satisfy the condition for the coefficient $\beta = \frac{\beta_o}{V_A^2} = 1$, used in the nonlinear equations (13) and (14) by simulation of $F_s(t)$, $\cot \Theta = \left(\frac{2\Omega_o^2 V_A}{c^2} \beta \right)^{1/2} = \sqrt{2} \Omega_H \sqrt{\frac{\beta}{V_A}}$. The data given below

in Table 1 show that this condition is in agreement with equation (24). The

second coefficients μ_o and $\mu = \frac{\mu_o}{\beta V_A}$ used in these equations and in the dispersion

equation (21), may be estimated from the following. Taking into account the

damping of the waves, the complex wave number is

$$k* = \frac{n}{c}\omega* = \frac{n}{c}(\omega + i\omega') = \frac{\omega}{c}(n_o + i\kappa_o) = k\left(1 + i\frac{\kappa_o}{n_o}\right) \ , \tag{25}$$

where κ and ω' are the imaginary parts of the complex coefficient of refraction

and of the complex frequency $\omega*$. The value of κ is the linear attenuation factor

of these Alfven waves in the magnetoplasma. Namely,

$$\kappa_o = \frac{n_o}{2} \cdot \frac{m}{M} \frac{\nu_{ei} \cdot \omega}{\Omega_H^2} \left(\cos^2 \Theta\right)^{-1} \tag{26}$$

[see *Alpert*, 1983]. As to the value of ω', it may be estimated by comparing the

imaginary parts of the dispersion equation (21) and of the equation (25). As a

result, it follows that

$$\mu_o = \frac{\kappa_o c}{k n_o^2} \ , \qquad \mu = \frac{\mu_o}{V_A} \ , \qquad \nu_{ei} = 2\frac{M}{m} \frac{\Omega_H^2 \cdot \mu}{V_A}\left(1 - \frac{\omega}{\Omega_H}\right)^{-1} \ . \tag{27}$$

It is seen from the analysis given above that the obtained formulas (20), (23),

(24), and (27) estimate the parameters of the magnetoplasma and of the hot

beam of ions. Thus, the result of simulation of the $F_s(t)$ signals, i.e., the finding

of the appropriate modulators $Z(t)$ and of the values $V_o = \frac{v_o}{\beta}$, $\mu = \frac{\mu_o}{V_A}$, and

$\beta = \frac{\beta_o}{V_A^2}$ may be used not only for the shaping of the envelopes of $F_s(t)$ but also

for diagnosis of the properties and of the behavior of the regions of the medium

where these packets of waves are produced.

3. RESULTS OF PROCESSING THE EXPERIMENTAL DATA

The trains of Pc-1 packets of waves $F_s(t)$, used in this study for simulation, were described in detail by *Alpert and Fligel* [1985, 1987], *Alpert* [1987a], and *Alpert et al.* [1989]. Some records of these signals are shown above in Figures 2 and 3. The main characteristics of these trains are given in Table 1, where $L = R/R_o$ estimates the position of the apogee of the Earth-magnetic field line, namely, the region of excitation of $F_s(t)$, where R_o is the radius of the Earth, τ_s is the duration of the signals, and N_{oi}, H_o, ω_o, ω_H, and n_A are the main parameters of the surrounding magnetoplasma. The velocity V_{bo} of the beam, given in the table, is estimated by formula (23).

The first step of the process of simulation of the signals $F_s(t)$ is to find the value of V_o [see eq. (13)]. Namely, the shaping of the oscillating character of $F_s(t)$ in the appropriate scale of time of the duration of the signal τ_s depends on V_o. The value V_o may be found after few attempts by first determining the behavior of these functions (see, for example, Fig. 5). If $F_s(t)$ can be described by one function $Z(t)$, as, for example, for the signals shown in Figures $2a, b$, it is very easy to estimate V_o. In this case, depending on the symmetry of $F_s(t)$, the position of the level $\alpha_o Z(t) = 1$ should be chosen. Then $Z(t)$ is a monotonic weak slow shock wave [see eqs. (7), (10), and (13)].

The second step of this process is the choice of the values of γ_{io} and of μ. For oscillatory function $F_s(t)$, V_o depends at most on the periodicity of this

Table 1
**Characteristics of the Trains of the Pc-1 Packets of Waves $F_s(t)$
and of the Magnetoplasma**

No.	Observation Station	No. of Signals	Duration of the Train (sec)	τ_s (sec)	f_{so} (Hz)	L	V_{bo} (m.s^{-1})
1	Sogra	9	950	25–58	1.0	4.2	6.9×10^6
2	Sogra	21	3520	50–130	0.5	6	4.46×10^6
3	Husafell	20	6980	170–280	0.5	6	4.89×10^6

No.	N_{oi} (cm^{-3})	Ω_o (Hz)	H_o (γ)	Ω_H (Hz)	$\dfrac{\omega_{so}}{\Omega_H}$	n_A	V_A (m.s^{-1})	$\dfrac{V_{bo}}{V_A}$
1	70	1.1×10^4	400	38	0.16	289	1.04×10^6	6.64
2	3–15	$(2\text{-}5) \times 10^3$	80–200	8–19	0.4–0.16	260	1.15×10^6	3.88
3	6	3.2×10^3	150	14	0.22	230	1.30×10^6	3.77

function. It is clear, by examination of equations (7), (12), and (13), that in this case V_o is larger, sometimes even much larger (see below Figs. 9 and 10), than the values V_o of the monotonic shock waves $Z(t)$. It also depends on the attenuation character of these functions, namely, on the appropriate value of μ.

Results of the simulation of a Pc-1 packet of waves of the simplest shape are given in Figure 6. This was the first attempt by the author to simulate

Table 2
Coefficients of Equations (1) and (2) Used to Calculate Signals $F_s(t)$
(see Table 1)

No.	V_o	α_o	γ_{io}	μ	α_{12}	α_{21}
1	0.05	18	0.14	0.3	–	–
2	0.03–0.055	12–21	0.1–0.16	0.02–0.4	–	–
3	0.0005 0.115,0.14	48,42 10	0.03,0.045	$\mu_1 = 0.1, 0.045$ $\mu_2 = 10^{-4}$	0.005	0
$s = 12$	0.07,0.16	8,2	0.1	10^{-4}	$\alpha_1 = 10$	$\alpha_2 = 1.5$

theoretically Pc-1 micropulsations. Because of the symmetrical shape of this signal, the function $Z(t)$ may immediately be found and then, by changing the value of γ_{io}[see eq. (1)], the suitable theoretical signal $Y(t)$ is estimated. Three steps on this process are shown in Figure 6. They characterize the influence of γ_{io} on the time scale of $Y(t)$, on the length τ_s of the signal.

Results of shaping one signal of a train of waves, also of simple shape (see Fig. 2b), are given in Figure 7. The main characteristics of this train of Pc-1 micropulsations are presented in Table 1, line no. 1. The parameters of equation (13), used by these calculations, are given in Table 2. The modulator $Z(t)$ of the theoretical function $Y(t)$ is a monotonic shock wave with one very small maximum.

The oscillatory and monotonic packets of waves $F_s(t)$ of another train of micropulsations (Table 1, line no. 2, see Fig. 2d above) were also shaped by one oscillatory shock wave $Z(t)$. Results of calculations of the signals $Y(t)$ of this train are given in Figure 8 and in Table 2, line no. 2. The shape of $F_s(t)$ of this train changed from signal to signal. They have two or three maxima. The last of them ($s = 16, 19$) become smooth signals with one maximum. It should be noted here that to simulate the signals $s = 16, 19$, it is necessary to use considerably larger values of μ. This means that the attenuation of the shock waves increased greatly at the end of this process. Certainly, the collision frequency between the electrons and ions and also the density N_{ib} of the beam of the hot protons are increased (see below Table 3, line no. 2). These results demonstrate the possibility and efficiency of the method used for diagnosis of the behavior of the regions of the magnetoplasma where the Pc-1 micropulsations are generated.

The last steps of this study were to simulate functions $F_s(t)$ of very complicated shape. Some of these Pc-1 packets of waves are shown above in Figure 3. The theoretical calculations of two of them ($s = 4$ and $s = 19$) have been completed (see Table 2, line no. 3). These signals may be simulated only by a modulator $Z(t) = \alpha_1 Z_1(t) + \alpha_2 Z_2(t)$ [see eqs. (13) and (14)]. The method of simulation of other functions $F_s(t)$ of this train should be modified. It appears that more than two shock waves $Z_i(t)$ should be used. Possibly, even some other processes should also be taken into account. However, results of approximate, incomplete, calculations of one other signal of this kind (see Table 2, line $s = 12$)

show that the principal property of this signal of very complicated shape, i.e., its oscillatory character, is reproduced well by these calculations. Moreover, the main property of the beam of the hot particles, the anisotropic coefficient $\frac{T_{i\perp}}{T_{i\parallel}}(t)$ is also well estimated. Results of simulation of these signals are given in Figures 9 to 12. The characteristics of their trains of Pc-1 micropulsations and of the parameters of the region of the magnetoplasma, where they were excited, are presented in Tables 1 and 2 on lines no. 3 and $s = 12$.

The calculations of $F_s(t)$, $s = 4$ and 19 were done by the alternative method [see eq. (14)]. The values used for the correlation coefficients α_{12} and α_{21} of equation (14) are given in Table 2, line no. 3. A weak interconnection between the functions $Z_1(t)$ and $Z_2(t)$ is used only in the equation of $Z_1(t)$. The functions $Z_1(t)$, $Z_2(t)$, of the first solution $Z(t)$ used for the system of equations (14) and the adequate coefficients of equation (14) are given in Figures 9–11 and in Table 2, line no. 3.

One may note that the results of calculations of the functions $Y_s(t)$ reproduce well the main properties of the experimental functions $F_s(t)$. Certainly, in some details they should be distinguished because of the "noise" – the background of oscillations of the magnetoplasma. Noise always exists in the magnetosphere, and its level was especially high during the period of excitation of this train of Pc-1 micropulsations [see *Alpert et al.*, 1989].

By these calculations, the sensitivity of the method we used towards the change of the values of the coefficients of the adequate equations was also studied. For this purpose, the system of equations (1), (13), or (14) was solved for different values of γ_{io}, μ, α_1, α_2, α_{12}, and α_{21}, and for different initial data $Z_{1,2}(t=0)$, $\dot{Z}_{1,2}(t=0)$. The values of these quantities were chosen to be different from those that were chosen as the optima for simulating the signals $F_s(t)$. For illustrative purposes, four calculated functions of $Y(t)$ are given in Figure 11. Three of them are almost similar. They simulate well the signal $F_s(t)$. They were calculated for $V_o = 0.115 = \text{const}$ and different values of other parameters. Namely, they are changed in the following limits:

$$\gamma_{io} = 0.03\text{--}0.035 \; , \qquad \alpha_{12} = 0.0045\text{--}0.005 \; , \qquad \mu = 0.1\text{--}0.015 \; ,$$

$$Z_1(0) = Z_2(0) = 0 \; , \qquad \dot{Z}_1(0) = 0.0005\text{--}0.003 \; , \qquad \dot{Z}_2(0) = 10^{-5}\text{--}10^{-10} \; . \tag{28}$$

But the theoretical signal $Y(t)$ does not exactly match the oscillatory character of the experimental $F_s(t)$ if the value $V_{02} = 0.1$ is used instead of $V_{02} = 0.15$. The general conclusions of many such calculations are the following. Good agreement between the functions $Y(t)$ and $F_s(t)$ is achieved in sufficiently small limits, especially of the values V_o. The results of the rough approximate calculations of $Y(t)$ to simulate the signals $F_s(t)$, $s = 12$ (see Tables 2 and 3) are given in Figure 12. The periodicity of $F_s(t)$ is also reproduced well by the theoretical function $Y_s(t)$ even in this case.

Table 3
Parameters of the Hot Beam Estimated by Simulation of $F_s(t)$

No.	N_{ib} (cm^{-3})	ν_{ei} (s^{-1})	$\left(\dfrac{T_{i\perp}}{T_{i\parallel}}\right)_{max}$	$\left(\dfrac{T_{i\perp}}{T_{i\parallel}}\right)_{min}$
1	0.08	10^{-3}	2.7	–
2	0.04–0.12	3×10^{-4}–6×10^{-3}	2.3	0.41
3	0.02	–	2.7	0.41
$s = 12$	0.03	–	3.2	0.45

by the formulas (17), (20), and (27) are brought together in Table 3. It should be noted here that by using in the calculations two or more functions $Z(t)$ for simulation of $F_s(t)$, it is not known how to estimate the effective value of the parameter μ. This problem demands a special discussion. Therefore, the collision frequency ν_{ei} cannot be estimated in this study for signals $F_s(t)$, $s = 4$, 19, or 12.

The values of the parameters of the magnetosphere given in Table 3 lie in the limits of the known experimental data estimated in different conditions, particularly those received by satellites. They are also in good agreement with results of other theoretical evaluations of these quantities [see *Alpert*, 1983; *Orayevsky et al.*, 1985].

4. CONCLUSIONS

The mechanism and the model of the process of generation and shaping of discrete packets of waves $F_s(t)$ of Pc-1 micropulsations used in this study describe well the basic features of these signals. The method used gives the possibility of diagnosing the properties of the region of the magnetoplasma where the signals are excited. The behavior of the parameters of this region in time, during the period of excitation of these signals, which can last for hours, may also be studied.

The following parameters of the source of this process, i.e., of the beam of the hot protons, were estimated in this study:

The velocity of the beam: $V_b \simeq (4.5\text{--}6.9)10^6\,\mathrm{s}^{-1}$

The density: $\qquad\qquad\qquad N_{ib} \simeq (0.04\text{--}0.12)\,\mathrm{cm}^{-3}$

The collision frequency: $\quad \nu_{ei} \simeq \left(3 \times 10^{-4}\text{--}6 \times 10^{-3}\right)\mathrm{s}^{-1}$

The anisotropy coefficient: $\left(\dfrac{T_{i\perp}}{T_{i\|}}\right)_{max} \simeq (2.3\text{--}3.2)\,,\ \left(\dfrac{T_{i\perp}}{T_{i\|}}\right)_{min} \simeq (0.41\text{--}0.45)\,.$

These values are in rather good agreement with the data known from the literature.

Further explanation of the theory and method used, and of the development and extension of this method for simulation of discrete packets of waves $F_s(t)$ of very complicated shape, is needed. These investigations would be interesting and useful not only for understanding and simulation of different processes, produced in the magnetosphere or in the deep space plasma, but also as a clue for understanding a variety of other similar discrete processes observed in nature.

ACKNOWLEDGMENTS

I would like to thank Dr. Kenneth Budden (Cavendish Laboratory, Cambridge) for useful discussions and for the computer program developed for this and the future study of the discussed problem. The author also thanks Arkadij Yaroslavtsev (IZMIRAN, Moscow) for help in performing some of the computer calculations and Carolann Barrett (Harvard–Smithsonian Center for Astrophysics) for her help in preparing this paper. The author is grateful to the the Smithsonian Institution for a Scholarly Studies Grant, which partially supported this work.

REFERENCES

Alpert, Ya. L., *The Near-Earth and Interplanetary Plasma, Vol. I*, Cambridge University Press, Cambridge, 1983.

Alpert, Ya. L., On an oscillating model of Pc-1 geomagnetic pulsations and on some properties of this process, *Planet. Space Sci., 34*, 537, 1986.

Alpert, Ya. L., On spectra and some properties of Pc-1 geomagnetic pulsations of a complex shape, *Planet. Space Sci., 35*, 1381, 1987*a*.

Alpert, Ya. L., On simulation of the process of the generation of geomagnetic pulsations in the magnetosphere, *Planet. Space Sci., 35*, 71, 1987*b*.

Alpert, Ya. L., and D. S. Fligel, On Fourier-analysis of geomagnetic pulsations Pc-1, "twofold" and "threefold" pulsations Pc-1 and fundamental frequencies of the magnetospheric cavity, *Planet. Space Sci., 33*, 993, 1985.

Alpert, Ya. L., and D. S. Fligel, On fundamental frequencies of the cavity of the magnetosphere, twofold and threefold Pc-1 pulsations, *Planet. Space Sci., 35*, 879, 1987.

Alpert, Ya. L., D. S. Fligel, and A. A. Yaroslavtsev, Spectra and geomagnetic pulsations of complex shape and simulation of their generation process, *Planet. Space Sci., 37*, 391, 1989.

Burgers, I. M., (Damping of nonlinear waves in dispersive media), *Proc. Ac. Sci.,* Amsterdam, *43*, 2, 1940.

Gendrin, R., S. Lacourly, A. Roux, J. Solomon, F. Z. Feigin, M. V. Gokhberg, V. A. Troitskaya, and V. Y. Yakimenko, Wave packet propagation in an amplifying medium and its application to the dispersion characteristics and to the generation mechanisms of Pc-1 events, *Planet. Space Sci.*, *19*, 165, 1971.

Gulelmi, A. V., and V. A. Troitskaya, *Geomagnetic Pulsations and Diagnosis of the Magnetosphere* (in Russian), Publ. House Nauka, Moscow, 1973.

Jacobs, J. A., *Geomagnetic Micropulsations*, Springer-Verlag, Berlin, 1970.

Kadomtsev, B. B., *Collective Phenomena in a Plasma*, Publ. House Nauka, Moscow, 1976.

Karpman, V. I., *Nonlinear Waves in Dispersing Media*, Publ. House Nauka, Moscow, 1970.

Kennel, C. F., and H. E. Petschek, Limit of stably trapped particle fluxes, *J. Geophys. Res.*, *71*, 1, 1966.

Korteweg, D. J., and G. de Vries, (Nonlinear waves in dispersive media), *Phil. Mag.*, *39*, 442, 1895.

Liemohn, H. B., ELF propagation and emission in the magnetosphere, *Boeing Sci. Res. Lab. Document DI-82-0890*, 1969.

Orayevsky, Y. N., *et al. Transport Processes in Anisotropic Near-Earth's Plasma*, Publ. House Nauka, Moscow, 1985.

Storey, L. R. O., An investigation of whistling atmospherics, *Phil. Trans. Roy. Soc., London, A246*, 113, 1957.

Tepley, L. R., Observation of hydromagnetic emissions, *J. Geophys. Res.*, *66*, 1651, 1961.

FIGURE CAPTIONS

Fig. 1. Dynamical spectrogram of a train of Pc-1 micropulsations.

Fig. 2. Packets of waves $F_s(t)$ of Pc-1 micropulsations of smooth shape, recorded in different conditions.

Fig. 3. Records of Pc-1 packets of waves $F_s(t)$ of very complicated shape.

Fig. 4. Schematic diagram of the method of simulation of $F(t)$ functions of different shape.

Fig. 5. Weak slow oscillatory and monotonic shock waves $Z(t)$. Solution of the Korteweg, de Vries, and Burgers equations for different values of the viscosity μ of the medium.

Fig. 6. Experimental $F(t)$ and theoretical $Y(t)$ signals of simple shape.

Fig. 7. Experimental packet of waves $F_s(t)$, theoretical signal $Y(t)$, and its modulator $Z(t)$. The functions $Y(t)$ and $Z(t)$ were calculated by equations (1) and (13).

Fig. 8. The same as in Fig. 7 for a train of oscillatory and smooth Pc-1 packets of waves.

Fig. 9. The same as in Fig. 8 of a packet of waves $F_s(t)$ of complicated shape ($s = 4$). Functions $Y_s(t)$ were calculated by joint solution of equations (1) and (14).

Fig. 10. The same as in Fig. 9, for $F_s(t)$, $s = 19$.

Fig. 11. Experimental Pc-1 packet of waves $F_s(t)$, $s = 4$ and the theoretical functions $Y(t)$ calculated for different values of the parameters of equations (1) and (14) [see eq. (26)].

Fig. 12. The same as in Figs. 9 and 10 for $s = 12$. The function $Y(t)$ was calculated by joint solution of equations (1) and (13).

Frequency (Hz)
0.60
0.55
0.50
0.45
(c)
Seattle
0600
UT
0640
23 Feb. 1967

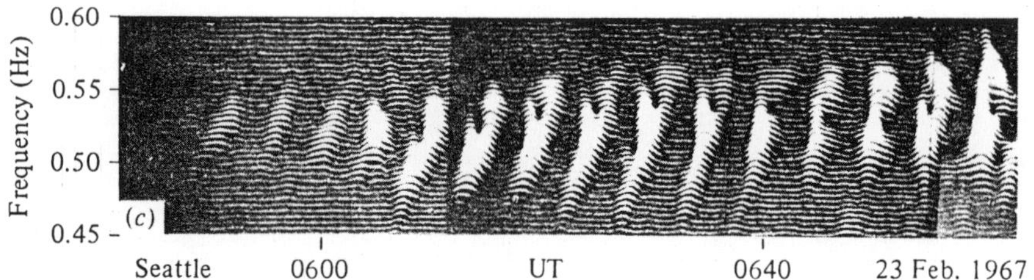

Fig. 1

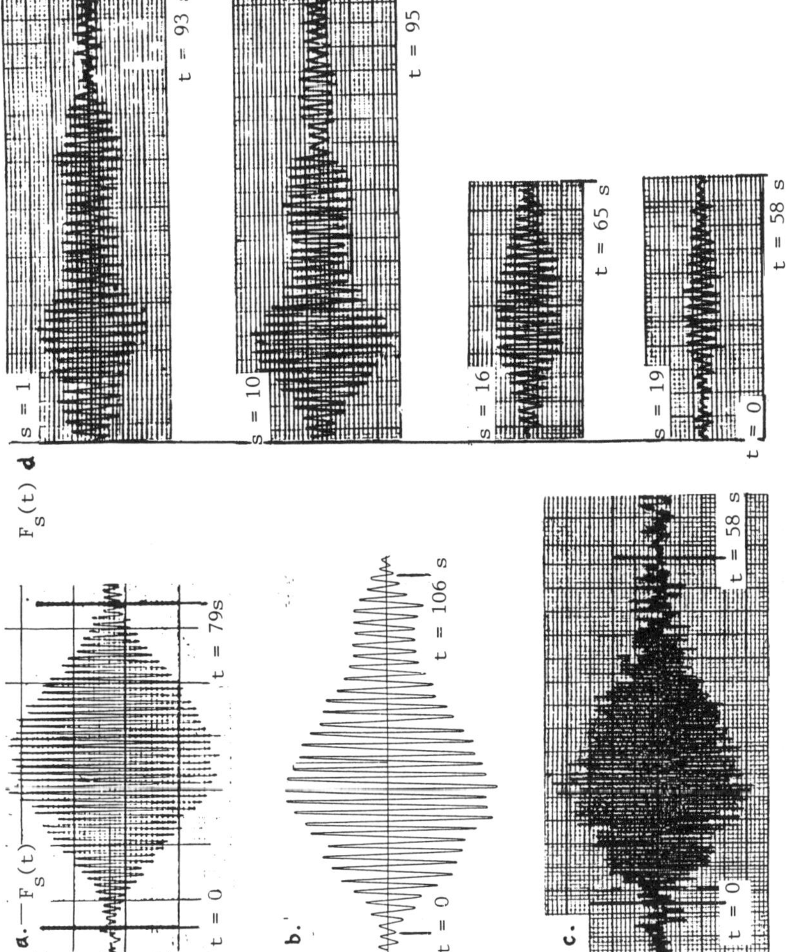

Fig. 2

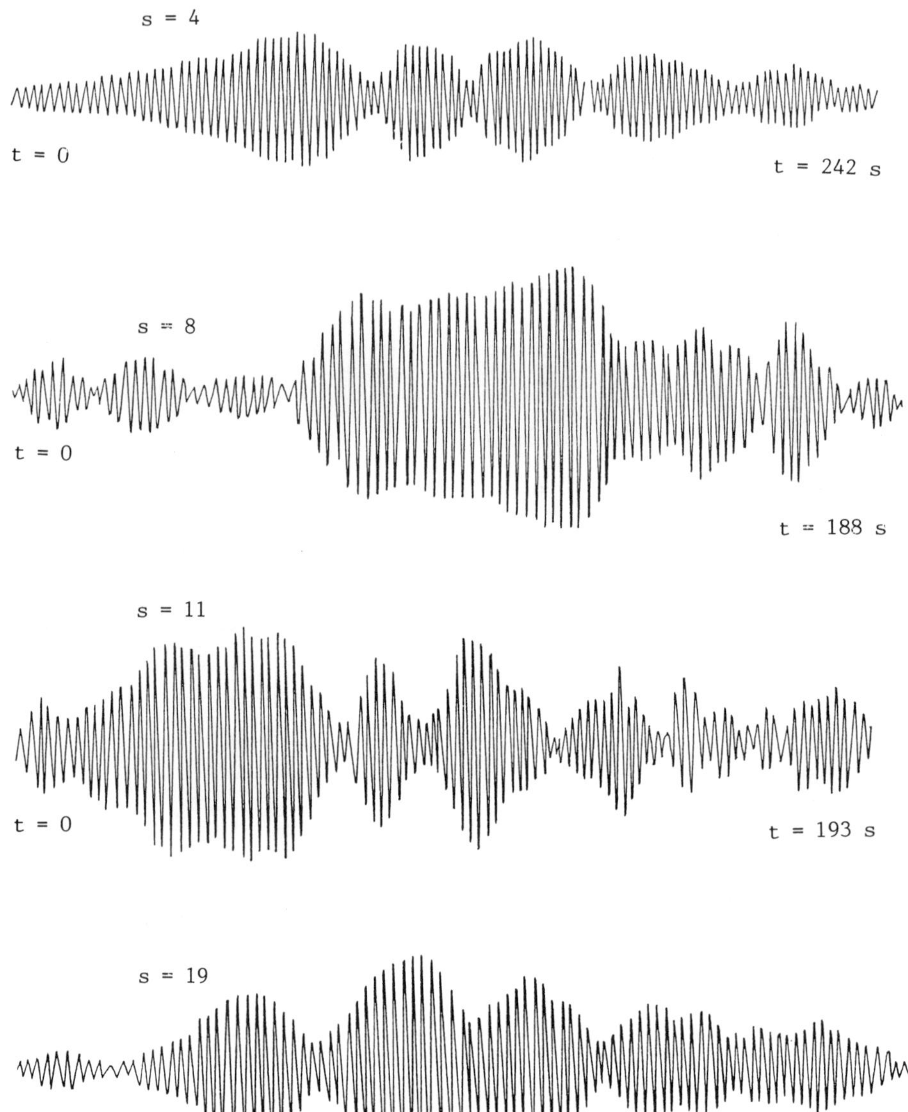

Fig. 3

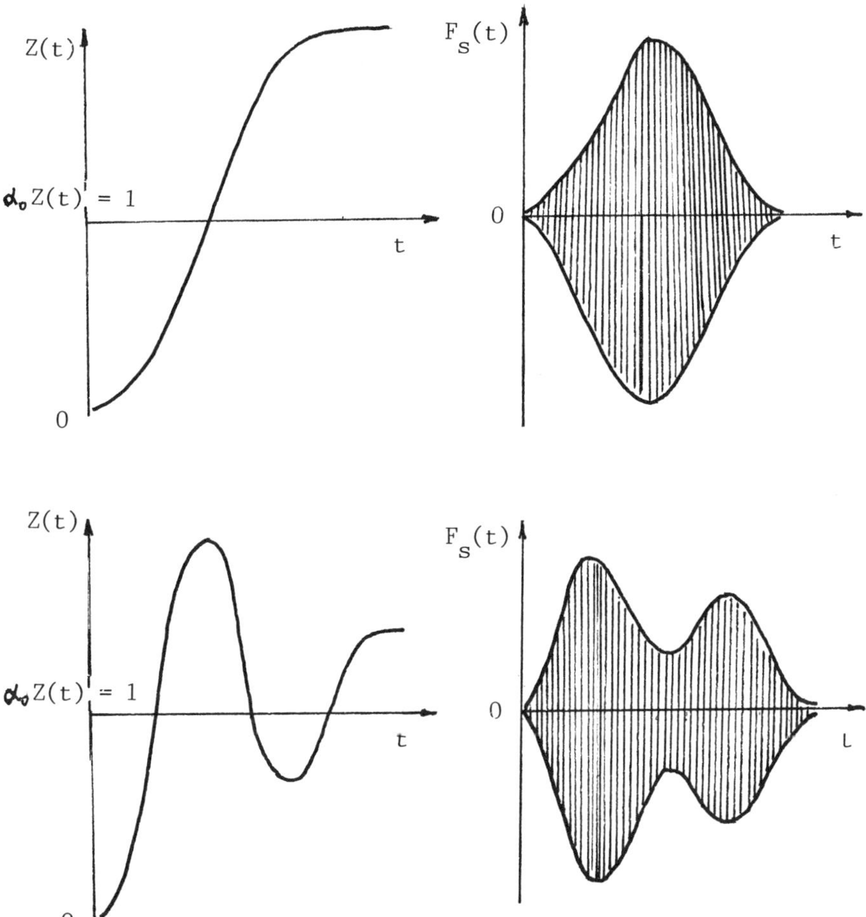

Fig. 4

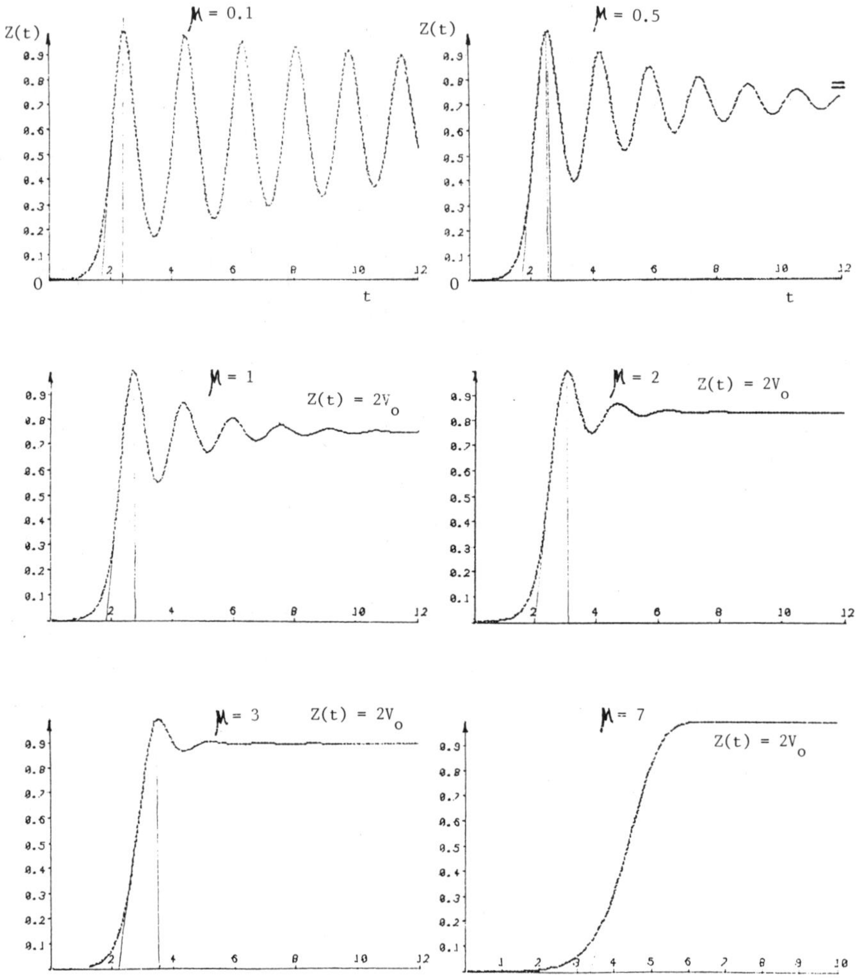

Fig. 5

Y(t)
t,s
15
45
60
$\gamma_{i0} = 3\,s^{-1}$
$= 1\,s^{-1}$
$= 0.3\,s^{-1}$
Experimental signal F(t)
$f = 0.8\,Hz$

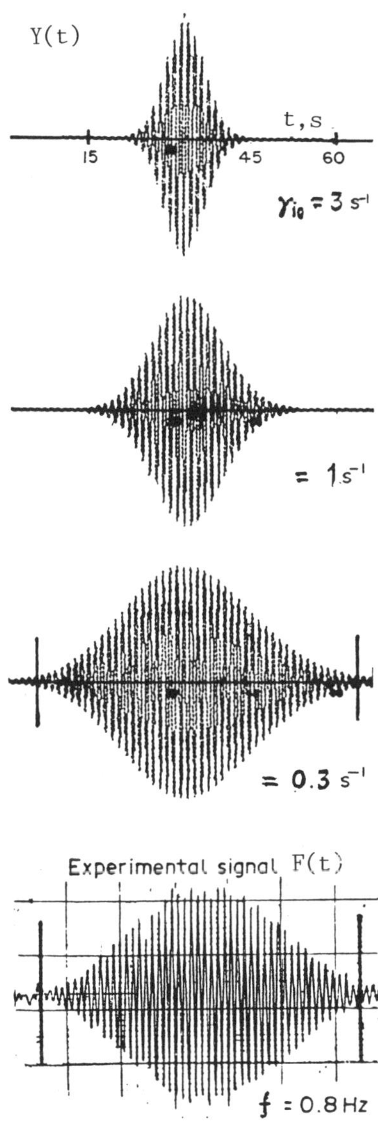

Fig. 6

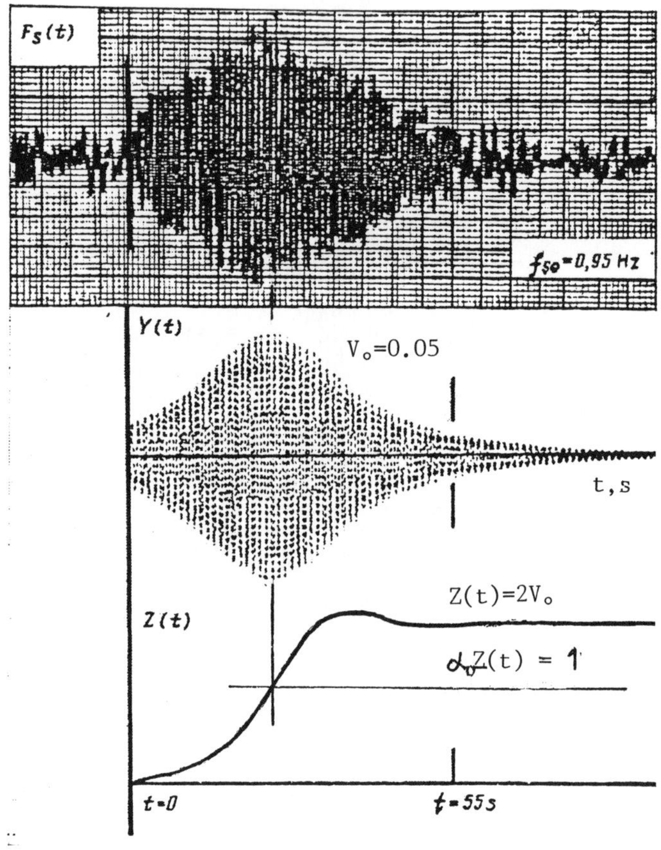

Fig. 7

Z(t)
V₀=0.03
μ=0.02
Y(t)
S=1
F_S(t)
t=0
t=92s
Z(t)
V₀=0.055
μ=0.02
Y(t)
S=10
F_S(t)
t=1464s

Z(t)
V₀=0.05
μ=0.4
Z(t)=2V₀
Y(t)
S=16
S=19
F_S(t)
t=2446s
t=2955s

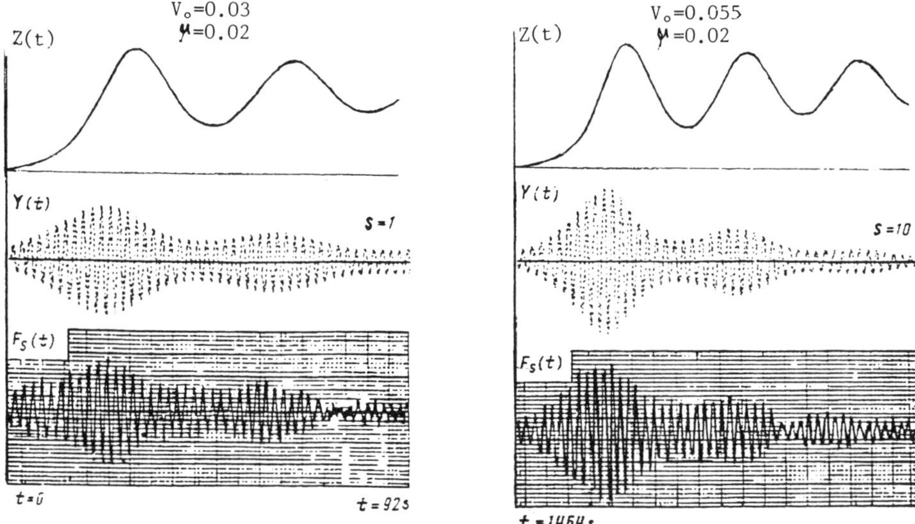

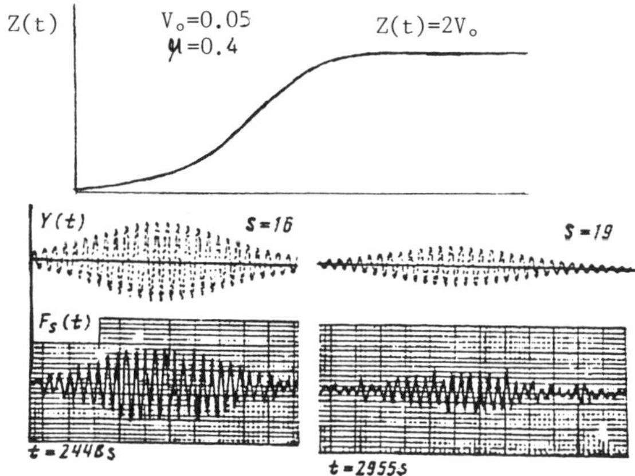

Fig. 8

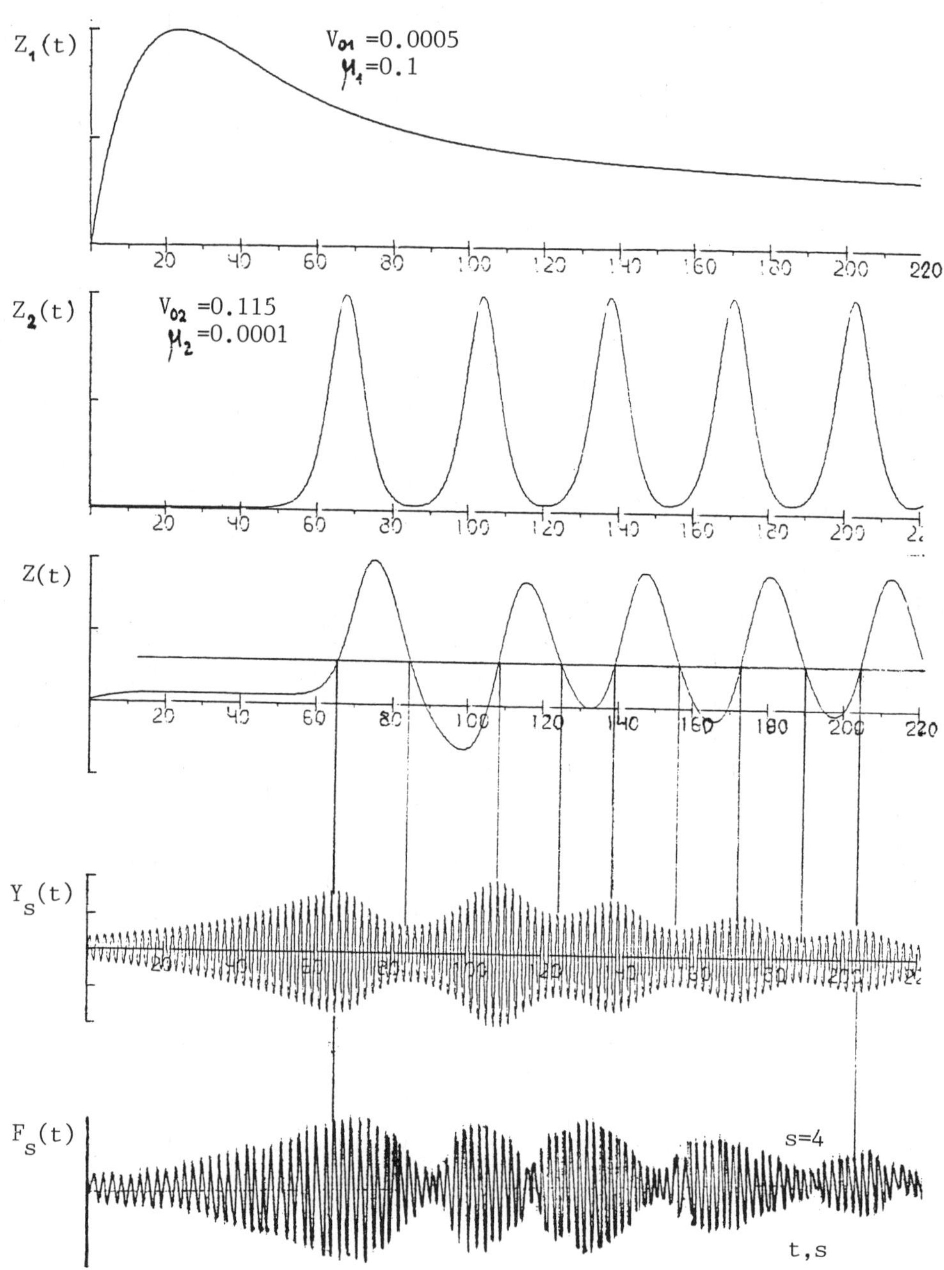

Fig. 9

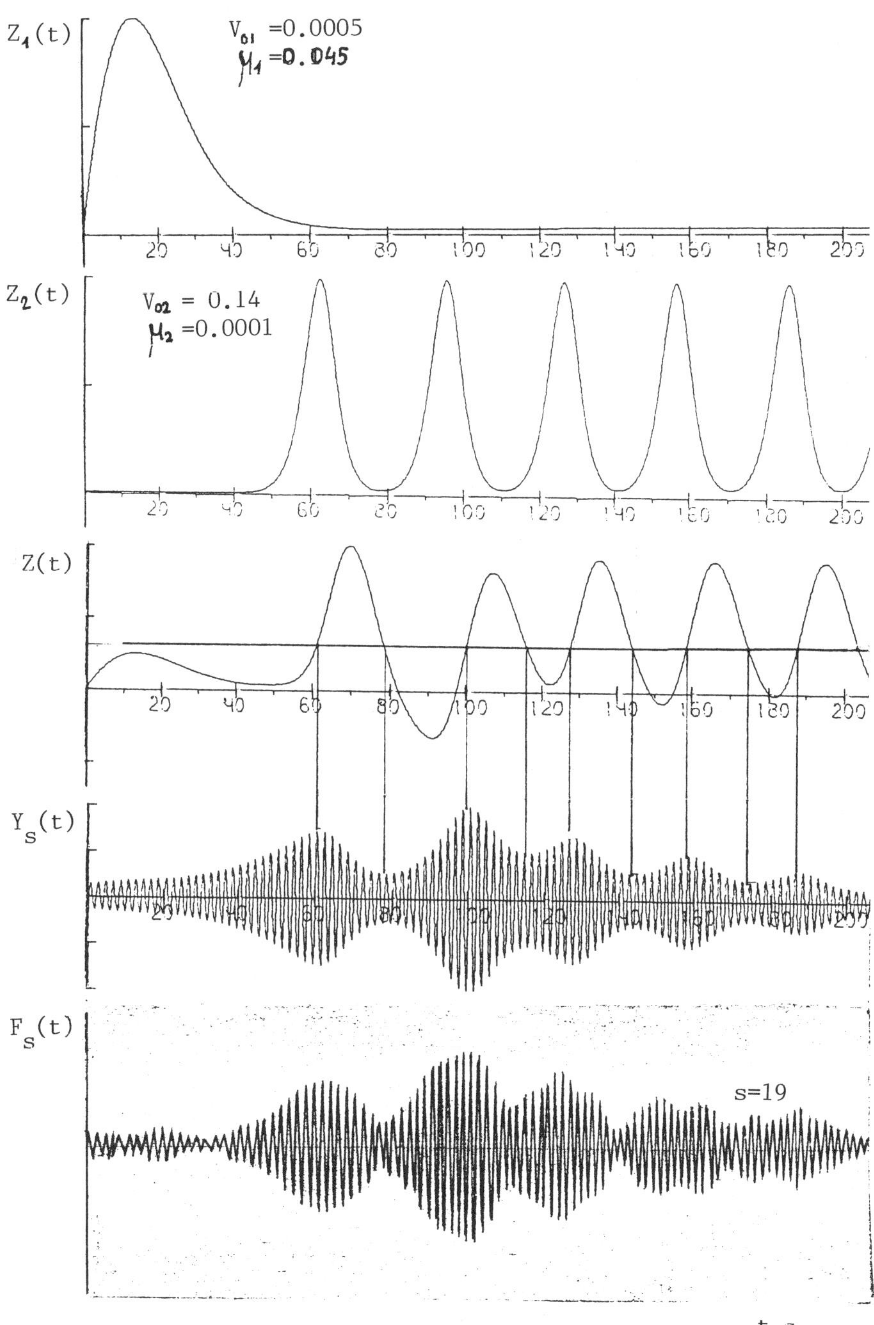

Fig. 10

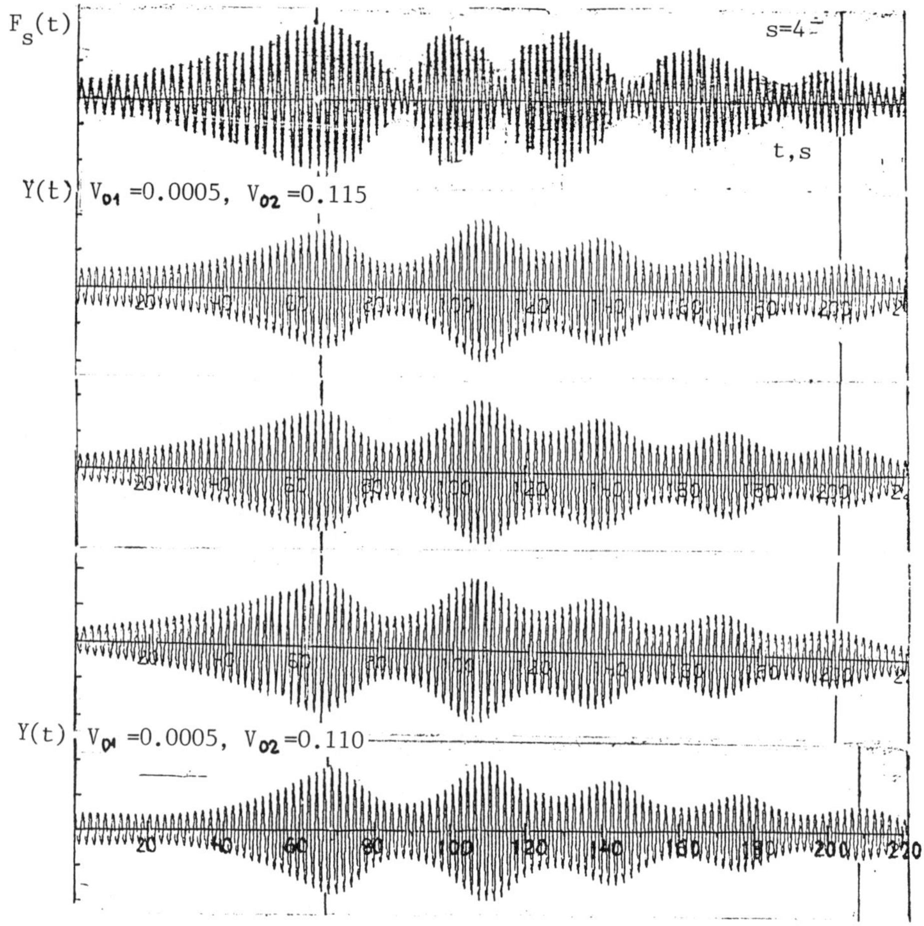
F_s(t)
s=4
t,s
Y(t) V01 =0.0005, V02 =0.115
20
40
60
80
100
120
140
160
180
200
Y(t) V01 =0.0005, V02 =0.110
220

Fig. 11

Z_1(t)
V_{01} =0.07, \mu_1 =10^{-4}
Z_2(t)
V_{02} =0.16 \mu_2 =10^{-4}
\alpha_1 Z_1(t)+\alpha_2 Z_2(t)
Y_s(t)
F_s(t)
s=12
f_s =0.5Hz
t,s

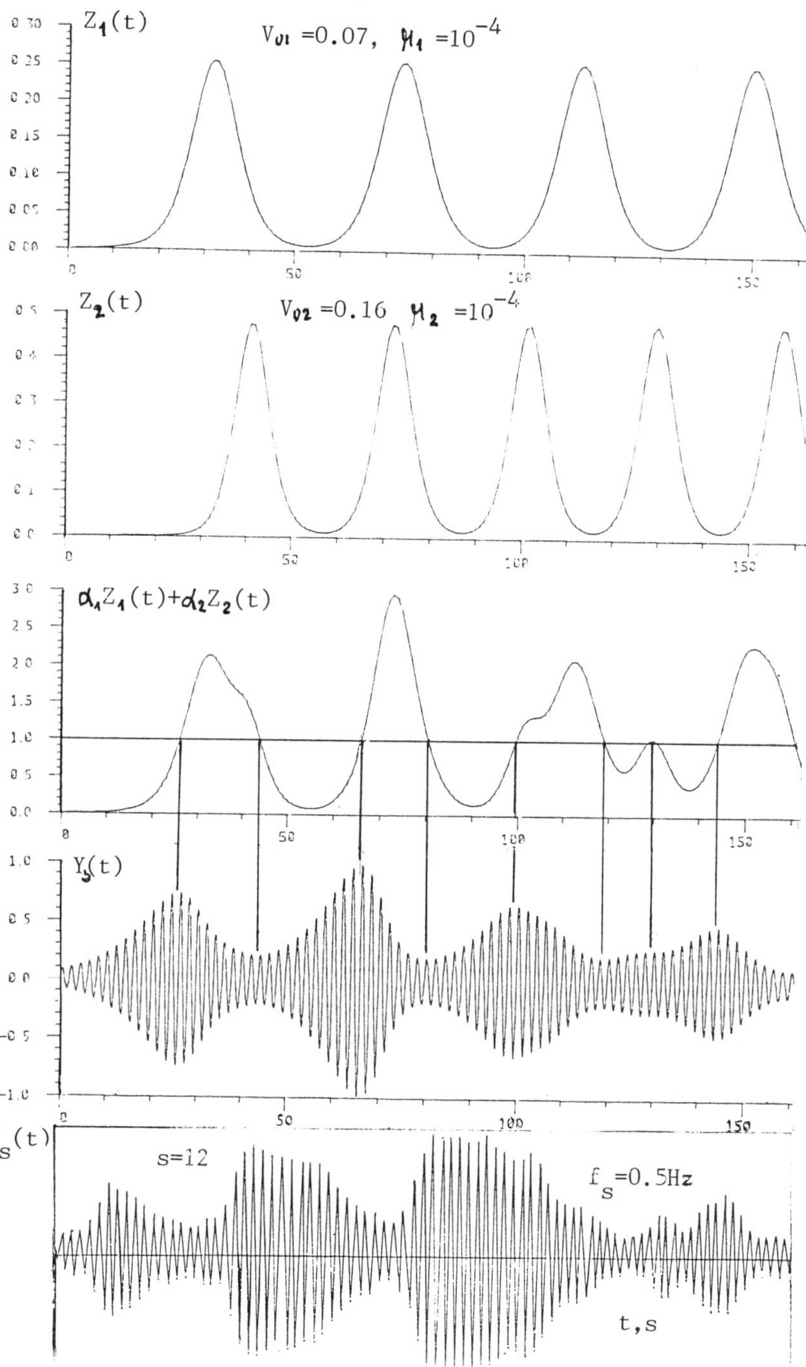

Fig. 12

HAMILTONIAN TEST PARTICLE DYNAMICS IN A PERTURBED MAGNETIC DIPOLE FIELD

Anthony A. Chan[1]

Princeton Plasma Physics Laboratory and Department of Astrophysical Sciences,
Princeton University, Princeton, NJ 08543

Alain Brizard

Lawrence Berkeley Laboratory, University of California, Berkeley, CA 94720

and Liu Chen

Princeton Plasma Physics Laboratory and Department of Astrophysical Sciences,
Princeton University, Princeton, NJ 08543

ABSTRACT

We have derived Hamiltonian, gyrophase-independent, test particle equations of motion which describe the nonrelativistic motion of charged particles in a perturbed magnetic dipole field. We assume low frequency ($\omega < \omega_c$), short perpendicular wavelength ($k_\perp^{-1} \sim \rho$), fully electromagnetic perturbations, where ω_c and ρ are the particle gyrofrequency and gyroradius, and ω and $k_\perp$ are the characteristic frequency and perpendicular wave number of the perturbations. The equations are derived using phase-space Lagrangian Lie transform perturbation methods [8]. The phase-space Lagrangian is derived to first order in the guiding center ordering parameter $\epsilon_0 = \rho|\nabla \ln B_0|$, where B_0 is the unperturbed magnetic field strength, and to first order in the perturbation ordering parameter ϵ. The corresponding equations of motion contain terms of order ϵ^2 which ensure that the Hamiltonian conservation properties of the unperturbed system are preserved. Numerical and analytic results for the interaction of energetic protons ($\gtrsim 100$ keV) with Alfvén waves in the Earth's magnetosphere [5], and for test particles in a dipole field fusion reactor [6], are presented. We find that large-scale radial diffusion can occur in both cases.

INTRODUCTION

It is well known that the motion of a charged particle confined in a magnetic field $\mathbf{B_0}$ takes place on three distinct time scales: gyration about a magnetic field line, characterized by the gyrofrequency $\omega_c = eB_0/Mc$; bounce motion along the field line, characterized by the bounce frequency $\omega_b \approx v_\parallel/S$, where $v_\parallel$ is the parallel velocity of the particle and S is the field line length between mirror points; and drift motion across the field line, characterized by the magnetic drift frequency $\omega_d \approx v_d/R_d$, where v_d is the ∇B and curvature drift velocity and R_d is the radius of the drift orbit. Each of these periodic motions is also associated with a corresponding phase angle: the gyrophase, the bounce phase, and the drift phase.

In guiding center theory, one uses the separation of the gyroperiod from the other time scales in the system to define a small parameter ϵ_0, where $|\partial \ln B_0/\partial t|/\omega_c \sim \rho|\nabla \ln B_0| \sim \epsilon_0 \ll 1$. The small parameter ϵ_0 is then used to remove the gyrophase

---

[1]Present address: Department of Physics and Astronomy, Dartmouth College, Hanover, NH 03755

dependence from the system order by order in ϵ_0. For lowest-order guiding center theory this may be accomplished by gyrophase averaging [11]. For higher-order calculations the gyrophase dependence can be removed systematically and efficiently using phase-space Lagrangian Lie transform perturbation methods [8, 3, 9]. Because the motion of a charged particle in a magnetic field is a Hamiltonian system, the removal of the gyrophase dependence to a given order in ϵ_0 results in the conservation of the conjugate quantity to the same order in ϵ_0, assuming that resonances involving the cyclotron motion can be neglected. In principle this procedure can be carried to arbitrary order in ϵ_0 and the conserved quantity, called the first adiabatic invariant, μ, is obtained as an asymptotic series in powers of ϵ_0. Littlejohn has shown that $\mu = MW^2/2B_0$, to all orders in ϵ_0, where W is the guiding center perpendicular velocity and B_0 is the magnetic field magnitude evaluated at the guiding center position [9]. The original system describing the motion of a charged particle in a magnetic field is reduced to a new system with one less degree of freedom which describes the motion of a *guiding center* in a magnetic field.

Similarly, the guiding center system can be reduced by defining a small parameter, ϵ_b, equal to the ratio of the bounce period to the next-higher time scale in the system, and removing the bounce phase dependence order by order in ϵ_b. Assuming that resonances involving the bounce motion can be neglected, the resulting adiabatic invariant is the longitudinal invariant $J_b = \oint U ds$, where U is the parallel velocity of the guiding center, s is the distance along the field line, and the integration is performed over one bounce period. Finally, if the drift frequency can be separated from any remaining frequency in the "bounce center" system, (for example, due to explicit time dependence with frequency $\omega \ll \omega_d$, or due to a spatial asymmetry with respect to the drift phase), and if resonances involving the drift motion can be neglected, the resulting adiabatic invariant is ψ, the magnetic flux linked by the bounce center during a drift orbit. We emphasize that the above hierarchy of adiabatic invariants exists only in the absence of resonances. In practice, however, plasmas are often subject to disturbances which can result in a resonance on a given time scale and in the destruction of the corresponding adiabatic invariant.

In this paper we consider the effect of low frequency ($\omega < \omega_c$), short perpendicular wavelength ($k_\perp^{-1} \sim \rho$), electromagnetic perturbations on a test particle trapped in a magnetic field. Motivated by applications to the Earth's magnetosphere and to Hasegawa's dipole field fusion reactor [6], we assume the unperturbed field to be a magnetic dipole field. We are particularly interested in processes which break the third adiabatic invariant ψ and lead to radial diffusion.

In order to study the perturbed motion on the bounce and drift time scales we remove the fast gyrophase dependence, as in guiding center theory, but now with finite gyroradius effects included. We accomplish this using the phase-space Lagrangian Lie transform methods. There are several advantages to these methods. First, the use of a Hamiltonian formulation provides a general picture of resonant wave-particle interactions in terms of breaking of the adiabatic invariants [7]. Second, the conservation properties of the unperturbed Hamiltonian system, such as conservation of energy and conservation of phase-space volume, are retained. These properties are very useful for checking the accuracy of numerical calculations and are essential for the use of Poincaré surface-of-section plots. Third, we are not restricted to using canonical variables. Canonical variables can be difficult to find and may not have a clear physical interpretation. Fourth, the methods are sufficiently general to allow straightforward extensions in a number of directions. For example, one could consider unperturbed (non-dipole) magnetic and electric fields which are slowly varying in time (compared to ω_c^{-1}), self-consistent kinetic theory, and relativistic motion. (See References [1], [9], and [10].) Finally, we note that Lie transform methods show their greatest advantage in the calculation of high-order nonlinear terms. Compared with conventional perturbation expansions, the computational effort in Lie transform expansions decreases rapidly as the order of the perturbation expansion increases [2].

The remainder of this paper is organized as follows. In the next section we outline the derivation of the gyrophase-independent phase-space Lagrangian and the corresponding equations of motion. Details of the derivation will be given in a future publication. In the third section we present calculations of test particle motion due to hydromagnetic waves in the Earth's magnetosphere and due to radio frequency waves in a dipole field fusion reactor. The final section gives a summary of our main results.

HAMILTONIAN GYROCENTER DYNAMICS

Consider a charged particle moving in a dipole magnetic field $\mathbf{B_0}(\mathbf{x})$ subject to electromagnetic perturbations $\delta\mathbf{B}(\mathbf{x},t)$ and $\delta\mathbf{E}(\mathbf{x},t)$. We define two small parameters: $\epsilon_0 = \mathcal{O}(\rho|\nabla \ln B_0|)$ and $\epsilon = \mathcal{O}(\delta B/B_0)$. For the ordering of $\delta\mathbf{E}$, we assume the $\delta\mathbf{E} \times \mathbf{B_0}$ drift velocity is $\mathcal{O}(\epsilon)$ compared to the unperturbed parallel velocity. Thus, ϵ_0 is the usual guiding center small parameter and ϵ orders the perturbation amplitudes. The length and time scales of the perturbation fields are assumed to be ordered as follows: the parallel scale length is the same order as the scale length of the background magnetic field, ie., $k_\parallel|\nabla \ln B_0|^{-1} = \mathcal{O}(1)$; the perpendicular scale length is comparable to the gyroradius, ie., $k_\perp\rho = \mathcal{O}(1)$; and $\omega/\omega_c = \mathcal{O}(\epsilon_0)$. Furthermore, we assume $\epsilon_0 \ll \epsilon$ so the effect of the perturbations can be treated as an $\mathcal{O}(\epsilon)$ correction to the guiding center motion in $\mathbf{B_0}$.

Except where otherwise noted the following equations are written in a dimensionless form which shows the order of each quantity in ϵ_0 and ϵ explicitly. The corresponding equations in terms of physical quantities may be obtained by setting $\epsilon_0 = \epsilon = 1$ and restoring factors of e, M, and c by dimensional analysis.

The unperturbed motion is described by the following guiding center phase-space Lagrangian [9]:

$$L_0 = (\epsilon_0^{-1}\mathbf{A_0} + \rho_\parallel\mathbf{B_0}) \cdot \dot{\mathbf{X}} + \epsilon_0 \mu\dot{\xi} - H_0, \tag{1}$$

$$H_0 = \tfrac{1}{2}\rho_\parallel^2 B_0^2 + \mu B_0 + \mathcal{O}(\epsilon_0^2). \tag{2}$$

Here $\mathbf{B_0} = \nabla\times\mathbf{A_0}$, $\mathbf{X} = \mathbf{x}-\epsilon_0\rho$ is the guiding center position, (where $\rho = (\hat{\mathbf{b}}\times\mathbf{v})/B_0$, with $\mathbf{x}$ and $\mathbf{v}$ denoting the particle position and velocity, respectively, and $\hat{\mathbf{b}} = \mathbf{B_0}/B_0$), ξ is the gyrophase, μ is the magnetic moment, $\rho_\parallel = U/B_0$ where U is the parallel velocity of the guiding center, and we have used $\hat{\mathbf{b}} \cdot \nabla \times \hat{\mathbf{b}} = 0$, which holds because the dipole field is curl-free. The quantities $\mathbf{A_0}$ and $\mathbf{B_0}$ in Equations (1) and (2) are to be evaluated at the guiding center position $\mathbf{X}$.

The perturbations $\delta\mathbf{B} = \nabla \times \delta\mathbf{A}$ and $\delta\mathbf{E} = -\nabla\delta\Phi - \partial\delta\mathbf{A}/\partial t$ are included by adding the perturbation phase-space Lagrangian, $\epsilon L_1 = \epsilon\delta\mathbf{A} \cdot \dot{\mathbf{x}} - \epsilon\delta\Phi$, to the unperturbed (guiding center) phase-space Lagrangian, L_0. In guiding center phase space coordinates $(\mathbf{X}, \rho_{\|}, \mu, \xi)$ we find

$$L_1 \;=\; \delta\mathbf{A} \cdot \dot{\mathbf{X}} + \epsilon_0 \left(\frac{\boldsymbol{\rho} \cdot \delta\mathbf{A}}{2\mu}\right)\dot{\mu} + \epsilon_0 \left(\frac{\mathbf{v}_\perp \cdot \delta\mathbf{A}}{B_0}\right)\dot{\xi} - H_1, \tag{3}$$

$$H_1 \;=\; \delta\Phi, \tag{4}$$

where $\delta\mathbf{A}$ and $\delta\Phi$ are functions of $(\mathbf{X} + \epsilon_0\boldsymbol{\rho}, t)$, and $\boldsymbol{\rho}$ and $\mathbf{v}_\perp$ are functions of $(\mathbf{X}, \mu, \xi)$. In Equations (3) and (4) we have kept only terms of leading order in ϵ_0.

The total Lagrangian, $L = L_0 + \epsilon L_1$, is now gyrophase dependent. We remove this gyrophase dependence by a Lie transformation from the guiding center coordinates $(\mathbf{X}, \rho_{\|}, \mu, \xi)$, to new coordinates $(\overline{\mathbf{X}}, \overline{\rho}_{\|}, \overline{\mu}, \overline{\xi})$, which we refer to as the *gyrocenter* coordinates [1]. By choosing the transformation such that the gyrocenter coordinates differ from the guiding center coordinates by terms of $O(\epsilon)$, and requiring that any terms secular in the gyrophase are eliminated, we obtain the following gyrocenter phase-space Lagrangian:

$$\overline{L} \;=\; (\epsilon_0^{-1}\mathbf{A_0} + \overline{\rho}_{\|}\mathbf{B_0} + \epsilon\langle\delta\mathbf{A}\rangle) \cdot \dot{\overline{\mathbf{X}}} + \epsilon_0\,\overline{\mu}\dot{\overline{\xi}} - \overline{H}, \tag{5}$$

$$\overline{H} \;=\; \tfrac{1}{2}\overline{\rho}_{\|}^2 B_0^2 + \overline{\mu}B_0 + \epsilon(\langle\delta\Phi\rangle - \langle\mathbf{v}_\perp \cdot \delta\mathbf{A}\rangle). \tag{6}$$

In Equations (5) and (6) $\langle\ldots\rangle = (1/2\pi)\oint \ldots d\overline{\xi}$ denotes gyrophase averaging, and all functions are now evaluated at the gyrocenter coordinates $(\overline{\mathbf{X}}, \overline{\mu}, \overline{\xi})$. The gyrocenter phase-space Lagrangian, $\overline{L}$, completely describes the gyrocenter motion. To obtain explicit equations of motion we first specify the form of the perturbation and then choose a set of spatial coordinates.

If we assume an eikonal form for the perturbations,

$$\delta\Psi(\mathbf{x}, t) = \widehat{\delta\Psi}(x_{\|}) \exp\left(i\epsilon_0^{-1}\int \mathbf{k}_\perp \cdot d\mathbf{x}_\perp - i\omega t\right), \tag{7}$$

where $\delta\Psi$ represents either $\delta\Phi$ or $\delta\mathbf{A}$, then

$$\langle\delta\Phi(\overline{\mathbf{X}} + \epsilon_0\boldsymbol{\rho}, t)\rangle \;=\; J_0\,\delta\Phi(\overline{\mathbf{X}}, t), \tag{8}$$

$$\langle\delta\mathbf{A}(\overline{\mathbf{X}} + \epsilon_0\boldsymbol{\rho}, t)\rangle \;=\; J_0\,\delta\mathbf{A}(\overline{\mathbf{X}}, t), \tag{9}$$

$$\text{and}\;\; \langle\mathbf{v}_\perp \cdot \delta\mathbf{A}(\overline{\mathbf{X}} + \epsilon_0\boldsymbol{\rho}, t)\rangle \;=\; -(2J_1/\lambda)\,\overline{\mu}\delta B_{\|}(\overline{\mathbf{X}}, t), \tag{10}$$

where we have shown the arguments of the functions $\delta\Phi$, $\delta\mathbf{A}$, and $\delta B_{\|} \equiv \hat{\mathbf{b}} \cdot \delta\mathbf{B}$ for clarity. In Equations (8) to (10) the Bessel functions J_0 and J_1, with argument $\lambda = k_\perp\rho$, show the reduction in the wave amplitudes due to finite gyroradius effects.

Next, we adopt magnetic dipole coordinates (χ, ψ, ζ) for the gyrocenter position $\overline{\mathbf{X}}$, where (χ, ψ, ζ) are defined by $\mathbf{B_0} = \nabla\chi = \nabla\psi \times \nabla\zeta$. In terms of spherical polar coordinates (r, θ, ϕ), we have $\chi = M_D \cos\theta/r^2$, $\psi = M_D \sin^2\theta/r$, and $\zeta = -\phi$, where M_D is the dipole magnetic moment. For the geomagnetic dipole: $M_D = B_E R_E^3$, with $B_E = 0.31$ Gauss and $R_E = 6380$ km. The coordinates (χ, ψ, ζ) form a right-handed, orthogonal system with scale factors $(h_\chi, h_\psi, h_\zeta) = (B_0^{-1}, R^{-1}B_0^{-1}, R)$ where $R = r\sin\theta$. As we will see, the use of magnetic coordinates explicitly separates the motion parallel and perpendicular to the unperturbed magnetic field [14].

Using the eikonal assumption, Eq. (7), the gyrocenter phase-space Lagrangian in magnetic coordinates takes the form

$$\mathcal{L} = (\epsilon_0^{-1}\psi + \epsilon J_0\gamma)\,\dot\zeta + (\bar\rho_\parallel + \epsilon J_0\alpha)\,\dot\chi + \epsilon J_0\beta\,\dot\psi + \epsilon_0\bar\mu\,\dot{\bar\xi} - \mathcal{H}, \tag{11}$$

$$\mathcal{H} = \tfrac{1}{2}\bar\rho_\parallel^2 B_0^2 + \bar\mu B_0 + \epsilon\left(J_0\delta\Phi + \frac{2J_1}{\lambda}\bar\mu\delta B_\parallel\right). \tag{12}$$

In Equations (11) and (12), we choose the gauge where $\mathbf{A_0} = \psi\nabla\zeta$, and we define $\boldsymbol{\delta A} = \alpha\nabla\chi + \beta\nabla\psi + \epsilon_0^{-1}\gamma\nabla\zeta$. Note that B_0 is now a function of (χ, ψ) and the perturbation quantities are functions of $(\chi, \psi, \epsilon_0^{-1}\zeta, t)$, where the factor ϵ_0^{-1} is due to the assumption $k_\zeta \gg k_\psi$ (k_ζ and k_ψ are the wave numbers in the ζ and ψ directions, respectively).

Finally, the equations of motion in magnetic dipole coordinates are obtained from Equation (11) using the Euler-Lagrange equations. Note that $\mathcal{L}$ is independent of $\bar\xi$, by construction, therefore $\dot{\bar\mu} = 0$, and $\bar\mu$ can be treated as a constant parameter in the other equations of motion:

$$\dot\chi = \bar\rho_\parallel B_0^2, \tag{13}$$

$$\dot\psi = \epsilon\left(\bar\rho_\parallel B_0^2\,\bar b_\psi + \bar e_\zeta - \bar\mu\,\partial_\zeta\overline{\delta B}_\parallel\right)a^*, \tag{14}$$

$$\dot\zeta = \epsilon_0\left[(\bar\rho_\parallel^2 B_0 + \bar\mu)\,\partial_\psi B_0 + \epsilon(\bar\rho_\parallel B_0^2\,\bar b_\zeta + \bar e_\psi - \bar\mu\,\partial_\psi\overline{\delta B}_\parallel)\right]a^*, \tag{15}$$

$$\dot{\bar\rho}_\parallel = -(\bar\rho_\parallel^2 B_0 + \bar\mu)\,\partial_\chi B_0 + \epsilon(\bar e_\parallel - \bar\mu\,\partial_\chi\overline{\delta B}_\parallel) + \epsilon\bar b_\zeta\dot\psi - \epsilon\bar b_\psi\dot\zeta. \tag{16}$$

In Equations (13) to (16) we have defined $\bar b_i = \varepsilon_{ijk}\,\partial_j\overline{\delta A}_k$ and $\bar e_i = -\partial_i\overline{\delta\Phi} - \partial_t\overline{\delta A}_i$, where $\overline{\delta A}_i = J_0\alpha$, $J_0\beta$, or $J_0\gamma$ for $i = \chi$, ψ, or ζ, respectively; $\overline{\delta\Phi} = J_0\,\delta\Phi$; and ε_{ijk} is the antisymmetric Levi-Civita tensor. Also, $\overline{\delta B}_\parallel \equiv (2J_1/\lambda)\delta B_\parallel$, and $a^* \equiv 1/(1+\epsilon\bar b_\parallel)$. Except for small terms containing derivatives of the Bessel functions, the quantities $\bar b$ and $\bar e$ are related to the perturbation fields $\boldsymbol{\delta B}$ and $\boldsymbol{\delta E}$ by the scale factors $(h_\chi, h_\psi, h_\zeta)$ as follows (with no implied summation): $\bar b_i = J_0|\varepsilon_{ijk}|\,h_j h_k\,\delta B_i$ and $\bar e_i = J_0 h_i\,\delta E_i$. Finally, the notation "∂_i" means "take the partial derivative with respect to the i argument". With this notation there are no hidden factors of ϵ_0; in particular, because the perturbation quantities are functions of $(\chi, \psi, \epsilon_0^{-1}\zeta, t)$, we have $\partial/\partial\zeta = \epsilon_0^{-1}\partial_\zeta$.

Equations (13) and (16) are the parallel velocity and parallel acceleration equations. Equations (14) and (15) describe the gyrocenter drifts in the radial (ψ) and azimuthal (ζ) directions. Physical interpretation of the leading-order terms in the equations of motion is straightforward. For example, in the $\dot\psi$ equation, Equation (14), the first term in the bracket arises because a particle tends to follow a bent field line, the second term is the $\boldsymbol{\delta E} \times \mathbf{B_0}$ drift, and the third term is the ∇B drift due to $\delta B_\parallel$.

Higher-order terms, such as $(\epsilon \bar{b}_\zeta \dot{\psi} - \epsilon \bar{b}_\psi \dot{\zeta})$ in the $\dot{\bar{p}}_{\parallel}$ equation and terms due to the factor a^*, ensure the Hamiltonian conservation properties of the system. Note that the ϵ_0 and ϵ factors explicitly show the order of the bounce and drift motions due to the background fields and due to the perturbations.

LOW FREQUENCY WAVE-PARTICLE INTERACTIONS

This section is a brief summary of the application of the preceding equations to two problems: the motion of energetic protons in magnetospheric hydromagnetic waves, and the drift-resonance-induced radial diffusion of ions and electrons in a dipole field fusion reactor.

(i) *Magnetic Drift-Bounce Resonance in Magnetospheric Hydromagnetic Waves*

Recently, Chen and Hasegawa [5] have shown that energetic protons ($\gtrsim 100$ keV) can excite hydromagnetic waves in the Earth's magnetosphere by a magnetic drift-bounce resonance. The magnetic drift-bounce resonance is characterized by the condition $\omega - m\bar{\omega}_d = \omega_b$, where $\bar{\omega}_d$ is the bounce-averaged magnetic drift frequency (positive westward for ions), m is the azimuthal mode number, ω_b is the bounce frequency, and $\omega \ll (\omega_b, m\bar{\omega}_d)$. We note that the drift-bounce resonance condition corresponds to $k_\perp \rho = \mathcal{O}(1)$, this was the original motivation for considering short perpendicular wave-length perturbations.

To investigate the effect of the waves on the particles we have solved the gyro-center equations of motion, Equations (13) to (16), numerically [4]. Guided by satellite observations, (see Reference [13] for a review), we assume the following model for the waves:

$$\delta B_{\parallel} = \sum_{m=m_0-\sigma_m}^{m_0+\sigma_m} \beta_m \sin k_{\parallel}s \cos(m\zeta - \omega t). \tag{17}$$

Here s is the length along the field line measured from the equator. This model describes a spectrum of westward-travelling compressional waves which are antisymmetric with respect to the equatorial plane. The azimuthal mode numbers are centered at $m = m_0$ with a spread given by σ_m. We use the following typical values: $m_0 = 70$, $\sigma_m = 10$, $\beta_m = \beta_0/2\sigma_m$, $\beta_0 = 10$ nT, and $k_{\parallel} = 2\pi/3R_E$. The wave amplitudes are assumed to be constant for $6 \lesssim L \lesssim 8$, and, since $\omega \ll (\omega_b, m\bar{\omega}_d)$, we will set $\omega = 0$ for simplicity.

First, consider the motion of resonant protons in one mode; i.e., one term in the sum in Equation (17). Figure 1 is a Poincaré surface-of-section plot of the flux coordinate ψ versus the azimuthal angle ζ whenever the gyrocenter trajectory crosses the equatorial plane travelling northward. The chain of "islands" corresponds to the primary resonance $m\bar{\omega}_d = \omega_b$. In the units used in this paper, $\psi = 1/L$, where L is the distance in Earth radii to the equatorial crossing point of the dipole field line. The resonant ψ-surface, denoted ψ_m, can be found by the solving the drift-bounce resonance condition, for a particle of a given energy and equatorial pitch angle, as discussed in Reference [4]. The island half-width, obtained using Equation (14), is given approximately by

$$\Delta \psi_m = 2\sqrt{\frac{2J_1(\lambda)\,\beta_m\,J_1(k_{\parallel}s_0)}{3\lambda\psi_m}}, \tag{18}$$

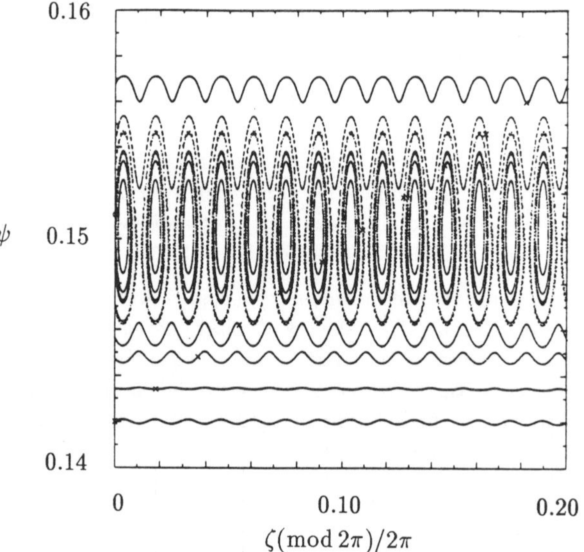

Figure 1: Poincaré surface-of-section plot for protons in a wave consisting of just the $m = 70$ term in Equation (17).

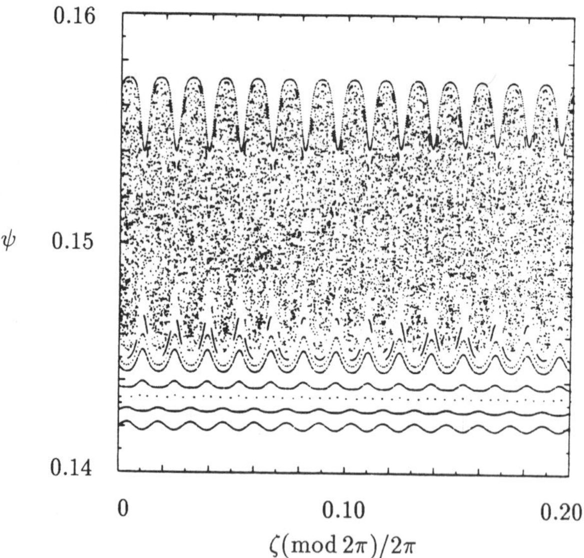

Figure 2: Same as Figure 1 except $m = 70$ and $m = 71$ modes are present together.

where s_0 is the unperturbed mirror-point distance, and the Bessel function $J_1(k_{\parallel}s_0)$ shows the reduction in the particle's response due to averaging over the bounce phase. We note that the drift-bounce resonance breaks the second and third adiabatic invariants, while the first adiabatic invariant and the particle energy are conserved.

The effect of adding a second mode is shown in Figure 2. The Chirikov overlap criterion [7] is exceeded for the observed wave amplitudes, as can be confirmed by comparing the island widths, given by Equation (18), with the spacing of the resonant flux surfaces [4].

Finally, numerical calculations show that the motion in several modes is well approximated by a quasilinear diffusion coefficient of the form

$$D_{\psi,\psi} = \frac{1}{4} N \tau_b m_0^2 \left[\frac{2J_1(\lambda)}{\lambda} \overline{\mu} \beta_m J_1(k_{\parallel}s_0) \right]^2 \tag{19}$$

In Equation (19,) $N \approx 2\sigma_m$ is the number of modes, and $\tau_b = 2\pi/\omega_b$ is the bounce period. Because the magnetic drift-bounce resonance instability is driven by radial gradients in the energetic proton pressure, this diffusion will contribute to the saturation of the growing modes.

(ii) *Magnetic Drift Resonance in a Dipole Fusion Reactor*

Hasegawa [6] has proposed a fusion reactor based on a magnetic dipole field configuration where the fueling and heating is supplied by an external radio-frequency (RF) source. To investigate the effect of the RF waves we consider particles in a dipole magnetic field subject to an azimuthal electric field perturbation of the form

$$\delta E_\zeta = \delta E_0 \cos(m\zeta - \omega t), \tag{20}$$

where $m = \mathcal{O}(1)$ and $\omega \simeq m\overline{\omega}_d \ll \omega_b$.

Both the electrons and the ions move radially by the $\delta E_\zeta \times \mathbf{B_0}$ drift. Also, because the ∇B and curvature drift frequency $\overline{\omega}_d$ is a function of energy, ψ, and equatorial pitch angle, *but not mass*, both the electrons and the ions will experience a drift resonant interaction with the RF waves when $\omega = m\overline{\omega}_d$. We have verified numerically the existence of primary drift-resonance islands in a surface-of-section plot of ψ versus ζ when $\omega t = \mathrm{mod}(2\pi)$. The resonant ψ-surface and the island half-width are given by

$$\psi_r = \sqrt{\frac{\omega}{3m\mu}}, \tag{21}$$

and

$$\Delta\psi_r = \sqrt{\frac{4m\,\delta E_0}{3\mu\psi_r^2}}, \tag{22}$$

respectively.

As in Section (i), we have shown numerically that, when the Chirikov criterion is exceeded for two or more modes, the particles will diffuse radially. In this case, however, the first two adiabatic invariants are conserved but the third adiabatic invariant and the particle energy are not conserved. In fact, because the dipole magnetic field magnitude is proportional to r^{-3}, particles which diffuse inward, while conserving the first two adiabatic invariants, increase their energy as $r^{-8/3}$ [12]. This is the basis for the RF-induced fueling and heating mechanism in the dipole fusion reactor.

SUMMARY

In this paper we have considered the effect of low frequency ($\omega < \omega_c$), short perpendicular wavelength ($k_\perp^{-1} \sim \rho$), electromagnetic perturbations on test particles trapped in a dipole magnetic field. In particular, we are interested in perturbations which break the third adiabatic invariant and result in radial diffusion.

Phase-space Lagrangian Lie transform perturbation methods provide a general and powerful formulation of the problem. Using these methods we have derived a gyrophase-independent phase-space Lagrangian and corresponding equations of motion which enable the efficient numerical evaluation of particle trajectories over many bounce and drift time scales. The Lagrangian is derived to first order in the guiding center small parameter ϵ_0, and to first order in the perturbation amplitude small parameter ϵ. The corresponding Euler-Lagrange equations contain higher-order terms which preserve the Hamiltonian properties of the unperturbed system. Assuming an eikonal form for the perturbations, and adopting magnetic coordinates (χ, ψ, ζ), we obtain equations of motion which explicitly separate the motion parallel and perpendicular to the unperturbed magnetic field and are convenient for numerical work.

For the two applications we have considered, we briefly summarize our results as follows: The motion of energetic ($\gtrsim 100$ keV) protons in magnetospheric hydromagnetic waves is diffusive, due to the overlap of primary drift-bounce resonances, and is well approximated by a quasilinear diffusion coefficient. Externally applied radio-frequency waves can cause drift-resonance-induced radial diffusion and heating in a dipole reactor configuration.

ACKNOWLEDGEMENTS

It is a pleasure to thank Roscoe White, Akira Hasegawa and Kazue Takahashi for useful discussions. This work was supported by NSF grant ATM-86-09585 and DOE contract no. DE-ACO2-76-CHO3073.

References

[1] A. Brizard. Nonlinear gyrokinetic Maxwell-Vlasov equations using magnetic coordinates. *J. Plasma Phys.*, 41(3):541–559, June 1989.

[2] J. R. Cary. Lie transform perturbation theory for Hamiltonian systems. *Phys. Rep.*, 79:129–159, 1981.

[3] J. R. Cary and R. G. Littlejohn. Noncanonical Hamiltonian mechanics and its application to magnetic field line flow. *Ann. Phys.*, 151:1–34, 1983.

[4] A. A. Chan, Liu Chen, and R. B. White. Nonlinear interaction of energetic ring current protons with magnetospheric hydromagnetic waves. *Geophys. Res. Lett.*, 16(10):1133–1136, October 1989.

[5] Liu Chen and A. Hasegawa. On magnetospheric hydromagnetic waves excited by energetic ring-current particles. *J. Geophys. Res.*, 93(A8):8763–8767, August 1988.

[6] A. Hasegawa. A dipole field fusion reactor. *Comments Plasma Phys. Controlled Fusion*, 11(3):147–151, 1987.

[7] A. J. Lichtenberg and M. A. Lieberman. *Regular and Stochastic Motion*, volume 38. Springer-Verlag, New York, 1983.

[8] R. G. Littlejohn. Hamiltonian perturbation theory in noncanonical coordinates. *J. Math. Phys.*, 23(5):742–747, May 1982.

[9] R. G. Littlejohn. Variational principles of guiding center motion. *J. Plasma Phys.*, 29(1):111–125, 1983.

[10] R. G. Littlejohn. Linear relativistic gyrokinetic equation. *Phys. Fluids*, 27(4):976–982, 1984.

[11] T. G. Northrop. *The Adiabatic Motion of Charged Particles*. Interscience Publishers, New York, 1963.

[12] M. Schulz and L. J. Lanzerotti. *Particle Diffusion in the Radiation Belts*, volume 7 of *Physics and Chemistry in Space*. Springer-Verlag, New York, 1974.

[13] K. Takahashi. Multisatellite studies of ULF waves. *Adv. Space Res.*, 8(9):427–436, 1988.

[14] R. B. White and M. S. Chance. Hamiltonian guiding center drift orbit calculation for plasmas of arbitrary cross section. *Phys. Fluids*, 27(10):2455–2467, October 1984.

Dynamo Activity and Turbulent Viscosity in Accretion Disks

P.H. Diamond
Department of Physics
University of California, San Diego

E.T. Vishniac, L. Jin
Department of Astronomy
University of Texas, Austin

Abstract

Accretion disks support low-m internal waves, which can be excited by accretion streams or gravitational effects, and which propagate inward to small radii. These waves, which are similar to oceanic internal waves, can in turn amplify small, "seed" magnetic perturbations, thus driving a dynamo in the disk. Using standard methods of mean-field electrodynamics the induced electric field is calculated and used to construct an α - ω dynamo scenario. As the dynamically generated magnetic field grows until $B_r B_\theta$ - $(H/r)\, P_{thermal}$, the magnetohydrodynamic stresses induced by the disk dynamo imply an effective value of α - H/r, which is sufficient to explain several features of the gross structure of accretion disks.

CHAOS IN DRIVEN ALFVÉN SYSTEMS

T. Hada and C. F. Kennel
Department of Physics and
Institute of Geophysics and Planetary Physics
University of California, Los Angeles
Los Angeles, CA 90024-1547

B. Buti
Jet Propulsion Laboratory
California Institute of Technology
Pasadena, CA 91125

and

E. Mjølhus
Institute of Mathematical and Physical Sciences
University of Tromso

Tromso, Norway

Abstract

The chaos in a one dimensional system that would be nonlinear stationary Alfvén waves in the absence of an external driver is characterized. The evolution equations are numerically integrated for the transverse wave magnetic field amplitude and phase using the derivative nonlinear Schrödinger equation (DNLS), including resistive wave damping and a long-wavelength monochromatic, circularly polarized driver. A Poincare map analysis shows that, for the non-dissipative (Hamiltonian) case, the solutions near the phase space (soliton) separatrices of this system become chaotic as the driver amplitude increases, and "strong" chaos appears when the driver amplitude is large. The dissipative system exhibits a wealth of dynamical behavior, including quasi-periodic orbits, period doubling

537

bifurcations leading to chaos, sudden transitions to chaos, and several types of strange attractors.

I.   *Extended Abstract*

The complex behavior of simple dynamical systems with a small number of degrees of freedom has attracted much recent interest in a variety of fields. New concepts of chaos stemming from the study of low-dimensional dynamical systems have impacted our approach to plasma turbulence. The traditional picture of the transition from order to turbulence in a fluid[1] views turbulence as a hierarchy of instabilities. As a controlling parameter--for example, the Reynolds or Rayleigh number--increases, there appears a succession of unstable modes with different frequencies that are not rationally related. Accordingly, the system will appear to be more and more complicated, though it is still not strictly chaotic since the correlation time is finite so long as the number of waves present is finite. On the other hand, recent studies show that chaos can appear in systems with as few as three wave modes[2,3]. Using a nonlinear Schrödinger equation that includes growth and damping, Russell and Ott[4] showed that a system of three interacting wave modes can be truly chaotic, in the sense that the autocorrelation of the wave amplitude vanishes in the limit of large correlation time. Their three wave system also has many well-known characteristics of other nonlinear dynamical systems, including quasi-periodic phase space orbits and period doubling bifurcations, and it manifests several different routes to chaos.

In this paper we discuss driven, one dimensional, finite amplitude, stationary Alfvén waves (both right-and left-hand polarized) using a dynamical system formulation. Our approach differs from Russell and Ott's[4] in the following way. Since the general nonlinear wave evolution equation has a continuous space dependence, it describes the time evolution of an infinite number of wave modes in wave number Fourier space. Russell and Ott[4] truncated the set of evolution equations for the Fourier modes, retaining only the three most dominant, so that the well-developed theory of dynamical systems with three degrees of freedom could be applied to it. Here, instead of limiting the number of Fourier modes, we will reduce the number of degrees of freedom by considering stationary nonlinear Alfvén waves. The number of free variables will again be three: two transverse components of the magnetic field and the wave phase.

A similar, but computationally more elaborate, numerical simulation study of Alfvén waves in a driven system was carried out by Ghosh and Papadopoulos[5], who integrated the derivative nonlinear Schrödinger equation (DNLS) including temporal growth and diffusive damping, keeping all the Fourier modes available to their periodic simulation system. One mode was chosen to grow, and the remainder were damped. Their results show that, as the growth rate increases, the saturation amplitude of the Alfvén waves manifests period doubling bifurcations; for a sufficiently large growth rate, their system becomes chaotic. The similarities between the multi-dimensional results obtained by Ghosh and Papadopoulos[5] and the behavior of the three-dimensional dynamical system discussed here, including the ways transitions to chaos take place, suggest that our simplified system may retain properties intrinsic to the evolution of driven Alfvén waves.

In short, starting from an extended DNLS equation that describes the evolution of one-dimensional, quasi-parallel, finite amplitude Alfvén waves subject to driven wave growth and diffusive damping, we derived a set of nonlinear ordinary differential equations with three degrees of freedom, by assuming that the nonlinear wave is stationary in the absence of driving. We then studied the behavior of this dynamical system by numerically integrating the equations.

When there is no driver, our system has three invariant points; among them are two attractors, which correspond to the fast and intermediate shock downstream stationary points. There are two solitons, a "dark" right-handed soliton, and a "bright" left-handed soliton. There are three nonlinear periodic travelling waves, pure right-handed, pure left-handed, and a mixed right-hand, left-hand mode. The periodic solutions may be visualized as following contours of constant "energy" in phase space; the soliton solutions follow contours of zero "energy".

When an external driver is applied in the absence of dissipation (the Hamiltonian system), the solutions near that soliton separatrix whose sense of polarization is the same as the driver's become chaotic first as the driver amplitude increases. "Global" or "strong" chaos appears when the driver amplitude is large. The global chaos tends not to invade those regions of phase space where the unperturbed travelling waves have polarizations opposite to that of the driver.

The left-hand driven dissipative system exhibits a wealth of dynamical behavior, including periodic limit cycles of different multiplicity, period doubling bifurcations and transitions to chaos, sudden transitions to chaos, reverse bifurcations from chaos, and several types of strange attractors with different fractal dimensions. The attractors appear in two sequences. The "main sequence" evolves from the absolute minimum of the potential through a sequence of limit cycles and chaotic regions that are by and large in the mixed polarization region of phase space; the chaos finally disappears only when the left-handed driver is strong enough to drive a limit cycle that is mostly in the pure left hand region of phase space. Since the potential contains no further structure with increasing field magnitude, the stable left-hand limit cycle appears to be the final large amplitude solution on the main sequence. By contrast, the sporadic sequence of attractors appears to evolve out of the saddle point of the potential, and to evolve through a sequence of limit cycles whose trajectories approach the saddle point.

We find it significant that the properties of the undriven, non-dissipative nonlinear travelling waves provide us a way to order the complex behavior of the driven, dissipative system.

Acknowledgements

We are pleased to acknowledge interesting discussions with R.D. Blandford, F.V. Coroniti, A.V. Gurevich, R. Hamilton, M. Nambu, R. Pellat, T. Sakaguchi, J. Greene, S. Putterman, V.D. Shapiro, and G.M. Zaslavskii, and, especially, R. Pellat. This research was supported by NASA NAGw-1624. The numerical computations were performed at the San Diego Supercomputer Center, which is supported by the National Science Foundation.

References

1. L.D. Landau, C. R. Acad. Sc. USSR, $\underline{44}$, 311 (1941).

2. J.M. Wersinger, J.M. Finn, and E. Ott, Phys. Fluids, $\underline{23}$, 1142 (1980).

3. M.I. Rabinovich and A.L. Fabrikang, Zh. Eksp. Teor. Fiz., $\underline{77}$, 617 (1979) [(Sov. Phys. JETP, $\underline{50}$, 311 (1979)].

4. D.A. Russell and E. Ott, Phys. Fluids, $\underline{24}$, 1976 (1981).

5. S. Ghosh and K. Papadopoulos, Phys. Fluids, $\underline{30}$, 1371 (1987).

Nonlinear Dynamic Phenomena in the Magnetosphere and Ionosphere Involving Interactions Between Large and Small Scale Processes

P.J. Palmadesso
Naval Research Laboratory
Washington, DC 20375

ABSTRACT

We describe research efforts aimed at developing an improved understanding of magnetospheric and ionospheric phenomena involving an interaction between large and small scale dynamic processes. This coupling between microscopic and macroscopic processes cannot be properly described by either local kinetic models or conventional ideal fluid descriptions, yet there is good reason to believe that such processes are important. Recent research at NRL has shown that multimoment, multispecies generalized fluid models with anomalous transport coefficients are a promising vehicle for this research, when supported by separate studies of relevant microprocesses using the methods of local and nonlocal kinetic theory and PIC computer simulations. We have seen and studied a number of interesting nonlinear dynamic phenomena which arise in this context.

Recent Developments in Ionospheric Modifications by Radio Waves

K. Papadopoulos

University of Maryland
Department of Physics and Astronomy
College Park, Maryland 20742

Abstract

A review of recent developments in two areas of ionospheric modifications by high power radiowaves is presented. The first deals with techniques to improve the power conversion efficiency of ground HF to ELF waves generated by utilizing the lower ionosphere as an active medium. The second explores the physics and power requirements for creating artificial ionospheric clouds in the atmosphere and lower ionosphere. Such clouds are of interest in studying in situ atmospheric chemistry, developing new diagnostic techniques and improving transionospheric communications and radars.

1 Overview

The development of high power phased arrays coupled with improved theoretical and experimental understanding of RF-plasma interactions in the weakly ionized collisionless regime opens up a new era in the exploration of the ionosphere as a strongly non-linear medium for fundamental and applied research. This review presents two applications of ionospheric heating utilizing the ionosphere as a non-linear active medium. The first one deals with the interaction of modulated HF radiowaves to produce with high efficiency ELF waves which couple to the earth ionosphere waveguide. The second discusses the application of UHF focused radiowaves to create artificial patches in the atmosphere and ionosphere. Such patches can be used as mirrors to reflect HF frequencies for communications and radar applications. Furthermore optical emissions from such patches can provide diagnostics of the atmospheric and ionospheric mixing ratios of minority species and radicals. The first topic is a mature one with a large amount of experimental data, while the second is exploratory. The review emphasizes physics scalings and power requirements rather than practical applications.

2 Electrojet Modulation ELF Generation

2.1 Introduction

The potential for using the ionosphere as an active medium to transform ground based HF power to ELF power and its significance to submarine communications was first noted in an internal report at the Naval Research Laboratory (Papadopoulos, 1973). It was shown that if the ionosphere is irradiated with an electromagnetic (em) signal that contains two carrier frequencies ω_1 and ω_2, the ionospheric plasma acts as an active, non-linear medium producing signals at the frequency difference $\omega \approx \omega_1 - \omega_2$ (Fig. 1). The effect was experimentally confirmed for the first time in Gorky USSR (Getmantsev et al., 1974). In the course of subsequent experimental and theoretical studies (Kotik et al., 1975) it was found that for the HF powers used the effect was strongest in the presence of ambient ionospheric currents, in which case the frequency conversion could be accomplished by either the two frequency technique or by simple amplitude modulation of the HF carrier at the desirable low frequency. The experimental and theoretical work in the seventies was conducted predominantly in the USSR (see Belyaev et al., 1987 for a comprehensive review). Following the success of the USSR experiments Stubbe and Kopka (1977) suggested that the high latitude ionosphere could be a much more efficient frequency converter due to the presence of strong electrojet currents. A powerful HF facility built by the Max-Planck Institut fur Aeronomie in Tromso Norway was the dominant site of experimentation in the 80's (Stubbe et al., 1981, 1982; Barr and Stubbe, 1984a,b; Barr et al., 1985). These experiments were extremely successful in producing large ELF/VLF signals on the ground by amplitude modulation of the HF power incident in the ionosphere. The Tromso HF transmitter operated in the frequency range of 2.5–5 MHz with an effective radiated power (ERP) of 100–300 MW. However, in the context of practical applications, the results were rather disappointing since the overall HF to ELF power conversion efficiency was found to be of the order of 10^{-8} while the conventional ground based ELF facilities have a conversion efficiency of 10^{-6}. The purpose of this paper is to examine the reasons for the low conversion efficiency and suggest techniques under which HF facilities similar to the Tromso facility can improve their conversion efficiency by two orders of magnitude. In addition to the theoretical results we present below preliminary experimental investigations using the HIPAS facility in Fairbanks, Alaska, which confirm the theoretical premise for the inefficiency of downconversion and the suggested improvements. This paper concentrates on frequency conversion in the low D-region of the ionosphere (< 80 km) where, in the low frequency ($f < 5$ MHz) and low power density $\left(< 1 - 2 \frac{mW}{m^2}\right)$ regime, the HF power is absorbed. Even larger efficiency gains can be achieved by using higher frequency ($f > 5$ MHz) higher power densities $\left(> 5 - 10 \frac{mW}{m^2}\right)$ transmitters, in which case the HF power is absorbed in the lower E region (85–95 km). This topic forms the subject of a separate paper (Papadopoulos et al., 1990a).

2.2 Baseline ELF Generation and Scaling

The process by which the HF power is converted to ELF in the lower D-region where the ambient electron neutral collision frequency for momentum transfer ν_0 is smaller than the electron cyclotron frequency Ω is the following (Chang et al., 1981). At high latitudes the solar wind interaction with earth's magnetosphere results in the creation of an electromotive force (emf). Since at high altitude the magnetic field lines are equipotential lines, the high altitude electric field $\underline{E}_o = E_o \hat{e}_x$ maps in the lower ionosphere where collisional processes allow for the generation of cross field currents. Two types of currents flow across the magnetic field $\underline{B} = B_o \hat{e}_z$. The Pedersen current

$$\underline{j}_p = \sigma_p E_o \hat{e}_x \tag{2.2.1}$$

in the direction of the electric field $\underline{E}_0$, and the Hall current

$$\underline{j}_H = \sigma_H E_o \hat{e}_y \tag{2.2.2}$$

perpendicular to $\underline{E}_0$ and $\underline{B}_0$. In eqs. (2.2.1) and (2.2.2) σ_p and σ_H are the Pedersen and Hall conductivities defined as

$$\sigma_p = \frac{ne^2}{m\Omega^2}\left(\frac{\nu_o}{1 + \nu_o^2/\Omega^2}\right) \tag{2.2.3a}$$

$$\sigma_H = \frac{ne^2}{m\Omega^2}\left(\frac{\Omega}{1 + \nu_o^2/\Omega^2}\right) \tag{2.2.3b}$$

where n is the ambient electron density. It is clear that

$$\frac{j_p}{j_H} = \frac{\nu_o}{\Omega}$$

Charge neutrality requires that the ambient ionospheric currents be divergent free, i.e.

$$\nabla \cdot \underline{J} = 0 \tag{2.2.4}$$

where $\underline{J}$ is the height integrated current density. Periodic electron heating at the HF modulation frequency ω, leads to a periodic conductivity modulation in the heated volume if $\omega \tau_c \lesssim 1$ where τ_c is the electron cooling time. In order to maintain the divergence free condition imposed by the quasineutrality requirement, a polarization electric field $\underline{E}_1$ builds in the modified region to maintain the divergence free condition, i.e.

$$\nabla \left[\underline{\underline{\sigma}} \cdot (\underline{E}_o + \underline{E}_1) \right] = 0 \tag{2.2.5}$$

Introducing the height integrated conductivity $\underline{\underline{\Sigma}}$, eq. (2.2.5) with $\underline{E}_1 = -\nabla \phi$ becomes (Chang et al., 1981)

$$\frac{\partial}{\partial x} \left(\Sigma_p \frac{\partial \phi}{\partial x} \right) + \frac{\partial}{\partial y} \left(\Sigma_p \frac{\partial \phi}{\partial y} \right) + \frac{\partial \Sigma_H}{\partial x} \frac{\partial \phi}{\partial y} - \frac{\partial \Sigma_H}{\partial y} \frac{\partial \phi}{\partial x}$$

$$= \left(\frac{\partial \Sigma_p}{\partial x} - \frac{\partial \Sigma_H}{\partial y} \right) E_o \tag{2.2.6}$$

Figure 2 shows the difference in the height integrated current $\underline{J}_1$ and the electric field $\underline{E}_1$ before and after the modification for a Gaussian modified region with radius R; $\underline{J}_1$ and $\underline{E}_1$ are the sources of radiation at the low frequency ω. From eqs. (2.2.3) and $\nu \gg \Omega$ we see that the fractional change in Hall conductivity is larger than the fractional change of the Pedersen conductivity. We, thus, examine the modification of the height integrated Hall current only. If the HF absorption occurs over a region $\triangle z \approx a$, to zero order the radiating current will be

$$J_{1H} \approx (E_o a)\frac{ne^2}{m\nu_o}\frac{\Omega}{\nu_o}\left(1 - \frac{\nu_o^2}{\nu_F^2}\right).$$

(2.2.7)

where ν_F is the modified collision frequency. Taking as L^2 the horizontal area of the heated region and assuming that the radiating source at ELF is an horizontal electric dipole we find its moment as

$$M \approx (E_o a)L^2\frac{ne^2}{m\nu_o}\frac{\Omega}{\nu_o}\left(1 - \frac{\nu_o^2}{\nu_F^2}\right)$$

(2.2.8)

The reason for the inefficiency of HF to ELF conversion can be identified by rewriting (2.2.8) as

$$M = (E_o a)L^2 \triangle\sigma_H$$

(2.2.9)

$$\triangle\sigma_H = \frac{ne^2}{m\nu_o}\frac{\Omega}{\nu_o}\left(1 - \frac{\nu_o^2}{\nu_F^2}\right)$$

(2.2.10)

Notice that $\nu_o \sim T_e$, where T_e is the electron temperature, which in its turn is proportional to HF power density

$$S = \frac{P_o G}{4\pi z_o^2}$$

(2.2.11)

where z_o is the absorption height, P_o the ground HF power and G the antenna gain. The efficiency arguments are the same if the radiating source is an horizontal magnetic dipole, although the scaling with ground power is different.

Figure 3 shows the saturated value of $\triangle \sigma_H$ computed using a kinetic code as a function of the power density S for an HF frequency of 2.8 MHz (Papadopoulos et al., 1990). It is clear that $\triangle \sigma_H$ after an initial linear increase with S saturates. Further increase of S or equivalently of the ERP does not produce additional modification. For the Tromso heater $S \approx 2\,mW/m^2$ is above the optimum value resulting in waste of the HF power. The behavior seen in Fig. 3 can be understood on the basis of eq. (2.2.10). For low values of S, we can write

$$\nu_F \approx \nu_o + \triangle \nu$$

with $\triangle \nu / \nu_o << 1$. As a result

$$\triangle \sigma_H \approx 2 \frac{ne^2}{m\nu_o} \frac{\Omega}{\nu_o} \frac{\triangle \nu}{\nu_o} \sim S \tag{2.2.12}$$

For higher values of S, $\triangle \sigma_H$ saturates at the value

$$\triangle \sigma_H \approx \frac{ne^2}{m\nu_o} \frac{\Omega}{\nu_o} \tag{2.2.13}$$

For the low power density regime from eqs. (2.2.9) and (2.2.12) we find

$$M \sim (E_o a) L^2 S$$
$$\sim (E_o a) \frac{1}{G} P_o G \sim (E_o a) P_o \tag{2.2.14}$$

where we used $L^2 \sim \frac{1}{G}$. Namely the ELF power scales as P_o^2. However, for the high power regime

$$M \sim (E_o a) \frac{1}{G} \frac{ne^2}{m\nu_o} \sim (E_o a) \frac{1}{G} \qquad (2.2.15)$$

Namely for values of S larger than saturation, efficient ELF production requires smaller values of G, which of course result in larger radiating moments. This effect was first noted in Chang et al. (1981). Returning to Fig. 3, we note that $\triangle\sigma_H$ deviates from linearily at $S_o \approx 5 \times 10^{-4} W/m^2$. For the Tromso facility $P_o \approx 1MW$, $P_o G \approx 150MW$ and $S \approx 2mW/m^2$. It is obvious that an HF to ELF efficiency will increase by a factor of 16 (i.e. $\frac{1}{G^2}$) if we decrease the gain of the antenna by a factor of 4 so that $S = S_o \approx .5mW/m^2$ at 70 km.

The final factor that could affect the conversion efficiency is the scaling of the absorption height z_o with power density. To examine its effect we rewrite M from eq. (2.2.9) as

$$M \simeq E_o L^2 \triangle\Sigma_H \qquad (2.2.16)$$

$$\triangle\Sigma_H = a(\triangle\sigma_H) \qquad (2.2.17)$$

For constant gain $L^2 \approx$ const. and

$$M \sim \triangle\Sigma_H \qquad (2.2.18)$$

Figure 4 shows the scaling of $\Delta\Sigma_H$ as a function of PG from a numerical computation at 2.8 MHz. It is clear that the scaling of the ELF to the HF power is weaker than linear.

2.3 Efficiency Increase by Antenna Scanning

Further efficiency increase can be achieved by relying on the following basic non-linearities of the electron heating in the ionosphere,

(i) The value of $\nu_o \sim T_e$ resulting in fast (exponential) electron heating.

(ii) The cooling rate decreases as a function of T_e, for T_e in the range of $600°K$ and $2000°K$ (Fig. 5).

(iii) For $S_o > .5\frac{mW}{m^2}$ the heating is of the runaway type for $T_e < 2000°K$ and thus very little energy is waisted in excitation of rotational levels.

This technique again relies on the fact that $M \sim L^2$ and major efficiency increase can be accomplished by increasing the modified area L^2 by electronically scanning the HF transmitter. The underlying physics of the efficiency increase can be understood by referring to Fig. 6a,b (Papadopoulos et al., 1989). Figure 6a shows the temperature variation for the case where a single spot in the ionosphere was continuously irradiated by the HF over a time $\frac{1}{2}\tau_{ELF}$ where $\tau_{ELF} = \frac{2\pi}{\omega}$. The periodic modulation in the temperature results in modulating the conductivity and the ionospheric current, which is the basis of the ionospheric ELF antenna.

Figure 6b shows the temperature variation if we send short pulses of duration $1.2 \times 10^{-3}\ \tau_{\mathrm{ELF}}$ with a repetition rate of $5 \times 10^{-2}\ \tau_{\mathrm{ELF}}$ over the half period instead of irradiating it continuously. The resultant temperature was only $\frac{2}{3}$ of the one for the standard case, resulting in a lower conductivity variation by a factor of 2. However, the HF energy expenditure over the τ_{ELF} time is 40 times lower, resulting in extreme increase in the efficiency of HF to ELF conversion.

The validity of this concept was demonstrated experimentally using the HIPAS facility last summer (Papadopoulos, 1990b). Figure 7 shows the relative field amplitude measured on the ground for pulses equal with $\tau_{\mathrm{p}} = .5\ \tau_{\mathrm{ELF}}$. For technical reasons only nonrepetitive single pulse irradiation could be performed at $f = \frac{1}{\tau_{ELF}} = 833 Hz$. For the cases where $\tau_{\mathrm{p}} > .25\ \tau_{\mathrm{ELF}}$ the value of B measured on the ground was independent of the HF energy per τ_{ELF}, indicative of efficiency improvements. The technique by which this concept can lead to much higher ELF powers without increasing the ground HF power or the antenna gain was discussed in detail in Papadopoulos et al. (1989). The scheme shown in Fig. 8 relies on scanning of the HF antenna very fast so that it can irradiate other spots during its off time τ_{OFF} increasing the effective ELF antenna area. The estimated increase in efficiency was of the order of 50–100.

It should be emphasized that the above efficiency increases do not include reductions in efficiency due to HF absorption at height lower than the desired modulation heights. These effects are presently under study.

3 Physics of Ionospheric Breakdown

3.1 Introduction

Generating artificial ionization in the atmosphere and ionosphere has a wide range of applications from communications and radar to novel ionospheric and atmospheric diagnostic instrumentation. In assessing the feasibility of such a concept as well as the engineering requirements and projected costs a thorough understanding of the physics of ionospheric breakdown is required. We review below the physical foundations for creating artificial ionospheric clouds by using focused high power ground based UHF facilities.

The most critical parameter in the description of breakdown processes in the value of the net ionization rate ν_{nei} as a function of the incident RF power density S, frequency ω, ambient neutral density N and gas composition. Although a large body of theoretical and experimental work exists on the subject of RF breakdown of air, a critical review of the subject (APTI Technical Report 5004) noted that most of the experimental results and their empirical extrapolations were not appropriate to the parameter range of interest here. In fact, erroneous application of the empirical breakdown thresholds and ionization rates results in major underestimates of the power requirements and the frequency optimization. On the other hand, analytic efforts by Gurevich et al. (1978) and Borisov et al. (1988)

resulted in substantial errors due to neglect of molecular dissociation processes. A comprehensive numerical kinetic computation was developed to calculate the ionization rates as a function of the local RF power density, frequency, and neutral density for ionospheric parameters. The code describes kinetically the electron energization by the incident RF and includes the most current values of cross-sections for elastic and inelastic processes. A brief description of the code is given in section 3.2 a more detailed discussion of the code and of its range of validity can be found in (APTI 1990).

3.2 Kinetic Modeling

The kinetic calculation is based on the numerical solution of a differential equation for the isotropic portion f_o of the distribution function $f(r,v,t)$ of a uniform weakly ionized plasma. When the quiver energy is below the energy required for ionization, the isotropic portion of the distribution dominates, while the anisotropic part can be neglected. The differential equations governing the time evolution of f_o are (Gurevich, 1978)

$$\frac{\partial f_o}{\partial t} - \frac{1}{2v^2}\frac{\partial}{\partial v}\{v^2[\delta\nu_m v f_o + (\delta\nu_m \frac{kT}{m} \mid$$

$$(3.2.1)$$

$$\frac{e^2 E_{pk}^2}{3m^2}(\frac{\nu_m}{\nu_m^2 + \omega^2}))\frac{\partial f_o}{\partial v}]\} - S_{oin.}$$

$$S_{oin} = S_{oin}^{rot} + S_{oin}^{vib} + S_{oin}^{opt} + S_{oin}^{dis} + S_{oin}^{att} + S_{oin}^{ion} \qquad (3.2.2)$$

$$S_{oin}^{rot} = -\frac{1}{2v^2}\frac{\partial}{\partial v}\left[v^2\left(\sum_k \frac{8B_k N_{nk}\sigma_k}{mv}\right)\left(vf_o + \frac{kT}{m}\frac{\partial f_o}{\partial v}\right)\right] \tag{3.2.3}$$

$$S_{oin}^{vib} = -\frac{2}{mv}\sum_{kj}[N(\epsilon + \epsilon_{kj})f_o(\epsilon + \epsilon_{kj})\sigma_{kj}^{vib}(\epsilon + \epsilon_{kj}) - \tag{3.2.4}$$

$$\epsilon f_o(\epsilon)\sigma_{kj}^{vib}(\epsilon)]$$

$$S_{oin}^{opt} = -\frac{2}{mv}\sum_{kj} N_{nk}[(\epsilon + \epsilon_{kj})f_o(\epsilon + \epsilon_{kj})\sigma_{kj}^{opt}(\epsilon + \epsilon_{kj}) - \tag{3.2.5}$$

$$\epsilon f_o(\epsilon)\sigma_{kj}^{opt}(\epsilon)]$$

$$S_{oin}^{dis} = -\frac{2}{mv}\sum_k N_{nk}[(\epsilon + \epsilon_k)f_o(\epsilon + \epsilon_k)\sigma_k^{dis}(\epsilon + \epsilon_k) - \tag{3.2.6}$$

$$-\epsilon f_o(\epsilon)\sigma_k^{dis}(\epsilon)]$$

$$S_{oin}^{att} = -\left(N_{o_2}\sigma_{att2}(v) + N_{o_2}^2\sigma_{att3}(v)\right)vf_o \tag{3.2.7}$$

$$S_{oin}^{ion} = \sum_k \int_{v_{ion}}^{\infty} N_k v'^3 F_k(v,v')f_o\sigma_{ion}(v')dv' \tag{3.2.8}$$

The set of equations (3.2.1–8) above are those which were solved numerically. The major assumptions required in the derivation of these equations are listed below:

- The quiver energy is below the threshold energy for the dominant loss processes, so rates of energy loss and electron production rates are determined by f_0.

- The plasma is locally uniform, and diffusion operates on much longer timescales than ionization, so transport may be neglected.

- The fractional ionization is low, so electron-electron and electron-ion elastic collisions, as well as electron-ion processes such as detachment and recombination, may be ignored.

- The number of molecules in excited states is low, so superelastic collisions are unimportant.

- The electric field is rapidly alternating, so the time-averaged field may be used.

In addition, in the implementation of the numerical solution, the neutral atmosphere was assumed to consist of molecular *oxygen* and *nitrogen* only, excluding other trace neutral constituents. To solve the equations for the distribution function numerically, it is necessary to transform them into an equivalent set of equations on discrete time and energy (or velocity) grids. One method for performing this transformation is the finite-difference method; it may be developed very simply from the definition of the derivative. Because the equations to be solved are first-order in time, this will yield an equation for the distribution function at a given timestep in terms of the function at the previous timestep.

Modeling of ionization in the atmosphere is accomplished by setting the distribution function to its initial value, a Maxwellian at the ambient electron temperature, and calculating successive distributions by applying the finite-difference equation repeatedly. This will result in a time history of the distribution function, which may be integrated over velocity to give the electron density, and which may also be used to calculate average collision rates for momentum-transfer, attachment, and ionization.

The set of equations (3.2.1–8) may be approximated by a finite-difference equation of the form

$$\frac{f_o^{1+\Delta t} - f_o^t}{\Delta t} = (T + U)f_o^{1+\Delta t}$$

Here the continuous function f_o has been replaced by the vector f_o, a one-dimensional array which approximation; and the effects of the electric field and of collisions have been combined into the matrices T and U. The matrix T is a tridiagonal matrix representing the effects of the electric field, elastic collisions, and rotational inelastic collisions, all of which appear in the original equations as terms containing the first and second derivatives of f_o with respect to ϵ. The other inelastic processes cause discontinuous changes in electron energies and cannot be represented in this way; their effects are combined in the upper triangular matrix U. This is a fully implicit equation because the operators representing the energy derivatives are applied to the value of f_o at the succeeding timestep;

it is unconditionally stable and convergent to the solution of the corresponding differential equation.

A valuable check on the accuracy of the numerical simulation is to compute the energy balance. Because the cross-section for each process is available and the energy loss due to each event is known, the energy absorbed by each process may be computed on each timeliest. The energy loss by the electric field may also be computed and compared to the sum of the energies absorbed. Agreement between the two totals was excellent, typically with errors less than .01%.

3.3 Comparison of Kinetic Computations and Experimental Data

Deriving ionization rates from the experimental observables is, with the exception of one recent experiment (Hayes et al., 1987), a convoluted process which involves a great degree of uncertainty. The most relevant data are derived from pulsed experiments [Sharfman and Morita (1964), Sharman et al. (1964), Herlin (1948), Gould (1956), Geballe and Harrison (1953)]. In these experiments, the ionization rate is computed by fixing the power density and increasing the pulse length until "breakdown" occurs. Two definitions of breakdown are used.

- when the ratio of the final to the original electron density is:

$$\frac{n_e(t_f)}{n_e(t_o)} = 10^8 \qquad (3.3.1)$$

- when the electron density reaches the critical density (i.e. the plasma frequency equals the wave frequency):

$$n_e(t_f) = 1.2 \times 10^{-8} f^2 \qquad (3.3.2)$$

The ionization rate ν_{net} is then computed by assuming that over the pulse length the electron transport or other losses are negligible and that the ionization rate is independent of time, so that

$$n_e(t_f) = n_e(t_o) exp(\nu_{neff}) \qquad (3.3.3a)$$

$$\tau = t_f - t_o \qquad (3.3.3b)$$

Equation (3.3.3) is often written in the form

$$\frac{\nu_{net}}{P} = \frac{1}{P_T} ln\left(\frac{n_e(t_f)}{n_e(t_o)}\right) \qquad (3.3.4)$$

where P is the gas pressure. The value of ν_{net}/P is determined by varying the RF power and pulse length until breakdown, as defined by either of equations (3.3.1) or (3.3.2) occurs. When definition (3.3.2) is used, the initial electron density is either assumed to be 1 cm^{-3} or measured in advance. It should be noted that for historical reasons the ionization rates determined from the microwave experiments were presented in a format that makes the interpretation and scaling of the results difficult. Namely, the plots give ν_{net}/P as a function of E_e/P, where E_e is an effective electric field defined by

$$E_e = E_{rms}\left(\frac{\nu_e^2}{\nu_e^2 + \omega^2}\right)^{1/2} \tag{3.3.5a}$$

$$\nu_e = 5.3 \times 10^9 P \tag{3.3.5b}$$

The confusion and possible pitfalls in using the ν_{net}/P vs. E_e/P format will be discussed later. In order to facilitate the comparison of the numerical results with the experimental results, the above format of presentation was maintained.

Figure 9a presents a summary of the early experiments along with the values determined in the numerical work. Given the uncertainties discussed above, the agreement is remarkable. Even more important is the comparison of the numerical results with the recent experiment by Hays et al. (1988) shown in Fig. 9b. This experiment was unique in that instead of using equations (3.3.1) — (3.3.3), the ionization rate was monitored as a function of time. It should be noted that the experiment was performed in the presence of a longitudinal magnetic field, and the microwave frequency was adjusted to resonate with the electron gyrofrequency. Under these conditions, the physics of the electron energization is similar to the one in the presence of a DC electric field with amplitude two times larger than the RF field (Hays et al., 1987).

3.4 RF Ionization at 70 km Altitude

The kinetic equation given in section 3.2 above was numerically solved over a wide range of frequencies, altitudes and power densities. For concreteness, we discuss here results relevant to air density of 2×10^{15} cm^{-3}, which corresponds to an altitude of approximately 70 km. Figures 9–11 show the time evolution of the electron distribution function $f(\epsilon)$ at intervals of 200 ns for an incident RF frequency of 300 MHz and powers of 300 W/m^2, 3 kW/m^2, and 30 kW/m^2, respectively.

For the lowest power density case, shown in Fig. 10 below, $f(\epsilon)$ reaches steady state after approximately 2 μs. In this case, the incident power density was below threshold and breakdown did not occur. The energy loss due to the vibrational excitation of molecular N_2 acted as a barrier that prevented generation of significant electron fluxes past 2–3 eV.

Figure 11 shows $f(\epsilon)$ for 3 kW/m^2 RF power density. Recent Soviet studies predict a threshold power density of 1.8 kW/m^2 for 300 Mhz at $N = 2 \times 10^{15}$ cm^{-3}. However, our results indicate that even at 3 kW/m^2, although there was significant electron flux above 5–6 eV, few electrons reached ionizing energies.

As will be discussed in a future publication, the optimistic results of Borisov et al. (1988) are due to their neglect of O_2 dissociation.

Increasing the power density to 30 kW/m^2 (Fig. 12) results insignificant electron fluxes reaching energies of 20–25 eV, ionizing both O_2 and N_2 molecules. The evolution of f(ϵ) is self similar in energy and only the total electron density n_e increases with time. The distribution reaches a self similar state in approximately 200 ns. Of primary interest is the rate of ionization as a function of RF frequency and power density. The results of the computations are shown in Fig. 13 for the 70 km altitude case. Figure 13 indicates a scaling of

$$\nu_{net} \propto \frac{S}{\omega^2} \tag{3.4.1}$$

This along with other scaling issues is the subject of the next section.

3.5 Ionization Rates and Efficiency Scaling Considerations

Based on the self similarity of f(ϵ), the results of section 3.4 can be generalized to produce universal relations for ionization rates for any combination of S, ω, and N. To accomplish this we rewrite equation (3.2.1) in the form

$$\frac{\partial}{\partial t} f_o(\epsilon) = \frac{1}{\sqrt{\epsilon}} \frac{\partial}{\partial \epsilon} \left[\epsilon^{3/2} D(\epsilon, \omega, \tilde{\epsilon}) \frac{\partial f_o(\epsilon)}{\partial \epsilon} \right] - NL(f_o(\epsilon)) \tag{3.5.1}$$

$$D(\epsilon, \omega, \tilde{\epsilon}) = \frac{2}{3} \tilde{\epsilon} \frac{\nu_e(\epsilon)}{1 + \nu_e^2(\epsilon)/\omega^2} \tag{3.5.2}$$

$$\tilde{\epsilon} = \frac{1}{2} m_e \frac{q_e^2 E_o^2}{m_e^2 \omega^2} \tag{3.5.3}$$

The elastic electron neutral collision frequency $\nu_e(\epsilon)$ has a maximum value ν_{max} given by

$$\nu_{max} = 3 \times 10^{-7} N \tag{3.5.4}$$

This value is reached when $\epsilon = 20\text{--}30$ eV. For the frequencies of interest here $\omega \gg \nu_{max}$, and equation (3.5.1) can be written as

$$\frac{\partial}{\partial \tau} f_o(\epsilon) = \frac{2}{3} \frac{\tilde{\epsilon}}{\sqrt{\epsilon}} \frac{\partial}{\partial \epsilon} \left[\epsilon^{3/2} g(\epsilon) \frac{\partial f_o(\epsilon)}{\partial \epsilon} \right] - L_1(f_o(\epsilon)) \tag{3.5.5}$$

$$\tau = \nu_e t \tag{3.5.6a}$$

$$\nu_e(\epsilon) = g(\epsilon) \nu_{max} \tag{3.5.6c}$$

$$L_1(f_o(\epsilon)) = \frac{1}{3} \times 10^7 L(f_o(\epsilon)) \tag{3.5.6c}$$

Notice that in equation (2.5.5), the altitude dependence enters through the normalized time τ, while the RF frequency and power density dependences enter through $\tilde{\epsilon}$. As a result the dimensionless ionization rate ν_{net}/ν_{max} is only a function of $\tilde{\epsilon}$. The computational results confirm this conjecture for $\omega \gg \nu_{max}$, but show a factor of two difference when $\omega \approx \nu_{max}$. We will return to this point later on.

3.6 Analytic Approximation — A Test Particle Approach

The physics underlying the ionization rates and scaling presented in section 3.5 can be understood by examining the RF acceleration of an electron in the presence of inelastic losses. The energization of a test electron in the presence of Rf waves with $\omega \gg \nu_{max}$ can be appproximated by

$$\frac{d(\epsilon)}{dt} = \nu_e(\epsilon)\tilde{\epsilon} \tag{3.6.1}$$

In the presence of only ionization losses, the effective ionization time is the sum of the energization time to ionizing energies τ_e and of the time τ_{ion} to make an ionizing collision once the electron energy $<\epsilon> > E_{ion}$. For the moderate values of $<\epsilon>$ (i.e. < 4–5 eV) of interest here it is easy to see that τ_e is the longest timescale. As a result

$$\nu_i = \frac{1}{\tau_e(\epsilon)} \tag{3.6.2}$$

We approximate the collision frequency by (Kroll and Watson, 1972)

$$\nu_e(\epsilon) = \nu_{max}g(\epsilon) = \nu_{max}\frac{\epsilon + .1}{\epsilon + 5} \tag{3.6.3}$$

where ϵ is in units of eV. Assuming that ionization occurs near 25 eV energy, as seen in the computations, we find from equations (3.6.1) and (3.6.3)

$$\nu_{max}\tau_e = \frac{35}{\tilde{\epsilon}} \tag{3.6.4}$$

From equations (3.6.2) and (3.6.4)

$$\frac{\nu_i}{\nu_{max}} \approx \frac{\tilde{\epsilon}}{35} \tag{3.6.5}$$

The presence of other inelastic losses, shown in Fig. 2.6.1, reduces the ionization rate to a great extent. We can compute the effect of inelastic collisions within the test particle theory by noting that there are two main barriers to electron energization. One is between 2–3 eV and is due to N_2 vibrational losses. The second one is between 10–20 eV and is due to molecular dissociation and optical emissions. Ionization losses become dominant near 25 eV. We can account for the inelastic losses by introducing a probability $P(\tilde{\epsilon})$ that under the RF action the electron will be accelerated through the loss barriers and undergo an ionization collision. As a result

$$\nu_{max}\tau_e \approx \frac{35}{\tilde{\epsilon}P(\tilde{\epsilon})} \tag{3.6.6}$$

or

$$\frac{\nu_i}{\nu_{max}} \approx \frac{\tilde{\epsilon}}{35}P(\tilde{\epsilon}) \tag{3.6.7}$$

We can write

$$P(\tilde{\epsilon}) = P_{vib}(\tilde{\epsilon})P_{opt}(\tilde{\epsilon}) \tag{3.6.8}$$

where $P_{vib}(\tilde{\epsilon})$ refers to the vibrational band and $P_{opt}(\tilde{\epsilon})$ to the dissociation-optical excitation band. We calculate first the N_2 vibration effect $P_{vib}(\tilde{\epsilon})$.

Since the band is relatively narrow we can approximate the vibrational excitation rate ν_{vib} by its value at the vibrational peak energy ϵ_{vib}. In this range the value of the diffusion coefficient in energy space $D_E =< \Delta_\epsilon^2/\tau >$ is given by

$$D_E(\epsilon_{vib}) = \epsilon_{vib}D(\epsilon_{vib}) \tag{3.6.9}$$

As a result, the energy diffusion time through the width Δ_{vib} of the vibrational barrier is

$$\tau_{D-vib} = \frac{\Delta_{vib}^2}{D_E(\epsilon_{vib})} \tag{3.6.10}$$

If $\tau_{vib} = 1/\nu_{vib}$ is the average time to excite vibrational states, then the probability $P_{vib}(\tilde{\epsilon})$ that an electron will cross the vibrational barrier is given by

$$P_{vib}(\tilde{\epsilon}) = exp\left[-\left(\frac{\tau_{D-vib}}{\tau_\nu}\right)^{1/2}\right] \tag{3.6.11}$$

From (3.6.9–11) we find

$$P_{vib}(\tilde{\epsilon}) = exp\left[-\left(\frac{3}{2}\frac{\nu_{vib}(\epsilon_{vib})}{\nu_m(\epsilon_{vib})}\frac{\Delta_{vib}^2}{\tilde{\epsilon}\epsilon_{vib}}\right)\right] \tag{3.6.12}$$

From cross-section data we can determine that $\nu_{\mathrm{vib}}(\epsilon_{\mathrm{vib}}) = \nu(\epsilon_{\mathrm{vib}})/2$, $\epsilon_{\mathrm{vib}} = 2.6$ eV, and $\triangle_{vib} = 1.0 eV$. Making these substitutions, we find that

$$P_{vib}(\tilde{\epsilon}) = exp\left[-\sqrt{\frac{29eV}{\tilde{\epsilon}}}\right] \tag{3.6.13}$$

The losses in the 3–8 eV range are much smaller and may be ignored. The remaining barrier is due to both optical excitations and dissociation; it occupies the range between 8–20 eV and reaches a maximum near 14 eV. We can find the probability $P_{opt}(\tilde{\epsilon})$ of penetrating this barrier by using an approach similar to the one above. In this case $\nu_{\mathrm{opt}}(\epsilon_{\mathrm{opt}}) = 0.2\ \nu_e(\epsilon_{\mathrm{opt}})$, $\epsilon_{opt} \approx 14eV$, and $\triangle_{vib} = 6eV$. For these values, we find that

$$P_{opt}(\tilde{\epsilon}) = exp\left[-\sqrt{\frac{.77eV}{\tilde{\epsilon}}}\right] \tag{3.6.14}$$

The total probability that an electron will cross both barriers and reach the energies necessary to ionize is given by

$$P(\tilde{\epsilon}) = P_{vib}(\tilde{\epsilon})P_{opt}(\tilde{\epsilon}) = exp\left[-\sqrt{\frac{.29eV}{\tilde{\epsilon}}}\right]exp\left[-\sqrt{\frac{/77eV}{\tilde{\epsilon}}}\right] \tag{3.6.15}$$

From equations (3.6.7) and (3.6.15) we find

$$\frac{\nu_i}{\nu_{max}} = 3.0 \times 10^{-2}\tilde{\epsilon}exp\left[-\sqrt{\frac{.29eV}{\tilde{\epsilon}}}\right]exp\left[-\sqrt{\frac{.77eV}{\tilde{\epsilon}}}\right] \tag{3.6.16}$$

Substituting the value of $\nu_{\max}$ from (3.5.4) and realizing that for quiver energies above the ionization threshold $\nu_{net} \approx \nu_i$, we arive at

$$\nu_{net} = 9.0 \times 10^{-7} N \tilde{\epsilon} \exp\left[-\sqrt{\frac{.29eV}{\tilde{\epsilon}}}\right] exp\left[-\sqrt{\frac{.77eV}{\tilde{\epsilon}}}\right] \tag{3.6.17}$$

Figure 3.6.14 compares the prediction of equation (3.6.16) with the kinetically calculated rates presented in section 3.6.5. The agreement is excellent over more than four orders of magnitude in the rate and two in power. The results deviate for $\tilde{\epsilon} > 4 - 5eV$. This is expected since at such power densities the energization time becomes smaller than the time for ionizing collisions.

In concluding this section, we should comment on the factor of two differences in the ionization rates between the $\omega \gg \nu_{\max}$ and $\omega = \nu_{\max}$ situations. To extend the range of validity of the equation toward the $\omega \approx \nu_{max}$ regime, observe that the effective quiver energy of an electron at energy ϵ is:

$$\tilde{\epsilon}_{eff}(\epsilon) = \frac{\tilde{\epsilon}}{1 + \nu_e^2/\omega^2} \tag{3.6.18}$$

rather than $\tilde{\epsilon}$. We can conjecture that equation (3.6.17) can be generalized from $\omega \gg \nu_{\max}$ range by substituting the appropriate value of effective quiver energy into each term:

$$\nu_{net} \approx 9.0 \times 10^{-7} N \tilde{\epsilon}_{eff}(25eV) exp\left[-\sqrt{\frac{.29eV}{\tilde{\epsilon}_{eff}(2.6eV)}}\right] exp\left[-\sqrt{\frac{.77eV}{\tilde{\epsilon}_{eff}(14eV)}}\right]$$

$$(3.6.19)$$

From (3.6.3) and (3.6.18):

$$\tilde{\epsilon}_{eff}(\epsilon) = \frac{\tilde{\epsilon}}{1 + h(\epsilon)\nu_{max}^2/\omega^2}$$

where

$$h(\epsilon) = \left(\frac{\epsilon + .1}{\epsilon + 5}\right)^2$$

This extension reproduces the behavior of the numerical results for $\omega \doteq \nu_{max}$. The case of $\omega < \nu_{max}$ will be presented elsewhere.

3.7 Formation of Artificial Ionospheric Clouds

The formation of artificial ionized clouds in the ionosphere by ground based RF was studied through the use of one–dimensional simulation. The code models the breakdown and formation of ionized clouds by solving the coupled propagation and ionization equations, including self-absorption, for a microwave pulse propagating upwards. The model is a fluid rather than a kinetic one, ignores transport, and uses the ionization rate equation (3.6.17). A detailed description of the mechanics of the one–dimensional code can be found in APTI (1990).

The need to control the altitude of the cloud requires the use of a focussed heater antenna. If the elements of the heater are spread out over a wider area, the desired location of the cloud is in the near field of the heater, where focusing through phase correction of individual elements is possible. The result is a vertically increasing electric field from the heater antenna at a controlled altitude. This method allows breakdown to occur at higher altitudes, thereby increasing lifetime and decreasing absorption effects. The resulting electron density profiles have a very sharp gradient, which further reduces absorption effects, and fine control over the position of the breakdown is possible.

One–dimensional simulations revealed a long pulse effect which increases the gradient of the plasma cloud. At each altitude there is a threshold power density S_c. As a long pulse (> 5 μsec) of microwave energy propagates upwards in the atmosphere and ionization takes place, self-absorption of the pulse decreases the field strength in the pulse. Moreover, the areas with the highest induced electron density will absorb more of the pulse. Eventually, the area of peak ionization will absorb so much of the pulse that the remaining power density will fall below S_c. At this point, further ionization can only occur below and in front of the point of peak electron density. This moves the electron density peak down vertically with time, bringing it closer to the point where the unattenuated power density is S_c. Figure 15 shows the formation of a one– dimensional profile made by a 10 μsec pulse at 1.5 μsec time samples.

The gradient of the electron density profile is dependent on the slope of the heater antenna's array factor near the initial breakdown point and the heater frequency. A steeper slope in power density creates a corresponding sharper electron density gradient along with a higher peak in electron density profile. Parameter studies show a linear dependence between power density slope and resulting electron density slope. Since self-absorption or attenuation goes down as the heater frequency is increased, a higher frequency will achieve a greater electron before the clamping phenomenon occurs, and subsequently a steeper slope as well.

Acknowledgments

The review is the result of many years of research with many collaborators. I would like to acknowledge the contributions of Drs. C.L. Chang, A. Drobot, P. Vitello and K. Tsang of SAIC; Drs. R. Shanny, P. Koert, R. Short, L. Sussman and T. Wallace of APTI and Drs. S. Ossakow and P. Palmadesso at NRL. Special thanks for the HIPAS experimental results go to T. Ferraro at Penn State and M. McCarrick, A. Wong and R.F. Wuerker at UCLA. The work was supported by AFGL, ONR and ARCO Power Technologies.

References

APTI, Physics Studies in Artifical Ionospheric Mirror (AIM) Related Phenomena, ARCO Power Technology Report 5004, Washington, DC, 1990.

Barr, R. and P. Stubbe, J. Atoms. Terr. Phys., 46, 315, 1984a.

Barr, R. and P. Stubbe, Radio Sci., 19, 1111, 1984.

Barr, R., M.T. Rietveld, P. Stubbe, and H. Kopka, J. Geophys. Res., 90, 2861, 1985.

Belyaev, P.P., D.S. Kotik, S.N. Mityakov, S.V. Polyakov, V.O. Rapoport and V. Yu. Trakhtengerts, Radiophysics 30, 248, 1987.

Borisov, N.D., A.V. Gurevich, and G.M. Milikh, Artificial Ionized Region in the Atmosphere, U.S. Air Force Foreign Technology Division, 1988. (Translation of Iskusstvennaya Ionizirovannaya Oblast' v Atmosfere, Moscow, 1986).

Chang, C.L., V. Tripathi, K. Papadopoulos, J. Fedder, P.J. Palmadesso, and S.L. Ossakow, Effect of the Ionosphere on Radiowave Systems, edited by J.M. Goodman, p. 91, U.S. Government Printing Office, Washington, D.C., 1981.

Geballe, R. and M. Harrison, Phys. Rev., 91, 1, 1953.

Getmantsev, G.G., N.A. Zuikov, D.S. Kotik, L.F. Mironenko, N.A. Mityakov, V.O. Rapoport, Yu. A. Sazonov, V. Yu. Trakhtengerts, and V. Ya. Eidman, JETP Lett., 20, 229, 1974.

Gould, L. and L. Roberts, J. Appl. Physics, 27, 1162, 1956.

Gurevich, A.V., Nonlinear Phenomena in the Ionosphere, Spring-Verlag, 1978.

Hays, G., L. Pitchford, J. Gerardo, J. Verdeyen, and Y. Li, Phys. Rev., A36, 2031, 1987.

Herlin, M.A. and S.C. Brown, Phys. Rev., 74, 291, 74, 1650, 1948.

Kotik, D.S. and V. Yu. Trakhtengerts, JETP Lett., 21, 51–52, 1975.

Kroll, N. and K.M. Watson, Phys. Rev., 5, 1883, 1972.

Papadopoulos, K. Naval Res. Lab., Plasma Physics Division (internal report), 1973.

Papadopoulos, K., K. Ko, A. Reiman and V. Tripathi, ESA SP-195, pp. 11–28, 1983.

Papadopoulos, K., A.S. Sharma and C.L. Chang, Commets Plasma Phys. Controlled Fusion, 13(1), 1–17, 1989.

Papadopoulos, K., C.L. Chang, P. Vitello and A. Drobot, Radio Science, (in press), 1990a.

Papadopoulos, K., Final report submitted to ONR, 1990b.

Sharfman, W. and T. Morita, J. Appl. Physics, 35, 2016, 1964.

Sharfman, W., W. Taylor, and T. Morita, IEEE Trans., AP-12, 709, 1964.

Stubbe, P. and H. Kopka, J. Geophys. Res., 82, 2319–2325, 1977.

Stubbe, P., H. Kopka and R.L. Dowden, J. Geophys. Res., 86, 9073, 1981.

Stubbe, P., H. Kopka, M.T. Reitveld and R.L. Dowden, J. Atmos. Terr. Phys., 44, 1123, 1982.

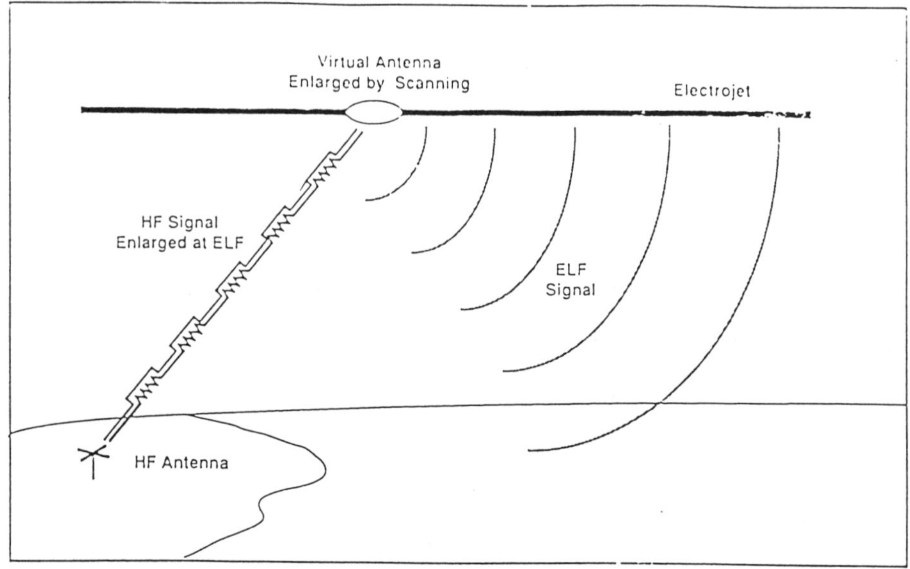

Figure 1. HF to ELF Conversion Process

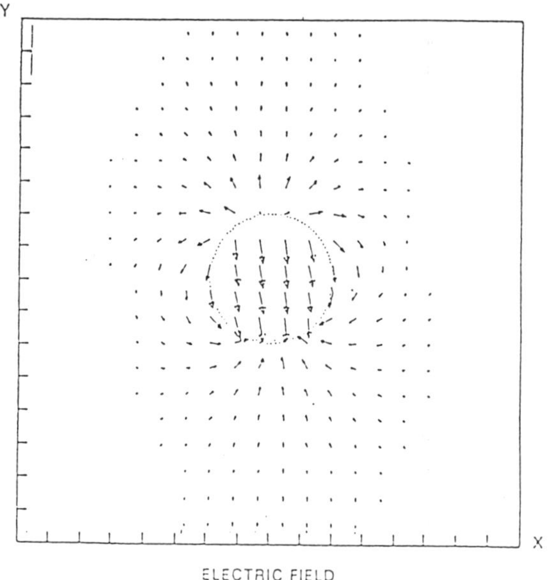

Figure 2 (a). Electric Field Established by Heating Modification

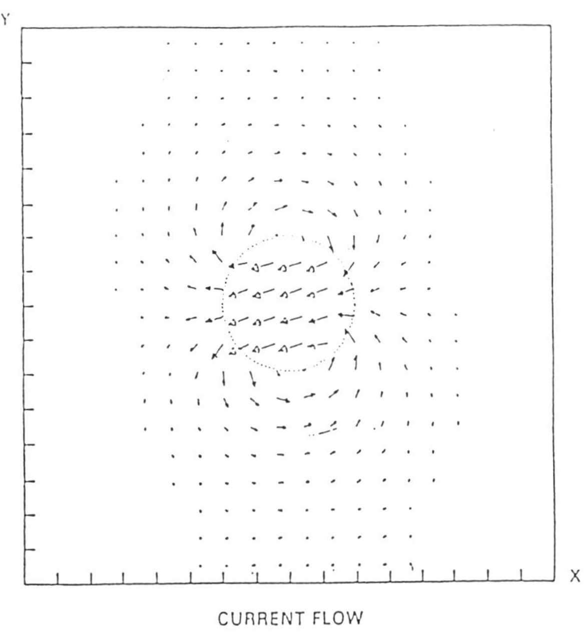

Figure 2(b).   Current Flow Established by Heating Modification

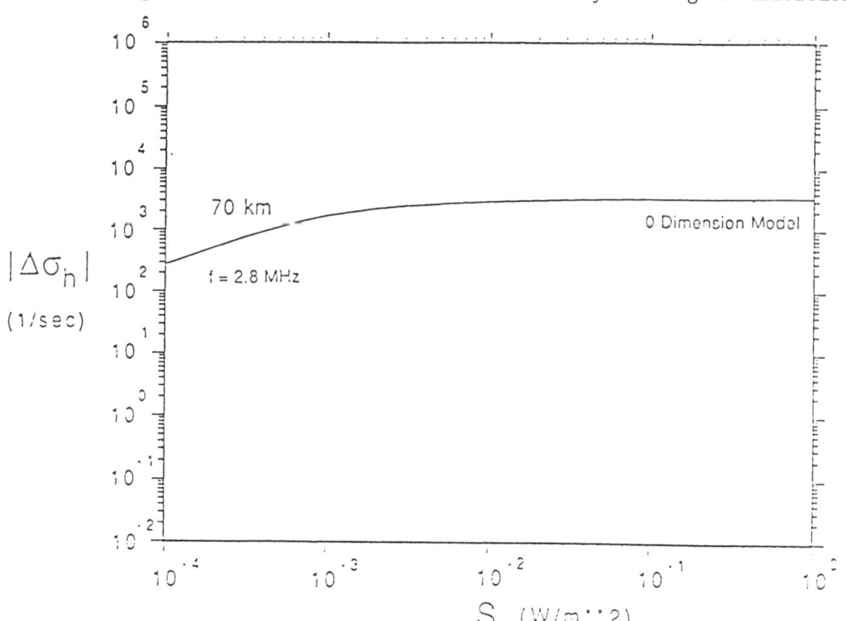

Figure 3. Modification of Hall Conductivity vs. Power Density

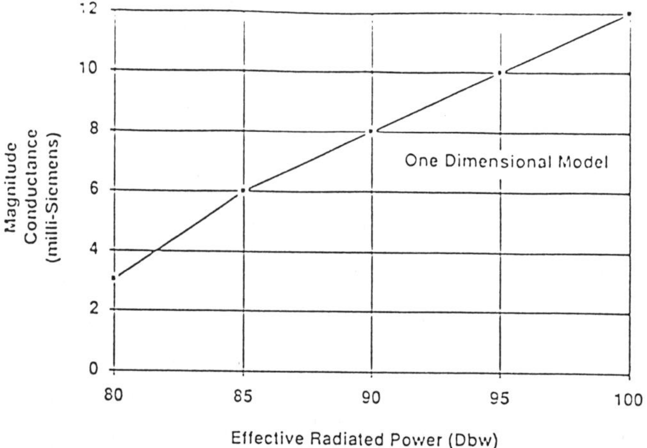

Figure 4. Modification of Hall Conductance at 2.8 MHz

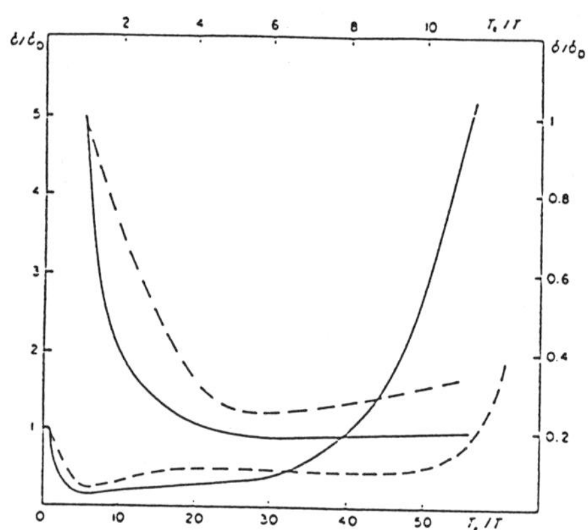

Figure 5. Electron Cooling Rate vs. Temperature

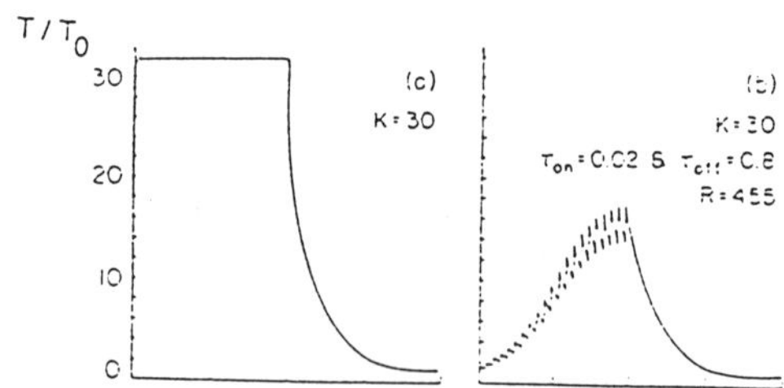

Figure 6. Ionospheric Electron Temperature

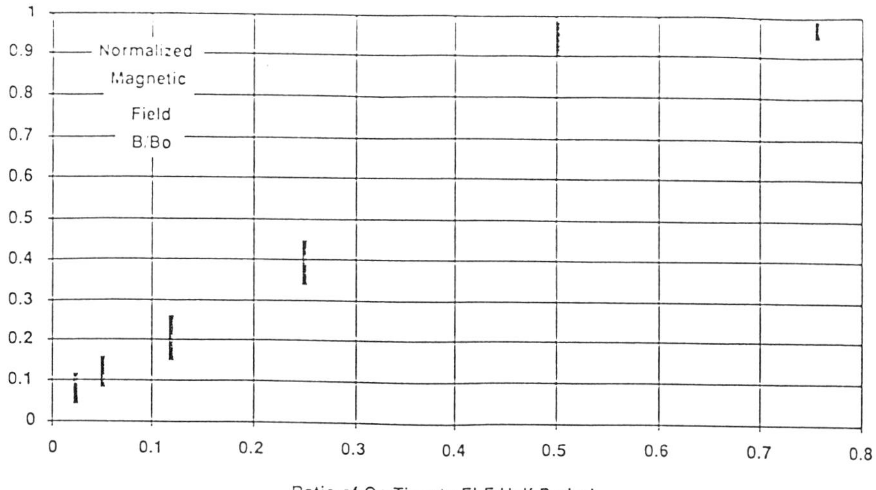

Figure 7.   HIPAS Experimental Results

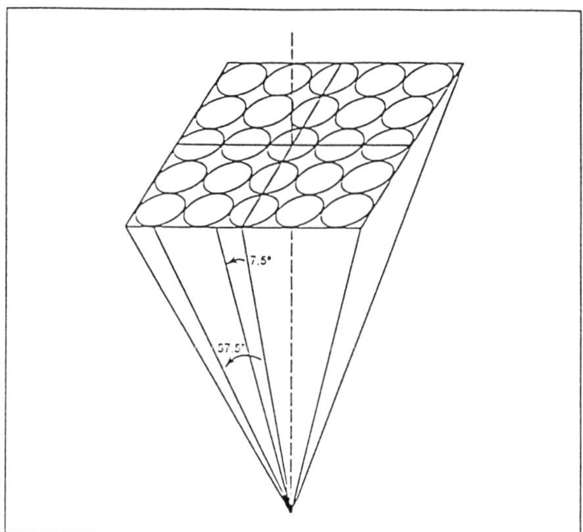

Figure 8.   ELF Power Enhanced by Factor of 50-100.

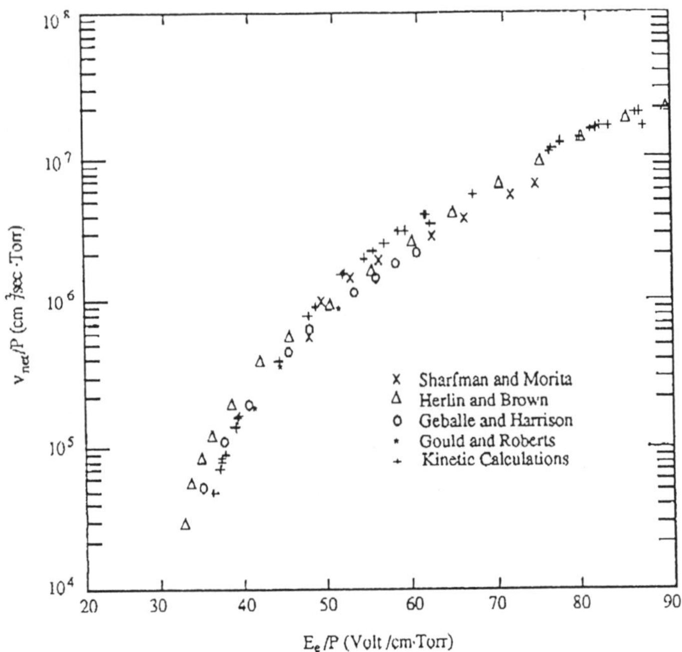

Figure 9a.   Comparison of Kinetic Calculations with Experiments

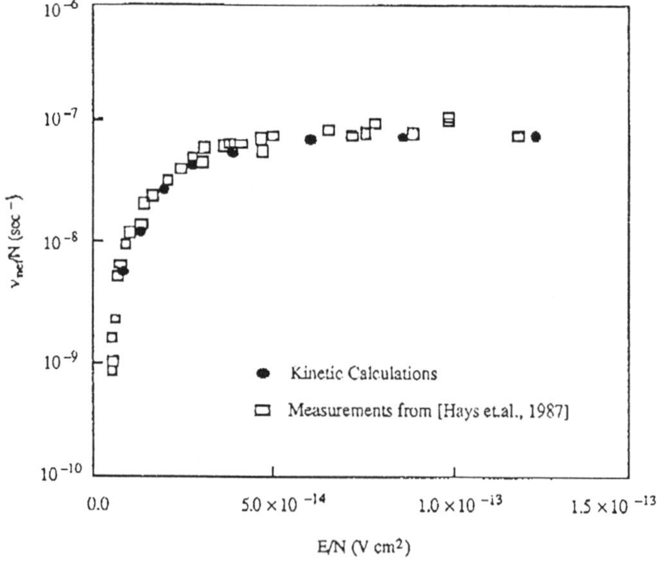

Figure 9b.   Comparison With Directly Measured Ionization Rates (Pure N_2)

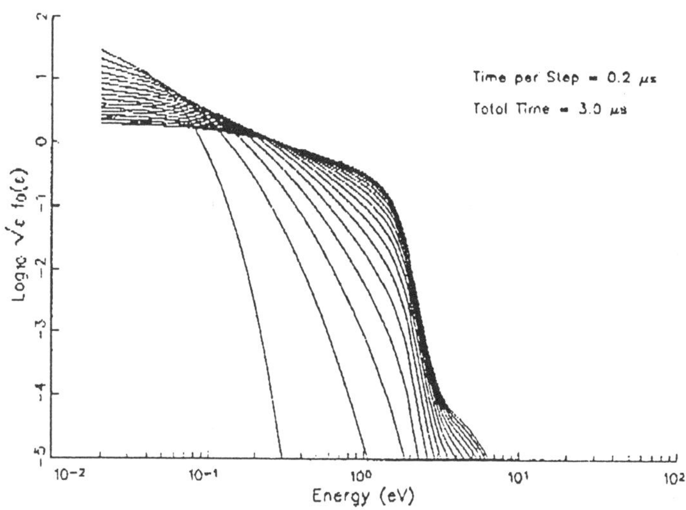

Figure 10.   Evolution of f_o for Power Density of 300 W/m^2

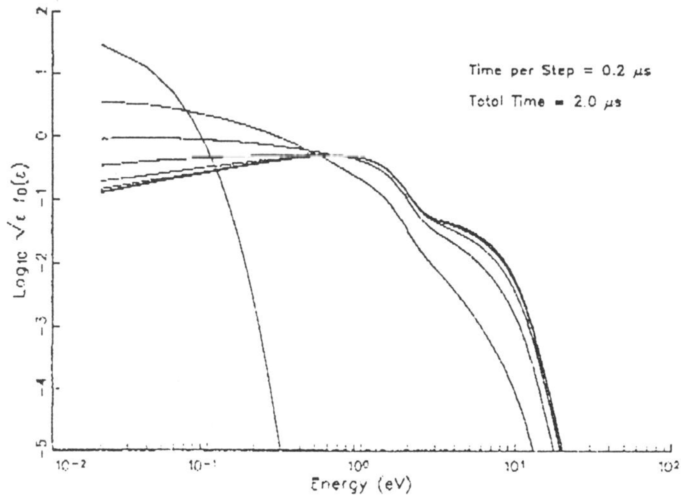

Figure 11.   Evolution of f_o for Power Density of 3 kW/m^2

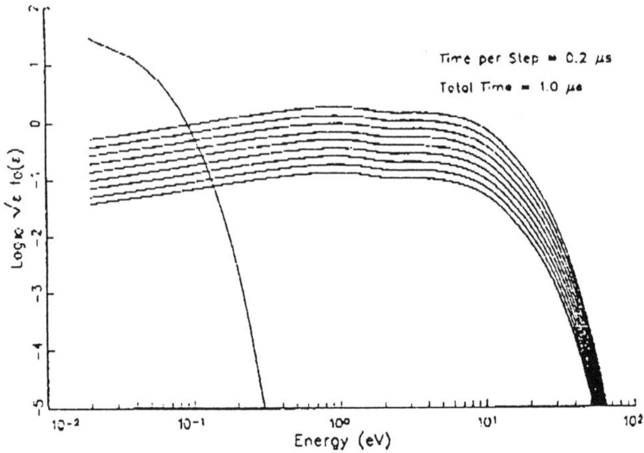

Figure 12. Evolution of f_o for Power Density of 30 kW/m^2

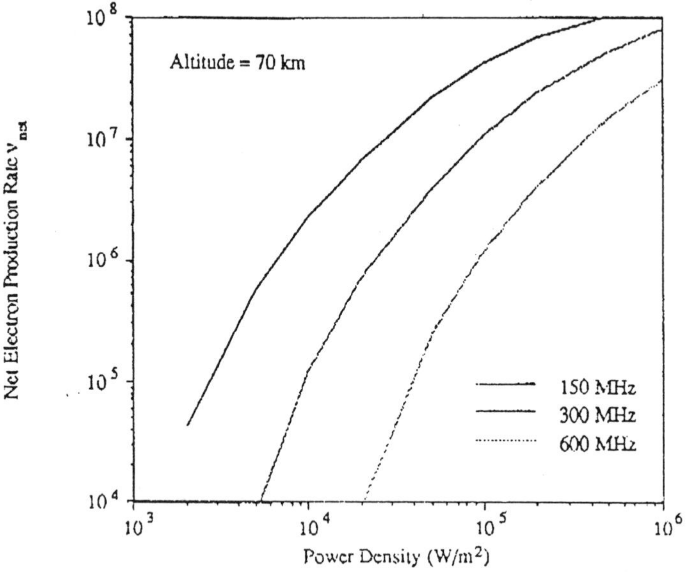

Figure 13. Net Ionization Rates at 70 km Altitude

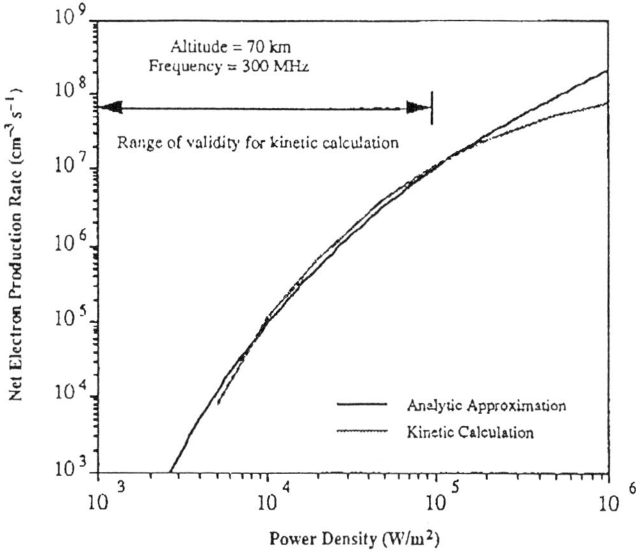

Figure 14.   Comparison of Analytic Approximation and Kinetic Calculations

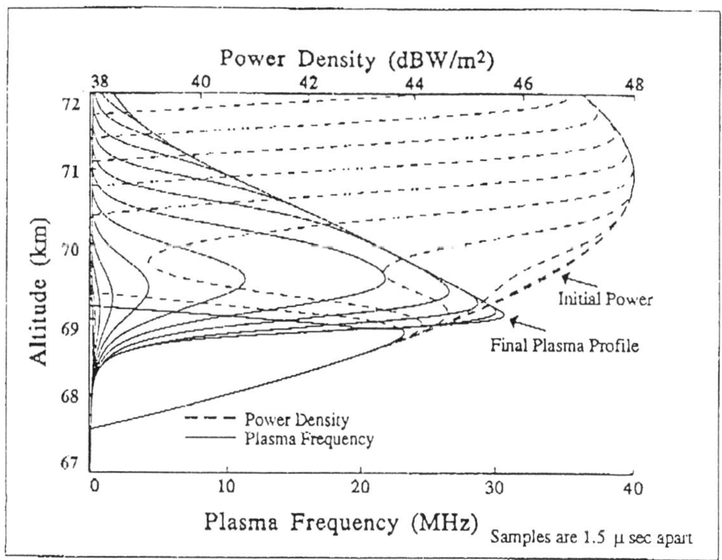

Figure 15.   Growth Through Time of AIM Cloud with f = 300 MHz, τ = 10 µsec

Ion Dynamics and Ozone

A. Y. Wong, , R. F. Wuerker, J. Sabutis, R. Suchannek
Plasma Physics Laboratory, Department of Physics, UCLA

C. D. Hendricks
Lawrence Livermore Laboratory, University of California

P. Gottlieb
TRW Systems, Redondo Beach
May ,1990.

Introduction.

The motivation of our research is the realization that charged particles have quite different reaction rates than uncharged ones. For example, chlorine atoms, which play a central role in the ozone destruction process, have a high affinity for electrons. Once a chlorine atom becomes negatively charged through acquiring an extra electron to fill its outer shell, it becomes inert in its reactions with certain species. Thermodynamic arguments using Gibbs' free energy on Cl_x^- and NO_x^- reactions with ozone show that reaction rates of negative ions with ozone are slower by several orders of magnitude.

Another fundamental point in considering electrons, ions and charged clusters in the atmosphere is that charged particles can be influenced remotely by externally applied electrostatic and electromagnetic fields. Since these charged particles interact with the neutral environment through collisions and dipole interactions we wish to explore such interactions and investigate the introduction of active perturbations in the atmosphere, such as electromagnetic fields, and electron deposition. Our research on ion dynamics, therefore, includes both chemical reactions among ions and neutrals as well as the transport and nucleating processes of charged particles in the atmosphere. Our interest in applying active perturbation techniques to the investigation of atmospheric sciences stemmed from our experience in making large-scale changes in the upper and lower ionosphere using ground-based transmitters.

We first present in section I results from computer modeling with the injection of negative charges into the atmosphere and observe the subsequent changes in the mixing ratios of key constituents of interest. We then present methods by which charged particles can be produced in section II. In section III a research program in laboratory and field experiments is discussed which will test the underlying assumptions in this concept. In section IV a method of transporting and accelerating ions is discussed.

<u>Theromodynamical Considerations</u>

We would like to compare the change in Gibbs' free energy for free atoms and negative ions for one of the important catalytic cycles which destroys ozone:

$$Cl + O_3 \rightarrow ClO + O_2 \qquad\qquad \Delta G = -168 \text{ kJ/mol}$$
$$ClO + O \rightarrow Cl + O_2$$

The change in Gibbs' free energy for the reaction involving Cl^- is nearly zero:

$$Cl^- + O_3 \rightarrow ClO^- + O_2 \qquad\qquad \Delta G \cong 13 \text{ kJ/mol}$$

The electron affinity (EA) for Cl is 3.7 ev, while the EA for ClO is 2 ev. The Cl^- reaction rate is less than 10^{-12} cm^3/sec. The low reaction rate implies that Cl^- will not be effective in the catalytic reaction.

In order to consider all the important species that participate in the atmospheric reactions we turn to computer modelling which is to be checked against laboratory modelling experiments.

<u>I.Computer Modeling</u>

A computer model of atmospheric transport and kinetic chemistry developed by J. Zinn and C. D. Sutherland at LANL was modified to include all known ion reactions related to ozone. The original code was used to simulate the atmosphere at a wide range of temperatures, as in J. Zinn. et al., 1982. It can model the atmosphere with a dimensionality ranging from 0 to 2, where a zero-dimensional atmosphere has no transport. The model predicts the densities of the chemical species in the atmosphere as a function of time, given the electron and neutral temperatures, the initial densities of the species, and the rate at which the species enter the atmosphere.

The present input consists of 62 species, including molecules and ions, and will expand as additional chemical processes are considered in the model. The computer program searches its library file of reactions for those reactions whose products and reactants are subsets of the input species. A smaller library file, formed by the computer program, contains only the reactions which meet this criterion. For the present input file, the smaller library file consists of 1104 reactions. The large library file contains reactions of the following types:

- ionizations and recombinations
- attachments and detachments
- charge transfer
- associations and dissociations
- excitations and de-excitations
- atomic interchange and rearrangements

These reactions are taken from D. Smith and M. J. Church, 1977 and a variety of other sources. The model will soon include cluster negative ion kinetics, as in G. Brasseur and A. Chatel, 1983. When reverse reaction rates are unknown, they are calculated by detailed balance.

All of the reactions included in the smaller library file combined with sources and sinks constitute a nonlinear first-order differential equation of the form for each species:

$$\frac{dN_i}{dt} = \overline{\sum_{j,k} K_{ijk} N_j N_k} + Q_i(t)$$

where N_i is the number density of species i, $Q_i(t)$ is due to external sources, and the sum is over all reactions involving species i, including two-body, three-body, and photo reactions. Using discrete time increments and matrix inversion, the program solves the many simultaneous equations of the above form as a function of time, thus yielding the time evolution of the density of each species. The time increment can be adjusted to accommodate processes with short and long time scales (from microseconds to millions of seconds).

Equilibrium is obtained by letting the system evolve for greater than 10^8 seconds, until the rate of change of N_i is less than 0.01 % per 10^6 seconds, for all i. The effect of a perturbation of the atmosphere, such as an influx of chlorine or electrons, can be determined by perturbing the equilibrium, and letting the system evolve to a new equilibrium. The equilibrium is perturbed through either an initial pulsed or continuous injection of various species, or a change in temperature.

In addition to listing the density as a function of time, the output gives the 12 most important reactions for each of the species, as a function of time. These are the reactions occurring at the highest rate per unit volume per unit time. The output is downloaded from the Cray to a PC for plotting.

A 40 km baseline atmosphere is established in the model starting with the actual observed concentrations for the major species, which remain fairly stable as the model is integrated for 10^5 seconds (over 1 day). The effects of increased active chlorine are best simulated by increasing the concentration of ClO, rather than increasing Cl. This is because the concentration of the former is more than 100 times the latter at 40 Km, and the rapid interchange between Cl and ClO assures that this ratio will be preserved for moderate increases in the concentration of either. Any modest increase in Cl will be immediately swallowed up in the much larger ClO, while a significant percent change in ClO will automatically be applied to Cl.

The increased catalytic destruction caused by additional active chlorine is demonstrated by re-running the simulation with an initial increase in concentration of ClO by 23%. After 10^5 seconds, both Cl and ClO concentrations are still increased by 23%, and increased catalytic destruction results in a decrease of 10% in the ozone concentration. The time dependence of this behavior is shown in figure 1. It should be noted that the reason the percentage decrease in ozone is much less than the increase in active chlorine is that NO is also a catalyst of ozone destruction, equal in importance to active chlorine at 40 Km.

The beneficial effect of the conversion of active chlorine to negative ions is demonstrated by re-running the 23% increased ClO simulation with an initial injection of 10^7 electrons/cc. or 300 electrons/cc sec This concentration of negative charges will remove just enough active chlorine to completely negate the effect of the 23% increase in ClO. This behavior of Cl and ozone is shown in figure 2.

II Experiments on Atmospheric Charging.

The concept of atmospheric charging can begin from the fact that the earth is negatively charged and possesses what is known as a "fair weather" electric field which varies diurnally between 100-130 volts meter at the earth's surface. This field intensity means that the earth carries a total electric charge of -500,000 coulombs. The ionosphere is oppositely charged, with the result that the earth-ionosphere combination can be thought as a huge spherical capacitor, leaking through the atmosphere. The potential difference between the earth and ionosphere is 300,000 volts, and the total leakage current is 1500-2000 amperes, giving a total atmospheric resistance of only 200 - 150 ohms. To maintain the "fair weather" field, the ionosphere has to be fed by other sources of current. These sources turn out to be the thunderstorms that occur daily all over the earth, primarily over the Amazon and African land masses.

Within thunderclouds charges are separated by the formation of water droplets, ice crystals, and gravity acting on these particles. The top of the thundercloud becomes positively charged, the middle negatively charged, and the bottom of the thundercloud slightly positively charged. These charged regions build up until there is a discharge to ground, leaving the top of the cloud charged positively. This excess charge redistributes itself by repulsion throughout the ionosphere, contributing to the maintenance of the "fairweather" electric field. Other thunderclouds make similar contributions. It is estimated that over the earth's surface there are 1,800 thunderstorms in process at any one time.

Charge separations are, thus, possible in the atmosphere; however, they require the formation of particulates (water droplets, ice crystals and larger ice particles referred to as "graupel"), and the action of gravity on the heavier particles to cause them to fall and separate away from the lighter less mobile oppositely charged particles. The properties of water also plays a prominent role which is beyond the present short summary; namely its electrical polarizability and generation of electrical potential during changes between the solid and liquid states.

Scheme for Artificially Electrifying the Atmosphere

Taking a clue from thundercloud phenomena, a scheme has been devised by Lawrence Livermore Laboratory (LLNL) and UCLA scientists for electrifying the atmosphere with either an airplane, a rocket or a balloon. The scheme is shown schematically in Figure 3. It is based on having a long wire or antenna biased at a sufficiently high voltage to emit electrons by field emission. In addition, the plane carries a water supply which releases water droplets with the opposite polarity. The charging of water droplets is carried out by sharp hollow needles at the bottomside. The release of charges of both polarities ensures that the airplane is not charged up and a steady increase of the ion population in the atmosphere is possible. The water droplets are large enough to fall under gravity, causing a separation between the opposite charges deposited by the antenna. Separation will occur when the emitted charges or ions around the antenna form small hydrated clusters or submicroscopic particles, of low mobility. The very processes that are set up by nature in thunderclouds are being attempted in the aforementioned scheme.

The atmospheric electrification scheme shown in fig 3 works in either polarity, the antenna can be biased positively and the neutalizing water-emitting needles biased negatively, creating a positively charged cloud, or the antenna can be biased negatively and the neutralizing needles biased positively as shown in Figure 3, creating a negatively charged cloud or region.

The neutralizing droplet size is determined by the size of the internal diameter of the needles, being approximately twice the needle diameter.

It can be shown that the maximum charge (q_{max}) on each individual water droplet is

$$q_{max} = 8\pi\sqrt{\varepsilon_0\, T\, r^3} \qquad\qquad 1)$$

where,

$\varepsilon_0 = 8.85 \times 10^{-12}$ Farads/meter

(permittivity of free space)

T = surface tension of water (0.075 Newton/meter),

and

r = droplet radius in meters.

For example, a droplet of 10 micrometers radius could be charged to 4 x 10^6 elementary charges or carry away 0.6 x 10^{-12} coulombs per particle. This equation is an upper bound since it expresses the fact that within the droplet the forces of electrical repulsion are just balanced by the forces of surface tension. Any more charge on the droplet would cause it to divide or emit the excess charge as a jet.

The charge to mass ratio of such a particle is easily computed and is the more significant number; namely,

$$\left[\frac{q}{m}\right]_{max} = \frac{6}{\rho}\left[\frac{2\,\varepsilon_o\,\Gamma}{r^3}\right]^{1/2} \tag{2}$$

where

ρ is the density of the droplet which
for water is 1000 kilograms/meter^3

Thus for our aforementioned 10 micron diameter water droplet ejected from the bottom of the airplane, the maximum charge to mass ratio would be 0.15 coulomb per kilograms. This value can also be interpreted as 0.15 coulomb of charge ejected by the antenna for each kilogram of water dropped away by the neutralizing needle.

Maximum charge values for water droplets of differnet sizes are included in Table I, which also includes values for the electric field strengths at the surface of the particle. The limiting charge values and charge to mass ratios apply to vacuum conditions where there are no collisions with neutral particles which might carry charge away . The limiting electric field for conductors in air at atmospheric pressure is 30 KV/cm.

TABLE I
Water Droplet Electrification

Drop Radius (microns)	Max Charge ($\mu\mu$Coulombs)	Max Charge/Mass Ratio (Coulombs/Kilogram)	Surface Field (KV/cm)
10	0.6	0.15	600
20	1.8	0.053	150
30	3.2	0.028	67
40	5.0	0.019	38
50	7.0	0.013	24
60	9.2	0.01	17
80	14.1	0.007	12
100	20	0.005	6
200	60	0.0017	1.5

Particles will fall with a limiting velocity (v_{term}) determined by the viscosity of air and the cross section of the particle as specified by Stokes law;

$$v_{term} = 2\rho g\ r^2/9\ \eta \qquad\qquad 3)$$

where

η is the viscosity of air

(1.83 x 10^{-5} km/m sec at 23 degrees C)

Stokes laws gives a terminal velocity of 1 cm/sec for water droplet of 10 micron radius, as shown in Table II. Since viscosity is approximately independent of pressure, the terminal velocity applies over a wide range of altitudes.

TABLE II
Terminal Velocities of Water Droplets in air at 23°C

Radius (microns)	Terminal Velocity (cm/second)
10	1
20	4
30	9
40	16
50	25
60	36
80	64
100	100

The neutralizing particles being rather highly charged would be expected to grow even larger in falling through conditions of saturated water vapor, increasing their rate of falls. Such processes are at work in thunderclouds.

Generation of electrons by photoelectric emission.

The most energy efficient method of injecting low energy electrons into the atmosphere is to use solar radiation which is approximately 1 KW/m^2. The photoelectric emission of electrons from a metal surface such as zinc has been found to be of the order of 1 μA / W with an equivalent injection flux of 10^{13} electrons/ sec. Assuming a surface area of 10 m^2 collecting 10 w of solar uv power injecting 10^{14} electrons/sec into a volume of 10 m x 10 m x 10 m , a rate of density change of 10^5 / cm^3sec will result. This is greater than the rate of density increase considered in the computer modeling in previous sections.

The rate of injection of electrons is determined by the rate at which positively charged droplets can be released by the space vehicle. Since each 10 μ droplet can carry a a maximum quantity of 4 x 10^6 elementary charges, a release of 10 μ particles at the rate of 2.5x 10^7/sec is required.

We have illustrated two methods of injecting charges into the atmosphere. The scheme is based upon releasing or forming ions by photoelectric effects or using a high voltage wire. The charging of the space vehicle is avoided by the release of oppositely charged water droplets. In both cases solar energy is used to provide the primary power.

Recombination and Lifetime of charges.

The lifetime of charges can be estimated from the densities of charged species, their relative velocity and the collisional cross section. For typical ion densities of 10^5/ cm^3 , relative velocity of 10^3 cm/sec and a collisional cross section of 10^{-14} cm^2 we have found a recombination rate of $v = n\ \sigma\ v$ /sec = 10^{-6} /sec.

When ions become attached to water molecules which have strong dipole moments their recombination rates can be considerably different. The binding energy of an ion to the surrounding dipoles is strong such that they have to overcome the binding potential in order to recombine with another ion of the opposite polarity. Experiments of releasing charges of opposite polarities in a cloud chamber have demonstrated the tendency of hydrated charges not recombining. The system has a lower energy state when oppositely charged ions are surrounded by dipoles than if they recombine. The dissociation of NaCl in water into hydrated Na^+ and Cl^- is a case in point. This leads to the possibility of accumulating positive and negative ions or making ion plasmas in the atmosphere without creating strong space potential. A plasma provides a source of electrons to be readily acquired by Cl neutrals.

III. Laboratory Experiments

The goal of these experiments is to reproduce in the laboratory the principal ozone-chlorine catalytic reaction:

$$Cl + O_3 \rightarrow ClO + O_2$$

$$ClO + O \rightarrow Cl + O_2$$

and then show that the above reaction is inhibited whenever the Cl atoms are replaced or converted into Cl^- ions.

A large laboratory device, (figure 4) was used for the modeling of the atmospheric physics and chemistry of the ozone depletion and conservation concept. Laser and optical diagnostics have been acquired and assembled. This 4m long chamber can be operated at different pressures corresponding to the various heights (10-100km) in the atmosphere. The ozone content in the chamber is monitored by the amount of UV light absorption at the 253 nm (Hg line). The lifetime of O_3 in the chamber (due to losses to the wall) is 4 hours at 1 Torr pressure of ozonized oxygen; the concentration of ozone is approximately 2%. The lifetime of ozone is long enough for us to perform experiments where controlled amounts of atmospheric species are introduced.

The following experiments are designed to study the feasibility in three regimes of the atmosphere. Regime 1 is from 10 to 25 km in altitude. Regime 2 is from 25 to 40 km, and Regime 3 from 40 to 80 km.

    a. Catalytic cycles with and without ion kinetics

    b. Surface interactions

    c. Ion reaction rates

    d. Recombination reactions among hydrates

    e. Recombination reactions of Cl and Cl^- with ozone

    f. Chamber charge spraying to electrify Cl and NO

    g. Ice crystals with charges as sites for $Cl_2 + 2e \rightarrow 2Cl^-$

      h. Formation of ion hydration clusters

      i. Measurements of ion kinetics reaction

      j. Chamber measurement of Cl^- reaction rates

      k. Comparison between optical and probe diagnostics

Field Experiments.

The vertical profile of ozone has been successfully detected by means of DIAL(Differential Absorption Lidar). The vertical resolution has been approximately 1Km while the horizontal resolution is 100 m . It is therefore possible to perturb a relatively small region of space and monitor the effects remotely. In a manner similar to laboratory experiments discussed above tracer gases can be injected into the atmosphere by space vehicles such as balloons or high-flying spacecrafts while the the ozone profile is monitored. Electrons are injected by the methods mentioned in section II and their effects on the atmospheric interactions observed.

IV. Ion acceleration and control.

The recent successful excitation of ELF waves(Extremely Low Frequency) in the ion cyclotron frequency range(McCarrick et al. 1990) has given rise to the possibility of performing selective acceleration of ions in the magnetic mirror field of the ionosphere. When an ion drifts upward or convects to the region where its cyclotron frequency exceeds the collisional frequency in the ionosphere it can be resonantly accelerated in the perpendicular direction by these wave fields. These ion cyclotron waves have long wavelengths(3-30 Megameters) and penetrate to the upper region of the atmosphere. Their group velocities are close to the earth's magnetic field. Using currently available transmitters and antenna array at the HIPAS Observatory (Wong et al, 1990) we have been able to excite waves of electric field of 200 μV/m. This is sufficient to accelerate ions resonantly in typical periods of tens of seconds to perpendicular energy of tens of ev , sufficient to escape the earth's gravitation field and be expelled to the magnetosphere or open magnetic field lines.

Summary.

We have shown how negative chlorine ions can be much less destructive than neutral chlorine and that there are practical methods

of injecting charges into the atmosphere to cause the production of
chlorine ions. These ideas are being presently tested in the
laboratory. When these ions drift to the upper region they can be
accelerated upward into the magnetosphere or open field lines.

REFERENCES

Brasseur, G. and Chatel, A., Annales Geophysicae, $\underline{1}$, 173-185 (1983).

Das Gupta, N. and Chosh, K., "A Report on the Wilson Cloud Chamber and
its Applications in Physics, Reviews of Modern Physics, $\underline{18}$, 225-290
(1946).
McCarrick M.J,, A.Y. Wong,R.F. Wuerker, B. Chouinard and D. Sentman,
to be published in special issue of Radio Science (1990).

Smith, D. and Church, M. J., Planetary and Space Science, $\underline{25}$, 433, (1977).

Wong A.Y., J. Carroll, R. Dickman, W. Harrison, W. Huhn, B. Lum,
M. McCarrick, J. Santoru, C. Schock, G. Wong and R.F. Wuerker,
to be published in special issue of Radio Science (1990).

Zinn, J. Sutherland, C. D., Stone, S. N., and Duncan, L. M., Journal
of Atmospheric and Terrestrial Physics, $\underline{44}$, 1143-1171 (1982).

Part VII

NONLINEAR ELECTROMAGNETIC WAVE-PLASMA INTERACTION

Recent applications of bifurcation theory to nonlinear waves in collisionless plasmas*

Mark Buchanan†
University of Virginia
Department of Physics
Charlottesville, VA 22903-2442

James Paul Holloway‡
University of Michigan
Department of Nuclear Engineering
Ann Arbor, MI 48109-2104

J. J. Dorning
University of Virginia
Department of Nuclear Engineering and Engineering Physics
Charlottesville, VA 22903-2442

ABSTRACT

We summarize some recent applications of bifurcation theory and nonlinear analysis to the study of collisionless plasmas, and in particular, use these techniques in the analysis of nonlinear wave phenomena described by the one-dimensional Vlasov-Maxwell or Vlasov-Poisson-Ampère equations. The analysis focuses on the nonlinear Poisson equation for the electric potential that results from using a BGK representation for the plasma species distribution functions and establishes that there exist undamped nonlinear spatially-periodic traveling waves arbitrarily close to spatially-uniform Vlasov equilibria. There exist, therefore, small perturbations of smooth plasma equilibria that in fact do not Landau damp. In addition, further analysis shows that there also exist solitary waves, close to smooth spatially uniform Vlasov equilibria, which are related to the spatially periodic traveling waves through a transcritical bifurcation. These results indicate the value of the direct, though sometimes technically detailed, application of the techniques of bifurcation theory, nonlinear dynamical systems analysis and nonlinear functional analysis to problems in plasma physics.

* Research Supported by NASA Grant No. NAGW-1669.
† Research Supported by NASA Grant No. NGT-53643.
‡ Research Supported by NASA Grant No. NGT-50183.

1. INTRODUCTION

Almost all really interesting problems in plasma physics are fundamentally nonlinear. Space plasmas—including low-density collisionless plasmas—fusion plasmas, charged particle beams, free electron lasers, etc. are for the most part dominated by nonlinearity, the most interesting physics often resulting from collective nonlinear effects. Nevertheless, these problems frequently have been approached using linearization or various perturbation methods, notwithstanding the fact that there have been available for some time rigorous applied methods of nonlinear functional analysis, nonlinear dynamic systems analysis, and bifurcation theory [see for example Chow & Hale (1982) or Guckenheimer and Holmes (1983)]. While linearization and naive perturbation methods can be useful and frequently add insight into many problems, they also can be incorrect and misleading. In cases where these *ad hoc* methods fail—and such cases are unfortunately difficult to identify—nonlinear analyses are required. The value of such rigorous nonlinear techniques has been demonstrated recently by their fruitful application to the study of undamped small amplitude nonlinear waves in plasma kinetic theory. It is these particular recent applications that will be summarized in this paper.

Low density multi-species plasmas that are essentially collisionless arise naturally in numerous areas of scientific and technological interest. The former include the interstellar medium, the interplanetary medium, and planetary foreshock regions, etc. in which there exist fairly expansive spatially homogeneous regions where one intuitively might expect to observe spatially-uniform homogeneous "quiescent" plasmas. Contrary to this intuition, elaborate plasma wave structure frequently is observed in the interstellar and interplanetary expanses, and especially in the earth's foreshock. Hence, a theoretical understanding of the origins of these waves in regions far from boundary structures, and of the reasons for their ubiquity is essential to the exploration and the understanding of the physics of these regions and the roles they play in connecting the magnetospheric regions of neighboring and distant celestial bodies.

Multi-species collisionless plasmas are well described by the Vlasov-Maxwell equations, and nonlinear wave motion along locally straight parallel magnetic field lines in these plasmas is well described by the one-dimensional Vlasov-Maxwell or Vlasov-Poisson-Ampère equations in which the nonlinear terms arise as a result of the interaction between the species distribution functions and the self-consistent electric field. These equations are

$$\frac{\partial f_\alpha}{\partial t}(x,u,t) + u\frac{\partial f_\alpha}{\partial x}(x,u,t) + \frac{q_\alpha}{m_\alpha}E(x,t)\frac{\partial f_\alpha}{\partial u}(x,u,t) = 0 \qquad \alpha = 1,2\ldots N \quad (1)$$

$$\frac{\partial E}{\partial x}(x,t) = 4\pi \sum_{\alpha=1}^{N} q_\alpha \int_{-\infty}^{\infty} f_\alpha(x,u,t)\,du \qquad (2)$$

$$\frac{\partial E}{\partial t}(x,t) + 4\pi \sum_{\alpha=1}^{N} q_\alpha \int_{-\infty}^{\infty} u f_\alpha(x,u,t)\, du = 0\,, \tag{3}$$

where m_α and q_α denote the mass and charge of particles of species α.

The spatially uniform solutions of these equations describe plasma oscillations [Holloway and Dorning (1989)], which include as a subset the spatially uniform Vlasov equilibria. These equilibria are characterized by spatially uniform distribution functions $F_\alpha(u)$, $\alpha = 1, 2 \dots N$ which must satisfy the conditions of charge neutrality

$$0 = \sum_{\alpha=1}^{N} q_\alpha \int_{-\infty}^{\infty} F_\alpha(u)\, du \tag{4}$$

and zero net current

$$0 = \sum_{\alpha=1}^{N} q_\alpha \int_{-\infty}^{\infty} u F_\alpha(u)\, du\,. \tag{5}$$

A problem of obvious physical interest is to understand how spatially non-uniform perturbations to any such equilibrium behave, and this problem has been the object of many analyses based on a linearized version of the equations. The classical analysis of this linear problem by Landau (1946) is well known, as are its predictions. Most importantly, Landau's analysis predicts that the electric potential created by a spatially periodic traveling wave perturbation to a thermal equilibrium plasma—described by Maxwellian distribution functions for example—will be damped away.

In this paper we summarize present results on spatially periodic traveling waves that do not Landau damp. These solutions are exact for the full nonlinear problem, having been obtained without making the assumptions inherent in the linear theory, but they are of small amplitude and therefore represent spatially periodic traveling wave perturbations to Vlasov equilibria. We then summarize results on solitary wave solutions that also can be small perturbations of Vlasov equilibria and which are related to the periodic waves through a transcritical bifurcation.

In order to study traveling wave solutions Eqs. (1)–(3) it is convenient to transform to the wave frame. This is accomplished by the straightforward transformation $f_\alpha(x,u,t) \to f_\alpha(x-vt, u-v, t)$ and $E(x,t) \to E(x-vt,t)$, where v is any phase velocity, to arrive at the Vlasov-Poisson-Ampère equations in the wave frame

$$u\frac{\partial f_\alpha}{\partial x}(x,u) + \frac{q_\alpha}{m_\alpha} E(x)\frac{\partial f_\alpha}{\partial u}(x,u) = 0 \qquad \alpha = 1, 2 \dots N \tag{6}$$

$$\frac{\partial E}{\partial x}(x) = 4\pi \sum_{\alpha=1}^{N} q_\alpha \int_{-\infty}^{\infty} f_\alpha(x,u)\, du \tag{7}$$

$$4\pi \sum_{\alpha=1}^{N} q_\alpha \int_{-\infty}^{\infty} u f_\alpha(x, u)\, du = 0 \tag{8}$$

which in fact are simply the steady-state form of the Vlasov-Poisson-Ampère equations. Hence, spatially periodic solutions to these equations are spatially periodic traveling wave solutions to Eqs. (1)–(3) in the original frame, and spatial 'bump' solutions to these equations are solitary wave solutions to Eqs. (1)–(3) in the original frame. In this paper we shall sketch the application of some nonlinear methods to the analysis of these stationary equations. The spirit of the analysis is to use the well-known BGK form for the distribution functions [Bernstein, Greene & Kruskal (1957)] developed in such a way that the BGK modes so constructed can be made small deviations from a given Vlasov equilibrium F_α as the electric potential goes to zero, and then to use bifurcation theory to analyze the nonlinear equation that results for the potential.

2. SMALL AMPLITUDE BGK MODES

The results described in this paper apply only to waves of the simple BGK form, in which the wave distribution function in the wave frame is given by

$$f_\alpha(x, u) = \begin{cases} g_\alpha^+ \left(\dfrac{u^2}{2} + \dfrac{q_\alpha}{m_\alpha} \phi(x) \right), & u \geq 0 \\[2ex] g_\alpha^- \left(\dfrac{u^2}{2} + \dfrac{q_\alpha}{m_\alpha} \phi(x) \right), & u \leq 0 \end{cases} \tag{9}$$

with $g_\alpha^+(\eta) = g_\alpha^-(\eta)$ for $\Phi_\alpha^{min} \leq \eta \leq \Phi_\alpha^{max}$, where

$$\Phi_\alpha^{min} = \inf_x \{(q_\alpha/m_\alpha)\phi(x)\} \tag{10}$$

and

$$\Phi_\alpha^{max} = \sup_x \{(q_\alpha/m_\alpha)\phi(x)\} . \tag{11}$$

As is well known, these distribution functions will exactly satisfy the stationary Vlasov equation—or the Vlasov equation in the wave frame—with the electric field given by $E(x) = -\partial\phi/\partial x$. Because the distribution function is constant along particle trajectories in the wave frame, it is forced to be an even function of velocity over that range of velocities which, at a given location, are sufficiently slow to allow particle reversal by the electric potential ϕ. We call the BGK form given in Eq. (9) simple because it implies that the distribution of particles over such intervals of velocity is determined only by the total particle energy; this means that trapped particles in spatial regions which are physically separated are artificially forced to have the same distribution, even though, in general, they need not.

Unfortunately, the particular BGK representation given in Eq. (9) is not particularly convenient. Indeed, its use obscures the real features of the analysis and encourages the retention of various unnecessary singularities at and near the phase velocity. Instead, it is better to write $f_\alpha = f_\alpha^e + f_\alpha^o$ where

$$f_\alpha^e(x, u) = g_\alpha^e\left(\frac{u^2}{2} + \frac{q_\alpha}{m_\alpha}\phi(x)\right) \tag{12}$$

gives the even part, and

$$f_\alpha^o(x, u) = \begin{cases} g_\alpha^o\left(\dfrac{u^2}{2} + \dfrac{q_\alpha}{m_\alpha}\phi(x)\right), & u \geq 0 \\[2ex] -g_\alpha^o\left(\dfrac{u^2}{2} + \dfrac{q_\alpha}{m_\alpha}\phi(x)\right), & u \leq 0, \end{cases} \tag{13}$$

gives the odd part, and $g_\alpha^o(\eta) = 0$ for $\Phi_\alpha^{min} \leq \eta \leq \Phi_\alpha^{max}$. Because the charge density in the Poisson equation for the electric potential that results from the BGK representation of the distributions $f_\alpha(x, u)$ depends only on the even parts of those distributions, it will contain only the functions g_α^e, $\alpha = 1, 2 \ldots N$, and these functions do not have any restriction imposed on their form by the presence of trapped particles; they can be smooth and even analytic functions of the particle energy. In contrast, the restriction $g_\alpha^o(\eta) = 0$ for $\Phi_\alpha^{min} \leq \eta \leq \Phi_\alpha^{max}$ implies a lack of analyticity in the functions g_α^o, but even these functions can be very smooth and have derivatives of all orders. There is no need for any of the functions g_α^e and g_α^o to be unbounded or to have any unbounded derivatives of any order, and the resulting distribution functions can be very smooth and well behaved, even in the limit of zero potential.

Thus far there has been nothing said which ties the BGK form to small amplitude waves near to any particular Vlasov equilibrium, described by spatially uniform distribution functions F_α, $\alpha = 1, 2 \ldots N$. To make this connection, let us shift the given Vlasov equilibrium into an arbitrary wave frame of phase velocity v and introduce the functions

$$F_\alpha^e(u) = \tfrac{1}{2}\left[F_\alpha(u + v) + F_\alpha(-u + v)\right] \tag{14}$$

$$F_\alpha^o(u) = \tfrac{1}{2}\left[F_\alpha(u + v) - F_\alpha(-u + v)\right], \tag{15}$$

which represent the even and odd parts of the spatially uniform distribution functions in the wave frame. (The phase velocity dependence of F_α^e and F_α^o has been suppressed to reduce the dreaded *notational clutter.*) Using the BGK form we can therefore represent this arbitrary equilibrium distribution as

$$F_\alpha^e(u) = G_\alpha^e\left(\frac{u^2}{2}\right) \tag{16}$$

and

$$F_\alpha^o(u) = \begin{cases} G_\alpha^o\left(\dfrac{u^2}{2}\right), & u \geq 0 \\[2mm] -G_\alpha^o\left(\dfrac{u^2}{2}\right), & u \leq 0 \end{cases} \tag{17}$$

where

$$G_\alpha^e(\eta) = F_\alpha^e\left(\sqrt{2\eta}\right) \tag{18}$$

and

$$G_\alpha^o(\eta) = F_\alpha^o\left(\sqrt{2\eta}\right). \tag{19}$$

We can now describe a wave which is a small amplitude perturbation of the original spatially uniform Vlasov equilibrium by writing the wave in the BGK form of Eqs. (12) and (13) using functions g_α^e and g_α^o which are close to the corresponding functions G_α^e and G_α^o defined by Eqs. (18) and (19).

The details of this construction begin by showing that the functions G_α^e, $\alpha = 1, 2 \ldots N$, are smooth; in fact, a careful analysis [Holloway (1989)] shows that the function G_α^e has at least half as many derivatives as the original distribution function F_α. With this knowledge, we can define an extension of each function G_α^e to negative values of its argument; essentially any extension will do, provided that it is non-negative everywhere; we call such an extension $\mathcal{G}_\alpha^e$. Using this extension, the BGK function g_α^e used in Eq. (12) will henceforth be written as

$$g_\alpha^e(\eta) = \mathcal{G}_\alpha^e(\eta) + h_\alpha^e(\eta), \tag{20}$$

where h_α^e is any smooth function which may subsequently be perturbed from zero. Thus, if the h_α^e, $\alpha = 1, 2 \ldots N$ are small and the potential ϕ is small the even part of the wave distribution function will be but a little different from that of the spatially uniform equilibrium distribution.

The treatment of the odd part is a bit more difficult because it requires the use of functions which are infinitely differentiable but not analytic. This difficulty is compounded by the need to construct a function g_α^o that satisfies four requirements: (1) g_α^o should be made close to G_α^o for non-negative values of its arguments, and made closer still as the potential becomes smaller, (2) $g_\alpha^o(\eta)$ should equal zero for $\eta \leq \Phi_\alpha^{max}$, (3) the net current of the wave should be kept equal to zero in the wave frame, and (4) the distribution function should never be negative. There is no unique way to develop a g_α^o satisfying these requirements, but the function

$$g_\alpha^o(\eta) = \left(1 - R(\eta/\Phi_\alpha^{max})\right)G_\alpha^o(\eta) + A_\alpha B\left(\frac{\eta - 2\Phi_\alpha^{max}}{\Phi_\alpha^{max}}\right) \tag{21}$$

will do the trick, provided that the potential is sufficiently small, the spatially uniform equilibrium distribution functions F_α are not zero at the phase velocity, and the functions h_α^e are non-negative [Holloway and Dorning (1990)]. In Eq. (21) R is any infinitely differentiable function which satisfies $0 \le R(\eta) \le 1$ for all η, $R(\eta) = 1$ for $\eta \le 1$, and $R(\eta) = 0$ for $\eta \ge 2$; B is any infinitely differentiable function which satisfies $0 \le B(\eta) \le 1$ for all η, $B(\eta) = 0$ for $|\eta| \ge 1$ and B not identically zero; and

$$A_\alpha = \left[\Phi_\alpha^{max} \int_{-2}^{\infty} B(\eta)\, d\eta \right]^{-1} \int_0^{\infty} R(\eta/\Phi_\alpha^{max}) G_\alpha^o(\eta)\, d\eta \,. \tag{22}$$

In this form for g_α^o the factor $\left(1 - R(\eta/\Phi_\alpha^{max})\right)$ ensures that condition (2) mentioned above is satisfied, while the expression for A_α ensures that condition (3), the net current in the wave frame is zero, is met. The positivity of the final distribution function, condition (4), is guaranteed by the facts that $|G_\alpha^o(\eta)| \le G_\alpha^e(\eta)$ for a non-negative spatially uniform equilibrium F_α, that $F_\alpha(v) \ne 0$ and that $A_\alpha \to 0$ as $\Phi_\alpha^{max} \to 0$. This last fact, along with the observation that $\left(1 - R(\eta/\Phi_\alpha^{max})\right)G_\alpha^o(\eta) \to G_\alpha^o(\eta)$ uniformly for all $\eta \ge 0$ guarantees that $g_\alpha^o(\eta) \to G_\alpha^o(\eta)$ as required in the condition (1).

It is important to note that the distribution functions f_α which can be constructed from the BGK form using these forms for g_α^e and g_α^o are close to the equilibrium distribution functions $F_\alpha(u + v)$ in the sense that, with I any closed and bounded interval of space, $\|\phi\| = \sup_{x \in I} |\phi(x)|$ and $\|h_\alpha^e\| = \sup_\eta |h_\alpha^e(\eta)|$,

$$\lim_{\substack{\|\phi\| \to 0 \\ \|h_\alpha^e\| \to 0}} \; \sup_{\substack{x \in I \\ u \in \mathbf{R}}} \left| f_\alpha(x, u) - F_\alpha(u + v) \right| = 0 \tag{23}$$

and, for any $p \ge 1$

$$\lim_{\substack{\|\phi\| \to 0 \\ \|h_\alpha^e\| \to 0}} \int_I \int_{\mathbf{R}} \left| f_\alpha(x, u) - F_\alpha(u - v) \right|^p du\, dx = 0 \,. \tag{24}$$

(Some technical assumptions concerning smoothness and decay at large velocities are required for this result, but these details are not important here.) Physically, this result, and the case $p = 1$ in particular, means that the wave distribution functions represent only a small rearrangement of the equilibrium distribution of particles. This, it seems to us, is all that it is physically relevant to require. A stronger claim, namely that the velocity gradients of the wave and equilibrium distribution functions be close to each other is not possible because of the influence of trapped particles; trapped particles (or particles which reverse direction) force the wave distribution function to be even over a range of velocities about the phase velocity, and this requires that the traveling wave distribution function satisfy $\partial f_\alpha / \partial u|_{u=v} = 0$ for all x, while in general $F_\alpha'(v) \ne 0$. In contrast the linear theory [Vlasov (1945), Landau (1946)] and the energy arguments for Landau damping [Dawson (1961)] require deviations from the equilibrium which satisfy $\partial f_\alpha/\partial u|_{u=v} \approx F'_\alpha(v)$. Thus, undamped waves are naturally in conflict with the essential assumptions of the linear theory, and the relative scarcity of undamped waves in the linear theory is therefore not surprising. But further, this lack should not be viewed as a physical result, but rather as a shortcoming of the linear approach. In the next five sections we shall show that a careful construction can in fact establish the existence of exact undamped nonlinear traveling wave solutions in a regime where the linear theory would predict their absence.

3. SMALL AMPLITUDE PERIODIC WAVES

In this section we shall give a sketch of the nonlinear functional analysis which can be used to understand the characteristics of periodic traveling waves which are small amplitude perturbations of an arbitrary Vlasov equilibrium. This analysis is of a rather general character, and allows for an understanding of the relationship between the wave number of the wave, its amplitude, and the wave distribution functions.

The analysis begins with the nonlinear ODE for the potential obtained when the BGK form of Eq. (12) is inserted into Poisson's equation, namely

$$\frac{d^2\phi}{dx^2} + 4\pi \sum_{\alpha=1}^{N} q_\alpha \int_{-\infty}^{\infty} \mathcal{G}_\alpha^e\left(\frac{u^2}{2} + \frac{q_\alpha}{m_\alpha}\phi(x)\right) + h_\alpha^e\left(\frac{u^2}{2} + \frac{q_\alpha}{m_\alpha}\phi(x)\right) du = 0 , \quad (25)$$

the task being to show that for *any* sufficiently small amplitude there is some choice of the functions h_α^e near zero for which this equation has periodic solutions ϕ of that amplitude or smaller. If we seek solutions of period $2\pi/k$ we can scale x so that our objective becomes that of finding small amplitude solutions of

$$\frac{d^2\Phi}{dy^2} + \mathcal{H}(\Phi, h, k) = 0 \tag{26}$$

where $\phi(x) = \Phi(xk)$ and

$$\mathcal{H}(\Phi, h, k) = \frac{4\pi}{k^2} \sum_{\alpha=1}^{N} q_\alpha \int_{-\infty}^{\infty} \mathcal{G}_\alpha^e\left(\frac{u^2}{2} + \frac{q_\alpha}{m_\alpha}\Phi(y)\right) + h_\alpha^e\left(\frac{u^2}{2} + \frac{q_\alpha}{m_\alpha}\Phi(y)\right) du \tag{27}$$

and $h = \left(h_1^e, \ldots, h_N^e\right)$ is used as a shorthand. Because

$$\sum_{\alpha=1}^{N} q_\alpha \int_{-\infty}^{\infty} \mathcal{G}_\alpha^e\left(\frac{u^2}{2}\right) du = \sum_{\alpha=1}^{N} q_\alpha \int_{-\infty}^{\infty} F_\alpha^e(u) \, du = 0 \tag{28}$$

we at once perceive that $\mathcal{H}(0, 0, k) = 0$. It is therefore useful to view the nonlinear ODE as a bifurcation problem depending on the parameters h_α^e, $\alpha = 1\ldots N$

and k. Of course the h_α^e are actually arbitrary functions, and we are therefore considering an infinite dimensional space of bifurcation parameters, but this is not a problem in either principle or practice.

Our desire then is to show that Eq. (26) has small amplitude, 2π periodic solutions, and to understand the dependence of these solutions on the parameters h_α^e, non-negative and near zero, and the parameter k near some specific value κ. It is from such solutions that we can construct, using the BGK form given by Eqs. (12), (13), (20) and (21), small amplitude nonlinear waves which, in contradiction with the Landau damping or growth of the linear theory, are of constant amplitude. In this scheme the parameters provide us with the control needed to ensure that the amplitude of the nonlinear waves can be made as small as desired, so that there can be no doubt that the undamped periodic nonlinear waves exist arbitrarily close to the original equilibrium and therefore in any possible linear regime, if in fact there is one.

One important caveat is called for: it is not enough to know that the 2π periodic solutions exist, they must also be non-constant. For a constant potential Φ, while it is 2π periodic, merely gives another spatially uniform Vlasov equilibrium, and not a plasma wave. Application of the implicit function theorem to Eq. (26) implies however, that if the functional derivative of the nonlinear operator with respect to the first argument

$$D_1 \mathcal{H}(0,0,\kappa) = \frac{1}{\kappa^2} 4\pi \sum_{\alpha=1}^{N} \frac{q_\alpha^2}{m_\alpha} \int_{-\infty}^{\infty} \frac{d\mathcal{G}_\alpha^e}{d\eta} \left(\frac{u^2}{2} \right) du \neq 1 \tag{29}$$

then the (locally) unique 2π periodic solution of Eq. (26) is spatially uniform; indeed, in this case Φ is a smooth function of the parameters h_α^e, $\alpha = 1, 2 \ldots N$, and k, for small functions h_α^e and wave numbers k near κ. Thus, in order to have the non-trivial periodic solutions which we seek it must be that

$$\kappa^2 = 4\pi \sum_{\alpha=1}^{N} \frac{q_\alpha^2}{m_\alpha} \int_{-\infty}^{\infty} \frac{d\mathcal{G}_\alpha^e}{d\eta} \left(\frac{u^2}{2} \right) du . \tag{30}$$

We will only be able to find non-trivial small amplitude waves that have a wave number k which is near κ as defined by this expression. Since k^2 must be positive for a spatially periodic solution, we can interpret the condition $\kappa^2 \geq 0$ as a necessary condition for the existence of undamped periodic BGK waves of arbitrarily small amplitude. While this result has been arrived at here by examining simple BGK modes, this necessary condition actually applies to all rigidly translating traveling waves [Holloway (1989), Holloway & Dorning (1989), Holloway & Dorning (1990)]. It is also interesting to note that from Eq. (30) we can derive a more familiar sort of expression for κ^2, namely

$$\kappa^2(v) = 4\pi \sum_{\alpha=1}^{N} \frac{q_\alpha^2}{m_\alpha} P \int_{-\infty}^{\infty} \frac{F_\alpha'(u)}{(u-v)} du \tag{31}$$

in which the explicit dependence of κ^2 on the phase velocity has been restored. [The equivalence of the integrals on the right of this expression with those on the right side of Eq. (30) follows easily from the dominated convergence theorem.] Equation (31) is the Vlasov dispersion relation originally derived (incorrectly) from the linear equations [Vlasov (1945)], yet now we see that nonlinear small amplitude plasma waves are actually governed by this relation and not by the more familiar Landau dispersion relation [Landau (1946)]. The reasons for the coincidence between Eq. (31), arrived at here through a full nonlinear analysis, and the Vlasov dispersion relation, incorrectly derived from the linear equations, are as yet unclear.

To proceed with a bifurcation analysis of the ODE we write $k = \kappa + \delta$ and use a standard technique from bifurcation theory, the Liapunov-Schmidt method [Cesari (1976), Chow & Hale (1982)], to discover that

$$\Phi(y) = A\cos(y) + B\sin(y) + \Psi(A, B, h, \delta; y). \tag{32}$$

Here Ψ is the demonstrably unique 2π periodic solution of the nonlinear equations

$$\frac{d^2\Psi}{dy^2} + Q\mathcal{H}(A\cos(y) + B\sin(y) + \Psi, h, \kappa + \delta) = 0, \tag{33}$$

$$0 = \int_0^{2\pi} \cos(y)\Psi(A, B, h, \delta; y)\, dy = \int_0^{2\pi} \sin(y)\Psi(A, B, h, \delta; y)\, dy \tag{34}$$

where Q is the projection onto the orthogonal complement of $\cos(y)$ and $\sin(y)$. The existence of a unique function Ψ which is a smooth function of all its arguments follows from the implicit function theorem; and because it is unique and is a smooth function of its arguments it can be approximated by a regular perturbation scheme.

Thus, if A and B were known as a function of the parameters h_α, $\alpha = 1, 2 \ldots N$, and δ we would have a complete understanding of the small amplitude 2π periodic solutions for Φ. The equations which A and B satisfy, called the bifurcation equations, are derived as part of the Liapunov-Schmidt procedure, and are

$$0 = \int_0^{2\pi} \cos(y)\mathcal{H}(A\cos(y) + B\sin(y) + \Psi(A, B, h, \delta; y), h, \kappa + \delta)\, dy - \pi A$$

$$\equiv H_c(A, B, h, \delta) \tag{35}$$

$$0 = \int_0^{2\pi} \sin(y)\mathcal{H}(A\cos(y) + B\sin(y) + \Psi(A, B, h, \delta; y), h, \kappa + \delta)\, dy - \pi B$$

$$\equiv H_s(A, B, h, \delta) \tag{36}$$

The functions H_c and H_s can be shown to satisfy

$$H_c(0, 0, h, \delta) = H_s(0, 0, h, \delta) = 0 \tag{37}$$

but $A = B = 0$ is not the only solution of Eqs. (35) and (36) close to zero.

The characterization of all solutions of Eqs. (35) and (36), and hence of Eq. (26), is made easier by a study of the symmetry properties of the equations. Equation (26) is invariant under phase shifts $y \mapsto y + \theta$ and spatial inversions $y \mapsto -y$, and these symmetries are reflected in Eqs. (35) and (36) as invariance under the group $O(2)$, generated by the transformations

$$\begin{bmatrix} A \\ B \end{bmatrix} \mapsto \begin{bmatrix} \cos(\theta) & -\sin(\theta) \\ \sin(\theta) & \cos(\theta) \end{bmatrix} \begin{bmatrix} A \\ B \end{bmatrix} \tag{38}$$

which rotates the point (A, B) by an angle θ, and

$$\begin{bmatrix} A \\ B \end{bmatrix} \mapsto \begin{bmatrix} 1 & 0 \\ 0 & -1 \end{bmatrix} \begin{bmatrix} A \\ B \end{bmatrix} \tag{39}$$

which reflects the point (A, B) about the A axis. In consequence, the functions H_c and H_s can be shown to have the properties

$$
\begin{aligned}
&(1a) \quad H_c(A, -B, h, \delta) &=& \quad H_c(A, B, h, \delta) \\
&(1b) \quad H_s(A, -B, h, \delta) &=& -H_s(A, B, h, \delta) \\
&(2a) \quad H_c(A', B', h, \delta) &=& \quad \cos(\theta) H_c(A, B, h, \delta) - \sin(\theta) H_s(A, B, h, \delta) \\
&(2b) \quad H_s(A', B', h, \delta) &=& \quad \sin(\theta) H_c(A, B, h, \delta) + \cos(\theta) H_s(A, B, h, \delta)
\end{aligned}
$$

when (A', B') are related to (A, B) by rotation through the angle θ. It follows from (2a) and (2b) that if (A, B) satisfy Eqs. (35) and (36) then so do (A', B') for all values of θ. It is therefore sufficient to consider only the cases with $B = 0$, since all other solutions can be found from this case. Further, from property (1b) we see that $H_s(A, 0, h, \delta) = 0$ for all A; therefore we only need to solve $H_c(A, 0, h, \delta) = 0$ in order to understand all (small amplitude) solutions of Eq. (26).

We now define a new function $H(A, h, \delta) = H_c(A, 0, h, \delta)$, and observe that it has three important properties:

$$(i) \quad H(0, h, \delta) = 0$$

$$(ii) \quad \frac{\partial H}{\partial A}(0, 0, 0) = 0$$

$$(iii) \quad H(A, h, \delta) = -H(-A, h, \delta).$$

The problem of small amplitude periodic BGK waves has now been rigorously reduced to finding the small A solutions of $H(A, h, \delta) = 0$, since from these we can reconstruct all small amplitude solutions of Eq. (26).

There are numerous approaches to studying the equation $H(A, h, \delta) = 0$: we can consider $\delta = 0$ and identify those select h_α^e in the infinite dimensional space of parameters for which the potential has a fixed wave number $k = \kappa$ [Holloway

(1989)]; we can consider $\delta = 0$ and introduce a one parameter family of functions h_α^e [Holloway (1989), Holloway & Dorning (1989)], thereby describing a simplified one-dimensional parameter space for the fixed wave number waves; or we can set $h_\alpha^e = 0$ and explore the δ-dependence [Holloway & Dorning (1990)]. Here we shall present a simplified combination of the latter two analyses, a combination which makes it easy to verify that solutions exist for ϕ corresponding to functions h_α^e that are both non-zero and non-negative, so that the expression given in Eq. (21) will guarantee a positive distribution function for sufficiently small potentials.

To be very specific let's choose a family of functions for each h_α^e of the form

$$h_\alpha^e(\eta) = \mu \mathcal{G}_\alpha^e(\eta), \tag{40}$$

which will be non-negative if $\mu \geq 0$, and show that, under some weak hypotheses, that we can find solutions of $H(A, h, \delta) = 0$ for all sufficiently small non-negative values of μ. While using this one parameter family of functions we shall write $M(A, \mu, \delta) = H(A, h, \delta)$. Besides $\kappa^2 > 0$ we need the hypothesis that

$$0 \neq \frac{24\pi^2}{\kappa} \left[\frac{1}{8} \sum_{\alpha=1}^{N} q_\alpha \left(\frac{q_\alpha}{m_\alpha} \right)^3 \int_{-\infty}^{\infty} \frac{d^3 \mathcal{G}_\alpha^e}{d\eta^3} \left(\frac{u^2}{2} \right) du \right.$$
$$\left. - \frac{5}{24\kappa^2} \left[\sum_{\alpha=1}^{N} q_\alpha \left(\frac{q_\alpha}{m_\alpha} \right)^2 \int_{-\infty}^{\infty} \frac{d^2 \mathcal{G}_\alpha^e}{d\eta^2} \left(\frac{u^2}{2} \right) du \right]^2 \right]. \tag{41}$$

This is a generic property of the Vlasov equilibrium under investigation, that is, it is satisfied by almost all Vlasov equilibria and almost all phase velocities. This hypothesis implies that

$$\frac{\partial^3 M}{\partial A^3}(0, 0, 0) \neq 0, \tag{42}$$

and we can also calculate that

$$\frac{\partial^2 M}{\partial \delta \partial A}(0, 0, 0) = \frac{-2\pi}{\kappa} \neq 0. \tag{43}$$

If there were no higher order terms then the equation $M(A, \mu, \delta) = 0$ would look like $\mu C(\delta) A + A^3 = 0$ and we therefore expect that the exact non-trivial solutions for A near $A = 0$ would look like $A = \pm \sqrt{-\mu C(\delta)}$. Application of the implicit function theorem confirms this expectation: all solutions of $M(A, \mu, \delta) = 0$ near $(A, \mu, \delta) = (0, 0, 0)$ are given by:

$$A = 0 \qquad \text{for all } (\mu, \delta) \text{ near zero}$$

$$A = \pm \sqrt{a(\mu, \delta)} \qquad \text{for all } (\mu, \delta) \text{ near zero such that } a(\mu, \delta) \geq 0, \tag{44}$$

and we can find $a(\mu, \delta)$ by a regular approximation scheme. In fact,

$$a(\mu, \delta) = -6 \left[\frac{\partial M}{\partial A}(0, \mu, \delta) \Big/ \frac{\partial^3 M}{\partial A^3}(0, \mu, \delta) \right] + o(|\mu|, |\delta|) \tag{45}$$

and $a(\mu, \delta) = 0$ near $(\mu, \delta) = (0, 0)$ if and only if

$$\frac{\partial M}{\partial A}(0, \mu, \delta) = 0. \tag{46}$$

The set of points at which this holds true, the set of points where the non-trivial solutions go to zero, is the bifurcation set. If this set makes a non-zero angle at $(0, 0)$ with respect to the δ axis in the (μ, δ) plane then there will be non-trivial solutions for A for some positive value of μ. But, by Eq. (43),

$$\hat{\delta} \cdot \nabla_{(\mu, \delta)} \frac{\partial M}{\partial A}(0, 0, 0) = \frac{\partial^2 M}{\partial \delta \partial A}(0, 0, 0) \neq 0 \tag{47}$$

and so this is in fact the case. Figure 1 shows a schematic of the bifurcation set and the solutions, thereby illustrating these points. In Fig. 1 we can schematically view A, δ and μ as representing the electric potential $\phi(x)$, the wave number k and the distribution function g_α^e respectively. The figure shows that for any wave number and distribution in the correct region—the region in which non-trivial solutions exist—a certain amplitude of the potential arises. But for a given amplitude of the potential there is a range of possible wave numbers and corresponding distribution functions which are consistent with that amplitude. Thus, small amplitude nonlinear plasma waves will exist with a range of wave numbers about κ and there will be no dispersion relation, no exact relation between wave number, phase velocity and amplitude.

This analysis proves that there are traveling wave solutions with phase velocity v of the one-dimensional Vlasov-Maxwell equations that are small amplitude perturbations of the Vlasov equilibrium given by $F_\alpha(u)$, $\alpha = 1, 2 \ldots N$, provided only that $\kappa^2(v)$ is positive and that Eq. (41) is satisfied; the wave distribution functions are smooth solutions of the Vlasov equation, are positive, and they do not become singular as the wave amplitude goes to zero. Other considerably more general analyses [Holloway (1989), Holloway & Dorning (1989), Holloway & Dorning (1990)] emphasize that Eq. (41) is not necessary, and that there is a tremendous non-uniqueness in these small amplitude waves: there is a range of wave numbers and distribution functions possible for a given wave amplitude, even close to a fixed equilibrium. The existence of this plethora of waves brings the physical intuition developed from the linear theory into question; even near the Maxwellian there are many nonlinear waves of arbitrarily small amplitude which trap particles and defy Landau.

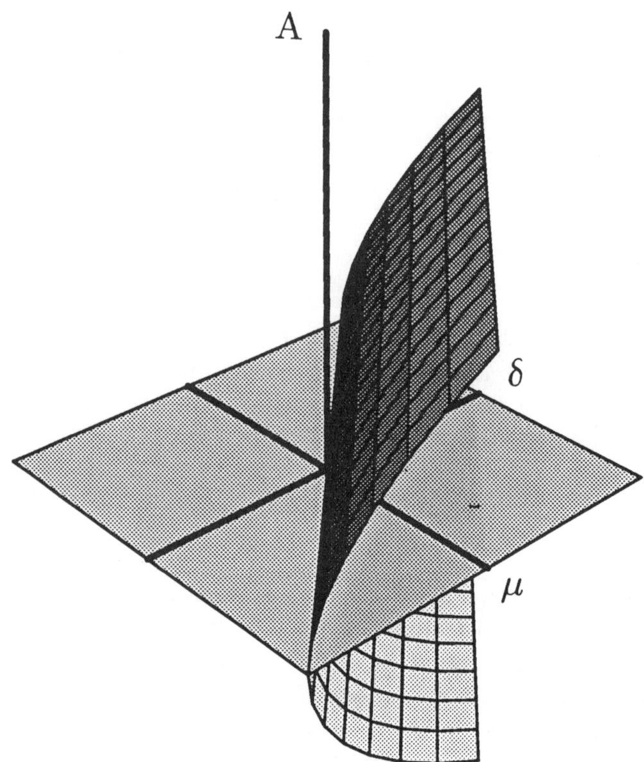

Figure 1. The solutions A versus the parameters, showing the non-trivial fold of solutions going to zero at the bifurcation set.

And there is good reason to doubt the linear theory: a general result of the theory of nonlinear dynamical systems is that linearization about an equilibrium provides a good local description of the nonlinear dynamics only when the spectrum of the corresponding linear evolution operator does not include any points on the imaginary axis [see for example Guckenheimer and Holmes (1983), pg. 13]. It has long been known that the spectrum of the linearized Vlasov–Poisson equations includes the entire imaginary axis (rotated to the real axis and described by the Van Kampen–Case singular eigenmodes in much of the literature) [Van Kampen (1955), Case (1959), Case (1977)]. The presence of the singular and undamped waves corresponding to the Van Kampen–Case modes could be interpreted as describing a large class of undamped plasma waves, but they are unlike the small amplitude waves just derived. The wave distributions developed here are smooth functions of velocity and remain so as their amplitude is decreased to zero, while the Van Kampen–Case singular eigenfunctions are first-order distributions in the sense of Schwartz and unbounded near the phase velocity; the Van Kampen–Case waves are thus highly nonphysical solutions of the unphysical linear equations while the waves described here are physically realizable solutions of the nonlinear equations.

An instance in which the linear theory does predict the presence of physically reasonable undamped plasma waves is when $\kappa^2(v) > 0$ and simultaneously $F_\alpha'(v) = 0$, $\alpha = 1, 2 \ldots N$. In this case the nonlinear waves described here do in fact approach the waves predicted by the linear theory in the zero amplitude limit, and in this rather special case the linear analysis does provide an adequate description of undamped plasma waves. But it can easily happen—in fact it is the rule rather than the exception—that $\kappa^2(v) > 0$ at phase velocities v where $F_\alpha'(v) \neq 0$. In this typical case the linear theory has completely missed the nonlinear waves because the assumptions for its validity are invalidated even at small amplitude.

The essential point in interpreting the results of this section as representing nonlinear waves which can be arbitrarily close to the arbitrary Vlasov equilibrium F_α is that we interpret close only in any of the senses implied in Eqs. (23) or (24). We do not require that the derivatives of the distribution functions be close. Frequently in physical problems there are processes—viscosity is an example—which tend to smooth the physical fields of interest, and then it can be that the closeness of the derivatives of the fields in fact follows from the closeness of the fields themselves. The Vlasov-Maxwell equations however, are an infinite dimensional Hamiltonian system [Morrison (1980); Marsden & Weinstein (1982)] and therefore lack any kind of smoothing; there is no physical basis for claiming that two distribution functions are "close" only when their velocity gradients are close.

Having established the existence of exact nonlinear undamped waves when $\kappa^2 \geq 0$, it is natural to ask the question, "What happens when κ^2 passes through zero and becomes negative?" Clearly the solutions to the linearized form of the nonlinear Poisson equation developed above for simple BGK waves are then exponential rather than trigonometric functions. However, remembering that this differential equation is merely the linearization of a nonlinear problem, we recognize that these growing and decaying exponential solutions in space represent motion toward and away from the point $(\phi, d\phi/dx) = (0, 0)$ in a two-dimensional phase space with coordinates ϕ and $\phi_x = d\phi/dx$ and where x plays the role of time. In fact, this suggests treating the nonlinear Poisson equation as a two-dimensional dynamical system with ϕ and ϕ_x as the phase variables. It is easy to show that this system is a Hamiltonian system and to study the transition of the $(\phi, \phi_x) = (0, 0)$ fixed point, as κ^2 passes through zero and becomes negative, from a center surrounded by solutions that are periodic (in space) to a saddle with "motion" given by $\exp(-|\kappa|x)$ toward it along its stable eigenspace and "motion" given by $\exp(+|\kappa|x)$ away from it along its unstable eigenspace. This local behavior, which results from a local bifurcation in which there is a transition from a center to a saddle point is interesting, but the important question is "What is the behavior on the stable and unstable manifolds of this $(0,0)$ fixed point, and do those manifolds go off to infinity, connect with one another forming a homoclinic orbit, or connect to another saddle point forming heteroclinic orbits?"

The remainder of this paper addresses these questions. However, before exploring the $\kappa^2 < 0$ regime through this Hamiltonian dynamical systems approach, it is both useful and instructive to use it to rederive the condition $\kappa^2 \geq 0$ as sufficient for the existence of undamped spatially periodic waves. These less complex techniques provide both an easy route to the $\kappa^2 \geq 0$ sufficient condition for periodic waves, and an easy bridge to the solitary wave solutions that are developed in Section 5. More importantly, however, they provide a glimpse at the bigger picture, or at least a less local picture, in the infinite-dimensional parameter space, of the undamped nonlinear waves that bifurcate from Vlasov equilibria. In particular, they establish the connection between the $\kappa^2 > 0$ region and the $\kappa^2 < 0$ region through a simple bifurcation that thereby relates the undamped periodic waves to the solitary waves.

4. THE SUFFICIENT CONDITION REVISITED

In the following two sections we investigate the nonlinear ordinary differential equation for the electric potential, Eq. (25), in the context of the theory of two-dimensional dynamical systems. First, in this section, we derive the sufficiency of the condition $\kappa^2(v) \geq 0$ for the existence of small amplitude undamped traveling waves nearby a particular Vlasov equilibrium—a result already demonstrated earlier in this paper. Here, however, we arrive at this result using a different procedure that exploits the analogy between Eq. (25) and the equation describing the one-dimensional classical mechanics of a single particle in a potential field. This alternative approach offers new perspective, and considerably simplifies the derivation. Then, in Sections 5 and 6, we establish general conditions that are sufficient for the existence of small amplitude solitary waves nearby any Vlasov equilibrium and show, in particular, that such solitary waves exist near a thermal equilibrium electron-proton plasma.

Setting $h_\alpha^e = 0$ in Eq. (25), we have an ordinary differential equation for the electric potential,

$$\frac{d^2\phi}{dx^2} = -4\pi \sum_{\alpha=1}^{N} q_\alpha \int_{-\infty}^{\infty} \mathcal{G}_\alpha^e \left(\frac{u^2}{2} + \frac{q_\alpha}{m_\alpha} \phi(x) \right) du \tag{48}$$

where, as discussed earlier, the $\mathcal{G}_\alpha^e(\eta)$ are smooth extensions to negative arguments of the functions $G_\alpha^e(\eta)$, defined by Eq. (18), and represent the even parts of the Vlasov equilibrium distributions. To derive the sufficient condition for small amplitude periodic waves we begin by rewriting Eq. (48) as

$$\frac{d^2\phi}{dx^2} = -\frac{dV_g}{d\phi}(\phi) \tag{49}$$

where

$$V_g(\phi) = 4\pi \sum_{\alpha=1}^{N} q_\alpha \int_0^\phi d\phi' \int_{-\infty}^\infty \mathcal{G}_\alpha^e\left(\frac{u^2}{2} + \frac{q_\alpha}{m_\alpha}\phi'\right) du. \tag{50}$$

The purpose of writing the nonlinear ODE in this form is that Eq. (49) is formally identical to the classical equation of motion of a single point particle in a one-dimensional potential field. In this instance, $\phi(x)$ plays the role of position, while x plays the role of the independent time variable. This analogy, first exploited by Bernstein, Green and Kruskal (1957), considerably simplifies the search for solutions since we may use familiar methods gleaned from the study of classical mechanics. We will call $V_g(\phi)$ the "pseudo-potential" to avoid confusion with the electric potential $\phi(x)$.

To look for small amplitude solutions we expand the pseudo-potential $V_g(\phi)$ in a Taylor series about $\phi = 0$. Since $\mathcal{G}_\alpha^e(\eta)$ has at least half as many derivatives as the equilibrium distribution function $F_\alpha(u)$, it suffices to assume that $F_\alpha(u)$ is twice differentiable in what follows. We find,

$$V_g(\phi) = \frac{1}{2}\kappa^2\phi^2 + o(\phi^2) \tag{51}$$

where κ^2 is defined in Eq. (30). The first two terms in the expansion vanish, the first because of our definition of $V_g(\phi)$, and the second because

$$\frac{dV_g}{d\phi}(0) = 4\pi \sum_{\alpha=1}^{N} q_\alpha \int_{-\infty}^\infty G_\alpha^e\left(\frac{u^2}{2}\right) du = 4\pi \sum_{\alpha=1}^{N} q_\alpha \int_{-\infty}^\infty F_\alpha^e(u) du = 0, \tag{52}$$

which expresses the fact that the Vlasov equilibrium is charge neutral.

As Eq. (51) indicates, the leading order term in $V_g(\phi)$ for small ϕ is quadratic with coefficient proportional to κ^2. This immediately implies that if κ^2 is positive, then there exists in $V_g(\phi)$ a local pseudo-potential well in the vicinity of $\phi = 0$. It follows then that there exist small amplitude periodic solutions with $\phi(x)$ confined to this well, i.e. there are solutions to Eq. (25) with $h_\alpha^e(\eta) = 0$, or equivalently to Eq. (49), for which $\phi(x)$ is of small amplitude and is periodic in x.

Thus, identifying Eq. (25) or Eq. (49) with the analogous equation of classical mechanics makes it very easy to demonstrate that $\kappa^2 > 0$ guarantees the existence of small amplitude periodic traveling waves near a Vlasov equilibrium. For this demonstration to be physically meaningful, however, we must of course also show that we can construct a corresponding distribution function which is positive and satisfies the zero current condition—which is just Ampère's law in the wave frame. But the results of Section 2 immediately provide us with an acceptable prescription for the odd part, using Eqs. (13) and (21), and the even part is just given by Eq. (12) with $g_\alpha^e = \mathcal{G}_\alpha^e$

To reiterate briefly, the Vlasov equation is satisfied by virtue of the BGK form of the distribution function; Poisson's equation has small amplitude periodic solutions whenever $\kappa^2 > 0$ as shown by the analogy with classical mechanics; and Ampère's equation is satisfied as a result of the construction of $g_\alpha^o(\eta)$ in Eq. (21). Thus, as shown previously [Holloway & Dorning (1990)], small amplitude undamped periodic traveling wave solutions to the Vlasov-Poisson-Ampère equations exist whenever $\kappa^2 > 0$, and in fact $\kappa^2 \geq 0$ is sufficient since arbitrarily close to any equilibrium with $\kappa^2 = 0$ is another equilibrium with $\kappa^2 > 0$.

5. SOLITARY WAVES

To investigate solitary wave solutions to the VPA equations it is instructive to again consider Eq.(25) with $h_\alpha^e(\eta) = 0$, or equivalently Eq. (49), as a two-dimensional dynamical system in the phase space $(\phi, \phi_x) = (\phi, d\phi/dx)$. To do this, we first rewrite the single equation as a system of two coupled first order equations,

$$\frac{d\phi}{dx} = \phi_x, \qquad \frac{d\phi_x}{dx} = -\frac{dV_g}{d\phi}(\phi). \tag{53}$$

Our goal is to study the bifurcation problem for this dynamical system with κ^2 as the bifurcation parameter by smoothly varying the BGK functions $\mathcal{G}_\alpha^e$—by changing the phase velocity for example—so as to change the value of κ^2 from positive to negative.

First, we note that the fixed points of the system satisfy the conditions,

$$\phi_x = 0, \qquad \frac{dV_g}{d\phi}(\phi) = 0. \tag{54}$$

Since $dV_g/d\phi|_{\phi=0} = 0$ as in Eq. (52), $(\phi, \phi_x) = (0,0)$ is clearly a fixed point of this system for all values of κ^2. By calculating the Jacobian on this branch of fixed points, which we regard as the basic branch, we find

$$\mathbf{J} = \begin{pmatrix} 0 & 1 \\ -d^2 V_g/d\phi^2|_{\phi=0} & 0 \end{pmatrix}, \quad \text{and} \quad |\mathbf{J}| = \frac{d^2 V_g}{d\phi^2}(0) = \kappa^2, \tag{55}$$

so that $|\mathbf{J}| = 0$ when $\kappa^2 = 0$, indicating that $(\phi, \phi_x, \kappa^2) = (0,0,0)$ is a bifurcation point.

To find other branches of fixed points in the neighborhood of the bifurcation point, we use the fixed point conditions, Eqs. (54). Clearly, any fixed point must lie on the $\phi_x = 0$ axis. To study the second condition, we again expand $dV_g(\phi)/d\phi$ about the trivial fixed point $(\phi, \phi_x) = (0,0)$, obtaining

$$\frac{dV_g}{d\phi}(\phi) = \kappa^2 \phi + \frac{1}{2}V^{(3)}(0)\phi^2 + o(\phi^2) \tag{56}$$

where

$$V^{(3)}(0) = 4\pi \sum_{\alpha=1}^{N} q_\alpha \left(\frac{q_\alpha}{m_\alpha}\right)^2 \int_{-\infty}^{\infty} \frac{d^2 G_\alpha^e}{d\eta^2}\left(\frac{u^2}{2}\right) du. \tag{57}$$

(For this expansion to be valid it is sufficient for the original Vlasov equilibrium distribution function $F_\alpha(u)$ to be four times differentiable.) For small ϕ then, the conditions for (ϕ, ϕ_x) to be a fixed point are

$$\phi_x = 0, \qquad \frac{dV_g}{d\phi}(\phi) = \phi\left(\kappa^2 + \frac{1}{2}V^3(0)\phi + o(\phi)\right) = 0. \tag{58}$$

Observing that the nature of the bifurcation at $\kappa^2 = 0$ depends upon the parameter $V^{(3)}(0)$, we consider the generic case in which $V^{(3)}(0) \neq 0$ when $\kappa^2 = 0$. We then find just one fixed point branch in addition to the basic branch $(0,0)$; to leading order in κ^2 near $\kappa^2 = 0$ it is

$$(\phi, \phi_x) = \left(-2\kappa^2/V^{(3)}(0), 0\right). \tag{59}$$

This shows that the generic case is a transcritical bifurcation. If, on the other hand, $V^{(3)}(0) = 0$ and $V^{(4)}(0) \neq 0$, then

$$\phi_x = 0, \qquad \frac{dV_g}{d\phi}(\phi) = \phi\left(\kappa^2 + \frac{1}{6}V^{(4)}(0)\phi^2 + o(\phi^2)\right) = 0 \tag{60}$$

are the conditions that (ϕ, ϕ_x) be a fixed point, and we find two other branches for $\kappa^2 < 0$ in addition to the basic branch; to leading order they are

$$(\phi, \phi_x) = \left(\pm\sqrt{\frac{-6\kappa^2}{V^{(4)}(0)}}, 0\right). \tag{61}$$

This non-generic case is a pitchfork bifurcation. Here we will consider further only the generic case $V^{(3)}(0) \neq 0$.

So, generically, as κ^2 passes through zero a transcritical bifurcation occurs at $(\phi, \phi_x, \kappa^2) = (0,0,0)$, and the two fixed point branches intersecting there are described locally by $(\phi, \phi_x)^1 = (0,0)$ and $(\phi, \phi_x)^2 = (-2\kappa^2/V^{(3)}(0), 0)$. For $\kappa^2 > 0$, the first (basic) branch is a branch of centers, around each of which there exists

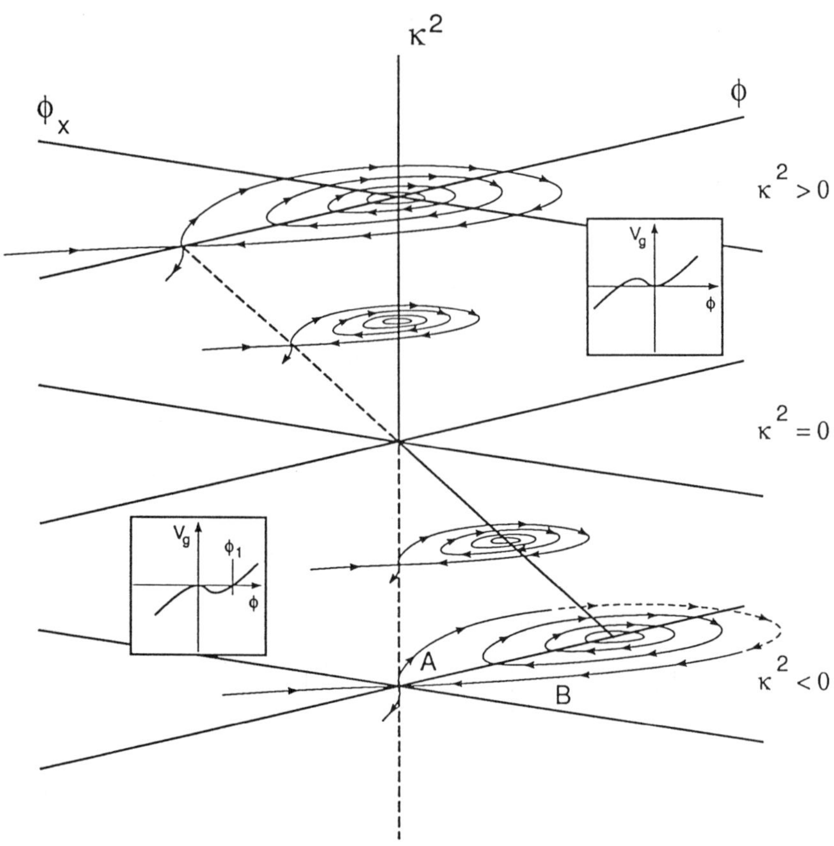

Figure 2. Bifurcation diagram (for the case $V^{(3)}(0) > 0$) showing the transcritical bifurcation at $\kappa^2(v) = 0$, and the corresponding phase space topology. The insets show the form of the pseudo-potential.

a family of closed orbits as indicated in Fig. 2. These are the small amplitude periodic solutions discussed previously. Also shown in Fig. 2 is the second fixed point branch, a branch of saddle points when $\kappa^2 > 0$. As κ^2 is decreased through zero, these branches exchange stability, so that $(\phi, \phi_x)^1$ becomes a saddle and $(\phi, \phi_x)^2$ becomes a center.

For $\kappa^2 < 0$ we find a new feature. As suggested by the dashed curve in Fig. 2, a solitary wave solution will exist if the unstable manifold (A) of the fixed point $(\phi, \phi_x)^1 = (0,0)$ connects with its stable manifold (B). For such a solution, $\phi(x) > 0$ for all x, and $\phi(x) \to 0$ as $x \to \pm\infty$. [Fig. 2 is drawn for the case $V^{(3)}(0) > 0$. The diagram for $V^{(3)}(0) < 0$ is obtained by reflection with respect to the ϕ_x axis.]

The bifurcation calculation also shows (Fig. 2) that in addition to the small amplitude periodic solutions for $\kappa^2 > 0$, there also exist similar solutions for small $\kappa^2 < 0$. This seems, at first, to contradict the necessity of $\kappa^2 \geq 0$ for the existence of small amplitude periodic solutions as derived in the first part of this paper. These solutions, however, are centered about a non-zero value of ϕ, $\phi_c = -2\kappa^2/V^{(3)}(0)$, and therefore are not arbitrarily close to the original Vlasov equilibrium except as κ^2 goes to zero. This can be seen easily from the distribution function written in the BGK form.

The information contained in the phase portraits of Fig. 2 is, of course, also contained in the pseudo-potential $V_g(\phi)$. (See insets in Fig. 2.) For $\kappa^2 < 0$ and small, $V_g(\phi)$ has a local maximum at $\phi = 0$, corresponding to the saddle at $(\phi, \phi_x)^1 = (0,0)$ in the phase portrait. Similarly, the center at $(\phi, \phi_x)^2 = (-2\kappa^2/V^{(3)}(0), 0)$ corresponds to a local well in the pseudo-potential at $\phi = -2\kappa^2/V^{(3)}(0)$. Thus, the pseudo-potential $V_g(\phi)$ decreases monotonically from $\phi = 0$ to $\phi = -2\kappa^2/V^{(3)}(0)$, and then begins to increase. The condition that must be satisfied in order for trajectory A to close with trajectory B is that $V_g(\phi)$ must be positive at some value of $\phi \neq 0$. From Eq. (56) it is clear that for $V^{(3)}(0) \neq 0$, this condition must be satisfied for $\kappa^2 < 0$ and sufficiently small (though still finite). This fact will be exploited in the next section to obtain explicit solitary wave solutions nearby a Maxwellian plasma of electrons and protons.

Finally, we note that these solitary waves are not isolated from the periodic waves. In fact, as is clear in Fig. 2, the family of small amplitude periodic waves undergoes a smooth transition as amplitude increases, to a solitary wave. In this transition, the wavelengths of the periodic waves (closed orbits in Fig. 2) increase, becoming infinite for the solitary wave (the homoclinic orbit A-B). But the solitary waves are not restricted to any finite amplitude; since the branch of solitary waves emanates from a transcritical bifurcation point at $\kappa^2 = 0$, the amplitudes of these waves go to zero as κ^2 goes to zero; hence, they also can be made arbitrarily close to spatially uniform Vlasov equilibria for which $\kappa^2 = 0$, and can in principle be initiated by small, albeit special, perturbations of these equilibria. This is somewhat different from the behavior of the periodic waves, which could be made arbitrarily close to a given equilibrium at a fixed value of the phase velocity and hence any fixed positive value of κ^2. In contrast, the solitary waves are reduced to zero amplitude by varying the phase velocity so that κ^2 goes to zero.

6. SOLITARY WAVES NEAR A MAXWELLIAN EQUILIBRIUM

In this section we consider a Maxwellian plasma consisting of electrons and protons at the same temperature. We will show that for such a plasma, small amplitude traveling solitary wave solutions exist at velocities $v \simeq 1.3v_e$ where v_e is the electron thermal velocity $v_e = \sqrt{kT/m_e}$.

To represent the even parts of the particle distribution functions, we choose BGK functions $\mathcal{G}_\alpha^e(\eta)$ as

$$
\mathcal{G}_\alpha^e(\eta) = \begin{cases} \dfrac{n}{2}\sqrt{\dfrac{m_\alpha}{\pi kT}}\left[e^{-\beta_\alpha\left(\sqrt{2\eta}+v\right)^2} + exp^{-\beta_\alpha\left(-\sqrt{2\eta}+v\right)^2}\right] & \eta \geq 0 \\[2em] \xi_\alpha(\eta) & \eta \leq 0 \end{cases}
$$

with $\beta_\alpha = m_\alpha/2kT$, and $\alpha = 1,2$ for electrons and protons and where each function $\xi_\alpha(\eta)$ is constructed to extend the definition of $G_\alpha^e(\eta)$ smoothly into negative arguments while still keeping it nonnegative. Since $G_\alpha^e(\eta)$ has finite derivatives of all orders at $\eta = 0$, and $G_\alpha^e(0) > 0$ as well, there is no difficulty in defining such an extension. When $\phi = 0$, these BGK functions represent the even parts of the spatially uniform Maxwellian equilibrium as seen in a frame of reference shifted by velocity v.

Calculated numerically, one finds that κ^2 depends upon phase velocity as shown in Fig. 3, where the critical feature in this diagram is that $\kappa^2 = 0$ at some velocity $v_c \simeq 1.3v_e$. For $v < v_c$, $\kappa^2 < 0$, although this is not shown in the diagram. A calculation of $V^{(3)}(0)$ shows that $V^{(3)}(0) > 0$ in a neighborhood of this critical velocity v_c. This information is sufficient to show that small amplitude solitary wave solutions exist in this plasma with phase velocities in an interval extending downward a finite distance from v_c.

Since $\kappa^2 < 0$ for $v < v_c$, the pseudo-potential has negative curvature at the origin in this velocity range, as shown in Fig. 4. Thus, $V_g(\phi)$ decreases as ϕ moves away from zero in either direction (locally). However, as ϕ becomes larger, the third order term becomes important. In fact, this term becomes increasingly influential as we consider v approaching v_c so that κ^2 approaches zero. For κ^2 very small the third order term dominates quickly over the second order term, bringing the pseudo-potential back into positive values. Since $V^{(3)}(0) > 0$, for $v \lesssim v_c$ a pseudo-potential well exists for positive ϕ; this is indicated by the small pseudo-potential well with turning point labeled ϕ_1 in Fig. 4.

By truncating the expansion of $V_g(\phi)$ in Eq. (56) after two terms and integrating the resulting non-linear Poisson equation, Eq. (49), one finds that an approximate analytical form of these small amplitude solitary waves with velocity v is

$$
\phi(x,t) = \phi_1 \operatorname{sech}^2\left[\frac{|\kappa|}{2}(x - vt)\right]
$$

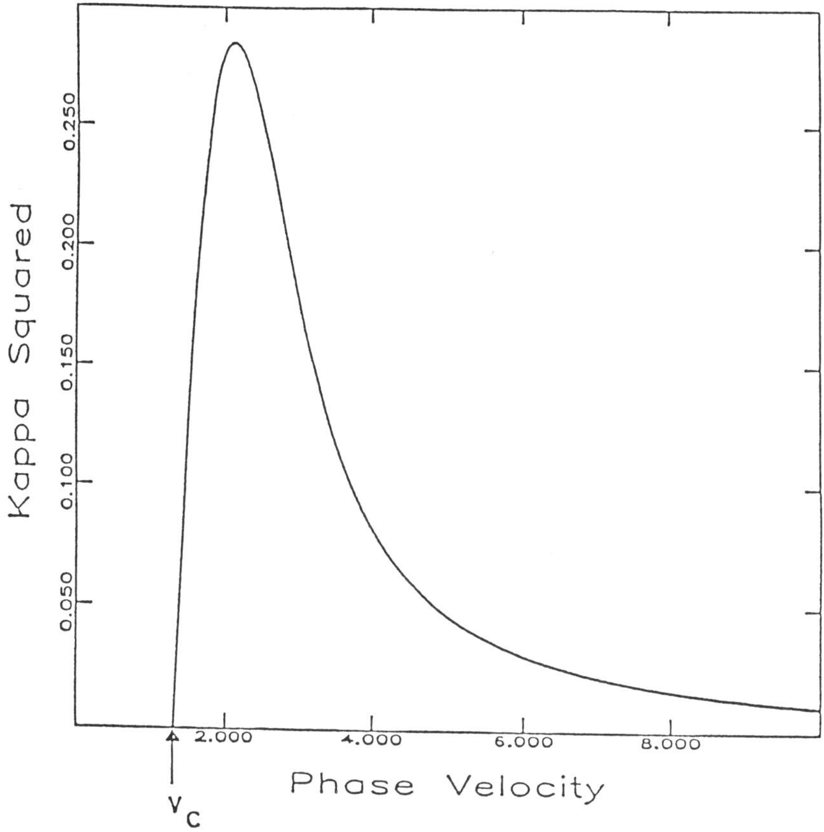

Figure 3. The function $\kappa^2(v)$ for an electron-proton plasma in thermal equilibrium with equal ion and electron temperatures. The scale for κ is the inverse electron Debye length and the scale for the phase velocity is the electron thermal velocity.

where $\phi_1 = -3\kappa^2/V^{(3)}(0)$. As is typical of solitary waves, the amplitude ϕ_1 of this wave depends on the phase velocity v through the phase velocity dependence of both κ^2 and $V^{(3)}(0)$, although this dependence has been suppressed in the notation.

While we have specifically considered here the Maxwellian two-species plasma, our arguments hold more generally. For an arbitrary plasma, small amplitude solitary wave solutions exist generically for phase velocities either slightly above or slightly below v^* where v^* is any root of $\kappa^2(v)$.

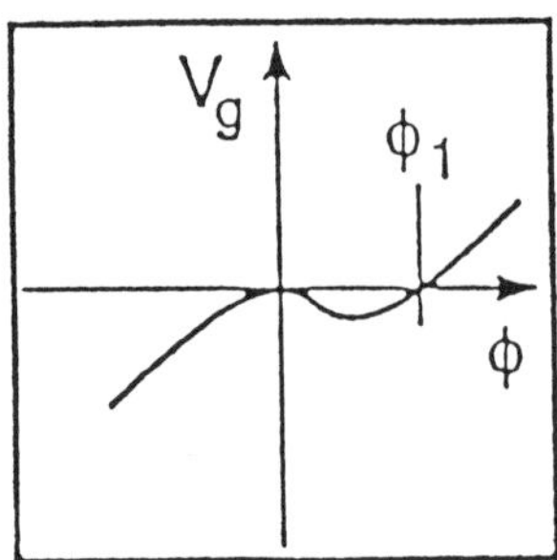
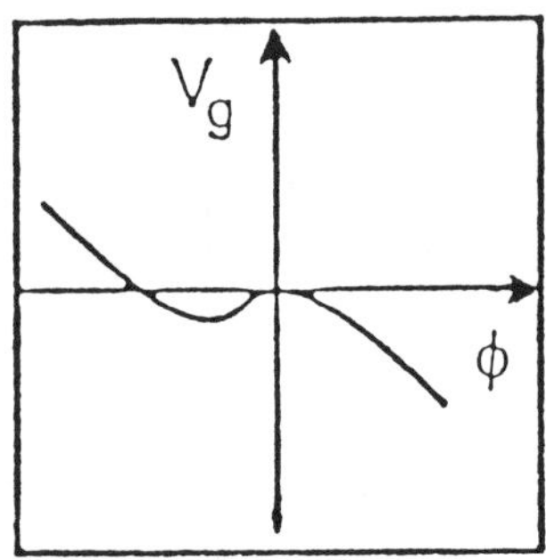

Figure 4. The local form of the pseudo-potential $V_g(\phi)$ for κ^2 small and negative for the two cases $V^{(3)}(0) > 0$ and $V^{(3)}(0) < 0$ respectively.

7. SUMMARY

In this paper we have summarized some of the results on exact nonlinear wave solutions to the Vlasov-Poisson-Ampère equations that we have obtained recently using modern methods of bifurcations theory, nonlinear dynamical systems analysis, and nonlinear functional analysis. The purpose was two-fold: first, to report the results and, second, to emphasize the usefulness of the methods of modern nonlinear analysis. The results reported—some previously published, others not—prove that, in addition to well known large amplitude periodic and solitary waves, there exist exact undamped spatially-periodic and solitary waves of arbitrarily small amplitude in collisionless plasmas. Because these small amplitude periodic and solitary waves were shown to exist arbitrarily close to spatially-uniform Vlasov equilibria they have implications concerning the relaxation—or more importantly the nonrelaxation—of certain classes of small perturbations of these equilibria. The fact that there are at least some such small amplitude spatially periodic nonlinear wave solutions that do not damp shows that the widely accepted implications of Landau's linear analysis [Landau (1945)] are incorrect for some cases where they previously were thought to apply. The nonlinear analysis also established the mathematical connection between the solitary waves and the periodic waves via a generic transcritical bifurcation, and the physical connection between these two classes of waves through the corresponding smooth transition between the types of equilibria from which each class bifurcates.

In the spirit of this conference on relativistic and nonlinear effects in plasmas, we have—through these applications—attempted to demonstrate the value of modern nonlinear methods in a direct frontal attack on some long-standing fundamentally nonlinear problems that arise in plasma kinetic theory. Using the simple BGK form for the traveling wave distribution function, the infinite dimensional Hamiltonian dynamical system was reduced to a one-dimensional bifurcation problem for the amplitudes of the periodic wave solutions and to a two-dimensional Hamiltonian dynamical system describing both the periodic and solitary wave solutions. The careful application of nonlinear techniques has been most efficacious for these problems, and has emphasized that the delicate Vlasov-Maxwell equations are fundamentally nonlinear and resistant to naive perturbation techniques based on the linearized form of the equations.

ACKNOWLEDGEMENTS

The first author (M. B.) gratefully acknowledges ongoing support from the NASA Graduate Student Researchers Program, Grant No. NGT-53643. The second author (J. P. H.) also thanks the NASA Graduate Student Researchers Program for past support under Grant No. NGT-50183.

REFERENCES

Bernstein, Ira B., John M. Greene and Martin D. Kruskal (1957). Exact Nonlinear Plasma Oscillations, *Physical Review*, **108**, 546–550.

Buchanan, M. L. and J. J. Dorning (1990). Bifurcation Theory of Solitary Waves in Vlasov Plasmas, *Transactions of the American Nuclear Society*, **61**, 166.

Case, K. M. (1959). Plasma Oscillations, *Annals of Physics*, **7**, 349–364.

Case, K. M. (1977). Plasma Oscillations, *Physics of Fluids*, **21**, 249–257.

Cesari, Lamberto (1976). "Functional Analysis, Nonlinear Differential Equations, and the Alternative Method" in *Nonlinear Functional Analysis and Differential Equations*, (L. Cesari, R. Kannan and J. D. Schuur, Eds.), Lecture Notes in Pure and Applied Mathematics, Vol. 19. Marcel Dekker, New York.

Chow, Shui-Nee and Jack K. Hale (1982). *Methods of Bifurcation Theory*. Springer Verlag, New York.

Dawson, J. (1961). On Landau Damping, *Physics of Fluids*, **4**, 869–874.

Guckenheimer, J. and P. Holmes (1983). *Nonlinear Oscillations, Dynamical Systems, and Bifurcations of Vector Fields*. Springer Verlag, New York.

Holloway, James Paul (1989). *Longitudinal Traveling Waves Bifurcating from Vlasov Plasma Equilibria*. PhD Dissertation in Engineering Physics, University of Virginia.

James Paul Holloway and J. J. Dorning (1989). Undamped Longitudinal Plasma Waves, *Physics Letters A*, **138**, 279–284.

Holloway, James Paul and J. J. Dorning (1990). Nonlinear but Small Amplitude Plasma Waves in *Modern Mathematical Methods in Transport Theory*, (W. Greenberg, Ed.), Birkhauser (in press).

Landau, L. (1946). On the Vibrations of the Electronic Plasma, *Journal of Physics*, **10**, 25–34.

Marsden, J. and A. Weinstein (1982). The Hamiltonian Structure of the Maxwell-Vlasov Equations, *Physica*, **4D**, 394–406.

Morrison, P. J. (1980). The Maxwell-Vlasov Equations as a Continuous Hamiltonina System, *Physics Letters*, **80A**, 383.

Van Kampen, N. G. (1955). On the Theory of Stationary Waves in Plasmas, *Physica*, **21**, 949–963.

Vlasov, A. (1945). On the Kinetic Theory of an Assembly of Particles with Collective Interaction, *Journal of Physics*, **9**, 25–40.

DYNAMICS OF CAVITONS IN STRONG LANGMUIR TURBULENCE*

D. F. DuBois, Harvey A. Rose and David Russell
Theoretical Division and Center for Nonlinear Studies**
Los Alamos National Laboratory, Los Alamos, NM, 87545

ABSTRACT

Recent studies of Langmuir turbulence as described by Zakharov's model
will be reviewed. For parameters of interest in laser-plasma experiments and for
ionospheric HF heating experiments a significant fraction of the turbulent energy
is in nonlinear "caviton" excitations which are localized in space and time. A
local caviton model will be presented which accounts for the nucleation-collapse-
burnout cycles of individual cavitons as well as their space-time correlations.
This model is in detailed agreement with many features of the electron density
fluctuation spectra in the ionosphere modified by powerful HF waves as measured by incoherent scatter radar. Recently such observations have verified a
prediction of the theory that "free" Langmiur waves are emitted in the caviton
collapse process. Observations and theoretical considerations also imply that
when the pump frequency is slightly lower than the ambient electron plasma
frequency cavitons may evolve to states in which they are ordered in space and
time. The sensitivity of the high frequency Langmuir field dynamics to the low
frequency ion density fluctuations and the related caviton nucleation process
will be discussed.

**Research supported by USDOE.

I. INTRODUCTION

There are many linear instabilities in plasmas which result in the excitation of intense Langmuir waves. An important subgroup of such instabilities are the radiation-induced parametric instabilities which have been studied extensively since 1965.[1,2] In the last decade important progress has been made in understanding the nonlinear, and usually turbulent state to which these instabilities evolve. It has become increasingly clear that the older approaches involving weak turbulence theory, as one extreme, and the wave-breaking of traveling Langmuir waves, as another limit, are not adequate, especially on the longer time scales associated with ion motion. Instead, a new paradigm is emerging involving the concepts of what is commonly called strong Langmuir turbulence (SLT) theory. This theory has its roots in the seminal work of V. E. Zakharov[3] who developed a compact mathematical model of SLT and concluded that localized collapsing Langmuir states could play a central role in the turbulent state.

In recent years it has been possible to carry out long time computer solutions[4−8] of Zakharov's model equations, suitably modified to treat the effect of various types of driving sources. This research has led to a "global" view of how the turbulent state is sustained by the balance of driving sources and the dissipation resulting from the transfer of energy from the collapsing electrostatic fields to accelerated electrons.

The new paradigm has received support from various experiments including beam driven laboratory experiments,[9] laser-plasma experiments[10,11] and recently from detailed ionospheric modification experiments.[12] The power spectra of turbulent fluctuations measured by Thomson scatter radars from the modified ionosphere provides detailed information concerning the dynamics of the elementary excitations, the "cavitons" and the "free modes" which are important in SLT. An important part of the SLT scenario is the controlling effect of low frequency density fluctuations on the localization of the Langmuir fields. Laser-plasma experiments,[10,11] which demonstrated various aspects of the nonlinear coupling of stimulated Raman scattering (SRS) and stimulated Brillouin scattering (SBS), provided useful tests of the predictions[13] that the ion sound waves from SBS would have a controlling effect on the Langmuir waves from SRS. The temporal signatures of the Thomson scatter signals from these two types of fluctuations where consistent with the evolution to collapse of the Langmuir waves in regimes where the levels of ion sound waves were low enough to permit the growth of intense Langmuir waves.

The research program at Los Alamos has concentrated on SLT driven by various radiation sources. These include the ponderomotive sources appropriate to SRS and SBS. Our most comprehensive studies have involved long wavelength electric field drivers with frequencies near the electron plasma frequency.[14] This method of driving is appropriate for the turbulence induced in the ionosphere, near the reflection density, by HF pump (or heater) waves of ordinary polarization. In a quiescent plasma such drivers can excite the well-known parametric decay instability[1] (PDI) or the modulational instability[15,16] (MI) (sometimes called the oscillating two stream instability.)

The results of this research on SLT driven near critical density can be summarized as follows:

1.) States of SLT can be excited for heater (pump) intensities only marginally above the threshold for parametric instabilities. Thus, for example, we expect the ionospheric heating experiments, which are estimated to be well above the threshold for these parametric instabilities will be in the SLT regime.

2.) In these states of SLT a significant part of the energy in high frequency density fluctuations is contained in <u>localized states</u> in the case of strong ion sound wave damping which is appropriate to the ionosphere.[4,5,14] These localized states, which we will call cavitons, consist of a high frequency Langmuir field trapped in a self-consistent density cavity (i.e., density depletion). The dynamics of these cavitons will be a major concern of this paper. It is important to emphasize that these localized states are <u>not</u> wavepackets of plane linear Langmuir waves, but new nonlinear Langmuir states and consequently cannot be described by perturbation arguments such as weak turbulence theory based on Langmuir wave states satisfying the linear dispersion relation.

3.) This state of SLT is sustained by a local nucleation process, (see 6. below), and not by linear parametric instabilities.[4,7,8] The developed turbulent state is <u>stable</u> to the excitation of these global parametric processes because of the level and localized nature of the turbulent fluctuations. Parametric instabilities may play a role in the transient excitation of the SLT state from <u>quiescent</u> initial conditions; for ionospheric parameters this would be the first ms following the turn-on of the heater.

4.) The localized states are trapped in self-consistently evolving density wells which collapse to small dimensions because of the dominance of the nonlinear ponderomotive force over the linear pressure force. The evolution from nucleation to collapse is discussed by Rose and Weinstein[17] and the collapse process follows the self-similar scaling discussed by several Soviet authors.[3,18]

5.) As the caviton's spatial dimension decreases to the order of 5-10 electron Debye lengths, λ_{De}, the electrostatic energy trapped in the caviton is rapidly given up in the acceleration of electrons resulting in the sudden dissipation or "burnout" of electrostatic energy. The interaction of cavitons by the exchange of hot electrons appears to be a weak effect. Even in regimes where there is significant energy in free modes (see (7.)) the burnout of cavitons is the dominant source of energy dissipation.

6.) The electrostatic burnout process leaves an empty density cavity, no longer supported by a ponderomotive force, which then evolves as a free, possibly nonlinear, ion sound pulse. These residual ion density wells provide <u>nucleation</u> centers for the excitation of new collapsing cavitons.[4,7,8] For strong ion sound damping, the burntout density wells relax in place. The radiation of ion sound waves following collapse is an important interaction mechanism for cavitons.

7.) The collapsing cavitons emit and absorb propagating Langmuir waves.[14] This provides another interaction mechanism for cavitons. Under some conditions the free mode-caviton interaction may be an important source driving caviton nucleation. The free modes generated by collapse have been observed in ionospheric heating.[12] These free modes can interact with one another by familiar wave-wave processes. They also can influence the nucleation of localized states.

8.) When the driving frequency exceeds the background plasma frequency, i.e. for overdense driving, the caviton cycles of nucleation-collapse-burnout can become stable limit cycles.[14] Under some conditions the cavitons can also settle into stable periodic spatial patterns. These ordered states of high spatio-temporal correlation have distinct signatures in the power spectra of the turbulent fluctuations.

Highlights of this research have been reported in several short articles.[4,5,7,8]

A longer article, ref. 14, is devoted mainly to the regime of ionospheric heating which greatly reduces the "volume" of the potentially very large parameter space of SLT which is considered, although what remains is still very rich in phenomena. Other applications such as laser-plasma interactions (e.g., see Rose, DuBois and Bezzerides[13,19]) involve many of the same or related phenomena and will be mentioned here briefly. The same conditions of excitation near the critical density in weak density gradients as considered for ionospheric heating might be approximated in long scale length laser produced plasmas with weak collisionality or in laboratory microwave-plasma experiments with sufficiently long-lived plasmas uneffected by boundaries. We will not treat such applications in detail here.

Here we will try to sketch out the major features of this strong turbulence scenario. The physical setting of HF heating of the ionosphere is ideal for observing SLT phenomena. We believe the SLT theory represents the most credible description of the experimental facts of ionospheric heating. This SLT approach represents a significant departure from the accepted or conventional ideas associated with parametric instabilities and weak turbulence cascades. Our most complete current understanding of SLT is based on simulations of a homogeneous, isothermal model described by Zakharov's equations.[3] This situation is best realized in ionospheric modification for early times (several ms) after heater turn on before large scale (several m) density and temperature fluctuations have had time to develop.

A more detailed comparison of the SLT theory to ionospheric modification experiments is given in reference 14.

2. ZAKHAROV'S MODEL OF NONLINEAR LANGMUIR WAVE-ION SOUND WAVE INTERACTIONS

The calculations to be reported here are based on solutions of Zakharov's model[3] of Langmuir wave-ion sound wave interactions. These are formulated in terms of the slowly time varying envelope field $\bar{E}$ (x,t) of the total electrostatic field $\underline{E}_{TOT}$ (x,t), where

$$\underline{E}_{TOT}(\underline{x}, t) = \frac{1}{2}\underline{\tilde{E}}(\underline{\tilde{x}}, \tilde{t})exp[-i\omega_p\tilde{t}] + c.c \tag{2.1}$$

where $\omega_p^2 = 4\pi e^2 n_0/m_e$ where n_0 is the mean plasma electron density. It is assumed that

$$|\tilde{\partial}_t\underline{\tilde{E}}| << |\omega_p\underline{\tilde{E}}| . \tag{2.2}$$

The total ion density is written as

$$n_{TOT} = n_0 + \tilde{n} \tag{2.3}$$

where $\tilde{n}$ is the fluctuation about the mean density; the spatial average of $\tilde{n}$ is then zero.

The equations of Zakharov's model are:

$$\underline{\tilde{\nabla}} \cdot \left[i(\tilde{\partial}_t + \tilde{v}_e\bullet) + \frac{3}{2}\,\omega_p\lambda_D^2\tilde{\nabla}^2 - \frac{1}{2}\,\frac{\tilde{n}}{n_o}\,\omega_p \right]\underline{\tilde{E}} = \underline{\tilde{E}}_0 \cdot \underline{\tilde{\nabla}}\left[\frac{\tilde{n}}{2n_0}\,\omega_p \right] \tag{2.4a}$$

$$\left[\tilde{\partial}_t^2 + 2\tilde{\nu}_i \bullet \tilde{\partial}_t - c_s^2 \widetilde{\nabla}^2\right] \tilde{n} = \frac{1}{16\pi m_i} \, \widetilde{\nabla}^2 |\underline{\tilde{E}} + \underline{\tilde{E}}_0|^2 \tag{2.4b}$$

where $\underline{\widetilde{\nabla}} \times \underline{\tilde{E}} = 0$. Here λ_D is the electron Debye length and $c_S = (\eta \, T_e/m_i)^{1/2}$ is the ion acoustic speed which is often expressed in terms of specific heat parameters. $\underline{\tilde{E}}_0$ is the possibly time-dependent pump which is assumed to be spatially uniform. This is the "heater" field in ionspheric modification experiments. Tildes are used to denote conventional dimensional quantities to distinguish them where necessary from dimensionless quantities introduced below.

The damping operators $\tilde{\nu}_e \bullet$ and $\tilde{\nu}_i \bullet$ which are nonlocal in coordinate space are local in Fourier space. In Fourier space it is also simple to include a weak background geomagnetic field $\underline{B}_0$ based on the modified Bohm-Gross dispersion relation for Langmuir waves.[14,20]

$$\omega(k)^2 = \omega_p^2 + 3k^2 v_e^2 + \omega_c^2 sin^2 \, \theta \tag{2.5}$$

where $v_e^2 = T_e/m_e$, $\omega_c = e \, B_o/mc$ and θ is the angle between $\underline{B}_o$ and $\underline{k}$.

In this paper we adopt the convention for spatial Fourier transforms:

$$E(\underline{k}) = (L)^{-D} \int d^D x \; exp[-i\underline{k} \cdot \underline{x}]E(\underline{x}) \tag{2.6}$$

where L is the linear dimension of the system and D is the dimensionality of space.

The ionospheric heater or "pump" field $\underline{\tilde{E}}_0(t)$ is included, ignoring pump depletion, by assuming that the spatially uniform, $k = 0$, Fourier component is a <u>given</u> function. We will generally take $\underline{\tilde{E}}_0(t)=\underline{\tilde{E}}_0 \exp\left[-i\tilde{\omega}_0 t\right]$ where $\tilde{\omega}_0 = \omega_H - \omega_p$ is the difference between the heater (or pump) frequency and the average plasma frequency.

The Langmuir wave dumping term is taken to be collisional damping plus Landau damping.

$$\tilde{\nu}_e(\tilde{k})/\omega_p = \tilde{\nu}_c/\omega_p + \sqrt{\pi/8}e^{-3/2}(k_D/\tilde{k})^3 exp - (k_D^2/2\tilde{k}^2) \tag{2.7}$$

for $\tilde{k} < 0.3 k_D$; this function is continued smoothly to increase as $\tilde{k}^2$ for large $\tilde{k}$. The latter step is necessary in order to arrest collapse at small scales as discussed by Zakharov and Shur[21] and Russell et al.;[4,8] it is essential for numerical resolution. This damping is an ad hoc addition to the model which is justified by comparing with particle in cell simulations[22–25] which show nearly complete dissipation of the trapped electrostatic field at the burnout stage of collapse. For the work reported here, where we treat heater intensities well above the collisional thresholds for parametric instabilities, we will take $\tilde{\nu}_c = 0$. This is valid provided all physically important rates are much larger than $\tilde{\nu}_c$.[14]

For ionospheric conditions we expect the ratio of electron to ion temperatures, T_e/T_i, to be of order unity for early times after the onset of heating. Fluid description of the ion density response is then expected to be quantitatively inaccurate because of the important role of Landau damping on ions.[26] Since kinetic simulations of the ion response are prohibitively expensive for the problems we treat here we have adopted the following strategy: We use the fluid description of (2.4b) but the sound velocity c_S and the ion Landau damping used in this equation are chosen to coincide with the least damped poles of the linear kinetic response. Using this procedure we find for $\tilde{k} \ll k_{D_e}$ that $\tilde{\nu}_i(\tilde{k})/\tilde{\omega}_i(\tilde{k}) = \nu_i$ where $\tilde{\omega}_i(k) = \tilde{k}(\eta\, T_e/m_i)^{1/2}$. The values of ν_i and η are found from the least damped roots of the full kinetic dispersion relation.[14]

We have found the qualitative features of the nucleation process to be unaffected by the values of ν_i in the regime $0.9 > \nu_i > 0.4$ for systems driven well above the nucleation threshold discussed in Sec. 3.

It is well-known,[3] that the <u>linearized</u> form of these equations contains the parametric decay instability (PDI)[1] and modulational instability[15,16] (MI or OTSI) of the pump wave. Furthermore, when weak turbulence analysis[27,28] is applied to these equations it yields the usual wave kinetic type of equations which lead to the weak turbulence cascade. However, the validity conditions for the weak turbulence approximations are very limiting.[28]

We have studied examples of the solution of these equations for parameters relevant to ionosphere heating in which the system is initially excited by a linear parametric instability and evolves to a state of SLT.[14] In this paper, however, we will consider only the developed turbulent state.

In carrying out numerical solutions of these equations it is convenient to use dimensionless <u>untilded</u> quantities which are related to dimensional tilded quantities in the following way:[27]

$$t \equiv \frac{2}{3}\left(\eta\frac{m_e}{m_i}\right)\omega_p\tilde{t}$$

$$x \equiv \frac{2}{3}\left(\eta\frac{m_e}{m_i}\right)^{1/2}\frac{\tilde{x}}{\lambda_D} \tag{2.8}$$

$$E = \frac{1}{\eta^{1/2}}\left(\frac{m_i}{\eta m_e}\right)^{1/2}\left(\frac{3}{16}\frac{\tilde{E}^2}{4\pi n_0 T_e}\right)^{1/2}$$

$$\eta = \frac{3}{4}\left(\frac{m_i}{\eta m_e}\right)^{1/2}\frac{\tilde{n}}{n_o}$$

The scaled equations then have the familiar simple form

$$\underline{\nabla}\cdot[i(\partial_t + \nu_e\bullet) + \nabla^2 - n]\underline{E} = \underline{E}_0\cdot\underline{\nabla}n + S_R \tag{2.9a}$$

$$[\partial_t^2 + 2\nu_i\bullet\partial_t - \nabla^2]n = \nabla^2|\underline{E} + \underline{E}_0|^2 + S_B \tag{2.9b}$$

In the scaled units there is a residual mass ratio dependence which occurs only in the scaled damping rate which is obtained from (2.7) as follows:

$$\nu_e(k) = (3/2)M\tilde{\nu}_e(2/3M^{-1/2}kk_D)\omega_p^{-1} \tag{2.10}$$

in terms of the scaled wavenumber k and $M = m_i/\eta m_e$. This residual mass ratio dependence reflects the ratio of the parametric instability space and time scales which increase with M and the mass ratio independent dissipation scale. A similar formula applies to $\nu_i(k)$.

Here we have added source terms S_R and S_B in (2.9a) and (2.9b) which arise when the turbulence is driven by the SRS interaction and the SBS interaction, respectively. We will not give the specific formulae here but refer the reader to ref. 13.

Note that in dimensional units Landau damping becomes significant for $\tilde{k}$ $> 0.2k_D(k_D = \lambda_D{}^{-1})$. Thus in dimensionless units this dissipation becomes significant for k greater than the dissipation scale k_d:

$$k > k_d \simeq (0.2) \cdot \frac{3}{2}\left(\frac{m_i}{\eta m_e}\right)^{1/2}$$

Since the dynamics of the decay instability involves $\tilde{k}$'s on the scale of $\tilde{k}_*$ $= (2/3)(\eta m_e/m_i)^{1/2} \, k_D$ we need Fourier components at <u>least</u> as small as this, if the parametric processes are important, and this sets the linear dimension of the simulation cell to be $\tilde{L}_x = \tilde{L}_y > 2\pi/k_*$. In dimensionless units $k_* = 1$ and $L_x = L_y > 2\pi$. The number of Fourier modes must be sufficient to probe deep within the dissipation range of $k_{max} >> k_d$ in order to resolve collapse. This sets a limit on the value of M which can be accommodated in a reasonably sized simulation of say 128×128 Fourier modes in two dimensions. In view of these limitations we have chosen M = 1836 for our simulations. In ref. 14, Section 3, we discussed the scaling of physical quantities with the mass ratio. This scaling allows us, at least roughly, to translate the simulation results to the larger mass ratios.

The validity conditions for Zakharov's model have been discussed elsewhere[3,27] and include the conditions.

$$\frac{|\tilde{E}|^2}{4\pi n_0 T_e} << 1 \tag{2.11a}$$

$$\frac{\tilde{n}}{n_0} << 1 \tag{2.11b}$$

A discussion of the degree to which these conditions are satisfied in our numerical simulations is given in reference 14.

Our simulations are carried out on a 128×128 square grid of sides $L_x = L_y = 2\pi$, with periodic boundary conditions in x and y. In physical units this implies $L_x = L_y = 404 \, \lambda_{D_e}$ and a grid-point spacing $\Delta x = \Delta y = 3.15 \, \lambda_{D_e}$. The Debye wavenumber in these units is 64.3 and the maximum wavenumber is 91. Spot checks with a dealiased code with a nominal 256×256 grid were used to confirm the validity of our simulations. Typically the spectrum $\langle |n(k)|^2 \rangle$ decreases by 4 orders of magnitude between the k values for which the spectrum peaks and the largest k values. The test of temporal and spatial resolution is energy conservation as expressed by the balance between the average dissipation and injection rates.[14]

3. THE LOCAL CAVITON MODEL

The accumulated evidence from many computer simulations of equations (2.4) shows that, at least for moderate to strong ion acoustic wave damping, ν_i $\gtrsim 0.1$, the strongly turbulent system is dominated by caviton "events" which are localized in space and time. Snapshots such as Figure 1 show the localized nature of $|\underline{E}(\underline{x},t)|^2$ and $n(\underline{x},t)$ as functions of $\underline{x}$ for given t. The power spectra $|\underline{E}(\underline{k},\omega)|^2$, which we will discuss in detail below, also have signatures of localized states. The envelope field $\underline{E}(x,t)$ in this case can be modeled by a sum over events i:

$$\underline{E}(x,t) = \sum_{i=0}^{N(t)} \hat{\underline{\varepsilon}}_i(\underline{x}-\underline{x}_i, t-t_i) + \underline{E}_{nonlocal}(\underline{x},t) \tag{3.1}$$

Here a caviton event i is localized at the space time point $\underline{x}_i$, t_i. The single event function $\hat{\underline{\varepsilon}}_i(\underline{x},t)$ has its maximum at $x=0$, $t=0$ with a spatial width $\underline{\delta}_i(t)$ and a temporal width or lifetime τ_i; from the simulations we find this lifetime to be of the order of 0.05 to 0.1 ms for ionospheric parameters. At a given time t, the number of events $N(t)$ which contribute to the sum in (3.1) are those for which $0< |t-t_i| < \tau_i$ which is clearly proportional to the volume of the system if the cavitons are roughly uniformly distributed. For example, if the portion of the heated volume observed by the radar is ionospheric heating experiments is $(200\ \text{m})^3$, the mean caviton spacing is 0.25 m which is about 50 λ_{De} as observed in our simulations, and accounting for the time scales of the caviton cycles as observed in simulations, we find $N(t) \sim 10^7$ which is crude but representative. In (3.1) the term $\underline{E}_{nonlocal}(\underline{x},t)$ represents the nonlocalized or free mode part of the envelope field which is relatively negligible for systems driven with heater (or pump) frequencies near or slightly below the ambient electron plasma frequency.

This local caviton model can be put into a more formal setting by introducing the instantaneous vector eigenfunctions $\underline{e}_\nu(\underline{x},t)$ of the operator on the left hand side of (2.9a). These satisfy (for $B_0 = 0$)

$$\underline{\nabla} \cdot \left[\lambda_\nu(t) + \nabla^2 - n(\underline{x},t)\right] \underline{e}_\nu(\underline{x},t) = 0 \tag{3.2}$$

where $\lambda_\nu(t)$ is the corresponding instantaneous eigenvalue and $\underline{\nabla} \times \underline{e}_\nu = 0$. These are nothing more than the Langmuir modes in a nonuniform density background. In ordinary units this can be written as

$$\underline{\nabla} \cdot \left[\omega_\nu - \omega_p(x,t) + \frac{3}{2} \lambda_D^2 \omega_{po} \nabla^2\right] \underline{e}_\nu(\underline{x},t) = 0 \tag{3.3}$$

where

$$\omega_p(x,t) = \omega_{po}\left(1 + \frac{1}{2}\frac{n(xt)}{n_0}\right)$$

is the spatially fluctuating plasma frequency. Thus we can relate λ_ν in scaled units to ω_ν in ordinary units:

$$\lambda_\nu(t) = \frac{\omega_\nu - \omega_{po}}{\omega_{po}} \cdot \frac{2}{3}\left(\frac{m_i}{\eta m_e}\right)^{1/2} \tag{3.4}$$

The complete description of these states for an arbitrary $n(\underline{x},t)$, especially for $D \geq 2$ is beyond our capability. In $D = 1$ it is relatively easy to compute these states from an arbitrary realization of $n(x,t)$ obtained from the complete numerical simulation of (2.9a,b).[7]

The form of the density fluctuation field $n(x,t)$ can have a profound effect on the eigenstates of the Langmuir field. Most importantly, some of these states are localized in a density minimum (i.e. depressions). Roughly the condition for a localized state to occur in a density depression ñ (ñ<o) and spatial extent $\tilde{\delta}$ is

$$\left|\frac{\tilde{n}}{n_o}\right|\left(\frac{\tilde{\delta}}{\lambda_D}\right)^2 \simeq O(1) \tag{3.5}$$

which can be satisfied for example for $|\tilde{n}/n_o| \sim 10^{-3}, (\delta/\lambda_D) \sim 30$. This condition is similar to the conditions on the depth and width of a potential well which can sustain a bound state of the Schroedinger equation of quantum mechanics. The density wells which trap these localized states might arise from initial background density fluctuations, from density wells remaining from earlier collapse events or from density fluctuations driven by some instability such as SBS.

In the case of SBS-generated ion sound waves, we can sometimes regard these density fluctuations as being periodic in space with a wavelength corresponding to the fastest growing SBS mode. The eigenstates $\underline{e}_\nu(x,t)$ in this case can be regarded as one dimensional Bloch waves with lattice wave vector $\underline{k}$ and with eigenvalues $\lambda_{\underline{k},\nu}(t)$ which lie in bands, labelled by the index ν, just as in solid state physics.[13,29] As the periodic density fluctuation grows exponentially in time due to the SBS instability the Langmuir mode eigenfuctions $\lambda_{\nu k}(t)$ change in time. If a stimulated Raman instability is simultaneously excited, the SRS frequency matching condition, $\Delta\omega = \omega_{\underline{k}_o}^{laser} - \lambda_{\underline{k},\nu}(t) - \omega_{\underline{k}_o-k}^{scatteredlight}$ $= 0$, can only be satisfied instantaneously for a given Langmuir Bloch mode with lattice wave vector $\underline{k}$. In fact if $\left|\frac{d\Delta\omega}{dt}\right| = \left|\frac{d}{dt}\lambda_{k,\nu}\right| > \gamma_R^2$, where γ_R is the instantaneous SRS growth rate, then it can be shown[13,29] that the SRS instability is <u>detuned</u> by the growing SBS ion sound wave and SRS is suppressed. Experiments carried out at the NRC Laboratory in Canada[10,11] appear to be consistent with this scenario. The experiment by Villeneuve et al.[11] verified the theoretical prediction[13] that a "seeded" SBS instability could suppress SRS. In other parameter regimes where SRS is not suppressed the (weaker) SBS ion sound wave may still impose its spatial periodicity on the SRS Langmuir eigenfunctions.[13] These envelope eigenfunctions have a periodic array of maxima of $|\underline{E}|^2$ (or $|\underline{e}_{k,\nu}|^2$) which have a finite ponderomotive force (PMF). This periodic PMF causes a periodic array of density wells to develop in which the Langmuir waves are trapped and can be driven to collapse. [Note the ponderomotive driven periodic density wells do not coincide in general with the density minima of the SBS sound waves but they have the same periodicity.] In Fig. 2, taken from ref. 13, we show typical spatial configurations of $|\underline{E}|^2$ and n, before and after collapse and burnout.

The impulsive time signature of the Thomson-scattering signal from Langmuir fluctuations in the experiment of Walsh et al.,[10] is consistent with the collapse of SRS driven Langmuuir fluctuations. The large ion density fluctuations remaining from the burntout cavitons then act as seeds for the subsequent strong SBS pulse. In Fig. 2 the time signatures of the Langmuir and ion sound fluctuations, obtained from numerical solutions of the SRS-SBS driven Zakharov equations, are shown. Further evidence consistent with the controlling effect of SBS generated ion sound waves on the SRS process is found in the experiments of Baldis et al.[30]

The electric field envelope of the Zakharov equations (2.9a,b) can be resolved in the complete set of states $\underline{e}_\nu(\underline{x},t)$:

$$\underline{E}(x,t) = \sum_\nu h_\nu(t)\underline{e}_\nu(\underline{x},t)exp(-i\omega_0 t) \tag{3.6}$$

(In an infinite system the sum may imply an integral over continuum states.) The equation of motion for the amplitudes $h_\nu(t)$ is readily found from (2.9a) to be

$$i\dot{h}_\nu(t)+(\omega_0-\lambda_\nu(t)h_\nu(t)+i\sum_{\nu\prime}\left[\langle\underline{e}_\nu \blacklozenge \nu_e\underline{e}_{\nu\prime}\rangle + \langle\underline{e}_\nu \blacklozenge \dot{\underline{e}}_{\nu\prime}\rangle\right]h_{\nu\prime} = \underline{E}_0\cdot\langle\underline{e}_\nu|n\rangle \tag{3.7}$$

Here we have taken the $\underline{e}_\nu$ to be a complete orthonormal set with

$$\langle\underline{e}_\nu \blacklozenge \underline{e}_{\nu\prime}\rangle = \int dx\underline{e}_\nu^*(\underline{x},t)\cdot\underline{e}_{\nu\prime}(\underline{x},t) = \delta_{\nu\nu\prime} \tag{3.8}$$

and have used the notation

$$\langle\underline{e}_\nu \blacklozenge \dot{\underline{e}}_{\nu\prime}\rangle = \int d\underline{x}\,\underline{e}_\nu^*(\underline{x},t)\cdot\frac{d}{dt}\,\underline{e}_{\nu\prime}(\underline{x},t) \tag{3.9a}$$

$$\langle\underline{e}_\nu \blacklozenge \nu_e\underline{e}_{\nu\prime}\rangle = \int d\underline{x}\int d\underline{x}'\underline{e}_\nu(\underline{x},t)\cdot\nu_e(\underline{x}-\underline{x}')\underline{e}_{\nu\prime}(\underline{x}',t) \tag{3.9b}$$

$$\underline{E}_0\cdot\langle\underline{e}_\nu|n\rangle = \underline{E}_0\cdot\int d\underline{x}\,\underline{e}_\nu^*(\underline{x},t)n(\underline{x},t) \equiv S_0 \tag{3.9c}$$

To understand the various terms in (3.7) first consider the case where $n(\underline{x})$ is independent of time and therefore $(d/dt)\,\underline{e}_\nu(\underline{x}) = 0$. Then the amplitude h_ν is driven directly by the source term $\underline{E}_0\cdot\langle\,\underline{e}_\nu|n\,\rangle$. If $\underline{e}_\nu$ were a plane wave state proportional to $exp\,i\,\underline{k}\cdot\underline{x}$; then this source term is proportional to $\underline{E}_0\cdot\underline{k}\,n(\underline{k})$ the so called direct conversion source term.[14,31] However, the important states are the localized states. The coefficient $\langle\,\underline{e}_\nu \blacklozenge \nu_e\,\underline{e}_{\nu\prime}\rangle$ couples states because of the nonlocal nature of the Landau damping. This term becomes important in the time dependent case only in the burnout phase. Note that by introducing the spatial Fourier transform of the eigenstates $\underline{e}_\nu(k,t)$ we can write

$$\langle\underline{e}_\nu \blacklozenge \nu_e\underline{e}_{\nu\prime}\rangle = \sum_k \underline{e}_\nu^*(\underline{k},t)\cdot\underline{e}_{\nu\prime}(\underline{k},t)\nu_e(\underline{k}) \tag{3.10}$$

For $\nu=\nu'$ this certainly becomes important in the burnout phase. For $\nu \neq \nu'$ this is less important if one of the states is not localized - e.g., noncollapsing - or is localized at a different space-time point.

In the time-dependent case of interest the coefficient $\langle e_\nu \blacklozenge \dot{e}_{\nu\prime}\rangle$ can provide a coupling between rapidly collapsing states, say ν' and a nonlocalized state ν. This is one of the mechanisms responsible for the excitation of the "free mode" states observed in the spectra. We will return to this detail below.

We note that a subset $\{i\}$ of the states $\{\nu\}$ are localized at $x=x_i$ in density depressions of $n(\underline{x},t)$ and some of these, which have the proper symmetry to couple to the pump $\underline{E}_0$, evolve to collapse. This subset of states can be viewed as local ground states of the "potential" $n(\underline{x},t)$.

We can now make a tentative connection between the localized event functions $\hat{\underline{\varepsilon}}_i$ ($\underline{x} - \underline{x}_i$, $t - t_i$) of (3.1) and the subset $\{i\}$ of localized eigenstates. It is reasonable to identify

$$\hat{\underline{\varepsilon}}_i(\underline{x} - \underline{x}_i, t - t_i) = \underline{\varepsilon}_i(\underline{x} - \underline{x}_i, t - t_i)exp(-i\omega_0 t) \qquad (3.11a)$$

where

$$\underline{\varepsilon}_i(\underline{x} - \underline{x}_i, t - t_i) = h_i(t)\underline{e}_i(\underline{x}, t) \qquad (3.11b)$$

The contributions from the remaining nonlocalized states in the set $\{\nu\}$ make up the term $\underline{E}_{nonlocal}$ (x,t) in (3.1).

The eigenstates $\underline{e}_i(\underline{x},t)$ are in a sense the natural basis or coordinates for describing the turbulent system. Unfortunately, they can only be obtained by first solving (2.9) for n($\underline{x}$,t). In spite of this they are conceptually useful and some observed properties of the turbulence can be related to general properties of these states. In effect the use of the states $\underline{e}_i(\underline{x}$,t) represents a huge reduction in the effective dimensionality of the problem. While we use $(128)^2$ Fourier modes for the simulation there may be of the order of 10 collapse sites in the cell and therefore roughly 10 localized states.

We have gained useful insight into the nature of these eigenstates and their connection to the observed turbulence by considering the scalar Zakharov model. In this model E(r,t) and E_0 are scalar fields and in place of (2.9) we have

$$\left[i(\partial_t + \nu_e\bullet) + \nabla^2 - n(\underline{x}, t)\right] E(\underline{x}, t) = E_0 n(\underline{x}, t) \qquad (3.12a)$$

$$(\partial_t^2 + 2\nu_i \bullet \partial_t - \nabla^2)n(\underline{x}, t) = \nabla^2|E_0 + E(\underline{x}, t)| \qquad (3.12b)$$

In this model only spherically symmetric collapsing cavitons are allowed and the three dimensional problem for an isolated collapse reduces to one in which E and n depend only on the radial coordinate r. This scalar model has several properties in common with the physical three-dimensional vector model (2.9): threshold and maximum growth rate for the modulational instability, collapse scaling exponents which are discussed below, no threshold energy for collapse and the possible failure of a density well to support a localized eigenstate.

Spherical symmetry is imposed by representing all fields in terms of the Fourier modes sin ($k_\ell r$), $k_\ell = \pi\ell/r_0$, $\ell = 1,2,--$, with r_o chosen large compared to a typical caviton size. In these scalar studies we have observed for $\nu_i(k)/k = 0.9$ that at the nucleation site, E(r,t) is dominated by its projection, $h_0(t)$, on the localized ground state $e_0(r,t)$. In nucleation $e_0(r,t)$ remains localized; at every time step $e_0(r,t)$ can be computed from n(r,t). Here we will adopt a simplified model in which $h_0(t)$ is evolved neglecting the excited state contributions $\nu' \neq 0$ in (3.7). The density evolves according to (3.12b) with the ponderomotive force replaced by $\nabla^2|E_0 + h_0(t)e_0(r,t)|^2$. The solution to this model is insensitive to boundary conditions (i.e., the choice of r_0).

Let us restrict our attention to the case where ν_i is large enough so that after burnout, the relaxing density fluctuation is essentially nonpropagating. Immediately after burnout, energy absorption is minimal because the eigenvalue is large and negative, implying a far from resonant coupling to S_0 (see (3.9c)). The ponderomotive force is negligible, and the density fluctuation evolves according to the acoustic Green's function. A simple model for this phase of the dynamics is obtained by replacing the rhs of (3.12b) by $I\delta(t) \nabla^2\delta^3(\underline{x})$, where the "impulse"

$I \equiv \int dt \int d\underline{x} |\varepsilon(\underline{x},t)|^2 = \int |h_0(t)|^2 \, dt \approx \langle |h_0|^2 \rangle \tau$. In three dimensions, the response of n is

$$n(x,t) = I \cdot G(|x|/t)/t^4 \qquad (3.13a)$$

where

$$G(\rho) = \frac{\nu_i}{\pi^2 \rho} \frac{d^2}{d\rho^2} \left[\frac{\rho}{(1+\rho^2)^2 - 4\rho^2(1-\nu_i^2)} \right]. \qquad (3.13b)$$

Even though n is evolving self similarly, $\underline{e}_0$ is not. The figure of merit, μ, for the ground state is simply expressed in terms of the width w, $w(t) \sim t$, of n, and its depth, d, $d \sim I/t^4$, as $\mu \sim dw^2 \sim I/t^2$. If μ is too small, there is no localized state. Since 3D solitons are unstable, as t increases, either enough energy will be accumulated so that another collapse follows, or the bound state will be lost. In the latter case, the bound state will be localized in the immediate vicinity of the expanding density fluctuation, until just before the bound state is lost. So that during the time when energy is being injected, one may be able to ignore the coupling between states localized at different collapse sites. At a particular collapse site there is a lowest lying localized state which has a nonzero source. Excited states at the same site typically have a smaller source term because they are oscillatory while $\underline{E}_0 \; n(\underline{x},t)$ is essentially uniform in direction. Also at a particular site there may only be a small number of localized states. This motivates the study of a model for the evolution of a caviton in a previously existing density fluctuation - a process we call caviton nucleation - in which only one localized state is present. For a given ion fluctuation the lowest lying state, $\underline{e}_0$, with nonvanishing source, S_0, (we shall call it the ground state) is calculated from (3.9c), h_0 is evolved according to (3.7) without the coupling to other amplitudes.

In Figure 3 we show some typical results from the scalar model driven by a spatially uniform field E_0 at the plasma frequency ($\omega_0 = 0$). For a range of E_0, a stable nucleation cycle is observed, with a complete cycle over the interval $0 < t < t_c$. We expect that in a turbulent environment of other such nucleation sites, the strict periodicity of this cycle may be lost, but at a given site there may be strong correlations over a few cycle times for strong ion acoustic damping.

In fact, for overdense drive where the drive frequency ω_H is less than the background plasma frequency ω_p, i.e. for $\omega_0 < o$, we have found[14] that over a range of driving amplitudes and ion wave damping strengths, ν_i, these cycles become stable limit cycles.

This is easily understood from the nucleation picture: For $\omega_0 < 0$ the relaxing density well remaining from a previous burnout comes earlier into resonance with the pump and therefore at a relatively deeper depletion compared to the $\omega_0 = 0$ case. Thus at the time of closest resonance $\lambda_\nu \sim \omega_0$ the eigenfunction $\underline{e}_\nu(\underline{x},t)$ is more confined. The caviton cycle presumably will be more stable and less effected by neighboring cavitons in this more confined caviton cycle. We expect more rapid caviton cycles, i.e., smaller τ_c, with less energy carried into collapse and this is verified by simulations. The overdense drive $\tilde{\omega}_0 < 0$ is much more efficient in the nucleation of cavitons.

An important observation of the scalar, local caviton model discussed in Section 3, is that the single event functions $\hat{\underline{\varepsilon}}_i(\underline{x} - \underline{x}_i, t - t_i)$ are phase locked to the pump. Thus it was more convenient to replace these functions in (3.1) by $\exp - i\omega_0 t \; \varepsilon_i(\underline{x} - \underline{x}_i, t - t_i)$; that is to explicitly separate out the pump phase.

[See also (3.11).] Another way to look at the problem is to rewrite (2.9) in terms of $\hat{E}(\underline{x},t = \exp(-i\omega_0 t)\,E(\underline{x},t)$, i.e., to envelope around the pump frequency. As an equation for $\hat{E}$ the equations are autonomous, i.e., the drive term has no explicit time dependence but an additional term, $\omega_0\,\underline{\hat{E}}(\underline{k},t)$, appears on the left hand side of (2.9a). The result of this is that the single caviton spectra for $\omega_0 \neq 0$, $|\underline{\hat{\varepsilon}}(\underline{k},\omega)|^2$, have their spectral energy mainly for $\omega < \omega_0$. The eigenvalue trajectories versus time have the property that they reverse near the resonance $\lambda(t)\leq\omega_0$ where the PMF reverses the relaxation of the density well.

In Figure 4 we show snapshots of the lowest two eigenstates $e_0(r,t)$ and $e_1(r,t)$ and the density $n(r,t)$ as they evolve during one of these cycles for a case where $\omega_0 < 0$. In Figure 3a we show the time evolution of $|E(r=0,t)|^2$ and $n(r=0,t)$, in Figure 3b the ground state eigenvalue $\lambda_0(t)$, the velocity Φ_0 of the phase of the amplitude $h_0(t) = |h_0(t)|\,\exp i\Phi_0(t)$, and in Figure 3c the electrostatic energy in the caviton $|h_o(t)|^2$. At the beginning of the cycle, t=0, the deep density well is relaxing from the previous burnout. From Fig. 3a we see that the peak $|E|^2$ occurs at about t=0.22 followed by its rapid burnout due to dissipation. The density well reaches its maximum depth shortly after at t $\simeq$ 0.235. The maximum spatial extent of the eigenfunction $\delta(t)$ occurs earlier at t $\simeq$ 0.15 which is also the time at which the well depth, $n(r=0,t)$, is shallowest. The well then deepens under the action of the PMF increasing the confinement of the eigenfunction. The eigenvalue $\lambda_o(t)$ approaches the pump frequency $\omega_0 = 0$, this causes a rapid increase in $|h_0(t)|^2$ as the mode frequency approaches resonance with the pump frequency. This rapidly increases the PMF and as the density well deepens again $\lambda_0(t)$ again decreases rapidly during collapse. This illustrates what we believe to be the typical behavior: As the relaxing density well becomes shallower and broader its eigenvalue approaches resonance with the pump causing a rapidly increasing PMF which initiates the next collapse. It is, of course, important that the state remain localized so that it maintains a significant PMF. Under some conditions for D$\geq$2 a localized bound state can be lost, i.e., λ_0 crosses zero before sufficient PMF is built up to initiate collapse. For the D=3 scalar model discussed here we find a finite nucleation threshold [Rose et al.[19]]. For E_0 below this value the cycle cannot be maintained even for an isolated caviton.

These scalar model calculations have been used to deduce scaling laws[19,14] for the dependence of quantitites such as the electrostatic energy taken into collapse, the caviton cycle time and the maximum caviton radius as a function of driver strength E_0. We refer the reader to references 19 and 14 for details. As an example, we find that the peak electric field in the caviton collapse process scales as $|E|^2_{peak} \sim E_0^{-2}$; the peak field in the caviton decreases with increased driver strength. The caviton cycle period scales as $\tau \sim E_0^{-1}$ and the maximum caviton radius, $\Delta \sim E_0^{-1}$. From these we estimate the space-time density of cavitons to scale as $(\tau\Delta^3)^{-1} \sim E_0^4$.

This isolated collapse model is oversimplified in several ways, one is its neglect of the turbulent environment of the collapse site. The density fluctuation $n(r,t)$ was constrained so that there was locally no net change in particle number - i.e., $\int dr\, r^2 n(r,t)=0$ - i.e., the local averaged plasma frequency is the same as the global average plasma frequency which is the zero of frequency in our envelope approximation. This constrains all bound or localized states to have $\lambda_\nu(t)<0$. However, locally on a scale larger than a single caviton but macroscopically small there can be fluctuations in the background plasma frequency away from the global average. If the local plasma frequency is different from zero this is equivalent to replacing n(r,t) in (5.12a) by δn_0 + n(r,t) and bound states can occur if $\lambda_\nu < \delta n_0$. Since there are local domains or "patches" of positive and negative δn_0 we conclude that in a large multicaviton system localized states can occur for $\lambda_\nu(t) < (\delta n_0)_{max}$ where $(\delta n_0)_{max} > 0$ and depends on the parameters E_0, ω_0, etc. which determine the turbulent state. Simulations with $\omega_0 > 0$ are consistent with this picture. The "localized" states for $\delta n_0 > 0$ are not strictly localized from the mathematical point of view; their eigenfunctions may have extended tails which are exponentially small but do not decay at large distances. Such states are better described as resonance states as discussed at the end of this section.

The self similar scaling of the parameters of the eigenstates during the collapse phase are well-known [e.g., Galeev et al.[18]]. A self-similar ansatz for $\underline{e}_i(\underline{x},t)$ can be written

$$\underline{e}_i(\underline{x},t) \;=\; \frac{1}{\delta^{D/2}(t)} \; \underline{\hat{e}}(\underline{x}\delta^{-1}(t)) \tag{3.18a}$$

where $\hat{e}(\zeta)$ is the normalized shape function of the collapsing state.

$$\int d\zeta |\hat{e}(\zeta)|^2 = 1 \tag{3.18b}$$

The collapses observed in our D = 2 simulations are not cylindrically symmetric as can be seen in Figure (4.2), but have a pancake shape with the narrow direction mainly aligned along the pump polarization (the x direction in Figure 1). The aspect ratio of the y dimension to the x dimensions appears to be in the range of 2 to 3. The ansatz of (3.18) implies that although the aspect ratio is not necessarily unity all dimensions scale with $\delta(t)$. Simulations of collapse in D = 3 for isolated cavitons[22] and for multicaviton states[6] display these pancake cavitons whose orientation arises either as a result of initial conditions or by coupling to a drive source such as we have used.

The scaling of the parameters of the self-similar supersonic collapse depend on the spatial dimension D as follows:

$$\delta(t) = (t_c - t)^{2/D} \tag{3.19}$$

$$\lambda(t) \sim \delta^{-2}(t) \sim (t_c - t)^{-4/D} \tag{3.20}$$

where t_c is the time of collapse. The self-consistent density behaves as

$$n(\underline{x},t) \;=\; -\frac{1}{\delta^2(t)} \; G(\underline{x}\delta^{-1}(t)) \tag{3.21}$$

where $G(\underline{x}\delta^{-1}(t))$ is a shape function related self-consistently to $\hat{e}(\underline{x}\delta^{-1}(t))$. The scalar model collapse behavior is consistent with these scalings for D = 3.

We can sometimes use the self-similar formula (3.18) for $\underline{e}_i(\underline{x},t)$ in other regimes -- e.g., nucleation but where $\delta(t)$ does not satisfy the scaling of (3.19). Examples of the evolution of $\delta(t)$ in the scalar model are shown in Figure 3.

In the scalar model results, presented above, the contributions from nonlocalized states or from localized excited states are neglected. Such localized excited eigenfunctions have one or more nodes in the region of the confining density well and would be expected to couple less efficiently to the pump in the overlap integral of (5.9c). Localized excited states which evolve to collapse are not observed in the simulations.

For an isolated density well the localized state eigenfrequencies lie below those of the nonlocal (or continuum) states. In the "patchy" model with fluctuating domains of differing mean plasma frequency, it appears that the eigenvalue ranges of localized and nonlocalized states may overlap.

The definition of a localized state as one of the localized eigenfunctions $\underline{e}_i(\underline{x},t)$ is actually too restrictive. A wavepacket of nonlocalized states whose λ_ν lie just above the localization limit can be a resonance state analogous to those known in quantum mechanical scattering theory if it is a superposition of states with a sharp peak in the density of states. Such a resonance state will appear spatially coherent and localized for a time $\Delta t \sim (\Delta \lambda)^{-1}$ where $\Delta \lambda$ is the frequency width of the resonance. The resonance state will then have a ponderomotive force over a time which may be sufficiently long to depress the density so that a strictly localized eigenstate can again appear. The effective source terms $\mathbf{E}_0 \bullet \langle e_\nu n \rangle$ are nearly the same for all the states comprising the resonance. A narrow resonance state therefore cannot be distinguished from a strictly localized eigenstate and so the definition of the states $\underline{e}_i(\mathrm{x},t)$ which are identified in (3.11) should be extended to include such narrow resonance states. It can be shown that for a sufficiently narrow resonance the equation of motion for its amplitude $h_i(t)$ is indistinguishable from the equation of motion discussed above for a localized state. The existence of such resonances is another reason why localized states appear to exist for $\lambda_i > 0$. As discussed above, the random density environment of a caviton can also raise the eigenvalue limit for localization to positive values. Such localized states are also best described as resonance states.

4. POWER SPECTRA OF TURBULENT FLUCTUATIONS IN THE LOCAL CAVITON MODEL

It is well-known that the power spectrum of electron density fluctuation, $\tilde{n}_e(\underline{x},t)$, contains information concerning the elementary excitations of a plasma and can be measured by incoherent Thomson scatter techniques. For frequencies ω near theelectron plasma frequency ω_p (or its negative) the ions cannot respond significantly and so one can relate the electron density fluctuation directly to the total electric field

$$4\pi e\,\tilde{n}_e(\tilde{\underline{x}},t) = \tilde{\underline{\nabla}} \cdot \tilde{\underline{E}}_{TOT}(\tilde{x},t) = \frac{1}{2}\tilde{\underline{\nabla}} \cdot \tilde{\underline{E}}(\tilde{x},t)e^{-i\omega_p t} + c.c. \qquad (4.1)$$

From this it is easy to see that the power spectrum of n_e is directly related to that of the envelope field. For $\tilde{\omega} \simeq \omega_p$ we have

$$\begin{aligned}
(4\pi e)^2 |\tilde{n}_e(\tilde{\underline{k}},\omega)|^2 &= (1/4)|\tilde{\underline{k}} \cdot \tilde{\underline{E}}(\tilde{\underline{k}},\tilde{\omega}-\omega_p)|^2 \\
&= (1/4)\lambda_D^2 |\tilde{\underline{k}} \cdot \tilde{\underline{E}}(\tilde{\underline{k}},\tilde{\omega}-\omega_p)|^2 (4\pi n_0 T_e)^{-1}
\end{aligned} \qquad (4.2)$$

To obtain the low-frequency spectrum associated with the ion line we note that for $\tilde{\omega} << \omega_p$ that quasineutrality is obtained so $n_e(\tilde{\underline{k}}\ \tilde{\omega}) \simeq n_i(\tilde{\underline{k}},\tilde{\omega}) = n(\tilde{\underline{k}},\tilde{\omega})$ and the low frequency electron density spectrum is obtained directly from the density fluctuation which appears in the Zakharov model:

$$|\tilde{n}_e(\underline{\tilde{k}},\tilde{\omega}|^2 = |\underline{\tilde{n}}(\underline{\tilde{k}},\tilde{\omega})|^2 \tag{4.3}$$

The power spectra can be found by taking the temporal Fourier transform of (3.1).

$$\underline{E}(\underline{k},\omega)_T = \sum_i^{N(T)} expi[\underline{k}\cdot\underline{x}_i - \omega t_i]\underline{\varepsilon}_i(\underline{k},\omega)_t + \underline{E}(\underline{k},\omega)_{nonlocal} \tag{4.4a}$$

where

$$\underline{\varepsilon}_i(\underline{k},\omega)_t = \int_{t-T/2}^{t+T/2} dt(t' - t_i)\underline{\varepsilon}_i(\underline{k},t' - t_i)\ exp[i\omega(t' - t_i)] \tag{4.4b}$$

is the single event Fourier coefficient. If we make the assumption, that all events are uncorrelated we obtain the power spectrum as a sum over single event spectra:

$$\langle|\underline{E}(\underline{k},\omega)_t|^2\rangle = \sum_{i=1}^{N(T)} |\underline{\varepsilon}_i(\underline{k},\omega)|^2 \equiv N(T)\langle|\underline{\varepsilon}(\underline{k},\omega)|^2\rangle \tag{4.5}$$

The important effect of correlations will be discussed below.

Simulation parameters can be chosen so that the collapse events are so well separated in time that we were able to compute the single event spectra $|\underline{\varepsilon}(\underline{k},\omega)|^2$ for this case. These spectra, shown in Figure 5, have a surprisingly rich structure including the following features: 1.) Essentially all the spectra energy occurs for $\omega < \omega_0$; in this case $\omega_0 = 0$; 2.) There are well defined peaks in the spectrum; 3.) For increasing k, i.e., increasing $k\lambda_D$, the peaks for more negative ω become relatively more important; 4.) The position of the maximum shifts in a step-wise fashion (see inset to Figure 6) where $-\omega_{max} \sim k$, i.e., $\omega_p - \tilde{\omega}_{max} \sim \tilde{k}\ c_S$; 5.) There is a weak "free mode" peak at $\omega \sim k^2$, i.e., roughly at the Bohm-Gross frequency. These single event spectral properties are similar to those shown in Figure 6 for the power spectra from multicaviton states in a magnetic field.[14] Caviton-caviton correlations also can have a strong effect on the spectral shape and will be discussed below.

We have obtained some insight into the sources of this structure from the scalar model discussed in Sec. 3. A realization of the single event spectrum $|\underline{\varepsilon}(\underline{k},\omega)|^2$ for the scalar model is constructed by taking the temporal transform of the function $f(t) = h_0(t)e_0(k,t)$ for $0<t< \tau_c$ and $f(t) = 0$ for $\tau_c <t<T$ where T is chosen to give the desired frequency resolution and $e_0(k,t)$ is the spatial Fourier dransform of the numerically obtained $e_0(r,t)$. The results are shown in Figure 7. These model spectra contain the features listed above for the $D=2$ Zakharov model (Fig. 6) except for 4.) and 5.).

The predominance of negative frequencies arises because the phase velocity $\dot{\Phi}$ in Figure 4 is predominantly negative; this in turn is related to the negativity of the eigenvalue $\lambda(t)$. In this model calculation Eq. 3.7 reduces to $ih_0 - \lambda_0 h_0 = S_0 \equiv E_0\langle e_0|n\rangle$ since we are neglecting coupling to excited states. Then if we write $h_0 = |h_0|\ exp\ i\Phi_0$ we see that $\dot{\Phi}_0$ is related to λ_0 by

$$-\dot{\Phi}_0 = \lambda_0(t) + S_0(t)|h_0|^{-1}cos\Phi_0 \qquad (4.6)$$

which is the equation used to compute Φ_0 in Figure 3.

The peaks arise from a modulation of the spectrum with an angular frequency $\Delta\omega = 2\pi/\tau_c$ where τ_c is the caviton lifetime as measured by the width in time of the total electrostatic energy pulse, $|h_0(t)|^2$, shown in Fig. (3). This is the same kind of modulation that arises in the spectrum of a single square wave pulse. Let us assume that the early-time spectrum from low-duty-cycle experiments can be identified with the incoherent average $\langle|\varepsilon(k,\omega)|^2\rangle$ of single-event spectra. In this averaged spectrum the individual spectral peaks may be smeared out but it is reasonable to assume that the half-power frequency width is approximately that of the first and strongest maximum of the single-event spectra. Application of this argument to the data of Djuth, Gonzales, and Ierkic[32] in this regime — e.g., their Figure 4 — leads also to a value $\tau_c \sim$ $0.1{\pm}0.05$ ms. The k dependence in this model arises from the k dependence of the eigenfunction, $e_0(k,t)$; for a localized state with $k\delta_0(t)$ we expect that $e_0(k,t) \sim \delta_0^{D/2}(t)$ exp $(-k\,\delta_0(t))$. For increasing k, smaller values of $\delta_0(t)$ are favored and these correspond to more tightly collapsed states with more negative frequencies.

The free-mode peak observed in the spectra Figures 5 and 6 is, of course, not seen in this simplified scalar model calculation since it neglects all of the excited states. The free mode excitation in the scalar model is discussed below. The behavior near $\omega = 0$, including the shift of the maximum peak with k, is also different in the scalar model than in the $D=2$ Zakharov simulation of Figure 5. The inclusion of excited states in the scalar model brings the results into closer qualitative agreement. In the case of overdense drive $\omega_0 < 0$, where free modes are only weakly excited, this single state model agrees well with the complete Zakharov simulations, such as those shown in Fig. 11.

The Fourier transform of E(r,t) is given by $E(k,\omega) \simeq \int$ dt exp i $(\omega t +$ $\Phi_0(t))$ $|h_0(t)|$ $e_0(k,t)$ where Φ_0 is the phase of $h_0(t)$. For large negative ω we can make asymptotic estimates based on a stationary phase evaluation of the time integral; the stationary phase points t = t_S occur approximately where ω $= -\dot{\Phi}_0(t_S)$. From Fig. 2 we see that the ground state has large negative phase velocities where $\dot{\Phi}_0(t) \to -\lambda_0(t)$ as t $\to t_c$ and can satisfy the stationary phase condition. In this temporal regime one comes closest to the self-similar scaling for the collapsing state: $e_0(rt) = \delta_0(t)^{-D/2}\,\Psi_0(r/\delta_0(t))$ with the spatial Fourier transform $e_0(k,t) = \delta_0(t)^{D/2} \cdot \int d\xi$ (exp $-\, ik\delta_0\xi)\Psi_0(\xi)$. The self similar behavior is $\delta_0(t) \sim (t_c - t)^{2/D} \sim \lambda_0(t)^{-1/2}$ where t_c is the collapse time. Using these behaviors in the stationary phase evaluation of the Fourier integral we find the asymptotic behavior $|E(k,\omega)|^2 \sim |\omega|^{-(1+3D/4)}$ as $\omega \to -\infty$. This asymptotic prediction is observed in the $D = 2$ vector Zakharov simulations and in the scalar simulations to an accuracy of 10%.

The spectrum obtained from incoherent scatter of radars (ISR) from the modified ionosphere is the result of about 10^8 events and the question of caviton-caviton correlations becomes important. In general if the events are correlated, (4.5) is replaced by

$$\langle |\underline{E}(\underline{k},\omega)_t|^2 \rangle = \sum_i^N |\underline{\varepsilon}_i(\underline{k},\omega)|^2$$

$$+ \sum_{i \neq j}^N \sum^N exp\; i[\underline{k}\cdot(\underline{x}_j - \underline{x}_i) - \omega(t_j - t_i)]$$

$$\cdot \underline{\varepsilon}_i(k,\omega)\cdot \underline{\varepsilon}_j^*(k,\omega) \tag{4.7}$$

The second or "coherent" term in this equation has N^2 potential contributions and so could have a potent effect on the spectrum if events are correlated.

A possible model of the effect of correlations is to assume that the dispersion in the single event transform $\underline{\varepsilon}_i(k,\omega)$ is small from event to event. This is true in the scalar model calculations and has been seen in the full vector simulation; especially in the case of overdense drive. Formally, this assumption is equivalent to writing $\underline{\varepsilon}_i(k,\omega) = \bar{\underline{\varepsilon}}(\underline{k},\omega) + \delta\underline{\varepsilon}_i(\underline{k},\omega)$ where $\bar{\underline{\varepsilon}}$ is the average over many events and $\delta\underline{\varepsilon}_i=0$. If we assume $|\delta\underline{\varepsilon}_i|^2 << |\underline{\bar{\varepsilon}}|^2$ we can write

$$\langle |\underline{E}(\underline{k},\omega)|_t^2 \simeq \langle|\rho(\underline{k},\omega)|^2\rangle|\bar{\underline{\varepsilon}}(\underline{k},\omega)|^2 \tag{4.8a}$$

where

$$\rho(\underline{k},\omega) = \sum_i^{N(T)} exp\; i[\omega\, t_i - \underline{k}\cdot\underline{x}_i] \tag{4.8b}$$

This quantity is just the space-time Fourier transform of the caviton event density

$$\rho(\underline{x},t) = \sum_i^{N(T)} \delta^D(\underline{x} - \underline{x}_i)\delta(t - t_i) \tag{4.9}$$

Eq. (4.8a) shows that in this approximation the single event spectrum $|\bar{\underline{\varepsilon}}(\underline{k},\omega)|^2$ is modulated by the correlation or structure factor: $\langle|\rho\,(\underline{k},\omega)|^2\rangle$.

5. COMPARISON WITH OBSERVED POWER SPECTRA

Recently Cheung et al.[12] have performed modification experiments at Arecibo which emphasized low duty cycle heating sequences; the heater was turned on periodically in pulses of duration up to 50 ms with an interpulse period (IPP) of 150 ms. The Thomson radar diagnostic pulses of duration 1.1 ms were also turned on and off with the same period and the delay time between the onset of the heater pulse and that of the radar pulse was varied. The comparisons between these observations and the SLT theory can be summarized as follows:

1.) For short radar delay times the many-pulse averaged observed spectra agree in detail with the smoothed simulation spectra. The main energy containing portion of these spectra occur for $\tilde{\omega} < \omega_H$ and there is a free mode peak for $\tilde{\omega} > \omega_H$.

Examples of experimental spectra from Cheung et al.[12] are shown in Figure 8. Examples of smoothed simulation spectra including geomagnetic field effects are shown in Figure 6 taken from DuBois et al.[14]

2.) The local caviton model accounts for the $\tilde{\omega} < \omega_H$ spectral features as arising from the nucleation-collapse-burnout caviton cycle as discussed in Section 4.

3.) Associated with each caviton cycle a nearly free Langmuir wave packet is radiated away from each caviton site. This is discussed in Section 6. The free mode peak occurs at a frequency $\tilde{\omega}_f = \omega_p[1 + (3/2)(\tilde{k}\lambda_p)^2 + (1/2) (\tilde{\omega}_c/\omega_p)^2 \sin^2\theta]$ associated with a free Langmuir wave for which $\tilde{\omega}_f > \omega_H$ yet is a distinct signature of the collapse process.

4.) Recently Djuth[33] et al. have presented evidence that these short delay-time spectra are produced in a thin turbulent layer within 100 m of the reflection altitude of the heater. This is consistent with the parameters of our simulations for which we assumed the altitude of the first standing wave maximum of the heater in a smooth ionosphere density profile with a scale length of about 50 km. [This determines the value of $\tilde{\omega}_0 = \omega_H - \omega_p$.]

It is important to realize that these short delay time observations, following the onset of the heating pulse, are completely at odds with the predictions of weak turbulence theory (WTT). WTT fails to predict the spectral shape, the altitude dependence of the turbulence or its angular dependence (on the direction of $\underline{k}$ observed by the radar relative to the geomagnetic field.) SLT on the other hand accounts for all of these observations qualitatively and at least semi-quantitatively.

The observations of Djuth[33] et al. indicate that the turbulent layer begins to spread downward from the reflection altitude at about 50 ms, following the onset of heating ultimately extending to an altitude 1 to 2 km below reflection after 100 ms or so. Changes in the spectrum occur on similar time scales.

In several sets of observations[34,31,32,12,33] sharp spectral peaks are observed to develop as the time delay of the radar pulse is increased following the onset of the heater pulse. In Figure 9, a 50-pulse average spectrum is shown in which the radar pulses occur 29 ms following the onset of a 30 ms heating pulse with a 150 ms IPP. This spectrum shows features observed in many previous long-time experiments[33] consisting of a main "decay line" peak lying about 3.0 ± 0.5 kHz below the heater frequency and two "cascade" peaks lying further below the heater frequency by 10.0 ± 0.5 kHz and 16.0 ± 0.5 kHz, respectively. This approximate "1:3:5" pattern of frequency displacements is sometimes associated with a weak turbulence cascade.[33]

6. CAVITON CORRELATIONS

The question to concern us next is whether such a spectral pattern can be explained in terms of caviton correlations? We have found that for overdense driving where $\tilde{\omega}_H < \omega_p$ (or $\omega_0 < 0$) stable cycles can be found with cavitons in ordered spatial arrays. In the case of overdense driving the turbulent state depends on initial conditions[14] or at least the memory of initial conditions decays more slowly in time than in cases where $\omega_0 > 0$.

We wish to present here an example of a correlated caviton state which has interesting properties and has led us to consider a class of perfectly correlated, alternating lattice models. This example is one of a class of simulations in which cavitons initially arranged in a regular array of sites persist at these sites, their cycles become very stable and become phase locked to one another in various temporal patterns.

In Figure 10 we show the initial locations of two density cavities which resulted from previous collapses and in which the initial electric field fluctuation is set to zero. This is the initial state for a simulation with $E_0 = 1.2$ and $\omega_0 = -25$ in scaled units. This value of ω_0 corresponds to a domain in physical units which is about 1% overdense. We note that for these parameters the system is stable to modulational instability but yet a energetic nonlinear state is sustained for long times because of these initial conditions. As time increases the cavitons on each of the two lattices evolve into a strict limit cycles with period $\tau = 0.59$ with the cycles of the two lattices becoming $\tau/2$ out of phase with one another.

In Figure 10 we also show the extended spatial periodicity implied by our periodic boundary conditions. This shows that the simulation is equivalent to two interpenetrating square lattices in which the $\bullet$ positions are all in phase but are $\tau/2$ out of phase with the $+$ positions which themselves are all in phase. Examples of the computed spectra are shown in Figure 11 for $\theta = 0$ and $\theta = 45°$.

In Figure 11 the single event spectra, the $|\bar{\varepsilon}(\underline{k},\omega)|^2$ in (4.8a), are also shown for the same values of $(\underline{k},\theta)$. These are easy to isolate from a single cycle at a given site. According to (4.8a) the single event spectrum modulates the spectrum of the structure factor. Comparison of the complete spectra and the single caviton spectra in Figure 11 verifies this.

In Figure 12 we show the ion line spectra obtained from the same simulations. Note, in the cases in which the plasma line spectra in Figure 12 has peaks at odd multiples of $2\pi/\tau$ (i.e. cases a and d) structure, the ion line consists of a symmetrical peak around $\omega = 0$ (corresponding to $\tilde{\omega} = \omega_H$) and two displaced peaks at $\omega = \pm 2\pi/\tau$. These two displaced peaks are shifted by exactly the same frequency as the "decay line" peak in the plasma line spectra in Figure 11 a and d. This correlation of frequency shifts in the plasma line and ion line spectra has often been offered as evidence of the parametric decay instability; here we see that the same spectral correlation can arise from completely different physics!

These considerations lead us to postulate models in which the cavitons tend to order themselves in a regular three-dimensional lattice in overdense regions of the ionosphere. In the ionosphere application, because the heater field varies within the radar observed region, due to the antenna pattern and the altitude dependent Airy pattern, the spacing, a, of cavitons in the lattice varies on a scale large compared to a. In regions of most intense E_0, a will be smallest. The orientation of the lattice may also vary within the observed region. Our scalar model simulations show that for $E_0 \simeq 0.5$ V/m the maximum isolated caviton size is about 40 λ_D so we might expect an intercaviton spacing a > 80-100 λ_D or about 40-50 cm. At Arecibo the radar wavenumber is about 2π (35 cm)$^{-1}$ which means that low order Bragg scattering from such a lattice is possible with small adjustments of the lattice spacing which could arise through variations in E_0. We imagine that the observed region of the heated ionosphere contains domains of ordered cavitons whose lattice spacing and orientation varies from domain to domain so that in some domain the Bragg resonance condition for the radar is satisfied. It is difficult to estimate how likely it is to have such a resonant domain present. These spatially and temporally ordered domains take some time to organize themselves after the onset of heating.

In the simplest interesting model the cavitons are arranged on two identical, interpenetrating lattices. In each lattice the cavitons undergo strict limit cycles with period τ but the cycles in the two lattices are displaced in time by $\tau/2$ as in the simulation of Figs. 10-12. The two lattices are symmetrically oriented with respect to each other in such a way that nearest-neighbor cavitons are members different lattices.

The structure factor $|\rho(\underline{k},\omega)|^2$ for this model was analyzed in detail in Ref. 14. The structure factor as a function of $\underline{k}=2\,\underline{k}_{radar}$, the wavevector observed by the backscatter radar, has resonances when $\underline{k}$ is equal to certain discrete vector values, $\underline{K}_\nu$, known as reciprocal lattice vectors. These satisfy the condition $\exp(i\underline{K}_\nu \cdot \underline{x}_n)=1$ for all caviton locations, $\underline{x}_n$, in the lattice. The magnitude of these vectors is determined by the lattice spacing and structure and can be written as $|\underline{K}|_\nu = (2\pi/a)|A_\nu|$ where a is the smallest lattice spacing and the numbers $|A_\nu|$ are determined by the lattice symmetry. The resonance conditions, $\underline{k}=\underline{K}_\nu$, are called Bragg resonance conditions. The lowest order Bragg condition corresponds to the case where the lattice spacing a is equal to half the radar wavelength; $a=(1/2)\lambda\text{radar}$. In this case it was shown[14] that the frequency spectrum has peaks at $\omega=m(2\pi/\tau)$ where m is an odd, positive or negative, integer. The strength of these peaks is proportional to N_c^2 where N_c is the number of caviton events observed in the time interval of the radar pulse. Higher order Bragg resonances arise when a is a larger multiple of $(1/2)\lambda\text{radar}$. These more loosely packed lattices may have peaks at $\omega=m(2\pi/\tau)$ where m is either odd (as for the primary resonance) or even. Regions of higher E_0 will produce more tightly packed lattices. (Recall that the local caviton model results, summarized in Section 3, predicted that the maximum caviton radius scaled like E_0^{-1}.) We expect stronger radar signals from the regions of stronger E_0 with the primary Bragg resonance. In this case we have the 1:3:5—type spectrum with resonances at $\omega=m(2\pi/\tau)$, m = -1, -3, -5,—and also with m=1, 3, 5—etc. Eq. (4.8) shows that the spectrum is the product of the structure factor, $|\rho(\underline{k},\omega)|^2$ and the single particle spectrum $|\varepsilon(k,\omega)|^2$.

The latter has most of its strength for $\omega < o$ ($\tilde{\omega} < \omega_H$) with only a small overlap of the rgion $\omega > o$, as in the examples in Fig. 11. Thus the complete spectrum, $|E(\underline{k},\omega)|^2$, has prominent peaks for m= -1, -3, -5—plus a weaker peak at m=+1. The latter can be identified with the often observed "anti-Stokes" linc.

In Ref. 14 it was noted that because of the scalings $\tau \sim E_0^{-1}$ and $\Delta \sim E_0^{-1}$ that if we identify the caviton lattice spacing, a, with the maximum caviton radius, Δ, we have $\tau \sim$ a. If the lattice domain is such as to satisfy the lowest Bragg resonance condition, $a = (1/2)\lambda\text{radar} = (\pi/k)$, we conclude that the downward frequency shift of the first decay line, $\Delta\omega = 2\pi/\tau \sim$ k. This has the same k dependence as the ion acoustic frequency and is found, numerically, to be close to the ion acoustic frequency, within a factor of 2. This scaling of the frequency interval of the 1:3:5 spectrum is qualitatively the same as predicted by weak turbulence theory! It is remarkable that the correlated caviton model seems to be able to account for all the observed features of the radar power spectra.

However, there are two problems with this explanation. First, it is observed[35,36] that when the heater is switched off the spectral peaks of the 1:3:5 pattern remain for several ms after switch-off, albeit with decaying amplitudes and altered relative amplitudes. We can not see how the correlations necessary to preserve this spectral structure can be maintained in the correlated caviton model <u>after</u> switch-off of the heater. A second problem with the correlated caviton explanation are the recent observations of Djuth et al.[33] that the 1:3:5 structured spectra arise from turbulence excited 50-100 ms following heater turn on, at altitudes up to 1 to 2 km below the reflection altitudes. At such underdense altitudes, with respect to the undisturbed ionosphere, it is harder to imagine that ionospheric irregularities can produce the <u>overdense</u> domains necessary for correlations.

Our theoretical understanding, based mainly on simulation results and on a new microscopic theory of caviton correlations[37] <u>for the case of overdense driving</u>, can be summarized as follows:

i.) Temporal correlations at a given site devleop, on the time of a cycle period, into stable limit cycles.

ii.) Temporal correlations <u>between</u> sites evolve more quickly than spatial correlations between sites.

iii.) Some spatial patterns appear to be stable equilibrium arrangements with definite temporal correlations between sites. Only temporal correlations where nearest-neighbor caviton cycles are in-phase or $\tau/2$ out of phase appear to be stable.

iv.) Those patterns which are not stable equilibria evolve to stable patterns on an experimentally relevant time scale - say tens of ms.

We have discussed (i) at some length above. An example of (ii) is the simulation discussed in relation to Figure 11 in which cavitons at neighboring sites became anti-correlated in time. Points (iii) and (iv) (as well as other examples of (i) and (ii) are based on simulations and theory which we will not present here. This work is part of our continuing research and will be published elsewhere.[37]

Near reflection altitude (critical density) it may not be difficult for ionospheric irregularities to produce the overdense domains necessary for correlations. If this is the case then why would the 1:3:5 spectral signature associated with caviton correlations <u>not</u> be observed? This could be due to a low probability of overdense domains which satisfy the spatial Bragg resonance condition for the particular radar wavelength.

Domains of coherent cavitons whose lattice spacing is not Bragg resonant for the given radar $\underline{k}$ are more likely to be present (near reflection altitude). Such domains will have weaker resonances at $\omega = 2\pi m\, \tau_d^{-1}$ for <u>all</u> m where the life time of the caviton cycle τ_d varies from domain to domain. The resonance at $\omega=0$ is common to all domains and is not smeared out by domain to domain variations of τ_d as are the resonances for m $\neq$ 0. The $\omega = 0$ resonance can be identified with what is conventionally called the "OTSI line" in ionospheric heating parlance. This line can be extremely narrow if the caviton cycles are long lived; we find $\Delta\omega\alpha\pi(M\tau_d)^{-1}$ where M is the number of cycles in the observation interval.[14] The caviton picture provides the only nonlinear description, which we know of, which is capable of understanding the very narrow width of the "OTSI line" observed in the experiments of Sulzer and Fejer.[38,39]

We believe, then, that it still remains a challenge to explain the 1:3:5 structured spectra which occur in some Arecibo observations 30-100 ms following the onset of heating. The results of Djuth et al.[33] show that the turbulence responsible for these spectral features occurs at significantly underdense altitudes with respect to the undisturbed ionosphere. We have carried out simulations[40] in such underdense regimes where $\omega_H - \omega_p > 3/2\ (k\lambda_D)^2\omega_p$. Again in such regimes caviton collapse effects are strong: the dominant sink of Langmuir dissipation is through caviton collapse and the dominant sources of ion density fluctuations are the burnout density cavities remaining after collapse. We cannot expect WTT to describe such a turbulent state. On the other hand the fraction of Langmuir energy in free modes is significantly higher at these underdense altitudes than at reflection density or in overdense domains. We are investigating the possibility that the beating of the pump with the collapse-enhanced ion density fluctuations can produce a source for free Langmuir modes which would excite the decay (or Stokes) line and the weaker anti-Stokes line. (Similar processes may be possible for other steps in the "cascade.") The theory must take into account the direct excitation of free modes by collapse - (the $M_{\lambda \lambda_0}$ term in (7.4) which follows) - and the scattering of free Langmuir modes from collapse-enhanced ion density fluctuations. The latter process produces an effective damping on free Langmuir modes which tends to counteract the parametric gain which in WTT is supposed to lead to the usual WTT decay cascade. The results of this study will be published elsewhere.[40]

7. RADIATION OF FREE LANGMUIR WAVES BY COLLAPSING CAVITONS

Our studies have shown that the collapse process invariably excites free (or propagating) Langmuir waves. These manifest themselves in the "free mode" peak which occurs in all the power spectra, $|E(\underline{k},\omega)|^2$, which we have computed. As discussed in Sec. 5 there is strong evidence that the free mode peak has been observed in the short time scale experiments of Cheung et al.[12]

These free mode states are extended states whose energy is not localized at a particular point in space. A single collapsing caviton will radiate a wave packet of free modes which spread out and whose amplitudes decay geometrically (as r^{-1}) away from the excitation center of the caviton.

Outside of the spatial region of an isolated collapsing caviton these radiated Langmuir waves are asymptotically free Langmuir waves obeying the dispersion relation of (2.5). In a many-caviton environment, the large density fluctuations generated by collapse distort their propagation. The free mode frequencies appear to approach the dispersion relation (2.5) as k increases. For lower k values the frequencies are shifted to somewhat higher values due to the perturbation of the density fluctuations.

The generation of free modes can be understood in terms of the coupled mode amplitude equations (4.7). For simplicity we ignore the dissipative coupling (4.9b) which is not important for the k values measured by most radars. The coupling between states is given in (4.7) involving the matrix (4.9a)

$$M_{\nu \nu \prime} = i \langle \underline{e}_\nu \, \dot{\underline{e}}_{\nu \prime} \rangle \tag{7.1}$$

By taking the time derivative of (3.2) and using the orthonormality condition (3.8) we can reexpress this as

$$M_{\nu \nu \prime} = i \, \frac{\langle \underline{e}_\nu | \dot{n} \, \underline{e}_{\nu \prime} \rangle}{\lambda_\nu - \lambda_{\nu \prime}} \tag{7.2}$$

where

$$\langle \underline{e}_\nu | \dot{n} | \underline{e}_{\nu \prime} \rangle \equiv \int d\underline{x} \, \underline{e}_\nu^*(xt) \bullet \underline{e}_{\nu \prime}(xt) \, \frac{dn}{dt}(x,t) \tag{7.3}$$

The free modes have a continuum of eigenvalues $\lambda_\nu(t)$ in an infinite space and so we can parameterize them directly in terms of their eigenvalue λ:i.e., $\underline{e}_\lambda$ (x,t). The free modes receive from or give energy to localized states and are driven directly by the heater E_0. In the following we will consider in detail the coupling to a unique collapsing state denoted by the subscript zero. The equation of motion for the amplitude h_λ of a given free mode then follows from (4.7) as

$$i\,\dot{h}_\lambda + (\omega_0 - \lambda)h_\lambda + \int d\lambda' \rho(\lambda')M_{\lambda\lambda'}h_{\lambda'} = -M_{\lambda\lambda_0}\,h_0 + \underline{E}_0 \bullet \langle \underline{e}_\lambda|n\rangle \qquad (7.4)$$

Here

$$M_{\lambda\lambda_0} = i\,\frac{\langle \underline{e}_\lambda|\dot{n}|\underline{e}_{\lambda_0}\rangle}{\lambda - \lambda_0(t)} \qquad (7.5)$$

and $\rho(\lambda')$ is the density of continuum states. The third term on the left hand side of (7.4) involving $M_{\lambda\lambda'}$ involves the scattering of one free mode from another. By considering the equation for

$$(d/dt)\sum_\lambda |h_\lambda(t)|^2$$

it is easy to see that the scattering terms do not change the total free mode energy while the terms on the right hand side of (7.4) do.

In Ref. 14 we have presented detailed numerical results for free mode radiation in the spherical scalar model discussed in Section 3. We found that, in the case of overdense driving, that the coupling to the time dependent collapsing Langmuir ground state, the $M_{\lambda\lambda_0}h_0$ term in (7.4), dominates the dynamic conversion term, $E_0 <e_\lambda|n>$, in that equation. During the early nucleation stage of the caviton cycle, there is relatively little energy, $|h_0|^2$, in the local ground state and the dynamic conversion term is the dominant source of free mode energy. At intermediate times during collapse the coupling to the time dependent ground state is dominant and produces the largest free mode energies of the entire cycle. Deep into collapse the coupling to the ground state is again unimportant because the coupling coefficient, $M_{\lambda\lambda_0}$, is going to zero as the coupling becomes more and more nonresonant. We also showed[14] that the hot electrons emitted during the burnout phase of collapse cannot produce a significant rate of production of free Langmuir waves by Cherenkov radiation.

As mentioned in Section 7, the energy in free modes is a higher fraction of the total Langmuir energy in the case of underdense driving ($\omega_H > \omega_p$). In the regime, $\omega_H - \omega_p > (3/2)(k\lambda_D)^2\omega_p$, the free modes occur at frequencies below the heater frequency, ω_H. The frequencies of localized states, which must be resonance states, apparently are in the range $\omega < \omega_H$ and overlap the free mode range. In these underdense regimes the emission and absorption of free modes by localized caviton states may be an important contribution to the nucleation process.[40]

In the experiments of Cheung et al.,[12] the ratio of the strength of the "collapse continuum" portion of the spectrum for $\omega < \omega_0$ to the strength of the free mode line increases as the time delay following the onset of the heating increases. It is observed that the strength of the free mode line does not change by as much as an order of magnitude while the strength of the collapse continuum increases by several orders of magnitude. This is consistent with the increase of the $\omega < \omega_0$ spectrum due to the onset of temporal correlations, i.e., a signal proportional to N_c^2 rather than N_c. The free mode line is not strengthened by correlations except in the unlikely case that the free mode peak at $\omega = \omega_f$ in the single caviton spectrum coincides with one of the correlation peaks at $\omega = 2\pi m/\tau_c$, which were discussed in Section 8. In addition, we know that free mode emission is weaker from overdense regions where $\omega_0 < 0$ which are likely to be correlated. Note, for example, that the free mode peaks in the single caviton spectra from the correlated simulations of Figure 11 where $\omega_0 = -25$ are not very strong and do not produce prominent peaks in the correlated spectra.

8. CONCLUSIONS

We have discussed some of the accumulating evidence that strong Langmuir turbulence theory explains phenomena observed in laser-plasma interaction experiments and in ionospheric modification experiments. The laser-plasma experiments[10,11] particularly illuminate the sensitive coupling between low frequency density fluctuations and high frequency Langmuir fluctuations. The theory in refs. 13 and 29 appears to describe the major feature of the experiments of refs. 10 and 11 which are transient in nature. In these experiments a short pulse of SRS activity, as in Fig. 2, representing only one generation of collapse, is observed before the level of SBS ion sound fluctuations has grown sufficiently to detune the SRS process. In other experiments, say with longer laser pulses, under some conditions the SBS ion sound wave level may saturate at a level which permits a more or less continuous SRS excitation. A theoretical description of such a long time scale, SRS-excited, state of Langmuir turbulence is very difficult. Such studies are complicated by the need to apply realistic boundary conditions on the Langmuir waves and scattered light waves generated by SRS in an inhomogeneous density profiles.

We have also explored in detail the implications of strong Langmuir turbulence theory for ionospheric heating experiments. The short time scale data from these experiments provide the best test of the theory available today. A major conclusion of our work is that weak turbulence theory (WTT) cannot be valid for the conditions of ionospheric heating. Our conclusion is based, first of all, on extensive numerical solutions of Zakharov's model encompassing many generations of collapsing cavitons. WTT follows under very special conditions from the Zakharov equations. The fact that the numerical solutions are dominated by coherent, collapsing cavitons proves that the nonlinear state is far from the regime of WTT. Recently Payne, Nicholson and Shen[28] have explored in detail the limit of WTT in numerical solutions of Zakharov's equations in one dimension and have established rough criteria for the validity of WTT. These stringent criteria are not satisfied for the conditions of ionospheric heating.

The strong Langmuir turbulence theory has developed on two levels. The first level is based on solutions of Zakharov's model equations. From the properties of these solutions we have proposed the local caviton model which is a more "phenomenological" level. The local caviton model is built on single caviton properties. Cavitons go through cycles of nucleation, collapse and burnout. Associated single caviton properties include their lifetimes (or cycle times) τ_c, the single caviton field fluctuation $\underline{\varepsilon}(\underline{x},t)$ and its power spectrum $|\underline{\varepsilon}(k,\omega)|^2$. These single caviton properties are not necessarily those of isolated cavitons, although the isolated caviton approximation is at least qualitatively useful in many cases. As the driving becomes increasingly overdense ($\omega_0 < 0$) we have evidence that caviton interactions decrease but the residual interactions can lead to coherent caviton states. It is a challenge to understand the mechanism(s) for self-organization of this weakly interacting caviton gas. For these highly correlated states, the name turbulence hardly seems appropriate.

We believe that the qualitative properties of the local caviton model will be those deduced from the Zakharov model. More complete and accurate descriptions of single caviton properties are needed to treat the end stages of collapse and the burnout processes, whereas the nucleation and early collapse stages should be accurately described by the model. We anticipate that these improvements will make quantitative but not qualitative changes in the picture developed in this paper.

We believe that the SLT model has at least three apparent successes in explaining the ionospheric heating observations for early times (<50 ms) following the onset of the heating pulse:

1.) The altitude dependence of the early time plasma line signal is easily explained because the localized caviton states are not tied to the linear dispersion relation, (2.5). Based on the sensitive dependence of the turbulence level on E_0 we concluded[14] that the strongest plasma line signal should occur near the altitude of strongest E_0, which is the first Airy maximum in an undisturbed profile. This is in agreement with the early time observations of Djuth et al.[33]

2.) At these early times following the onset of heating the broad featureless spectrum for $\tilde{\omega} < \tilde{\omega}_H$, observed in many experiments,[12,31,32,33] is explained by the dynamics of local caviton states. This part of the spectrum arises from the caviton cycle of nucleation-collapse-burnout. It is not consistent with WTT.

3.) A new prediction of SLT theory, the free mode peak, has been unambiguously observed by Cheung et al.[12] This arises because of the radiation of free Langmuir waves by collapsing cavitons. Again. this feature is not consistent with WTT.

There is no experimental evidence concerning the dependence of this early time turbulence on the angle θ between the radar $\underline{k}$ and the geomagnetic field. The theory predicts that the same qualitative features would be seen at Tromsø where $\theta \sim 0°$ as observed at Arecibo where $\theta \sim 45°$, but the plasma line signal should be several orders of magnitude stronger at $\theta \sim 0°$. Short time scale experiments have not been carried out at Tromsø.

The SLT theory appears to be able to predict, at least qualitatively, many properties of the sharp spectral features observed at longer delay times following the onset of heating. These predictions depend on the existence of overdense domains of temporally and spatially correlated cavitons. Structures similar to the "decay line," the 3:5:7--"cascade lines" and the anti-Stokes line appear in the SLT spectrum of correlated cavitons provided Bragg resonance conditions, depending on the radar wavelength, are satisfied by some correlated domains. In addition we have new theoretical insight into the caviton-caviton interaction mechanisms which establish these correlations.[37] However, inspite of the attractiveness of this scenario, we believe this is probably not the mechanism which produces the above mentioned sharp spectral features. The reasons for this conclusion were given in Section 6.

The probability of finding Bragg resonant domains in the observed region may be small. However, there may be overdense domains of correlated cavitons, near reflection altitude, whose long-lived temporal correlations produce the narrow "OTSI" lines observed by Sulzer and Fejer.[38,39] We know of no other satisfactory explanation of such lines which takes into account the nonlinear evolution of the OTSI instability. Single pulse radar taken by Cheung et al.[12] and more recently by Sulzer et al.[41] may be consistent with temporal correlations at relatively early times following the onset of heating.

An alternative SLT scenario which might account for the sharp spectral features at later delay times was mentioned in Section 6. This theory must be consistent with the new information from Djuth et al.[33] concerning the altitude of the turbulence producing these features. The theory must also be consistent with our observations that simulations of Zakharov's equations[40] for these underdense altidues show that a significant source of Langmuir dissipation and of ion density fluctuations is caviton collapse. Clearly WTT does not satisfy these requirements.

These questions are a challenge for future work. We believe that the new strong Langmuir turbulence model presented here is considerably more successful in describing the early-time behavior of the heated ionosphere than the conventional theory. We hope this paper will stimulate new experimental and theoretical tests of these ideas.

ACKNOWLEDGMENTS

We are grateful to Drs. P. Y. Cheung, T. Tanikawa, J. Santoru and A. Y. Wong for important collaborations and discussion. We are also grateful to Dr. F. T. Djuth and Dr. M. P. Sulzer for permission to discuss their unpublished experimental results. Continuuing discussions and collaborations with Bandel Bezzerides are gratefully acknowledged. This research was supported in part by the U.S. Department of Energy, the Office of Naval Research, the Los Alamos Center for Nonlinear Studies and the Institute for Geophysics and Planetary Physics.

REFERENCES

1. D. F. DuBois and M. V. Goldman, Radiation induced instability of electron plasma oscillations, Phys. Rev. Lett. **14**, 544, 1965. V. P. Silin, Parametric Resonance in a plasma, Soc. Phys. JETP, Engl. Tranl. **21**, 1127, 1965.
2. M. V. Goldman and D. F. DuBois, Stimulated Scattering of Light from Plasmas, Physics of Fluids, **8**, 1966.
3. V. E. Zakharov, Collapse of Langmuir waves, Sov. Phys. JETP, 35, 908, 1972.
4. D. Russell, D. F. DuBois and H. A. Rose, Nucleation in Two-Dimensional Langmuir Turbulence, Phys. Rev. Lett. **60**, 581-584.
5. D. F. DuBois, Harvey A. Rose and David Russell, Power spectra of fluctuations in strong Langmuir turbulence, Phys. Rev. Lett. **61**, 2209, 1988.
6. P. A. Robinson, D. L. Newman and M. V. Goldman, Three dimensional strong Langmuir turbulence and wave collapse, Phys. Rev. Lett. **61**, 702, 1988.
7. G. D. Doolen, D. F. DuBois and H. A. Rose, Nucleation of cavitons in strong Langmuir turbulence, Phys. Rev. Lett. **54**, 804-807, 1985.

8. D. Russell, D. F. DuBois and H. A. Rose, <u>Collapsing caviton turbulence in one dimension</u>, Phys. Rev. Lett. **56**, 838-842, 1986.

9. P. Y. Cheung and A. Y. Wong, <u>Three dimensional self-collapse of Langmuir waves</u> Phys. Rev. Lett. **52**, 1222 (1984).

10. C. J. Walsh, D. M. villeneuve, and M. A. Baldis, <u>Electron Plasma-Wave Production by Stimulated Raman Scattering: Competition with Stimulated Brillouin Scattering</u>, Phys. Rev. Lett. **53**, 1445 (1984).

11. D. M. Villeneuve, H. A. Baldis, and J. E. Bernard, <u>Suppression and Stimulated Raman Scattering by Seeding of Stimulated Brillouin Scattering in a Laser-Produced Plasma</u>, Phys. Rev. Lett. **59**, 2547 (1987).

12. P. Y. Cheung, A. Y. Wong, T. Tanikawa, J. Santoru, D. F. DuBois, Harvey A. Rose, and David Russell, <u>Short time scale evidence for strong Langmuir turbulence in H-F heating of the ionosphere</u>, Phys. Rev. Letters **62**, 2676 (1989).

13. H. A. Rose, D. F. DuBois and B. Bezzerides, <u>Nonlinear Coupling of Stimulated Raman and Brillouin Scattering in Laser-Plasma Interactions</u>, Phys. Rev. Lett. **58**, 2547-2550, 1987.

14. D. F. DuBois, H. A. Rose and David Russell, <u>Excitation of Strong Langmuir Turbulence Near Critical Density; Application to HF Heating of the Ionosphere</u>, Los Alamos National Laboratory Report LA-UR-89-1419 submitted to Journal of Geophysical Research.

15. A. A. Vedenov and L. I. Rudakov, <u>Interactions of waves in continuous media</u>, Sov. Phys. Dokl. **9**, 1073 (1965).

16. K. Nishikawa, <u>Parametric excitation of coupled waves. I. General formulation</u>, J. Phys. Soc. Japan, 24, 916-922, 1968a, and <u>Parametric excitation of coupled waves. II. Parametric plasma-photon interactions</u>, J. Phys. Soc, Jap. 24, 1152-1158, 1968b.

17. H. A. Rose and M. I. Weinstein, <u>On the bound states of the nonlinear Schrödinger equation with a linear potential</u>, Physica D, 30, 207-218, 1988a.

18. A. A. Galeev, R. Z. Sagdeev, Yu S. Sigov, V. D. Shapiro, and V. I. Schevchenko, <u>Nonlinear theory for the modulational instability of plasma waves</u>, Fiz. Plazmy 1, p 10, 1975 [Sov. J. Plasma Phys. 1 p 5, 1975].

19. H. A. Rose, D. F. DuBois, David Russell, and B. Bezzerides, <u>Experimental signatures of localization in Langmuir wave turbulence</u>, Physica D 1988b.

20. J. C. Weatherall, J. P. Sheerin, D. R. Nicholson, G. L. Payne, M. V. Goldman, and P. J. Hansen, <u>Solitons and ionospheric heating</u>, J. Geophys. Res. 87, 823-832, 1982.

21. V. E. Zakharov and L. N. Shur, <u>Self similar regimes of wave collapse</u>, Sov. Phys. JETP **54**, 1064, 1981.

22. A. I. Dyachenko, V. E. Zakharov, A. M. Rubenchik, R. Z. Sagdeev and V. F. Shvets, <u>Two dimensional Langmuir collapse and two dimensional Langmuir cavitons</u>, Pis'ma v ZhETF **44**, 504 (1986) [JETP Letters, 44, 648, 1986].

23. A. I. Dyachenko, A. M. Rubenchik, R. Z. Sagdeev, V. F. Shvets, and V. E. Zakharov, <u>Computer simulation of the Langmuir collapse of the isolated cavity in Plasma Theory and Nonlinear and Turbulent Processes in Physics</u>, Vol. 2, ed. by V. G. Bar'yakhtar et al., World Scientific 1987.

24. C. Aldrich, B. Bezzerides, D. F. DuBois and H. A. Rose [1984 unpublished].

25. V. E. Zakharov, A. N. Pushkarev, A. M. Rubenchik, R. Z. Sagdeev and V. F. Shvets, <u>Final stage of 3D Langmuir collapse</u>, JETP Lett. **45**, #8, 287 (1988) [Pis'ma Zh. Eksp. Teor. Fiz. **47**, #5, 239 (1988)].

26. B. D. Fried, and R. W. Gould, <u>Longitudinal ion oscillations in a hot plasma</u>, Phys. Fluids 4, 139 (1961).
27. D. R. Nicholson, <u>Plasma Theory</u>, John Wiley & Sons, N.Y., 1983.
28. G. L. Payne, D. R. Nicholson, and M. M. Shen, <u>Numerical test of weak turbulence theory</u>, to be published in Physics of Fluids.
29. A. Y. Wong and P. V. Cheung, <u>Three dimensional self-collapse of Langmuir waves</u>, Phys. Rev. Lett. **52**, 1222, 1984.
30. H. A. Baldis, P. E. Young, R. P. Drake, W. L. Kruer, Kent Estabrook, E. A. Williams, and T. W. Johnston, <u>Competition between the stimulated Raman and Brillouin scattering instabilities in 0.35-μm irradiated CM foil targets</u>, Phys. Rev. Lett. **62**, 2829 (1989).
31. A. Y. Wong, G. J. Morales, D. Eggleston, J. Santoru, and R. Behnke, <u>Rapid conversion of electromagnetic waves to electrostatic waves in the ionosphere</u>, Phys. Rev. Lett. **47**, 1340, 1981.
32. F. T. Djuth, C. A. Gonzales and H. M. Ierkic, <u>Temporal evolution of the H-F-enhanced plasma line in the Arecibo F region</u>, J. Geophys. Res. 91, p 12,089-12,107, 1986.
33. F. T. Djuth (private communication).
34. J. A. Fejer, C. A. Gonzales, H. M. Ierkic, M. P. Sulzer, C. A. Tepley, L. M. Duncan, F. T. Djuth, S. Gangulys, and W. E. Gordon, <u>Ionospheric modification experiments with the Arecobø Heating Facility</u>, J. Atmos. Terr. Phys. 47, p. 1165, 1985.
35. A. Y. Wong, <u>Nonlinear Phenomena in laboratory and space plasmas</u>, Physica Scripta T2/1, 262-270, 1982.
36. P. Y. Cheung (private communication).
37. Harvey A. Rose, D. F. DuBois and David Russell (to be published).
38. M. P. Sulzer, H. M. Ierkic, J. A. Fejer and R. L. Showen, <u>HF-induced ion and plasma line spectra with two pumps</u>, J. Geophys. Res., **89**, 6804, 1984.
39. M. P. Sulzer, H. M. Ierkic, and J. A. Fejer, <u>Observational limitations on the role of Langmuir cavitons in ionospheric modification experiments at Arecibo</u>, J. Geophys. Res. (1988).
40. D. F. DuBois, David Russell and Harvey Rose (to be published).
41. M. P. Sulzer (private communication).

Figure captions.

Fig. 1. Two-dimensional plots of $|E|^2$ (upper surface) and n (lower surface) at two different times. Parameters in scaled units $E_0 = 1.2$, $\nu_i = 0.9$, $\omega_c/\omega_p = 0$, ν_e = Landau damping continued smoothly as k^2 at large k, $m_i/\eta m_e = 1836$, $L_x = L_y = 2\pi$ and a 128×128 spatial grid. The collapses are anistropic with the narrow dimension along the x axis, the drive direction.

Fig. 2. (a),(b) Spatial profiles of $|E|^2$, n and $|A_R|^2$ for two times for $n_o/n_c = 0.045$, $L = 15\ \lambda_o$ and $V_{osc}|c = 0.035$. Here E, n and A_R are, respectively, the Langmuir field envelope, the ion density fluctuation and the envelope of backscattered light wave; n_o is the electron density in the simulation slab of length L and n_c is the critical (reflection) density of the light which is incident on the slab from the left and has an intensity expressed in terms of the oscillating velocity V_{osc} - $(e/m\omega_o)E_{light}$. The units are $4\pi n_o T_e/439$ for $|E|^2$, $n_o/2754$ for n, arbitiary units for $|A_R|^2$, 91 λ_{De} is the spatial unit, and 140 psec is the temporal unit. We show only the leftmost portion of the slab, $0 \leq x \leq 3\lambda_o$. (c) Temporal history of the total Langmuir energy, total ion wave energy, and SBS backscatter reflection coefficient (in arbitiary units) for the case above. The unit of time is 150 ps. (d) Same but for $V_{osc}/c = 0.07$, $n_o/n_c = 0.055$. Here the unit of time is 132 ps. The mean square thermal fluctuations of the initial undisturbed plasma were $\langle |E|^2 \rangle_{thermal} \simeq 0.1$ and $\langle n^2 \rangle_{thermal} \sim 900$. In (c) and (d) note the initial pulse of Langmuir energy associated with a single generation of collapse and burnout followed by the growth of SBS backscattered light.

Fig. 3. The temporal evolution of (a) $|E(r = 0,t)|^2$ and $n(r = 0,t)$, (b) $\lambda_0(t)$ and the phase velocity $-d\Phi/dt$ and (c) the caviton width $\delta(t)$ and the electrostatic energy in the caviton $|h_0(t)|^2$ in the scalar model; $E_0 = 1.8$, $m_i/\eta m_e = 2 \times 10^4$, $\nu_i = 0.9$ and $\omega_0 = 0$.

Fig. 4. Scalar model for $E_0 = 1.8$, $m_i/\eta m_e = 2 \times 10^4$, $\nu_i = 0.9$, $\omega_0 = -80$. Showing shape of n(r,t) and $|E(r,t)|^2$ as functions of r for two times in the caviton cycle evolution and the profiles of the two lowest eignefunctions $e_0(r,t)$ and $e_1(r,t)$ vs r at the same times. The qualitative behavior is the same for the case $\omega_0 = 0$.

Fig. 5. Spectra, $|\Psi(k,\omega)^2 = k^{-4}|E(k,\omega)|^2$, for D = 2 isolated collapse events. $E_0 = 0.8$, $\nu_i = 0.9$, $m_i/\eta m_e = 1836$, $\omega_0 = 0$. $k_x = 8$, $k_y = 0$. Inset Solid line, negative frequency at maximum of spectrum vs $k = k_x$; dashed line, ion acoustic frequency shift, $|\omega| = k$.

Fig. 6. Power spectra $|E(\underline{k},\omega)|^2$ $E_0 = 1.2$, $\omega_0 = 5$, $\omega_c/\omega_p = 0.2$, $L_x = L_y = 2\pi$, M = 1836. Spectra are smoothed over an angular frequency interval $\Delta\omega \simeq \pi$. Spectra intensity scales are arbitrary. The spectra are for various values of (k,θ). Because of a numerical coincidence, when these results are scaled to the more realistic value $M = 9 \times 1836$, the frequency scales can also be read as kHz of frequency (not angular frequency).

Fig. 7. Power spectra $|E(\underline{k},\omega|^2$ for the scalar model parameters of Fig. 2. Spectrum for k = 40.0.

Fig. 8. Experimental spectra. (a) Heater pulse width 10 ms, Interpulse period (IPP) 150 ms, $f_H = 7.3$ MHz spectra taken in 1.1 ms internals delayed 4 ms from onset of heating pulse. Note the free mode peak at ~ 72 kHZ above the heater frequency which is 256 on the scale, (b) Heater pulse width 10 ms, IPP 150 ms, $f_H = 7.3$ MHz spectral delayed by 1.5 ms from onset of heating. Free mode peak at 52 kHz. From Cheung et al. [1989].

Fig. 9. Plasma line spectra for 30 ms, heater pulses with 150 IPP at 7.3 Mhz and radar pulse delayed by 29 ms after onset of heater pulse, averages over 10 and 50 radar pulses, respectively.

Fig. 10. Two dimension simulation of alternating lattice model. Caviton locations marked $\bullet$ are $\tau/2$ out of phase with locations marked $+$. The caviton cycle period is τ. The central square is the simulation cell. The other cells are implied by the periodic boundary conditions. The $\bullet$'s and $+$'s denote the initial caviton locations which do not change for $\omega_0 = -25$ or $\omega_0 = -15$. Other parameters: $E_0 = 1.2$, $v_i = 0.9$, $m_i/\eta m_e = 1846$.

Fig. 11. Total correlated power spectra $|E(k,\omega)|^2$ (sharp line spectra) and single caviton spectra $|\varepsilon(k,\omega)|^2$ (smooth spectra) for the coherent alternating lattice for various values of (k,θ). Showing spectra for even and odd values of $v = (1/2\pi)\,\underline{k} \cdot$ where $\underline{d}$ is the displacement vector between the $\cdot$ and $+$ lattices (see ref. 14) d. Here $|\psi(\underline{k},\omega)|^2 = k^{-4}|E(k,\omega)|^2$ or $k^{-4}|\varepsilon(k,\omega)|^2$ for the two classes of spectra, respectively.

Fig. 12. Ion line spectra $|n(\underline{k},\omega)|^2$ for the coherent alternating lattice simulation. Both the total spectrum and the single caviton spectrum are again shown.

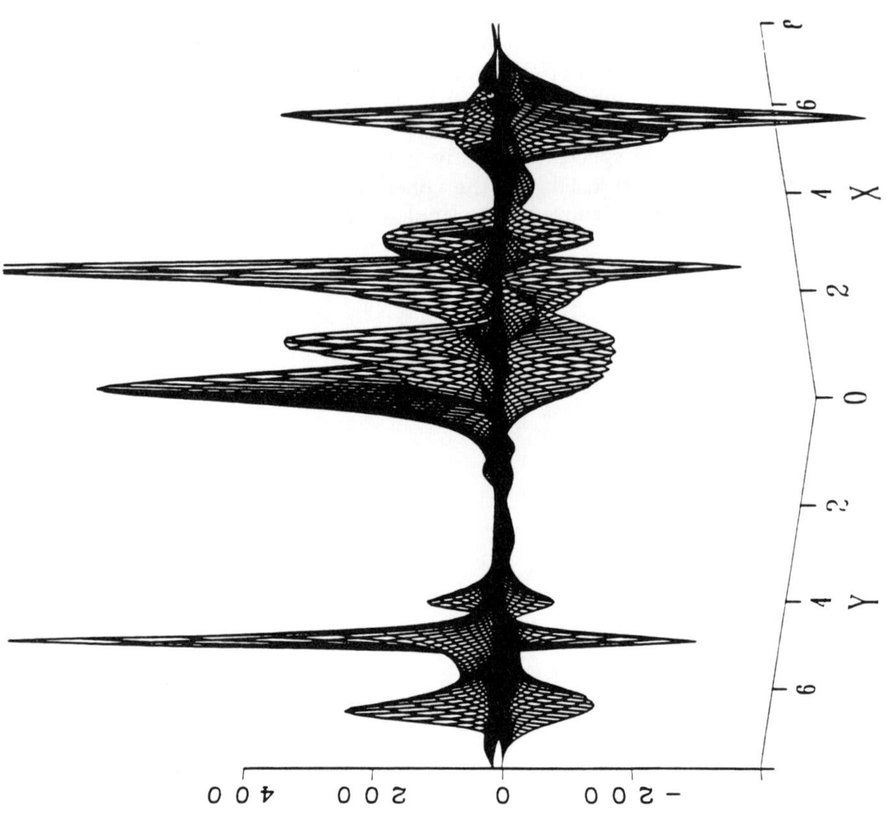

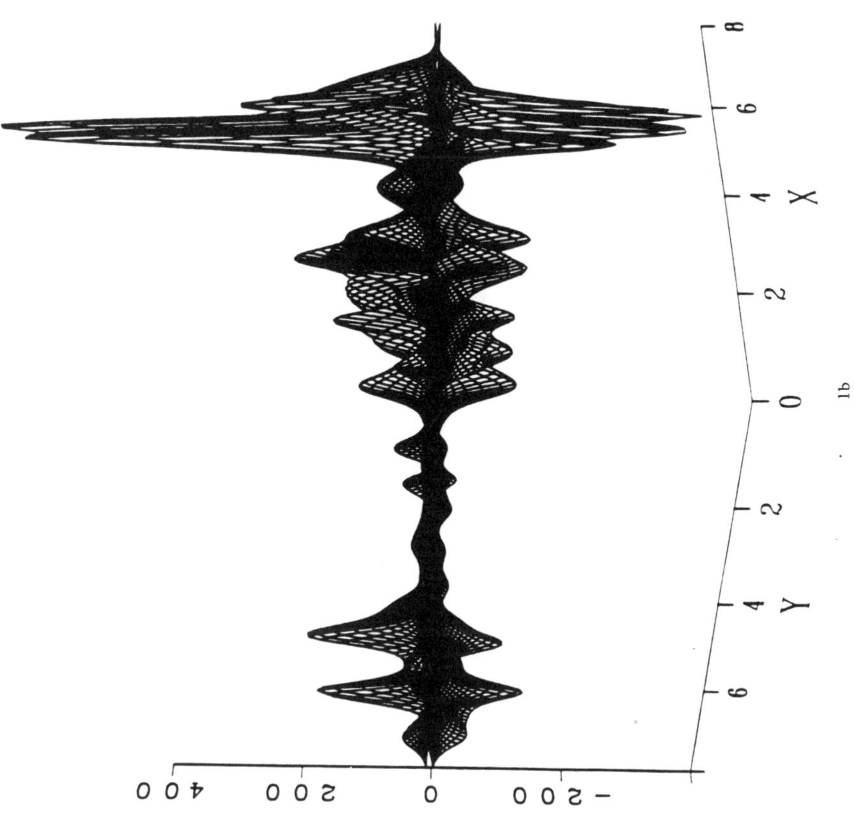
400
200
0
-200
-200
8
6
4
2
0
2
4
6
X
Y
1b

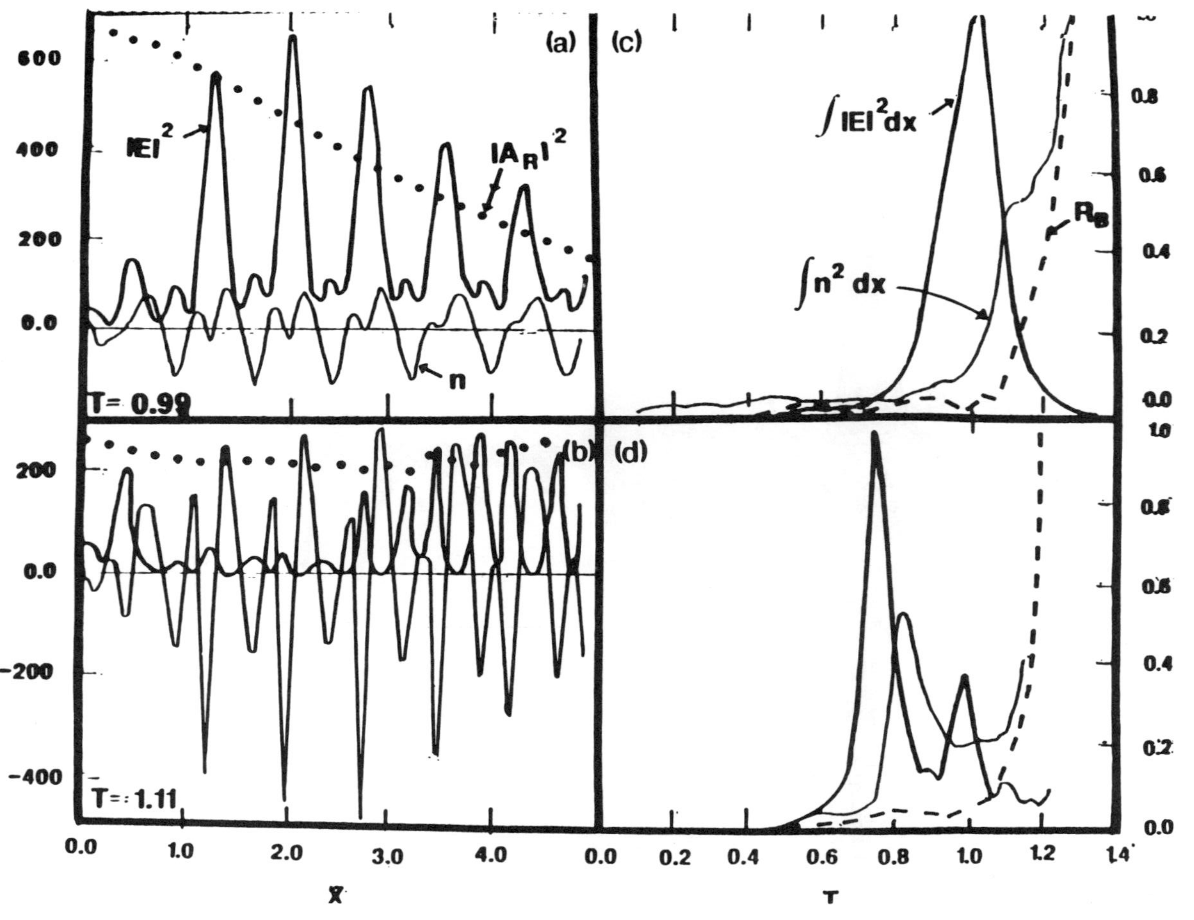
(a)
(c)
|E|²
|A_R|²
n
T= 0.99
∫|E|²dx
∫n²dx
R_B
600
400
200
0.0
0.8
0.6
0.4
0.2
0.0
(b)
(d)
T= 1.11
200
0.0
-200
-400
1.0
0.8
0.6
0.4
0.2
0.0
0.0   1.0   2.0   3.0   4.0
X
0.0 0.2 0.4 0.6 0.8 1.0 1.2 1.4
T

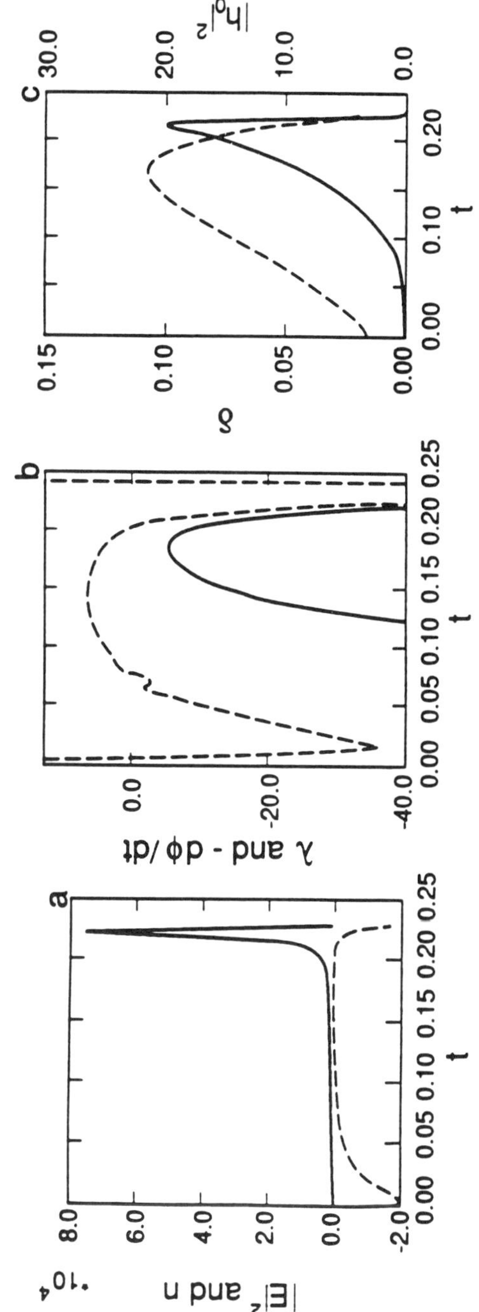

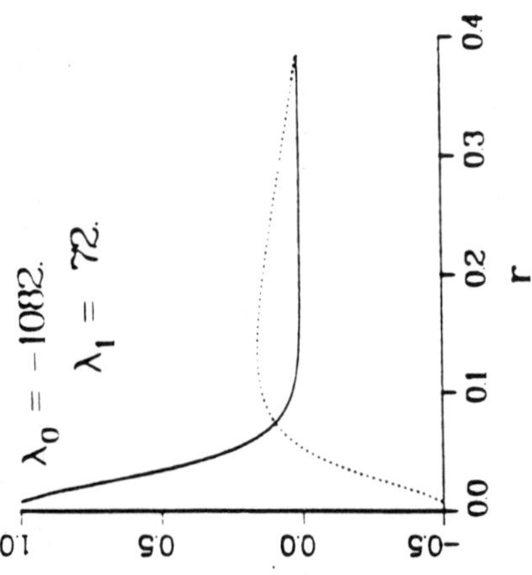

$\lambda_0 = -1082.$
$\lambda_1 = 72.$
r

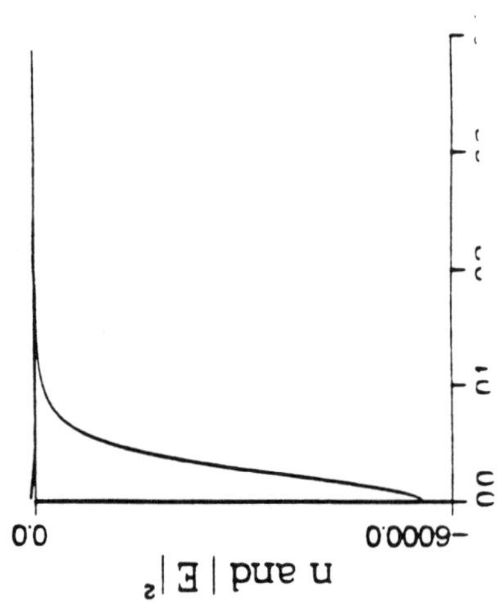

n and $|E|^2$
-6000.0
0.0

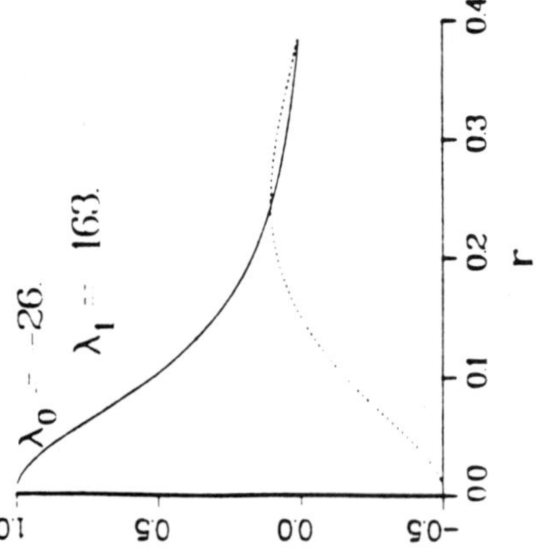

$\lambda_0 = -26.$
$\lambda_1 = 163.$
r

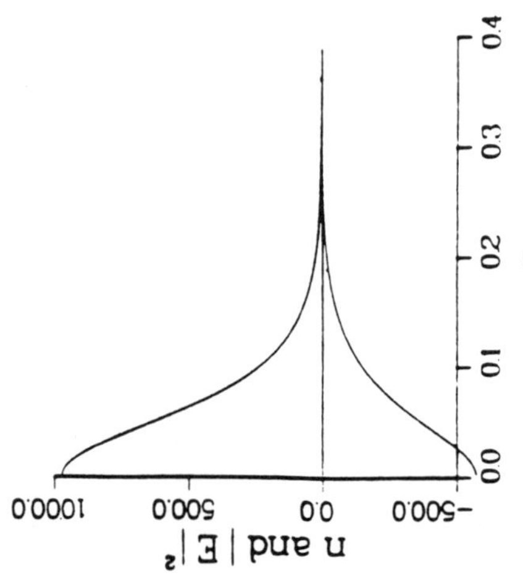

n and $|E|^2$
-500.0
0.0
500.0
10000.0

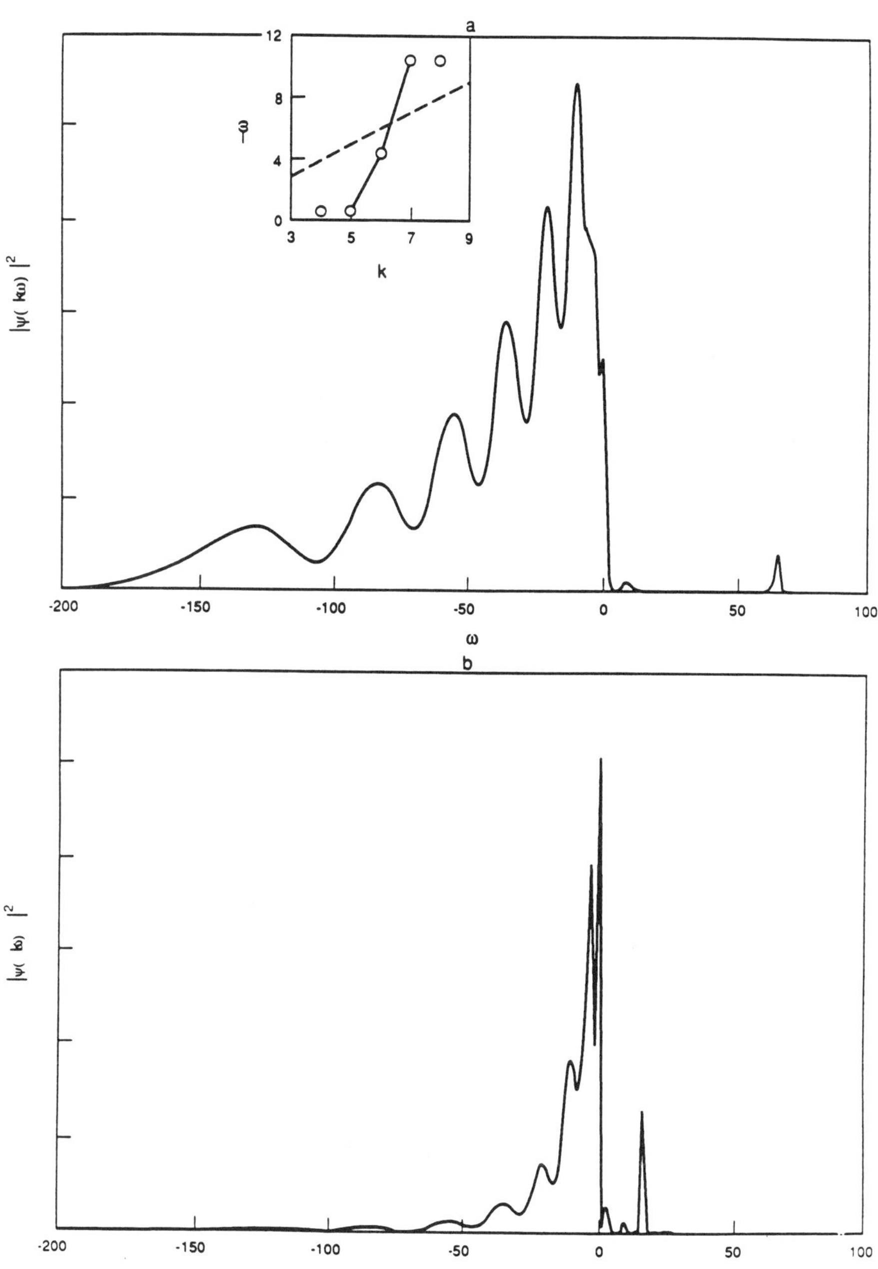

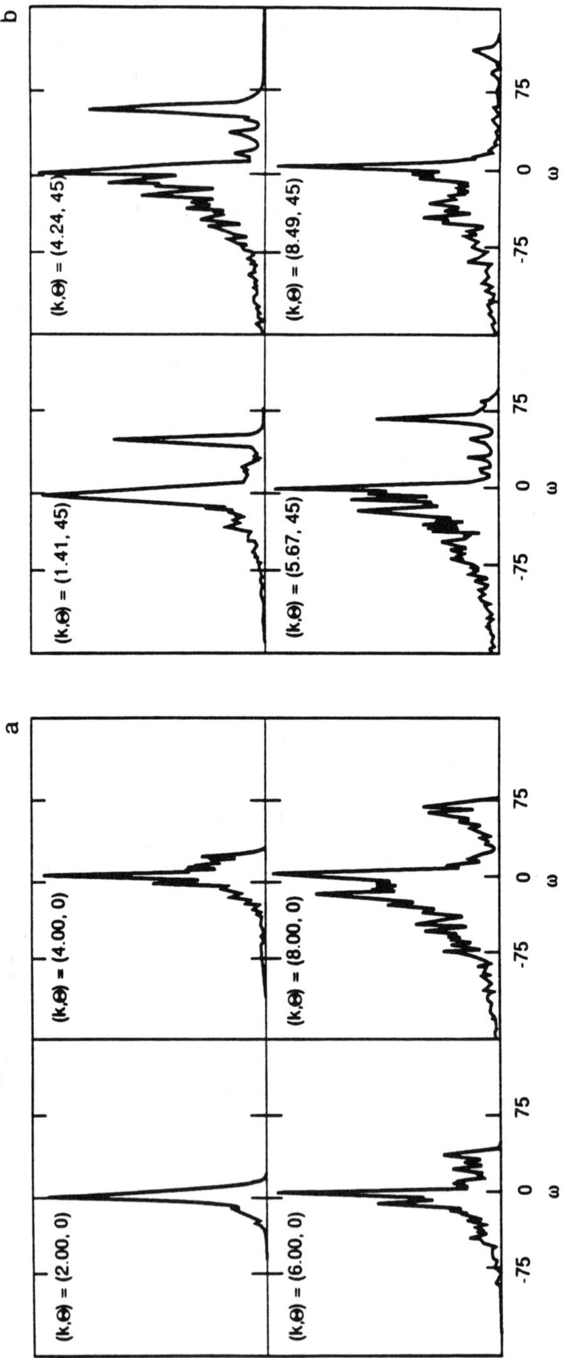
a
b
(k,θ) = (2.00, 0)
(k,θ) = (4.00, 0)
(k,θ) = (6.00, 0)
(k,θ) = (8.00, 0)
(k,θ) = (1.41, 45)
(k,θ) = (4.24, 45)
(k,θ) = (5.67, 45)
(k,θ) = (8.49, 45)
ω
ω
-75
0
75
-75
0
75

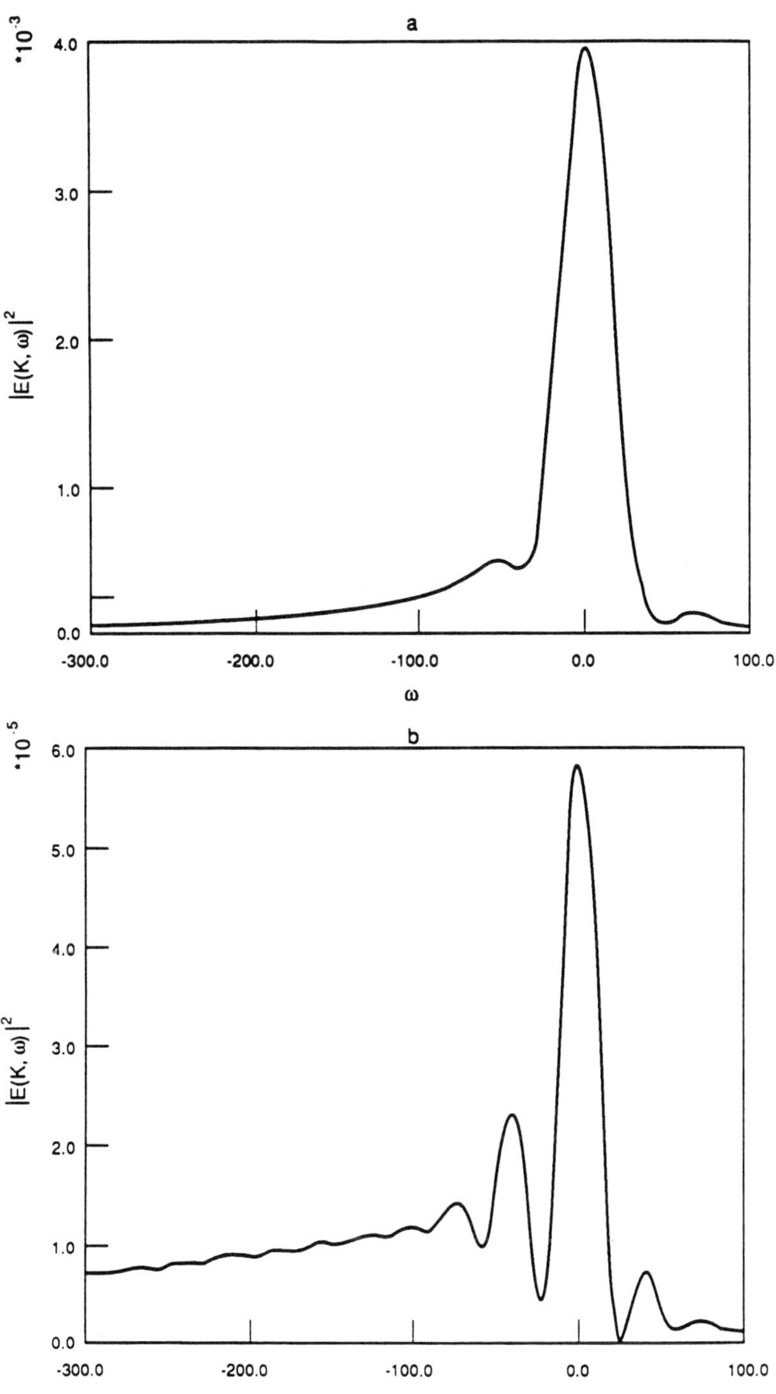

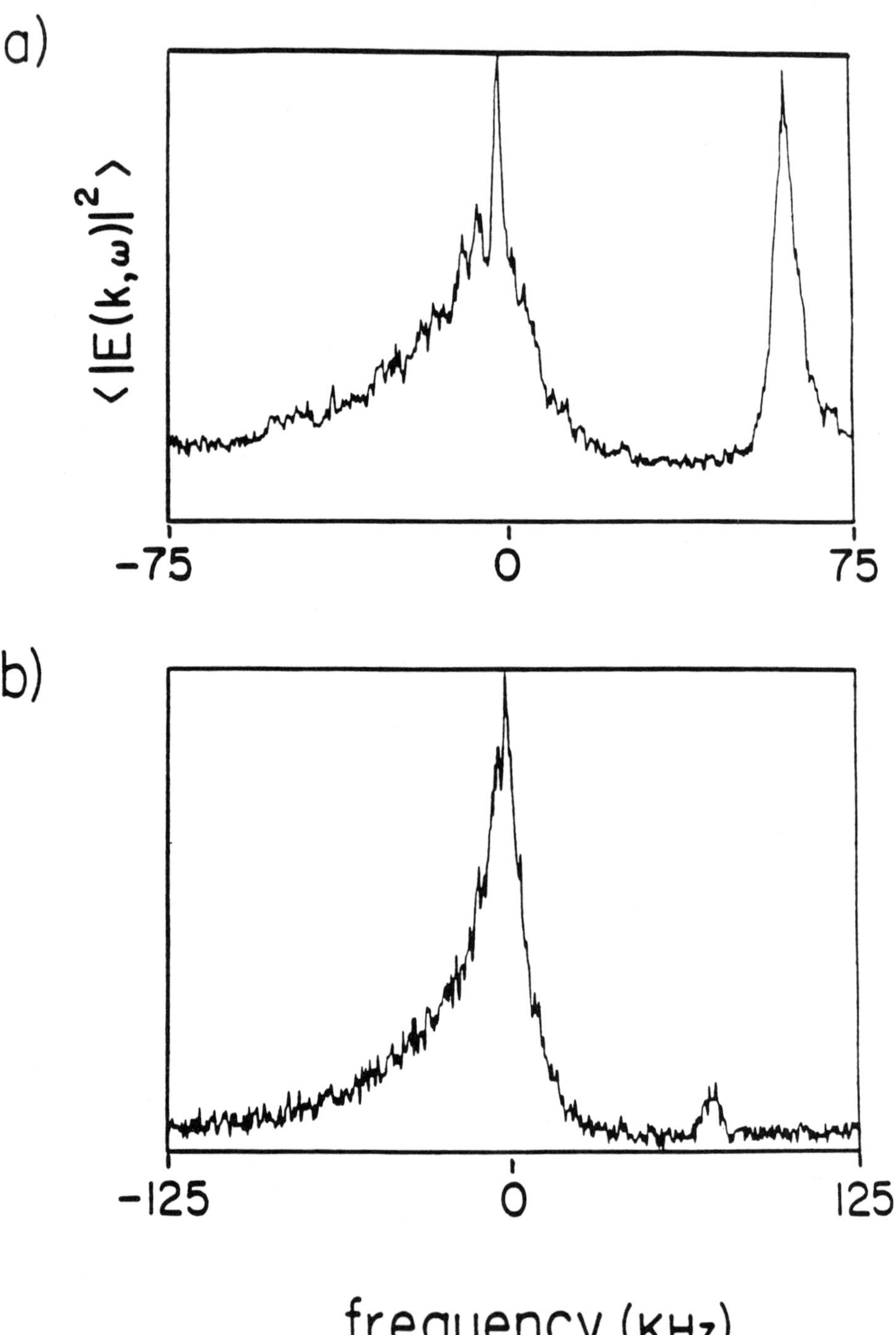
a)
⟨|E(k,ω)|²⟩
-75
0
75
b)
-125
0
125
frequency (KHz)

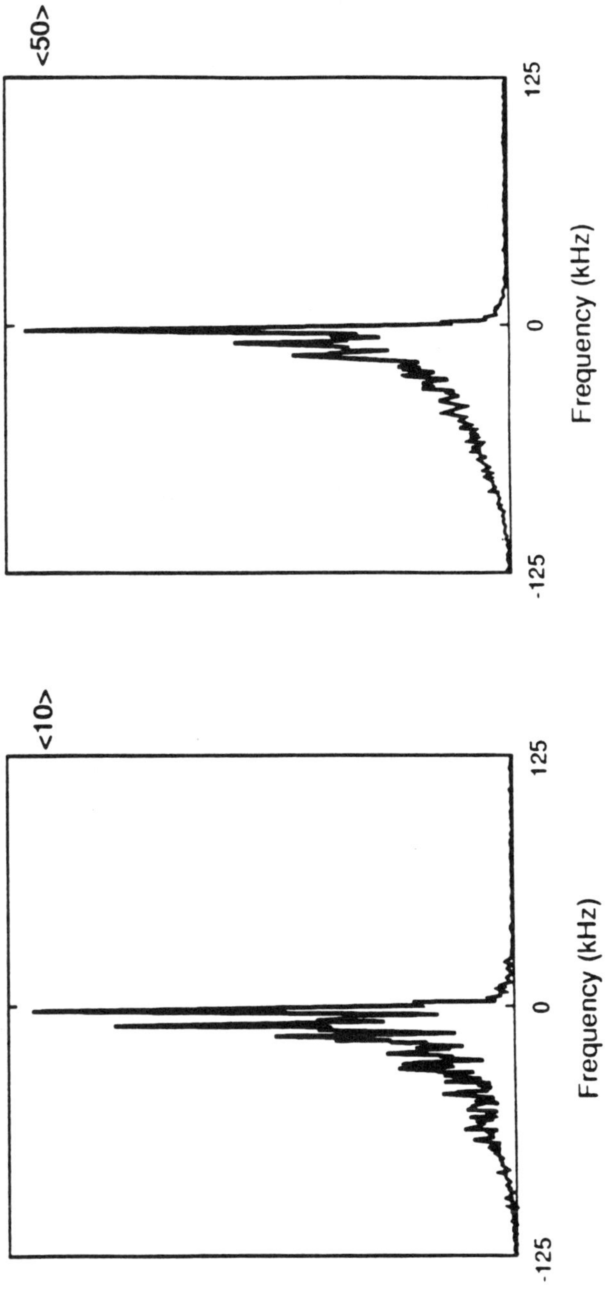

<50>
<10>
Power
Frequency (kHz)
Frequency (kHz)
125
0
-125
125
0
-125

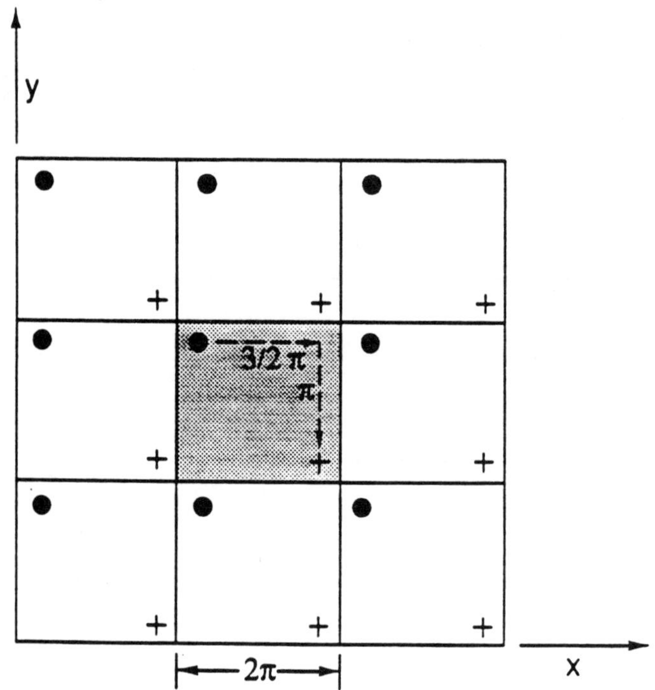

y
3/2 π
π
2π
x

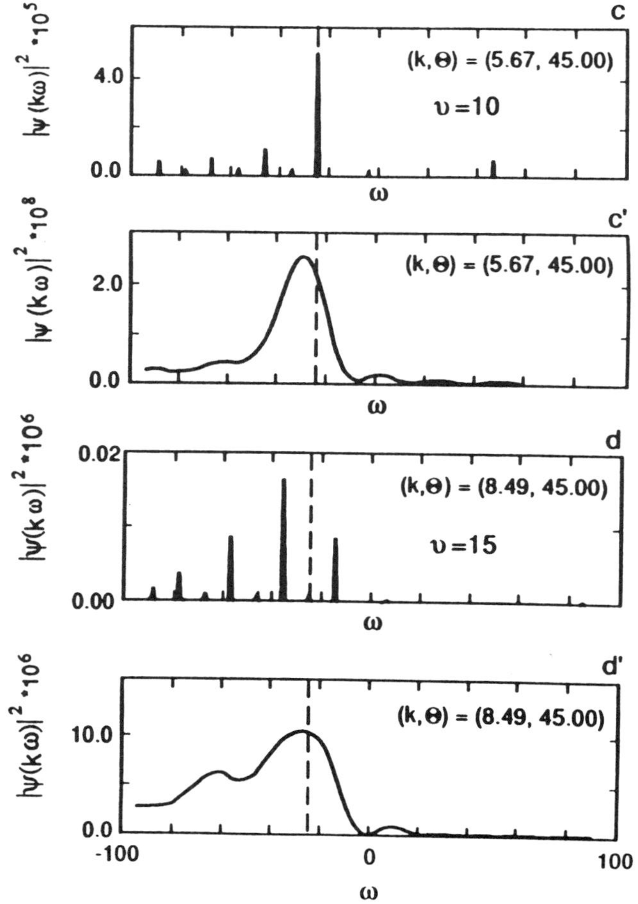
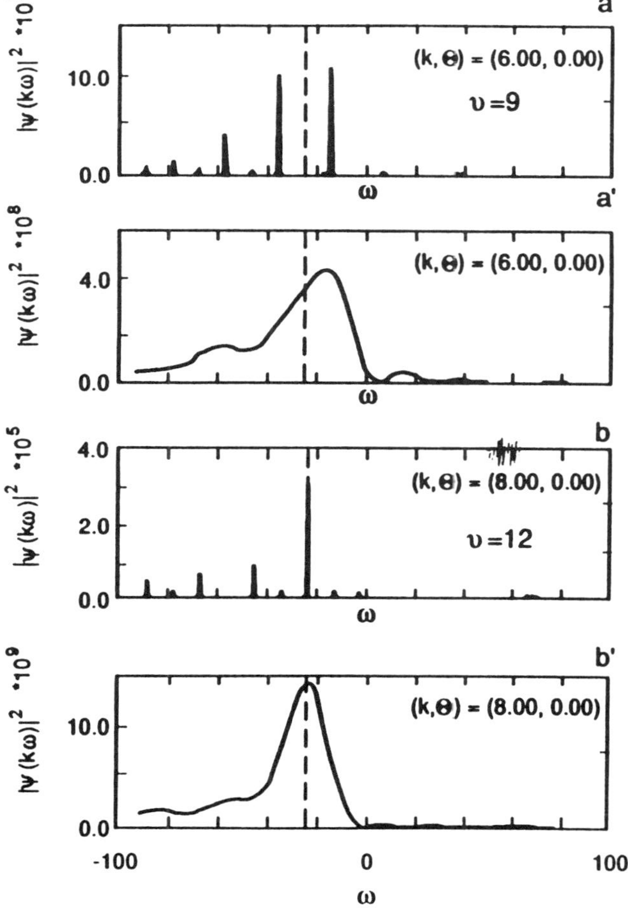

a
$|\psi(k\omega)|^2 \cdot 10^5$
10.0
0.0
$(k,\Theta) = (6.00, 0.00)$
$\upsilon = 9$
ω
a'
$|\psi(k\omega)|^2 \cdot 10^8$
4.0
0.0
$(k,\Theta) = (6.00, 0.00)$
ω
b
$|\psi(k\omega)|^2 \cdot 10^5$
4.0
2.0
0.0
$(k,\Theta) = (8.00, 0.00)$
$\upsilon = 12$
ω
b'
$|\psi(k\omega)|^2 \cdot 10^9$
10.0
0.0
$(k,\Theta) = (8.00, 0.00)$
-100
0
100
ω
c
$|\psi(k\omega)|^2 \cdot 10^5$
4.0
0.0
$(k,\Theta) = (5.67, 45.00)$
$\upsilon = 10$
ω
c'
$|\psi(k\omega)|^2 \cdot 10^8$
2.0
0.0
$(k,\Theta) = (5.67, 45.00)$
ω
d
$|\psi(k\omega)|^2 \cdot 10^6$
0.02
0.00
$(k,\Theta) = (8.49, 45.00)$
$\upsilon = 15$
ω
d'
$|\psi(k\omega)|^2 \cdot 10^6$
10.0
0.0
$(k,\Theta) = (8.49, 45.00)$
-100
0
100
ω

Guiding Center Stochasticity and Transport Induced by Electrostatic Waves

K. Kupfer, A. Bers, and A.K. Ram

Plasma Fusion Center and Research Laboratory of Electronics

Massachusetts Institute of Technology

Cambridge, Massachusetts 02139

Abstract

We present a study of plasma transport driven by externally imposed wave fields; the transport arises from the induced chaotic particle dynamics. In particular, we examine the guiding center motion of charged particles interacting with electrostatic waves in a tokamak. This motion is described by a time-dependent Hamiltonian in a four-dimensional phase space composed of the guiding center's three position coordinates and its velocity along the magnetic field. As a specific focus, the work is applied to electrons in lower-hybrid current drive (LHCD) fields. We examine the dynamics for two cases: (i) a large ensemble of randomly phased waves, and (ii) a few large amplitude waves. For case (i), we calculate the local quasilinear diffusion tensor in the two-dimensional action space of the guiding center and compare it to numerical observations of the stochastic diffusion induced by the same spectrum of waves. An average radial diffusion coefficient is estimated from the quasilinear tensor and we discuss its scaling for typical LHCD parameters. In case (ii), we find enhanced radial transport in the presence of waves whose corresponding wave-particle resonance conditions are simultaneously satisfied at a single point in action space. In the vicinity of this point the radial excursions are not limited by the trapping width of the resonances. The mean square radial deviation of an orbit from its initial average flux surface, $\sigma(t)$, is found to obey a power law for long times, i.e. $\sigma(t) \approx \sigma_o t^\gamma$. When the amplitude of the waves is small enough that their primary resonances retain stable fixed points (even though the overlap criteria is satisfied), we find that γ is below unity, indicating an anomalous diffusive process; at larger amplitudes $\gamma \approx 1$. When the shear of the magnetic field lines is taken to be zero, resonances can coalesce throughout action space. In the presence of such a degenerate resonance we find that orbits will exhibit a "stochastic streaming" in the radial direction.

I Introduction

In this paper we are studying the guiding center motion of magnetically confined charged particles interacting with electrostatic waves in a plasma. As a specific focus, the work is applied towards describing the wave induced radial transport of electrons during lower-hybrid current drive (LHCD) in a tokamak. During LHCD, electrons interact with the wave field and experience stochastic accelerations along magnetic field lines. Radial transport follows because the toroidal ∇B and curvature drifts are coupled to the parallel stochastic acceleration. In addition, the perpendicular component of the wave field gives rise to a fluctuating $E \times B$ drift which also leads to radial transport. This intrinsic wave induced transport may be partially responsible for the observed anomalous confinement of superthermal electrons during LHCD [1]. It is also important to consider wave induced transport when attempting to control the wave induced current profile.

Guiding center motion is described by a time-dependent Hamiltonian in a four-dimensional phase space composed of the guiding center's three position coordinates and its parallel velocity (along the magnetic field). The stochastic motion of guiding centers through prescribed wave fields has been studied previously in two limiting cases: motion along the magnetic field with no drifts [2, 3, 4], and drift motion without parallel wave driven accelerations [5, 6]. In each of these cases one can view the dynamics in a two-dimensional surface of section and the relevant diffusive process is one-dimensional. In our recent work, the radial component of the guiding center drift was coupled to a stochastic two-dimensional map which determined the parallel motion [7]. We are generalizing these approaches by considering the dynamics in a four-dimensional phase space. We use Littlejohn's Lagrangian formulation of guiding center motion [8] to derive canonical coordinates for a tokamak equilibrium with an imposed electrostatic wave field. Using the standard approach of classical mechanics, we obtain the generating function which relates the toroidal coordinates to the action-angle coordinates of a particle's unperturbed orbit (the case when the imposed wave field vanishes). One can derive an explicit form for the action-angle coordinate transformation for circulating particles in a low β circular tokamak. This is done by expanding the generating function to exploit the fact that the deviation of an electron's orbit from its average flux surface is small compared to the scale length of the magnetic inhomogeneity. Upon transforming the wave field from toroidal coordinates to action-angle coordinates, it is written as a Fourier series in the angles and each component is associated with a discrete curve in action space along which the wave-particle resonance condition is satisfied. Ignoring drift motion leads to the usual resonance overlap criterion for the formation of a thick stochastic layer in a two-dimensional phase space which describes the parallel motion. When the drift motion is included, the stochastic process leads to radial transport. Generically, wave-particle resonance curves will intersect at points in action space where the ratio of the unperturbed frequencies is a rational number (i.e. rational unperturbed KAM tori). We show that in the vicinity of these intersections, particles can make large excursions in the radial direction. This leads to enhanced diffusion in the presence of a few large amplitude waves. Conventional quasilinear treatments of transport in RF heated plasmas [9, 10] cannot account for such effects.

For LHCD, determining the wave field in the plasma from that launched at the edge is a complex problem which has not yet been solved. Here we assume that there exists in the core of the plasma some spectrum of electrostatic travelling waves which approximately satisfy the local dispersion relation for frequencies above the lower-hybrid frequency. We consider two examples. In the first example we calculate the quasilinear diffusion tensor in action space [11] for the case of a broad spectrum of randomly phased waves; we compare this to the diffusion observed by numerically following the exact, stochastic orbits. In the second example we consider the stochasticity and radial transport induced by a few large amplitude waves.

The paper is divided into six sections, organized in the following manner. Canonical coordinates for the guiding center are derived in Section II. In Section III an explicit form for the action-angle coordinate transformation is developed for circulating particles. In Section IV general features of the guiding center's interaction with the wave field are discussed. In Section V we consider quasi-linear diffusion induced by a broad spectrum of waves and in Section VI we consider the case of a few large amplitude waves. Our conclusions are in Section VII. In Appendix A we derive the canonical coordinates for a magnetized slab with shear. Appendix B contains the guiding center equations of motion in toroidal coordinates.

II   Hamiltonian Guiding Center Formalism

Starting from the Lagrangian formulation of guiding center motion developed by Littlejohn [8], we will derive canonical coordinates for an axisymmetric, time-independent equilibrium under the effect of electrostatic waves. The Lagrangian is

$$L(\mathbf{x}, \dot{\mathbf{x}}, u, \dot{u}) = [\frac{e}{m}\mathbf{A}(\mathbf{x}) + u\mathbf{b}(\mathbf{x})] \cdot \dot{\mathbf{x}} - H(u, \mathbf{x}, t) \quad , \tag{1}$$

where

$$H = \frac{1}{2}u^2 + MB(\mathbf{x}) + \frac{e}{m}\Phi(\mathbf{x}, t) \quad , \tag{2}$$

$\mathbf{x}$ is the three-dimensional spatial position vector of the guiding center, u is its parallel velocity, and M is its magnetic moment divided by its mass. The vector potential is $\mathbf{A}(\mathbf{x})$, $\mathbf{B}(\mathbf{x})$ is the magnetic field, and $\Phi(\mathbf{x}, t)$ is the scalar potential of the electrostatic wave-field. Also $\mathbf{b}(\mathbf{x}) = \mathbf{B}/B$ and $B(\mathbf{x}) = |\mathbf{B}|$. If we let z_i denote one of the four guiding center coordinates, $\mathbf{x}$ and u, then the guiding center's motion is determined by the four Euler-Lagrange equations,

$$\frac{d}{dt}\frac{\partial L}{\partial \dot{z}_i} = \frac{\partial L}{\partial z_i} \quad . \tag{3}$$

Since L is independent of $\dot{u}$, one finds that $\partial L/\partial u = 0$, so that u is in fact the parallel velocity, i.e. $u = \mathbf{b} \cdot \dot{\mathbf{x}}$. The remaining three Euler-Lagrange equations determine the perpendicular guiding center drift and the parallel acceleration. With a little algebra one obtains the following equations of motion:

$$\dot{\mathbf{x}} = (B_{\parallel}^{*})^{-1}[u\mathbf{B}^{*} + \mathbf{b} \times (\frac{m}{e}M\nabla B + \nabla\Phi)] \quad , \tag{4}$$

$$\dot{u} = -\frac{\mathbf{B}^{*}}{B_{\parallel}^{*}} \cdot (\frac{e}{m}\nabla\Phi + M\nabla B) \quad , \tag{5}$$

where $\mathbf{B}^{*} = \mathbf{B} + (mu/e)\nabla \times \mathbf{b}$ and $B_{\parallel}^{*} = \mathbf{b} \cdot \mathbf{B}^{*}$. Since the magnetic moment is conserved, we consider it a parameter and not a dynamical variable. This restricts us to the case when the frequency of the wave field is much smaller than the gyro-frequency of the particles under consideration, a condition which is well satisfied by electrons in LHCD fields. It is assumed that the electron Larmor radius is small compared to the perpendicular wavelength of the electrostatic field. The amplitude of the wave field is restricted by the usual limitations of guiding center analysis, which require both the parallel acceleration and the $E \times B$ drift during a gyro-period to be small in the following sense: $|k_{\parallel}^2\Phi/(\omega_{rf}B)| \ll 1$

and $|mk_\perp^2 \Phi/(eB^2)| \ll 1$, where $k_\parallel$ and $k_\perp$ are characteristic parallel and perpendicular wave numbers and ω_{rf} is the frequency of the wave field ($\omega_{rf} = 2\pi f_{rf}$).

We now transform the Lagrangian into the toroidal coordinates (ψ, θ, ϕ), where ψ is the usual poloidal flux function and ϕ is the toroidal angle. The coordinate θ is chosen so that $\nabla\theta$ is perpendicular to both $\nabla\psi$ and $\nabla\phi$. Also θ increases by 2π upon one rotation around a flux surface at fixed toroidal angle. Using these coordinates the magnetic field for a scalar pressure MHD equilibrium can be written as [12]

$$\mathbf{B} = g(\psi)\nabla\phi + F(\psi,\theta)\nabla\theta \quad , \tag{6}$$

where the vector potential is chosen to be

$$\mathbf{A} = \psi\nabla\phi - G(\psi,\theta)\nabla\theta \quad . \tag{7}$$

These relations define the functions $g(\psi)$, $F(\psi,\theta)$, and $G(\psi,\theta)$. It is straightforward to show that the θ average of $\partial G(\psi,\theta)/\partial\psi$ is the tokamak safety factor $q(\psi)$ [13]. Substituting (6) and (7) into (1) the Lagrangian becomes

$$L = p_\phi\dot\phi + p_\theta\dot\theta - H \quad , \tag{8}$$

where the momenta conjugate to ϕ and θ are

$$p_\phi = \psi + g(\psi)u/B \quad , \tag{9}$$

and

$$p_\theta = -G(\psi,\theta) + F(\psi,\theta)u/B \quad . \tag{10}$$

The system has been normalized in the following manner: the units of velocity are v_o (chosen to make the parallel velocity of order unity, i.e. $v_o \sim \omega_{rf}/k_\parallel$) and the units of time are R_o/v_o, where R_o is the major radius. The normalized Hamiltonian is

$$H = u^2/2 + MB(\psi,\theta) + \Phi(\psi,\theta,\phi,t) \quad , \tag{11}$$

where the electrostatic potential is normalized to mv_o^2/e and the magnetic field is normalized to B_o (the toroidal field on axis). The constant M is now the magnetic moment of the guiding center normalized to mv_o^2/B_o. Both ψ and G are normalized to $mR_o v_o/e$, whereas F and g are normalized to $R_o B_o$. It is convenient to introduce the function $G'(\psi,\theta) \equiv \partial G(\psi,\theta)/\partial\psi$. With the above normalization the functions $g(\psi)$, $F(\psi,\theta)$, $B(\psi,\theta)$, and $G'(\psi,\theta)$ are all slowly varying in ψ. Generally one cannot invert (9) and (10) to obtain ψ and u as explicit functions of the canonical momenta. It is easy, however, to obtain the equations of motion in the coordinates $(\rho_\parallel, \psi, \theta, \phi)$, where $\rho_\parallel \equiv u/B$. The results are shown in Appendix B. The formalism is most similar to that of White [14] except that we have used an orthogonal coordinate system to describe the tokamak geometry. In general, these results differ from those obtained by Kaufman [11] because the parallel velocity in our formulation is not strictly proportional to the time rate of change of the toroidal angle (as is the case when the guiding center is following a magnetic field line).

In the absence of the electrostatic potential, the toroidal angular momentum, p_ϕ, and the unperturbed Hamiltonian, H_o, are both conserved. (Note, $H_o = u^2/2 + MB(\psi, \theta)$.) Transforming to the action-angle coordinates of the unperturbed Hamiltonian is done by following the standard procedure of classical mechanics [15]. The definitions of H_o and p_ϕ are used to obtain p_θ as a function of θ, H_o, and p_ϕ. The poloidal action, I_θ, is defined as the area (divided by 2π) in the (p_θ, θ) plane enclosed by, or beneath a curve of constant H_o and p_ϕ, depending on whether the orbit is trapped, or circulating. This is written as

$$I_\theta = \oint \frac{d\theta}{2\pi} p_\theta(\theta; H_o, p_\phi) \quad . \tag{12}$$

The toroidal action, I_ϕ, is identical to p_ϕ. The canonically conjugate angle variables, ζ_θ and ζ_ϕ, are obtained from the generating function,

$$S(\theta, \phi, I_\theta, I_\phi) = \int^\theta d\theta' p_\theta(\theta'; H_o(I_\theta, I_\phi), I_\phi) + I_\phi \phi \quad , \tag{13}$$

by differentiating with respect to the actions,

$$\zeta_i = \frac{\partial S}{\partial I_i} \quad , \tag{14}$$

where the subscript i denotes either θ, or ϕ. Also, the identities, $p_\theta = \partial S/\partial\theta$ and $p_\phi = \partial S/\partial\phi$, are recovered from the generating function. Note, that H_o is a function of the actions as given implicitly by (12). The unperturbed frequencies for both the θ and ϕ motions are

$$\dot{\zeta}_i = \frac{\partial H_o}{\partial I_i} \equiv \Omega_i(I_\theta, I_\phi) \quad . \tag{15}$$

Physically, Ω_θ, is the bounce (transit) frequency of a trapped (circulating) particle. Ω_ϕ is the bounce-average of $\dot\phi$, which is equivalent to $\Omega_\theta \Delta\phi/2\pi$, where $\Delta\phi$ is the amount of toroidal rotation during one poloidal orbit. In action-angle coordinates the Hamiltonian takes on the form

$$H = H_o(\mathbf{I}) + \Phi(\mathbf{I}, \boldsymbol{\zeta}, t) \quad , \tag{16}$$

where $\mathbf{I} = (I_\theta, I_\phi)$ and $\boldsymbol{\zeta} = (\zeta_\theta, \zeta_\phi)$.

III Action-Angle Coordinates For Circulating Electrons

In this section we consider an ensemble of electrons in the vicinity of a specific flux surface in the core of the plasma. Furthermore, we consider only circulating electrons since they are primarily responsible for the absorption of the RF power during LHCD. We write the functions $g(\psi)$, $F(\psi,\theta)$, $B(\psi,\theta)$, and $G'(\psi,\theta)$ as power series in the inverse aspect ratio, ϵ, defined as r/R_o, where r is the radial distance from the magnetic axis:

$$
\begin{aligned}
g(\psi) &= 1 + \mathcal{O}\epsilon^2 \\
F(\psi,\theta) &= \mathcal{O}\epsilon^2 \\
B(\psi,\theta) &= 1 - \epsilon(\psi)\cos\theta + \mathcal{O}\epsilon^2 \\
G'(\psi,\theta) &= q(\psi) + q_1\cos\theta + \mathcal{O}\epsilon^2 \quad .
\end{aligned}
\tag{17}
$$

The ordering in (17) assumes a low β circular equilibrium [13] and reflects the fact that the ratio of the poloidal to toroidal field is of order ϵ and modifications to the vacuum toroidal field are of order ϵ^2. The safety factor $q(\psi)$ is the poloidal average of $G'(\psi,\theta)$ and $q_1(\psi)$ is of order ϵ, as needed to satisfy the identity $G' = (\mathbf{B}\cdot\nabla\phi)/(\mathbf{B}\cdot\nabla\theta)$. For example, if one ignores the Shafranov shift of the flux surfaces, so that $\psi = \psi(r)$, then $q_1(\psi) = -\epsilon(\psi)q(\psi)$. In a self consistent toroidal equilibrium the shift of the surfaces must be included to correctly determine $q_1(\psi)$ through order ϵ. We ignore all terms of order ϵ^2 and higher.

To obtain an explicit form for the action-angle transformation we exploit the fact that both $B(\psi,\theta)$ and $G'(\psi,\theta)$ are slowly varying functions of ψ; i.e $B = B(\lambda\psi,\theta)$ and $G' = G'(\lambda\psi,\theta)$, where $\lambda \ll 1$. Physically, the size of λ is obtained by considering the relation between radial distance and the normalized poloidal flux, $\Delta r = \rho_\theta\Delta\psi$, where $\rho_\theta = mv_o/eB_\theta$ and B_θ is the poloidal component of the magnetic field at $r = r_o$. Typically ρ_θ is small compared to the radial distance over which B and G' vary. We now consider a two parameter expansion, in which both ϵ and λ are small. We consider λ to be of order ϵ, so that we keep terms of first order in λ, while ignoring terms of order $\epsilon\lambda$ and λ^2. For fast electrons in the core of the plasma, $\lambda \sim \rho_\theta/r_o \sim 10^{-2}$, whereas $\epsilon_o \sim 10^{-1}$. Technically our expansion should go to second order in ϵ to consistently retain terms of first order in λ. However, we shall see that going to first order in λ allows us to treat the important physical effect of magnetic shear; going to second order in ϵ would make things unnecessarily complicated. By ignoring terms of order $\epsilon\lambda$, we take $\epsilon(\psi) \approx \epsilon_o$ and $q_1(\psi) \approx q_{1o}$, where the subscript denotes that each quantity is evaluated at the surface r_o. The safety factor is of order unity so we expand it to first order in λ, i.e.

$$q(\psi) \approx q_o + q_o s_o \psi \quad , \tag{18}$$

where s_o is of order λ and represents the shear of the magnetic field. Note that we have chosen $\psi = 0$ at $r = r_o$. Using these expansions in (9) and (10) yields the canonical momenta,

$$
\begin{aligned}
p_\phi &= \psi + u(1 + \epsilon_o \cos\theta) \tag{19} \\
p_\theta &= -\int^\psi d\psi' G'(\psi', \theta) \\
&= -q_o(\psi + \frac{s_o}{2}\psi^2) - q_{1o}\psi \cos\theta \quad . \tag{20}
\end{aligned}
$$

To the same order the unperturbed Hamiltonian is $H_o = u^2/2 - M\epsilon_o \cos\theta$, where we have dropped the additive constant, M. For well circulating particles, which satisfy M/H_o of order unity or smaller, one obtains the following relations for $u(\theta, H_o)$ and $\psi(\theta, H_o, p_\phi)$:

$$u \approx \hat{u} + (M\epsilon_o/\hat{u}) \cos\theta \quad , \tag{21}$$

$$\psi \approx \hat{\psi} - \epsilon_o(\hat{u} + M/\hat{u}) \cos\theta \quad , \tag{22}$$

where $\hat{u} \equiv \sqrt{2H_o}$ and $\hat{\psi} \equiv p_\phi - \hat{u}$. The poloidal action is obtained by substituting (22) into (20) and averaging in θ; ignoring terms of order ϵ^2 and $\epsilon\lambda$ one finds that

$$I_\theta \approx -q_o\hat{\psi} - \frac{q_o s_o}{2}\hat{\psi}^2 \quad . \tag{23}$$

Noting that $p_\phi = I_\phi$, we can write the unperturbed Hamiltonian as

$$H_o = \hat{u}^2/2 = (1/2)[I_\phi - \hat{\psi}(I_\theta)]^2 \quad . \tag{24}$$

From H_o it follows that $\Omega_\phi = \hat{u} = q(\hat{\psi})\Omega_\theta$; ignoring terms of first order in λ would result in $q(\hat{\psi}) \approx q_o$, so that the ratio of the unperturbed frequencies would be a constant. The generating function in (13) can now be written explicitly to the required order

$$S(\theta, \phi, I_\theta, I_\phi) = I_\theta\theta + I_\phi\phi + \epsilon_o S_T(I_\theta, I_\phi)\sin\theta \quad , \tag{25}$$

where

$$S_T = q_o I_\phi + (1 + \frac{q_{1o}}{q_o\epsilon_o})I_\theta + \frac{q_o^2 M}{(q_o I_\phi + I_\theta)} \quad . \tag{26}$$

One obtains $\zeta_\theta = \theta + \epsilon_o(\partial S_T/\partial I_\theta)\sin\theta$ which is easily inverted through first order in ϵ to obtain $\theta \approx \zeta_\theta - \epsilon_o(\partial S_T/\partial I_\theta)\sin\zeta_\theta$. Similarly one obtains the relation $\phi \approx \zeta_\phi - \epsilon_o(\partial S_T/\partial I_\phi)\sin\zeta_\theta$. We thus have the following transformation equations:

$$\theta \approx \zeta_\theta - [1 + \frac{q_{1o}}{q_o\epsilon_o} - \frac{q_o^2 M}{(q_o I_\phi + I_\theta)^2}]\epsilon_o \sin\zeta_\theta \tag{27}$$

$$\phi \approx \zeta_\phi - [q_o - \frac{q_o^3 M}{(q_o I_\phi + I_\theta)^2}]\epsilon_o \sin \zeta_\theta \tag{28}$$

$$\psi \approx \hat{\psi}(I_\theta) - [I_\phi + \frac{I_\theta}{q_o} + \frac{q_o M}{(q_o I_\phi + I_\theta)}]\epsilon_o \cos \zeta_\theta \quad . \tag{29}$$

The equation for ψ was obtained by substituting ζ_θ for θ in the order ϵ term of (22). Using these equations, the electrostatic potential can be transformed from the toroidal coordinates, ψ, θ, and ϕ, to action-angle coordinates.

Let us briefly discuss the physics of various terms that arise in the above transformation. The order ϵ terms signify the toroidal effects, the parallel magnetic gradient and the curvature of the field lines. These give rise to oscillations in both $\theta(t)$ and $\phi(t)$, as well as in the radial coordinate $\psi(t)$, the latter coming from the combination of the ∇B and curvature drifts. The order ϵ oscillation in ϕ vanishes when $M \approx I_\phi + I_\theta/q_o$, in which case the oscillation along the field line from the magnetic mirror is just cancelled by the shrinking of the pathlength as the particle rotates from the outside to the inside of the tokamak. For fast electrons during LHCD the curvature effects will dominate since M will typically be much smaller than the average parallel energy, H_o. The only order λ contribution comes from the shear of the magnetic field lines, i.e. q is not a constant so that the proportionality between Ω_θ and Ω_ϕ depends on the poloidal action. We note that the order $\epsilon\lambda$ terms, which have been neglected here, would be needed to properly treat trapped particles since it is these terms that give rise to the toroidal precession of the banana orbit.

IV   Stochasticity In Action-Angle Coordinates

During LHCD the RF wave-field is driven at a single frequency, ω, by a source at the edge of the plasma. We write the potential as $\Phi = \text{Re}\{\Phi_c(\psi,\theta,\phi)e^{-i\omega t}\}$, where Φ_c is a complex function of the original toroidal coordinates. The electron dynamics is obtained by transforming Φ_c into the above action-angle coordinates. Since the transformed potential is a periodic function of the angles, ζ_θ and ζ_ϕ, we can write it as the following Fourier series:

$$\Phi = \text{Re}\{\sum_{n,l} C_{n,l}(\mathbf{I})e^{i(n\zeta_\phi + l\zeta_\theta - \omega t)}\} \quad , \tag{30}$$

where $C_{n,l}(\mathbf{I})$ is a complex function of the actions. We represent the transformation to action-angle coordinates as follows:

$$\phi = \zeta_\phi + \Lambda(\zeta_\theta,\mathbf{I}) \tag{31}$$

$$\theta = \Theta(\zeta_\theta, \mathbf{I}) \tag{32}$$

$$\psi = \Psi(\zeta_\theta, \mathbf{I}) \quad , \tag{33}$$

where explicit forms for Λ, Θ, and Ψ can be obtained by direct comparison with equations (27)-(29). The Fourier coefficients are

$$C_{n,l}(\mathbf{I}) = \int \frac{d\zeta_\theta}{2\pi} e^{-il\zeta_\theta + in\Lambda(\zeta_\theta, \mathbf{I})} \Phi_n(\Theta(\zeta_\theta, \mathbf{I}), \Psi(\zeta_\theta, \mathbf{I})) \quad , \tag{34}$$

where

$$\Phi_n(\theta, \psi) = \int \frac{d\phi}{2\pi} e^{-in\phi} \Phi_c(\theta, \psi, \phi) \quad . \tag{35}$$

The wave-particle resonance condition is simply $n\Omega_\phi(\mathbf{I}) + l\Omega_\theta(\mathbf{I}) = \omega$, which defines a curve in the two-dimensional action space. For a single mode (i.e. a distinct n and l) the actions will no longer be conserved, but there will still exist two independent constants of motion. These are easily found to be $nI_\theta - lI_\phi$, $\omega I_\phi - nH$, and $\omega I_\theta - lH$, where two of the three are independent (note H is the complete Hamiltonian). The orbits in this case are regular everywhere in phase space. For two modes it is easily shown that there will be only one constant of motion, in which case the dynamics can be obtained from an autonomous Hamiltonian with two degrees of freedom. Generally, there will be three or more modes and no constants of motion. In this case there exists three types of motion[16]: regular motion on a KAM torus, stochastic motion across resonances in the region where they overlap, and stochastic motion along resonances which are interconnected in action space (i.e. Arnold diffusion).

Physically the unperturbed guiding center motion has two well distinguished time scales, fast motion in the parallel direction (along the magnetic field) and slow drifts across the field. In the case of LHCD, the basic mechanism for induced stochastic dynamics arises from parallel accelerations caused by the waves. Ignoring the drift motion altogether will lead to the usual resonance overlap criterion for the formation of a thick stochastic layer in a two dimensional phase plane which describes the parallel motion alone. On a slower time scale this is coupled to the drift motion enabling particles to shift their average radial position. After a long time, particles can make large excursions perpendicular to the magnetic field, which in turn effects a gradual change in the structure of their parallel dynamics. For example, KAM curves in the parallel portion of phase space will gradually shift allowing particles to move from one resonance to another, even when such a transition is prohibited by motion in the parallel phase plane alone. The latter effect is Arnold diffusion.

To separate the fast motion along the magnetic field from the slow perpendicular motion, we introduce the following linear transformation of the action-angle coordinates:

$$
\begin{aligned}
I_\phi &= p_1 + p_2 \\
I_\theta &= -q_o p_2 \\
z_1 &= \zeta_\phi \\
z_2 &= \zeta_\phi - q_o \zeta_\theta \quad .
\end{aligned}
\tag{36}
$$

The new coordinates are z_1 and z_2, and their respective canonical momenta are p_1 and p_2. Note, the generating function for this transformation is $S(\zeta, \mathbf{p}) = (p_1 + p_2)\zeta_\phi - q_o p_2 \zeta_\theta$. Using equation (24) and (23) and neglecting terms of order s_o^2 (since s_o is of order λ), one can write the Hamiltonian as

$$
H = \frac{p_1^2}{2} + s_o p_1 \frac{p_2^2}{2} + \Phi(\mathbf{p}, \mathbf{z}, t) \quad .
\tag{37}
$$

Following from (30) the potential is

$$
\Phi = \mathrm{Re}\Big\{ \sum_{\mathbf{k}} C_{\mathbf{k}}(\mathbf{p}) e^{i(k_1 z_1 + k_2 z_2 - \omega t)} \Big\} \quad ,
\tag{38}
$$

where $k_1 = n + l/q_o$, $k_2 = -l/q_o$, $\mathbf{k} = (k_1, k_2)$, and $C_{\mathbf{k}}(\mathbf{p}) = C_{n,l}(\mathbf{I})$. If we take q_o to be a rational value then Φ is periodic in both z_1 and z_2 (the periodicity length for z_1 is an integer multiple of 2π depending on q_o, and for z_2 it is $2\pi q_o$). In Appendix A we derive this Hamiltonian for a magnetized slab with shear; the difference between a slab and a torus is contained in the coefficients $C_{\mathbf{k}}(\mathbf{p})$ of the wave spectrum, which are determined by equation (34) for a torus. From the Hamiltonian (37), we define the unperturbed frequencies $\mathbf{\Omega}(\mathbf{p}) = (\Omega_1, \Omega_2)$, where

$$
\Omega_1 = p_1 + s_o p_2^2/2 \quad ,
\tag{39}
$$

$$
\Omega_2 = s_o p_1 p_2 \quad .
\tag{40}
$$

Because $s_o \ll 1$, the frequencies are generally well separated, $\Omega_1 \gg \Omega_2$, with Ω_2 vanishing at the rational surface, $p_2 = 0$. The resonance curves in action space are $\mathbf{\Omega}(\mathbf{p}) \cdot \mathbf{k} = \omega$, which are nearly parallel lines given by $p_1 = \omega/k_1$. Two modes whose wave vectors, $\mathbf{k}$, correspond to the same k_1, will have resonance curves which intersect in action space at the point $p_1 = \omega/k_1$ and $p_2 = 0$. When $s_o = 0$, Ω_2 vanishes everywhere and these resonances are degenerate, becoming a single straight line throughout action space. We will show below that the intersection of resonance curves in action space and their degeneracy when s_o vanishes, plays an important role in a particle's radial motion.

Since the motion is fast in the (p_1, z_1) plane compared to the (p_2, z_2) plane, it is convenient to define the "parallel Hamiltonian" as

$$H_{\parallel} = \frac{p_1^2}{2} + \mathrm{Re}\{\sum_{k_1} A_{k_1} e^{i(k_1 z_1 - \omega t)}\} \quad , \tag{41}$$

$$A_{k_1} = \sum_{k_2} C_{\mathbf{k}}(\mathbf{p}) e^{ik_2 z_2} \quad , \tag{42}$$

where we consider z_2 and p_2 held fixed in the determination of the resonant amplitudes, A_{k_1}. (To obtain $H_{\parallel}$ from H, one sets the slow frequency, Ω_2, identically to zero.) For given values of z_2 and p_2 one can obtain a two dimensional surface of section for the parallel dynamics. Since the motion is both periodic in z_1 and in time, t, one can either view the dynamics in the (p_1, z_1) phase plane at successive values in time spaced by $2\pi/\omega$, or in the $(H_{\parallel}, t)$ plane at successive values of z_1 spaced by the periodicity length. One finds that the motion defined by $H_{\parallel}$ becomes stochastic in the region of overlapping resonances given by the usual criterion

$$2(\sqrt{|A_\nu|} + \sqrt{|A_\mu|}) > \omega|\frac{1}{\nu} - \frac{1}{\mu}| \quad , \tag{43}$$

where the ν and μ are neighboring components of the k_1 spectrum and $4\sqrt{|A_{k_1}|}$ is the full trapping width of the resonance at $p_1 = \omega/k_1$.

We now consider the radial motion obtained from the Hamiltonian in (37) by differentiating with respect to the coordinate z_2:

$$\dot{p}_2 = -\mathrm{Re}\{\sum_{k_1} B_{k_1} e^{i(k_1 z_1 - \omega t)}\} \quad , \tag{44}$$

where

$$B_{k_1} = \sum_{k_2} i k_2 C_{\mathbf{k}}(\mathbf{p}) e^{ik_2 z_2} \quad . \tag{45}$$

We may also consider the quasi-static limit where z_2 and p_2 are held fixed in B_{k_1}, so that the time rate of change of p_2, as given by (44), is completely determined by the orbit $z_1(t)$, which in turn is determined by the following:

$$\dot{z}_1 = p_1$$
$$\dot{p}_1 = -\mathrm{Re}\{\sum_{k_1} i k_1 A_{k_1} e^{i(k_1 z_1 - \omega t)}\} \quad . \tag{46}$$

Let us look at the radial motion of a particle trapped in the vicinity of a specific resonance. Ignoring all the non-resonant components of the spectrum, we consider only a single value of k_1. We chose to write $B_{k_1} = ik_1 A_{k_1}(\beta e^{i\xi})$, where β and ξ are both real. Substituting this into (45) and using (42), one obtains

$$\beta e^{i\xi} = \frac{\sum_{k_2} k_2 C_{\mathbf{k}}(\mathbf{p}) e^{ik_2 z_2}}{k_1 \sum_{k_2} C_{\mathbf{k}}(\mathbf{p}) e^{ik_2 z_2}} \quad . \tag{47}$$

If there is only one finite term in each of the sums over k_2 in (47), then $\beta = k_2/k_1$ and $\xi = 0$. In the generic case, the spectrum will contain more than one Fourier component, $C_{\mathbf{k}}(\mathbf{p})$, at a given k_1, so that ξ will generally be non-zero; this corresponds to the intersection of two or more resonances in action space. Without loss of generality, we can chose $k_1 A_{k_1} = -\alpha$, where α is real and positive. Ignoring the sums over k_1 in equations (44) and (46), they can be written in the following form:

$$\dot{p}_2 = -\alpha\beta \sin{(\varphi + \xi)} \tag{48}$$

$$\ddot{\varphi} + \alpha \sin\varphi = 0 \quad , \tag{49}$$

where $\varphi = k_1 z_1 - \omega t$. The solution for $p_2(t)$ in terms of $\varphi(t)$ and $p_1(t)$ can be written as follows:

$$p_2(t) = p_2(0) + (\beta \cos\xi)\delta p_1(t) - (\alpha\beta \sin\xi)\int_0^t dt' \cos\varphi(t') \quad , \tag{50}$$

where $\delta p_1(t) = p_1(t) - p_1(0)$. Although $\delta p_1(t)$ is bounded by the trapping width of the resonance, the integral over $\cos\varphi(t)$ is clearly not bounded since for deeply trapped particles $|\varphi(t)| \leq \pi/2$. In this case the equations predict a secular drift in $p_2(t)$, i.e. $p_2(t) \sim -(\alpha\beta \sin\xi)t$ for times long compared to the bounce frequency of motion in the resonance. As particles approach the separatrix surrounding the resonance, $\cos\varphi(t)$ is negative for most of the time, so the drift changes direction. Generally, this secular drift cannot persist indefinitely, since after some time the quasi-static approximation is invalidated by the secularity, i.e. we no longer can consider z_2 and p_2 to be fixed in (47). In the specific case that $\sin\xi = 0$, the secular term in (50) vanishes and the excursion in $p_2(t)$ is proportional to $\delta p_1(t)$. Although the solution in (50) is not valid for long times when the secularity is present, it does show that the excursions in $p_2(t)$ are not generally restricted by the trapping width of the resonance. This effect gives rise to Arnold diffusion along an interconnected "web" of resonances when the amplitudes are too small to allow stochastic motion across the resonances [16].

V   Quasilinear Diffusion During LHCD

When the spectral amplitudes of the wave field are large enough to satisfy the overlap criterion (43), yet small enough that their corresponding trapping widths are narrow compared to the width of the connected stochastic region, electrons in this region of phase space may be expected to diffuse quasilinearly. The quasilinear diffusion tensor in action space is [11]:

$$D_{i,j}(\mathbf{p}) = \sum_{\mathbf{k}} \frac{\pi}{2} |C_{\mathbf{k}}(\mathbf{p})|^2 \delta\left(\mathbf{\Omega}(\mathbf{p}) \cdot \mathbf{k} - \omega\right) k_i k_j \quad , \tag{51}$$

where the unperturbed frequencies, $\mathbf{\Omega}(\mathbf{p})$, are given in equations (39) and (40). (Note, that we have written the scalar potential as the real part of a complex Fourier series (38), so that its spectral energy is $|C_{\mathbf{k}}|^2/2$.) Although $D_{i,j}(\mathbf{p})$ is a singular tensor field it should be smoothed out by coarse-graining action space because the resonances are broadened by their trapping widths. Neglecting Ω_2 compared to Ω_1, we approximate the resonance condition as $p_1 \approx \omega/k_1$ and coarse grain in p_1 by averaging over an interval whose width is larger than the distance between resonances, but much smaller than the width of the stochastic region. The coarse-grained diffusion tensor is a function of $\mathbf{p}$; we will refer to this as the local diffusion tensor. The off diagonal elements of the local diffusion tensor are necessary in the Fokker-Planck operator to give the correct diffusion paths in $\mathbf{p}$ space. From (51) we can obtain a global estimate of the radial diffusion coefficient, $D_r = \langle D_{2,2}(\mathbf{p})\rangle_{\mathbf{p}}$, where the averaging is over the entire stochastic region in $\mathbf{p}$ space. This global quantity is simply an indicator of the magnitude of the radial diffusion process which is described in detail by the local diffusion tensor. The quasilinear autocorrelation time, τ_{ql}, is

$$\tau_{ql} = \frac{\pi \langle k_1 \rangle}{\omega \Delta k_1} \quad , \tag{52}$$

where $\langle k_1 \rangle$ is the average value of k_1 over the spectrum whose width is Δk_1. Assuming that most of the spectral energy is located within the stochastic region, we may write D_r in the following way

$$D_r \approx \tau_{ql} \langle \sum_{\mathbf{k}} \frac{1}{2} k_2^2 |C_{\mathbf{k}}(\mathbf{p})|^2 \rangle_{p_2} \quad , \tag{53}$$

where $p_1 = \omega/k_1$ when it appears in $|C_{\mathbf{k}}(\mathbf{p})|^2$ and the final averaging is only over p_2. The global radial diffusion coefficient is converted into physical units by multiplying the expression in (53) by $(v_o/R_o)\rho_\theta^2$.

To evaluate the local diffusion tensor (51), or the global radial diffusion coefficient (53), we need to have an expression for the wave field inside the plasma. We consider that in the core of the plasma, in the vicinity of r_o, there exists an ensemble of waves propagating in both the toroidal and poloidal directions and across flux surfaces. The potential is assumed to be of the following form:

$$\Phi = \mathrm{Re}\{\sum_{m,n} a_{m,n} e^{i(n\phi+m\theta+\kappa_{m,n} r-\omega t)}\} \quad , \tag{54}$$

where $\kappa_{m,n}$ is determined by assuming that ratio of parallel to perpendicular wave numbers is fixed by the local dispersion relation for high frequency (above the lower hybrid frequency) electrostatic waves. In particular, we take

$$k_{\parallel} \approx (n + m/q_o)R_o^{-1}$$
$$k_{\perp} \approx [\kappa_{m,n}^2 + (\frac{m}{r_o})^2]^{\frac{1}{2}} \quad , \tag{55}$$

with $k_{\perp}/k_{\parallel}$ equal to the ratio of the local electron plasma frequency to the RF driving frequency [17]. Typically the spectrum launched into the plasma from the wave guide array is broad in n and vanishes outside the range $n_1 \leq n \leq n_2$; it is comparatively narrow in m and centered about $m = 0$. Although the results of toroidal ray tracing [18] and toroidal normal mode analysis [19, 20] show a significant upshift in the poloidal mode numbers of the spectrum inside the plasma, we will initially assume that the spectrum in the core is approximately the same as the spectrum launched into the plasma. In this case $k_{\parallel} \approx n/R_o$ and we can estimate n_2 from the condition that the lowest phase velocity in the spectrum, $\omega_{rf} R_o/n_2$, should intersect the tail of the central electron distribution function at about $4 \sqrt{kT/m_e}$, in order to drive sufficient current in the core of the plasma [21, 22]. For a frequency (f_{rf}) of about 2 GHz, in a 2.5 kev plasma with a major radius of 1 meter, this leads to $n_2 \approx 150$. In the following example we take $n_2 = 150$ and $n_1 = 100$, with $a_{m,n}$ identically vanishing outside of this range. Furthermore, we assume $a_{m,n}$ is finite only for $m = 0$. We now transform the potential (54) into action-angle coordinates using the transformation equations (27)-(29). We first note that $r = r_o + \rho_\theta \psi$ and rewrite (54) as

$$\Phi = \mathrm{Re}\{\sum_{n} a_n e^{i[n(\phi+\alpha\psi)-\omega t]}\} \quad , \tag{56}$$

where $\alpha = (k_{\perp}/k_{\parallel})(\rho_\theta/R_o)$. Since the RF frequency is on the order of the ion plasma frequency, we have $k_{\perp}/k_{\parallel} \sim \sqrt{m_i/m_e}$, so that $\alpha \sim \sqrt{m_i/m_e}\lambda\epsilon_o$. When using (28) and (29) to transform the combination $(\phi + \alpha\psi)$, there will be a term of order ϵ coming from the transfomation of ϕ and a term of order $\epsilon\alpha$ coming from the transformation of ψ. Since α is typically small we need only retain the order ϵ term which comes from transforming ϕ, so we evaluate the potential at $\psi \approx -I_\theta/q_o$ (i.e. at an electron's average radial position). We find that

$$\phi + \alpha\psi \approx \zeta_\phi + \alpha p_2 - q_o\epsilon_o(1 - \frac{M}{p_1^2})\sin\zeta_\theta \quad , \tag{57}$$

where we have used the definitions of p_1 and p_2 given in (36). Substituting (57) into (56), one obtains the following relation for the Fourier components:

$$\begin{aligned}
C_{n,l}(\mathbf{p}) &= a_n e^{in\alpha p_2}\left[\int \frac{d\zeta_\theta}{2\pi} e^{-il\zeta_\theta - inq_o\epsilon_o(1-M/p_1^2)\sin\zeta_\theta}\right] \\
&= a_n(-1)^l J_l\left(nq_o\epsilon_o(1-M/p_1^2)\right)e^{in\alpha p_2} \quad ,
\end{aligned} \tag{58}$$

where J_l is the ordinary Bessel function of order l. Using the definition of $\mathbf{k}$ following (38) we can change notation from $C_{n,l}(\mathbf{p})$ to $C_{\mathbf{k}}(\mathbf{p})$.

First consider $\epsilon_o = 0$, so that only the $l = 0$ coefficients in (58) are finite. Suppose we can write $|a_n|$ in terms of a smooth function for $n_1 \leq n \leq n_2$, i.e.

$$a_n = f(n)e^{i\xi_n} \quad , \tag{59}$$

where $f(n)$ is smooth and the phases, ξ_n, are randomly distributed. The coarse-grained diffusion coefficient, $D_{1,1}$, which follows from (51) can now be written as

$$D_{1,1}(p_1) = \frac{\pi\omega^2}{2p_1^3}[f(\frac{\omega}{p_1})]^2 \quad . \tag{60}$$

When $\epsilon_o = 0$ the other elements of the diffusion tensor vanish because p_2 is a conserved quantity. It is convenient to chose

$$f(n) = \Phi_o(\frac{n_1}{n})^{3/2} \quad , \tag{61}$$

so that $D_{1,1}$ in (60) is a constant. Chosing the velocity scale, $v_o = \omega_{rf}R_o/n_1$, so that $\omega = n_1$ (i.e. t is physical time normalized to $n_1\omega_{rf}^{-1}$) we find that $D_{1,1}(p_1) = (\pi/2)\Phi_o^2 n_1^2 \equiv D_o$. Note, that when $\Phi_o > (4n_1)^{-2}$, the overlap criterion is satisfied for $n_1/n_2 \leq p_1 \leq 1$, outside of which the diffusion coefficient drops from D_o to zero.

Assuming (61) and (59), and using (58), the quasilinear diffusion tensor for finite ϵ_o may be written as follows:

$$D_{i,j}(p_1) = D_o \sum_{l=l_1}^{l_2} \frac{k_1 k_i k_j}{(k_1 + k_2)^3} J_l^2(X) \tag{62}$$

where $k_1 = n_1/p_1$, $k_2 = -l/q_o$, $X = \epsilon_o q_o(k_1 + k_2)(1 - M/p_1^2)$, $l_2 = n_1 q_o(1 - p_1)/p_1$ and $l_1 = q_o(n_1 - n_2 p_1)/p_1$. The limits on the l sum arise because $k_1 = n + l/q_o = n_1/p_1$ and a_n

vanishes for $n < n_1$ and $n > n_2$. Also the diffusion tensor is not a function of p_2, because we approximated the radial dependence of the RF field as $e^{i\kappa_{m,n}r}$. The diffusion tensor (62) should be considered a local quantity defined in the vicinity of $p_2 = 0$, where $\epsilon = \epsilon_o$ and $q = q_o$. In Figure 1 we have plotted the elements of $D_{i,j}/D_o$ as functions of p_1 by numerically computing the sum in (62) for the case when $n_1 = 100$, $n_2 = 150$, $q_o = 2$, $\epsilon_o = .1$, and $M = 0$. It can be shown analytically that the central flat region of $D_{2,2}$ scales as $D_o\epsilon_o^2/2$. Figure 2 shows $D_{i,j}$ for the same parameters, except that $M = .2$. We see that although $D_{1,1}$ is nearly the same, $D_{1,2}$ and $D_{2,2}$ are significantly altered, since the factor $(1 - M/p_1^2)$ tends to reduce the argument of the Bessel functions in equation (58). Physically, the oscillation along a field line from the magnetic mirror tends to compensate for the shrinking of the parallel pathlength as particles rotate from the outside to the inside of the torus. For these examples, the basic scaling of the global radial diffusion coefficient, D_r, is obtained by substituting $k_2 \sim k_1\epsilon_o$ into (53). The resulting expression yields $D_r \sim D_o\epsilon_o^2$. Clearly, if the poloidal mode numbers, m, are large enough (as when they upshift due to toroidal effects on wave propagation) there will be an enhancement in the radial diffusion coefficient over the above example with m identically zero. One finds that for $|m| \gg q_o\epsilon_o n$ the dominant terms in the radial diffusion are obtained by ignoring the order ϵ effects altogether; in this case D_r becomes the usual expression for $E \times B$ diffusion induced by a poloidal electric wave field. We note that the $E_\phi \times B_\theta$ drift is one order of ϵ smaller than the drift induced via the parallel electric field by coupling to the equilibrium ∇B and curvature drifts. Consequently, the $E_\phi \times B_\theta$ drift is ignored because we have truncated the action-angle transformation to first order in ϵ.

The quasilinear diffusion can be compared to that observed by numerically following the exact orbits of an ensemble of particles whose initial conditions are at the same point in action space and uniformly distributed throughout the range of both angles. We introduce the following diagnostic quantity:

$$d_{i,j}(t; \mathbf{p_o}) = \frac{\langle p_i(t)p_j(t)\rangle - \langle p_i(t)\rangle\langle p_j(t)\rangle}{2tD_o} \tag{63}$$

where $\mathbf{p}(0) = \mathbf{p_o}$ and the angular brakets indicate an ensemble average. In Figures 3 and 4 we show $d_{i,j}(t; \mathbf{p_o})$ for $\Phi_o = 10^{-5}$ and two different values of $p_1(0)$. The other parameters are the same as given above for Figure 1. We can identify two time scales on which $d_{i,j}(t)$ varies: the fastest is τ_{ac}, the time required for the autocorrelation function of $\dot{p}_1(t)$ to decay, the slow time scale is τ_d, which is the time required for the ensemble to spread across the stochastic region, or the diffusion time. The time τ_{ac} can be distinguished in Figures 3 and 4 as the time it takes for the initial transient in $d_{i,j}(t)$ to decay, i.e. $\tau_{ac} \lesssim 1$. The quasilinear estimate of τ_{ac} is given by τ_{ql} in (52) (for this example, $\omega \sim \langle k_1 \rangle$ and $\Delta k_1 \sim 50$, giving $\tau_{ql} \sim .1$). For times long compared to τ_{ac}, but short compared to τ_d, Figures 3 and 4 show that $d_{i,j}(t; \mathbf{p_o}) \approx D_{i,j}(p_1(0))/D_o$, where the quasilinear diffusion tensor is given by equation (62) and is shown in Figure 1. The diffusion time for these parameters is quite long, on the order of 10^4, because D_o is so small (approximately 10^{-6}). In Figure 3, one sees that the transient portion of $d_{1,2}(t)$ persists much longer than those of the diagonal components of $d_{i,j}(t)$. We note that in the case of Figure 3, we chose $p_1(0) = .8$, which corresponds to the central flat region of $D_{i,j}(p_1)$ where $D_{1,2}$ nearly vanishes. When $p_1(0)$ is chosen towards the outer portion of the stochastic region, where $D_{1,2}(p_1)$ is large, the transient portion of $d_{1,2}(t)$ decays faster, as shown in Figure 4. From the Figures one can also identify smaller transients in $d_{i,j}(t)$ which take place on a longer time scale than τ_{ac}, but are still fast compared to τ_d. This intermediate time scale, which we define as τ_r, can be identified as the time required for the ensemble to spread in p_1 across the width between two neighboring resonances. The time scale, τ_r, arises generically when analyzing diffusion in two-dimensional stochastic maps with periodicity in only one direction, such as the Fermi map [23].

Having established the connection between the diffusion predicted by the local quasilinear tensor and that observed numerically, it is interesting to go back and look at the scaling for radial diffusion given by D_r in (53). In terms of the average parallel electric wave field

$$E_{\parallel}^2 = (\frac{mv_o^2}{eR_o})^2 \langle \sum \frac{1}{2} k_1^2 |C_{\mathbf{k}}(\mathbf{p})|^2 \rangle_{p_2} \quad , \tag{64}$$

we can write the radial diffusion coefficient in physical units as

$$D_{rad} \sim \frac{\langle k_1 \rangle E_{\parallel}^2}{2 f_{rf} \Delta k_1 B_o^2} \langle \langle (\frac{k_2 B_o}{k_1 B_\theta})^2 \rangle \rangle \quad , \tag{65}$$

where $D_{rad} \equiv D_r(v_o/R_o)\rho_\theta^2$ and the double brackets indicate an averaging over both the spectrum (weighted by the spectral energy) and the radial coordinate. For $\langle k_1 \rangle \approx \Delta k_1$, $f_{rf} \approx 2$ GHz and $B_o \approx 5$ Tesla, one finds that $D_{rad} \sim (.1 \text{m}^2/\text{sec})E_{\parallel}^2 \langle \langle (k_2 B_o/k_1 B_\theta)^2 \rangle \rangle$, where $E_{\parallel}$ is expressed in kV/cm. In the example problem, where we assumed that the poloidal mode numbers were identically zero, the term enclosed by double brackets is of order unity. For upshifted poloidal mode numbers, this term could be as large as $(B_o/B_\theta)^2$. Thus with large wave fields, $E_{\parallel} \gtrsim .1$kV/cm, quasilinear scaling can yield an appreciable RF induced radial diffusion. However at such large amplitudes, the typical trapping widths of individual resonances are on the order of the width of the stochastic region and the quasilinear approximation is invalid. For example, in Figure 3, $\Phi_o = 10^{-5}$ corresponding to an $E_{\parallel}$ of about 10^{-3}kV/cm and a trapping width $(4\sqrt{\Phi_o})$ of about 1.3×10^{-2}. Increasing Φ_o by two orders of magnitude results in a trapping width that is about one fourth of the width of the stochastic interval in p_1.

VI   Stochasticity And Diffusion From A Few Large Amplitude Waves

In this section we assume that the potential is given by (54), where m and n are the same order of magnitude. Ignoring the order ϵ piece of the action-angle transformation we write the potential as

$$\Phi = \sum \Phi_{\mathbf{k}} \cos\left(\mathbf{k} \cdot \mathbf{z} + \alpha_{\mathbf{k}} p_2 - t\right) \quad , \tag{66}$$

where $k_1 = (n + m/q_o)/n_o$, $k_2 = -m/(n_o q_o)$, $\alpha_{\mathbf{k}} = \rho_\theta \kappa_{m,n}$, and n_o is some typical toroidal mode number of the wave spectrum. The Hamiltonian has the same form as in equation (37), where we have rescaled the coordinates z_1 and z_2 by the factor n_o; t is physical time normalized to ω_{rf}^{-1} and the velocity scale is $v_o = \omega_{rf} R_o / n_o$. Noting that $\rho_\theta = v_o B_\phi / \omega_{ce} B_\theta$ and $q_o R_o / r_o = B_\phi / B_\theta$, and using (55), one obtains

$$\alpha_{\mathbf{k}} = \frac{k_\perp \omega_{rf} B_\phi}{k_\parallel \omega_{ce} B_\theta} [k_1^2 - (\frac{B_\phi k_\parallel}{B_\theta k_\perp})^2 k_2^2]^{\frac{1}{2}} \quad , \tag{67}$$

Typical parameter values are: $(k_\perp / k_\parallel)^2 \sim 10^3$, $B_\phi / B_\theta \sim 10$, and $\omega_{ce}/\omega_{rf} \sim 10^2$. Using these one finds that $\alpha_{\mathbf{k}} \sim \sqrt{10 k_1^2 - k_2^2}$. In the following we consider the stochasticity and radial transport induced by just a few such modes, where the spacing between the k_1 mode numbers is on the same order as k_1. In this situation the amplitude of the waves must be large to satisfy the overlap criterion and the width of the stochastic interval in p_1 is on the order of a few resonant trapping widths.

We first consider the case of just three modes, since for fewer modes than this there will always exist at least one constant of the motion. In particular we take: $\Phi_k = \Phi_o$ for $\mathbf{k} = (1.5, -.5)$, $(2, -1)$, and $(3, -1)$, where Φ_o is real and positive. Recall that the resonance condition is $\Omega(\mathbf{p}) \cdot \mathbf{k} = 1$, where the unperturbed frequencies are given in equations (39) and (40). Since $\Omega_2 \ll \Omega_1$, the resonance condition is $p_1 \approx 1/k_1$, so that the resonances for these modes occur at $p_1 = 1/3, 1/2$, and $2/3$, respectively. Each point in the (p_2, z_2) plane can be associated with a two-dimensional parallel Hamiltonian as defined in (41). We see that $|A_{k_1}| = \Phi_o$, where A_{k_1} is defined in (42), so that the overlap criterion in (43) is satisfied when $\Phi_o > 1.7 \times 10^{-3}$. The (p_1, z_1) surface of section for the parallel Hamiltonian is shown in Figure 5 for the case when $\Phi_o = 2 \times 10^{-3}$. Since $|A_{k_1}|$ is independent of p_2 and z_2, the parallel surface of section is similar at all points in the (p_2, z_2). We now let the coordinates p_2 and z_2 evolve according to the complete Hamiltonian dynamics, as opposed to being held constant in the above two-dimensional surface of section. For example, Figure 6 shows the time series, $p_1(t)$ and $p_2(t)$, of a typical stochastic orbit when $\Phi_o = 10^{-2}$. In this case $p_2(t)$ shows a fast stochastic variation superimposed upon a slow meandering. Focusing in on the fast stochastic motion, one can see that it is directly correlated to the p_1 time series. We showed in Section IV, following equation (50), that for motion in the vicinity of any non-degenerate resonance the excursions in p_1 and p_2 are proportional by the ratio of k_1/k_2. Since this ratio varies as the orbit jumps stochastically from one resonance to another, the overall time series for the two actions will not be proportional. Averaging over the fast stochastic variations one sees that the mean of p_2 tends to wander, whereas p_1 is confined to the region of overlapping resonances. Varying the shear parameter, s_o, does not have a significant effect on either the p_1 or p_2 time series in this example.

We now add a fourth mode to the spectrum in the above example. The additional mode is chosen in such a way as to create a degenerate resonance. In particular we take: $\Phi_k = \Phi_o$ for $\mathbf{k} = (1.5, -.5)$, $(2, -1)$, $(2, 0)$ and $(3, -1)$, where Φ_o is real and positive. For the set of modes in the first example, the resonance curves were non-degenerate, in the sense that each A_{k_1} was composed of only a single mode in the sum over k_2. On the other hand, in the present example, the resonance curves for the two modes with equal k_1 components, $(2,0)$ and $(2,-1)$, intersect at the point $p_1 = 1/2$ and $p_2 = 0$. In the limit of vanishing Ω_2, the resonances for these two modes coalesce throughout action space creating a degenerate resonance at $p_1 = 1/2$. We define Φ_d as the potential obtained by explicitly adding the two degenerate modes. Using the notation $\mathbf{k} = (k_1, k_2)$ and $\mathbf{k}' = (k_1, k_2')$ for the two degenerate modes, one can write Φ_d as follows:

$$\Phi_d = 2\Phi_o \cos \vartheta \cos \left(k_1 z_1 + \bar{\alpha} p_2 + \bar{k}_2 z_2 - t \right) \quad , \tag{68}$$

where

$$\vartheta = \frac{(k_2' - k_2)}{2} z_2 + \frac{(\alpha_{\mathbf{k}'} - \alpha_{\mathbf{k}})}{2} p_2 \quad , \tag{69}$$

$\bar{k}_2 = (k_2 + k_2')/2$, and $\bar{\alpha} = (\alpha_{\mathbf{k}} + \alpha_{\mathbf{k}'})/2$. The amplitude of the degenerate resonance is $2\Phi_o |\cos \vartheta|$, which depends on a particles position coordinates in the (p_2, z_2) plane as given by ϑ in (69). For the four mode example, the (p_1, z_1) surface of section of the parallel

Hamiltonian is shown in Figure 7 for two different positions in z_2 at $p_2 = 0$ and with $\Phi_o = 2 \times 10^{-3}$. In Figure 7a, the amplitude of the degenerate resonance vanishes because $\vartheta = \pi/2$ and the phase space is dominated by the remaining two resonances at $p_1 = 1/3$ and $2/3$. At this value of Φ_o the trapping widths of these resonances do not overlap and we see that they are seperated by phase spanning KAM curves in the surface of section. In Figure 7b, the degenerate resonance has nearly the same amplitude as the other two and the overlap criterion is satisfied forming a thick layer of stochasticity. At larger values of Φ_o, greater than about 7×10^{-3}, the trapping widths of the $p_1 = 1/3$ and $2/3$ resonances overlap. In this case a stochastic layer in parallel phase space is seen at all values of p_2 and z_2.

We continue our analysis of the four wave example by looking at the time series determined from the four-dimensional Hamiltonian. Figure 8 shows a typical stochastic orbit for $\Phi_o = 10^{-2}$ and $s_o = 0$; we see that $p_2(t)$ and $z_2(t)$ exhibit fluctuations about a uniform average drift. The origin of this streaming motion is related to the degenerate resonance as explained in Section IV following equation (50). The direction of the stream depends on the initial conditions in the (p_2,z_2) plane. It is possible to show that the streaming motion progresses along a line in the (p_2, z_2) plane, because ϑ in (69) is approximately conserved by the Hamiltonian dynamics with zero shear. This can be determined from the equations of motion by using the potential Φ_d given in (68) and ignoring the additional contributions of the two non-degenerate waves to the total potential; one finds that ϑ is proportional to $p_1(t)$, so that with $p_1(t)$ bounded to the region of overlapping resonances the streaming must approximately conserve ϑ. In addition the streaming can move in either direction along a line of constant ϑ depending on the the initial value of ϑ in Φ_d. When shear is included, the streaming motion cannot persist since ϑ is no longer approximately conserved. Figure 9 shows the time series for the same initial condition as in Figure 8, but with $s_o = 10^{-2}$. We see that $p_2(t)$ moves stochastically in a random walk fashion. As soon as p_2 becomes appreciable, the time dependence of z_2 is dominated by the linear frequency (i.e. $\dot{z}_2 \approx \Omega_2$, where $\Omega_2 = s_o p_1 p_2$). We note that the excursions in p_2 are no longer approximately proportional to the excursions in p_1, as was the case in the example with three non-degenerate resonances (see Figure 6). In fact we can see directly by comparing the $p_2(t)$ time series in Figures 6 and 9, that the radial excursions are significantly enhanced by the presence of intersecting resonances.

For stochastic orbits, the power spectra of the p_1 time series obtained from the four-dimensional Hamiltonian are the same as those obtained from the two-dimensional parallel Hamiltonian. The parallel Hamiltonian provides a useful way of visualizing the actual parallel dynamics which is really a projection onto the (p_1, z_1) plane of motion in a four-dimensional phase space. Regular orbits, or "sticky" orbits initially near invariant surfaces of the parallel Hamiltonian are significantly altered by the four-dimensional dynamics. In the four-dimensional case the invariant surfaces seen in the Figure 7 no longer isolate portions of the (p_1, z_1) plane. For example, chosing an initial condition in the small primary island of the degenerate resonance at $p_1 = 1/2$ (see Figure 7b), we know that in the two-dimensional case the orbit remains regular and is confined within this structure. Figure 10 shows the time series obtained from the four dimansional Hamiltonian for the same initial condition. We see that after many oscillations the orbit becomes stochastic. Although the orbit appears quasi-periodic while it is trapped in the island structure, only a small amount of Arnold diffusion is required for the orbit to escape into the stochastic region of the (p_1, z_1) plane. Consider the same initial condition, but now with $s_o = 0$, as in Figure 11. In this case, although we see aperiodic behavior in $p_1(t)$, the orbit never escapes the island, during which time it streams uniformly in the (p_2, z_2) plane. It is also interesting to compare the time series in $p_2(t)$ for two stochastic orbits, one at higher amplitude, as previously shown in Figure 9, and one at lower amplitude, as in Figure 12 (both with finite shear). In the lower amplitude case the motion is characterized by sharp jumps in p_2 intermingled with smaller scale fluctuations; basically the orbit spends long periods of time in the vicinity of each of the resonances as it stochastically moves from one to the other; when stuck in the vicinity of the intersecting resonances (i.e. the degenerate resonance when s_o vanishes), the orbit makes a large excursion in p_2. At larger amplitude the relative sharpness of these excursions fades, since the overall stickyness to any particular resonance is reduced.

To assess the radial transport induced by the waves, we consider the evolution of an ensemble of initial conditions. The ensemble is initially located at a specific p_2, z_2, and p_1, but uniformly spread in z_1 across a region of stochastic phase space. In particular we have chosen 1280 initial conditions at $p_2 = 0$, $z_2 = \pi/2$, $p_1 = .43$, and $0 < z_1 \le 4\pi$ corresponding to a stochastic strip in the (p_1, z_1) surface of section. The ensemble is allowed to evolve in time according to the equations of motion. The mean of the distribution, $\bar{p}_2(t)$, and the variance, $\sigma(t)$, are defined in the usual way: $\bar{p}_2(t) = \langle p_2(t) \rangle$ and $\sigma(t) = \langle p_2(t)^2 \rangle - \langle p_2(t) \rangle^2$, where the angular brackets denote ensemble averages. We will first discuss the results with finite shear ($s_o = 10^{-2}$). For long times, the mean is stationary and the variance grows

SPECTRUM $\mathbf{k} = (k_1, k_2)$	Φ_o	γ	σ_o
Three Waves			
$\mathbf{k} = (1.5, -.5), (2, -1), (3, -1)$	10^{-2}	1.0	5×10^{-6}
Four Waves			
$\mathbf{k} = (1.5, -.5), (2, -1), (2, 0), (3, -1)$	10^{-2}	1.0	1×10^{-4}
	2×10^{-3}	0.9	5×10^{-5}

Table 1: Results for three wave and four wave examples; $\sigma(t) \approx \sigma_o t^\gamma$, where $\sigma(t) = \langle p_2(t)^2 \rangle - \langle p_2(t) \rangle^2$. When $\gamma = 1$ the radial motion is a random walk and the diffusion coefficient in physical units is $\sigma_o \pi f_{rf} \rho_\theta^2$. Additional parameters are $s_o = 10^{-2}$ and $\alpha_{\mathbf{k}} = (10k_1^2 - k_2^2)^{1/2}$ following equation (67).

proportional to a power of time, i.e. $\sigma(t) \approx \sigma_o t^\gamma$. By "long times" we mean long compared to the time required for the ensemble to spread over the stochastic interval in p_1 and long compared to the time required for the ensemble to mix throughout the range of z_2. On the other hand, the time cannot be so long that the ensemble has spread beyond the range of the local shear, i.e. $s_o \langle |p_2| \rangle \ll 1$. In this sense, the observed power law behavior for $\sigma(t)$ is not truely asymptotic, but rather it is applicative only over a range of times. Note, that $\alpha_{\mathbf{k}} \gg s_o$ so that for "long times" the ensemble has typically spread over a region large enough to sample the fast variation of the wave field with respect to p_2. For "long times" the results are independent of the ensemble's initial conditions, as long as they are located in the stochastic region of the (p_1, z_1) plane. For the four wave spectrum at large amplitude ($\Phi_o = 10^{-2}$), we find that $\gamma = 1.0$ and $\sigma_o = 1 \times 10^{-4}$; $\sigma(t)$ is shown in Figure 13, where the linear regime (i.e. $\sigma = \sigma_o t^\gamma$) is for $t/2\pi \gtrsim 10^3$. When γ is unity, the motion in p_2 is a classical random walk with a diffusion coefficient of $\sigma_o/2$. At smaller amplitude ($\Phi_o = 2 \times 10^{-3}$), the exponent in the power law for $\sigma(t)$ decreases; $\gamma = 0.9$ and $\sigma_o = 5 \times 10^{-5}$. Comparing this to the initial three wave spectrum, where there was no degenerate resonance, one finds a sharp decrease in the radial transport: for example at $\Phi_o = 10^{-2}$, $\gamma = 1.0$ and $\sigma_o = 5 \times 10^{-6}$. The results are summarized in Table 1. For the four wave example with $\Phi_o = 10^{-2}$, if we take $\rho_\theta \sim 10^{-3}$m for energetic electrons and $f_{rf} \sim 2GHz$, the radial diffusion in physical units ($\sigma_o \pi f_{rf} \rho_\theta^2$) is about 1 m^2/sec.

In the four wave example, the results with s_o zero are dramatically different because of the streaming orbits. In general, we observe either the entire ensemble, or part of it, streaming along a line of constant ϑ in the (p_2, z_2) plane. For example, at $\Phi_o = 10^{-2}$, in the case with zero shear we see $\bar{p}_2(t) \approx \mu t$, where the magnitude and sign of μ depend on the ensemble's initial value of ϑ. For an ensemble initially at $\vartheta = \pi/4$, we find that

$\mu \approx -3 \times 10^{-4}$. Compare this to the case with finite shear shown in Figure 14; for a short time $\bar{p}_2(t)$ streams negatively at a relatively uniform rate, at $t/2\pi \approx 250$ the slope of $\bar{p}_2(t)$ abruptly changes sign and it streams positively until $t/2\pi \approx 500$, after which it breaks into a fluctuating state with no average time rate of change. The effect of shear is to make $\Omega_2(\mathbf{p})$ finite for orbits which move away from the rational surface at $p_2 = 0$. In Figure 15, we show the ensemble averaged variance in z_2 as a function of time for the case with $s_o = 10^{-2}$. (The variance in z_2 is defined as $\langle z_2(t)^2 \rangle - \langle z_2(t) \rangle^2$). The long time dependence of the z_2 variance is a power law with an exponent of 3. This power law is predicted by integrating $\dot{z}_2 \approx s_o p_1 p_2$ with $p_1 \approx$ constant and with $p_2 \approx \pm \sqrt{\sigma_o t}$, as given by the effective random walk in p_2, and then squaring to form $\langle z_2(t)^2 \rangle$. When the shear is zero, the variance in z_2 ceases to spread after an initial transient which corresponds to the spreading of the ensemble throughout the stochastic region in p_1. In the lower amplitude case, $\Phi_o = 2 \times 10^{-3}$, Figure 16 shows two distributions in p_2 of ensembles initially started at $p_2 = 0$ and allowed to evolve until $t/2\pi = 10^4$: one is for $s_o = 10^{-2}$ and the other for $s_o = 0$. While the case with shear shows an evenly spread distribution function (whose variance is growing as $t^{0.9}$), the case with zero shear shows a long negative tail. This tail forms because of the fast streaming motion of particles which get stuck in the primary island of the degenerate resonance in the (p_1, z_1) plane, since at this amplitude the primary resonance still retains a stable fixed point as shown in Figure 7b. In the four-dimensional dynamics, a particle's motion in the (p_2, z_2) plane, allows it to pass in and out of such islands. Thus, even though all initial conditions are chosen in the stochastic region, the island gradually acquires a small population of steaming particles.

VII   Conclusions

The Lagrangian formulation of guiding center motion provides an elegant and powerful method for deriving canonical coordinates in toroidal geometry. Although an explicit transformation between toroidal coordinates and action-angle coordinates cannot generally be obtained, it is possible to obtain an approximate transformation by exploiting the fact that the deviation of an electron's orbit from its average flux surface is small compared to the scale length of the equilibrium magnetic inhomogeneity. The straight forward expansions used in the text provide an explicit representation for the Hamiltonian in action-angle coordinates that is valid for circulating particles in the vicinity of a given flux surface. This is given by the unperturbed Hamiltonian in (24) and the transformation equations, (27) through (29), which are needed to obtain the wave field in action-angle coordinates. In the action-angle representation both the wave induced $E \times B$ drift and the radial motion that arises from the coupling of the wave induced parallel acceleration to the equilibrium toroidal drift, occur at the same level of description, i.e. they simply result in different components of the wave field's action-angle spectrum. Once the poloidal electric field of the wave is large enough (approximately $|m| > \epsilon q n$), the dominant mechanism for radial transport is the wave induced $E \times B$ drift.

The unperturbed frequencies, $\Omega_\phi(\mathbf{I})$, and $\Omega_\theta(\mathbf{I})$ of the action-angle coordinates are nearly proportional except for a small term which arises due to the shear of the magnetic field lines. A simple linear transformation defined by (36) yields a new canonical coordinate system, $(\mathbf{p}, \mathbf{z})$, whose unperturbed frequencies satisfy the ordering $\Omega_1(\mathbf{p}) \gg \Omega_2(\mathbf{p})$. The frequency $\Omega_2(\mathbf{p})$ vanishes at $p_2 = 0$, which corresponds to some rational q surface in the plasma. The Hamiltonian expressed in $(\mathbf{p}, \mathbf{z})$ coordinates, as given in (37), describes the dynamics in the vicinity of this rational surface. Because of the seperation between Ω_1 and Ω_2, the four-dimensional phase space naturally divides into two coupled two dimensional phase planes: (p_1, z_1) and (p_2, z_2), where the motion in the first plane is fast compared to the second. Holding the coordinates p_2 and z_2 fixed, we define an effective two dimensional Hamiltonian, $H_\parallel$, as in (41). The stochastic threshold determined by $H_\parallel$ occurs when the wave spectrum satisfies the usual overlap criterion, as given by (43).

For a broad spectrum of randomly phased waves, with closely spaced resonances whose trapping widths overlap, we have measured the stochastic diffusion in $\mathbf{p}$ space and compared it with the local quasilinear diffusion tensor evaluated for the same spectrum. For wave fields large enough that the quasilinear radial diffusion becomes appreciable, the trapping widths of individual resonances extend over a large portion of the stochastic region. Our initial investigations of this large amplitude regime indicate a dramatic reduction of the radial diffusion in comparison to the quasilinear value; more numerical work is needed to determine the scaling of the radial diffusion with respect to the wave amplitude. We are currently considering the case when the wave modes are not randomly phased and the field is represented by a spatially localized wave packet.

For the case when stochasticity is induced by a few large amplitude waves, we find that the radial transport is greatly enhanced when the wave-particle resonance curves intersect in action space. When resonances do not intersect, the excursions in p_2 are always nearly proportional to the excursions in p_1, where the latter are bounded by the trapping widths of the resonances. When resonances intersect, excursions in p_2 are greatly enhanced, even though the stochastic motion in p_1 looks very similar. Without shear, the intersecting resonances become degenerate and we find that stochastic trajectories stream in the (p_2, z_2) plane. At large amplitudes, an ensemble of trajectories, initially at the same point in the (p_2, z_2) plane and spread throughout the stochastic region in the (p_1, z_1) plane, will stream at a uniform rate, i.e. $\langle p_2(t) \rangle \approx \mu t$, where μ depends on the initial condition in the (p_2, z_2) plane. At lower amplitudes, where the primary resonances of the parallel Hamiltonian still retain stable fixed points, some trajectories in the ensemble get trapped by the degenerate resonance; their corresponding streaming motion creates a long tail to the ensemble's radial distribution, as shown in Figure 16. With small but finite shear, the streaming motion does not persist and the resulting transport in p_2 is diffusive; in the large amplitude case we find a diffusion coefficient of about 1 m^2/sec (see Table 1). At lower amplitudes the transport is slower than would be given by a classical random walk. In general we find that after an initial transient, the growth in the mean square of p_2 is proportional to a power of time; at lower amplitudes the exponent is below unity. We are currently in the process of obtaining the scaling of this relation with respect to wave amplitude.

VIII   Acknowledgements

We wish to acknowledge fruitful discussions with Dr. Dieter J. Sigmar. This work was supported in part by National Science Foundation Grant No. ECS-88-2475 and in part by U.S. Department of Energy Contract No. DE-AC02-78ET-51013.

References

[1] R.K. Kirkwoood, PhD Thesis Department of Nuclear Engineering, Massachusetts Institute of Technology, Cambridge, MA (September 1989); Phys. Fluids B (to appear in June, 1990).

[2] V. Fuchs, V. Krapchev, A.K. Ram, and A. Bers, Physica **14D** 141 (1985).

[3] G.M. Zaslavskii, Sov. Phys. JETP 61 (1985) 1176.

[4] A.K. Ram, A. Bers, K. Kupfer, Phys. Lett. A **138**, 288 (1989).

[5] R.G. Kleva, J.F. Drake, Phys. Fluids **27**, 1686 (1984).

[6] W. Horton, D.I. Choi, Plasma Phys. and Contr. Fusion **29**, 901 (1987).

[7] K. Kupfer, A. Bers, A.K. Ram, *Proceedings of the Eighth Topical Conference on RF Plasma Heating*, May 1-3, 1989, Irvine, CA, A.I.P. Conf. Proc. 190 (R. McWilliams, ed.) New York, pp. 434-437.

[8] R.G. Littlejohn, J. Plasma Phys. **29** 111 (1983).

[9] T.M. Antonsen, Jr., K. Yoshioka, Phys. Fluids **29**, 2235 (1986).

[10] K.C. Shaing, Phys. Fluids **31**, 2249 (1988).

[11] A.N. Kaufman, Phys. Fluids **15**, 1063 (1972).

[12] A.H. Boozer, Phys. Fluids **27**, 2441 (1984).

[13] J.P. Freidberg, *Ideal Magnetohydrodynamics*, Plenum Press (New York, 1987).

[14] R.B. White, M.S. Chance, Phys. Fluids **27**, 2455 (1984).

[15] L.D. Landau, E.M. Lifshitz, *Mechanics*, Pergamon Press (New York, 1976).

[16] A.J. Lichtenberg, M.A. Lieberman, *Regular and Stochastic Motion*, Springer-Verlag (New York, 1983).

[17] A. Bers, *Proceedings of the Third Topical Conference on RF Plasma Heating*, Jan. 11-13, 1978, California Institute of Technology, Pasadena, CA, pp. A1:1-10.

[18] P.T. Bonoli, R.C. Englade, Phys. Fluids **29**, 2937 (1986).

[19] D. Moreau, Y. Peysson, J.M. Rax, A. Samain, J.C. Dumas, Nucl. Fusion **30**, 97 (1990).

[20] D. Moreau, J.M. Rax, A. Samain, Plasma Phys. and Contr. Fusion **31**, 1895 (1989).

[21] N.J. Fisch, Phys. Rev. Lett. **41**, 873 (1978)

[22] V. Fuchs, R.A. Cairns, M.M. Shoucri, K. Hizanidis, A. Bers, Phys. Fluids **28**, 3619 (1985).

[23] A.J. Lichtenberg, M.A. Lieberman, *Diffusion In Two Dimensional Mappings*, Electronics Research Laboratory Memorandum No. UCB/ERL M88/5, University of California, Berkeley (1988).

Appendices

A   The Sheared Slab

Here we derive canonical coordinates for a magnetized slab with shear. The vector potential is chosen to be

$$\mathbf{A} = B_o(x\hat{y} - \frac{x^2}{2L_s}\hat{z}) \quad , \tag{70}$$

so that the magnetic field is

$$\mathbf{B} = B_o(\hat{z} + \frac{x}{L_s}\hat{y}) \quad , \tag{71}$$

where L_s is the length scale of the shear. Substituting these into the Lagrangian in equation (1) one obtains the canonical momenta:

$$p_z = u\frac{B_o}{B(x)} - s\frac{x^2}{2} \quad , \tag{72}$$

$$p_y = x(1 + su\frac{B_o}{B(x)}) \quad , \tag{73}$$

where the Hamiltonian is

$$H = \frac{u^2}{2} + MB(x) + \Phi(x,y,z,t) \quad . \tag{74}$$

The units of velocity are v_o and the units of time are ω_c^{-1}, where $\omega_{ce} = eB_o/m$. The units of distance are ρ_o, defined as v_o/ω_{ce}, and $s \equiv \rho_o/L_s \ll 1$. Noting that $u = p_z + sp_y^2/2 + \mathcal{O}s^2$, and $B(x) = 1 + \mathcal{O}s^2$, by ignoring terms of order s^2 one obtains

$$H_o = \frac{p_z^2}{2} + sp_z\frac{p_y^2}{2} \quad , \tag{75}$$

which is the unperturbed Hamiltonian in equation (37). The potential $\Phi(x,y,z,t)$ can be written in the canonical coordinates most simply by exchanging x for p_y and ignoring terms of order $s\partial\Phi/\partial x$, which arise through equation (73).

Note, the unperturbed Hamiltonian system is invariant under the following scale transformation: $z = aZ$, $y = bY$, $t = aT$, $p_y = (a/b)P_y$, and $p_z = P_z$. The equations of motion in the lower case and upper case variables are identical if one rescales the shear parameter as $S = (a/b)^2 s$. The scale used in the text is obtained by making the assignment $b = q_o r_o/\rho_o$ and $a = R_o/\rho_o$, where upon we may interpret Z as z_1, Y as z_2, and the parameter S as s_o.

B   Guiding Center Equations in Toroidal Coordinates

Here we obtain the equations of motion in the coordinates $(\rho_{\parallel}, \psi, \theta, \phi)$, where $\rho_{\parallel} \equiv u/B$. By directly applying the Euler-Lagrange equations to the Lagrangian in (8), where p_{θ} and p_{ϕ} are functions of ψ, θ, and $\rho_{\parallel}$, one obtains the four equations of motion:

$$\dot{\phi} = \mathcal{J}^{-1}[(G' - \rho_{\parallel}F')\frac{\partial H}{\partial \rho_{\parallel}} + F\frac{\partial H}{\partial \psi}] \tag{76}$$

$$\dot{\theta} = \mathcal{J}^{-1}[(1 + \rho_{\parallel}g')\frac{\partial H}{\partial \rho_{\parallel}} - g\frac{\partial H}{\partial \psi}] \tag{77}$$

$$\dot{\rho}_{\parallel} = -\mathcal{J}^{-1}[(1 + \rho_{\parallel}g')\frac{\partial H}{\partial \theta} + (G' - \rho_{\parallel}F')\frac{\partial H}{\partial \phi}] \tag{78}$$

$$\dot{\psi} = \mathcal{J}^{-1}[g\frac{\partial H}{\partial \theta} - F\frac{\partial H}{\partial \phi}] \quad , \tag{79}$$

where

$$\mathcal{J} = F(1 + \rho_{\parallel}g') + g(G' - \rho_{\parallel}F') \quad . \tag{80}$$

The prime superscript indicates differentiation with respect to ψ and the Hamiltonian, H, is considered a function of $\rho_{\parallel}$, ψ, θ, and ϕ, as given by

$$H = \frac{1}{2}\rho_{\parallel}^2 B^2(\psi, \theta) + MB(\psi, \theta) + \Phi(\psi, \theta, \phi, t) \quad . \tag{81}$$

$\mathcal{J}$ is the Jacobian of the coordinate transformation from the canonical momenta p_{θ} and p_{ϕ} to the non-canonical coordinates ψ and $\rho_{\parallel}$.

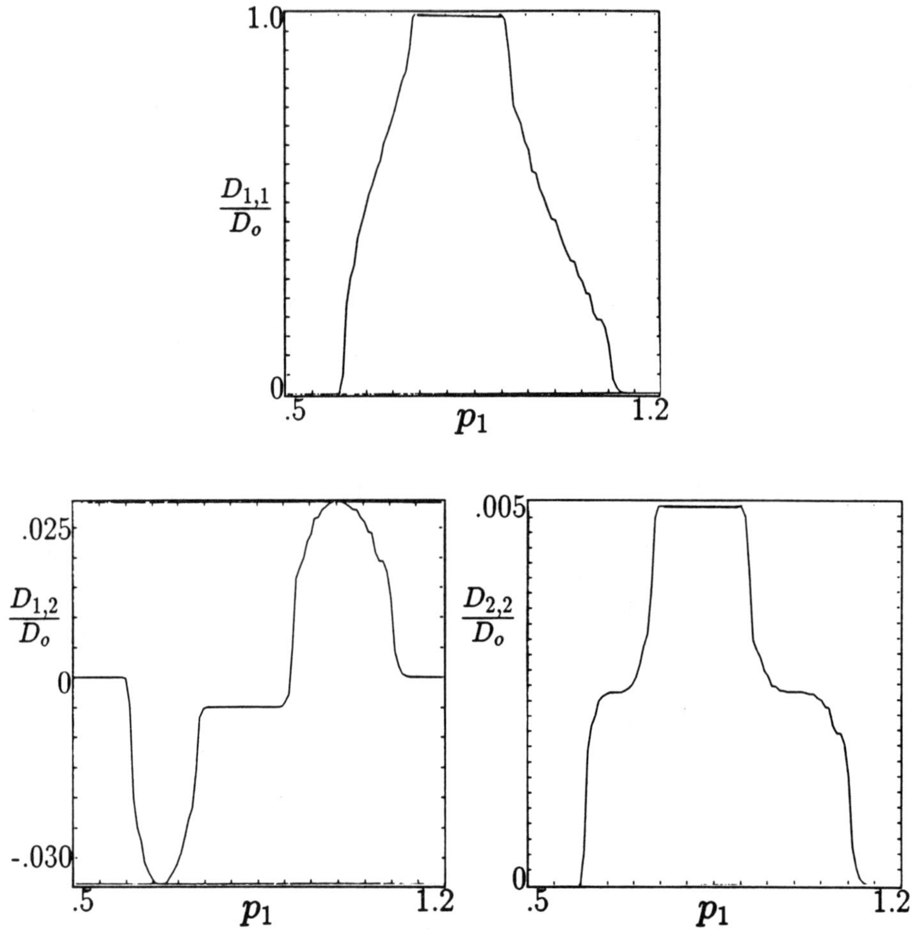

Figure 1: Elements of of quasilinear diffusion tensor, $D_{i,j}(p_1)$, determined from equation (62) for the following parameters: $n_1 = 100$, $n_2 = 150$, $q_o = 2$, $\epsilon_o = .1$, and $M = 0$. For $\epsilon_o = 0$, $D_{1,1} = D_o$ for a range of p_1 ($n_1/n_2 \leq p_1 \leq 1$) outside of which it vanishes; in this case the other elements of $D_{i,j}$ are zero for all p_1.

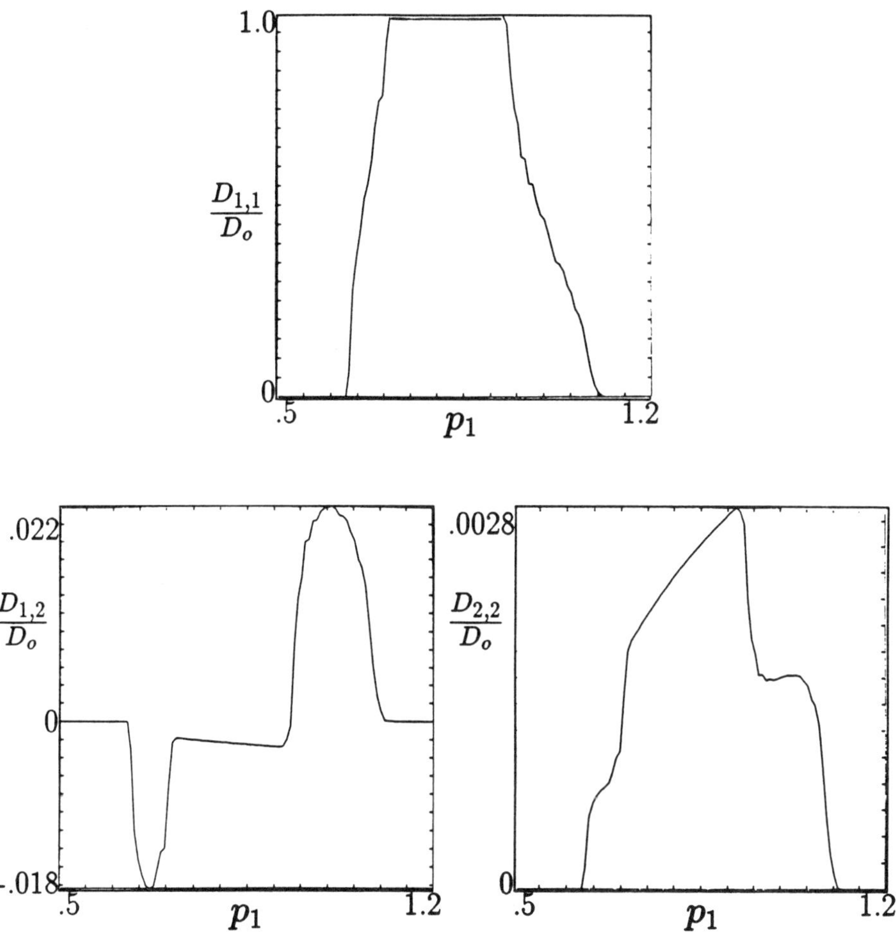

Figure 2: Elements of $D_{i,j}(p_1)$ for the same parameters as figure 1 except that $M = .2$.

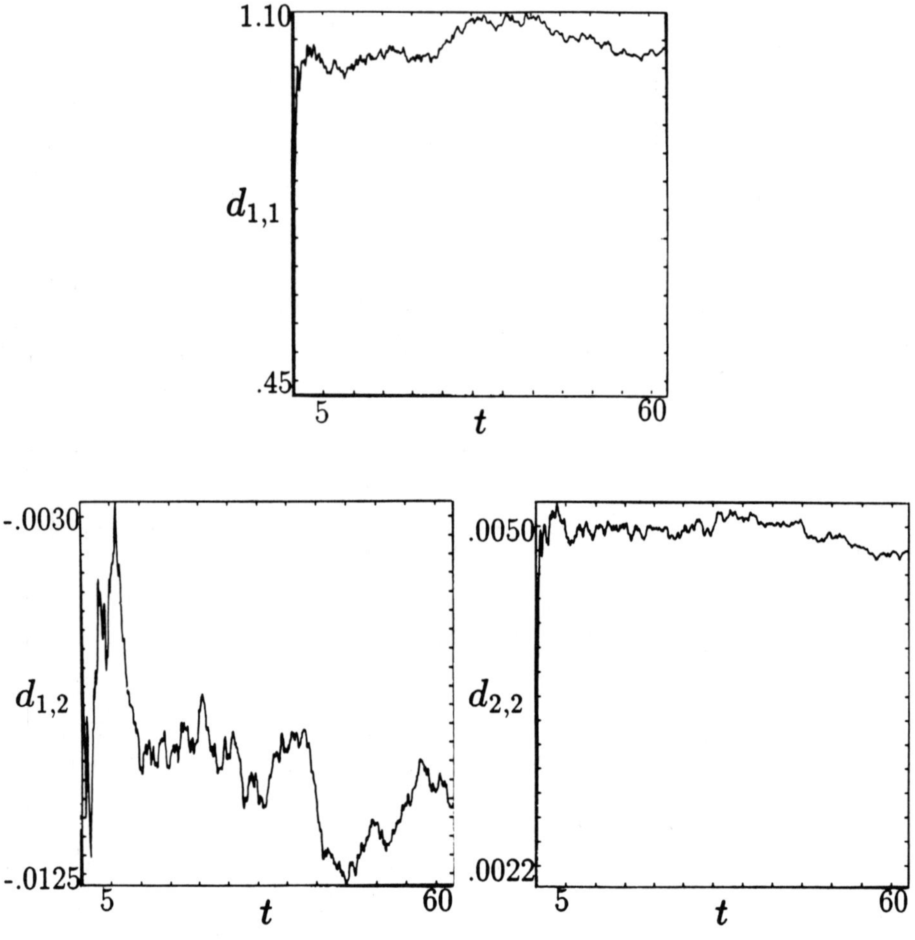

Figure 3: Elements of the diffusion diagnostic, $d_{i,j}(t; \mathbf{p}_o)$, defined in equation (63), where the parameters are the same as in figure 1 and $\Phi_o = 10^{-5}$. The ensemble of 1280 particles was initially at $p_1(0) = .8$ and $p_2(0) = 0$. (t is physical time normalized to $n_1\omega_{rf}^{-1}$).

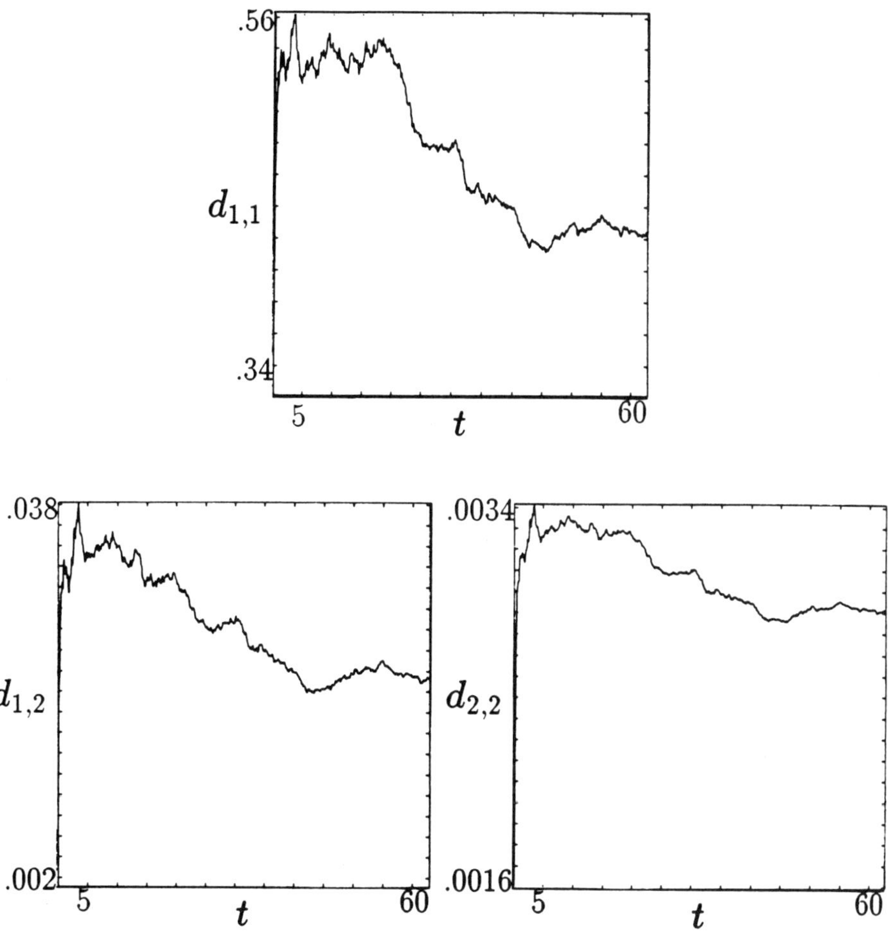

Figure 4: Elements of $d_{i,j}(t; \mathbf{p}_o)$ for the same parameters as in figure 3 except that $p_1(0) = 1.0$.

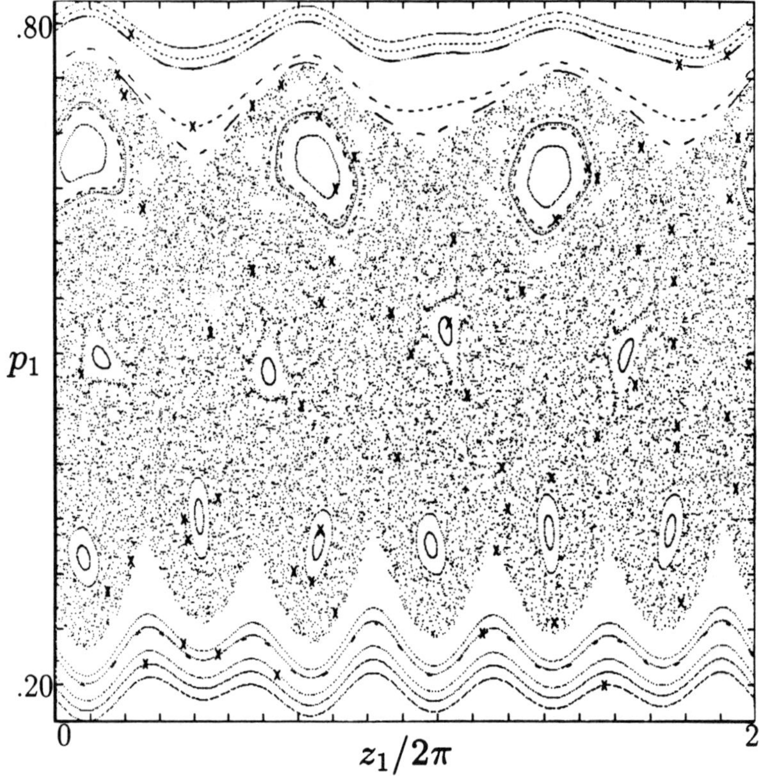

Figure 5: Surface of section, (p_1, z_1), of the parallel Hamiltonian for three waves with $\Phi_o = 2 \times 10^{-3}$, $p_2 = 0$ and $z_2 = \pi/2$. The wave-particle resonance conditions are satisfied at $p_1 = 1/3$, $1/2$, and $2/3$. Each initial condition is marked by an "X".

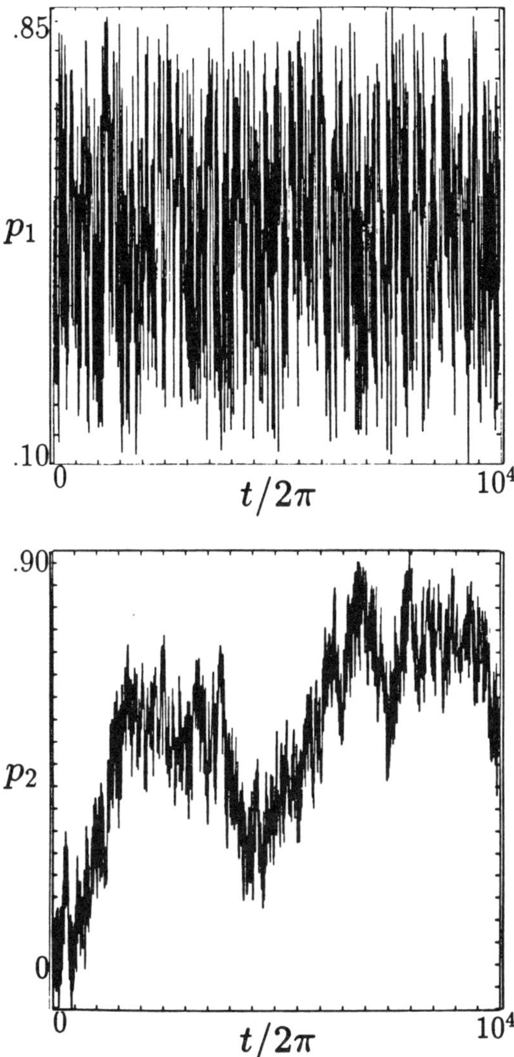

Figure 6: Stochastic time series, $p_1(t)$ and $p_2(t)$, for three waves with $\Phi_o = 10^{-2}$ and $s_o = 0$. (t is physical time normalized to ω_{rf}^{-1}).

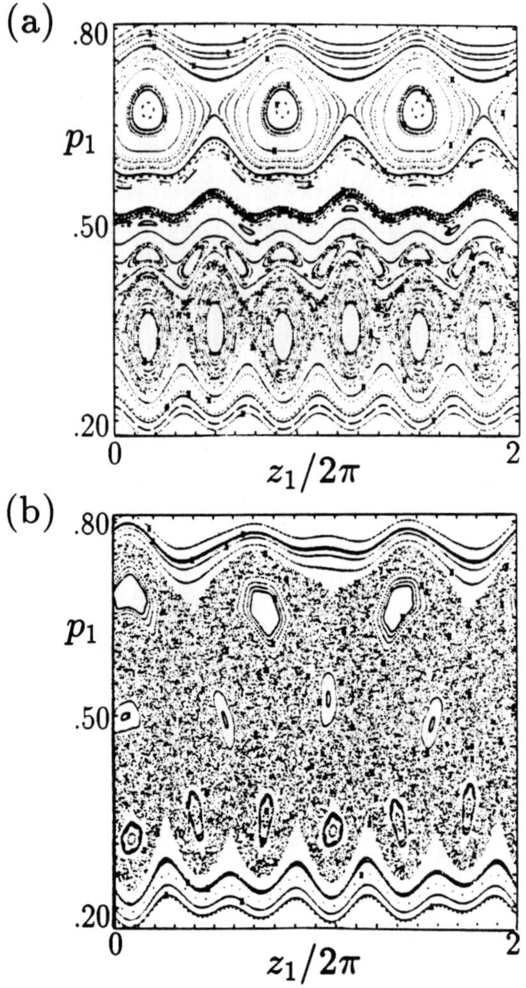

Figure 7: Surface of section, (p_1, z_1), of the parallel Hamiltonian for the four wave spectrum with degeneracy. The surface of section is shown at two values of z_2: (a) $z_2 = \pi$ and (b) $z_2 = \pi/2$, where in both cases $p_2 = 0$ and $\Phi_o = 2 \times 10^{-3}$. The wave-particle resonance conditions are satisfied at $p_1 = 1/3$, $1/2$, and $2/3$, where the $p_1 = 1/2$ resonance is degenerate.

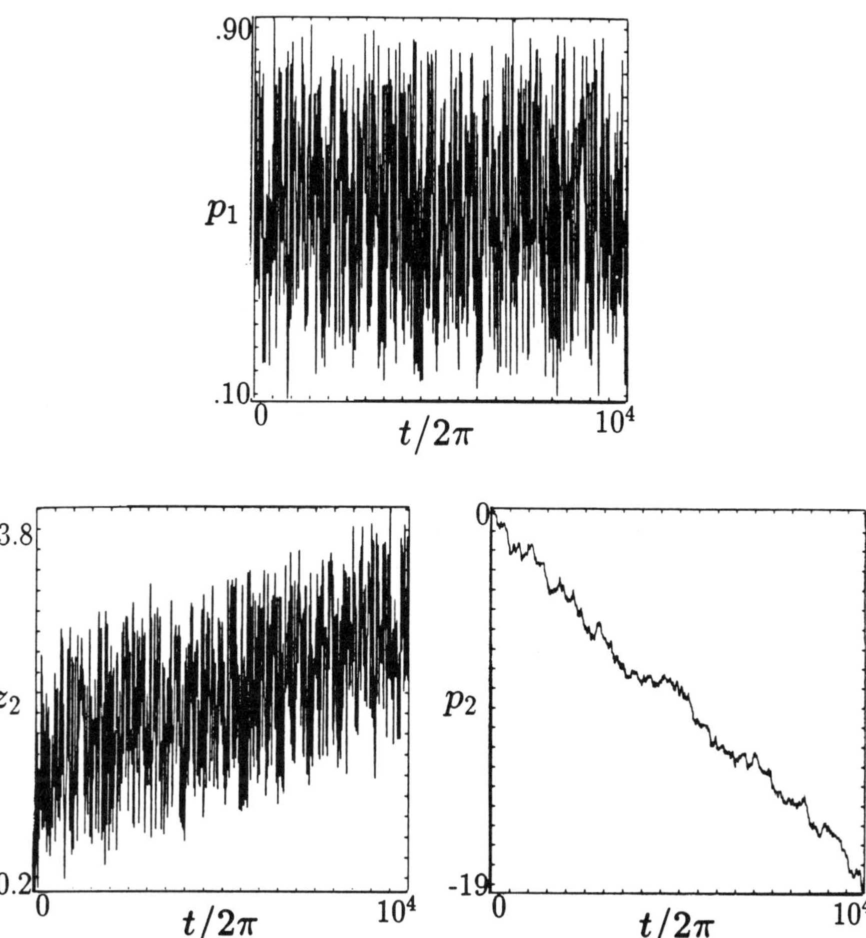

Figure 8: Stochastic time series, $p_1(t)$, $p_2(t)$, and $z_2(t)$, for four waves with $\Phi_o = 10^{-2}$ and $s_o = 0$. (t is physical time normalized to ω_{rf}^{-1}).

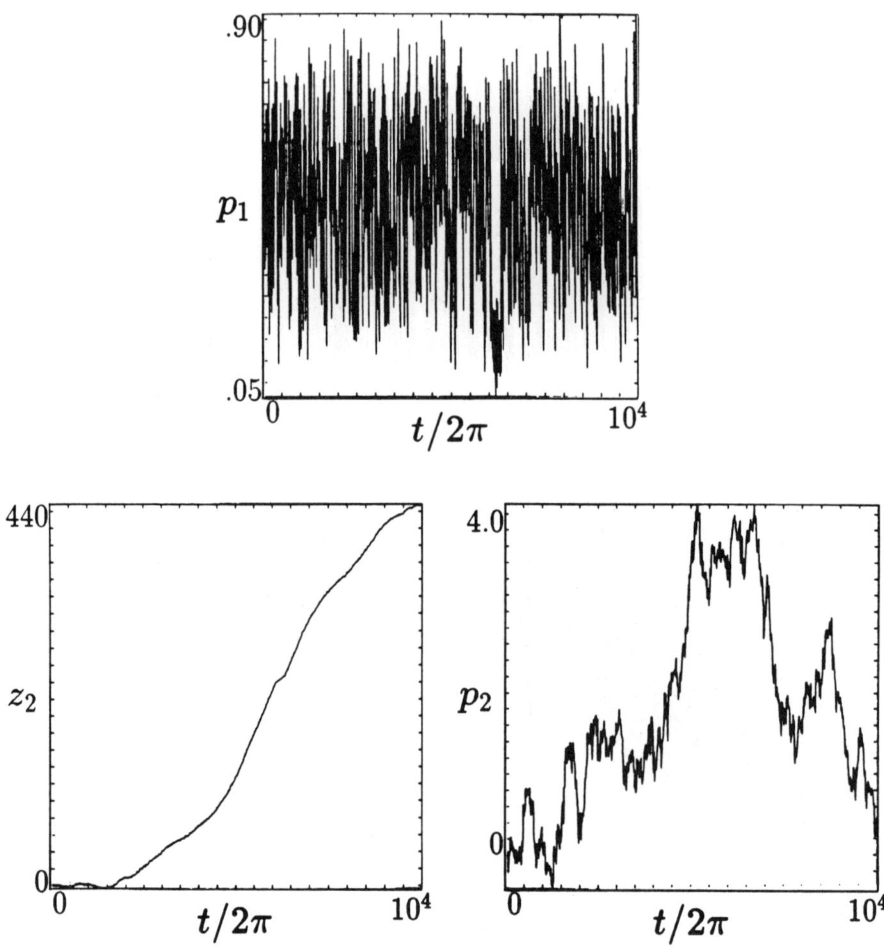

Figure 9: Stochastic time series, $p_1(t)$, $p_2(t)$, and $z_2(t)$, for four waves with $\Phi_o = 10^{-2}$ and $s_o = 10^{-2}$. (t is physical time normalized to ω_{rf}^{-1}).

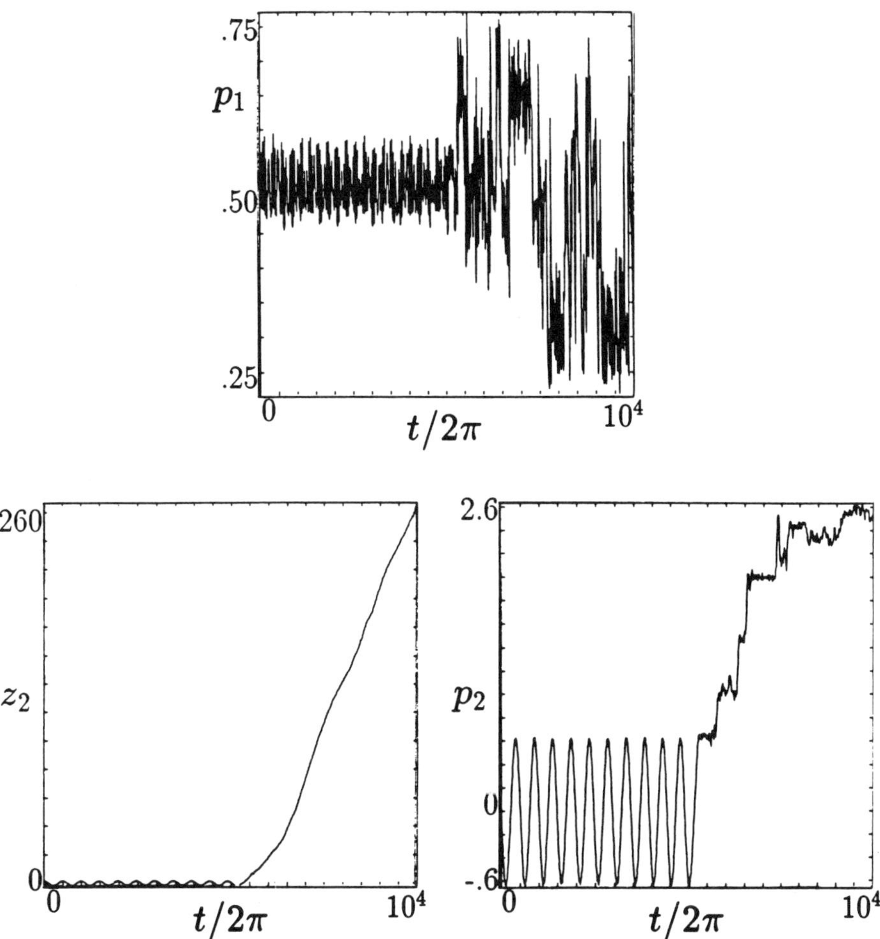

Figure 10: Time series, $p_1(t)$, $p_2(t)$, and $z_2(t)$, for initial condition in primary island of figure 7b with $s_o = 10^{-2}$. (t is physical time normalized to ω_{rf}^{-1}).

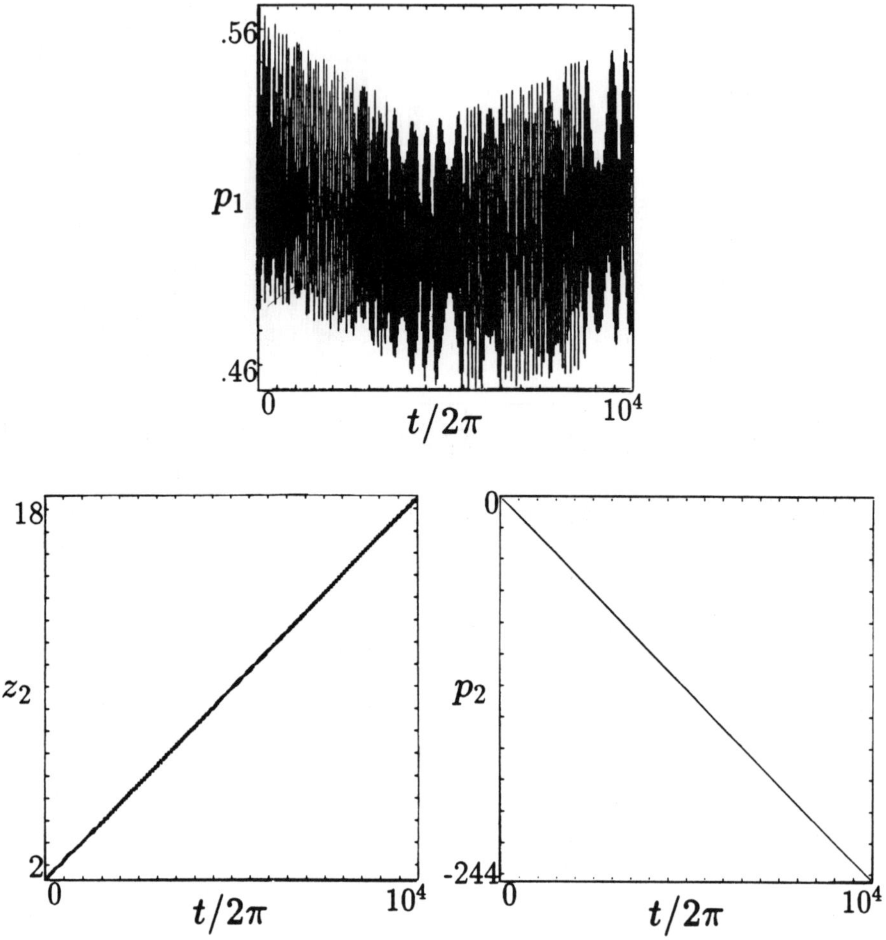

Figure 11: Time series, $p_1(t)$, $p_2(t)$, and $z_2(t)$, for initial condition in primary island of figure 7b with $s_o = 0$. (t is physical time normalized to ω_{rf}^{-1}).

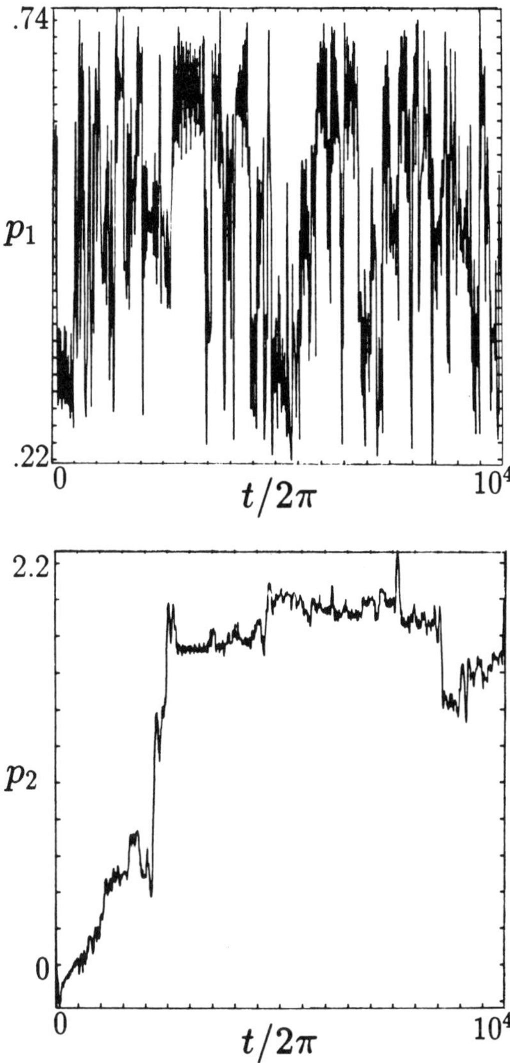

Figure 12: Stochastic time series, $p_1(t)$ and $p_2(t)$, for four waves with $\Phi_o = 2 \times 10^{-3}$ and $s_o = 10^{-2}$. (t is physical time normalized to ω_{rf}^{-1}).

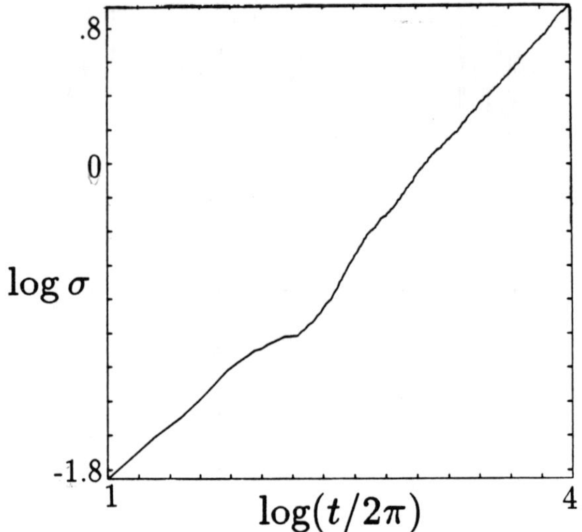

Figure 13: Log-log plot of the radial variance, $\sigma(t)$, versus time for four waves with $\Phi_o = 10^{-2}$ and $s_o = 10^{-2}$. The ensemble was orginally localized at $p_2 = 0$. (t is physical time normalized to ω_{rf}^{-1}).

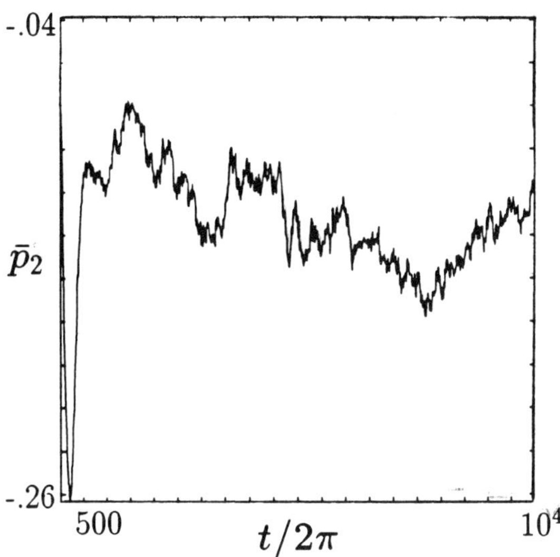

Figure 14: The ensemble average of p_2, $\bar{p}_2(t)$, versus time for the same parameters as in figure 13. (t is physical time normalized to ω_{rf}^{-1}).

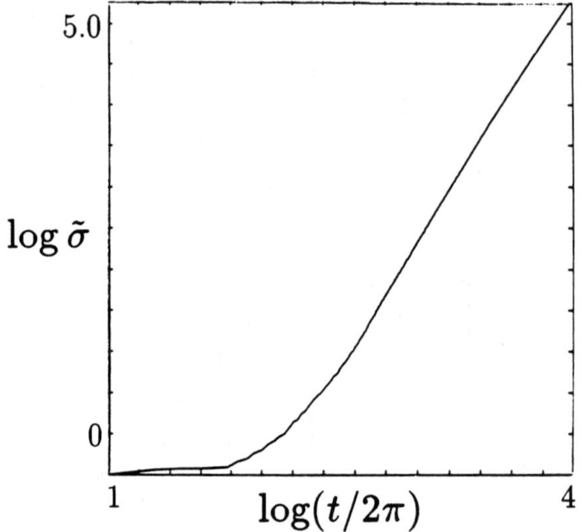

Figure 15: Log-log plot of the ensemble averaged variance in z_2, $\tilde{\sigma} \equiv \langle z_2(t)^2 \rangle - \langle z_2(t) \rangle^2$, versus time for the same parameters as in figure 13. (t is physical time normalized to ω_{rf}^{-1}).

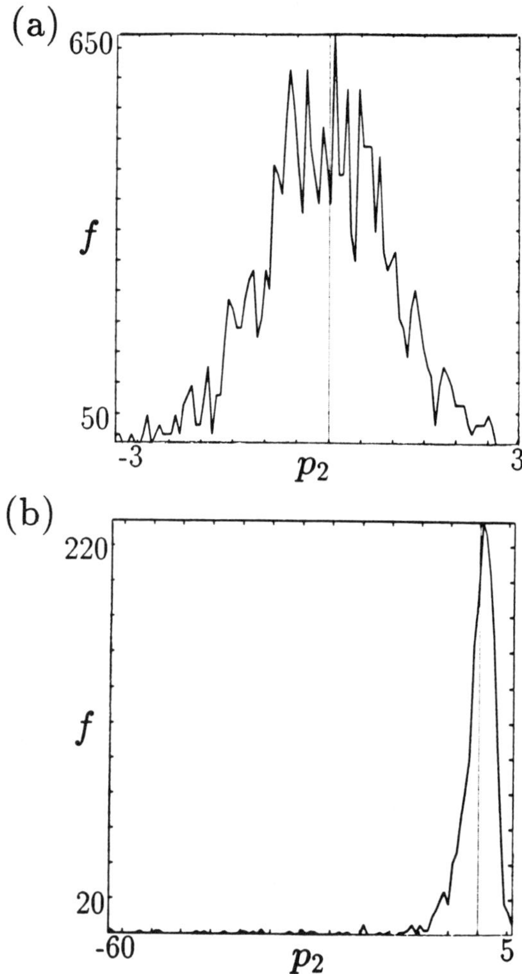

Figure 16: An ensemble of 1280 particles is initially localized at $p_2 = 0$ and spread in the stochastic region of the (p_1, z_1) plane, given by the surface of section in figure 7b. The distribution of the ensemble, $f(p_2, t)$, is shown at $t = 2\pi \times 10^4$ for two different cases: (a) $s_o = 10^{-2}$, and (b) $s_o = 0$. The dashed line marks $p_2 = 0$ and the area under each curve is equal to the total number of particles in the ensemble.

Solitons and Chaos in Resonance Absorption of EM Waves in Inhomogeneous Plasmas

C. S. Liu. W. Shyu, P. N. Guzdar, H. H. Chen, and Y. C. Lee

Laboratory for Plasma Research

University of Maryland

College Park, MD 20742-3511

ABSTRACT

The process of resonance absorption at the critical layer of an inhomogeneous unmagnetized plasma can be modelled by the nonlinear Schrödinger equation with additional inhomogeneity and driving terms. The transition from solitons to chaos is studied for this equation as the driving amplitude is increased.

In laser irradiated, the plasma density rises from zero to hundred times the liquid density in a very short distance, typically a few hundred microns. Inhomogeneity is therefore an important factor in considering various mechanisms in laser-plasma coupling. At high intensities the laser couples with the plasma through collective modes like Langmuir waves, ion acoustic waves and other electromagnetic waves. These mode coupling processes are resonant in nature. For example, parametric instability requires the frequency and wave number matching conditions

$$
\begin{aligned}
\omega_0 &= \omega_\ell + \omega_a, \\
\vec{k}_0 &= \vec{k}_\ell + \vec{k}_a,
\end{aligned}
$$

(1)

where $\omega_0, \vec{k}_0$ is the incident laser light frequency and wave number, respectively, $\omega_\ell, \vec{k}_\ell$, the Langmuir wave frequency and wave vector and $(\omega_a, \vec{k}_a)$ the ion acoustic wave frequency and wave vector. Also from the linear dispersion relation of these waves

$$\omega_\ell^2 = \omega_p^2 + k_\ell^2 v_e^2$$

and $\qquad\qquad\qquad\qquad\qquad\qquad\qquad\qquad\qquad\qquad\qquad\qquad\qquad$ (2)

$$\omega_a = k_a c_s$$

with $\omega_p = (4\pi n e^2/m)^{1/2}$, $v_e = T_e/m$, and $c_s = T_e/M_i$. n is the electron density, T_e and T_i are the electron and ion temperatures, respectively, and m and M_i the electron and ion masses, respectively. In an inhomogeneous plasma the frequency matching condition demands that this interaction take place near the critical surface where $\omega_0 \simeq \omega_p(x)$. Furthermore, the $\vec{k}$ matching condition limits the width of the resonance zone. Hence the instability is strongly localized.

In an inhomogeneous plasma the wave number of the plasma wave has spatial variations determined by the dispersion relation

$$\omega_\ell^2 = \omega_p^2(x) + k_\ell^2(x)v_e^2 = \text{const.}$$

Thus the width of the interaction region x_I near the resonance is given by the detuning condition $\kappa' x_I^2 = |\frac{d}{dx}[\vec{k}_0 - \vec{k}_\ell - \vec{k}_a]x_I^2| = 1$. The propagation of the decay waves (Langmuir and ion acoustic) with group velocities v_1 and v_2 out of this resonance zone thus gives rise to an effective damping rate $\gamma_d = \frac{\sqrt{v_1 v_2}}{x_I} \simeq (\kappa' v_1 v_2)^{1/2}$ for the parametric process. To overcome this damping due to inhomogeneity the laser intensity must exceed a certain threshold, i.e.,

$$\gamma_0 > \gamma_d = (\kappa' v_1 v_2)^{1/2},$$

where γ_0 is the growth rate of the parametric instability in a homogeneous plasma.[1]

Because the inhomogeneity causes strong localization of the instability, as the wave amplitude increases it can give rise to ponderomotive force which can in the nonlinear phase lead to the formation of solitons. Furthermore with the increase of laser intensity the soliton formation can occur in a chaotic manner. Here we will discuss the formation of solitons and transitions to chaos for a simple one spatial dimensional plasma. Two and three dimensions are clearly important considerations for nonlinear plasma waves since plasma wave collapse occurs. Nevertheless the 1D model is interesting in its own right.

We shall consider the simple laser plasma interaction in which the incident light suffers resonance absorption at the critical density. An obliquely incident EM wave has an electric field component E_{0x} along the density gradient in the x direction. The oscillation of the electrons in the x direction gives rise to a density perturbation given by

$$\frac{\delta n}{n_0} \propto \frac{v_{0x}}{\omega_0} \frac{1}{n_0} \frac{dn_0}{dx}$$

where

$$v_{0x} = -\frac{eE_{0x}}{m\omega_0}$$

is the oscillation velocity, ω_0 is the frequency of the incident wave and $\frac{1}{n_0}\frac{dn_0}{dx}$ the background density scale length. At the critical density where $\omega_0 = \omega_p(x)$, the density oscillations excites plasma waves resonantly.

Using the electron continuity equation, 1D equation of motion and Poisson's equation,

$$\frac{\partial}{\partial t}\delta n = -v_x\frac{dn_0}{dx} - n_0\frac{\partial v_x}{\partial x}, \tag{3}$$

$$\frac{\partial}{\partial t}v_x = -\frac{e}{m}(E_{0x}e^{-i\omega_0 t} + E_x) - \frac{1}{n_0 m}\frac{\partial(nT_e)}{\partial x}, \tag{4}$$

$$\frac{\partial E_x}{\partial x} = -4\pi e\delta n, \tag{5}$$

we can derive the equation for the driven linear plasma wave equation.

$$\frac{\partial^2 E_x}{\partial t^2} - v_e^2\frac{\partial^2 E_x}{\partial x^2} + \omega_p^2(x)E_x = -\omega_p^2(0)E_{0x}e^{-i\omega_0 t}, \tag{6}$$

where

$$\omega_p^2(x) = \omega_p^2(0)(1 + x/L). \tag{7}$$

Furthermore, writing $E_x(x,t)e^{-i\omega_p(0)t} + c.c.$ where $\omega_p(0)$ is the fast time scale $\simeq \omega_0$ and the time dependence in the $E(x,t)$ is slow, we get

$$i\frac{\partial E}{\partial t} - xE + \frac{\partial^2 E}{\partial x^2} = E_d, \tag{8}$$

where x has been normalized to the Airy length $x_0 = [v_e^2 L/\omega_p^2(0)]^{1/3}$, time has been normalized to $2L/(x_0\omega_p(0))$ and $E_d = LE_{0x}/x_0$.

In the steady state ($\partial/\partial t = 0$) the solution of this equation is shown in fig. 1 for $E_d = 1$. Near the origin the amplitude is large. For $x > 0$ the solution is evanescent, whereas for $x < 0$ it is outward propagating. The localized large amplitude near $x = 0$ can therefore give rise to a ponderomotive force which can push the plasma out of the region of intense field. Thus to account for this effect the local plasma frequency given by equation (5) can be rewritten as

$$\omega_p^2(x) = \omega_p^2(0)(1 + \frac{x}{L}) + \frac{\delta n}{n_0}, \tag{9}$$

where the additional term $\delta n/n_0$ is a measure of the density perturbation caused by the ponderomotive force. Furthermore since this $\delta n/n_0$ causes a nonlinear shift in the frequency this leads to detuning of the resonance and hence saturation of the instability.

To compute the density perturbation we use the slow time equations for the ions

$$n_0 M \frac{\partial v_i}{\partial t} = -\frac{\partial}{\partial x}(\delta n(T_i + T_e)) - \frac{\omega_p^2(0)}{2\omega_0^2} \frac{\partial}{\partial x} \frac{|E|^2}{4\pi}. \tag{10}$$

The first term on the R.H.S. is from the ion pressure gradient and gradient of the electrostatic potential. However, since for electrons $\frac{\delta n}{n_0} = \frac{e\phi}{T_e}$, the $-n_0 e \partial \phi/\partial x$ term is equated to $-\frac{\partial}{\partial x}(\delta n T_e)$. The second term is the ponderomotive force arising from the high frequency plasma wave. In the quasistatic approximation, the two terms on the R.H.S. balance and therefore give us the desired density perturbation (for $T_i = T_e$ and $\omega_0 = \omega_p(0)$)

$$\frac{\delta n}{n_0} = -\frac{|E|^2}{16\pi n_0 T_e}. \tag{11}$$

Using (7) and (9) in Eq. (4), together with the slow time ansatz, Eq. (6) now becomes

$$\frac{\partial A}{\partial t} - xA + \frac{\partial^2 A}{\partial x^2} + P|A^2|A = 1, \tag{12}$$

where $A = E/E_d$ and $P = \frac{L^2 k_D^2 v_0^2}{2v_e^2}$ and the space and time normalizations are the same as those used in Eq. (5). Furthermore, $k_D^2 = \omega_p^2(0)/v_e^2$, $v_0^2 = e^2 E_{0x}^2/m^2\omega_p^2(0)$.

Thus the process of resonance absorption in an inhomogeneous plasma can be represented by a nonlinear Schrödinger equation with two additional terms representing the effects of inhomogeneity ($-xA$) and driving (the term on the right-hand side).

With the right-hand side set equal to zero, this equation has been shown to be exactly solvable with multi-soliton solutions [4,5]. Also the threshold condition for single soliton and multi-soliton production for this case have been derived [5,6]. The inhomogeneity $-xA$ term in Eq. (2) leads to acceleration of the solitons into the underdense region [6] just like a wavepacket

$$\frac{dV_g}{dt} = \left(\frac{\partial^2 \omega}{\partial \kappa^2}\right)\dot{\kappa} = -\left(\frac{\partial^2 \omega}{\partial \kappa^2}\right)\left(\frac{\partial \omega}{\partial x}\right).$$
(13)

With the driving term on the right-hand side, this equation was first numerically integrated by Morales and Lee [2,3]. For $P < 0.8$, they found steady state solutions which could be represented by modified Airy functions. For $P > 1.0$ they found periodic emission of solitons. Adam et al. [7] later re-examined this problem and found that for $P < 0.3$, the solution is asymptotically stationary and behaves like an Airy function. For $0.3 \leq P \leq 1$, however they observed periodic growth of soliton-like structures near the critical density ($x = 0$), which subsequently convected down the density gradient, very much like the accelerated solitons found by Chen and Liu [4]. With still higher P, the soliton emission appeared to be chaotic. Various analytic attempts were also made to evaluate this transition process (Laedke and Spatschek, Anderson, Bussac et. al., Larroche et. al.) [9,10,11,12]. For the homogeneous case with dissipation, Moon and Goldman [8] have shown that the driven NSE displays the intermittency route to chaos. However, a definite route to chaos for the inhomogeneous case was not identified in any of the studies. In this paper, we wish to examine numerically the details of transitions from the stationary state to the chaotic attractor for increasing values of P.

Before discussing the solutions of NSE, we should mention the boundary conditions and numerical scheme. We solve Eq. (12) numerically for different P by a splitting method [13]. To implement outward propagation boundary conditions, we introduce an artificial spatial damping which increases towards the boundary. This absorbs the outward going waves and prevents any reflection at the boundary.

(i) In the range $0 \leq P \leq 0.575$, the solution is a steady state much like the Airy function for $P = 0$. This solution is well known [2,3] and is shown in Fig. 1.

(ii) $0.575 < P < 1.088$. Periodic emission of solitons to the underdense region is observed (Figs. (2), Figs. (3a)). Figure 2a displays the amplitude as a function of x at a fixed time. The solitons are produced near $x = 0$ and propagate downstream ($x < 0$). The corresponding density profile is shown in Fig. 2b. A three-dimensional plot of the amplitude as a function of x and t in Fig. 2c displays the periodic emission of solitons and the subsequent propagation. The amplitude at $x = 0$ varies periodically as shown in Fig. 3a. A typical frequency spectrum is displayed in Fig. 4a. $|A|_{max}$ and the emission rate increase slowly with P in this regime. Our numerical results show that the emission frequency ω_1 increases from 2.7 to 5.4 (Figs. (4a)). Chen and Liu [6] first calculated the emission rate as the inverse of the time for the soliton to propagate out of the resonance.

(iii) $1.088 < P < 1.155$. The time evolution of the wave amplitude at $x = 0$ undergoes period doubling (Figs. (3b)), the typical period doubled frequency spectrum is shown in Figs. (4b). The cause of the period doubling is the competition between convection and nonlinear detuning of the resonance. The nonlinear propagation tries to push the resonance point inward into the overdense region ($x > 0$), whilst the convective effect tries to shift the linear resonance point to the underdense region ($x < 0$). The emission point therefore oscillates between two resonance points and this causes the period doubling.

(iv) $1.1575 < P < 1.255$. A small frequency modulation beyond the period doubling phase is observed (Figs. (3c)). The frequency spectrum shows a second frequency ω_2 (Figs. (4c)) with $\Delta\omega = \omega_2 - \omega_1 \sim 0.15$. Other peaks are observed at the linear combination of these two frequencies and their harmonics.

We have been able to show analytically [15] that for $P > 1.25$ the two soliton bound-state solution appears. We observe that the outgoing waves have an internal oscillation characteristic of a breather. Numerical results show that the amplitude of the two solitons are $\eta_1 = 3.6$ and $\eta_2 = 0.3$. The breather frequency is given by [14]

$$\omega_2 = (\eta_1^2 - \eta_2^2)\frac{P}{2}.$$ (14)

The calculated difference between ω_1 and ω_2 is $\Delta\omega \sim 0.2$, which compares well with the numerically obtained $\Delta\omega \sim 0.15$ (Figs. (4c)). Thus at this point the system has evolved to a two-torus.

(v) For $P > 1.255$, the emission of soliton shows chaotic behavior (Figs. (5)). The

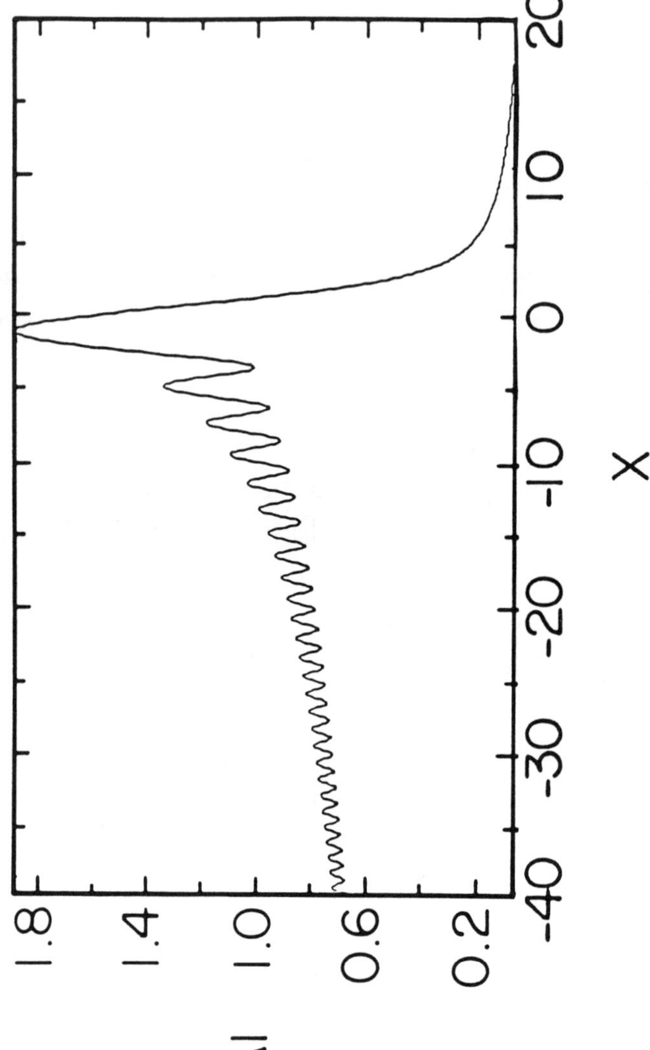

Figure 1.   Wave amplitude *vs.* x for steady state solution of Eq. (6) with *Ed* = 1.

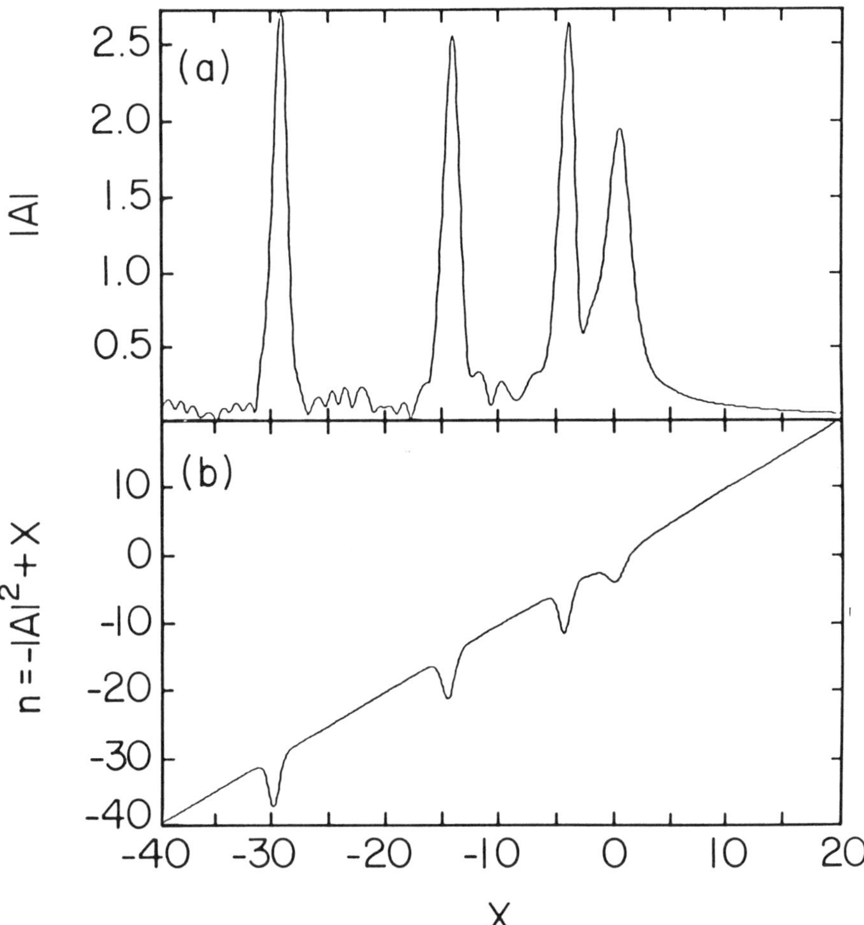

Figure 2.   (a) Wave amplitude vs. x. (b) Density profile $n = -|A|^2 + x$ vs. x.

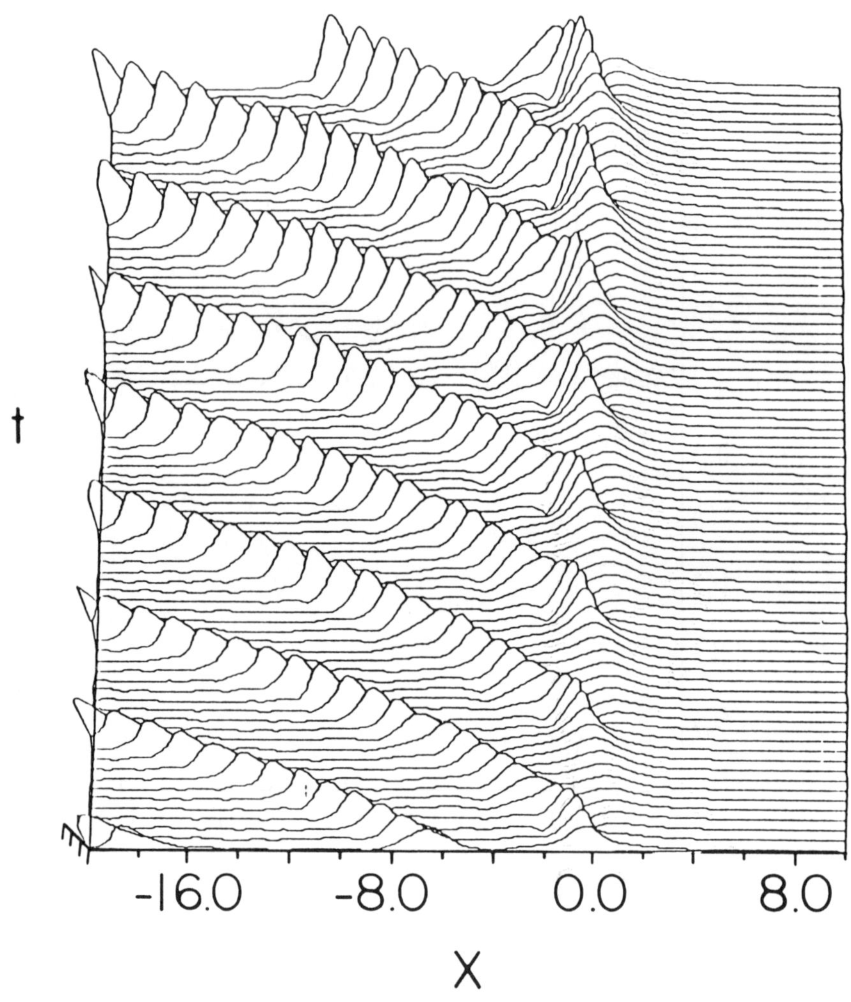

Figure 2.   (c) 3-D plot of wave amplitude as a function of x and t for P = 0.8.

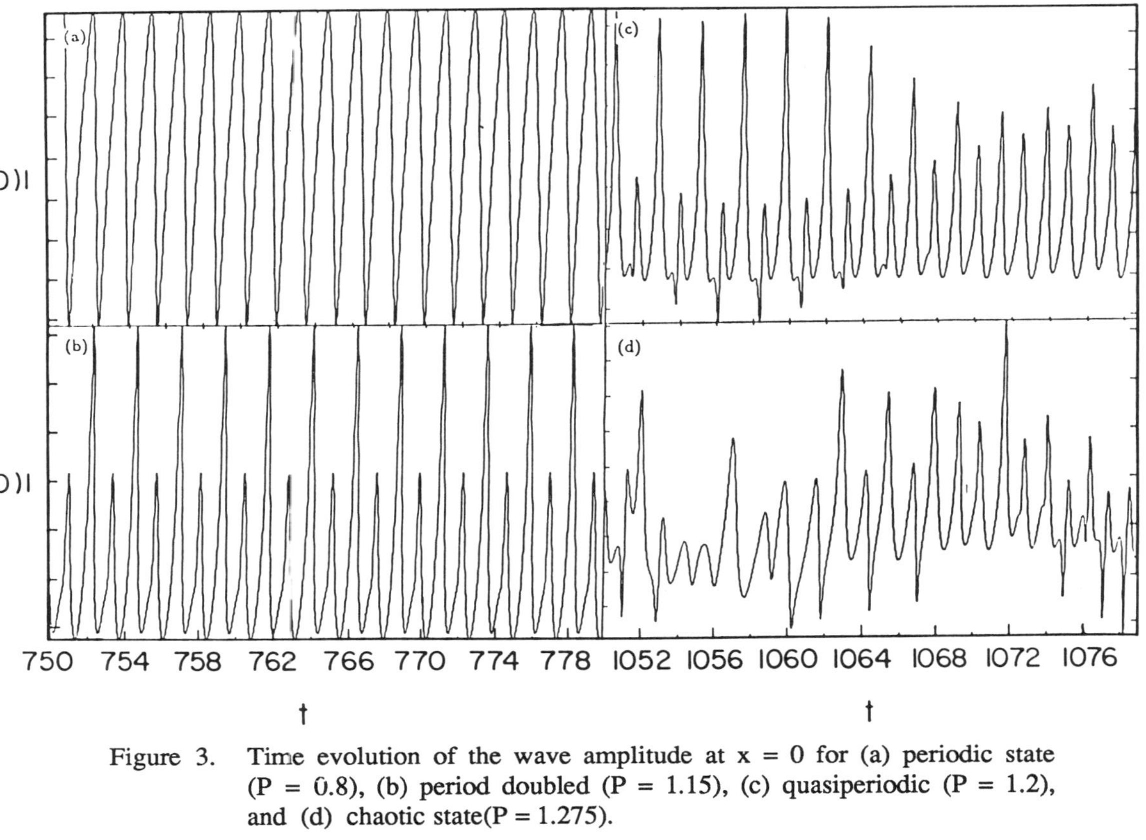

Figure 3.   Time evolution of the wave amplitude at x = 0 for (a) periodic state (P = 0.8), (b) period doubled (P = 1.15), (c) quasiperiodic (P = 1.2), and (d) chaotic state(P = 1.275).

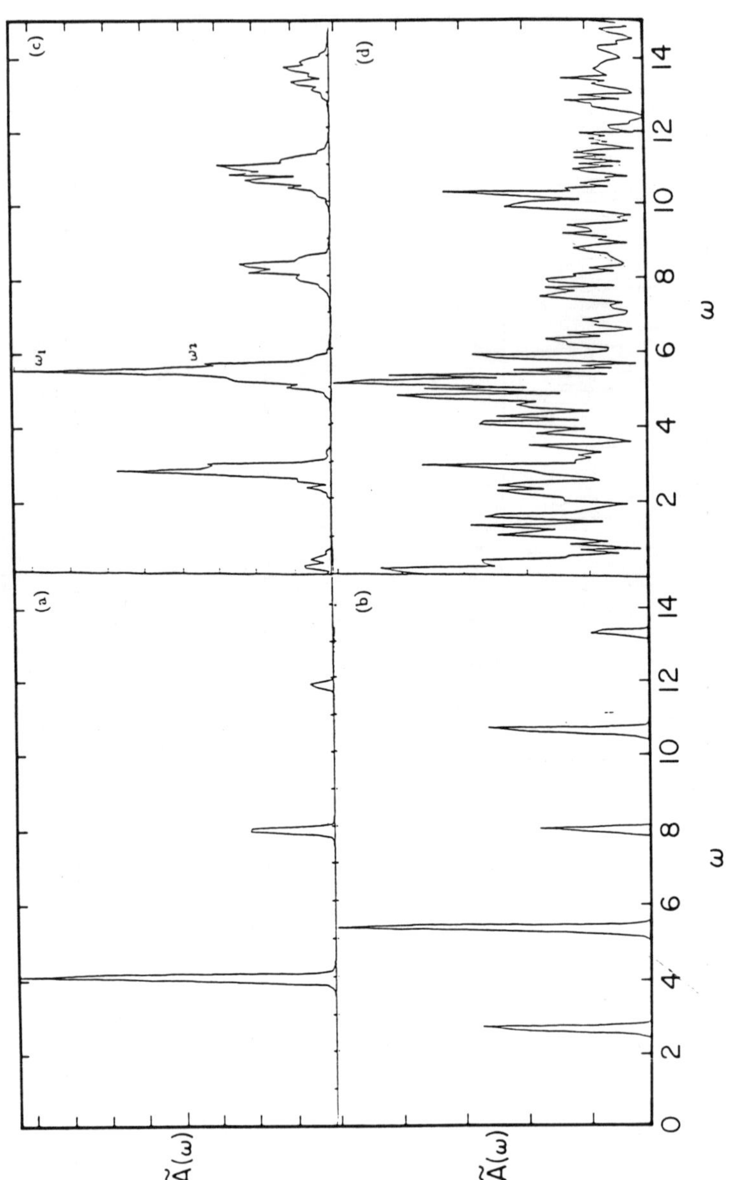

Figure 4.   The frequency spectrum at $x = 0$. (a) Periodic state ($P = 0.8$), $\omega_1 \sim 4.0$, (b) period doubled ($P = 1.15$), $\omega_1 \sim 5.2$, $\omega_1/2 \sim 2.6$, (c) quasiperiodic ($P = 1.2$), $\omega_1 \sim 5.4$, $\omega_2/2 \sim 5.6$ (d) chaotic state ($P = 1.275$).

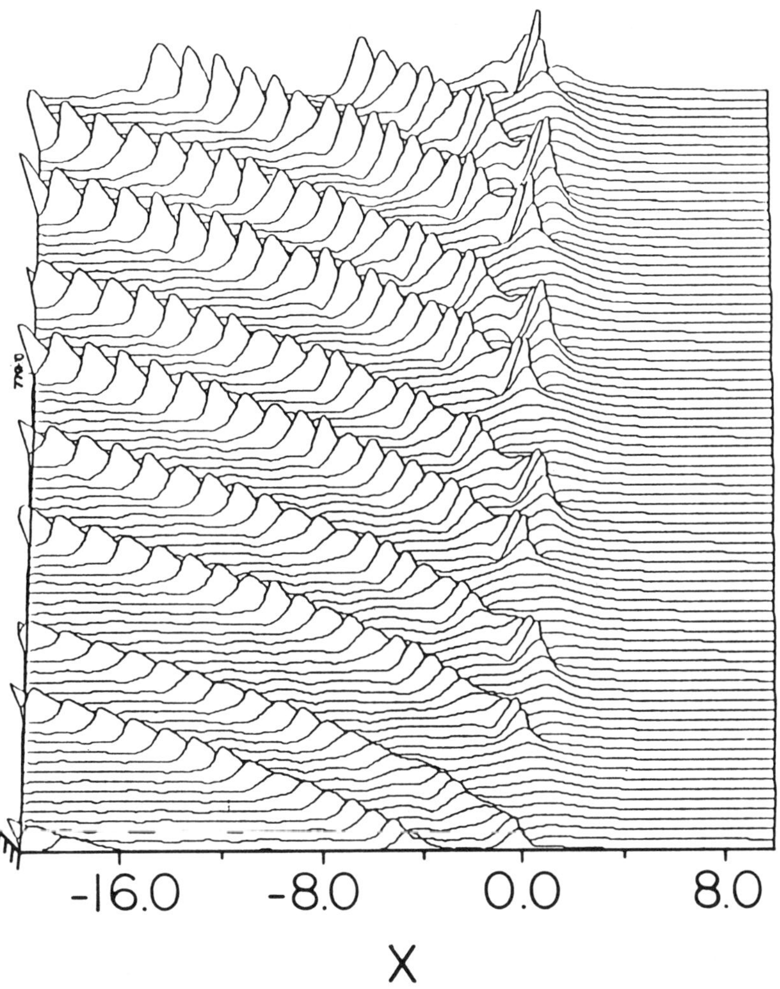

Figure 5.   3D-plot of chaotic emission of solitons for P = 1.275 as a function of x and t.

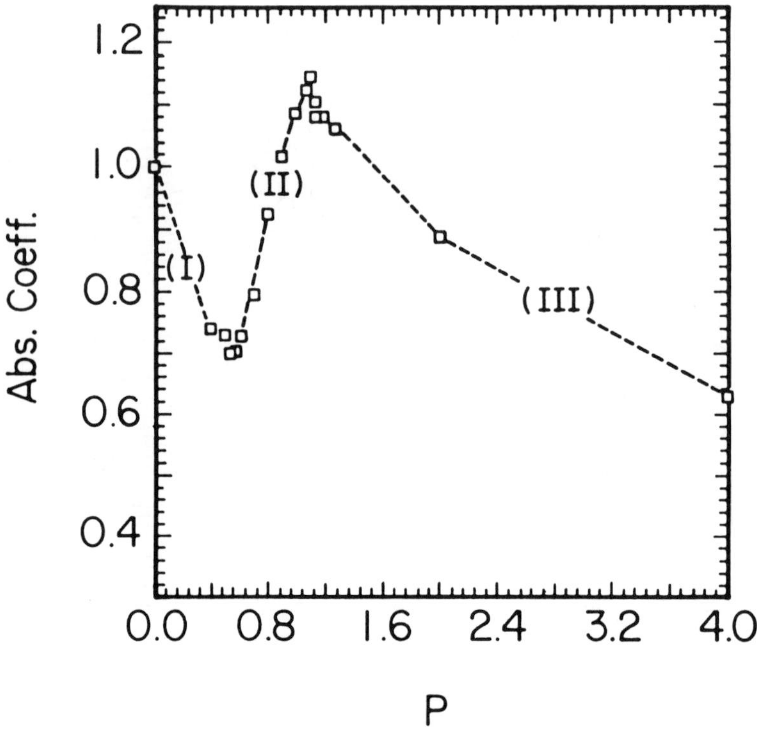

Figure 6.   Absorption coefficient vs. P. Region (I): steady state (0 < P < 0.575), Region (II): periodic state (0.575 < P < 1.09), Region (III): quasiperiodic and chaotic state 1.09 < P < 4.0.

time evolution of wave amplitude becomes chaotic (Figs. (3d)). The frequency spectrum shows a very broad distribution (Figs. (4d)). Thus the chaotic state occurs due to the breakup of the two-torus. We have calculated the correlation dimension D of the system at $P \sim 1.3$ and have found $D = 2.6$. The Lyapunov exponent $\lambda \simeq 0.3$ for this value of P.

We now consider the absorption of the incident radiation. The absorption coefficient a is defined as [7],

$$a = \frac{\int_t^{t+T} \int_{-\infty}^{\infty} \Im A(z',t')dz'dt'}{T\pi} \tag{15}$$

where T is the period for time averaging and $\Im A$ is the imaginary part of wave function. The numerically computed absorption coefficient for different P is shown in Fig. (6). We see that for the steady state, increasing P leads to steepening of the density gradient at the critical surface. Hence the absorption decreases. During the periodic emission phase, for increasing nonlinearity, the increased trapping and higher rate of soliton production results in the increase of absorption rate. In the chaotic phase, the detuning effect is so strong that the density profile always remains steep. Hence as P increases the absorption coefficient decreases.

In conclusion, we have studied the soliton formation process for a driven nonlinear Schrödinger equation and investigated the transition process to chaos. With the increase of P, the nonlinearity parameter, the system makes transitions from a fixed point to a periodic attractor, followed by period doubling. Further increase in P leads to quasiperiodic, two torus behavior followed by a break up of the torus. The two frequencies involved are the convection frequency and the breather oscillation frequency of the bound state of two solitons. Finally the absorption coefficient increases during the periodic phase unlike the steady state and chaotic phase where the absorption coefficient decreases.

NOTE: We recently received a preprint from Spatschek et al. [16] which addresses the same problem as discussed in our work.

ACKNOWLEDGMENTS

This work was supported in part by the Plasma Physics Division, NRL, Washington, D.C. 20375-5000 under contract No. N00014-89-k-2013.

REFERENCES

1. C. S. Liu, *Advances in Plasma Physics*, Vol. 6, Eds. A. Simon and W. B. Thompson (Wiley, New York, 1976) p. 121.

2. G. J. Morales and Y. C. Lee, Phys. Rev. Lett. **33**, 1016 (1974).

3. G. J. Morales and Y. C. Lee, Phys. Fluids **20**, 1135 (1977).

4. H. H. Chen, C. S. Liu, Phys. Rev. Lett. **37**, 693 (1976).

5. M. Ablowitz, A. Newell, D. Kaup and H. Segur, Study App. Math. **53**, 249 (1974).

6. H. H. Chen and C. S. Liu, Plasma Physics, Wilhelmsson, Eds. New York and London; Plenum Press, 1977, pp. 211-221.

7. J. C. Adam, A. Gourdin Serveniere, and G. Laval, Phys. Fluids **25**, 376 (1982).

8. E. W. Laedke and K. H. Spatschek, Phys. Rev. Lett. **45**, 993 (1980).

9. D. Anderson, Phys. Rev. A **27**, 3135 (1983).

10. H. T. Moon and M. V. Goldman, Phys. Rev. Lett. **53**, 1821 (1984).

11. M. N. Bussac, P. Lochak, C. Meunier and A. Heron-Gourdin, Physica **17D**, 313 (1985).

12. O. Larroche, M. Casanova, D. Pesme, M. N. Bussac, Laser and Particle Beams, Vol. 4, pt. 3, p. 545 (1986).

13. S. Qian, Ph. D. Thesis, University of Maryland, 1986.

14. Alex P. K. Wai, Ph. D. Thesis, University of Maryland, 1988.

15. W. Shyu, to be published.

16. K. H. Spatschek, H. Pietsch and Th. Eickermann, "Chaotic Behavior in Time of Langmuir Solitons", to appear in the Proceedings of the International Conference on Plasma Physics (Delhi 1989).

KINETIC NONLINEARITIES IN NONUNIFORM PLASMAS*

G. J. Morales and J. E. Maggs

Physics Department, University of California, Los Angeles
Los Angeles, CA   90024

INTRODUCTION

The properties of basic kinetic nonlinearities (e.g., particle trapping, nonlinear Landau damping, etc.) in uniform plasmas are well understood, and an extensive literature exists on the subject.[1-3] However, in many problems of current interest, plasma nonuniformity plays an important and fundamental role. Although helpful insight can often be obtained using local, or WKB generalizations of nonlinear phenomena in uniform media, such extrapolations often miss important features. Consequently it is of interest to develop a systematic study of nonlinear phenomena in the context of a well-defined problem in which the intrinsic plasma nonuniformities are retained without undue complexity. Clearly there are numerous topics to be considered in this area, and in recent years we have analytically and numerically investigated a few key issues motivated by broader research interests. This manuscript presents an abbreviated summary of the highlights of these studies.

Our studies of kinetic nonlinearities in nonuniform plasmas have been motivated, in large measure, by several experimental observations which have not yet been fully explained quantitatively from first-principles calculations. Prominent among these are issues related to wave generation, possibly by distortions in the electron distribution function. In laser-plasma experiments, an ongoing assessment is being made[4,5] of anomalies observed in the spectral features of Raman scattering which can not be explained by conventional parametric instability theories. In ionospheric modification experiments that use powerful ground-based HF transmitters, anomalies are also observed[6,7] in the wave spectrum sampled by radar backscattering. Typically, frequencies different than that of the HF wave are observed, and are not trivially related to parametric instabilities. In such

---

*Work supported by Office of Naval Research and NASA

experiments, stimulated electromagnetic emissions[8] are also observed over a broad frequency band.

Another topic that has motivated our investigations is electron acceleration by waves in nonuniform plasmas. In the topside ionosphere, spontaneous generation of fast field-aligned electron bursts have been observed.[9] We speculate that the underlying mechanism involves whistler wave mode conversion in low density plasma irregularities. Also, various aspects of beat-wave excitation are currently being considered in regards to features introduced by plasma nonuniformity.

PHYSICAL MODEL

In surveying the literature of wave-particle interactions in nonuniform media, many studies are found to be deficient because the source of wave generation is ill-defined, and the boundary conditions are not well-posed. Regrettably, some otherwise fine studies confuse the meaning of temporal and spatial nonlinearities and are thus unable to make concrete predictions in spite of significant analysis. To overcome these shortcomings, without introducing unnecessary complexity, we use a simple model in our studies in which wave generation and the essential role of plasma nonuniformity are retained. We consider the model to be a "theoretical laboratory" in which one can systematically study various kinetic nonlinearities in a universal scaled form, so that meaningful comparisons to a variety of experimental scenarios can be established.

The essence of the model is illustrated in Figure 1. It consists of a plasma with a linear density profile $n_0(z)$, having scale length L in the neighborhood of the spatial point $z = 0$, where the local plasma frequency ω_p matches the frequency ω of an external source, $S(z) \exp(-i\omega t)$. We envision the density gradient to be supported by low energy electrons because, conceptually, they have a relatively short collision mean free path. To extract interesting effects related to wave-particle interactions we assume the presence of energetic electrons (fast tail electrons) whose density n_t is spatially uniform and small compared to the low-velocity electron density, $n_t << n_0$. The assumption of uniform density for fast electrons rests upon their collision lengths being much longer than the scale lengths of the wave structures generated in the plasma. The spirit of the model is to treat the response of the slow electrons through a warm-fluid description, involving a differential equation in configuration space, while the tail electrons are treated kinetically in Fourier transform space.

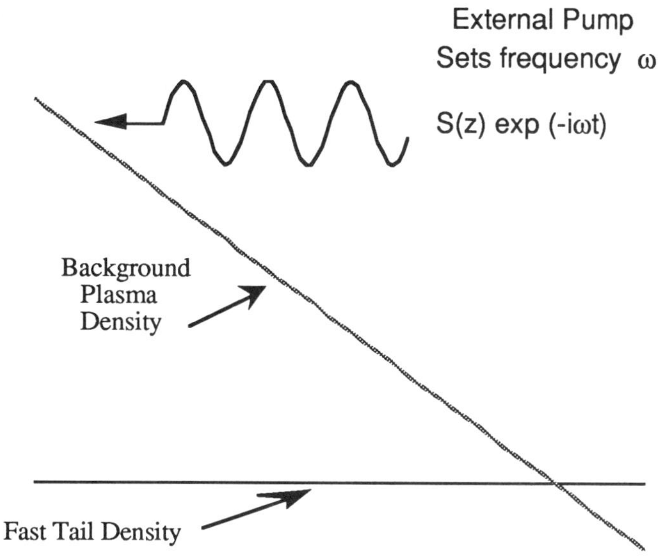

Figure 1.   The plasma model consists of a cold plasma with a density gradient and a hot uniform tail.

The spatial dependence of the source function $S(z)$ used to resonantly excite plasma waves can model several physically interesting situations of experimental relevance.   By choosing $S(z)$ constant, the model rigorously describes the process of linear mode conversion of an oblique electromagnetic wave[10] into a Langmuir wave (as in laser-plasma studies) in an unmagnetized plasma, or the mode conversion of an oblique electrostatic whistler wave[11] into a Langmuir wave in a magnetized plasma. The choice of a sinusoidal $S(z)$ can describe beat excitation by transparent electromagnetic waves (up or down the density profile) as well as direct conversion[12] on density ripples.

SLOPE REVERSAL

The self-consistent resonant electric fields excited in the

presence of a small population of tail electrons is obtained from the solution of a differential equation in Fourier transform space (k-space),

$$\left[\frac{i}{L}\frac{d}{dk} + i\Gamma - \frac{3k^2}{k_D^2} + i4\pi\,\mathrm{Im}\,\tilde{\chi}_t \right] \tilde{E}(k,\omega) = E_0\,\tilde{S}(k) \ , \tag{1}$$

where Γ represents collisional damping, and k_D is the Debye wavenumber, $\tilde{\chi}_t$ is the susceptibility of the tail electrons and its effect in (1) is to describe Landau damping. Equation (1) can be solved analytically and then inverted numerically to obtain the spatial pattern $E(\xi)$, where the relevant scaled spatial coordinate is $\xi = (k_D L/\sqrt{3})^{2/3}\, z/L$. Figure 2.a exhibits the spatial dependence of the field amplitude for the case corresponding to linear mode conversion (constant S). The top curve corresponds to the undamped situation (no collisions, no tail), the middle curve contains collisional damping (but no tail), and the bottom curve contains collisional and Landau damping. It is evident collisions cause a reduction in the peak response near resonance where the effective phase velocity is large. Wave-particle interactions with the tail electrons are responsible for damping the mode-converted Langmuir wave as its local phase velocity decreases with distance away from the resonance point. The parameter α is the scaled tail density, $\alpha \equiv (n_t/n_0)\,(\pi/2)^{1/2}\,(k_D L)(\,\bar{v}/v_t)$, with $\bar{v}$, v_t the thermal velocity of the slow and tail electrons respectively.

Figure 2.b exhibits the electric field pattern of a Langmuir wave generated through beat excitation with a wavenumber vector k_b that points in the direction of decreasing density. In this case, local enhancement of the field occurs in the neighborhood of the WKB matching point, and tail Landau damping eventually destroys the unidirectional plasma wave. This wave-particle interaction results in a distortion of the tail distribution and gives rise to a heat-flux to be described later.

When the beat wave number vector points in the direction of increasing density, the excited Langmuir wave initially propagates to its cutoff point (i.e., $\omega_p(z) = \omega$) where it reflects, and then proceeds to propagate in the direction of decreasing density. The corresponding partially-standing wave pattern is shown in Fig. 2.c. In this case wave-particle interactions occur with fast

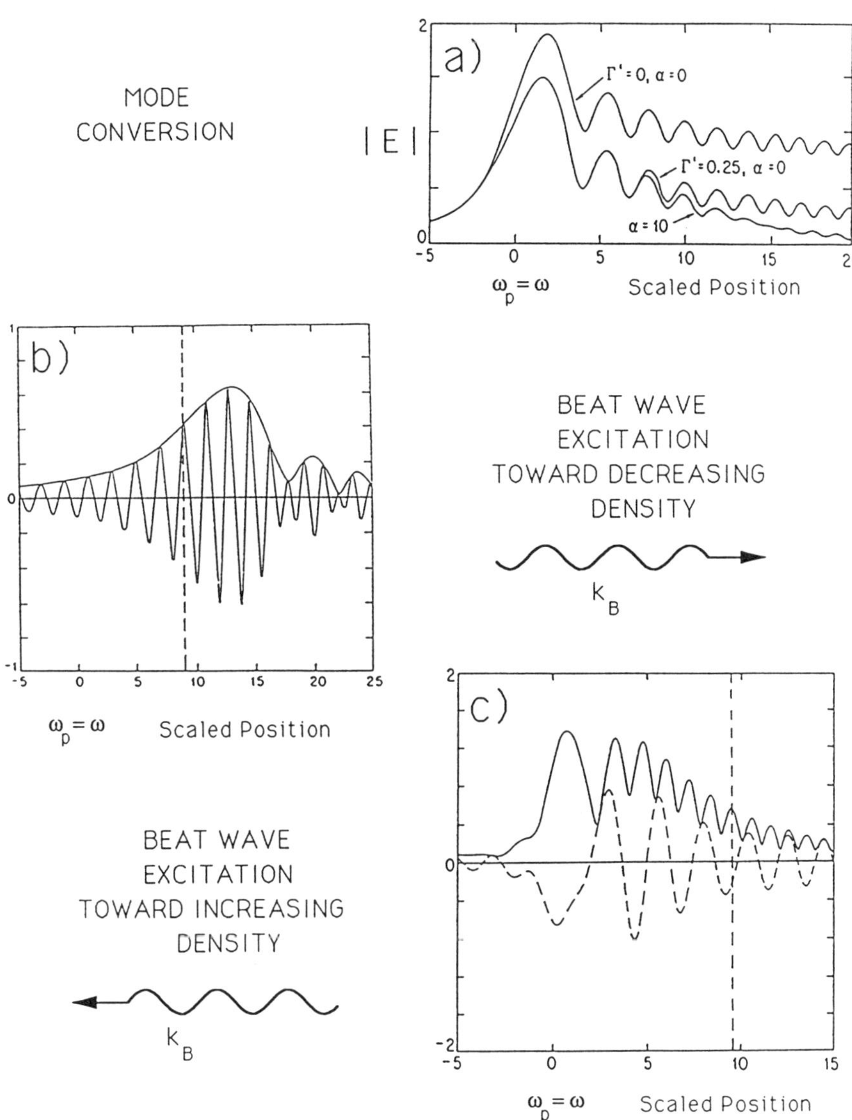

Figure 2. a) The effects of collisional damping ($\Gamma=0.25$, $\alpha=0$) and collisionless damping ($\alpha=10$.) on the mode converted field amplitude. b) the field excited by a beat wave propagating towards decreasing density, and c) towards increasing density.

electrons moving in both directions so that a bi-directional heat flux develops.

Once the self-consistent, resonantly excited fields have been determined it is possible to calculate the second order (in pump-field amplitude E_0) modifications[13] in the tail distribution function from

$$< \delta f_t(z \to \pm \infty \, , v) > = (\frac{e}{2\,m})^2 \frac{1}{v} \frac{\partial}{\partial v} \left\{ \, | \tilde{E}(k = \frac{\omega}{v})|^2 \, \frac{1}{v} \frac{\partial}{\partial v} \, f_{0t} \right\} \, . \quad (2)$$

An interesting feature is the ability to generate a slope reversal in the tail distribution once a threshold field amplitude is exceeded. Slope reversal can occur because the increase in the spectral power density at $k = \omega/v$ can be much faster than the rate of decrease in the zero order tail distribution.   Consequently, in a given velocity bin, the number of electrons that are promoted to higher velocities can be larger than the number entering the bin from lower velocities, thus a region of positive slope (bump-in-tail) can develop in the velocity distribution.   The exact value of the   threshold   field   is   determined   from   a   transcendental equation[13], but roughly its magnitude is $E_0 > p(mv_t^2/\pi eL)$, with p ~ 2-5.   A detailed analysis[14] related to ionospheric HF-modification experiments   shows   that   the   presently   available   HF   facilities operate at power levels large enough to exceed this threshold, so that bump-in-tail distributions can be expected to develop and contribute to the noise spectrum observed in such experiments. The   corresponding   observation   of   this   phenomenon   in   laser-plasma experiments is presently being debated[4,5] by experts.

Figure 3.a shows an example of bump formation associated with mode-conversion (the field pattern is like that shown in Fig. 2.a), while Fig. 3.b shows the corresponding slope reversal obtained from beat-excitation in the direction of decreasing density.   As can be seen, modifications occur in the velocity interval $v < \omega/k_b$ because the local phase velocity always remains below this value.   Figure 3.c shows a composite distribution function ($v < 0$ corresponds to $z \to - \infty$, $v > 0$ to $z \to + \infty$) in which bump formation results from beat excitation in the direction of increasing density.

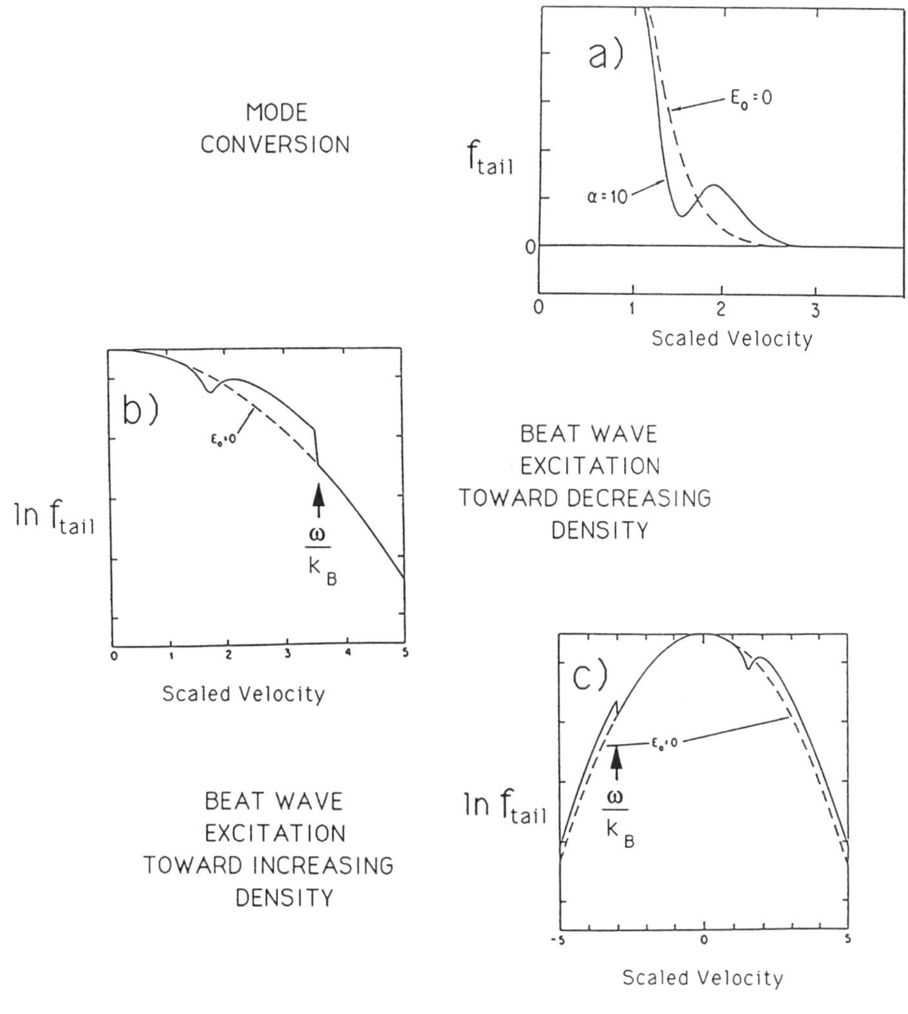

Figure 3.   a) The electron distribution for $\alpha=10$, b) for a beat wave propagating towards decreasing density, and c) towards increasing density.

EFFICIENCY OF BEAT EXCITATION

Since the physical model describes accurately the self-consistent excitation of Langmuir waves in a density gradient, we have considered the case in which the source $S(z)$ corresponds to the beat-ponderomotive force generated by two transparent ($\omega_j \gg$

ω_p, j = 1,2) electromagnetic waves having frequencies ω_j. The resulting beat wave vector k_b can point in either the direction of decreasing (+, for positive z-direction) or increasing (-) density. The modified electron distribution function can be calculated from Eq. (2) to yield[15] the heat flux δQ induced in the plasma by the beat wave process in the (±) directions

$$\delta Q_{\pm} = \pm \frac{e^2}{4\,m} \int_0^{\pm\infty} dv \mid \tilde{E}(k = \frac{\omega}{v})\mid^2 \frac{\partial}{\partial v} f_{0t}^{\pm}(v) \quad . \tag{3}$$

Using the solution of Eq. (1) in Eq. (3) the plasma heating efficiency η for beat wave excitation in a nonuniform plasma can be determined analytically.   The compact simple expression for (±) excitation is

$$\eta_{\pm} = \eta_0 \left\{ 1 - \exp[-2\alpha(1 \mp \exp(-w^2))] \right\} \quad , \tag{4}$$

where w = $(\omega_1 - \omega_2)/\sqrt{2}k_D v_t$, α is given in the discussion of Fig. 2, and

$$\eta_0 = 2\pi \left(\frac{\omega_1 - \omega_2}{\omega_j} \right)^5 (k_{0j}L) \frac{P_2}{n_0 mc^2} \quad , \tag{5}$$

with k_{0j} the wavenumber of the jth wave, P_2 the power density carried by ω_2, and c the speed of light.   In the limit of large α, i.e., long scale length L, the efficiency predicted by Eq. (4) reduces to the expression previously found by Rosenbluth and Liu[16], which scales linearly with L.   However, in the limit of small α, i.e., short scale length, Eq. (4) predicts a transition to a scaling proportional to L^2.

BUMP-ON-TAIL RESONANT EXCITATION

Motivated by the finding that slope reversal can be produced in the fast-tail electron distribution by resonant excitation, we have investigated the subsequent modifications to the mode conversion process (constant S, Fig. 2.a) when a bump-on-tail is present.   The issues of interest are: 1) Determination of the net amplification resulting from the convective instability driven by the slope reversal; 2) Modifications in the fast-tail distribution due to changes in the pump field amplitude, E_0.   For

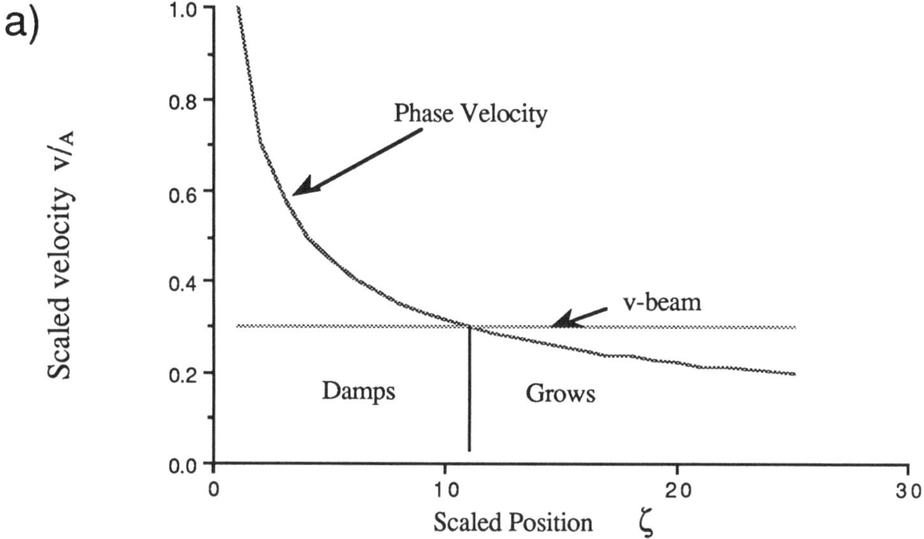

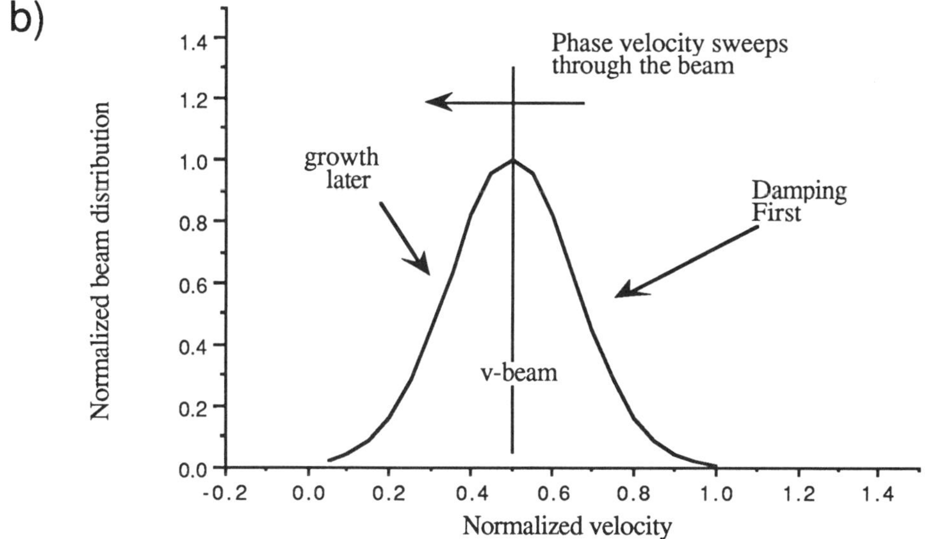

Figure 4.   The spatial varying phase velocity results in first damping and then growth from a beam distribution.

concreteness we present the results produced by a beam distribution of the form,

$$f_b(v) = \frac{n_b}{(2\pi\bar{v}_b^2)^{1/2}} \ \exp\{-(v-v_b)^2/2\bar{v}_b^2)\} \ , \qquad (6)$$

with $n_b \ll n_0$. Mathematically, the beam effects are introduced through the susceptibility $\tilde{\chi}_t$ in Eq. (1).

The underlying physics of this problem is illustrated in Fig. 4. Figure 4.a shows the spatial dependence of the phase velocity of the wave excited by mode conversion, i.e., $\omega/k(z) \sim z^{-1/2}$, and the unperturbed velocity v_b of the peak of the beam, i.e., where $\partial f_b/\partial v = 0$. In the spatial region near plasma resonance (i.e., $z = 0$), $\omega/k > v_b$ and the wave is damped by the beam because it samples the negative slope region of the distribution. However, at some positive value of z, $\omega/k < v_b$, and in this region wave amplification occurs because the wave samples the positive slope region of the beam distribution. A complementary view is illustrated in Fig. 4.b, which emphasizes that the phase velocity of

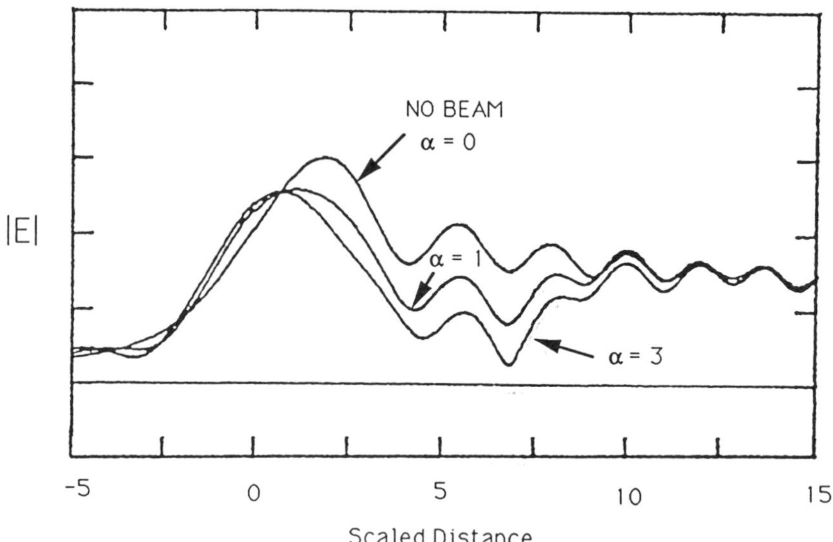

Figure 5.   Effect of beam on the mode converted field amplitude for densities corresponding to α = 1., and α = 3.

the wave continuously sweeps through the beam velocity distribution, first starting in the damping region so that the fast part of the beam is accelerated, and then proceeding to the growth region in which the slower beam electrons are decelerated.

The effect of the beam on the amplitude of the resonantly excited field is illustrated in Fig. 5 for two values of the scaled beam density α (as defined in the discussion of Fig. 2) with $v_b/\bar{v}_b = 5$, and $v_b = 2v_A$, where $v_A = \omega L/(k_D L/\sqrt{3})^{2/3}$ is the maximum phase velocity of the field (determined by the width of the Airy pattern at $z = 0$).   It is evident that as the beam density increases, the peak amplitude of the resonance decreases and the mode converted wave initially experiences strong damping, as anticipated from Fig. 4.   The regrowth of the wave indeed occurs (around $\xi \approx 7$), but with the remarkable signature that asymptotically the amplitude returns to the value obtained in the absence of the beam.   This implies that, although there is spatial rearrangement of the wave energy, there is no net gain produced by the beam.   This property can be demonstrated analytically to be rigorously correct and independent of the beam distribution parameters for symmetric (about the beam drift velocity v_b) beam distributions.   Selective alteration of beam symmetry can

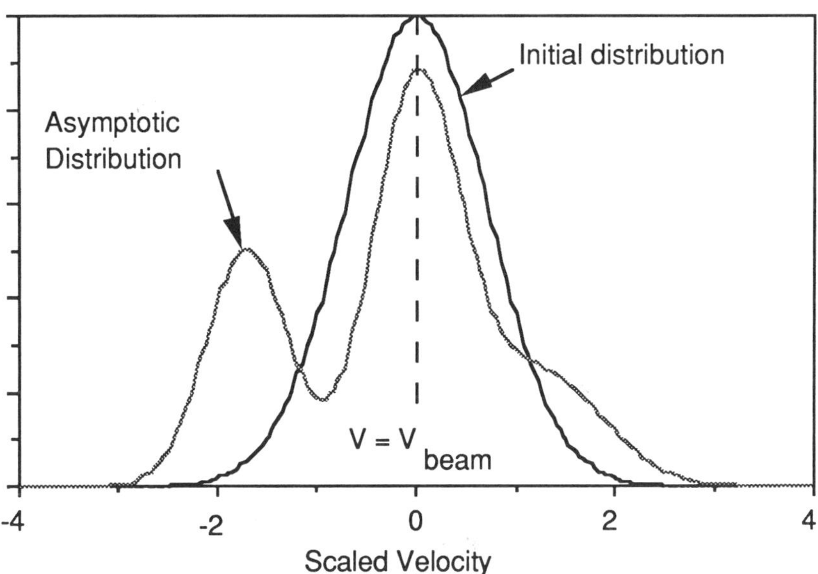

Figure 6.   The beam velocity distribution changes as a result of its interaction with the resonant field.

lead to either enhanced damping or growth of the resonant field (these results are not shown here).

Although a symmetric beam does not yield net wave amplification, the resonantly excited wave does modify the beam distribution function as illustrated in Fig. 6 for a scaled pump amplitude $(\pi L e E_0 / \sqrt{2m} \; \bar{v}_b^2) = 5$. It is seen that a few of the beam electrons are accelerated to higher velocities, while a significant number are decelerated during the process of wave regrowth. Beyond some threshold value of E_0 (equivalent to that found for slope reversal) which is exceeded in this example, a multiple-beam distribution is formed.

NONLINEAR LANDAU DAMPING OF RESONANT FIELDS

Since the effective phase velocity of a resonantly excited electric field near plasma resonance scales as $\bar{v}_e/(k_D L)^{1/3}$, the mode converted wave is essentially undamped in plasmas having

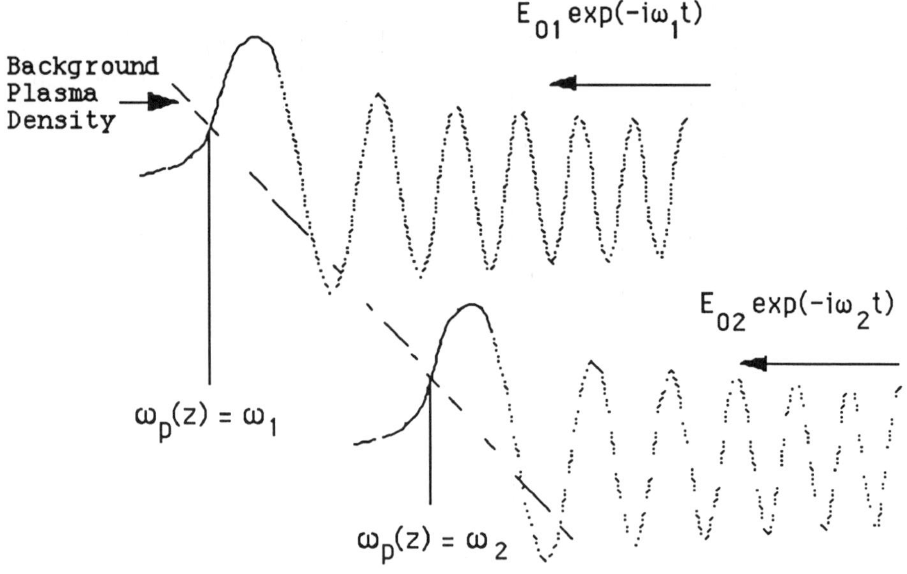

Figure 7.   Two external sources having amplitudes E_{01} and E_{02} and frequencies ω_1 and ω_2 drive in resonant field patterns at two separate spatial locations.

large scale lengths L, as illustrated by the top curve in Fig. 2.a.   It is therefore of interest to consider the role of nonlinear (beat) Landau damping in such an environment.   Figure 7 illustrates the geometry of a case in which two external sources (uniform pumps) are present with closely spaced frequencies, $\omega_1 > \omega_2$, and excite individual resonant electric fields of the type shown in Fig. 2.a.   The two resonant field structures are linearly weakly damped (by collisions), and propagate with phase velocities $\omega_1/k_1(z)$, $\omega_2/k_2(z)$ which decrease as $z^{-1/2}$, as sketched in Fig. 8.

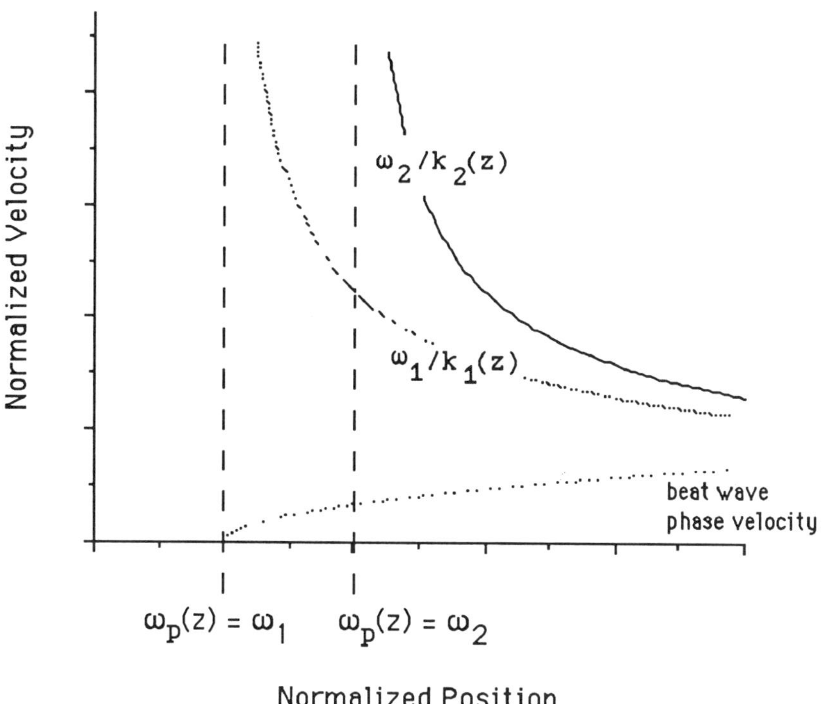

Figure 8.   The phase velocity of the beat wave varies as $z^{1/2}$ while the phase velocities of the individual drivers vary as $z^{-1/2}$.

Nonlinearly the waves generate a beat idler whose local phase velocity is $(\omega_1 - \omega_2)/(k_1(z) - k_2(z))$.   For small frequency separation the idler phase velocity equals the local group velocity and increases as $z^{1/2}$ and scales as $\bar{v}_e/(k_D L)^{1/3}$.   Thus, the idler interacts strongly with slow electrons and background ions.

This problem is presently being investigated by S. Srivastava in his Ph.D. dissertation, and the mathematical details are too lengthy to be presented here. It suffices to say that the formulation consists of three coupled equations to be solved self-consistently for the fields E_1, E_2 oscillating at frequencies ω_1, ω_2, respectively, and the idler field E_3 at frequency $\omega_1 - \omega_2$. The equations for the plasma waves ($j = 1,2$) are differential equations of the form

$$\frac{3}{k_D^2} \frac{d^2}{dz^2} E_j + \left(\frac{z-z_j}{L} \right) E_j = E_{0j} + S_j(z) , \qquad (7)$$

with z_j the cutoff point, E_{0j} the external pump amplitude, and S_j the nonlinear (beat) source arising from coupling to the other two modes. The idler response must be calculated in k-space because of its low effective phase velocity which requires that the response of the ions and slow electrons be described kinetically, giving

$$\frac{i}{L} [\, 1 - \varepsilon(\omega_1 - \omega_2, k) \,] \frac{d}{dk} \tilde{E}_3(k) + \varepsilon(\omega_1 - \omega_2, k) \tilde{E}_3(k) = \tilde{S}_3(k) , \qquad (8)$$

where ε is the kinetic dielectric and $\tilde{S}_3(k)$ the source in k-space (obtained from a convolution over $\tilde{E}_1 \tilde{E}_2{}^*$). The relevant nonlinear parameter in the system is $(eE_0L/\sqrt{2}T_e)^2$, where T_e is the electron temperature and, for simplicity, $E_{01} = E_{02} = E_0$. The spatial pattern of nonlinear Landau damping is illustrated in Fig. 9. The top panel shows the amplitude of the individual fields (non-interacting) for $(\omega_1 - \omega_2)/\omega_1 = 0.2$. The middle panel shows the linear and nonlinear (for $(eE_0L/\sqrt{2}T_e)^2 = .15$) amplitudes for the highest frequency wave (i.e., ω_1). It is evident that the nonlinear interaction causes strong damping. The corresponding behavior for the wave at frequency ω_2 is shown in the bottom panel where clear enhancements in the wave amplitude are observed. As expected, the highest frequency wave transfers part of its energy to the lower frequency wave and simultaneously accelerates slow electrons and ions through the idler field. Figure 10 displays the idler field and the power dissipation to both particle species for $T_e/T_i = 1.0$.

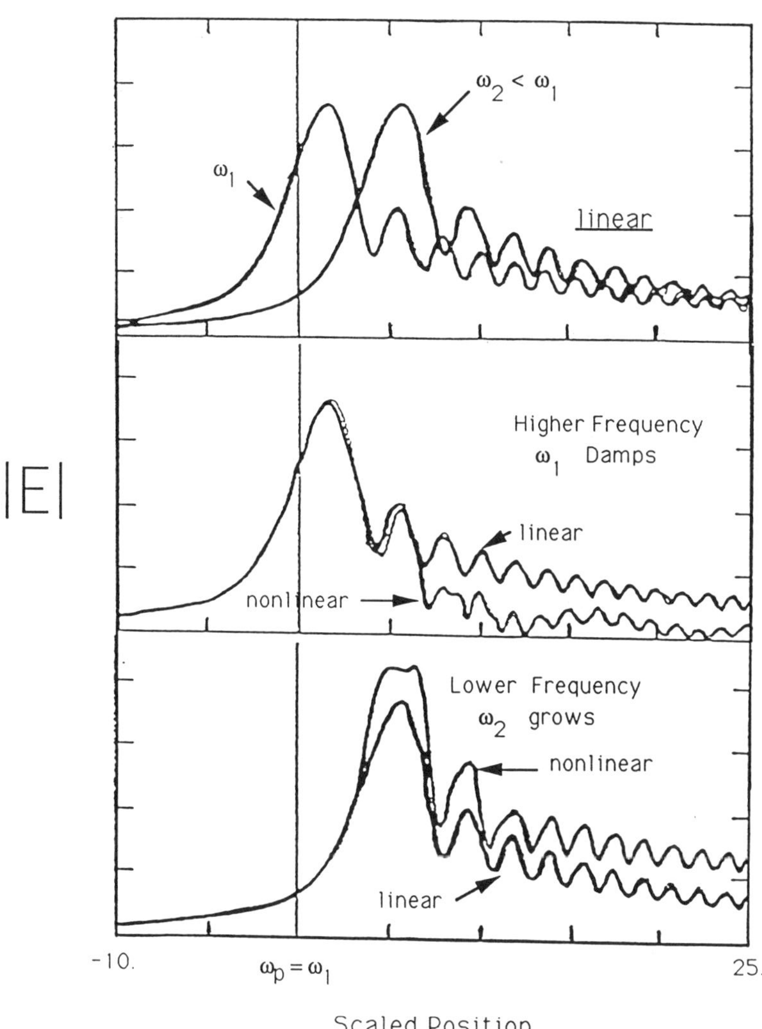

Figure 9.   The field amplitude of the two high frequency waves   in the linear (top panel) and nonlinear cases (bottom panels).

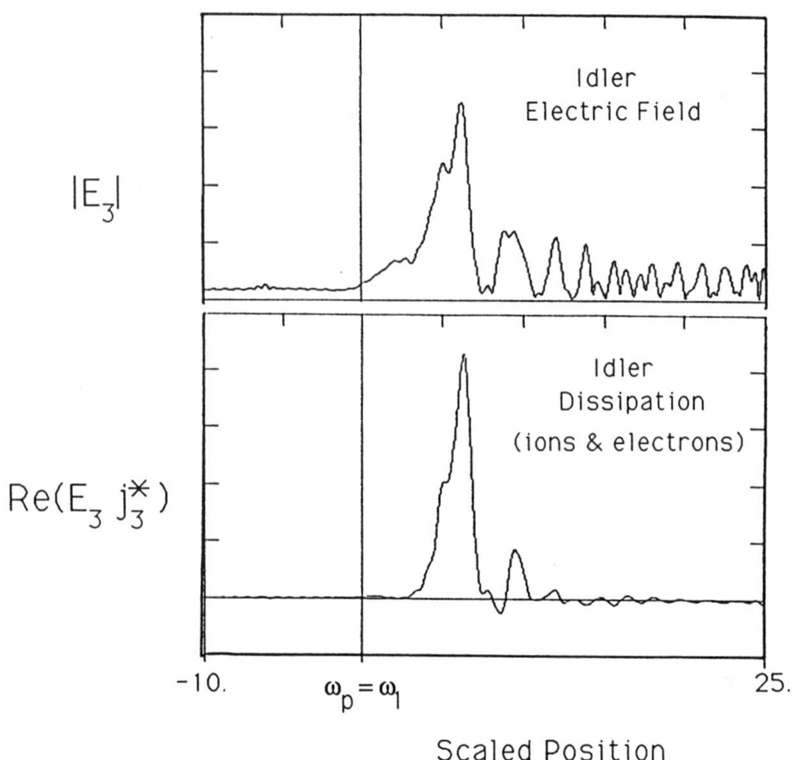

Figure 10.   The idler field and spatial pattern of dissipation are shown for a frequency separation of 2% in the driving fields.

NONLINEAR DYNAMICS OF ELECTRONS ACCELERATED BY RESONANT FIELDS

In the analysis of the formation of slope reversal in the electron velocity distribution discussed in Sec. 3, strong wave-particle nonlinearities, such as particle trapping, were not included.   To obtain a better understanding of the role strong nonlinearities play in this process, we have performed an extensive numerical and analytical study[17] of individual electron orbits associated with an undamped, driven-Airy pattern whose amplitude corresponds to the top curve in Fig. 2.a.   We have found that the nonlinear interactions can be grouped in three categories depending upon the initial velocity of the particle as illustrated in Fig. 11.

   Although the boundaries between the different categories of interaction can be precisely determined the expressions for these boundaries are complicated, so it is convenient to think of the categories as corresponding to fast ($v \geq v_A$), intermediate ($v \approx .5$ v_A), and slow ($v \leq .2v_A$), where the comparison velocity v_A is the

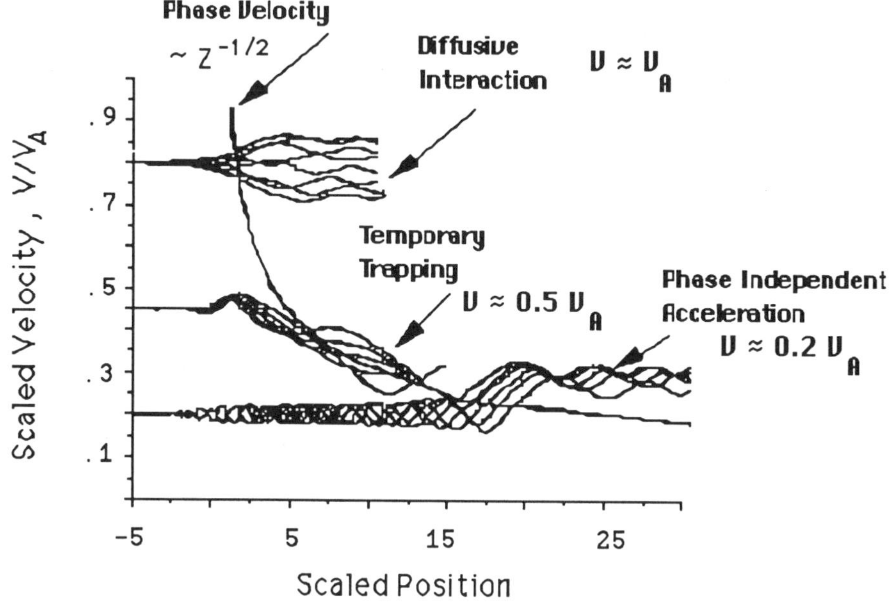

Figure 11.   The interactions of particles with a resonant field can be divided into three categories depending upon velocity: fast $V \approx V_A$, intermediate $V \approx 0.5 V_A$, and slow $V \approx 0.2 V_A$.

fastest phase velocity of the driven-Airy pattern, $v_A \equiv \omega L/(k_D L/\sqrt{3})^{2/3}$. The fast particle trajectory in phase space, as shown in Fig. 11, intersects the phase velocity curve, $v_p \sim z^{-1/2}$, at a very large angle, hence the time of interaction is relatively short, and results in a diffusive interaction in which particles are both accelerated and decelerated as in the second-order theory discussed in Sec. 3. For intermediate velocities it is possible for some particles to be temporarily entrained by the wave. The conditions for this trapping behavior are rather restrictive, occurring only for a narrow velocity and wave-phase range. The

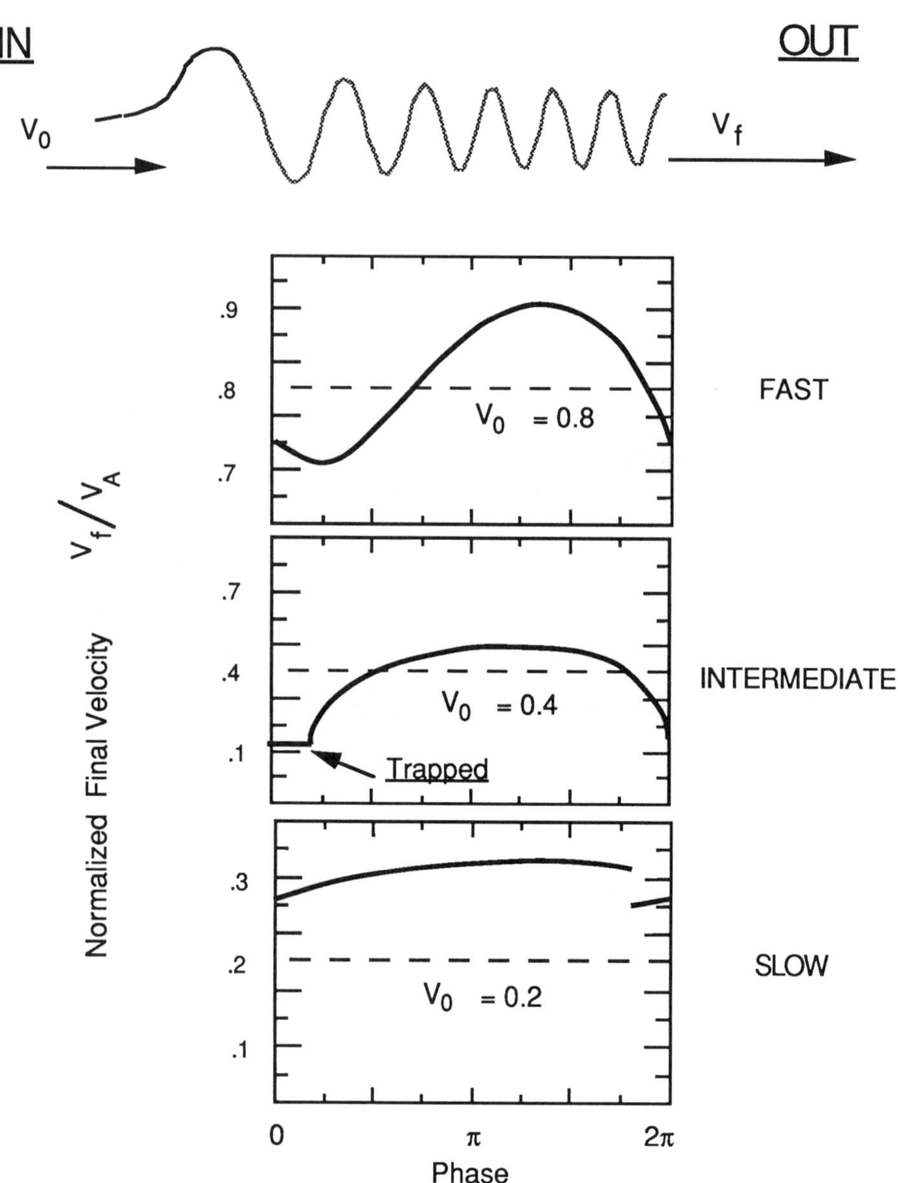

Figure 12.   The behavior of the velocity change as a function of initial velocity and initial wave phase depends strongly upon the initial velocity of the particle.

few particles that can be trapped, however, exhibit deceleration as they adiabatically follow the decreasing wave phase velocity.

Eventually, at some value of z, the adiabatic invariant associated with the oscillatory trapped motion[18] is violated and the particle escapes.   Perhaps the most interesting and important category of interaction is experienced by the slow particles.   Every one of these particles is found to be accelerated, independently of the initial wave phase.   Clearly, energizing the entire population of slow particles should cause catastrophic damping of the wave.

The dependence of the asymptotic velocity v_f on the wave phase of a particle injected on the overdense side of plasma resonance with velocity v_0 is shown in Fig. 12.   The top panel corresponds to fast particles ($v_0 = 0.8\ v_A$) and exhibits a nearly sinusoidal behavior, characteristic of diffusive behavior (i.e., some phases result in energy gain and others energy reduction).   The middle panel shows the behavior of intermediate velocity ($v_0 = 0.4\ v_A$) particles.   The constant step on the left side corresponds to the trapped particles that have been decelerated by the wave. The bottom panel shows clearly that slow particles ($v_0 = 0.2\ v_A$)

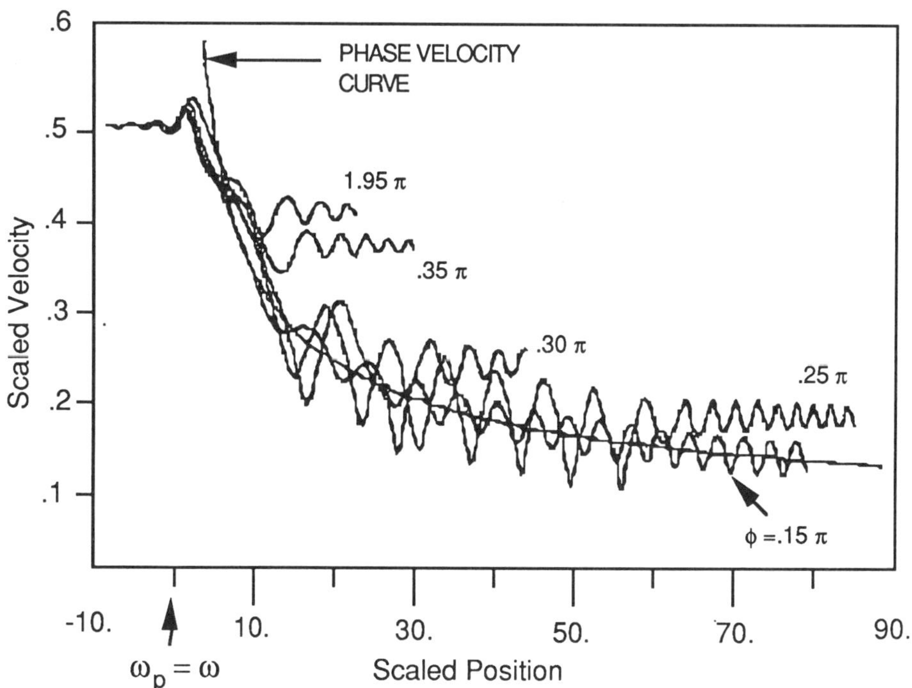

Figure 13.   The trapping of particles by the resonant field is only temporary and occurs only over a narrow range of intitial phases.

experience a velocity increase for all wave phases.

The transient nature of the trapping process is exhibited in Fig. 13 which shows the phase trajectories of particles with initial velocity $0.5v_A$ for those wave phases which result in entrainment. It is evident that a small phase change results in significantly different detrapping positions.

The physical process responsible for the phase-independent acceleration of slow particles is illustrated in Fig. 14. Essentially, if the particle has the wrong phase to experience acceleration when first encountering the wave it is slowed down at a rate

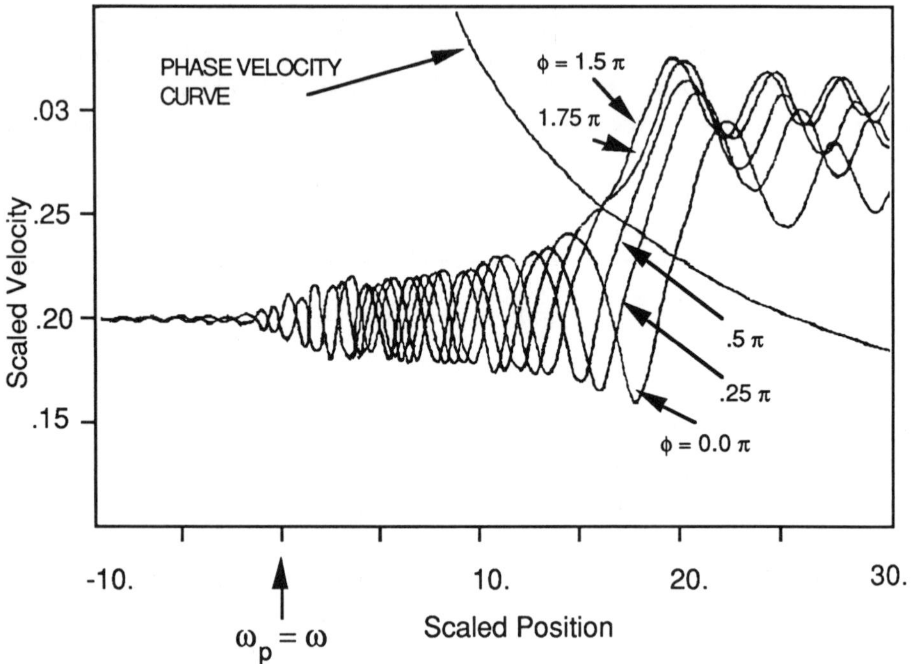

Figure 14.   All slow particles are accelerated independently of the initial wave phase.

faster than the decrease in wave phase velocity, and thus immediately re-encounters the wave with a phase that results in acceleration.   Once acceleration occurs, the particle does not come into resonance with the wave again because the wave phase velocity continues to decrease with position.

REFERENCES

1. R. C. Davidson, "Methods in Nonlinear Plasma Theory", W. A. Benjamin, Inc. (1974).
2. "Reviews of Plasma Physics", M. A. Leontovich, Ed., Vol. 7 Consultants Bureau (1979).
3. "Handbook of Plasma Physics", M. N. Rosenbluth and R. Z. Sagdeev, Eds., Vol. 2, North Holland Publishing Co. (1983).
4. R. P. Drake, Phys. Fluids B, 2, 225 (1990).
5. A. Simon and R. W. Short, Phys. Fluids B, 2, 227 (1990).
6. R. L. Showen and D. M. Kim, J. Geophys. Res., 83, 623 (1978).
7. P. Stubbe, H. Kopka, H. Lauche, M. T. Rietveld, A. Brekke, O. Holt, T. B. Jones, T. Robinson, A. Hedberg, B. Thide, M. Crochet and H. J. Lotz, J. Atmos. Terr. Phys., 44, 1025 (1982).
8. B. Thide, H. Derblom, A. Hedberg, H. Kopka and P. Stubbe, Radio Science, 18, 851 (1983).
9. A. D. Johnstone and J. D. Winningham, J. Geophys Res, 87, 2321 (1982).
10. D. E. Hinkel-Lipsker, B. D. Fried and G. J. Morales., Phys. Rev. Lett., 62, 2680 (1989).
11. J. E. Maggs and G. J. Morales, J. Plasma Phys., 41, 301 (1989).
12. A. Y. Wong, G. J. Morales, D. E. Eggleston, J. Santoru and R. Behnke, Phys. Rev. Lett., 47, 1340 (1981).
13. G. J. Morales, M. M. Shoucri and J. E. Maggs, Phys. Fluids, 31, 1471 (1988).
14. M. M. Shoucri, G. J. Morales and J. E. Maggs, J. Geophys. Res., 92, 246 (19870.
15. G. J. Morales and J. E. Maggs, Phys. Fluids, 31, 3807 (1988).
16. M. N. Rosenbluth and C. S. Liu, Phys. Rev. Lett., 29, 701 (1972).
17. J. E. Maggs and G. J. Morales, UCLA PPG-Report 1283, Feb. (1990).
18. G. Laval and R. Pellat, J. Geophys. Res., 73, 3255 (1968).